ESSENTIALS OF
ANATOMY AND
PHYSIOLOGY
FOR NURSING PRACTICE

Applying your learning alongside the person-centred nursing framework

NEAL COOK

ANDREA SHEPHERD

JENNIFER BOORE

FOREWORD BY
BRENDAN MCCORMACK
AND TANYA MCCANCE

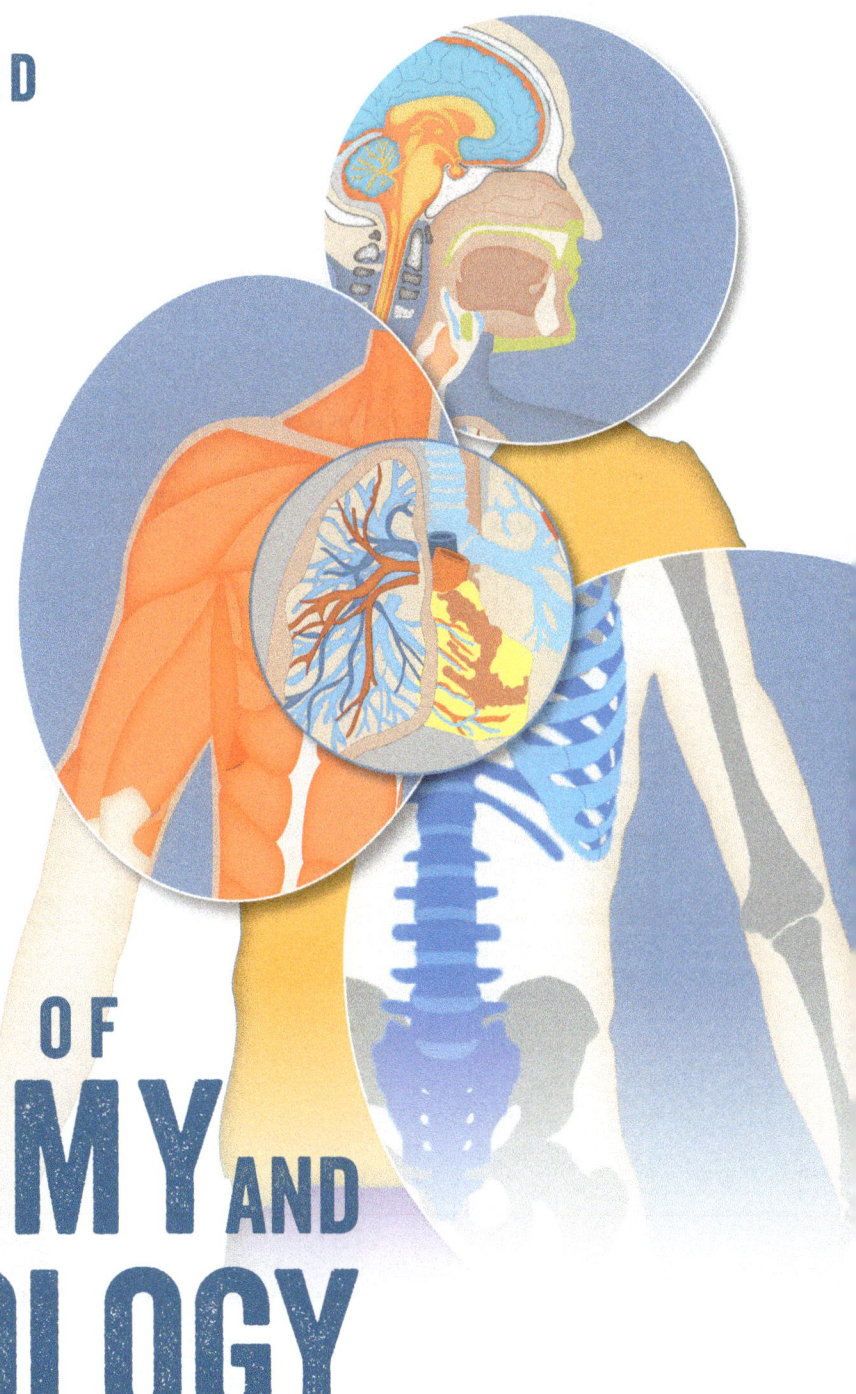

2nd
Edition

ESSENTIALS OF
ANATOMY AND
PHYSIOLOGY
FOR NURSING PRACTICE

SAGE

Los Angeles | London | New Delhi
Singapore | Washington DC | Melbourne

Los Angeles | London | New Delhi
Singapore | Washington DC | Melbourne

SAGE Publications Ltd
1 Oliver's Yard
55 City Road
London EC1Y 1SP

SAGE Publications Inc.
2455 Teller Road
Thousand Oaks, California 91320

SAGE Publications India Pvt Ltd
B 1/I 1 Mohan Cooperative Industrial Area
Mathura Road
New Delhi 110 044

SAGE Publications Asia-Pacific Pte Ltd
3 Church Street
#10-04 Samsung Hub
Singapore 049483

Editor: Alex Clabburn
Development editor: Martha Cunneen
Editorial assistant: Ozlem Merakli
Production editor: Tanya Szwarnowska
Copyeditor: Clare Weaver
Proofreader: Katie Forsythe
Marketing manager: George Kimble
Cover design: Shaun Mercier
Typeset by: C&M Digitals (P) Ltd, Chennai, India

Library of Congress Control Number: 2020933931

British Library Cataloguing in Publication data

A catalogue record for this book is available from the British Library

ISBN 978-1-5264-6031-8
ISBN 978-1-5264-6032-5 (pbk)

CONTENTS

FOR LECTURERS

There is a wealth of online resources, including a new teaching and video guide, test bank and illustration bank, to support your teaching in class.

All resources have been designed and formatted to upload easily into your LMS or VLE.

To access the resources visit **https://study.sagepub.com/essentialap2e** for more details.

ABOUT THE AUTHORS

Dr Neal Cook is a Reader and the Associate Head of School at the School of Nursing, Ulster University and a Principal Fellow of the Higher Education Academy. Neal is also President of the European Association of Neuroscience Nurses and an Executive Board Member of the British Association of Neuroscience Nurses. Neal has taught anatomy, physiology and pathophysiology to undergraduate and post-graduate nursing students across a number of courses since he commenced working in higher education. Neal is also an Advanced Life Support Instructor, teaching life support courses in Health and Social Care Trusts and in the University. He has worked in the fields of neurosciences and critical care since registering as a nurse, becoming a specialist practitioner and subsequently moving into education and research. Neal has published clinical, research and education papers in the fields of education and neurosciences and remains very active in these endeavours. He remains clinically active in neurosciences and remains a Registered Nurse with the Nursing and Midwifery Council (UK).

Andrea Shepherd is a Lecturer in Nursing at the School of Nursing, Ulster University and a Fellow of the Higher Education Academy. She has taught anatomy, physiology and pathophysiology to undergraduate nursing students across a number of courses since she commenced working in higher education. Andrea is also an Advanced Life Support Instructor, teaching life support courses in Health and Social Care Trusts and in the University. She has worked in the fields of critical care and orthopaedics since registering as a nurse, becoming a specialist practitioner and subsequently moving into education. She currently takes a lead role in adult pre-registration nursing, is clinically active in critical care and remains a Registered Nurse with the Nursing and Midwifery Council (UK).

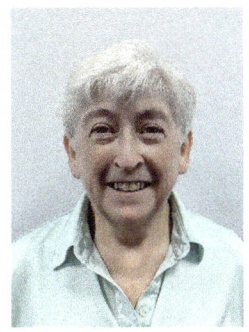

Professor Jennifer Boore is Emeritus Professor of Nursing at the School of Nursing, Ulster University. Jenny started her career as a Registered Nurse, followed by becoming a midwife. She practised as a nurse and midwife in the UK and Australia for some years before returning and beginning her first degree in human biology. After working as a clinical teacher with degree students she obtained a Research Fellowship at the University of Manchester and completed her PhD on pre-operative preparation of patients. From 1977 to 1984 Jenny worked as a Lecturer in Nursing at the Universities of Edinburgh and Hull and was then appointed as Professor of Nursing at the University of Ulster in 1984 (the first Professor of Nursing in Ireland). Jenny has an extensive background in education, research and professional regulation. She has taught anatomy, physiology and pathophysiology to undergraduate and post-graduate nursing students across a number of courses throughout her career. Her contributions to nursing have been recognised in achieving the honours of Fellow of the Royal College of Nursing in 1993 and Officer of the Order of the British Empire in 1996.

The School of Nursing at Ulster provides pre-registration and post-registration nursing education across two campuses in Northern Ireland and internationally. The Person-Centred Nursing Framework (McCormack and McCance, 2010[1], 2019[2]) and the Person-Centred Practice Framework (McCormack and McCance, 2017[3]) inform the curricular framework for pre-registration nursing courses at the School and influence a wide variety of programmes and research activity within the School and the Institute of Nursing and Health Research. Both the School and the Institute of Nursing and Health Research are recognised as excellent and leading in their field nationally and internationally.

---

[1]McCormack, B. and McCance, T. (2010) *Person-Centred Nursing – Theory and Practice*. Chichester: Wiley-Blackwell.

[2]McCormack, B. and McCance, T. (2019) *The Person-Centred Nursing Framework 2010 Revised*. Belfast: Ulster University.

[3]McCormack, B. and McCance, T. (eds) (2017) *Person-Centred Practice in Nursing and Health Care: Theory and Practice*, 2nd edn. Chichester: Wiley-Blackwell.

FOREWORD

Person-centredness and person-centred nursing have become key agendas in nursing and health/social care policy and strategy. This is no surprise really, given the times we live in and the ever-increasing desire to pay more attention to our 'humanness'. Healthcare increasingly recognises the need to keep the person at the heart of decision-making at all levels of the system, whilst nurses are re-engaging with what has been the core of our identity, i.e. providing care in a way that is respectful of persons. However, to the best of our knowledge, this is the first book to take a person-centred approach to the facilitation of learning about anatomy and physiology (A&P). It is odd that this topic has not been considered in this way before, as without the consideration of our physical makeup from a person-centred perspective, then there is no real foundation upon which to build.

The teaching of A&P in nursing has historically been a subject that has been treated like many of the other 'ologies', i.e. it is a subject that has been given to experts in the subject area to teach. This has resulted in significant problems for many nursing students who do not have A&P as a key strength but are able to understand key principles in the context of their daily practice. Many student nurses appear to struggle with understanding the fundamentals of A&P in the curriculum and it is our experience that this struggle sometimes exists because of a difficulty in applying principles rather than a struggle with understanding. Nursing is a practice-based profession and so nursing curricula should reflect that fundamental principle. It is therefore exciting that this book is written by nurse academics who are experts in A&P in a nursing context and who also are well versed in person-centredness and person-centred nursing.

This book takes the reader by the hand as they go on a journey of discovery of the body as it manifests itself in persons. In the book *The Little Prince* the author is holding the Little Prince in his arms and realises that 'What I see here is nothing but a shell. What is most important is invisible' (*The Little Prince*, Antoine de Saint-Exupéry). The teaching of A&P in nursing can sometimes be about the 'shell' rather than the invisible qualities of persons. In this book the authors locate the reader in a family context and so engage the reader in the world of real people, to see beyond the human shell and understand the invisible qualities of persons from a bio-science perspective. This approach offers a real opportunity for students to transcend issues such as mind–body dualism and socio-cognitive dissonance that have plagued healthcare practices historically. Like a precious gift, the body is unwrapped so that the multiple and complex layers of personhood can be engaged with and the invisible qualities of persons exposed.

We are delighted to support this approach and to extol its virtues to student nurses and nurse educators everywhere.

Professor Brendan McCormack, Queen Margaret University, Edinburgh, Scotland

Professor Tanya McCance, Ulster University, Northern Ireland

ACKNOWLEDGEMENTS

The authors would like to thank all who have provided support, direction and inspiration in the development of this book. The book was inspired by student nurses in anticipation that it will support them in their learning and development. In particular, we would like to thank:

- Our families, friends and colleagues who have supported, encouraged and motivated us throughout. This book was a great undertaking for us and we could not have succeeded without you all.
- The team at SAGE who have been enthusiastic and professional critical friends throughout. We are very fortunate to have had the pleasure of working with you all.
- Professor Brendan McCormack and Professor Tanya McCance. Thank you for the inspiration and drive for person-centred nursing and for the eloquent Foreword. We are delighted with your support and contribution. It is exciting to see the framework continue to develop.
- Those who provided their photographs for the Bodie Family, you make a lovely family!:

  - Annalee Cook
  - David Freeman
  - Deirdre Ward
  - Derek Shepherd
  - Ian Munnerley
  - Joseph Friel
  - Kevin Holly
  - Leslie Smith
  - Maureen Smith
  - Niamh O'Doherty
  - Peter Monaghan
  - Stephaine Cook
  - Tracy Munnerley
  - Trevor Adams

- To all who have reviewed and provided invaluable feedback in the development of this book. Your time and contributions have shaped the book throughout its development.

THANKS FROM THE PUBLISHERS

The publisher would like to thank all of the students and lecturers who spoke to us and helped to review this book's content, design, and online resources to ensure it is as useful as possible.

STUDENTS

Firstly, we would like to say a huge thank you to all of the students at Ulster University and the University of the West of Scotland who provided invaluable feedback throughout the process of creating this book.

With special thanks to those who reviewed chapters and trialled the QR codes:

Dominik Bruch
Nicki Dunn
Kelly Gould

Katy McWilliams
Louise Ormond
Emily Sharp

Trudy Walls

LECTURERS

Sarah Ashelford, University of Bradford
Stuart Baker, University of South Wales
Michael Barbagallo, Federation University
Maria Bennallick, University of Plymouth
John Campbell, University of Cumbria
Donal Deehan, Liverpool John Moores University
David Gallimore, Swansea University
Laura Ginesi, University of East Anglia
Helen Godfrey, University of the West of England
Sarah Greenwood, City, University of London
Ray Hayes, Northumbria University
Gay James, Coventry University
Moira Lewitt, University of the West of Scotland
Leo Mayers, University of Worcester

Damion McCormick, University of Nottingham
Debbie McCraw, University of the West of Scotland
Brenda Reeve, University of Hull
Sheila Sobrany, Middlesex University
David Tait, Edinburgh Napier University
Vanessa Taylor, The University of Manchester
Donald Todd, Robert Gordon University
Jane Turner, University of Derby
Ann Williams, Queen Margaret University
Marjorie Wilson, Teesside University
Nicola Witton, Keele University
Debbie Wyatt, University of Chester

The publisher would like to extend a special thanks to Professor Harry Chumman at the University of Greenwich and Matthew Barton at Griffith University for their enormous contributions in compiling the multiple-choice questions and the flashcard decks for the online student resources, and the video teaching questions for the online instructor resources.

Lastly, the publishers would like to thank Shaun Mercier, our in-house designer for drawing so many wonderful illustrations for this book.

PRAISE FOR THE FIRST EDITION

'The inclusion of the fictional family makes the rather dry subject of anatomy and physiology come alive. As this is not hospital but patient based, it also gives the book a health focus which is often missing. The interactive nature of the book and the four stages of learning are also features that I particularly like – it recognises the different ways in which students now learn.'

David Gallimore, Senior Lecturer in Nursing,
Swansea University

'Nursing students will benefit enormously from the clear explanations, informative videos, colourful diagrams and simple and easy to remember definitions of key concepts.'

Harry Chumman, Senior Lecturer,
University of Greenwich

'A well-presented and clear book that should enable student nurses at all levels to establish a sound understanding of A&P related to clinical practice.'

Donal Deehan, Senior Lecturer, School of Nursing and Allied Health,
Liverpool John Moores University

'The use of patient case studies and links to practice is something that has been less clearly focused upon in other A&P textbooks and will be extremely helpful in making the concepts within A&P "real" for student nurses.'

Jane Turner, Senior Lecturer in Nursing, Department of Healthcare Practice,
University of Derby

'A clear strength of this book is how it situates anatomy and physiology within a philosophical framework of nursing care (namely the Person-Centred Practice Framework of McCormack and McCance, 2016). Anatomy and physiology are often taught separately from the psycho-social aspects of care. This can make it difficult for students to apply and integrate their knowledge of human physiology and anatomy to patient care. The philosophical framework developed by the authors of this book will go a long way towards developing a truly holistic nursing care. Students will develop a clear understanding of how human anatomy and physiology underpin safe and effective nursing care and clinical decision-making.'

Sarah Ashelford, Lecturer, School of Nursing,
University of Bradford

HOW TO USE YOUR BOOK

Essentials of Anatomy and Physiology for Nursing Practice has been developed with a number of features and online resources to help you succeed in your study.

FOUR STAGE LEARNING JOURNEY

Each chapter guides you through your learning journey in four stages, using video, activities, key concepts and examples to support you along the way

2. APPLY: Put your knowledge and understanding into practice

1. UNDERSTAND: Get to grips with the 'need to know' essentials

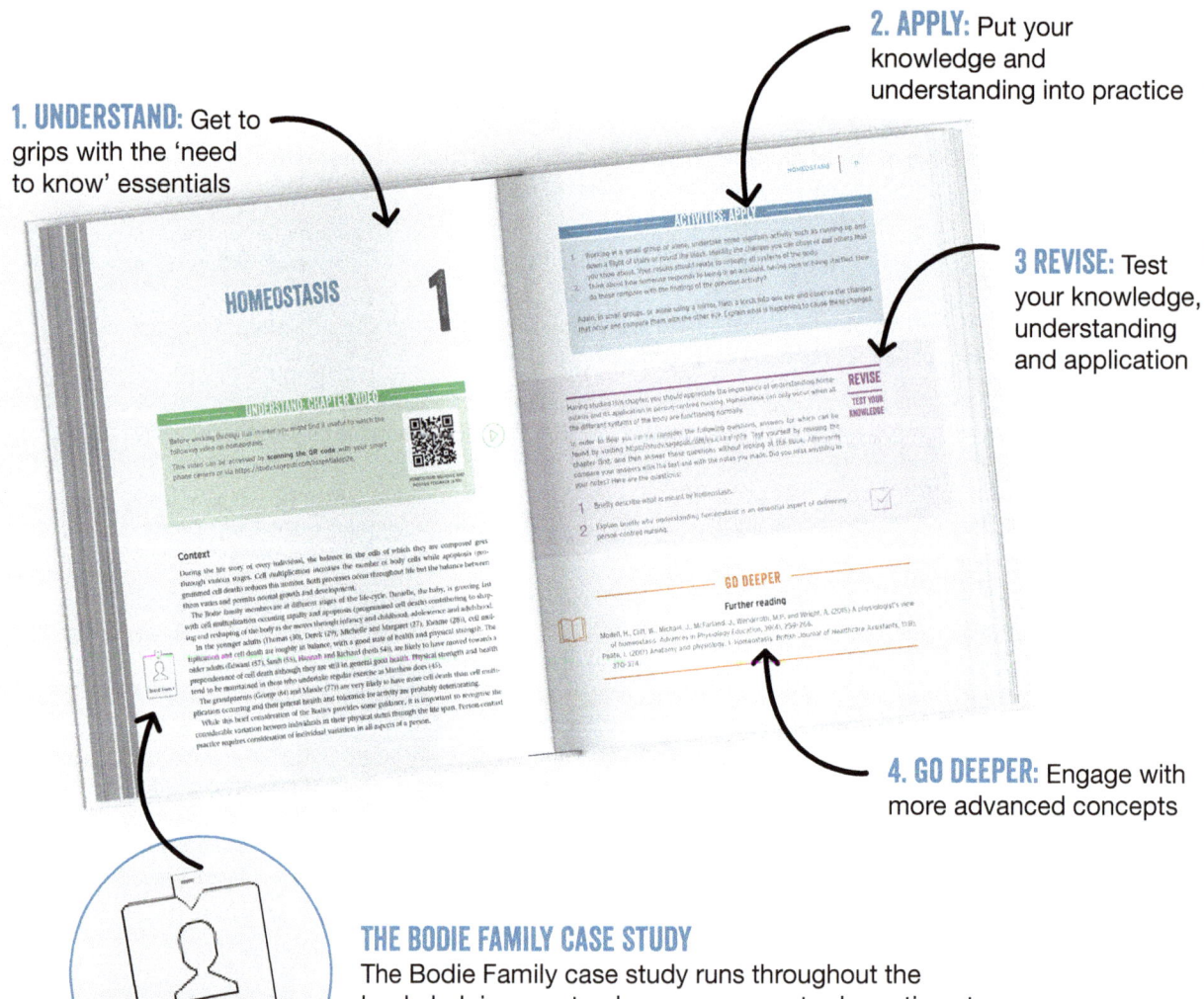

3 REVISE: Test your knowledge, understanding and application

4. GO DEEPER: Engage with more advanced concepts

THE BODIE FAMILY CASE STUDY

The Bodie Family case study runs throughout the book, helping you to place person-centred practice at the heart of your learning. Meet the family on pp xxx.

CURATED ONLINE RESOURCES

Online resources help to improve your understanding and bring each chapter to life. Access the online resources by visiting https://study.sagepub.com/essentialap2e.

 Visualise essential concepts, theory and practice with curated video links. *See below for how to access the videos.*

 Test your knowledge with quizzes and check your answers to the end of chapter Revise questions.

 Check your understanding of new vocabulary with audio flashcards for each chapter.

 Explore further with recommended readings, websites and apps.

"The QR codes in this book are great and very easy to use, and give immediate online access to the material that you are studying."

– First year Nursing student

HOW TO ACCESS THE VIDEOS

Scan the QR codes throughout the chapters to be taken **directly** to each video. Just follow the simple steps below

Got an iPhone?

- Open the camera app on your phone
- Hold the camera over the QR code for the video you want to view (but don't take a photo)
- Tap the notification when it appears

Got an Android?

- Open the camera app on your phone or download a QR scanner app
- Hold the camera or open the app over the QR code for the video you want to view
- Tap the notification when it appears

If you're not a fan of QR codes, you can also view the videos at

https://study.sagepub.com/essentialap2e

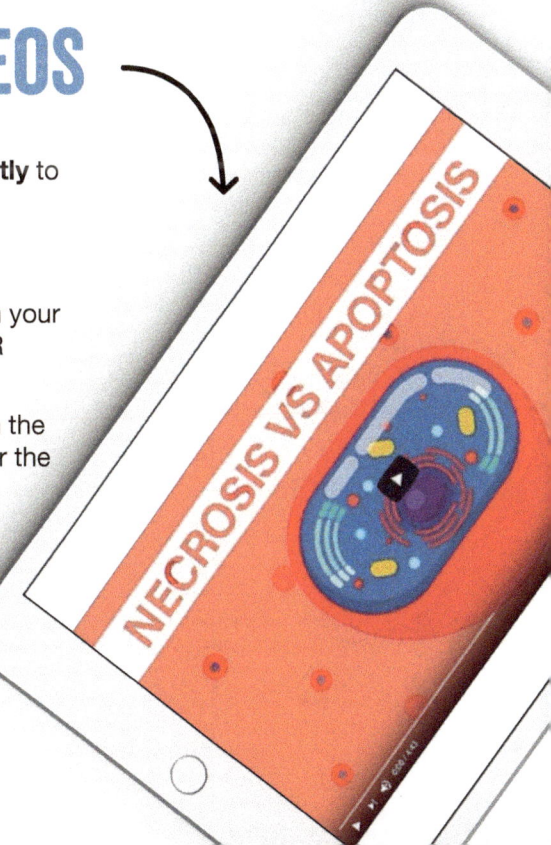

INTRODUCTION
OVERVIEW OF THIS BOOK

ANATOMY AND PHYSIOLOGY FOR PERSON-CENTRED NURSING (PCN)

Person-centred nursing (PCN) is an approach to care which is growing in importance and use world-wide (McCormack and McCance, 2010). As the title says, it emphasises the individuality of each person when you are assessing their needs and planning and implementing interventions. Providing person-centred care requires us to ensure that the person we are caring for remains at the forefront of what we do and how we do it.

Within this book we want to help you develop your understanding of the structure and function of the human body so that you are able to provide the most appropriate care. The main focus is on the healthy body throughout the lifespan and promoting normal physiology during illness. It includes examples of health promotion activities, an important aspect of nursing, and briefly refers to some common disorders to enhance understanding.

UNDERSTANDING PERSON-CENTRED NURSING

The Person-Centred Nursing (PCN) Framework (McCormack and McCance, 2010, 2019) focuses on placing people at the centre of their care so that meaningful and effective outcomes are realised (McCormack, 2018). The Framework emanated from McCormack's (2001, 2003) conceptual framework focusing on person-centred practice with older people, and McCance et al.'s (2001) framework focusing on people's experiences of caring in nursing. It emphasises the importance of focusing on the unique characteristics of the individual in assessing need and planning and providing care. It is based on the primary concept of personhood which emphasises the importance of the individual with their own characteristics, values, beliefs and attitudes, their own life story and future plans. Person-centred nursing necessitates knowledge and understanding of the individuality of the person. However, the focus is not solely on the provision of care, but the culture within which that care is provided. Person-centredness is only realised when a person-centred culture exists in practice as this culture is necessary to facilitate people to experience person-centredness and work in a person-centred way (McCormack, 2018).

Biological characteristics vary between individuals and understanding how the human body functions is equally essential for providing high-quality care. However, the approach used within the PCN Framework focuses on individuals' unique psychological, social, spiritual and environmental characteristics but within a social context. In providing person-centred care, it is essential to integrate and coordinate all the different components of care for the person whilst creating and maintaining a culture of person-centredness. The biological functioning of an individual influences psychological, social and spiritual responses and vice versa and so one component of person-centredness is recognising and responding to this interrelationship in practice.

McCormack and McCance (2010, 2019) describe the PCN Framework and Figure I1 shows the key elements within this model starting from the outside layer and working inwards.

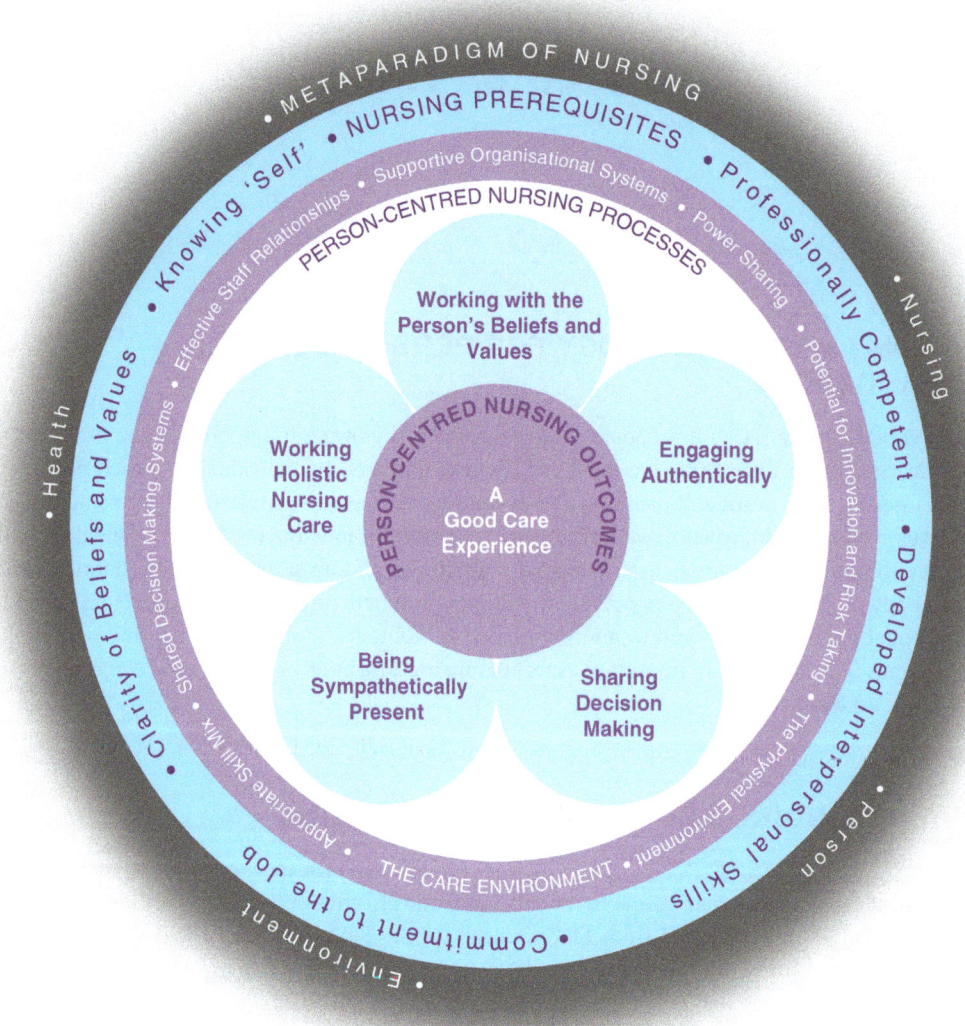

Figure I1 Person-Centred Nursing Framework

McCormack, B., McCance, T., Bulley. C., Brown. D., McMillan. A., Martin. S. (eds) *Fundamentals of Person-Centred Healthcare Practice*, Wiley-Blackwell. © 2021 John Wiley and Sons. Reprinted with kind permission of John Wiley and Sons.

Metaparadigm of nursing

A metaparadigm is the set of concepts central to a discipline, in this case nursing. Within the PCN Framework, this refers to *nursing*, the *person*, the *environment* and *health*. The *person* relates to the 'recipient' of care (including family and community groups), recognising that they are equal partners in a collaborative, therapeutic relationship. Understanding a person, including what is occurring biologically but certainly not limited to it, is central to being able to meet their needs. When we think of the person in this context, it reflects that they are whole and exist within a cultural context of a family, their culture, spirituality and their life experiences. The goal of nursing, within this context, is that the person is empowered to manage their health and well-being with dignity and positive life experiences. *Environment* considers the physical setting of care but also the culture within that setting and the philosophies of practice; this environment influences how a person responds to their health and so has to provide the ideal conditions for health to be realised by the person and those caring for them. *Health* refers to the health and health-related quality of life of that person. Here, *health* is considered within its widest sense and not solely on a biomedical model; the intellectual social and spiritual aspects, for example, are also central. *Nursing* within the metaparadigm refers to the profession, its values, beliefs and philosophies around practice and how we facilitate people to achieve optimal outcomes but in a way that is holistic and person-centred.

Nursing prerequisites

The nursing prerequisites are those attributes of the nurse that are required before being able to provide high-quality person-centred care and are shown in the second circle of Figure I1. Four of the attributes identified are central for interaction, professional behaviour and role performance while the Professionally Competent prerequisite has been defined as 'the knowledge and skills of the nurse to make decisions and prioritize care, and includes competence in relation to physical or technical aspects of care' (McCormack and McCance, 2006: 475).

To achieve this prerequisite, you need to learn and understand how the human body works and be able to apply this in planning and providing care, in combination with the other prerequisites identified. For example, professional competence requires nurses to understand that the signs and symptoms of illness represent an impairment in achieving homeostasis. More specifically, nurses must be able to pinpoint where that imbalance is and recognise that understanding how the body maintains homeostasis usually is the key to interventions to assist the return to homeostasis. How we achieve that knowledge, retain it and apply it to practice will be underpinned by our self-awareness (Knowing Self, Clarity of Beliefs and Values), dedication to compassionate care (Commitment to the Job), and our ability to work with others through respectful cognisance of values and beliefs (Developed Interpersonal Skills).

The care environment

The third layer in this model identifies a number of factors within the care environment that influence the quality of care provided and the experiences and reactions of the staff, persons being cared for and visitors. Many of these factors are determined by the organisational structure and the model of the team working within the clinical setting. In a practice team that prides itself on a person-centred approach to care and effective team working, the members are likely to make decisions based on the research evidence and an understanding that the social environment can minimise stress.

To comprehend the evidence supporting the physical care environment requires an understanding of the biological sciences. Stress is often understood as a psychological state of being distressed (regulated by the nervous and endocrine systems) but its physiological changes can have a deleterious effect on an individual's physical health status. It is important to be able to recognise and minimise these effects.

In considering this layer of the model, we must recognise that in order to work effectively with others, we must develop therapeutic and professional relationships that enable us to draw on the strength of others to find effective solutions. For example, we may recognise that a person's immunity is compromised and that they are showing signs of infection. Resolving this challenge requires nurses to work effectively with doctors and pharmacists, at the very least, in order to utilise their expertise for the common good of the person in their care. However, they all must also respect that the person being cared for has choice in deciding whether to have the infection treated or not – their particular life circumstances may result in them deciding that treating the infection is detrimental to their health in other ways (e.g. risk to a fetus for a pregnant woman). In this way, a sharing of power is vital. Each member of the care team does not oppress the other, including the person being cared for. This leads neatly into the next level of the model, processes – in person-centred nursing processes.

Person-centred nursing processes

These processes are identified in the five 'petals' of the Framework (Figure I1). These all interact and are essential in the provision of care but, in the context of this book, providing holistic care includes the care required to support physiological function. However, psychological and spiritual aspects of care interact with physiologically based care in achieving effective care outcomes.

In the previous layer of the model we gave the example of someone with an infection. Person-centred nursing processes recognise the need to be authentic in the way we engage with people, showing compassion and understanding of their perspective on their health in a wider, holistic context, and working with their beliefs in a shared decision-making process. Taking the example on a different track, this could mean arranging for someone to receive the antibiotic therapy at home as they feel a need to be cared for in their own personal space. This could include recognising that their immune response to the antibiotics could be monitored effectively through blood samples taken by the community nurse, enabling the wider context of the person's health to be taken into consideration.

Person-centred outcomes

This component lies at the centre of the Framework (Figure I1) and centres around having a good care experience, incorporating all aspects of well-being, and is the end result of the other layers of the Framework integrating together. The physiological parameters that nurses measure may relate to the physical care but are equally applicable to the range of care processes and person-centred outcomes. For example, a raised heart and respiratory rate may indicate a person being in a state of anxiety or pain: understanding how the human body works will enable you to understand the implications of these measurements. In working with the example used earlier regarding caring for a person with an infection, if we have followed through the model, the person may feel their care experience was positive as they managed to have the care needed in their own home, they were involved in that decision, and this led to an overall feeling of personal well-being that was achieved, in part, by restoring a state of physiological and psychological homeostasis.

The importance of this model of care is its focus on the uniqueness of the individual. It emphasises the importance of integrating all aspects of the person in ensuring high quality of care for the individual person and their family.

ACTIVITY I.1: GO DEEPER

Person-centred practice

Read the following web resources to broaden your knowledge of person-centred practice.

The weblinks can be accessed by visiting https://study.sagepub.com/essentialap2e.

Websites

The health foundation

Integrated care for patients and populations

Person-centred care

Journal article

See also: Nolan, M.R., Davies, S., Brown, J., Keady, J. and Nolan, J. (2004) 'Beyond "person-centred" care: a new vision for gerontological nursing', *International Journal of Older People Nursing in association with Journal of Clinical Nursing*, 13(3a): 45–53.

THE STRUCTURE OF THIS BOOK

This book is written in five sections to provide a logical structure for the content:

- *Introductory Concepts (Chapters 1–4):* these four chapters consider core concepts that relate to the broad area of the book as a whole. It studies the structure of the human body and how it functions, including homeostasis, the genetic control of the human structure and function, and the microbes in and on us, which influence our function.
- *Control and Coordination (Chapters 5–7):* these three chapters examine in detail the way in which body function is regulated through the nervous and endocrine systems. These are introduced early as both systems are necessary in the regulation of all systems in the body.
- *Preservation of the Internal Environment (Chapters 8–12):* these five chapters consider in some depth the major physiological systems that primarily maintain homeostasis (the balance of chemicals) within the body, recognising that all systems play a role. These systems chiefly deal with the delivery of substances necessary to keep cells living and working, removing products that are considered waste, and balancing the biochemistry of the body for optimal cellular and organ function.
- *Support and Protection of the Internal Environment (Chapters 13–15):* these three chapters consider those systems and mechanisms which play major roles in protection of the body.
- *The Next Generation (Chapters 16 and 17):* these final chapters consider maintenance of the species through reproduction and how the body develops through the lifespan.

We have endeavoured to write this book so that each chapter can be studied alone and make sense on its own, thus allowing chapters to be studied in the order preferred by the lecturer and/or students. However, many chapters also relate to others in the book, helping to demonstrate the interlinking of

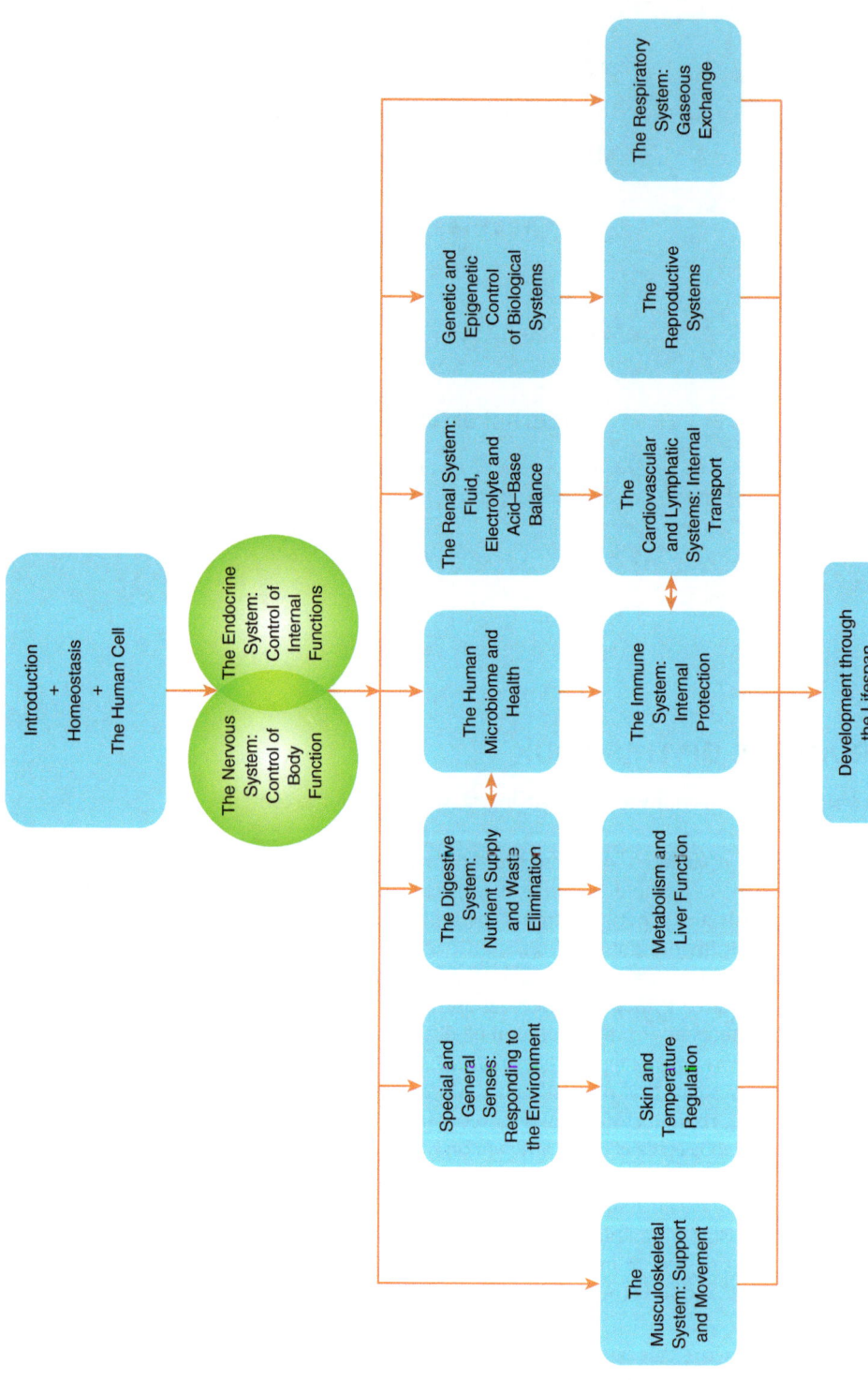

Figure I2   Relationships between the chapters

the different components of the human body. Additionally, we recognise that there are some chapters that must be read first in order for other chapters to make sense. Figure I2 details how each chapter inter-relates in terms of working through this book, illustrating the core chapters that underpin all chapters, and other chapters that are key to read before progressing to a related chapter.

HOW THIS BOOK WORKS

Whether you are an undergraduate student of nursing/midwifery or a Registered Nurse seeking to improve your knowledge and understanding of the human body, this book aims to support you in that journey. In Chapter 1 the systems of the body are described in brief but are then discussed in more detail in the following chapters. It is hoped that this will enable the exploration of the different systems of the body in the order that fits with the individual thinking and teaching approaches experienced by you as a student. You can also move to any relevant chapter in any order that fits with your studies (Figure I2).

This book is further supported by a range of online resources for both students and lecturers (https://study.sagepub.com/essentialap2e) to aid learning and teaching. The website will be referred to throughout the book and it contains a wealth of resources that will help you during your initial learning and also as you consolidate your knowledge and prepare for tests and exams.

1 Understand	• Learners are often unsure about how much or how little they need to know. This book will identify the material that is fundamental to understand each particular system or component of the human body in order that you know that this is what you really need to learn. There will be some activities to support this process. Guidance to specific materials, including animations, will be provided which will aid understanding of certain body actions.
2 Apply	• Understanding the human body is not enough as a nurse caring for people. Being effective as a nurse requires you to be able to apply the knowledge and understanding that you have of the human body. In this book, content that enables you to apply what you are learning to the care of people in society is highlighted.
3 Go Deeper	• For some of you, curiosity and only knowing the fundamentals of the workings of the human body will not be enough for your personal needs. You may also want to improve and advance your knowledge for professional reasons. This section will highlight content that goes deeper than the essential knowledge you need to have. Some of you will find this helpful or just interesting, others may choose not to study these areas.
4 Revise	• At this stage, you have learned the essential knowledge, applied it to practice and gone deeper if you wanted or needed to. You may have had a gap in your learning, or you may wish to refresh your knowledge or prepare for an exam. This stage will guide you on how you can revise through keypoints, and direct you to interactive activities to test your knowledge, understanding and application.

Figure I3 The four stages of learning

The four stages of learning

In addition, this book is set out in a staged process that takes you through your learning journey, using a colour-coded sequence. Four main approaches are used within each of the different topic areas covered: see Figure I3.

Throughout these four stages, this book will relate the content to a fictional family living in society. Health is determined by many factors, and it is within this context that nursing care is provided. Additionally, people must always remain at the centre of your nursing practice, and this book will focus your learning to do this.

Learning features

There are a number of features included in the book that will support your learning.

ACTIVITIES

You will be referred to activities that you can undertake to further your understanding and encourage you to reflect on applying your new knowledge and understanding to person-centred practice.

As this book is about the healthy functioning of the body, the content is focused on promoting health and preventing ill health, enabling you to make links with a holistic approach to nursing practice. You will be challenged to think about the content you are learning and how it relates to health in this context, enabling you to use reflection and other study skills to find answers in order to have a wider perspective on health and its determinants. This will be achieved through the use of case studies and practice scenarios focusing on the following:

- changes which occur through the stages of life and the implications for person-centred practice,
- implications for the health of people and their families of public health interventions and patient education strategies,
- reflective moments where you can pause and place what you have learned within the context of caring for people within the person-centred practice framework.

 Look out for the video icon which means that there is a useful video link available as part of the online resources or **by scanning the QR code**.

KEY CONCEPTS

Anatomy and physiology is a vast, fascinating and complex subject. There may be times where you find certain parts of the book challenging, so at the start of each chapter you will find a short selection of links to videos and reading material that provide a simple overview of the core chapter concepts. All materials indicated by a video icon are available to access online at https://study.sagepub.com/essentialap2e or **by scanning the QR code**. This material will be insufficient on its own but will hopefully make the subsequent information easier to digest. It will also be a place to go if you are struggling at any stage and need to revisit some of the fundamentals. There is a large amount for you to learn and it is important to remember that your knowledge will grow steadily with time and experience; so do not be alarmed if there are points where the content of the book feels difficult or challenging. By the end of your studies you will be amazed at how much you have learned in a short space of time.

The Bodie family

To help you grasp the application of anatomy and physiology to person-centred practice, we are using a fictional, multi-generational family throughout this book to provide examples of the approach we hope you will use through your practice as a nurse.

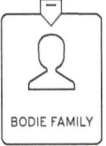
BODIE FAMILY

You can find the details of this family at the end of this introduction. Look out for the Clipboard icon, which will remind you to refer back to the Bodie family tree.

 **REVISE**

We have also provided a range of revision materials that can be used to check your progress and understanding of the material, or to prepare for upcoming tests and exams. Each chapter ends with a series of revision questions with the answers available online via **https://study. sagepub.com/essentialap2e**. You will also be able to access flashcards for each chapter to help you break down the task of learning and memorising key facts and information, as well as more general revision materials, articles and websites that cover the bigger picture - ideal for exam preparation.

Reference material

We recognise that while some of you will have a good level of background knowledge and experience of bioscience, others may have difficulties with terminology, SI units and some of the underpinning science. We have prepared some quick reference material which we hope will help you. This covers:

- The Person-Centred Nursing Framework
- Introductory Science (Appendix 1).
- Units and Numerals (Appendix 2).
- Descriptors of the Body (Appendix 3, p. 553).
- Terminology and Abbreviations (pp. xxxix–xliii).
- Spelling Variations (p. xlv).

Glossary

The Glossary at the back of this book defines specialist terms and is a tool you should access regularly. Students have often said that studying anatomy and physiology can feel like learning a new language and you are likely to encounter a lot of unfamiliar words and phrases as you work through the book; the Glossary should be your first port of call whenever this happens. There is also an interactive version of the book's Glossary in the online resources that comes with a range of useful features including audio pronunciations so you can hear as well as read through the terminology. We hope that this will help you feel more confident when engaging with colleagues on your practice learning experiences, although it is worth noting that individual pronunciation can vary greatly, so please bear this in mind if you hear something said differently to how it sounds in the Glossary.

INTERNATIONAL STANDARDS

This book provides the knowledge of anatomy and physiology necessary to meet the standards for registration as a nurse within a range of countries, including those who meet the requirements of the European Union (EU). The EU Directive on the recognition of professional qualifications (EU, 2005) lays out the principal requirements for general nurses, one of which is acquisition of:

> adequate knowledge of the sciences on which general nursing is based, including sufficient understanding of the structure, physiological functions and behaviour of healthy and sick persons, and of the relationship between the state of health and the physical and social environment of the human being (EU, 2005, Article 31: 41).

This statement clearly demonstrates the importance of understanding the biological sciences for nursing and currently applies to the 27 countries of the EU and four countries of the EEA (European Economic Area).

The Nursing and Midwifery Council (NMC) in the UK identifies the proficiencies for the four fields of nursing (Adult, Children's, Learning Disabilities, and Mental Health Nursing) across seven platforms (NMC, 2018):

- Being an accountable professional
- Promoting health and preventing ill health
- Assessing needs and planning care
- Providing and evaluating care
- Leading and managing nursing care and working in teams
- Improving safety and quality of care
- Coordinating care

These platforms are synergistic with the key concepts of the Person-Centred Nursing Framework (McCormack and McCance, 2019).

While the proficiencies and platforms are applied to the particular field of nursing, there is considerable overlap in the content. For example, one of the proficiencies particularly relevant to this book states:

> demonstrate and apply knowledge of body systems and homeostasis, human anatomy and physiology, biology, genomics, pharmacology and social and behavioural sciences when undertaking full and accurate person-centred nursing assessments and developing appropriate care plans (NMC, 2018: 14).

This book aims to provide an understanding of anatomy and physiology suitable for those undertaking any field of nursing or midwifery including the necessary knowledge of development through infancy, childhood and adolescence. However, student midwives will, almost certainly, use additional texts to develop their specialist knowledge of pregnancy, labour and the post-natal period including lactation.

Anatomy and physiology are mandatory requirements for nursing education in all countries, as far as can be determined. This is illustrated by the American Association of Colleges of Nursing (AACN) statement, requiring a good understanding of anatomy and physiology, of what the graduate of the Baccalaureate nursing qualification should achieve:

> Implement holistic, patient-centered care that reflects an understanding of human growth and development, pathophysiology, pharmacology, medical management, and nursing management across the health–illness continuum, across the lifespan, and in all healthcare settings. (AACN, 2008: 31)

THE BODIE FAMILY

In order to develop your application of human sciences to person-centredness you need to be able to reflect on how the experience of health in society relates to a multi-generational family. Through this book we will develop your understanding of how anatomy and physiology influences health and nursing practice. This can be through applying the content to a variety of situations and conditions that can impact on what we do as nurses. Thinking about caring in the context of a family will help you to do this effectively. You may also be able to relate these situations to people you have met in practice or indeed your own family. In order to achieve this, we have created the fictional Bodie family and you will come across different members of the family in the various chapters of this book. So now it is time to meet the Bodie family, outlined in Figure I4.

The grandparents are:

George Bodie, an 84-year-old retired engineer, worked for a major rail company network for around 50 years. He remains very active and has a strong interest in cars. George met his wife, Maud, now 77 years of age, at the Young Peoples' Guild at their local church. Maud was a dinner lady at a primary school before her marriage. They have been married for 59 years and have three children, four grandchildren and one great-grandchild and a Jack Russell dog which they take for walks together. They now live in a bungalow in a small housing development on the outskirts of a small town. Their children all live within a 20-mile radius.

The three children and their families are:

Edward Bodie, a 57-year-old accountant who is married to Sarah, a 55-year-old community midwife. Their 30-year-old son, Thomas, is a pilot with an international airline. He lives primarily in New York with his partner Jack.

Hannah Jones, a 54-year-old social worker who is married to Richard Jones, the 54-year-old vice-principal of a primary school. They have three children. The eldest, Derek, is a 29-year-old anaesthetist in a regional hospital and they also have 27-year-old identical twin daughters, Michelle and Margaret. Michelle is a linguist who works as a translator for the EU in Brussels where she met her husband Kwame Zuma, a 28-year-old South African civil servant. They have a two-month-old baby girl, Danielle. Margaret is in Australia undertaking post-graduate research in marine biology of the Barrier Reef.

Matthew Bodie is a 45-year-old unemployed electrician with a keen interest in competing in triathlons.

More information about each family member can be found in their individual profiles that follow. Now that you know how this book works and you have met the Bodie family, it is time to start your learning journey.

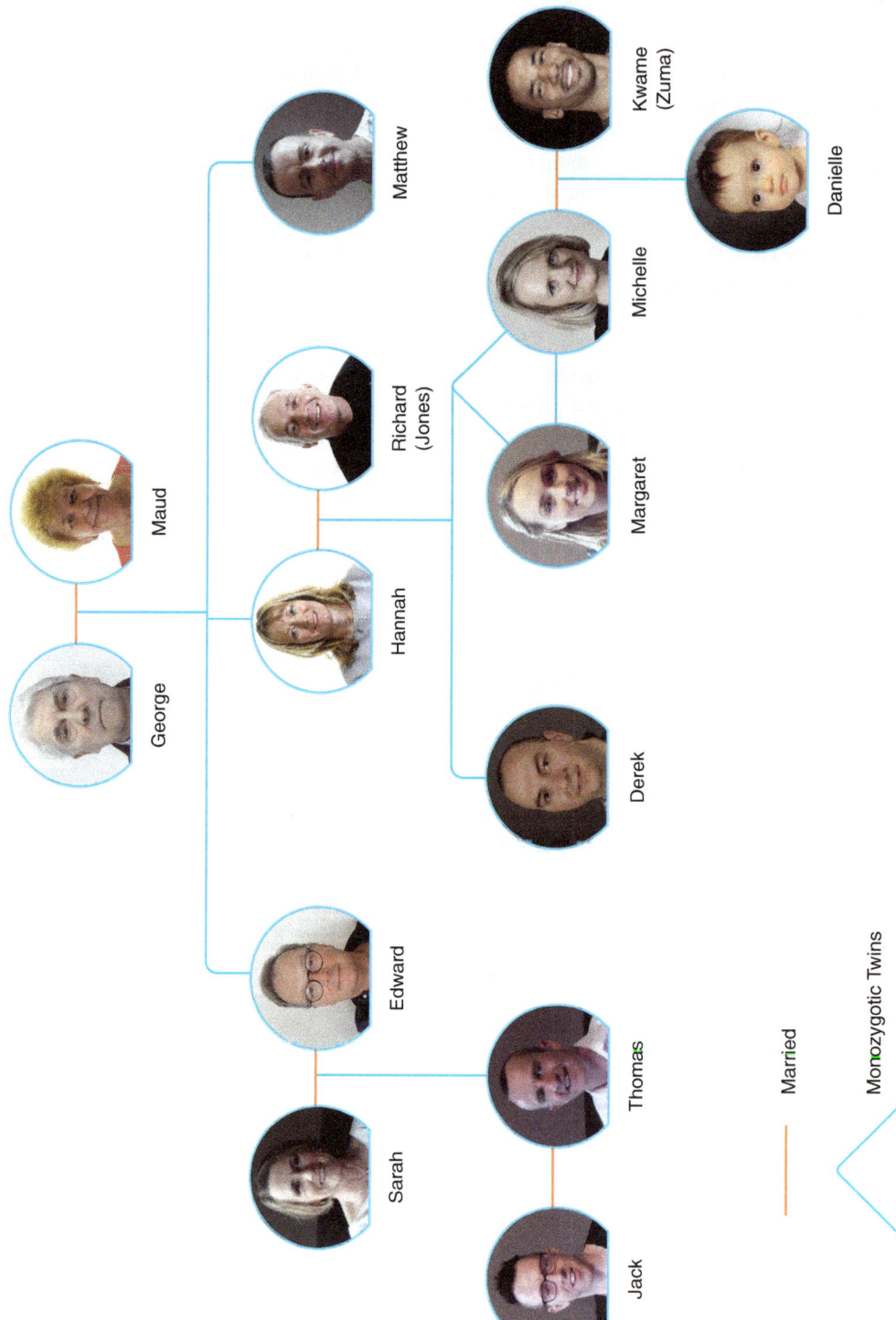

Figure 14 The Bodie family

George — Maud
Hannah — Richard (Jones)
Matthew
Edward — Sarah
Jack — Thomas
Derek
Margaret — Michelle
Kwame (Zuma) — Danielle

Married

Monozygotic Twins

George Bodie

Social

Marital Status:	Married to Maud Bodie for 59 years.
Children:	Father to Edward, Hannah and Matthew.
Occupation:	Retired engineer, having worked for a major rail network for 50 years.
Hobbies:	George is very active, having a strong interest in cars, walking their dog and spending time with the children, grandchildren and great-grandchild.
Housing:	Lives in a bungalow in a small housing development with Maud. Their children live within a 20-mile radius.

Health

Age:	84	**Weight:**	84 kg
Height:	1.8 m (5'11")	**Eye Colour:**	Brown
BMI:	25.9 (overweight category (lower end))		
History:	Appendix removed at the age of 15. Has had raised cholesterol (hypercholesterolaemia) since he was 75, since when he has taken statins to lower it successfully. George gets the flu vaccine every year from his GP. George is not consciously aware of being thirsty.		

Maud Bodie

Social

Marital Status:	Married to George Bodie for 59 years.
Children:	Mother to Edward, Hannah and Matthew.
Occupation:	Dinner lady until she married George.
Hobbies:	Maud likes to walk their dog and spend time with the children, grandchildren and great-grandchild.
Housing:	Lives in a bungalow in a small housing development with George. Their children live within a 20-mile radius.

Health

Age:	77	**Weight:**	73 kg
Height:	1.68 m (5'6")	**Eye Colour:**	Blue
BMI:	25.8 (overweight category (lower end))		
History:	Maud was diagnosed with heart failure three years ago (right ventricular failure) following a heart attack (myocardial infarction). She is prescribed digoxin and warfarin to improve the contractility of her heart and to thin her blood for better perfusion. She also was diagnosed with hypothyroidism at the age of 53, for which she takes thyroxine. Maud gets the flu vaccine every year from her GP.		

Edward Bodie

Social

Marital Status:	Married to Sarah Bodie for 30 years.
Children:	Father to Thomas Bodie.
Occupation:	Accountant.
Hobbies:	Edward enjoys wine and likes to make homemade wine in his spare time. He also likes to read modern fiction and is close to his son, Thomas.
Housing:	Lives in semi-detached house with a garden, located down a country lane.

Health

Age:	57	**Weight:**	90 kg
Height:	1.8 m (5'11")	**Eye Colour:**	Brown
BMI:	27.7 (overweight category)		
History:	Edward suffers from chronic lower back pain.		

Sarah Bodie

Social

Marital Status:	Married to Edward Bodie for 30 years.
Children:	Mother to Thomas Bodie.
Occupation:	Community Midwife.
Hobbies:	Sarah is an active painter, favouring watercolours. She also loves to play the guitar, having learned it as a child.
Housing:	Lives in semi-detached house with a garden, located down a country lane.

Health

Age:	55	**Weight:**	65 kg
Height:	1.73 m (5'8")	**Eye Colour:**	Brown
BMI:	21 (healthy weight category)		
History:	Sarah had childhood chickenpox. Sarah is currently going through the menopause. She has recently stopped taking hormone replacement therapy after doing so for two-and-a-half years.		

Thomas Bodie

Social

Marital Status:	Partner to Jack Garcia for six years.
Children:	None, although he and Jack are keen to adopt in the future.
Occupation:	Pilot for an international airline.
Hobbies:	Thomas loves to ski and snowboard. He is very passionate about music, having learned to play the guitar from his mother. Thomas also loves to sing. Thomas attends the gym three times weekly to stay fit.
Housing:	His primary home is with Jack in New York, although he often stays with his parents when his work brings him in their direction. He and Jack live in a city-centre apartment.

Health

Age:	30	**Weight:**	72 kg
Height:	1.83 m (6')	**Eye Colour:**	Brown
BMI:	21.4 (healthy weight category)		
History:	Thomas is very healthy, eating a balanced diet. He has regular health checks as part of his occupation. Because of irregular hours of work, he has to ensure adequate sleep to maintain circadian rhythm.		

Jack Garcia

Social

Marital Status:	Partner to Thomas Bodie for six years.
Children:	None, although he and Thomas are keen to adopt in the future.
Occupation:	Self-employed sculptor with his own studio in New York.
Hobbies:	Jack is very active in the gym, attending most mornings. He also collects art and is a volunteer at a local pet rescue centre.
Housing:	Jack lives in a city-centre apartment with Thomas.

Health

Age:	28	**Weight:**	73 kg
Height:	1.85 m (6'1")	**Eye Colour:**	Blue
BMI:	21.3 (healthy weight category)		
History:	Jack is very healthy, eating a balanced diet. His father has a history of testicular cancer and so Jack has been vigilant in performing Testicular Self-Examination (TSE) monthly since he was a teenager.		

Hannah Jones

Social

Marital Status:	Married to Richard Jones for 28 years.
Children:	Mother to Derek, and Michelle and Margaret (identical twins).
Occupation:	Social worker.
Hobbies:	Hannah likes to read and has an interest in hobby crafts.
Housing:	Hannah lives with her husband in a detached Victorian home in the countryside.

Health

Age:	54	**Weight:**	57 kg
Height:	1.7 m (5′7″)	**Eye Colour:**	Brown
BMI:	19.7 (healthy weight category)		
History:	Hannah likes to eat healthily, being fond of wholemeal bread and plenty of fruit and vegetables (five-a-day). Hannah smoked in her early 20s, but has successfully given up. She is currently going through the menopause. Hannah currently takes hormone replacement therapy using patches which are reportedly effective.		

Richard Jones

Social

Marital Status:	Married to Hannah Jones for 28 years.
Children:	Father to Derek, and Michelle and Margaret (identical twins).
Occupation:	Primary school teacher.
Hobbies:	Richard is a keen cyclist.
Housing:	He lives with his wife in a detached Victorian home in the countryside.

Health

Age:	54	**Weight:**	64 kg
Height:	1.78 m (5′10″)	**Eye Colour:**	Brown
BMI:	20.1 (healthy weight category)		
History:	Richard, like his wife, is fond of wholemeal bread and plenty of fruit and vegetables (five-a-day). He has suffered from stress in the past that was successfully managed through six weeks of cognitive behavioural therapy. Richard was diagnosed with type 2 diabetes five years ago.		

Derek Jones

Social

Marital Status:	Single.
Children:	None.
Occupation:	Anaesthetist in a regional hospital.
Hobbies:	Derek is a keen chef and often takes cooking course holidays in different countries to widen his culinary skills and knowledge.
Housing:	Lives alone in a new build riverside apartment.

Health

Age:	29	**Weight:**	69 kg
Height:	1.85 m (6′1″)	**Eye Colour:**	Blue
BMI:	20.1 (healthy weight category)		
History:	Derek has had mild persistent asthma since childhood for which he has been prescribed a β_2 agonist and a low-dose corticosteroid inhaler to reduce inflammation and dilate his airways. He is particular about having spare inhalers.		

Margaret Jones

Social

Marital Status:	Single.
Children:	None.
Occupation:	Post-graduate researcher in marine biology in Australia.
Hobbies:	Margaret likes to go horse-riding and goes running four times weekly.
Housing:	Lives with friends in a shared city-centre home.

Health

Age:	27	**Weight:**	69 kg
Height:	1.73 m (5′8″)	**Eye Colour:**	Brown
BMI:	23 (healthy weight category)		
History:	Margaret has hay fever, which she manages well with antihistamines when it flares up.		

Michelle Zuma

Social

Marital Status:	Married to Kwame Zuma for three years.
Children:	Mother to Danielle.
Occupation:	Linguist, working as a translator for the EU in Brussels.
Hobbies:	Michelle has a love for the theatre and has pursued both dance and acting interests.
Housing:	Lives with Kwame and Danielle in their house in Brussels.

Health

Age:	27	**Weight:**	64 kg
Height:	1.68 m (5'6")	**Eye Colour:**	Brown
BMI:	21.3 (healthy weight category)		
History:	Michelle broke (fractured) her right ankle during dance classes when she was 22. This healed uneventfully. She was severely ill when she was 10 and her growth was slowed: she is still slightly shorter than her twin sister.		
	Michelle is currently breast-feeding Danielle after some initial difficulties.		

Kwame Zuma

Social

Marital Status:	Married to Michelle Zuma for three years.
Children:	Father to Danielle.
Occupation:	South African civil servant working in Brussels.
Hobbies:	Kwame loves growing plants and vegetables. He is also a keen chess player.
Housing:	Lives with Michelle and Danielle in their house in Brussels.

Health

Age:	28	**Weight:**	85 kg
Height:	1.90 m (6'3")	**Eye Colour:**	Brown
BMI:	23.5 (healthy weight category)		
History:	Kwame has no significant medical history.		

Danielle Zuma

Social

Marital Status:	Single.
Children:	None.
Occupation:	Sleeping, eating and developing.
Hobbies:	Bath toys, plastic keys and being read stories.
Housing:	Lives with her parents in their house in Brussels.

Health

Age:	2 months	**Weight:**	5 kg (11 lb)
Height:	0.58 m (23″)	**Eye Colour:**	Brown
History:	Danielle is a healthy baby and is still being breast-fed by her mother. She is undergoing routine vaccinations.		

Matthew Bodie

Social

Marital Status:	Single.
Children:	None.
Occupation:	Electrician (currently unemployed).
Hobbies:	Matthew has a keen interest in undertaking triathlons. He also likes to restore vintage cars, which his father enjoys helping with. Matthew also enjoys DIY activities.
Housing:	Lives alone in a cottage.

Health

Age:	45	**Weight:**	82 kg
Height:	1.83 m (6′)	**Eye Colour:**	Brown
BMI:	24.4 (healthy weight category)		
History:	Matthew has suffered from depression since his early 30s. This is successfully treated with antidepressant therapy.		

REFERENCES

AACN (American Association of Colleges of Nursing) (2008) *The Essentials of Baccalaureate Education for Professional Nursing Practice*. Washington: AACN.

EU (2005) Directive 2005/36/EC of the European Parliament and of the Council of 7 September 2005 on the recognition of professional qualifications, Article 31. Available at: http://eur-lex.europa.eu/LexUriServ/LexUriServ.do?uri=OJ:L:2005:255:0022:0142:EN:PDF (accessed 5 January 2015).

McCance, T.V., McKenna, H.P. and Boore, J.R.P. (2001) 'Exploring caring using narrative methodology: an analysis of the approach', *Journal of Advanced Nursing*, 33: 350–56.

McCormack, B. (2001) *Negotiating Partnerships with Older People – A Person-Centred Approach*. Basingstoke: Ashgate.

McCormack, B. (2003) 'A conceptual framework for person-centred practice with older people', *International Journal of Nursing Practice*, 9: 202–9.

McCormack, B. (2018) 'The need to prioritize the person in nursing and healthcare: considering "Healthfulness"', *Obzornik Zdravstvene Nege*, 52(4): 220–4.

McCormack, B. and McCance, T. V. (2006) 'Development of a framework for person-centred nursing', *Journal of Advanced Nursing*, 56 (5): 472–479.

McCormack, B. and McCance, T. (2010) *Person-Centred Nursing: Theory and Practice*. Oxford: Wiley-Blackwell.

McCormack, B. and McCance, T. (2019) *The Person-Centred Nursing Framework 2010 Revised*. Belfast: Ulster University. Available at: www.ulster.ac.uk/nursingframework (accessed 25 November 2019).

NMC (Nursing and Midwifery Council) (2018) *Future Nurse: Standards of Proficiency for Registered Nurses*. London: NMC.

TERMINOLOGY

Prefixes	Meaning	Prefixes	Meaning
adip/o	fat	glyc/o-; glycos/o-	glucose; sugar
aer/o	air	haem-	blood
af-	toward	hapl/o-	simple; single
all/o-	other	hepat/o-	liver
a/an-	without	homeo-	sameness; unchanging; constant
ana-	up; apart; backward; again; anew	hydr/o-	water
angi/o-	vessel (blood)	hyper-	above; excessive
ante-	before; forward	hypo-	deficient; below; under; less than normal
anti-	against	inter-	between
apo-	off; away	intra-	within; into
ather/o-	plaque (fatty substance)	is/o-	same; equal
aut/auto-	self; own	juxta-	near
bar/o-	pressure; weight	kary/o-	nucleus
bas/o-	base; opposite of acid	kerat/o-	hard; horny tissue
chondr/o-	cartilage	kines/o-; kinesi/o-	movement
chrom/o-	colour	leuk/o-	white
de-	lack of; down; less; removal of	lip/o-	fat; lipid
dipl/o	double	ly/o-	to dissolve, loosen
dis-	apart, to separate	macro-	large
dys-	bad; painful; difficult; abnormal	mal-	bad
		meso-	middle
ecto-	out; outside	meta-	change; beyond
eff-	out; out of; from;	micro-	small
end/o-	in; within	multi-	many
eosin/o-	red; rosy; dawn-coloured	my/o-	muscle
epi-	above; upon; on	neo-	new
erythr/o-	red	neur/o-	nerve
eu-	good; normal	neutr/o-	neutral
ex/o-	out; away from	noci-	to cause harm, injury or pain
extra-	outside	oste/o-	bone

Prefixes	Meaning	Prefixes	Meaning
para-	near; beside; abnormal; apart from; along the side of	proxim/o	near
		sarc/o-	flesh (connective tissue)
		somat/o-	body
patho-	disease	sub-	under; below
peri-	surrounding	supra-	above; upper
pneum/o-; pneumon/o-	lung; air; gas	syn-; sym-	together; with
		tel/o-	complete
poly-	many; much	thromb/o	clot
post-	after; behind	trans-	across; through
pre-	before; in front of	viscer/o-	internal organs
pro-	before; forward		

Suffixes	Meaning	Suffixes	Meaning
-aemia	blood	-oid	resembling; derived from
-ase	enzyme	-ole	little; small
-blast	embryonic; immature cell	-phage	eat; swallow
-crine	to secrete; separate	-phil/o	like; love; attraction to
-erg/o	work	-phob/o	fear
-genesis; -gen	producing; forming	-poiesis	formation
-ism	process; condition	-sclerosis	hardening
-itis	inflammation	-stasis	to stop; control; place
-kinesis	movement	-tax/o	order; coordination
-lysis	breakdown; separation; destruction; loosening	-ton/o	tension
		-trophy	nourishment; development (condition of)
-metry	process of measuring	-zyme	enzyme

Parts of Words	Meaning	Parts of Words	Meaning
aemia	blood	ket/o	ketones; acetones
cost/o	rib	ox/oxia	oxygen
cyt/o, -cyte	cell		

ABBREVIATIONS

Abbreviation	Expansion	Abbreviation	Expansion
ACE	Angiotensin Converting Enzyme	CHP	Capsular Hydrostatic Pressure
Acetyl CoA	Acetyl Coenzyme A	CNS	Central Nervous System
ACh	Acetylcholine		
AChE	Acetylcholinesterase	COAD	Chronic Obstructive Airway Disease
ACSM	American College of Sports Medicine	COPD	Chronic Obstructive Pulmonary Disease
ACTH	Adrenocorticotrophic Hormone	COX-1	Cyclo-Oxygenase 1
		COX-2	Cyclo-Oxygenase 2
ADH	Anti-diuretic Hormone	CRH	Corticotrophin Releasing Hormone
ADP	Adenosine Diphosphate		
ALD	Assistive Listening Devices	CRP	C-Reactive Protein
ANP	Atrial Natriuretic Peptide	CS	Caesarean Section
ANS	Autonomic Nervous System	CSF	Cerebrospinal Fluid
		CT	Computed Tomography (scan)
APP	Acute Phase Protein		
APR	Acute Phase Response	CVS	Cardiovascular System
ATP	Adenosine Triphosphate	DCT	Distal Convoluted Tubule
AV	Atrioventricular	DHA	Omega-3 Docosahexaenoic Acid
BBB	Blood–Brain Barrier		
BCG	Bacillus Calmette–Guérin (vaccine)	DIT	Diiodotyrosine
		DNA	Deoxyribonucleic Acid
BCOP	Blood Colloid Osmotic Pressure	ECF	Extracellular Fluid
		ECG	Electrocardiogram
BMI	Body Mass Index	EDSS	Expanded Disability Status Scale
BMR	Basal Metabolic Rate		
bpm	beats per minute	EPO	Erythropoietin
CAC	Citric Acid Cycle	ER	Endoplasmic Reticulum
CHD	Coronary Heart Disease	ETC	Electron Transport Chain

Abbreviation	Expansion	Abbreviation	Expansion
EU	European Union	ICS	International Continence Society
EWAS	Epigenome-Wide Association Studies	ICSH	Interstitial Cell-Stimulating Hormone
$FADH_2$	Flavin adenine dinucleotide	Ig	Immunoglobulin
FG	Fast Glycolytic	IGF-1	Insulin-Like Growth Factor-1
FOG	Fast Oxidative–Glycolytic	IQ	Intelligence Quotient
FP	Filtration Pressure	IU	International Unit
FSH	Follicle Stimulating Hormone	IVF	In Vitro Fertilisation
GA	General Anaesthetic	JGA	Juxtaglomerular Apparatus
GABA	Gamma-Aminobutyric Acid	LCFA	Long Chain Fatty Acid
GFR	Glomerular Filtration Rate	LCPUFA	Long Chain Polyunsaturated Fatty Acid
GHP	Glomerular Hydrostatic Pressure	LDL	Low Density Lipoprotein
GIT	Gastrointestinal Tract	LH	Luteinising Hormone
GnRH	Gonadotrophin-Releasing Hormone	LRT	Lower Respiratory Tract
GTP	Guanosine triphosphate (Guanosine-5'-triphosphate)	MHC	Major Histocompatibility Complex
hCG	Human Chorionic Gonadotrophin	MIF	Müllerian-Inhibiting Factor
HD	Huntington's disease	MIT	Monoiodotyrosine
HDL	High Density Lipoprotein	MMR	Measles, Mumps and Rubella
hGH	Human Growth Hormone	MRI	Magnetic Resonance Imaging
HLA	Human Leucocyte Antigen	mRNA	Messenger RNA
HMP	Human Microbiome Project	MRSA	Methicillin Resistant *Staphylococcus aureus*
hPL	Human Placental Lactogen	MSH	Melanocyte Stimulating Hormones
HR_{max}	Maximal Heart Rate	MV	Minute Volume
HRT	Hormone Replacement Therapy	NADH	Nicotinamide adenine dinucleotide hydride
HTM	High Threshold Mechanoreceptors	NEC	Necrotising Enterocolitis
IBD	Inflammatory Bowel Disease	NFP	Net Filtration Pressure
ICF	Intracellular Fluid	NHP	Net Hydrostatic Pressure

Abbreviation	Expansion
NHS	National Health Service (UK)
NICE	National Institute for Health and Care Excellence (UK)
NIH	National Institutes of Health (USA)
NMC	Nursing and Midwifery Council (UK)
NMJ	Neuromuscular Junction
NO	Nitric Oxide
NREM	Non-Rapid Eye Movement
NRT	Nicotine Replacement Therapy
NS	Nociceptive Specific
NSAID	Non-Steroidal Anti-Inflammatory Drug
PAMPs	Pathogen-Associated Molecular Patterns
PCN	Person-Centred Nursing
PCT	Proximal Convoluted Tube
PCV	Pneumococcal conjugate vaccine
PF_3	Platelet Factor 3
PKU	Phenylketonuria
PMN	Polymodal Nociceptors
PNS	Peripheral Nervous System
PPV	Pneumococcal polysaccharide vaccine
PTH	Parathyroid Hormone
RAAS	Renin Angiotensin Aldosterone System
RAS	Reticular Activating System
RBC	Red Blood Cell
RCPCH	Royal College of Paediatrics and Child Health
REM	Rapid Eye Movement

Abbreviation	Expansion
RNA	Ribonucleic Acid
SA	Sinoatrial
SCFA	Short Chain Fatty Acid
SG	Specific Gravity
SNS	Sympathetic Nervous System
SO	Slow Oxidative
SR	Sarcoplasmic Reticulum
SRY	Sex-Determining Region of the Y
T_3	Triiodothyronine
T_4	Thyroxine
TDF	Testis-Determining Factor
TEF	Thermic Effect of Food
TENS	Transcutaneous Electrical Nerve Stimulation
TNF	Tumour Necrosis Factor
TRH	Thyrotrophin-Releasing Hormone
TSE	Testicular Self-Examination
TSH	Thyroid Stimulating Hormone
U3A	University of the Third Age
URT	Upper Respiratory Tract
UV	Ultraviolet
VLDL	Very Low Density Lipoprotein
V_T	Tidal Volume
VTE	Venous Thromboembolism
WCC	White Cell Count
WDR	Wide Dynamic Range
WHO	World Health Organization
WHR	Waist–Hip Ratio

SPELLING VARIATIONS

UK English	International alternative
Adrenaline/noradrenaline	Epinephrine/norepinephrine
Adrenocorticotrophic	Adrenocorticotropic
Ageing	Aging
Anaemia	Anemia
Caesarean	Cesarean
Catalyse	Catalyze
Colonise	Colonize
Colour	Color
Diarrhoea	Diarrhea
Dyspnoea	Dyspnea
Fertilisation	Fertilization
Fibre	Fiber
Gonadotrophin	Gonadotropin
Haemoglobin	Hemoglobin
Haemorrhage	Hemorrhage
Hydrolyse	Hydrolyze
Ischaemic	Ischemic
Labour	Labor
Leucocyte	Leukocyte
Litre	Liter
Neuron	Neurone
Oedema	Edema
Oesophagus	Esophagus
Oestrogen	Estrogen
Sulphate	Sulfate
Thyrotrophin	Thyrotropin
Tumour	Tumor

PART 1

INTRODUCTORY CONCEPTS

This section of the book considers core issues which underpin the functioning of most of the human body. It consists of the following four chapters:

- *Chapter 1. Homeostasis*

  Within this chapter we present the concept of homeostasis, the stability of the conditions of the internal environment for the cells of the body, and how the different systems of the body contribute to this. These are all considered in greater depth in later parts of the book.

- *Chapter 2. The Human Cell*

  In this chapter the structure and function of human cells and the different components within the cell are described. The two types of cell division are examined: meiosis for the formation of gametes for reproduction; and mitosis involved in growth and development. Adaptation of the different cell types and tissues comprising the systems of the body are considered.

- *Chapter 3. Genetic and Epigenetic Control of Biological Systems*

  The basic blueprint for life is carried as a code within the nucleic acids, DNA (Deoxyribonucleic acid) and RNA (Ribonucleic acid) and is virtually identical for all life on earth. The way in which this information is used to create the numerous proteins making up the body structure and regulating its function, and the influence of environmental factors in the presentation, is introduced. The different ways in which characteristics are transmitted from generation to generation, and the importance for nurses of this understanding, are considered.

- *Chapter 4. The Human Microbiome and Health*

  The human body carries at least ten times as many microbial cells as human ones and their importance in health and disease is considerable. Their distribution and function are examined.

HOMEOSTASIS

1

LEARNING OUTCOMES

When you have finished studying this chapter you will be able to:

1. Understand what is meant by homeostasis
2. Explain briefly why understanding homeostasis is an essential aspect of person-centred nursing practice
3. Identify the major functions of the body in contributing to the overall functioning of the human body

INTRODUCTION: THE HUMAN BODY AND HOMEOSTASIS

Homeostasis is a core concept in relation to the physiological aspects of the individual, and all of the biological systems of the body contribute to it. So, what is meant by homeostasis? In brief, homeostasis is the property of maintaining equilibrium – a stable (or nearly stable) condition of the different properties in the body (for example, body temperature, blood glucose level) through the action of the different bodily systems.

In this chapter we are going to explore this concept in some detail as an introduction to the book as a whole. In addition, please note that all biological works use the same terms to describe the position and orientation of different components of the body – these are presented in the Appendix section and can be used for reference.

Context

BODIE FAMILY

Within the Bodie family, there are people of four generations with different health needs and varying abilities to maintain homeostasis. Danielle, the youngest, is only two months old and at this age her homeostatic mechanisms are limited in ability. She requires appropriate interaction, nutrition and care to ensure that her body systems maintain homeostasis as she grows into a healthy child, mature adult and well older woman. The adults of working age have to balance meaningful work, family life, social activities and physical exercise in aiming to maintain homeostasis demonstrated through health and well-being. The great-grandparents' abilities to maintain homeostasis will be declining, and good nutrition (not too much salt or sugar), physical exercise and mental activity are necessary to maintain optimum health and, at least as important, to reduce the risk of developing cognitive impairment. The care provided during illness or following accidents aims to support physiological function and maintain homeostasis to promote recovery and healing. The PCN Framework (McCormack and McCance, 2010, 2019) provides guidance on all the different components of healthy living – psychological, social, environmental and physiological – and is applicable to all generations of this family.

Within the context of person-centred practice, we need to consider how the human body is composed and how it works. Starting at the beginning, all living things are composed of cells. Some are single-celled (unicellular) (Chapter 4), others, including human beings, are multicellular consisting of millions of cells (considered in more detail in Chapter 2). In all such organisms every cell of the body performs a range of functions to maintain its own existence and also contributes to homeostasis of the whole through its special contribution as part of tissues, organs and systems of the body.

UNDERSTAND

An organism is an individual living thing that responds to stimuli, grows and maintains homeostasis, and reproduces. It can be plant or animal, uni- or multicellular.

The human body consists of a number of body systems which contribute to the maintenance of homeostasis through interacting to undertake the functions above. A good understanding of the body, its functions and organisation will enable you to provide appropriate care and to help people to maintain their own health. Figure 1.1 identifies the different body systems and their contributions to homeostasis and continuation of the species, which are outlined in Table 1.1. In later chapters you will learn about the different systems in greater detail.

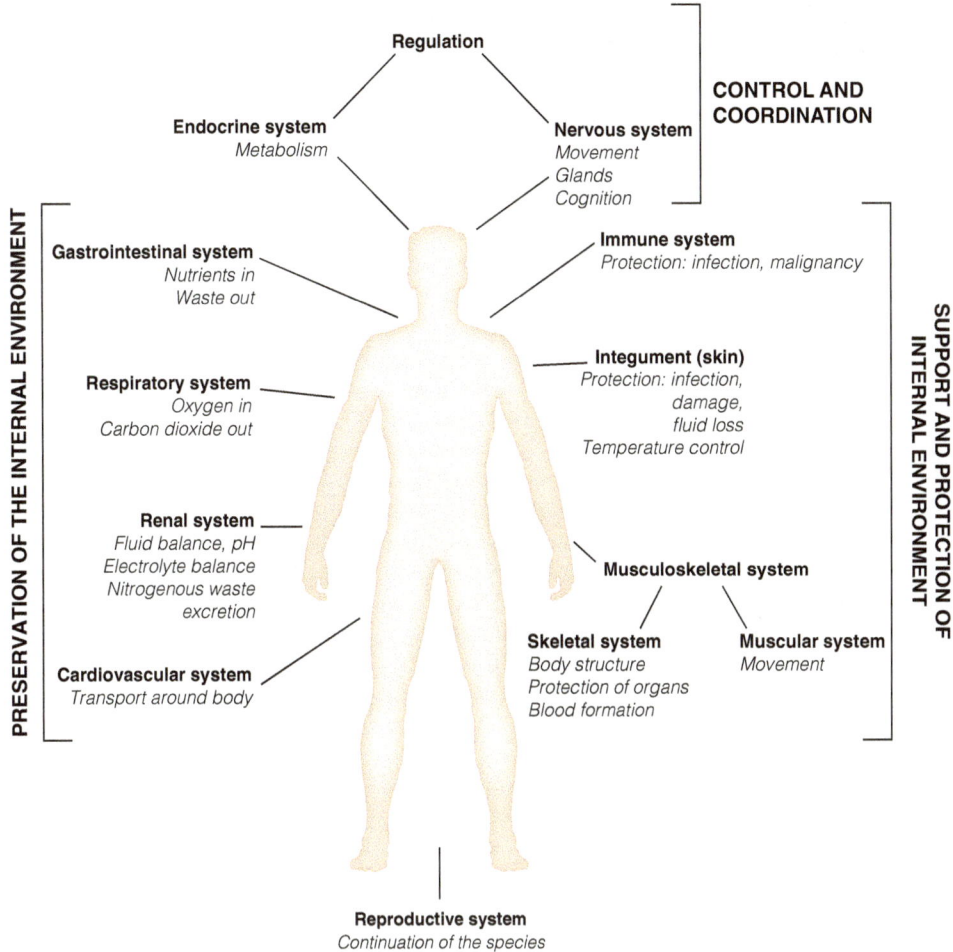

Figure 1.1 Body systems in homeostasis

Regulation through feedback

In order to understand how the body functions to maintain homeostasis, it is necessary to understand feedback and how it balances the levels of the different parameters so important for life (see Figures 1.2a and 1.2b).

There are three parts to any feedback system identified in the figures:

- Sensory receptors, which monitor the level of the parameter being regulated;
- The control centre, which receives information about changes and sends orders to effectors;
- Effectors: muscles, glands, etc., which alter the parameter in whichever direction is needed.

The control centre is often the Central Nervous System (CNS) or the hypothalamus and/or pituitary gland (part of the CNS and also of the endocrine system) but in some situations the sensory receptor, control centre and the effector organs are all in the same cell. An example of this is in the control of blood glucose levels, which is managed completely by the specialist cells of the islets of Langerhans (in the pancreas).

For each parameter controlled though feedback, there are set points of upper and lower levels between which the blood constituent (e.g. blood glucose, different electrolytes or blood oxygen concentration) or physical measurement (e.g. blood pressure or ventilation rate) varies. At the bottom of each of the two figures (Figures 1.2a and 1.2b) the way in which the parameter varies is illustrated.

Table 1.1 Contributions to homeostasis and species continuation

	Chapters	Notes
Part 2: Control and Coordination	5. The Nervous System: Control of Body Function	Rapid regulation of bodily activities through speedy nerve impulses carried through central and peripheral nervous systems (PNS): Central Nervous System (CNS) is brain and spinal cord; PNS is all of the nervous system outside the CNS
	6. Special and General Senses: Responding to the Environment	• Sensory division carries information to CNS from special and general sense organs and information is interpreted in the CNS • Motor division controls glandular secretions and muscular movements: ○ somatic (voluntary) actions ○ autonomic (involuntary) activities: ▫ sympathetic division: fight, flight, fright response increases nutrient supply ▫ parasympathetic division: reduced activity, but increased availability of nutrients
	7. The Endocrine System: Control of Internal Functions	Slower regulation of bodily activity through action of numerous endocrine glands, which regulate metabolism, and thus level of activity, of many organs of body through secretion of hormones (chemical messengers) carried in body fluids to target cells
Part 3: Preservation of the Internal Environment	8. The Digestive System: Nutrient Supply and Waste Elimination	Consists of gastrointestinal tract (GIT) and accessory organs. Enables food to enter the body and be used for body function GIT: essentially external to internal body environment, hollow tube runs from mouth (where food enters) to anus (where waste exits)
	9. Metabolism and Liver Function	Accessory organs: liver, gallbladder and pancreas contribute to digestion and storage of nutrients. Liver has major role in metabolism
	10. The Respiratory System: Gaseous Exchange	The respiratory system in combination with the cardiovascular system (see below): enables oxygen (O_2) to reach body cells, eliminates carbon dioxide (CO_2)
	11. The Renal System: Fluid, Electrolyte and Acid–Base Balance	The renal system is crucial in homeostasis through maintaining chemical composition of body fluids and eliminating waste products
	Appendix 1: Introductory Science	CO_2 from cell activity dissolves in water to form carbonic acid which, in excess, decreases pH (acid-base balance), i.e. increases the acidity of body fluids. Maintaining pH within its normal range (7.35–7.45) is undertaken by renal and respiratory systems combined
	12. The Cardiovascular and Lymphatic Systems: Internal Transport	Cardiovascular system: heart pumps blood with requirements for cell function through blood vessels around the body Contains: erythrocytes (red blood cells), carry oxygen; lymphocytes (white blood cells), involved in immunity; thrombocytes (platelets), control bleeding

Chapters	Notes
Part 4: Support and Protection of the Internal Environment	
	The circulation has two main parts:
	• Systemic: carries blood around upper body, including head and arms, and lower body containing major organs and legs
	• Pulmonary: carries blood between heart and lungs, permits exchange of blood gases
13. The Immune System: Internal Protection	The lymphatic system is functionally part of cardiovascular and immune systems; considered in relation to the immune system
	Immune system consists of the lymphatic system, lymph nodes, parts of GI tract, bone marrow, thymus gland, spleen
	It defends the body against disease-causing microorganisms and abnormal cells (e.g. cancer cells) through:
	• Innate immunity (non-specific defence mechanisms): common to all, activated by any foreign bodies. Reduces entry of microorganisms and other foreign bodies
	• Adaptive immunity: highly specific response to particular foreign body dealing with infective (pathogenic) organisms (used in immunisations)
	The foreign body is an antigen: different types of white blood cells (leucocytes) create antibodies
	Antigen–antibody reaction neutralises antigen effect. The immune response takes time to achieve full level of activity
14. Skin and Temperature Regulation	The skin, or integument, provides protection in different ways:
	• Flexible waterproof tissue prevents loss of fluid, nutrients and electrolytes from body, prevents damage from minor trauma and is physical barrier to pathogenic microorganisms
	• Sensory receptors enable the skin to send information about heat, trauma, etc. and CNS initiates evasive action
	• Prevents the body getting too hot, protects against UV light, production of vitamin D for bone formation
	Temperature regulated by hypothalamus, skin plays an important role
15. The Musculoskeletal System: Support and Movement	Consists of skeleton and muscular system
	Skeleton:
	• provides framework and, with muscular system, permits movement
	• provides protection for major organs within skull, chest (thorax), pelvis
	Bone marrow has major role in formation of blood cells
	Muscles have power of contraction. Three groups shown in Table 1.2
Part 5: The Next Generation	
16. The Reproductive Systems	Male and female reproductive systems together enable continuation of the species through creation of the embryo, which develops in the female. The mother's physiology adapts to safely carry baby through pregnancy and to nurture infant through breast-feeding
17. Development through the Lifespan	Body systems develop through the stages of life including fetal development, childhood, puberty, maturity and ageing. Functional ability also alters through these stages. Understanding these enhances the quality of person-centred care

Table 1.2 Types of muscle

Skeletal muscle (voluntary, striated or striped)	Responsible for voluntary movement of the body, thus providing protection by enabling movement away from danger; controlled by somatic (voluntary) nervous system	Chapter 15
Smooth muscle (involuntary or non-striated)	Different in appearance from skeletal muscle, controlled largely by autonomic nervous system. Examples: gastrointestinal tract and contraction of the blood vessels	Chapter 8
Cardiac muscle	Forms the heart and beats regularly throughout life, controlled by autonomic nervous system	Chapter 12

There are two types of feedback:

- Negative feedback is the most common type of feedback system (Figure 1.2a). This aims to maintain the particular parameter (e.g. body temperature) within normal limits. If the particular characteristic being controlled moves away from normal, this is sensed and information interpreted by the control system. This then initiates changes to return the parameter back towards its original level.

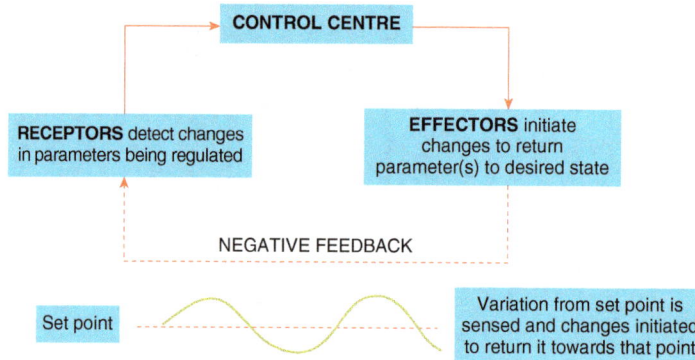

Figure 1.2a Negative feedback

- Positive feedback works to increase the change occurring (Figure 1.2b). For example, the initial changes that occur in blood clotting stimulate other reactions, which increase blood clotting to prevent loss of blood.

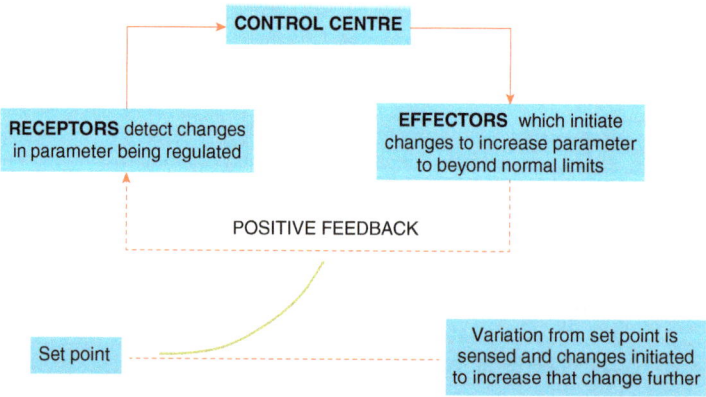

Figure 1.2b Positive feedback

APPLY

Negative feedback is the most common form of feedback used to maintain homeostasis but there are also a number of important positive feedback systems. Below is an example of each so that you can apply the detail you have learned to feedback systems in homeostasis.

Negative feedback

Body temperature is a classic example of where negative feedback is used to maintain the temperature of the internal environment around 36.8° C (or 1°C either side of this). When the body temperature falls, this is detected by receptors, called thermoreceptors, who relay this information to the control centre. In this case, this is the hypothalamus, a structure within the brain (Chapter 5). The hypothalamus then activates the effectors, such as muscles, glands and cells, to engage in processes to bring the temperature back up. For example, metabolism increases in cells and skeletal muscle shivers to generate heat (Figures 1.3a and 14.7). This negative feedback system has therefore brought the parameter being monitored and regulated back towards the level needed to maintain the internal environment.

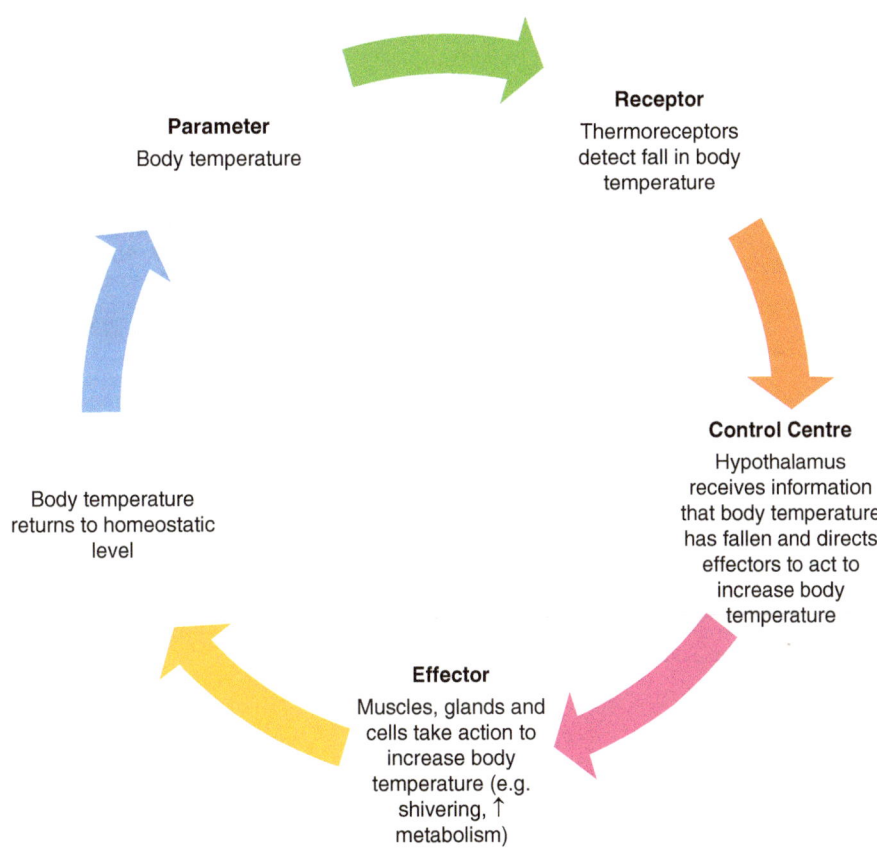

Parameter
Body temperature

Receptor
Thermoreceptors detect fall in body temperature

Control Centre
Hypothalamus receives information that body temperature has fallen and directs effectors to act to increase body temperature

Effector
Muscles, glands and cells take action to increase body temperature (e.g. shivering, ↑ metabolism)

Body temperature returns to homeostatic level

Figure 1.3a Body temperature regulation through negative feedback

(Continued)

Positive feedback

Coagulation (blood clotting) is an example of when the body uses positive feedback. This is a complex process and is discussed in detail in Chapter 12 (Figure 12.2). Another example is the contraction of the uterus (womb) during labour (Figure 1.3b). The pressure of the fetus on the cervix activates stretch receptors; these receptors carry this information to the hypothalamus in the brain (control centre) which triggers the pituitary gland, a small endocrine gland in the brain, to release oxytocin into the blood. Oxytocin is carried into the blood and when it reaches the uterine muscles, it stimulates them to contract. The more stretch detected by the stretch receptors in the cervix, the more oxytocin is released and this increases the frequency and intensity of contractions through this positive feedback mechanism. Once the baby has been born, there is no pressure on the cervix and so the feedback mechanism stops.

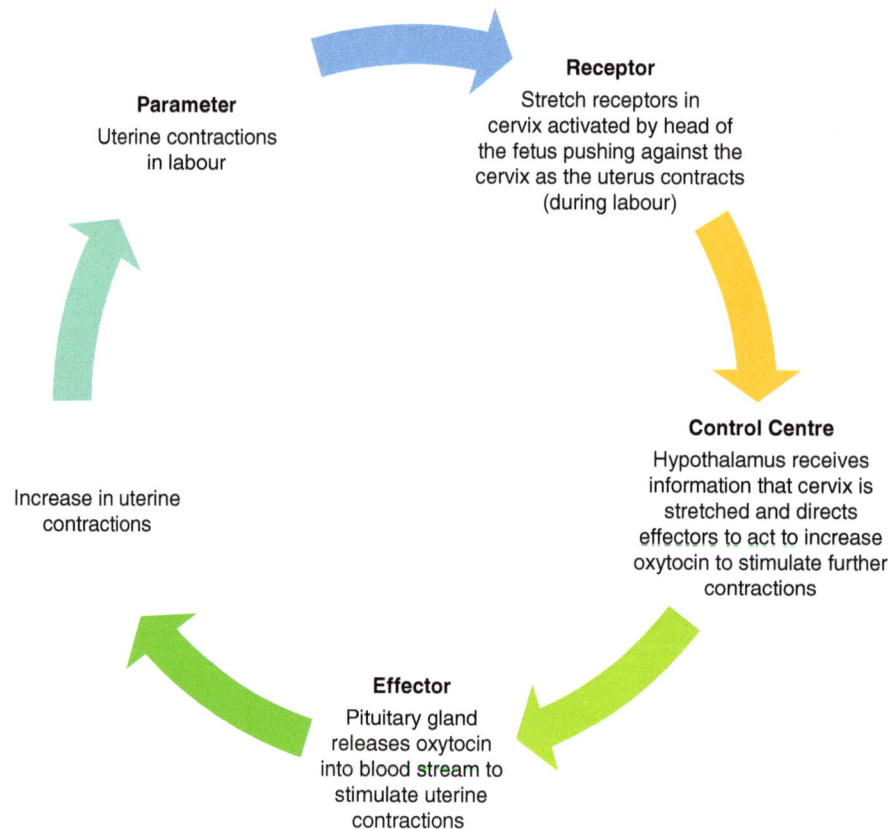

Parameter
Uterine contractions
in labour

Receptor
Stretch receptors in
cervix activated by head of
the fetus pushing against the
cervix as the uterus contracts
(during labour)

Control Centre
Hypothalamus receives
information that cervix is
stretched and directs
effectors to act to increase
oxytocin to stimulate further
contractions

Increase in uterine
contractions

Effector
Pituitary gland
releases oxytocin
into blood stream to
stimulate uterine
contractions

Figure 1.3b Uterine contraction regulation through positive feedback

Integration of function

In considering homeostasis, it is important to emphasise the integration of functions of all the different systems of the body. The nervous and endocrine systems carry out the main functions of control and coordination and are studied early in this book (Chapters 5, 6 and 7). These two systems are closely linked both anatomically and functionally and there are multiple feedback loops in operation at any one time.

The brain is the main controlling point of the nervous system, sending nervous impulses rapidly along nerves transmitting information to and from different parts of the body and dealing with both voluntary and involuntary activity. The hypothalamus and pituitary gland are the main control centres for the endocrine system of numerous endocrine glands, but they are also part of the brain.

UNDERSTAND

Endocrine glands release hormones directly into the blood stream and are thus known as ductless glands.

Exocrine glands release secretions (e.g. sweat, digestive enzymes) through ducts to the surface of an organ (e.g. the skin) or into a cavity (such as the digestive tract) and are ducted glands.

As we will see in studying the other systems of the body, the nervous and endocrine systems have links to organs and systems to control and integrate their functions. All the other systems also have links with each other's functions. Various bodily systems ensure that nutrients are provided, waste products removed, chemical composition of the blood maintained, the body protected and moved, and other activities performed. All these systems are supplied with the necessary oxygen and nutrients, and waste products removed, by the cardiovascular system. What is clear is that all systems work together for full body function.

In delivering person-centred care, it is also important to understand the changes that take place in the human body through fetal development, maturation, adult life and ageing. As these changes occur, the ability of each person to care for themselves and have normal social interactions and behaviour require adjustment to cope with these changes.

HOMEOSTASIS AND THE PERSON

Now that you have some understanding of how the body maintains homeostasis through the collective work of all the systems, where does this fit with person-centred practice? Holism is a concept central to person-centred practice. It refers to all of the concepts that make up personhood, for example, the biological, psychological, sociological and spiritual aspects of a person. So, let's make some links.

A person's state of mind, for example, fundamentally originates in their nervous system where experiences and learning integrate with emotion. The strength of someone's muscles and bones, including how they are coordinated by the nervous system, are central to the ability of a person to undertake their particular job; for example, a builder would be dependent on strength and coordination. In some cultures, someone's career may also be an important part of their identity, how they provide for themselves and their family, and in turn how this enables them to pursue activities that maintain well-being as well as to afford food that supports optimal health.

It is essential as a nurse to consider the social context of a person's health and well-being. While you will explore the anatomy and physiology of the human body in this text, it is necessary to apply this within the context of people living in society. For example, we know that excessive alcohol consumption inhibits the function of phagocytes (a type of white blood cell) and also suppresses the development of T-cells that form part of the immune response (Williamson and Williamson, 2010). Understanding how this inhibits homeostasis provides the nurse with the opportunity to engage in health education and promotion activities to prevent people from becoming immunocompromised through excessive alcohol use. However, the person-centred nurse must consider the social factors that influence such behaviours – it is not merely a case of educating people to drink alcohol sensibly, it is about considering what has caused the excessive alcohol intake and addressing those factors. For example, this could be low stress tolerance, low self-esteem, addiction, homelessness or adverse childhood experiences. Considering that alcohol consumption is highest in those living in low socio-economic circumstances (WHO, 2011), the challenge for nurses and healthcare services is to truly examine those relationships in order to access the real potential to improve health and well-being.

When you think of all of these issues, they map directly into the PCN Framework – central to the person-centred outcomes are: a good care experience where involvement with care, feeling of well-being and the existence of a healthful culture are all central to that experience. Identifying a good care experience will be influenced by how holistic care is – this relates to the person's experiences (think of the nervous system and its control of emotions, experiences and learning) and how they are able to fulfil their social roles (such as their job, providing for their family, engaging in rewarding activities that are healthy). The latter also contributes to their sense of well-being. A nurse should focus on facilitating empowerment to support the return to optimal health whereby all of this is possible, thus working within the context of person-centred practice.

ACTIVITY 1.1: APPLY

1. Working in a small group or alone, undertake some vigorous activity such as running up and down a flight of stairs or round the block. Identify the changes you can observe and others that you know about. Your results should relate to virtually all systems of the body.
2. Think about how someone responds to being in an accident, having pain or being startled. How do these compare with the findings of the previous activity?

Again, in small groups, or alone using a mirror, flash a torch into one eye and observe the changes that occur and compare them with the other eye. Explain what is happening to cause these changes.

CONCLUSION

You should now have an understanding of homeostasis and how all of the body systems work in harmony to maintain it. Additionally, you should also now understand how homeostasis links with the PCN Framework through the functions of the human body. While this chapter has provided an overview of these concepts, we will explore each system and other biological concepts in more detail later in the book to enhance your understanding and the relevance to practice. However, before that, in the next chapter we will examine the individual building blocks of the human body, the cells.

GO DEEPER

Further reading

Modell, H., Cliff, W., Michael, J., McFarland, J., Wenderoth, M.P. and Wright, A. (2015) 'A physiologist's view of homeostasis.', *Advances in Physiology Education*, 39(4): 259-266.

Peate, I. (2017) 'Anatomy and physiology, 1. Homeostasis', *British Journal of Healthcare Assistants*, 11 (8): 370-374.

- Homeostasis is essential for maintaining the function of individual cells, bodily systems and the human body as a whole. The body responds to the internal and external environments to maintain health and safety.

- The function of all body systems is integrated and regulated by the two regulatory systems – the nervous and endocrine systems.

- Specific terms are used in describing the human body to enable everyone to understand it in the same way.

- Homeostasis is a key component of personhood. Personhood is the focus of person-centred practice.

Having studied this chapter, you should appreciate the importance of understanding homeostasis and its application in person-centred nursing. Homeostasis can only occur when all the different systems of the body are functioning normally.

In order to help you revise, consider the following questions, answers for which can be found by visiting **https://study.sagepub.com/essentialap2e**. Test yourself by revising the chapter first, and then answer these questions without looking at the book. Afterwards compare your answers with the text and with the notes you made. Did you miss anything in your notes? Here are the questions:

1 Briefly describe what is meant by homeostasis.

2 Explain briefly why understanding homeostasis is an essential aspect of delivering person-centred nursing.

For additional revision resources visit: **https://study.sagepub.com/essentialap2e**.

- Revise key terms with interactive flashcards.

- Test yourself with multiple-choice questions.

- Access the glossary with audio to hear how complex terms are pronounced.

- Explore recommended websites suitable for revision.

REFERENCES

McCormack, B. and McCance, T. (2010) *Person-Centred Nursing: Theory and Practice*. Oxford: Wiley-Blackwell.

McCormack, B. and McCance, T. (2019) *The Person-Centred Nursing Framework 2010 Revised*. Belfast: Ulster University. Available at: www.ulster.ac.uk/nursingframework (accessed 25 November 2019).

McCormack, B. and McCance, T.V. (2006) 'Development of a framework for person-centred nursing', *Journal of Advanced Nursing*, 56(5): 472–9.

Williamson, J. and Williamson, R. (2010) 'Alcohol and the pancreas', *British Journal of Hospital Medicine*, 71(10): 556–61.

WHO (World Health Organization) (2011) *Environmental Burden of Disease Associated with Inadequate Housing*. Copenhagen: WHO.

THE HUMAN CELL

UNDERSTAND: CHAPTER VIDEOS

Before working through this chapter, you might find it useful to watch these external video clips on the human cell.

These videos can be accessed by **scanning the QR codes** with your smart phone camera or via https://study.sagepub.com/essentialap2e.

INTRODUCTION TO CELLS (2:54)

CELL STRUCTURE AND FUNCTION (9:06)

LEARNING OUTCOMES

When you have finished studying this chapter you will be able to:

1. Describe the structure and functions of the different components of human body cells
2. Describe the two types of cell division: meiosis for the formation of gametes for reproduction, and mitosis involved in growth and development
3. Identify how different types of body cells are adapted for their different functions and how they interact synergistically

INTRODUCTION

In the previous chapter you were introduced to the systems of the multicellular body, which supply the necessary conditions for life. It will make understanding the human body easier if you know rather more about how each cell works and this chapter focuses on the structure and function of the different types of cell in the human body.

The two types of cell division will also be examined:

- **mitosis** for growth and repair of tissues
- **meiosis** for creation of sperm or ova (gametes) containing half the normal genetic material for formation of the next generation.

Context

During the life story of every individual, the balance in the cells of which they are composed goes through various stages. Cell multiplication increases the number of body cells while apoptosis (programmed cell death) reduces this number. Both processes occur throughout life but the balance between them varies and permits normal growth and development.

The Bodie family members are at different stages of the life-cycle. Danielle, the baby, is growing fast with cell multiplication occurring rapidly and apoptosis (programmed cell death) contributing to shaping and reshaping of the body as she moves through infancy and childhood, adolescence and adulthood.

In the younger adults (Thomas (30), Derek (29), Michelle and Margaret (27), Kwame (28)), cell multiplication and cell death are roughly in balance, with a good state of health and physical strength. The older adults (Edward (57), Sarah (55), Hannah and Richard (both 54)), are likely to have moved towards a preponderance of cell death although they are still in general good health. Physical strength and health tend to be maintained in those who undertake regular exercise as Matthew does (45).

The grandparents (George (84) and Maude (77)) are very likely to have more cell death than cell multiplication occurring and their general health and tolerance for activity are probably deteriorating.

While this brief consideration of the Bodies provides some guidance, it is important to recognise the considerable variation between individuals in their physical status through the lifespan. Person-centred practice requires consideration of individual variation in all aspects of a person.

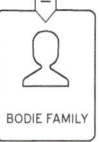

BODIE FAMILY

THE CELL

Introduction

This chapter focuses on the human cell and how it works. However, although not essential, it is interesting to understand where this fits in the wider biological context.

—————————————— **GO DEEPER** ——————————————

Types of living cells

There are three types of living cell, two prokaryotes and the eukaryotes:

Prokaryotes (before nucleus): there are two of these – bacteria and archaea, both of which are single-celled organisms without membrane-bound organelles (small organs) with specialist functions such as a nucleus, mitochondria, etc.:

- **Bacteria** (*bacterium* – singular) very similar to archaea in size and shape;
- **Archaea** (*archaeon* – singular) much of its metabolism is more similar to eukaryotes than bacteria.

Eukaryotes: have well-defined membrane-bound nuclei and organelles and form all the multicellular organisms on earth. Lane (2015) has proposed that these developed through endosymbiosis in which these complex cells arose from a unique merger of a bacterium into an archaeon resulting in the first eukaryote from which all others evolved. The bacterium replicated, transferred much of the surplus DNA (genetic material – see later in this chapter) to the host archaeal chromosomes, and developed into the mitochondria. The major benefit is the large amount of Adenosine Triphosphate (ATP) available from the mitochondria thus increasing the energy available and quantity of protein formed.

The human (mammalian) cell is a complex structure able to carry out all the functions required to maintain cell life by making its contribution to homeostasis through the activities of the different organelles (small organs) within the cell. Figure 2.1 illustrates a generic cell filled with liquid cytoplasm and the range of organelles. These are surrounded by the cell (plasma) membrane.

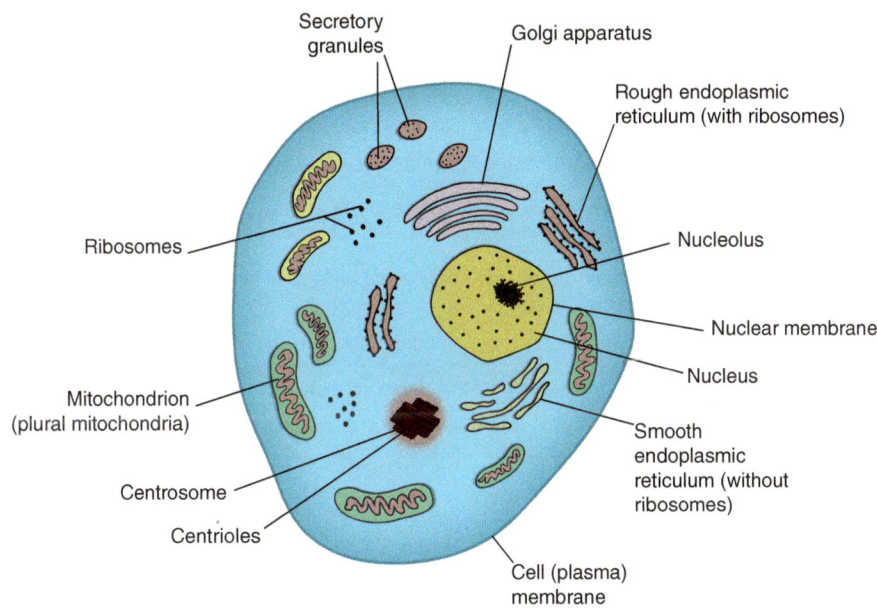

Figure 2.1 A human cell

Cell survival

Each individual cell needs the right conditions in order to continue to survive. It requires a constant supply of energy (primarily glucose), oxygen and water to enter the cell. However, it is also necessary for waste products to leave the cell; these include carbon dioxide, ammonia and heat. Alongside the substances that enter and leave the cell, cell survival depends on a number of key conditions, namely, core temperature, correct pH (7.35–7.45) and appropriate electrolyte balance.

Table 2.1 identifies the different organelles and their functions within the cells of the body.

The nucleus of the cell is unique to eukaryotes and contains the genetic material of the individual, which determines their characteristics. We are going to consider the nucleus first, including how it carries genetic information and how the cells divide.

Table 2.1 The functions of cell organelles

Organelle	Functions
Cell (plasma) membrane (double membrane)	Controls movement of substances, ions and nutrients into, and waste products out of, the cell thus determining composition of cytoplasm (cell contents excluding organelles)
	Response to external stimuli
Nucleus (double membrane)	Contains genetic information within DNA of chromosomes of the cell; provides the template (RNA) for protein formation. Controls activities of the cell
Nucleolus	Made of protein and RNA and is in the nucleus. Synthesises and assembles ribosomes, which leave the nucleus and enter the cytoplasm
Mitochondria (double membrane)	The 'power houses' of the cell. Glucose used as fuel to create ATP for storage of energy, which is released when required
Ribosomes	The protein factories of the cell, translate RNA into protein. When loose in cytoplasm, forms proteins for use within the cell
Endoplasmic reticulum	Directs movements of lipids and proteins through the cell
Smooth	Synthesises lipids and steroid hormones
Rough	Combined with some ribosomes – synthesises proteins for 'export' from the cell
Golgi body (apparatus)	'Packages' proteins within membrane as vesicles, stored, then exported through cell (plasma) membrane
Lysosomes	Contain enzymes that break down unneeded large molecules which are recycled or excreted from cell
	One type digests foreign bodies such as microbes
Cytoskeleton	Microfilaments and microtubules form an internal framework of the cell allowing movement
	Centrosome and centrioles: play important role in cell division

THE NUCLEUS

The nucleus is surrounded by a double membrane similar in structure to the cell membrane (Table 2.1). The nucleus contains the genetic information that determines the constitution of the body in the 46 chromosomes (23 pairs). The nucleus also contains the nucleolus, which is involved in the formation of ribosomes in the nucleus that are then moved out into the cytoplasm of the cell through the nuclear pores.

Chromosomes

Humans have 23 pairs of chromosomes, which makes 46 chromosomes in total. The nucleus contains 44 autosomes (i.e. not sex chromosomes) and two sex chromosomes. Men have an X and a Y sex chromosome and women have two X sex chromosomes. Normally, each body cell contains the complete (diploid) number of 46 chromosomes (23 from each parent). However, gametes, which are the sex cells (i.e. ova and sperm) are composed of a haploid set of 23 chromosomes.

UNDERSTAND

Diploid: two complete sets of chromosomes.

Haploid: one complete set of chromosomes present in gametes.

UNDERSTAND

The gametes are the specialist cells formed in the reproductive organs that combine at fertilisation to form the zygote (the fertilised egg). These differ in the two genders:

Female: Ovum (pl. ova);

Male: Spermatozoon (or sperm) (pl. spermatozoa).

Chromosomes differ in size and, when being examined, a cell is prepared so that the chromosomes are spread out in a smear (Figure 2.2a). The individual chromosomes are then cut out, paired up and laid out in order of size as a karyotype (Figure 2.2b) with the 22 pairs of autosomes followed by the two sex chromosomes at the end. As explained earlier, the female karyotype has two X chromosomes, while the male has one X and one Y chromosome in each cell. Each pair of autosomes are called homologous chromosomes and are the same size and carry comparable genes for the same characteristics. The genes may not be identical and allele is the term used for an alternative form of the same gene located at the corresponding position on homologous chromosomes.

(a)

(b)

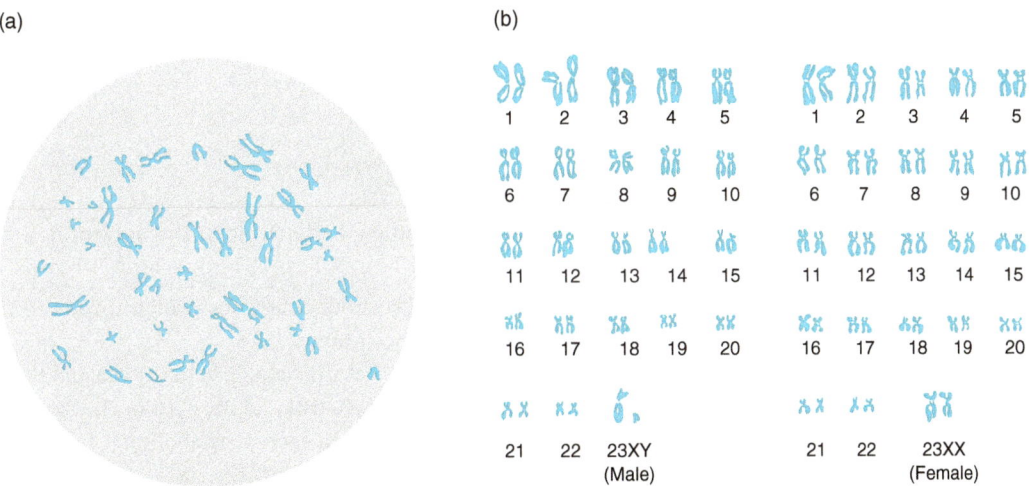

Figure 2.2 The human chromosome (a) Chromosome smear (b) Human karyotype

Deoxyribonucleic acid (DNA)

Chromosomes are composed of Deoxyribonucleic Acid (DNA). DNA acts as the genetic code that provides instructions for the formation of RNA (Ribonucleic Acid), which is transported out of the nucleus and provides a code for the formation of proteins (in collaboration with ribosomes). The proteins may form part of the structure of the cell or the matrix surrounding it, or are enzymes, which act as catalysts for the range of chemical reactions in the cell.

Chromosomes are formed of two DNA strands with backbones of alternating sugar (deoxyribose) and phosphate molecules connected by pairs of nitrogen-containing bases (Figure 2.3a), two purine (adenine and guanine) and two pyrimidine (thymine and cytosine); a purine always connects with a pyrimidine. Adenine joins with thymine by two chemical bonds, and guanine connects with cytosine by three chemical bonds. The two chains form a double helix (Figure 2.3b) described by Watson and Crick in 1953.

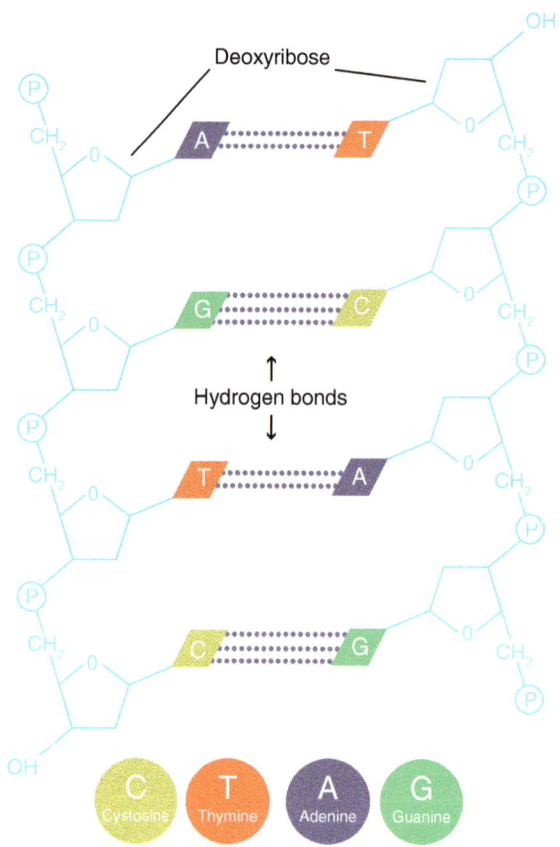

Figure 2.3 (a) DNA structure – The sugar-phosphate backbone and pairing of the nucleotides

In forming the chromosomes, the DNA helices are further coiled around histones (proteins) and coiled and folded yet further to create the demonstrable thickness of the chromosomes.

The chromosomes in the nucleus of the cell contain about 99.9% of the total DNA in the cell and hold information, half from each parent, determining the individual characteristics of the person. The remaining 0.1% of DNA is in the mitochondria (see below) and is concerned with energy metabolism (Chapter 9). Under all normal circumstances this is handed down only from the mother as the spermatozoon (male gamete) loses its tail and mid-piece (in which the mitochondria are arranged) in the process of entering and fertilising the ovum. Thus, all mitochondria in the body develop from those in the ovum.

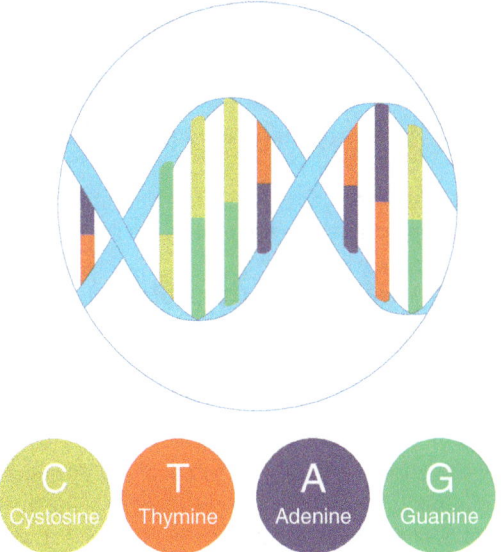

Figure 2.3 (b) DNA structure –DNA helix

Protein formation

We have already indicated that the DNA in the chromosomes carries the genetic material that determines the individual's characteristics. It does this by acting as a template for the formation of the proteins of the body from the amino acids, the nutrients absorbed into the body after the breakdown of proteins in the diet (Chapter 8). This determines both structure and function of the components of the body. The process occurs in the two main stages illustrated in Figure 2.4.

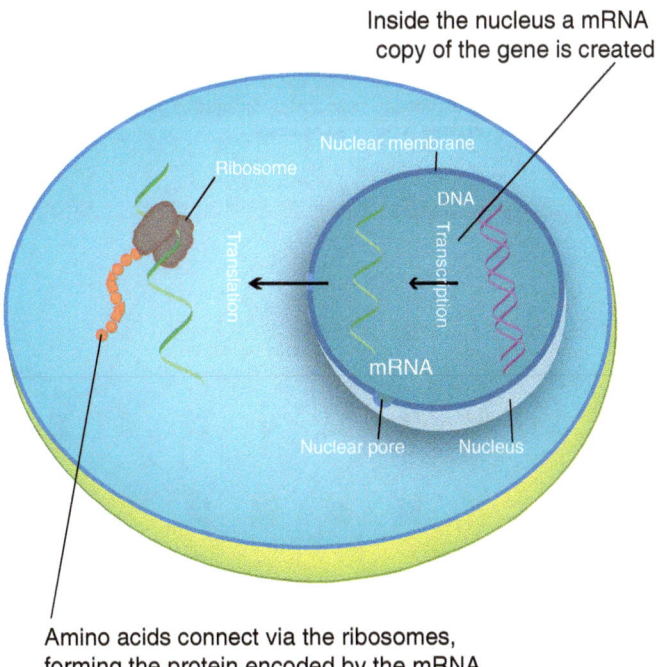

Figure 2.4 Transcription (to RNA) and translation (to proteins)

— **GO DEEPER** —

Transcription of DNA to Ribonucleic Acid (RNA)

The bases within DNA are linked by weak bonds, which allow separation of the DNA strands to permit formation of messenger RNA (mRNA) by transcription (Figure 2.5). The RNA is a single-strand molecule similar in structure to DNA except that the deoxyribose is replaced by ribose, and thymine is replaced by uracil. Each base in the RNA is complementary to the base in the DNA.

Translation (to proteins)

The bases in the DNA transcribed into RNA act as a code in which each group of three bases is code for a specific amino acid. The strand of messenger RNA (mRNA) is the template for translation of this code into a chain of amino acids forming a polypeptide chain which becomes a protein. The mRNA is carried out of the nucleus and joins with a ribosome (see below) in the cytoplasm where the translation takes place (Figure 2.6). Each group of three bases (called a codon) is translated into one amino acid picked up by transfer RNA (tRNA). Figure 2.7 shows the genetic code and illustrates how a number of codons can code for a single amino acid, or can code to stop or start the formation of the chain of amino acids.

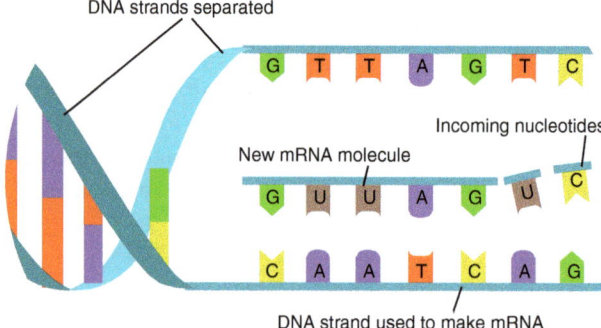

Figure 2.5 Formation of mRNA from DNA

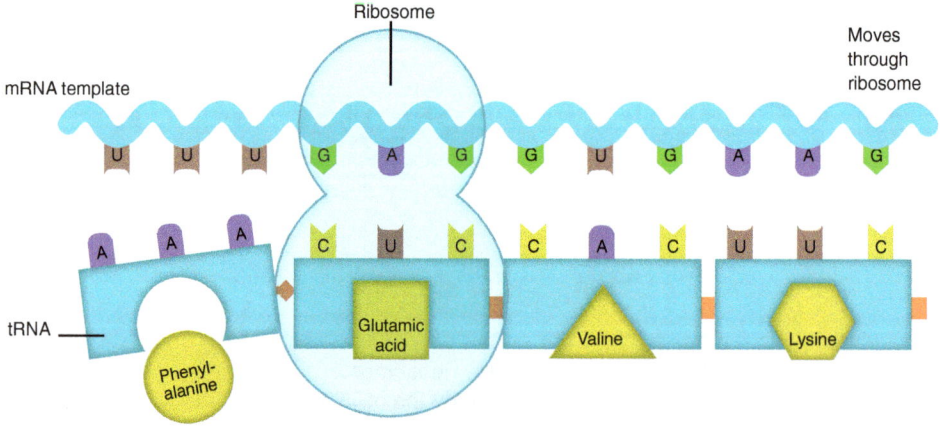

Figure 2.6 Translation of mRNA into polypeptide chain

	U	C	A	G	
U	UUU Phenylalanine	UCU Serine	UAU Tyrosine	UGU Cysteine	U
	UUC Phenylalanine	UCC Serine	UAC Tyrosine	UGC Cysteine	C
	UUA Leucine	UCA Serine	UAA** Stop Codon	UGA** Stop Codon	A
	UUG Leucine	UCG Serine	UAG** Stop Codon	UGG Tryptophan	G
C	CUU Leucine	CCU Proline	CAU Histidine	CGU Arginine*	U
	CUC Leucine	CCC Proline	CAC Histidine	CGC Arginine*	C
	CUA Leucine	CCA Proline	CAA Glutamine	CGA Arginine*	A
	CUG Leucine	CCG Proline	CAG Glutamine	CGG Arginine*	G
A	AUU Isoleucine	ACU Threonine	AAU Asparagine	AGU Serine	U
	AUC Isoleucine	ACC Threonine	AAC Asparagine	AGC Serine	C
	AUA Isoleucine	ACA Threonine	AAA Lysine	AGA Arginine	A
	AUG Methionine (start codon)	ACG Threonine	AAG Lysine	AGG Arginine	G
G	GUU Valine	GCU Alanine	GAU Aspartate	GGU Glycine	U
	GUC Valine	GCC Alanine	GAC Aspartate	GGC Glycine	C
	GUA Valine	GCA Alanine	GAA Glutamate	GGA Glycine	A
	GUG Valine	GCG Alanine	GAG Glutamate	GGG Glycine	G

Figure 2.7   The genetic code

*Arginine is one example of many amino acids identified by more than one codon

**Codons where amino acids not coded for and translation stops

The nucleolus

The nucleolus is the largest structure in the nucleus of eukaryotic cells. It is composed of proteins and RNA and its main function is to synthesise ribosomes and assemble them for export into the cytoplasm of the cell. It is also involved in how the cell responds to stress.

OTHER ORGANELLES

The cell or plasma membrane is double-layered and the nucleus and mitochondria are covered with similar membranes. Most of the other organelles are surrounded or formed from single-layered membranes.

Cell or plasma membrane

The boundary of each cell is a double-layered lipid membrane, the cell or plasma membrane, composed of phospholipids (fatty molecules with a phosphate group), proteins and carbohydrates arranged in a mosaic structure (Figure 2.8) (Chapter 8 includes the structure of different nutrients). The phosphate ends of phospholipids are attracted to water (hydrophilic) and they face outwards from the cell membrane while the fatty acid tails are water repellent (hydrophobic) and face each other in the centre of the membrane, preventing passage of all but very small molecules.

Some protein molecules are incorporated into one layer of the membrane while some pass through both layers and facilitate transport across the membrane. Proteins compose about 50% of the structure of these membranes and are particularly important in transport of substances across the cell membrane. Proteins in combination with carbohydrates on the outside of the cell membranes can act as receptors by having a specific binding site where hormones or other substances can link. This initiates other actions in the cell.

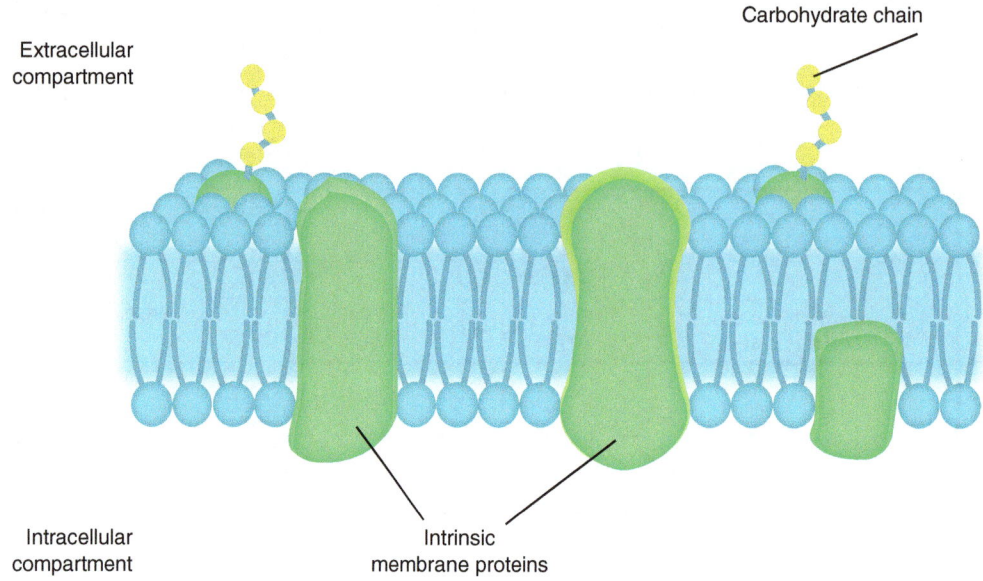

Figure 2.8 Cell membrane

The composition of the cytoplasm (Intracellular Fluid (ICF)) is very different from that of the Extracellular Fluid (ECF) surrounding it (Table 2.2). The concentration gradient of a substance sets the initial parameter, with substances moving from high to low concentration (Chapter 11). However, the plasma membrane plays a vital role in maintaining the differences between the interior and exterior of the cell through regulating the movement of substances in and out of the cell.

Table 2.2 Ionic composition of cytoplasm (ICF) compared to extracellular fluid (ECF) (these values can vary depending on which type of cell is being looked at and results can vary somewhat in different laboratories)

Element	Ion	ICF	ECF
Sodium	Na⁺	15	141
Potassium	K⁺	140	4
Calcium	Ca²⁺	0.0001	2.5
Chloride	Cl⁻	8	103
Bicarbonate	HCO₃⁻	15	25

Transport across the cell membrane

The cell membrane controls how substances can move in and out of the cell and these are discussed in detail in Chapter 11. They include passive movement, in which molecules pass down a concentration gradient and no energy is required, and active movement, in which energy is required to move molecules against resistance.

- **Carrier proteins** facilitate the movement of specific molecules. These are proteins which pass through the cell membrane and provide a site recognised by specific molecules that link to it. The protein then changes conformation and releases the molecule on the other side of the membrane (Figure 2.9). The proteins assist movement of substances by what is known as carrier-mediated transport requiring energy.

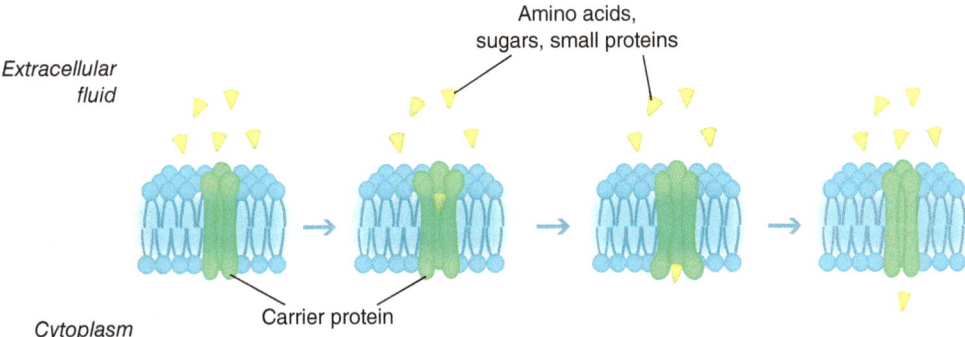

Figure 2.9 Carrier proteins

- **Ion channels** (Figure 2.10) determine the concentration of fluid and inorganic ions (which have a positive or negative charge) within the cytoplasm, including sodium (Na$^+$), potassium (K$^+$), chloride (Cl$^-$), bicarbonate (HCO$_3^-$) and calcium (Ca^{2+}). The distribution of ions within the Intracellular Fluid (ICF) and outside the cell in the ECF results in an electrical difference across the cell membrane (the electrical potential) with the inside of the cell being more negative than the outside.

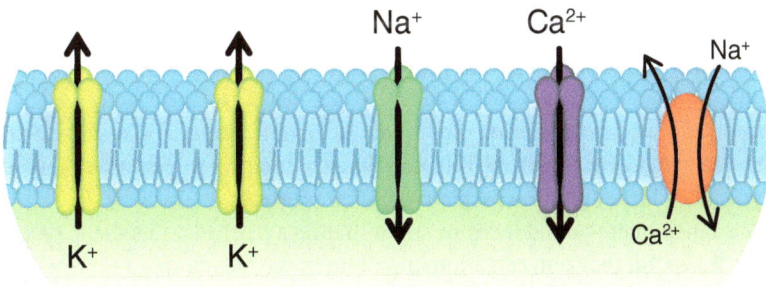

Figure 2.10 Ion channels

- **Exocytosis and endocytosis:** these processes enable large molecules that cannot pass though the cell membrane to move between the ICF and ECF. Endocytosis enables molecules to enter the cell by engulfing it (Figure 2.11a). Exocytosis enables the contents of a vesicle formed from the Golgi apparatus to fuse with the cell membrane and the contents are released from the cell (Figure 2.11b).

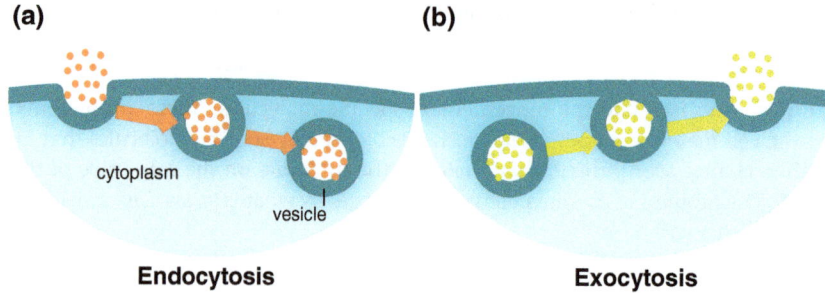

(a)   **(b)**

cytoplasm

vesicle

Endocytosis   **Exocytosis**

Figure 2.11 Endocytosis and exocytosis

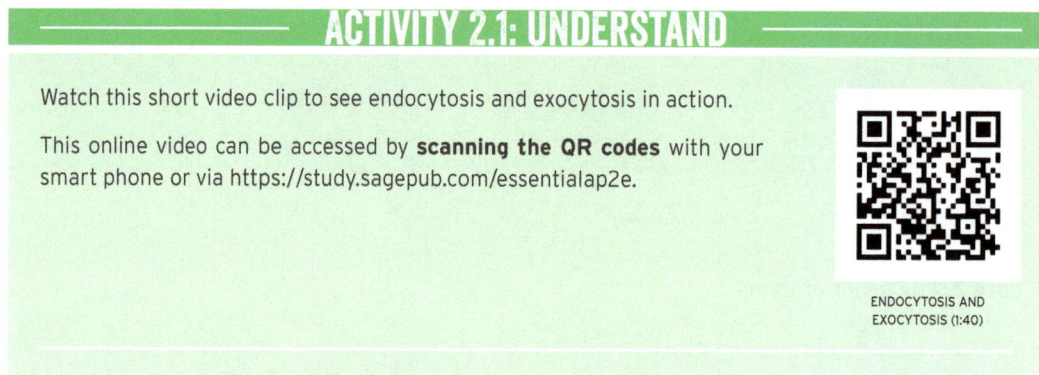

ACTIVITY 2.1: UNDERSTAND

Watch this short video clip to see endocytosis and exocytosis in action.

This online video can be accessed by **scanning the QR codes** with your smart phone or via https://study.sagepub.com/essentialap2e.

ENDOCYTOSIS AND
EXOCYTOSIS (1:40)

Receptors and the cell membrane

The cell membrane also plays an important role in signalling between cells through the action of receptors. These are proteins embedded in the plasma membrane surrounding the cell, within the cell or in the nucleus that bind to chemicals from outside the cell called ligands. Receptors detect the specific molecules (ligands) to which they are sensitive, for instance a hormone (Chapter 7), neurotransmitter (Chapter 5), small protein, a drug or part of an infectious agent (bacterium or virus) and modify the activity of the target cell.

The receptors modify the activity of the target cells in one of four ways:

1. By opening an ion channel and thus modifying the action potential (i.e. the electrical difference across the membrane) which in nerve cells cause nerve impulses to pass along the cell (Chapter 5). In other types of cells they activate particular cellular processes.
2. By activating a membrane-bound receptor and initiating a particular metabolic pathway.
3. By activating a receptor that activates another protein in the membrane (a G protein). The G protein can have effects on the cell in one of two ways: either through influencing activity of the ion channels or by influencing the concentration of second messengers.
4. By activating an intracellular receptor to adjust the transcription of specific genes.

Receptors are discussed again in Chapters 5 and 7.

Major Histocompatibility Complex (MHC)/Human Leucocyte Antigen (HLA)

The Major Histocompatibility Complex (MHC) proteins (known as Human Leucocyte Antigens (HLAs) in humans) are based in the cell membrane and play an important role in protecting against external agents entering the body (e.g. bacteria or viruses) or abnormalities developing in cells (e.g. cancer). The defence of the body depends on the immune system (Chapter 13) being able to recognise the difference between 'self' and 'non-self' cells. The specific body cell proteins involved in this recognition are known as antigens.

During development the fetus develops antigens on its cells which act as markers for the 'self'. However, each cell modifies itself in response to infection or by changes in the cell, becoming, for example, malignant, and binds fragments derived from the breakdown of these pathogens (infective agents). The cells of the body's immune system recognise these not-self antigens and destroy these cells (Chapter 13).

Mitochondria

The mitochondria (singular mitochondrion) are surrounded by a bilipid layer similar to the plasma membrane and they act as the power houses of the cell. They metabolise nutrients to produce ATP (Adenosine Triphosphate) – the energy store of the cell – which is used to power the various cell activities (Chapter 9).

Mitochondria are positioned in cells according to the particular cell function and where energy is required. For example, in a cardiac muscle cell where energy is required all the time, the mitochondria are clustered near the muscle fibres.

Ribosomes

These granules are formed by the nucleolus within the nucleus of the cell and pass through the nuclear pores into the cytoplasm. They are composed of RNA and protein and those that are either loose or in groups within the cytoplasm form proteins for use within the cell. Other ribosomes combine with Endoplasmic Reticulum (ER).

Endoplasmic reticulum (ER)

ER is a cell-wide network of membrane which provides a surface on which lipids and proteins can be formed and transported round the cell.

GO DEEPER

Cellular transport

ER forms transport vesicles to carry the substances formed between the ER, plasma membrane, Golgi apparatus and lysosomes. Special transfer proteins carry the same substances to mitochondria and lysosomes.

- **Smooth ER** is involved in the synthesis of steroid molecules (a type of lipid) in cells where these are produced, and in calcium storage. This is particularly important in muscle cells in which smooth ER is known as sarcoplasmic reticulum.
- **Rough ER** is combined with ribosomes and thus appears rough under the microscope. The ribosomes with the ER form proteins such as enzymes or hormones which are exported from the cell for use elsewhere in the body.

Golgi apparatus or body

The Golgi apparatus is more important and larger in cells that have a secretory role. It is composed of stacks of flattened sacs of membrane that receive proteins and lipids from the ER and package them into secretory vesicles. These are stored and moved to the plasma membrane when needed and exported by exocytosis.

Lysosomes

These are one of the types of secretory vesicles created by the Golgi apparatus and essentially are bags of enzymes. These break down large molecules that are no longer required into smaller fragments, which may be reused or eliminated from the cell. Those in white blood cells contain enzymes which break down microbes.

Cytoskeleton

As indicated in Table 2.1 this forms a framework for the cell and enables movement. The centrosome is the core from which single filaments radiate and enable movement of vesicles and organelles. The centrosome is near the nucleus and has two cylinders, centrioles, at right angles to each other. During cell division these separate and move to opposite poles of the nucleus (see Cell Division below).

CELL DIVISION

From the formation of the zygote, cell division continues to occur throughout life. There are two types of cell division: one which forms the gametes, called meiosis, and the other is mitosis, the cell division involved in growth and development.

Meiosis

Meiosis occurs in relation to reproduction and is the division that occurs to form the gametes – sperm or ova in preparation for fertilisation and formation of the zygote which develops into the fetus. As already stated, the gamete from each parent normally contains half the full number of chromosomes (i.e. 22 autosomes and one sex hormone – 23 in total).

Figure 2.12 shows the stages involved in meiosis. Initially the DNA replicates so that each chromosome in a pair of homologous chromosomes consists of two chromatids. Some of the DNA often swaps over (chromosomal crossover) between the two chromosomes as shown in the different colours in the chromatids in meiosis I. The pairs of chromosomes separate into two daughter cells each with half the original number of chromosomes, but with the sister chromatids remaining together. During meiosis II the chromatids separate into two separate cells so that from the original single cell, there are now four daughter cells. These can all be somewhat different in their DNA structure. These cells then mature into the male sperm or a female ovum with three polar bodies (Figure 2.13), which usually die.

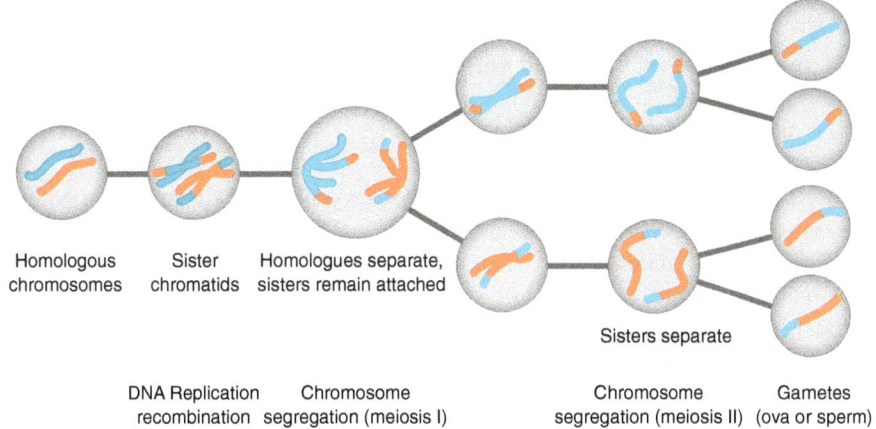

Homologous chromosomes   Sister chromatids   Homologues separate, sisters remain attached   Sisters separate

DNA Replication recombination   Chromosome segregation (meiosis I)   Chromosome segregation (meiosis II)   Gametes (ova or sperm)

Figure 2.12   Meiosis in formation of gametes

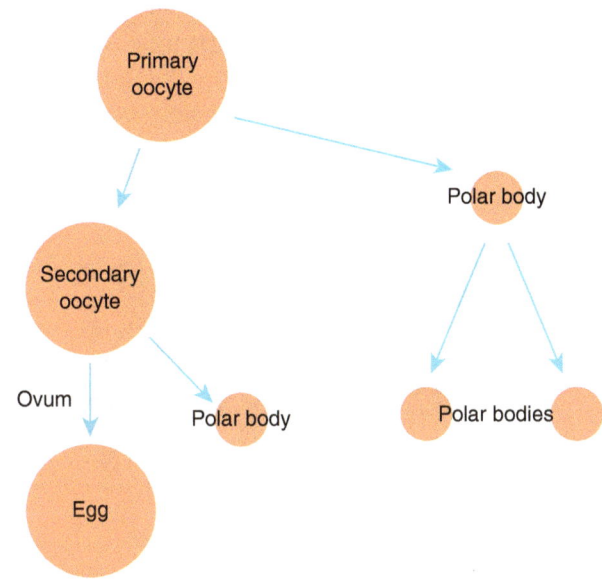

Figure 2.13   Formation of polar bodies

ACTIVITY 2.2: UNDERSTAND

Go online and watch the following video clip to see meiosis in action.

This external video link can be accessed by **scanning the QR code** with your smart phone camera or via https://study.sagepub.com/essentialap2e.

MEIOSIS (1:48)

Cell cycle

The cell cycle consists of a series of events in which cellular components are doubled and then accurately segregated into daughter cells. It consists of two phases, interphase and mitosis (M) (Figure 2.14).

Interphase is divided into three phases:

1. Gap phase 1 (G_1) – the cell grows in size, organelles are copied and the molecular building blocks required for later in the cycle are made.
2. Synthesis phase (S) – a complete copy of the DNA in the nucleus is synthesised and the centrosome is also duplicated.
3. Gap phase 2 (G_2) – cell continues to grow in size, makes proteins and the organelles begin to organise their contents in preparation for mitosis.

DNA replication is confined to the synthesis phase. G_1 and G_2 separate the synthesis and mitosis phases. G_1 and G_2 are periods where cells obtain mass, integrate growth signals, organise a replicated genome and prepare for chromosome segregation (Barnum and O'Connell, 2014).

After each gap phase there are checkpoints that act as a surveillance mechanism to detect that the cell is the appropriate size, that replication has occurred, and that there is integrity of the chromosomes, through accurate segregation at mitosis. Checkpoints are observed in G_1 (transition from G_1 to S), G_2 (transition from G_2 to M) and mitotic spindle.

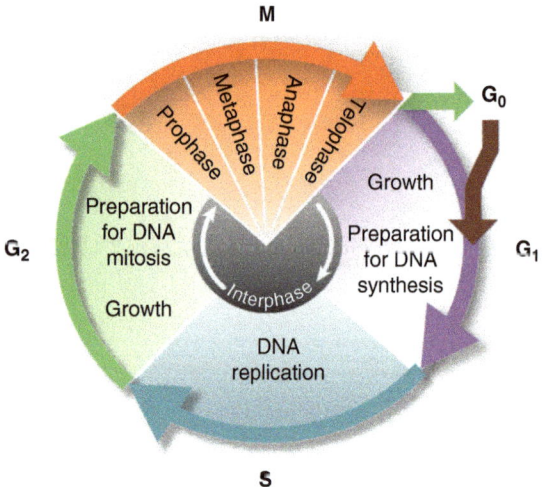

Figure 2.14 Cell cycle

GO DEEPER

Cell cycle checkpoints

- **G_1 checkpoint** - at this point the cell must decide whether to divide or not. The cell is checked for size, nutrients, molecular signals (e.g. growth factors) and DNA damage. If the cell does not receive the appropriate information at this checkpoint it may leave the cell cycle and enter a resting phase known as G_0. Some cells may stay in this phase permanently while others remain in this phase until the conditions required improve and then they continue to divide.

- **G₂ checkpoint** – this checkpoint occurs before M phase to ensure that the cell produces healthy daughter cells. The checkpoint will look for any damaged DNA (DNA integrity) and whether the DNA was copied completely during S phase (DNA replication). If any errors or damage are identified the cell will pause at G₂ phase checkpoint allowing the cell to repair damaged DNA or to complete DNA replication. If the cell is unable to repair the DNA or complete the replication it may undergo apoptosis, (programmed cell death) thereby ensuring damaged DNA is not passed on to daughter cells.
- **Spindle checkpoint** – this occurs at the M checkpoint and examines the cell to ensure that all sister chromatids are attached to the spindle microtubules correctly. The chromosomes must be firmly attached to at least two spindle fibres from opposite poles otherwise the cell cycle cannot proceed.

Mitosis

After the merger of the ovum and the sperm in fertilisation, mitosis, the alternative type of cell division, occurs as the embryo grows and differentiates into the different cells and tissues making up the body. Mitosis is when a cell divides into two genetically identical daughter cells. Continued cell division occurs as the body develops through the stages of development discussed in Chapter 17. It also enables tissue repair.

Mitosis consists of five main stages shown in Figure 2.15 and described below.

- **Interphase:** occurs between cell divisions and for most of this time the cells continue to function. During part of this phase, each chromosome duplicates to form two chromatids tightly coiled around each other.
- **Prophase:** the two chromatids become visible. The centrioles separate and go to each end of the cell with the mitotic spindle of microtubules between. The nuclear membrane disappears.
- **Metaphase:** the chromatids line up along the equator of the spindle attached by their centromeres (where the DNA is constricted and the two chromatids join).
- **Anaphase:** the microtubules forming the spindle begin to contract and draw the two chromatids of a chromosome apart to the ends of the cell.
- **Telophase:** The mitotic spindle disappears, the chromosomes reform and the nuclear membrane reforms.

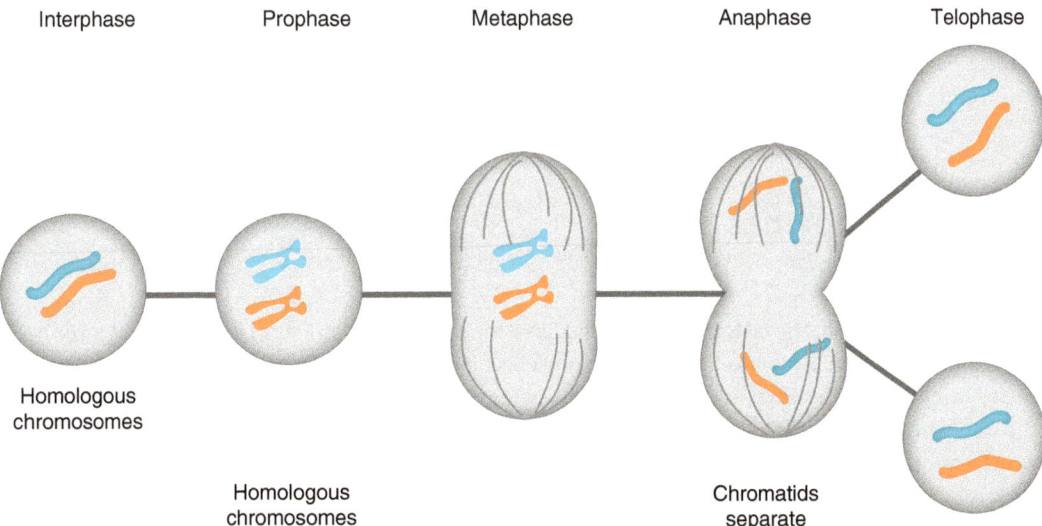

Interphase   Prophase   Metaphase   Anaphase   Telophase

Homologous chromosomes

Homologous chromosomes

Chromatids separate

Figure 2.15 Mitosis

The original cell then divides into two daughter cells by the division of the cytoplasm and cell membrane and may re-enter the cell cycle. However, many specialised cells remain in interphase and undergo no further cell division.

ACTIVITY 2.3: UNDERSTAND

Go online and watch the following video clip to see mitosis in action.

This external video link can be accessed by **scanning the QR code** with your smart phone or via https://study.sagepub.com/essentialap2e.

MITOSIS (6:10)

GO DEEPER

Cell division and ageing

At the end of each DNA chain is a section called a telomere – these protect the ends from damage and consist of a repeated chain of nucleotides (in vertebrates this sequence is TTAGGG). These shorten at each cell division, placing a limit on the number of cell divisions that can occur, resulting in ageing (Allsopp et al., 1995). Human fetal cells divide between 40 and 60 times before the telomeres become too short for further division and the cells become quiescent (dormant) and eventually die. This ageing of cells correlates with overall ageing of the body but there are certain conditions of premature ageing in which this takes place more rapidly than normal (e.g. progeria, a rare genetic condition that causes a child's body to age rapidly).

The enzyme telomerase can lengthen the telomere and result in continued division. This is linked with development of cancer; metastases (i.e. secondary growths) often show telomerase activity (Jafri et al., 2016).

Although rare, errors in DNA during cell division can occur and, without repair, it is suggested that 16,000 nucleotide errors could occur in a single cell (Westman, 2006). A number of different repair mechanisms exist to minimise the deleterious effects of such errors.

TYPES OF BODY CELLS

Introduction

Following fertilisation, the cells initially created have the potential to differentiate into any type of body cell – they are known as stem cells and, while most become fixed as a particular type of cell, some retain their flexibility and continue in certain tissues as stem cells. Cells with the ability to develop into any of the cells which make up the body are known as pluripotent stem cells (e.g. embryonic stem cells); those which are more limited but able to form more than one cell type are known as multipotent (e.g. adult stem cells and cord blood cells).

All body cells contain the same set of instructions, the DNA, but the expression of those genes is altered to produce the different types of cell. Differentiation is not thought to involve loss of DNA but occurs by modifying the expression of genes either through epigenetic changes to the DNA (Chapter 3) or by environmental factors extrinsic to the cell such as specific small molecules, secreted proteins from other cells, temperature and oxygen. The changes in gene expression can turn genes on or off or adjust the level of transcription of RNA and formation of proteins (Ralston and Shaw, 2008).

Cell differentiation produces five main types of body cell within the different tissues of the body:

- Blood and lymph
- Connective tissue
- Nervous tissue
- Muscle tissue
- Epithelial tissue.

Nervous and muscle tissue are both excitable tissues.

Blood and lymph

These differ from the other tissues in that the cells are not combined to form solid structures but are dispersed and transported in liquid. Table 2.3 outlines the major information about formation and functions of these blood and lymph tissues, which are sometimes included as connective tissues.

The cells within these systems are formed from pluripotent stem cells, originally present in the embryo and capable of forming any type of body cell, which create the two multipotent stem cells from which the range of blood cells are formed (Figure 2.16).

Table 2.3 Components of blood and lymph

	Constituent	Formation	Function
Blood		Carried in blood vessels: arteries, capillaries and veins	Transport medium
	Plasma	Water containing ions, plasma proteins, nutrients and waste products	Supplies all organs and tissues Carries nutrients, waste products, blood gases, blood cells
	Erythrocytes (red blood cells)	Formed in bone marrow, require iron and vitamin B_{12}. Have no nucleus. Lifespan 120 days	Transport O_2 to cells of the body, and some CO_2 to the lungs in haemoglobin of red blood cells (RBC)
	Platelets (thrombocytes)	Cell fragments, formed in bone marrow	Initiate haemostasis (blood clotting)
	Leucocytes (white blood cells)	Formed in bone marrow or lymphoid tissue	See below
Lymph		From tissue fluid from plasma, carried in lymphatic vessels. Returns fluid to blood circulation	
	Lymphatic fluid	Clear fluid formed in the tissues	Transports substances between blood and body cells Carries leucocytes
	Leucocytes		Combat infection, foreign bodies, malignant cells

(Continued)

Table 2.3   (Continued)

Constituent	Formation	Function
Agranulocytes:		
Monocytes	Formed in bone marrow	Phagocytic - engulf bacteria
Lymphocytes	Formed in bone marrow, lymphoid tissue, thymus, spleen	Produce antibodies (B Lymphocytes)
Granulocytes:		Produce histamine and heparin
Basophils		
Eosinophils	Formed in bone marrow	Congregate at inflammation, antihistamine properties
Neutrophils		Phagocytic, become macrophages when migrate to tissues

All organs of the body are supplied with blood through the circulation (Chapter 12). Fluid from the blood enters the tissue spaces to provide the cells with the necessary nutrients and remove waste products. Some of the tissue fluid re-enters the circulation via the capillaries. Excess tissue fluid enters the lymph channels, passes through lymph nodes where it acquires additional lymphocytes, and returns to the circulation at the subclavian vein (Chapters 12 and 13).

Connective tissue

These tissues are also known as structural tissues as they provide structural support for the organs of the body. They vary considerably in appearance, but all consist of cells, inter-cellular substance (matrix) and fibres, the last two being manufactured by the cells. The cells in connective tissue are much more spread out than in other tissues and there is considerably more matrix than in other tissues. The matrix and fibres vary according to the type of tissue and its function, major functions being structural support, protection, transport and insulation. The types of cells in connective tissue are given in Table 2.4.

Table 2.4   Cells in connective tissue

Cells	Description	Function
Fibroblasts: large flat cells	Produce extracellular matrix and fibres: • Collagen - minimal stretch coarse fibres in wavy bundles • Elastin - fine branching elastic fibres	Major role in tissue repair
Adipocytes: vary in size and shape	Singly or grouped in most connective tissues abundant in adipose tissue	Major energy store as fat in adipose cells
Macrophages: irregular shape	Some fixed, some mobile. Phagocytic - engulf and digest cell debris, bacteria, foreign bodies	Part of immune system
Leucocytes: white blood cells	Small numbers in healthy connective tissue	Raised numbers in infection
Mast cells: like basophils	In loose connective tissues and fibrous capsule of some organs, around blood vessels	Histamine, heparin, etc., released after damage

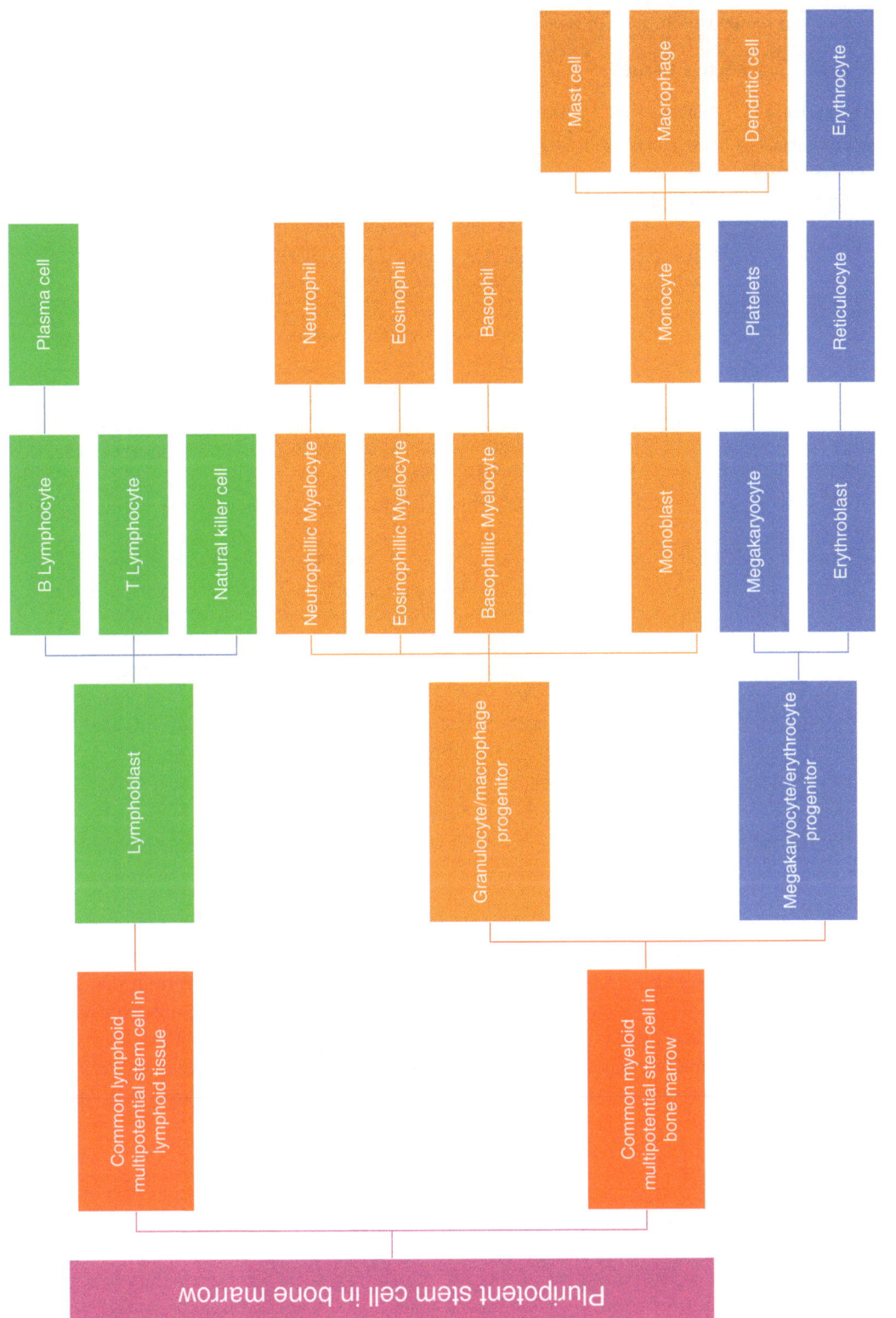

Figure 2.16 Formation of blood cells

The different tissues which these cells can form are specified in Table 2.5.

Table 2.5   Connective tissues

Type of tissue	Subtypes	Description
Loose (areolar) connective tissue	Most generalised type	Connects and supports other tissues through elasticity (elastin fibres) and tensile strength (collagen fibres)
Adipose tissue	White adipose tissue	Is 20-25% of (non-obese) body under skin, stores fat around kidneys and other organs, acts as insulation and energy store
	Brown adipose tissue	Has substantial blood supply. Maintains body temperature by producing considerable heat
Dense connective tissue	Fibrous tissue	Bundles of collagen fibres with little matrix. Forms ligaments to bind bones together. Protective covering for: bones, some organs (e.g. kidney, brain)
		Muscle sheaths (or fascia) become tendons (Chapter 15)
	Elastic tissue	Can stretch and recoil, few cells, much elastic tissue secreted by fibroblasts in organs which need to stretch or change shape (e.g. blood vessels, lungs, trachea)
Lymphoid tissue	Or reticular tissue	See Chapter 13. Internal protection: immune system
Cartilage	Hyaline cartilage	Solid and smooth bluish-white matrix, cells in small groups
		Flexible, supportive and smooth for movement at joints: ends of long bones, costal cartilages join ribs to sternum, parts of airway
	Fibrocartilage	Contains dense collagen fibres (see above) within matrix like hyaline cartilage. Tough and flexible: intervertebral discs, in knee, hip and shoulder joints, forms ligaments joining bones
	Elastic fibrocartilage	Consists of yellow elastic fibres in solid matrix. Supports and maintains shape (e.g. lobe of ear)
Bone	Compact bone	A very dense hard tissue (Chapter 15)
	Cancellous (spongy) bone	A spongy bone tissue (Chapter 15)

Adipose tissue

Adipose tissue has long been known as an energy source for the body and as having an important role in maintaining body temperature. Adipose tissue is composed of adipocytes, specialist cells that store fat. There are two types of adipose tissue, white and brown. White fat contributes to temperature regulation by reducing heat loss, and brown fat, with a considerably greater blood supply, produces heat through metabolism. The recognisable deposits of brown fat are particularly important in maintaining body temperature in babies as they generate heat and until fairly recently were not thought to exist in adults.

GO DEEPER

Brown adipose tissue

It has now been confirmed that precursor cells for brown adipose tissue are present in the supraclavicular regions (and possibly in other areas) of adults and that, under cold conditions, these can convert into active brown adipose tissue. Lean men or those working in cold conditions have more brown fat than those not working in such conditions, or those who are obese or overweight (Lee et al., 2011, 2014).

White adipose tissue is more prevalent in adults and also acts as an endocrine organ secreting the hormone leptin which influences food intake by reducing appetite and fat storage (Chapters 7 and 9). Interestingly, we never lose adipocytes, we only regulate the amount of fat stored in them. If there is too much fat for the number of adipocytes, we replicate adipocytes to increase available storage.

Cartilage

Cartilage is firmer than other connective tissues with fewer cells (chondrocytes) within a substantial matrix. Details of the three types are given in Table 2.5 and they vary due to the presence or absence of particular fibres (Figure 2.17).

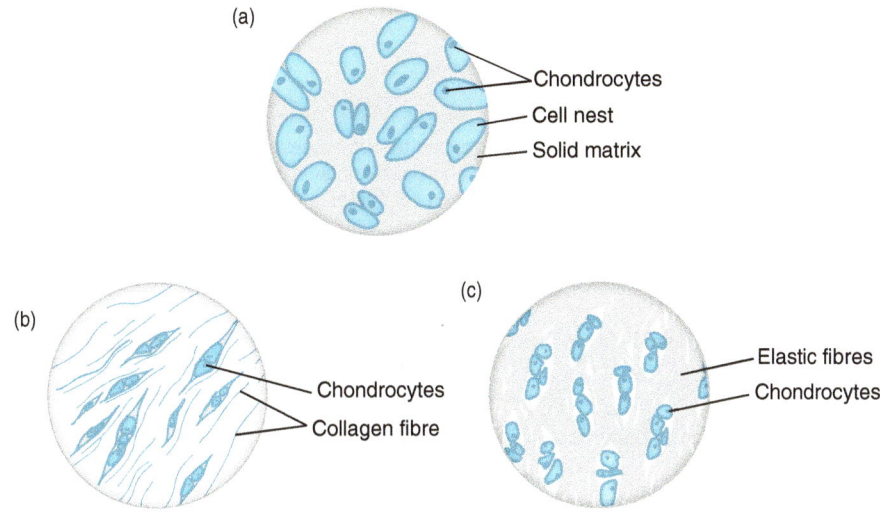

Figure 2.17 Cartilage: (a) hyaline, (b) fibrocartilage and (c) elastic fibrocartilage

Bone

Bone is the main component of the skeleton and the detailed structure of bone and its repair following injury are discussed in Chapter 15. The main functions are to provide the basic framework of the body, protection of organs within spaces in the skeleton, and formation of blood cells.

Nervous tissues

Nerve and muscle are both excitable tissues meaning that the cells respond to chemical, electrical or mechanical stimulation. Nerve cells respond by transmitting a nerve impulse to carry instructions to or from the central nervous system (Chapter 5). Muscle cells respond by contracting (Chapter 15).

Muscle tissue

There are three types of muscle tissue – details are given in Table 2.6. They are discussed in relation to the systems with which they are concerned in the following chapters:

- **Striated:** Chapter 15 – musculoskeletal system
- **Non-striated:** Chapter 8 – gastrointestinal tract
- **Cardiac:** Chapter 12 – cardiovascular and lymphatic systems.

Table 2.6   Types of muscle tissue

	Skeletal muscle	Smooth muscle	Cardiac muscle
Location	Attached to the bones or the skin (facial muscles only)	Found in the walls of hollow visceral organs and blood vessels	Located in the walls of the heart
Appearance	Single, long cylindrical cells	Single, narrow, rod-shaped cells	Branching chains of cylindrical cells
Nucleus	Multiple nuclei; peripherally located	Single nucleus, centrally located	Single nucleus, centrally located
Striations	Yes	No	Yes
Neural control	Voluntary (and involuntary reflexes)	Involuntary	Involuntary
Function	Body movements	Moving food through digestive tract, emptying urinary bladder, changing blood vessel diameter, etc.	Pumping blood through the blood vessels

Epithelial tissue

These tissues cover the body and line cavities, organs and hollow organs, and line glands. Their main functions are:

- protection of organs from damage such as dehydration or the effect of chemical agents or trauma;
- secretion;
- absorption.

Table 2.7   Epithelial tissues of the body

	Type of tissue	Position in body/function
Simple epithelia	Squamous (pavement): exchange of small molecules between compartments	Walls of lung alveoli
		Endothelium of blood vessels
	Cuboidal (cubical)	Small collecting ducts in kidney
	Columnar: secretion or absorption	Large collecting ducts of kidney
		Lining of small intestine
	Ciliated: waft materials along surface	Lining of fallopian tubes and respiratory tract
Stratified epithelium	Pseudostratified columnar ciliated	Lines upper airways
	Stratified squamous: waterproof and keratin barrier to bacteria	Epidermis of skin
	Transitional	Bladder lining
Glandular epithelium	Goblet cells: secrete mucus	Epithelial surface of airways and gut
	Endocrine: secretions enter blood stream	Hormones (e.g. thyroxine, insulin)
	Exocrine: secretions via duct onto surface	E.g. salivary and sweat glands and intestinal lining
Serous membranes	Double layer loose areolar connective tissue lined with simple squamous epithelium	Pleura, pericardium, peritoneum
	Serous fluid between two layers	

There are numerous types of epithelium (Table 2.7) but they share certain characteristics:

- Cells are tightly joined together by specialist cell-to-cell junctions to form a continuous sheet called epithelium (pl. epithelia);
- These cells lie on a basement membrane of connective tissue fibres which supports and separates the epithelium from the underlying tissue;
- Continual cell division takes place to replace any dead and damaged cells which occur as skin and gut lining are continually abraded;
- The epithelial sheets enable transport of substances in a particular direction either into or out of the compartment as functionally necessary.

A range of epithelial tissues and glands are illustrated in Figure 2.18. These include single layer and multiple layer tissues including transitional epithelium, which lines the bladder and can stretch to allow increased storage of urine. The figure of the stratified epithelium of the skin shows how the outer layers become thinned and then fall away. The glandular epithelium consists of unicellular or multicellular glands. Goblet cells secrete mucus, a slippery substance which lubricates the tissues of mucous membranes. The multicellular glands can secrete a number of different substances including sebaceous or intestinal secretions.

UNDERSTAND

Mucus (noun): viscous, slippery substance secreted as a protective lubricant coating cells and glands of the mucous membranes.

Mucous (adjective): means containing, producing, or secreting mucus.

CONCLUSION

This chapter has provided the foundation about human cells and how they work, multiply and differentiate into different types of cell that you can refer back to while you learn about the systems in which these cells function in later chapters. In relation to person-centred practice, this chapter provides some of the basic knowledge for you to understand how the human body works, and how that relates to psychological well-being.

GO DEEPER

Further reading

Peate, I. (2017) 'Anatomy and physiology, 2. The cell and tissues', *British Journal of Healthcare Assistants*, 11(9): 422-26.

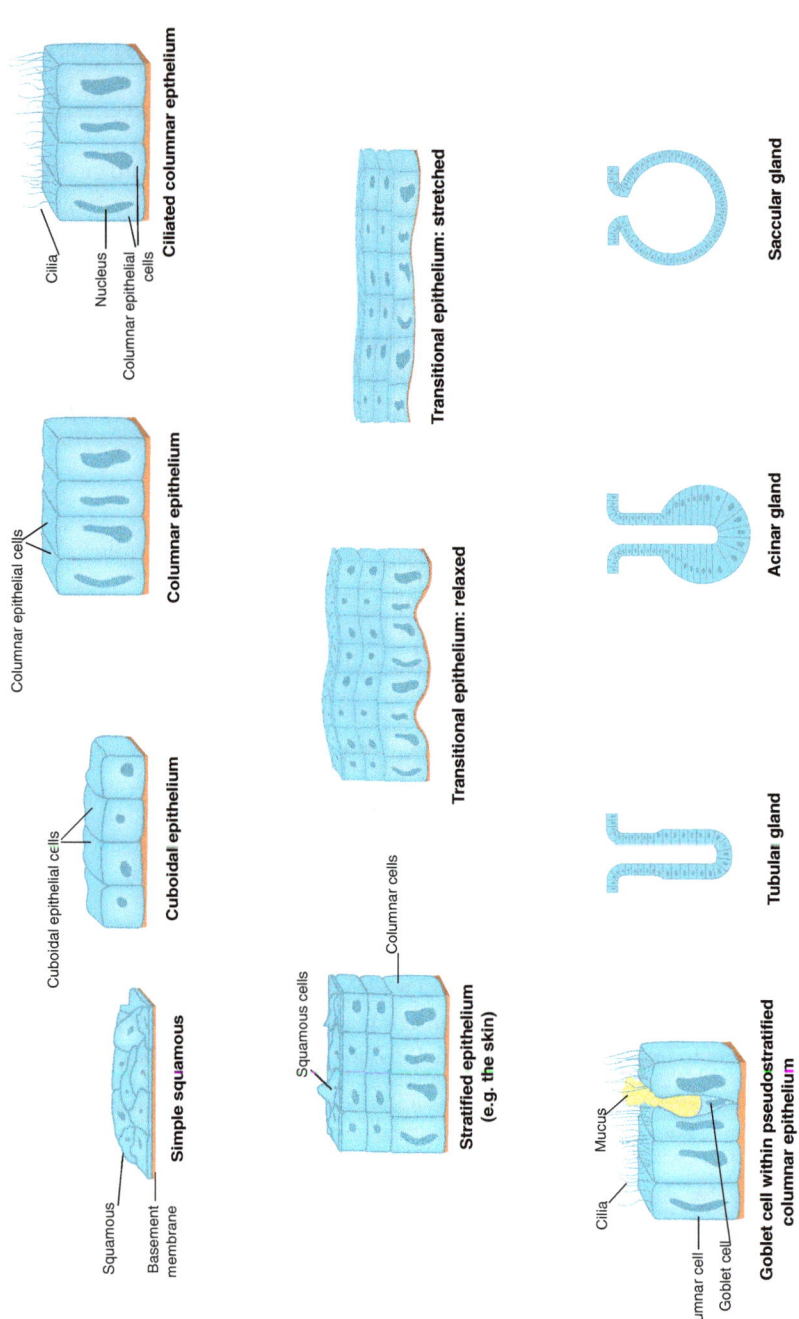

Figure 2.18 Epithelial tissues

- Understanding the structure of the human cell and how it works will help you to understand the human body as a whole.

- There are two types of cell division:

    o Meiosis occurs to prepare the gametes with the haploid number of chromosomes. When a sperm fertilises the egg the zygote then has the diploid number of chromosomes and further cell division is mitosis.

    o Mitosis forms two identical daughter cells and is the key to growth and development. It is linked with cell differentiation to form the different types of cells and tissues forming the body: blood and lymph, connective tissue, nervous and muscle cells, epithelial cells.

- Understanding the function of the different cells of the body and how they interact synergistically will enable you to understand their contribution to homeostasis.

This chapter and the previous one provide an underpinning for the remaining chapters in this book, so it is important to ensure that you have a good understanding of the content. In revising this chapter, it is useful to work through from the beginning. The areas to revise are as follows:

1 All the cell organelles and their functions, including transport across the cell membrane, DNA and RNA.

2 Meiosis – cell division for gamete formation; Mitosis – cell division for growth, development and repair.

3 The different types of cells and tissues: blood and lymph, connective, nervous and muscle (excitable) tissues, epithelial tissues.

4 The role of apoptosis in human development and functioning.

5 How this knowledge can be used in achieving person-centred practice.

In order to help you revise, consider the following questions, answers for which can be found by visiting **https://study.sagepub.com/essentialap2e**.

Test yourself by revising the chapter first, and then answering these questions without looking at the book. Afterwards compare your answers with the text and with the notes you made. Did you miss anything in your notes? Here are the questions:

1 Identify the different organelles in the human body and state their functions.

2 Outline the stages of mitosis and meiosis.

3 Identify the five types of human cells, outline the different variants within each group and clarify their functions.

REVISE

For additional revision resources visit: **https://study.sagepub.com/essentialap2e**.

ACE YOUR ASSESSMENT

- Revise key terms with interactive flashcards.
- Test yourself with multiple-choice questions.
- Access the glossary with audio to hear how complex terms are pronounced.
- Explore recommended websites suitable for revision.

REFERENCES

Allsopp, R.C., Chang, E., Kashefi-Aazam, M., Rogaev, E.I., Piatyszek et al. (1995) 'Telomere shortening is associated with cell division in vitro and in vivo', *Experimental Cell Research*, 220(1): 194–200.

Barnum K.J. and O'Connell M.J. (2014) 'Cell cycle regulation by checkpoints', in E. Noguchi and M. Gadaleta (eds) *Cell Cycle Control. Methods in Molecular Biology (Methods and Protocols)*, vol. 1170. New York, NY: Humana Press.

Jafri, M.A., Ansari, S.A., Alqahtani, M.H. and Shay, J.W. (2016) 'Roles of telomeres and telomerase in cancer, and advances in telomerase-targeted therapies', *Genome Medicine*, 8(1):69, 1–18.

Lane, N. (2015) *The Vital Question: Why Is Life the Way It Is?* London: Profile Books.

Lee, P., Linderman, J.D., Smith, S., Brychta, R.J., Wang, J. et al. (2014) 'Irisin and FGF21 are cold-induced endocrine activators of brown fat function in humans', *Cell Metabolism*, 19(2): 302–9.

Lee, P., Swarbrick, M.M., Zhao, J.T. and Ho, K.K.Y. (2011) 'Inducible brown adipogenesis of supraclavicular fat in adult humans', *Endocrinology*, 152(10): 3597–602.

Ralston, A. and Shaw, K. (2008) 'Gene expression regulates cell differentiation', *Nature Education*, 1(1): 127.

Watson, J.D. and Crick, F.H.C. (1953) 'Molecular structure of nucleic acids: a structure for deoxyribose nucleic acid', *Nature*, 171(4356): 737–8.

Westman, J.A. (2006) *Medical Genetics for the Modern Clinician*. Philadelphia: Lippincott Williams Wilkins.

GENETIC AND EPIGENETIC CONTROL OF BIOLOGICAL SYSTEMS

LEARNING OUTCOMES

When you have finished studying this chapter you will be able to:

1. Describe the major modes of inheritance: dominant, recessive, sex-linked, mitochondrial and multifactorial
2. Understand the influence of environmental factors on DNA (epigenetic changes) and the influence on individuals affected and later generations
3. Demonstrate awareness of the importance of understanding genetic and genomic aspects of disease causation and their impact on families

INTRODUCTION

We have already introduced the 46 chromosomes in the nucleus of the cell that contain the code for the formation of proteins in the cells of the body (Chapter 2). In this chapter we are examining in more detail how this system of control works to create us as individuals varying in many characteristics. We will also consider how such characteristics are transmitted to future generations and the influence of environmental conditions on genetically determined characteristics.

Two terms are used, often interchangeably but incorrectly, in this area of study: genetics and genomics.

- **Genetics:** refers to specific individual genes and how they are inherited. It has been particularly valuable in identifying the genes involved in numerous rare genetic disorders and enabling early diagnosis and treatment of many. Although these disorders are rare, there are a wide range of different conditions.
- **Genomics:** is considerably more complex. It refers to the complete genetic makeup of an individual and how it interacts with environmental and lifestyle factors. Many complex disorders, including cancer, diabetes and cardiovascular disease, occur because of variation in several genes, and genes interacting with each other and the environment. Genomic research is a relatively new and rapidly developing area arising from the Human Genome Project, which in 2003 announced completion of the mapping of the genes of the total human genetic structure (NHGRI, 2003).

The greater understanding will enable more effective diagnosis, and implementation of personalised intervention and nursing care.

Context

In the Bodie family we can see some examples of the transmission of characteristics. Many physical traits are transmitted by the genes carried in the DNA of the chromosomes including eye and hair colour. There are a number of different colours of eye that can occur.

BODIE FAMILY

Derek, the eldest child of Hannah and Richard, has blue eyes, although his sisters both have brown eyes. Of his grandparents, George has brown eyes and Maud blue ones and none of their children or other grandchildren have blue eyes. So how did Derek get his blue eyes?

The genetics of eye colour is fairly complicated with several genes now known to be involved in determination of the various different colours and patterns of eyes (Seddon et al., 1990). Even so, it appears that 74% of the blue–brown variation in human eye colour is explained by one gene on chromosome 15 (Sturm and Larsson, 2009). The easiest explanation is to focus on that one gene site with the gene for brown eyes dominant while that for blue eyes is recessive.

George has brown eyes, and thus has one or two of the paired genes for brown eyes, while Maud must have carried two genes for blue eyes. Each of their children received one gene for eye colour from each parent, blue from Maud and brown from George, and thus have brown eyes. Hannah married Richard Jones who was also brown eyed (the commonest eye colour in the world). However, as Derek has blue eyes, he must have received a gene for blue eyes from both parents, so Richard must be heterozygous for eye colour.

The twins Margaret and Michelle are monozygotic, that is identical, derived from the splitting of a fertilised egg into two, each carrying identical genetic makeup. However, epigenetic changes (discussed later in this chapter) result in the development of some differences in how their genes are expressed. Margaret is working in Australia and has developed a tan and her hair has become somewhat bleached. In addition, Michelle was seriously ill during childhood during which her growth was reduced and she is still slightly shorter than her sister.

GENETIC STRUCTURES

Genes

You have already seen in the previous chapter that the 46 chromosomes are formed of DNA and are paired.

A gene is a section of DNA which codes for a comparable RNA chain which leaves the cell and is translated into a specific polypeptide chain, with each being shaped into a specific protein that performs a particular structural or functional role. Many of these proteins are the enzymes that facilitate the numerous chemical reactions involved in cell metabolism (Introductory Science and Chapter 9). Each chromosome carries numerous genes. The estimated number of human genes within the nucleus has varied considerably but, through the Human Genome Project (NHGRI, 2003) work, it is now thought that there are 20,000–25,000 human genes on the 46 chromosomes.

The mitochondria of the cell also contain DNA, but this is much more limited in scope and size. Each mitochondrial genome contains 37 genes, of which 24 provide the code for the various components of RNA which are necessary for synthesis of proteins, which are coded for by the remaining 13 genes.

Each gene on a chromosome codes for a specific characteristic. Sometimes the gene on the two chromosomes of the pair is the same (homozygous), but sometimes these can vary (i.e. heterozygous); for example, the gene on one chromosome may code for brown hair, the gene on the paired chromosome may code for fair hair. Alleles are the variant genes for a specific characteristic on the paired chromosomes.

Non-coding DNA

The genes in the human chromosomes are only about 1–1.5% of the total DNA in the human chromosomes. The remaining 98–99% has been called non-coding (sometimes junk) DNA, which does not appear to code for the formation of any of the proteins arising from the genes in the DNA. However, The ENCODE Project Consortium (2011) (following the Human Genome Project) is aiming to identify all the DNA elements (genes and non-coding DNA) and their functions, in order to enhance human health.

--- **GO DEEPER** ---

DNA functioning

The Consortium estimated that 80.4% of DNA was involved in a DNA or RNA associated activity in at least one type of cell and that there were several sites for regulation of gene translation, both near to and distant from the gene. However, other work has found that only about 10% of our genome has been actively selected for some function involved in control of DNA transcription and, thus, the majority of our genome has no useful function (Strachan et al., 2015). Research is ongoing to try to elucidate the function, or lack of, of the non-coding DNA: it is hoped that agreement will eventually be reached.

A number of genetically determined conditions are caused by single point abnormalities, that is, an alteration in one of the nucleotides forming the DNA (Chapter 2, Figures 2.3 and 2.5). However, many disorders are determined by the combined influence of a number of different genes and other factors through multifactorial inheritance. The modes of inheritance are discussed below.

INHERITANCE

The overall structure of an individual's genetic makeup is known as the genotype. How this is expressed (shown) in the individual is the phenotype. The basic principles determining how genetic inheritance occurs are determined by the Laws (or Principles) of Inheritance. These laws help us to understand the patterns of inheritance seen in family trees which identify where a particular condition occurs in the family.

APPLY

Family tree

One of the key tools for diagnosis in this area is the family tree: the meanings of the symbols used are shown in Figure 3.1.

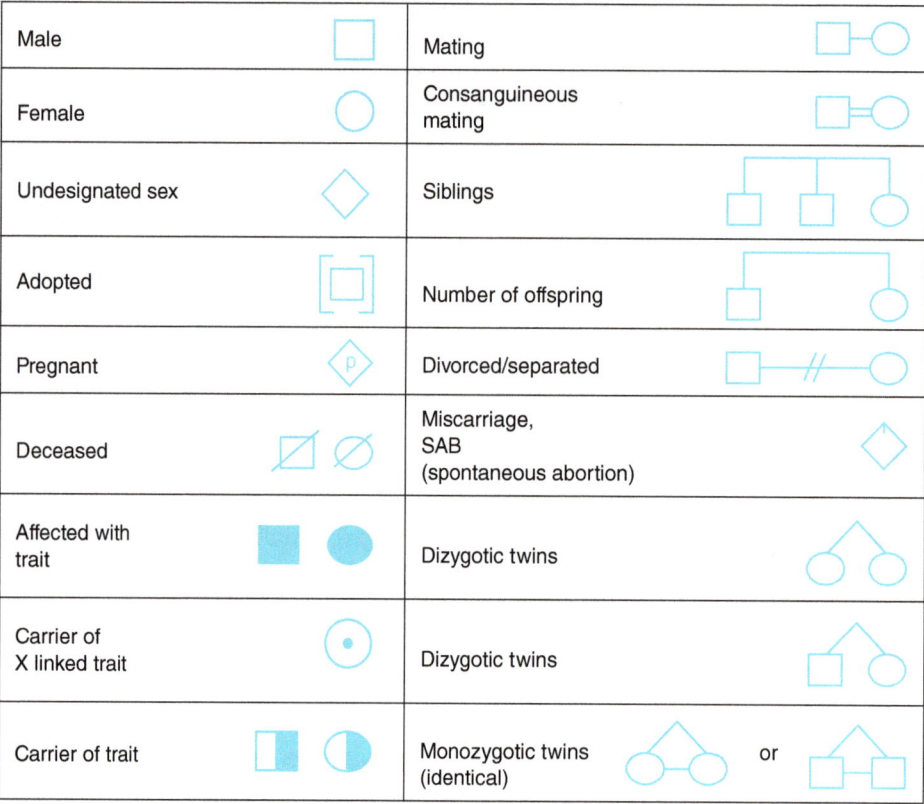

Male	□	Mating	□—○
Female	○	Consanguineous mating	□=○
Undesignated sex	◇	Siblings	
Adopted	⬚□⬚	Number of offspring	
Pregnant	◇p	Divorced/separated	□—//—○
Deceased	⬜̸ ⊘	Miscarriage, SAB (spontaneous abortion)	◇
Affected with trait	■ ●	Dizygotic twins	
Carrier of X linked trait	⊙	Dizygotic twins	
Carrier of trait	◫ ◑	Monozygotic twins (identical)	or

Figure 3.1 Meaning of symbols in family trees

 Go to http://study.sagepub.com/essentialap2e to access the online reading **Genetic Family History**. The weblink will help you understand how to take a genetic family history, including taking and drawing a family history.

The Laws of Inheritance

Much of the fundamental work on how characteristics were passed on from generation to generation was carried out by an Augustinian monk, Gregor Mendel, in what is now the Czech Republic (cited by Turnpenny and Ellard, 2007). The three (although some writers only quote two) Laws of Inheritance are based on Mendel's work and described in different ways. The following related section on Modes of Inheritance describes the Mendelian Patterns of Inheritance (dominant, recessive and X-linked).

GO DEEPER

Patterns of inheritance

Mendel worked in the monastery garden and carried out many experiments to identify how a number of characteristics in sweet peas were inherited. His work (published 1866) was not recognised as important until 1900 when three European scientists re-discovered it. It provided an explanation for the processes behind Darwin's work on evolution first published in 1859 with the 6th edition published as *The Origin of Species by Means of Natural Selection* in 1872. This is often considered the definitive edition.

The Laws of Inheritance are based on the pattern of the paired chromosomes (normally 23 pairs) within each cell. The pair of chromosomes carry equivalent (but not necessarily identical) genes on each chromosome within the pair. Each of the paired genes is an allele with the paired alleles coding for a matched characteristic.

Law of Dominance

Some alleles are dominant (i.e. demonstrated in the cell and body) while others are recessive (i.e. masked by the dominant gene); an organism with at least one dominant allele will display the effect of the dominant allele. A dominant allele is shown as a capital letter (e.g. A) while the recessive allele at the same site is in small print (e.g. a). An example of a condition determined by a dominant gene is achondroplasia in which the long bones are shortened.

A = dominant gene   a = recessive gene
Probable offspring from two heterozygous parents:

1: homozygous dominant (AA) – shows dominant characteristic
2: heterozygous (Aa) – both show dominant characteristic
1: homozygous recessive (aa) – does not show dominant characteristic

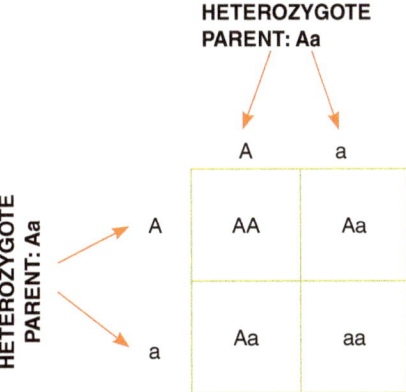

Figure 3.2 A 'Punnett Square'

Law of Segregation

This law makes it clear that the comparable alleles on the paired chromosomes are separated at cell division as the two members of the chromosome pair separate to different parts of the cell. During gamete formation, the alleles for each gene segregate from each other so that each gamete carries only one allele for each gene.

Law of Independent Assortment

Genes for different traits can segregate independently during the formation of gametes. However, there is some variability in this because some genes are on the same chromosome and, if fairly close, are increasingly likely to be inherited together.

Table 3.1 Examples of single point genetic conditions

Autosomal dominant	Achondroplasia	A type of dwarfism[1] with shortened long bones
	Polycystic kidney disease	Multiple cysts in both kidneys, may damage the liver, pancreas and, rarely, the heart and brain
	Marfan syndrome	Disorder of connective tissue, tall, long limbs, other defects
	Huntington's disease	Neurodegenerative disorder → cognitive decline and behavioural symptoms, common onset in midlife
	Familial hypercholesterolaemia	High blood cholesterol levels (particularly low-density lipoproteins - 'bad' cholesterol), early cardiovascular disease
	Hereditary spherocytosis	Haemolytic anaemia with sphere-shaped (instead of normal biconcave disk) red blood cells, prone to rupture
Autosomal recessive	Polycystic kidney disease	Presentation as above, but autosomal recessive condition
	Sickle cell anaemia	Abnormal, rigid, sickle-shaped red blood cells → reduced flexibility. Heterozygote protects against malaria
	Cystic fibrosis	Mucoviscidosis (thick viscid mucus), affects the lungs, pancreas, liver, intestine. Diagnosed by sweat test
	Tay-Sachs disease	Causes progressive deterioration of nerve cells and mental and physical abilities. Usually death by four years
	Phenylketonuria	Phenylalanine cannot be broken down. Build up causes intellectual disability, seizures, etc. Special diet needed
	Niemann-Pick disease	Metabolic disorders when sphingomyelin accumulates in lysosomes. Varies in severity with some dying in early years with others living to adulthood
	Spinal muscular atrophy	Defect in the SMN1 gene (protein necessary for motor neurons). Neuronal death and muscle atrophy
X-linked recessive	Duchenne muscular dystrophy	Progressive skeletal muscle weakness, defects in muscle proteins, death of muscle cells and tissue. Early death
	Haemophilia A	Deficiency in clotting factor VIII, causes increased bleeding and usually affects males
	Lesch-Nyhan syndrome	Uric acid in body fluids, with severe gout and kidney problems. Neurological signs, moderate intellectual disability
	Male pattern baldness	Androgenic alopecia: susceptibility of hair follicles to miniaturisation → hair loss in up to 70% of men, 40% of women
	Colour blindness	Colour vision deficiency, most usual cause is fault in development of some of retinal cones that perceive colour

---

[1]This term is not politically correct but is medically correct. It is considered more sensitive to refer to people of short stature or little people.

The application of these last two laws is illustrated in the Punnett Square in Figure 3.2 showing the probability of the offspring of two heterozygous parents having a particular genotype.

Modes of inheritance

Single gene inheritance

In these conditions the specific gene involved is altered by just one nucleotide in the DNA, which is transcribed into RNA in the nucleus that then alters one amino acid in the transcribed polypeptide chain. This is called a Single Point Genetic Condition. Table 3.1 lists a number of such conditions.

In the following section, a case study is included in relation to each of the main types of single gene inheritance. Drawing out the family histories and Punnett Squares will help you to answer the questions.

Dominant inheritance

This type of inheritance illustrates clearly the Law of Dominance. In the simplest example, the dominant gene is carried by one parent on one chromosome and thus there is, on average, a 50% chance at each pregnancy that the offspring will inherit the dominant gene from that parent (Figure 3.3a). As the other parent does not carry this gene, the offspring has (effectively) a 'single dose' of this characteristic. In dominant inheritance, the characteristic is demonstrated in this situation; in each pregnancy there is a 50/50 chance that the abnormal dominant gene will be inherited from the parent who carries it. The family tree illustrates the condition passing through the generations (Figure 3.3b).

GO DEEPER

Dominant conditions

There are numerous dominant conditions (Table 3.1): some are evident at birth, some appear later in life, and some have reduced penetrance which means that not everyone carrying the gene demonstrates the disorder. An example of a dominant condition is achondroplasia in which the gene may be handed down from a parent with the condition or may be due to a new mutation (a spontaneous change in the DNA structure). In this condition the person affected is of short stature with short limbs, a large head and a prominent forehead. These abnormalities are visible with a single gene, but occasionally both parents carry the condition and a child may be born who is homozygous for the dominant condition. Babies born with two genes for this condition have severe skeletal disorders and usually die shortly after birth.

Dominant conditions can have variable penetrance, ranging from complete when everyone who has the abnormal gene will show the condition, through varying degrees of incomplete penetrance when a proportion of those with the gene do not show the signs and symptoms. In addition, some conditions are late-onset disorders which arise later in life.

(a) Punnett Square for dominant inheritance

One parent carries a dominant gene, e.g. for Huntington's Disease

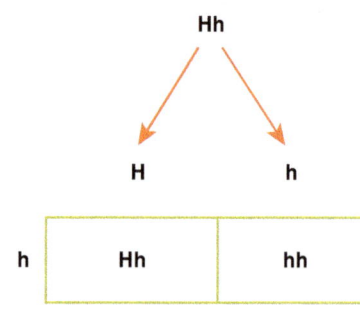

HETEROZYGOTE PARENT:

Hh

H   h

PARENT HOMOZYGOUS
FOR NORMAL GENE : hh

	H	h
h	Hh	hh

H = dominant gene

On average: 50% of offspring will carry one dominant gene

50% will carry two normal genes

(b) Family tree for dominant condition

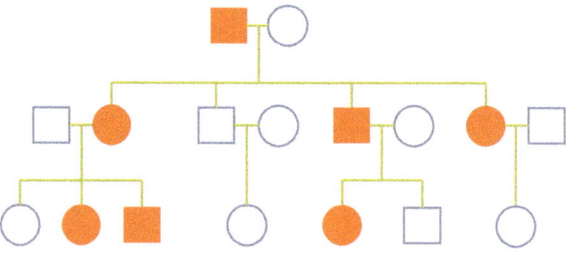

Figures 3.3a and 3.3b   Dominant Inheritance

APPLY

Genetics and inheritance

Huntington's disease (HD) is a dominant condition which manifests later in life usually with physical symptoms occurring between 35 and 44, although it may occur earlier in succeeding generations. It often starts with changes in mood and mental ability but progresses to disturbed coordination and unsteadiness. As it progresses, coordinated movement becomes more difficult, mental abilities and behaviour become more disturbed, and dementia develops. Life expectancy averages 20 years after diagnosis (Paulsen et al., 2005). The presence of the gene for HD can now be identified in carriers before signs and symptoms appear.

This is a condition which raises numerous issues to be taken into account in realising person-centred practice for the person with the condition and the family. These will include:

- Fore-knowledge at diagnosis of what the progress of the disease will be like, and planning for future care.

- Implications for children of the person with the condition: should they be/do they want to be tested for the presence of the gene?

  o Do they want to know what may (or may not) lie before them?
  o How will they cope with uncertainty if they do not get tested?
  o If they are positive, what will be the implications for their own life and having children?

Understanding genetics and inheritance will enable you to provide appropriate support to individuals and families with such conditions, although it is important to work within the limits of one's knowledge. Genetic Nurse Specialists have high levels of knowledge of genetics and skills in counselling to enable them to work effectively with individuals and families with such concerns (Jenkins and Lea, 2005). (Note: people with HD and their relatives will commonly refer to the condition as Huntington's.)

If a couple, one of whom carries the gene for HD, wish to have children, it is now possible to have pre-implantation genetic diagnosis. In vitro fertilisation (IVF) is used to prepare an embryo which is genetically tested to ensure that it does not have the HD gene before implantation in the uterus for gestation.

ACTIVITY 3.1: APPLY

Review the case study below and then answer the questions:

Case study: Huntington's disease (HD)

My father, John, is 45 and is beginning to show signs of forgetfulness and clumsiness with some muscle twitching. Huntington's disease runs in his family; his mother died of it and I am worried that he is developing it. His older brother was tested for it and does not have it, but his younger sister with two children does carry the gene for the disease although she has no symptoms. John has not been tested for the condition.

Questions

1. How is Huntington's disease inherited?
2. If Dad has the gene for HD, what is the risk that I also carry it and the risk that I could pass it on to my daughter?
3. How can a family history be used with genetic testing to identify the risk of my brother's children and extended family members developing HD? What are the potential difficulties involved for families?

Recessive inheritance

In recessive conditions both parents carry a gene for the disorder and create gametes (sperm or ova) in which half from each parent are likely to carry the recessive gene. Figure 3.4a shows the distribution of the recessive gene in the gametes and the average proportion of offspring with the different genetic makeups. Each pregnancy arising from two parents both of whom are recessive for a condition will have a one in four chance of the offspring of each pregnancy having the condition. Figure 3.4b is an illustrative family tree showing the carrier status running through the generations on both sides of the family tree and the individual with the disease showing up in the third generation.

(a) Punnett Square for recessive inheritance

Both parents are heterozygous for the abnormal gene, i.e. each carries one copy of the gene (small c), e.g. for cystic fibrosis.

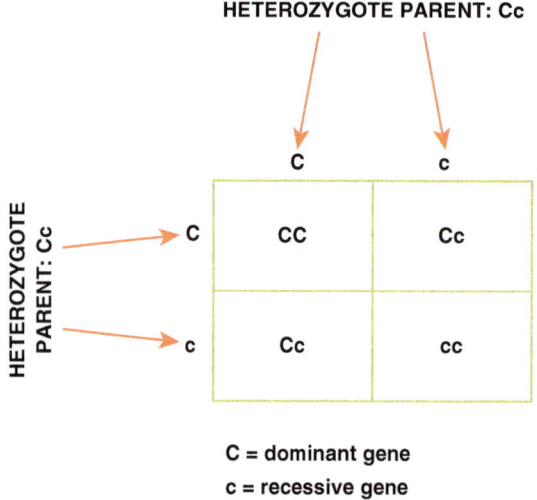

HETEROZYGOTE PARENT: Cc

C = dominant gene

c = recessive gene

Possible offspring from two heterozygous parents:

    1: homozygous for normal gene (CC): no disease occurs

    2: heterozygous (Cc): one normal and one abnormal gene, carriers of disease

    1: homozygous recessive (cc): two abnormal genes, has the disease

That is: there is a 1 in 4 chance of each child having the condition.

(b) Family tree for recessive condition

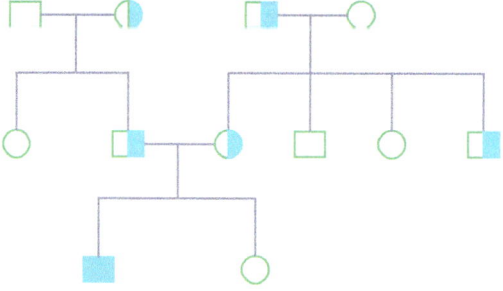

Figures 3.4a and 3.4b   Recessive Inheritance

    Recessive conditions are more likely to develop when there are close relationships between partners, or within a closely in-bred community (Table 3.1).

APPLY

Phenylketonuria

Phenylketonuria (PKU) is a recessive condition in which the amino acid phenylalanine builds up in the body because it cannot be metabolised normally. It causes brain damage and can lead to learning/intellectual disabilities, behavioural difficulties and epilepsy if not treated early. Danielle, the baby in the Bodie family, had a heel prick test shortly after birth (as all babies in the UK have) to test for PKU and a number of other rare but serious conditions. Her parents were reassured that none of these genetic disorders were identified in Danielle.

BODIE FAMILY

Those identified with PKU need a low protein and specialised diet with supplements to ensure all necessary nutrients are obtained, but phenylalanine does not accumulate. The diet with regular blood monitoring is needed throughout childhood to ensure normal intelligence. In adults, the specialised diet is not essential, but some people function less well if they stop taking it.

GO DEEPER

Recessive conditions

Some recessive genes cause disease when present on both chromosomes (i.e. homozygous for the condition) but provide protection against certain other diseases when only one gene is present (i.e. heterozygous). One important example is associated with disorders of erythrocytes including: sickle cell trait, thalassaemia trait, and glucose-6-phosphate dehydrogenase (G6PD – an enzyme in glucose metabolism) deficiency. These conditions are common in regions with a high incidence of malaria, and research has demonstrated that carrying a single gene for these conditions provides some protection against malaria (Ayi et al., 2004).

ACTIVITY 3.2: APPLY

Review the case study below and then answer the questions:

Case study: Cystic fibrosis – Jill and Roger

My uncle Alan (my mother's brother) and his wife had a daughter 20 years ago who was diagnosed with cystic fibrosis and has recently had a heart and lung transplant to resolve her respiratory symptoms. Neither of my parents nor my two sisters have this disorder. My partner (Roger) and I (Jill) are considering having a baby but I am concerned about the risk of the child also having cystic fibrosis.

Questions

1. What is the mode of inheritance of cystic fibrosis and what proportion of the population carry the relevant gene?
2. What is the risk that Jill carries the gene for cystic fibrosis?
3. How can Roger and Jill minimise the risk of having a child with cystic fibrosis?

Sex-linked inheritance

The commonest form of sex-linked inheritance is X-linked recessive, which appears most often in men. Figure 3.5 provides some detail about this form of inheritance.

X-linked dominant and Y-linked conditions are both much less common than other forms of inheritance and X-linked conditions are more common than Y-linked because the X chromosomes are much bigger than the Y chromosomes and, thus, carry more genes. However, infertility has been related to Y chromosome inheritance.

(a) Punnett Square for X-linked inheritance

The father has one X chromosome and the mother has two, one of which carries the gene for colour blindness (X^B).

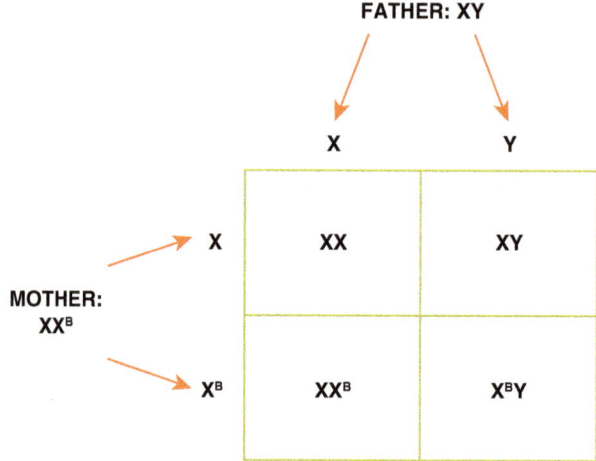

50% of the boy babies and 50% of the girl babies will carry the abnormal X chromosome.

The girls will be carriers for colour blindness

The boys will be colour blind

(b) Family Tree for X-linked recessive condition

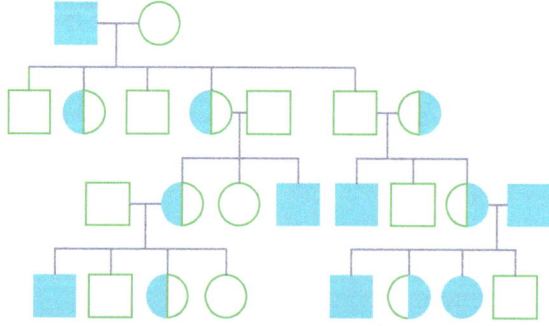

Figure 3.5 X-linked recessive inheritance

GO DEEPER

Haemophilia

One notable instance of X-linked inheritance was haemophilia in the British Royal Family. The allele for this was carried by Queen Victoria who passed it through her daughters to many of the royal families of Europe. It had not been present in any of her forebears and it is possible that this was a new mutation in Queen Victoria or her parents. It has also been suggested that she may have been illegitimate and thus acquired this gene (Potts, 1996).

ACTIVITY 3.3: APPLY

Review the case study below and then answer the questions:

Case study: Duchenne muscular dystrophy

Jane and David have been married for two years and 18 months ago had their first child, a little boy named Michael (Mike for short). Mike seemed to be doing well at first, but in his second year his parents became worried as his walking was not developing as expected. He has been referred for genetic testing.

Jane's mother had three siblings, including one boy several years older who she remembers as badly disabled and who died in his teens. She does not know what was wrong with him. Jane's younger sister has not yet had any children, her older one has a boy (six years) and a girl of three years without any symptoms. David's family has no history of any serious disorders.

Jane and David have just been told by the paediatric consultant that Mike has Duchenne muscular dystrophy.

Questions

1. What is the mode of inheritance of Duchenne muscular dystrophy?
2. How likely is that each further child by Jane and David will have the same condition?
3. Briefly discuss the chances that Jane's younger sister will have a boy or girl affected with this condition.
4. How does this condition affect the individual concerned?

Mitochondrial inheritance

In the previous chapter we have looked at the function of mitochondria. Their primary function is converting nutrients into useable energy: they are the 'powerhouses of the cell'. They are unique within the cell in having multiple copies of DNA in the form of circular chromosomes. Mitochondria are inherited only from the mother's egg and are passed on to all of her offspring. The mitochondria also contain about 1% of total cellular DNA (Chapter 2).

It has been estimated that at least 1 in 8,500 of the general population suffer from mitochondrial disease (Chinnery and Turnbull, 2001). These are a varied group of disorders that occur due to dysfunction of the respiratory chain in the mitochondria (see Chapter 9 and Table 3.2) and those organs most dependent on aerobic metabolism, including nervous tissues, are most likely to develop such disorders. These disorders are variable in severity and their complexity presents a challenge in diagnosis (Chinnery, 2000). Figure 3.6 demonstrates the type of inheritance pattern found with these conditions.

Table 3.2 Examples of mitochondrial disorders

Various cancers
Leukaemia
Intellectual disabilities: a range of conditions including, e.g. fragile X syndrome, Rett syndrome, Angelman syndrome
Nervous disorders
Muscular disorders

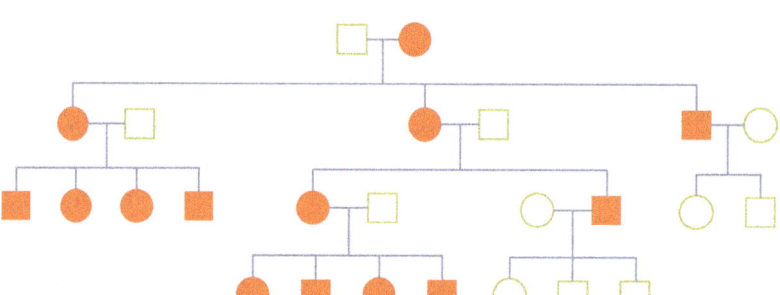

Figure 3.6 Mitochondrial inheritance family tree

GO DEEPER

Mitochondrial DNA

Each individual is formed from the fusion of an egg (from the mother) and a sperm (from the father) and gains half the DNA in the nucleus from each parent. However, all the mitochondria, with the mitochondrial DNA, in the zygote are provided from the mother's ova and are passed on to all her children. It has been reported that more than one in 4,000 babies are born with severe mitochondrial diseases, which occur because of abnormalities within the mitochondrial DNA (MMDC, 2014). An approach to prevention of such conditions approved by the UK Parliament in 2015 and with licences now being applied for involves forma-tion of a three-parent fetus (Devlin, 2015). The nucleus of the ovum (either before or after fertilisation) is transplanted into an ovum from a healthy woman, replacing the original nucleus which has been removed. The mitochondria of the mother are thus replaced by those from a healthy woman. The baby will have about 99% of their DNA from the original parents but will not suffer from a mitochondrial disorder.

Polygenic inheritance

Polygenic and multifactorial inheritance need to be differentiated. Polygenic inheritance occurs when a characteristic is determined by several or many genes with each having a small effect. Characteristics involving numerous genes include: height, weight, intelligence, skin colour (Figure 3.7) and eye colour, although some of these may also be influenced by other factors.

Multifactorial inheritance

Multifactorial inheritance involves both more than one gene and environmental factors.

Multifactorial inheritance is responsible for the presentation of many of the personal characteristics which make us all different; for example, skin colour, height, personality, intelligence.

Characteristics determined by single point inheritance are visible as discontinuous or discrete aspects of the phenotype. However, multifactorial conditions are continuous in nature, that is, they present along a continuum. For example, the height of children in a family may show considerable variation. Figure 3.7 illustrates the polygenic inheritance of skin colour but, in addition, exposure to UV light from the sun can stimulate the formation of additional melanin in the skin (Chapter 14) and further increases the darkness of the skin.

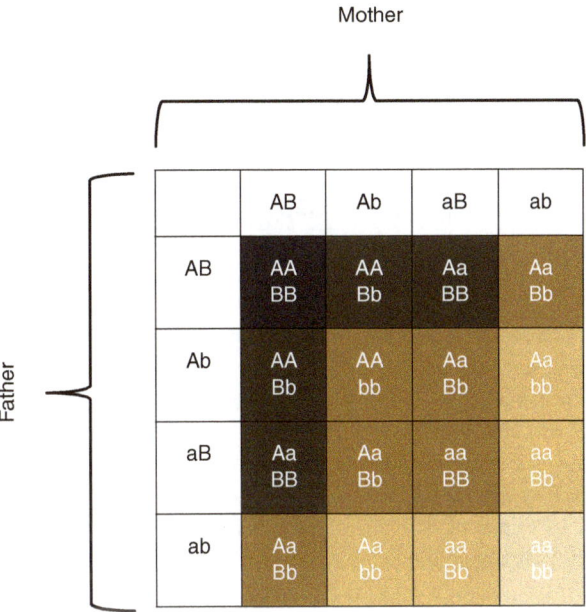

Figure 3.7 Polygenic inheritance of skin colour

Multifactorial inheritance and disease

Multifactorial inheritance involves the transmission of genetic information, but its presentation is also influenced by environmental factors such as diet or disease. A number of characteristics of this type of inheritance have been described (Lobo, 2008):

- No Mendelian pattern of inheritance is shown and, although numerous cases may occur within the one family, it can also occur in isolation.
- Environmental factors can alter the risk of the disease occurring.
- The disorder is not sex-linked, but may occur more commonly in one gender than another (the sex hormones may be an environmental factor influencing presentation of the disorder).
- The concordance (i.e. the rate at which both twins carry a disorder) does not match the expected Mendelian proportions in mono- (identical) and dizygotic (non-identical) twins.
- Some of these diseases occur more frequently in particular ethnic groups. For example, 'African Americans have the highest overall mortality rate from coronary heart disease (CHD) of any ethnic

group in the United States' (Clark et al., 2001: 97). However, while this is determined by genes, diet and exercise, it is highly likely that this is also related to access to healthcare – a classic example of multifactorial inheritance.

A number of disorders are inherited in this way and examples are given in Table 3.3.

Table 3.3 Examples of multifactorial inheritance

Alzheimer's disease
Heart disease
Some cancers
Neural tube defects
Schizophrenia
Type 2 diabetes
Narcolepsy

APPLY

Multifactorial inheritance

BODIE FAMILY

In the Bodie family, Richard Jones (married to Hannah) was diagnosed with type 2 diabetes five years ago (at 49 years). This is a disease in which the body is unable to metabolise glucose normally (Chapter 7 and Chapter 9) and the high blood glucose levels increase the risk of cardiovascular disease, renal disease, eye disorders and various other conditions.

The genetic background of an individual influences the risk of developing the condition and Richard's father also has diabetes. Type 2 diabetes occurs when a mix of genetic and environmental factors interact and tends to run in families. Someone with diabetes in their family is two to six times more likely to develop it than someone without (Diabetes in the UK, 2012).

However, development of obesity and lack of exercise also increase the risk of developing this disease (Cheng, 2005). Richard was overweight for a time in his early 40s when he stopped coaching the school team in football. After he was diagnosed with type 2 diabetes he worked successfully on reducing his weight and has made sure that his children understand the risks and are minding their own weights.

EPIGENETICS

What is epigenetics?

Some of the factors in multifactorial inheritance are probably included under epigenetics, described as:

> (it) literally means 'above' or 'on top of' genetics. It refers to external modifications to DNA that turn genes 'on' or 'off.' These modifications do not change the DNA sequence, but instead, they affect how cells 'read' genes. (Rettner, 2013)

This has become an area of growing importance in the understanding and, we hope, treatment of disease. The Human Epigenome Project (2014) 'aims to identify, catalogue and interpret genome-wide DNA methylation patterns of all human genes in all major tissues'.

ACTIVITY 3.4: UNDERSTAND

Watch these two video clips to enhance your understanding of epigenetics. These online videos can be accessed by **scanning the QR codes** with your smart phone or via https://study.sagepub.com/essentialap2e.

WHAT IS EPIGENETICS? (5:02)

EPIGENETICS AND THE INFLUENCE OF OUR GENES (18.40)

GO DEEPER

Inheritance of acquired characteristics

Lamarck (1744-1829) was a French biologist who is now remembered for his theory of the *inheritance of acquired characteristics*, which proposed that changes acquired over the life of an organism (such as the lengthening of the giraffe's neck through repeated stretching) may be transmitted to the offspring. His theory was accepted in the USSR (Union of Soviet Socialist Republics, dissolved 1991) for many years but was totally rejected by Western scientists and the mechanisms elucidated by Mendel's work have been accepted. Genetic Mendelian inheritance has been described as 'hard', that is, 'mediated by the transmission of gene alleles that are impervious to environmental influence' (Bonduriansky, 2012: 330). Many geneticists considered that DNA transfer was the only possible mechanism for transfer of characteristics. The converse, 'soft' inheritance as described by Lamarck, had been totally rejected. However, empirical evidence is now pointing to mechanisms for inheritance that enable acquired traits to be transmitted to succeeding generations without altering the DNA sequence.

Broadly, epigenetics can be described 'as a bridge between genotype and phenotype; a phenomenon that changes the final outcome of a locus or chromosome without changing the underlying DNA sequence' (Korkmaz et al., 2011: 139). It allows a fine control over the expression of genes.

Mechanisms of epigenetics

Three distinct mechanisms have been identified in achieving these changes and determining whether a gene is expressed or silenced: DNA methylation, histone modification, and post-transcriptional gene regulation by non-coding microRNAs.

GO DEEPER

Mechanisms of epigenetics

DNA methylation

In this process specific enzymes add a methyl group to a cytosine nucleotide of the DNA (Figure 3.8). This changes the shape/structure of DNA and alters its response to the machinery for transcribing the gene. In some situations, this determines whether the gene inherited from the mother or the father is expressed (Korkmaz et al., 2011).

(Continued)

Histone modification

Histones are the proteins around which DNA is wound and together these form the chromosomes. Chemical modification of histones can determine whether the DNA is compact (condensed) and cannot be transcribed, or not compact and active (Figure 3.9).

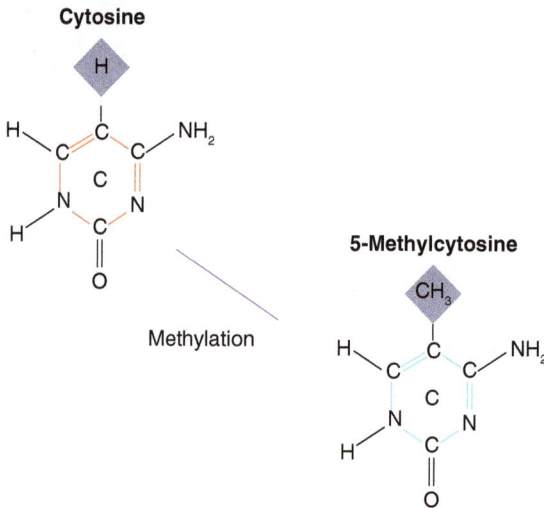

Figure 3.8 Methylation of DNA

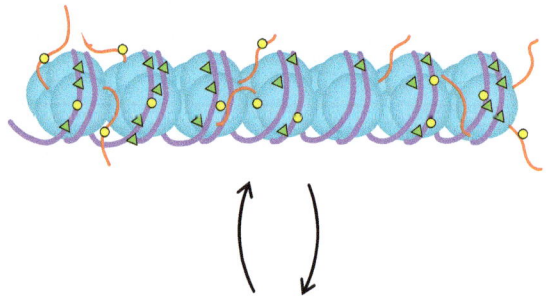

Figure 3.9 Histone modification of DNA activity

Post-transcriptional gene regulation by non-coding microRNAs

These microRNAs (miRNAs) are about 18–25 nucleotides in length but have a key role in regulating messenger RNA (mRNA) activity in the translation of particular proteins, or in degrading mRNA. They appear to have various effects on cell function and are a focus of work in biomedical research (Kala et al., 2013).

Epigenetics implications

Epigenetic changes have been associated with some cancers and rare developmental syndromes but study of the effects on more common and non-malignant disease is yet to produce verifiable results through epigenome-wide association studies (EWAS) (Murphy and Mill, 2014).

The potential for epigenetic changes to be transmitted through the generations is of considerable interest. There is evidence of transgenerational inheritance in plants and in some animals (e.g. nematodes or worms), but the evidence for such inheritance in man is, as yet, unverified (Morgan and Whitelaw, 2008).

--- **GO DEEPER** ---

Genomic science

The NHS in England is aiming to develop personalised healthcare through a set of proposals that will provide 'better, safer, sustainable care for all' (NHS, 2014). This will be achieved largely through improving the use of data and technology by health and social care professionals. Genomic science is a key area for research and development of resources to identify epigenetic changes as well as the genetic structure and presence of abnormal genes. This is likely to enhance the opportunities for diagnosis, treatment and care, which is planned to provide individualised (person-centred) practice.

It is anticipated that this work will be able to identify epigenetic changes as well as the genetic structure and presence of abnormal genes.

Genetic data from the Finnish population has been used extensively in examining single-gene and polygenic conditions (Wang et al., 2014) and this approach is being extended to other countries. It is anticipated that such data will be able to identify genetic factors enabling early screening for disease and selection of the most appropriate treatments.

CONCLUSION

This chapter has looked at genetics and inheritance in health and disease. At present this is little emphasised in nursing, but the growing concept of personalised healthcare depends to a large extent on characteristics of the individual transmitted through genetics. As our understanding of the effect of environmental factors and disease grows, epigenetics is also likely to become more important. During your career these areas are certain to become of growing relevance to your practice.

--- **GO DEEPER** ---

Further reading

Atkinson, P., Featherstone, K. and Gregory, M. (2013) 'Kinscapes, timescapes and genescapes: families living with genetic risk', *Sociology of Health & Illness*, 35(8): 1227–41.

Jacobs, C. and Patch, C. (2013) 'Identifying individuals who might benefit from genetic services and information', *Nursing Standard*, 28(9): 37–42.

Tonkin, E. and Skirton, H. (2013) 'The role of genetic/genomic factors in health, illness and care provision', *Nursing Standard*, 28(12): 39–46.

- Being able to create family trees will help you to recognise the different forms of inheritance and thus enable you to provide appropriate explanations to individuals and families with inherited conditions.

- Knowing about the influence of environmental factors on DNA (epigenetic changes) will help you with educational programmes for many people with long-term conditions.

REVISE

TEST YOUR KNOWLEDGE

This chapter contains some complex ideas which you need to understand to cope with the changes you are likely to see introduced into clinical practice during your career and to be able to communicate effectively with persons with these conditions for whom you are caring. The major areas you need to revise are:

1 Application of the Laws of Inheritance in dominant, recessive and sex-linked inheritance.

2 Mitochondrial and multifactorial inheritance and the importance of multifactorial inheritance in disease.

3 The importance of epigenetic changes in disease.

In order to help you revise, consider the following questions, answers for which can be found by visiting **https://study.sagepub.com/essentialap2e.**.

Test yourself by revising the chapter first, and then answer these questions without looking at the book (the following website might help you: http://geneticseducation.nhs.uk/genetic-testing-for-health). Afterwards compare your answers with the text and with the notes you made. Did you miss anything in your notes? Here are the questions:

1 What do the following terms mean?
- Karyotype.
- Autosomes.
- Homologous chromosomes.
- Allele.
- Gene.

2 Outline how the information carried on DNA is transcribed into RNA and translated into proteins.

3 Describe the Three Laws of Inheritance.

4 Draw a Punnett Square to illustrate the genetic outcome from parents one of whom carries a dominant gene for a particular characteristic. What is the common pattern of descent through generations in such conditions?

5 Draw a Punnett Square to illustrate the genetic outcome from parents both of whom carry a recessive gene for a particular characteristic. What is the common pattern of descent through generations in such conditions?

REFERENCES

Ayi, K., Turrini, F., Piga, A. and Arese, P. (2004) 'Enhanced phagocytosis of ring-parasitized mutant erythrocytes: a common mechanism that may explain protection against falciparum malaria in sickle trait and beta-thalassemia trait', *Blood*: 104(10): 3364–71.

Bonduriansky, R. (2012) 'Rethinking heredity, again', *Trends in Ecology and Evolution*, 27(6): 330–6.

Cheng, D. (2005) 'Prevalence, predisposition and prevention of type II diabetes', *Nutrition & Metabolism*, 2: 29.

Chinnery, P.F. (2000) 'Mitochondrial disorders overview', *GeneReviews* (Online. Last update 14 August 2014). Available at: www.ncbi.nlm.nih.gov/books/NBK1224/ (accessed 4 May 2020).

Chinnery, P.F. and Turnbull, D.M. (2001) 'Epidemiology and treatment of mitochondrial disorders', *American Journal of Medical Genetics (Semin. Med. Genet.)*, 106: 94–101.

Clark, L.T., Ferdinand, K.C., Flack, J.M. 3rd, Hall, W.D., Kumanyika, S.K. et al. (2001) 'Coronary heart disease in African Americans', *Heart Disease*, 3(2): 97–108.

Darwin, C.R. (1872) *The Origin of Species by Means of Natural Selection; Or The Preservation Of Favoured Races in the Struggle for Life*, 6th edn. Available at: www.gutenberg.org/files/2009/2009-h/2009-h.htm (accessed 4 May 2020).

Devlin, H. (2015) 'Britain's House of Lords approves conception of three-person babies', *The Guardian*, 24 February.

Diabetes in the UK (2012) *Key Statistics on Diabetes*. London: Diabetes UK. Available at: www.diabetes.org.uk/Documents/Reports/Diabetes-in-the-UK-2012.pdf (accessed 4 May 2020).

Human Epigenome Project (2014) Available at: www.epigenome.org (accessed 13 September 2014).

Jenkins, J.F. and Lea, D.H. (2005) *Nursing Care in the Genomic Era: A Case Based Approach*. Sudbury, MA: Jones and Bartlett.

Kala, R., Peek, G.W., Hardy, T.M. and Tollefsbol, T.O. (2013) 'MicroRNAs: an emerging science in cancer epigenetics', *Journal of Clinical Bioinformatics*, 3: 1–9.

Korkmaz, A., Manchester, L.C., Topal, T., Ma, S., Tan et al. (2011) 'Epigenetic mechanisms in human physiology and diseases', *Journal of Experimental and Integrative Medicine*, 1(3): 139–47.

Lobo, I. (2008) 'Multifactorial inheritance and genetic disease', *Nature Education*, 1(1): 5.

Mendel, G. (1866) Versuche über Plflanzenhybriden. *Verhandlungen des naturforschenden Vereines in Brünn, Bd. IV für das Jahr 1865, Abhandlungen*, 3–47. (Experiments in Plant Hybridization, translated into English by William Bateson in 1901.) Available at: www.esp.org/foundations/genetics/classical/gm-65.pdf (accessed 4 May 2020).

MMDC (Mitochondrial and Metabolic Disease Center) (2014) *Specializing in Mitochondrial Disease Treatment and Research*. MMDC, University of California San Diego.

Morgan, D.K. and Whitelaw, E. (2008) 'The case for transgenerational epigenetic inheritance in humans', *Mammalian Genome*, 19: 394–7.

Murphy, T.M. and Mill, J. (2014) 'Epigenetics in health and disease: heralding the EWAS era', *The Lancet*, 383(9933): 1952–4.

NHGRI (National Human Genome Research Institute) (2003) *The Human Genome Project Completion: Frequently Asked Questions*. NHGRI, National Institutes of Health, USA.

NHS (2014) *Personalised Health and Care 2020: Using Data and Technology to Transform Outcomes for Patients and Citizens, a Framework for Action*. National Information Board, HM Government, UK.

Paulsen, J.S., Hoth, K.F., Nehl, C., Stierman, L., The Huntington Study Group (2005) 'Critical periods of suicide risk in Huntington's disease', *The American Journal of Psychiatry*, 162(4): 725–31.

Potts, W.T.W. (1996) 'Royal haemophilia', *Journal of Biological Education*, 30(3): 207–17.

Rettner, R. (2013) 'Epigenetics: definition & examples', *LiveScience*, 24 June. Available at: www.livescience.com/37703-epigenetics.html (accessed 4 May 2020).

Seddon, J.M, Sahagian, C.R., Glynn, R.J., Sperduto, R.D. and Gragoudas, E.S. (1990) 'Evaluation of an iris color classification system. The Eye Disorders Case-Control Study Group', *Investigative Ophthalmology and Visual Science*, 31(8): 1592–8.

Strachan, T., Goodship, J. and Chinnery, P. (2015) *Genetics and Genomics in Medicine*. New York: Garland Science (Taylor & Francis Group).

Sturm, R.A. and Larsson, M. (2009) 'Genetics of human iris colour and patterns', *Pigment Cell and Melanoma Research*, 22: 544–62.

The ENCODE Project Consortium (2011) 'A user's guide to the Encyclopedia of DNA Elements (ENCODE)', *PLoS Biology*, 9(4): e1001046.

Turnpenny, P.D. and Ellard, S. (2007) *Emery's Elements of Medical Genetics*, 13th edn. Philadelphia: Churchill Livingstone (Elsevier).

Wang, S.R., Agarwala, V., Flannick, J., Chiang, C.W.K., Altshuler, D. et al. (2014) 'Simulation of Finnish population history, guided by empirical genetic data, to assess power of rare-variant tests in Finland', *The American Journal of Human Genetics*, 94(5): 710–20.

THE HUMAN MICROBIOME AND HEALTH

4

UNDERSTAND: CHAPTER VIDEO

In working through this chapter, you might find it useful to have an overview of the human microbiome.

The following weblink may be useful in enhancing your understanding. The URL for this weblink can be easily accessed via **http://study.sagepub.com/ essentialap2e**

Activity 4.1: Understand has further resources to enhance your understanding.

THE HUMAN MICROBIOME
(11:31)

LEARNING OUTCOMES

When you have finished studying this chapter you will be able to:

1. Identify the major groups of microbes found in different parts of the body
2. Describe the normal structure of the prokaryotes (single-celled microbes – bacteria)
3. Understand how the body acquires its microbiota (the microorganisms of a particular site) and how this varies in different parts of the body (intrapersonal) and between different people (interpersonal)
4. Discuss the importance of the normal flora in maintaining health of the human body and factors that may alter this flora
5. Consider potential implications of variations in the normal flora
6. Outline the implications for health of pathogenic microorganisms and the importance of appropriate use of antibiotics

INTRODUCTION

You have learned quite a lot about the human cell, but more than 90% of the cells in the human body are not human in origin but microbial, that is, only visible under a microscope. In a book about the human body they deserve some attention. They form the microbiome, which has been defined as the 'ecological community of commensal,[1] symbiotic,[2] and pathogenic[3] microorganisms that literally share our body space and have been all but ignored as determinants of health and disease' (Lederberg and McCray, 2001: 8) and have been described as the 'forgotten inner organ represented by the enteric microflora' (O'Hara and Shanahan, 2006: 692). The individual organisms comprising the normal flora (microbiota) interact with one another and with the host (Clemente et al., 2012). That is, they are an integral part of the human body.

Following the Human Genome Project, The National Institutes of Health (USA) are funding the Human Microbiome Project (HMP) involving around 80 research institutions to study the role of microbes in human health and disease (NIH HMP Working Group, 2009); this is a relatively recent area of work and its importance is growing. The human microbiome has been described as 'our second genome' (Grice and Segre, 2012: 151).

In this chapter we are going to look at the major groups of microbes within and on the human body, their structure and requirements for metabolism and life, and the importance of the normal flora (also called the 'indigenous microbiota') of the body. It is now clear that the microbiota, both internal and external, play a major role in maintaining health (NIH, 2012). This chapter will include an outline of the microbiological content of the gastrointestinal tract, skin, mouth, upper respiratory tract and reproductive tract, and the development and factors which alter the nature of the flora. The importance of the human microbiome in health will be the main focus. However, the potential implications for health of some pathogenic organisms and approaches to minimising risk will be introduced.

ACTIVITY 4.1: UNDERSTAND

Watch the following video clips online to help develop your understanding of the human microbiome.

These online videos can be accessed by **scanning the QR codes** with your smart phone or via https://study.sagepub.com/essentialap2e.

THE INVISIBLE UNIVERSE
(5:28)

YOU ARE MAINLY MICROBE
(3:53)

---

[1]Living on or within another organism, deriving benefit without harming or benefiting the host.

[2]A relationship between different species where both organisms benefit from the presence of the other.

[3]An agent causing disease or illness to its host.

Context

Within the Bodie family, the different individuals will have different microbiota. Danielle, the baby, is still being breast-fed and will have acquired most of her microbes during and since birth from her mother. The microbiota of the adults will have developed over time, influenced by the different environments, foods and drink ingested, individuals with whom they have contact and hygiene practices. While close family members are likely to have a degree of similarity in their microbiota, they will also have considerable variation.

BODIE FAMILY

MICROBIAL STRUCTURE AND FUNCTION

Introduction

The number of microbial cells normally living on and in the human body is in the region of 10^{14} (100 trillion) bacteria and 10^{15} (a quadrillion) viruses compared to about 10^{13} (10 trillion) human cells comprising the body.[4] Other microbes are present in smaller numbers. Each person's microbiome is unique, varying between individuals, between different parts of the body and over time. Most of the focus on microbes in healthcare education is on those that cause disease, which is not the emphasis here: in this chapter we are considering microbes and human health. However, the types of organisms involved in both contexts are similar in structure and much of their function: this section provides an overall introduction.

There are a number of different types of microbes that may be found in the human body: most are prokaryotes, both bacteria and some archaea, a few eukaryotes – fungi, protozoa (protists) and helminths (worms), and many viruses (mostly bacteriophages – i.e. that attack bacteria) (Todar, 2012). Table 4.1 indicates the main types of microorganisms found in humans. Their metabolism may be:

- Obligate aerobes: can only grow in presence of oxygen.
- Facultative aerobes: can grow if oxygen is available or not.
- Obligate anaerobes: cannot grow if any oxygen is available.

Relationship between microbes and host

Microbes have differing relationships with their hosts. In this section, we will look at some of these relationships, specifically for microbes that are transients, commensals, opportunists and pathogens.

---

[4]Understanding the numbering of microbial cells is complicated by the different systems seen in various publications. Until fairly recently British and German publications used the long scale (each new term a *million* times larger than the previous one) and American and French ones used the short scale (over one million, each new term one *thousand* times larger than the previous one). Modern British use is normally the short scale and is used in this text.

Long Scale			Short Scale		
$10^{15} =$	1,000,000,000,000,000	= 1000 billion	$10^{15} =$	1,000,000,000,000,000	= 1 quadrillion
$10^{14} =$	100,000,000,000,000	= 100 billion	$10^{14} =$	100,000,000,000,000	= 100 trillion
$10^{13} =$	10,000,000,000,000	= 10 billion	$10^{13} =$	10,000,000,000,000	= 10 trillion

Transients

These microbes get incorporated briefly into the microbiota from the diet or environment. However, they do not remain long and are unable to colonise the body because:

- the normal flora have already taken up the available sites,
- the immune system eliminates them,
- physical or chemical characteristics of the body prevent their growth.

Table 4.1   Main types of microbial organisms

Microbe type	Class of organism	Structure	Metabolism	Additional information
Bacteria	Prokaryotes	DNA or RNA within cell without membrane surrounding organelles. No nucleus	Varied oxygen requirements	(Figures 4.2–4.4)
Archaea	Prokaryotes			Much research still required
Viruses	'At the edge of life'	RNA or DNA (genetic material), surrounded by a protein coat, sometimes with a lipid coat	Dependent on host cell metabolism for replication	Non-active outside host Replicate only inside living cells of host using materials of cell to produce proteins encoded in RNA or DNA and form viruses Each virus restricted to specific number/type of cell. Those that infect bacteria are known as 'phages' (Figures 4.7 and 4.8)
Fungi	Eukaryotes	Separate from animals and plants with some characteristics of both Cell nuclei surrounded by membrane	Rely on carbon fixed by other organisms	Most grow from their tips as tubular filaments (hyphae), which branch producing a network of hyphae. Some grow by budding or binary fission as single cells: includes moulds, yeasts Some produce spores dispersed for reproduction. Further research to clarify role in microbiota needed (Figure 4.9)
Protozoa/ protists	Eukaryotes	Eukaryote that is not animal plant or true fungus Independent cells or non-differentiated into tissues	Some use sunlight, others rely on organic compounds	Some are pathogens of animals, some of plants, some non-pathogenic (Figure 4.10)
Helminths	Eukaryotes	Worm-like parasites living in animal hosts	Feed on living hosts	Can cause weakness and disease but can also reduce incidence of allergy and autoimmune conditions (Johnston et al., 2014). Uncommon in developed countries (Figure 4.11)

Commensals

Commensals are the normal flora of the body in symbiosis with their host. They live harmlessly in or on their selected site and gain nutrients from their host. They may be beneficial to the individual carrying them as they prevent potentially harmful microbes becoming established. In addition, many of them have a positive effect by, for example, forming particular vitamins or influencing the immune system (discussed later) though these may be described as symbiotic.

Opportunists

These are microbes which are normally harmless within the body but can cause disease when the host's immune system defences are lowered by other disease or by therapy (e.g. radiotherapy, chemotherapy or various drugs). They can also cause infection when they gain access to a part of the body where they are not usually found (such as *Escherichia coli* normally found in the gut, causing urinary tract infection if it gains access to the bladder). This is endogenous (self) infection when the organisms originate from the same person. Exogenous (or cross) infection is when the microbes come from someone or somewhere else.

Pathogens

Pathogens can cause disease, although the virulence (severity of the disease it causes) depends on both the host and the organisms. Some microbes are highly virulent and even in small doses will almost always cause disease (e.g. plague or the Ebola virus), while others are less virulent and will only cause disease in large quantities or when the host is susceptible (see opportunists).

GO DEEPER

Classification of microbes

It will be easier to understand this chapter with some knowledge of the classification of microbes. In the 18th century, the Swedish botanist, physician and zoologist, Carl Linnaeus, developed the scientific taxonomy still used today, with some modifications (Figure 4.1). However, growing understanding of microorganisms leads to frequent changes in their classification. Staining (particularly Gram staining), used for visual recognition of bacteria, and distinctive aspects of metabolism have been used in classification. More recently, DNA and RNA fingerprinting have become important in microbial classification. At present there are 11,940 different species names for bacteria but only 451 for archaea - an area in which there is much research yet to be completed.

In this chapter phylum and genera (plural of genus) will be mainly used when discussing microbes in the human microbiome.

(Continued)

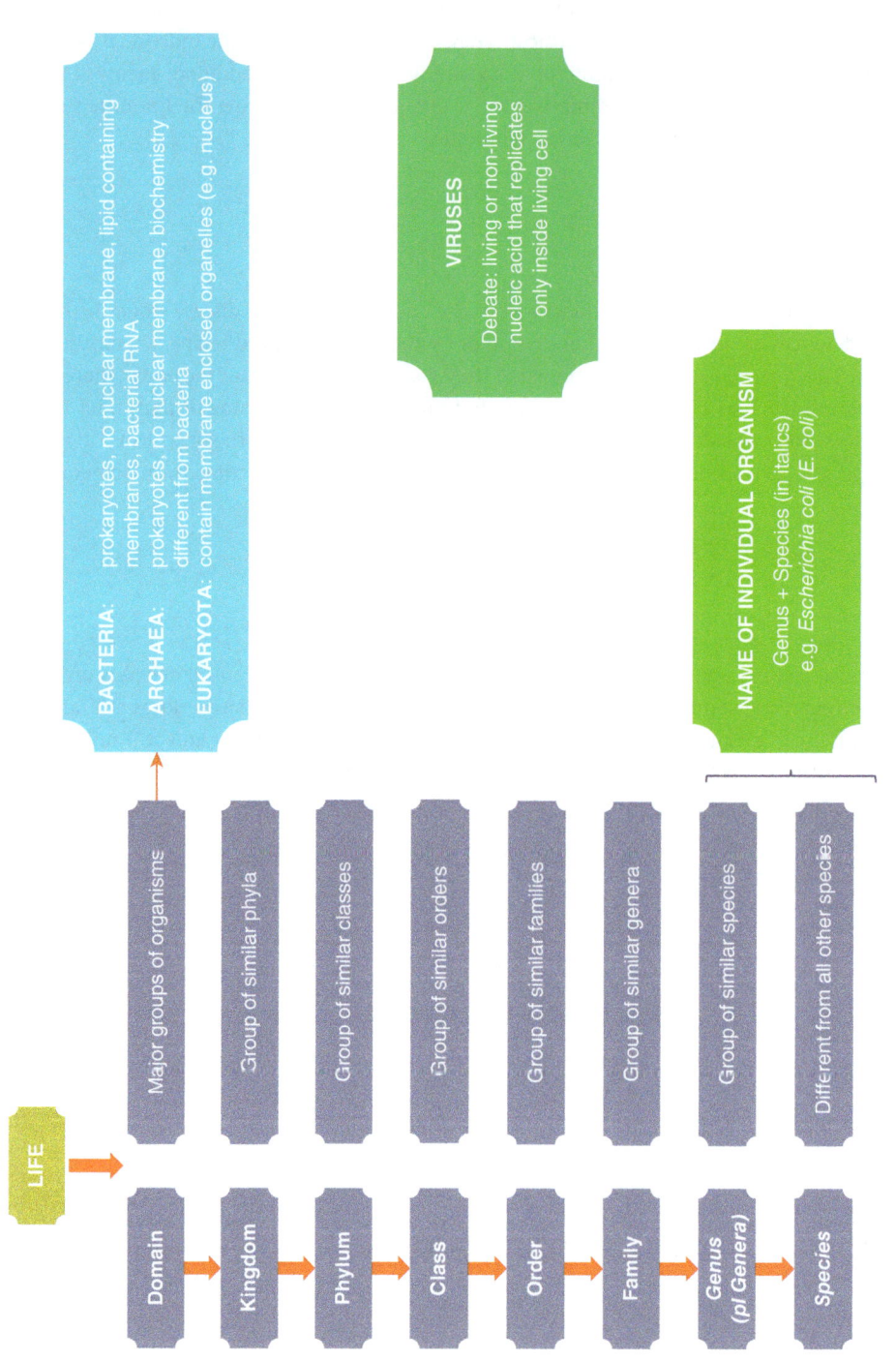

Figure 4.1 Classification of life

Types of microbes

Prokaryotes

Bacteria

We will start by looking at bacteria which are the most important microbes in the context of the human microbiome. They are single-celled prokaryotes and a typical bacterium is shown in Figure 4.2, although they differ in size and shape. The major difference from eukaryotes is that the cell organelles in bacteria are not enclosed by cell membrane but lie directly within the cytoplasm: they do not have an enclosed nucleus. It is considered that the mitochondria in eukaryotes are bacteria that, during evolution, took up residence and a major role in energy metabolism within the organism (archaea) that developed into the ancestor of present eukaryotes (Chapter 2).

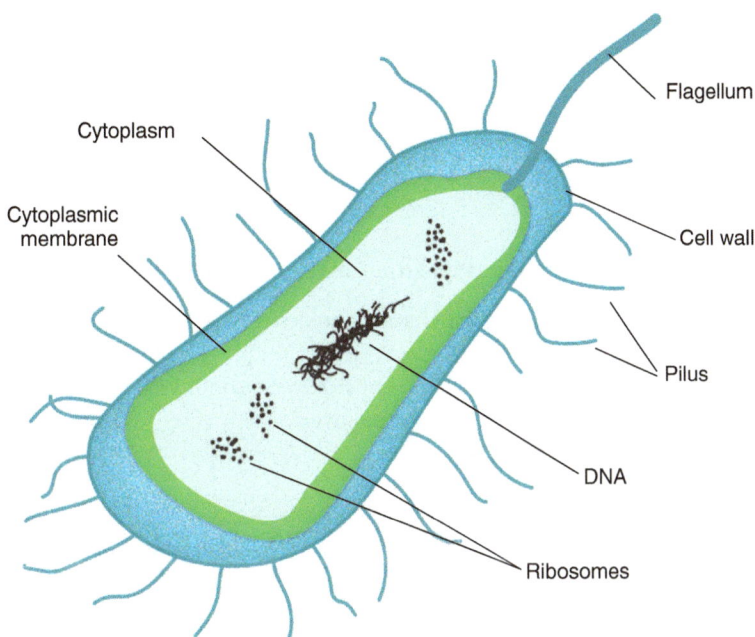

Figure 4.2 A bacterium

The following characteristics are used to describe bacteria:

1. Shape and organisation (Figure 4.3):

   i. Round (coccus) which exist individually and in pairs, chains and clusters.
   ii. Rod (bacillus) which also exist individually and in chains.
   iii. Curved (vibrio) individuals.
   iv. Spiral (spirochaete) individuals.

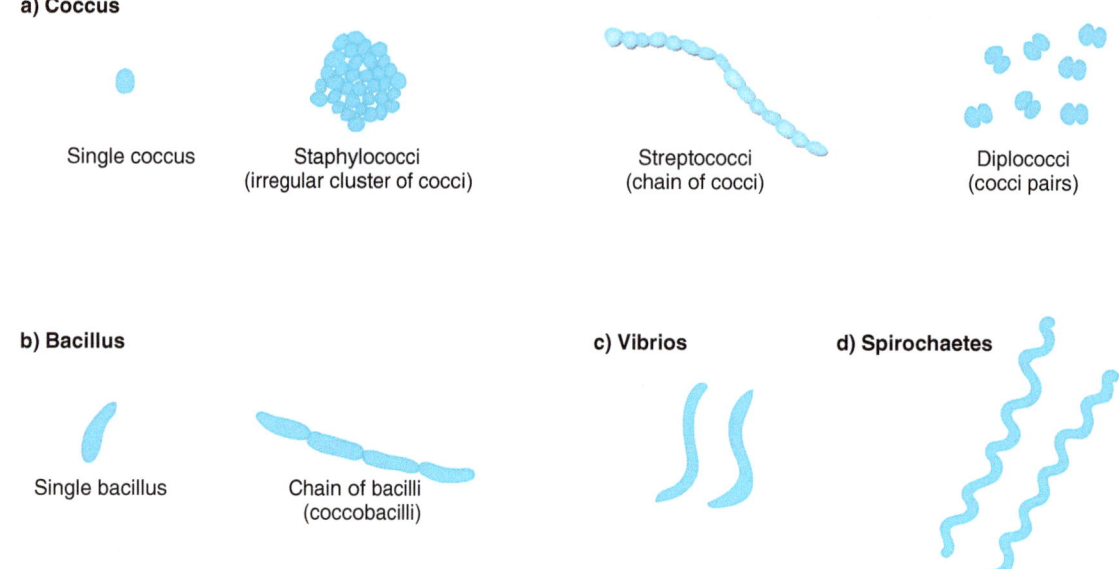

Figure 4.3 Shapes of bacteria

2. Structure: presence or absence of specialised structures in different bacteria (Figure 4.4):

   i. Mucous capsule protects against dehydration and desiccation in dry conditions.
   ii. Flagella enable bacteria to move.
   iii. Spore formation occurs in some bacteria under adverse conditions with germination and cell division recommencing when conditions improve.
   iv. Pili provide attachment to the host.

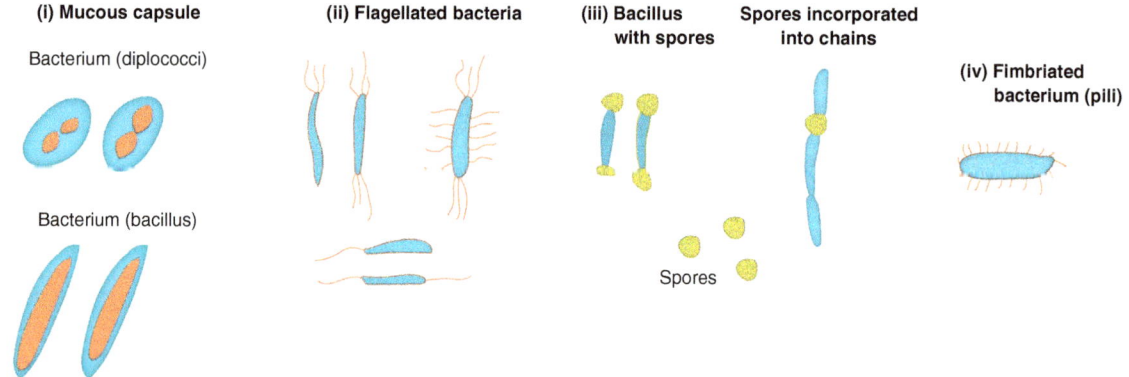

Figure 4.4 Specialised structures in bacteria

3. Effect of dyes on microbial specimens fixed on microscope slides. Bacteria are colourless and thus need to be stained in order to be viewed. There are a variety of stains but the Gram stain is a frequently used one that differentiates bacteria according to the physical and chemical structure of the cell wall. Bacteria are either Gram-positive (stained by the crystal violet dye) or Gram-negative

(stained red or pink by a counter stain) when the dark Gram stain is not absorbed by the cell wall. Some bacteria are Gram-variable due to variation in the thickness and structure of the cell wall.

4. Conditions of growth. Within the human microbiome there are a large number of commensal bacteria, some of which are opportunists. Some species of bacteria can be pathogenic or non-pathogenic depending on the environment, and some pathogenic species are similar to those within the normal flora (Gould and Brooker, 2008). The conditions for growth of bacteria vary with individual species but must meet their needs in relation to:

- Temperature – 35–40°Celsius.
- Moisture.
- pH level – Neutrophiles – optimum pH 7; Acidophiles – optimum below pH 5; Alkophiles – optimum above pH 8.
- Oxygen – Obligate aerobes – absolute requirement for oxygen in aerobic respiration, e.g. most environmental bacteria; obligate anaerobes – cannot use O_2 and can be killed by O_2 metabolites. They produce energy by anaerobic respiration/fermentation, e.g. inhabitants of large intestine. Facultative anaerobes – can grow in either the presence or absence of O_2. Use aerobic respiration if O_2 available but use fermentation or anaerobic respiration if O_2 unavailable (growth more rapid in O_2), e.g. *E.coli*.
- Energy source – Phototrophs – use sunlight as energy source, e.g. plants, algae, photosynthetic bacteria; chemotrophs – use chemical compounds as energy source (metabolism), e.g. mammalian cells, fungi, many types of bacteria.

Bacterial division

Different organisms will grow at different speeds and demonstrate exponential growth through binary fission as the DNA replicates and separates, and the cell divides into two identical daughter cells (Figure 4.5). The time taken for this process to occur is known as the generation time and varies with different species. *Escherichia (E.) coli* has a generation time of 20 minutes, thus the number of bacteria double in each 20 minutes.

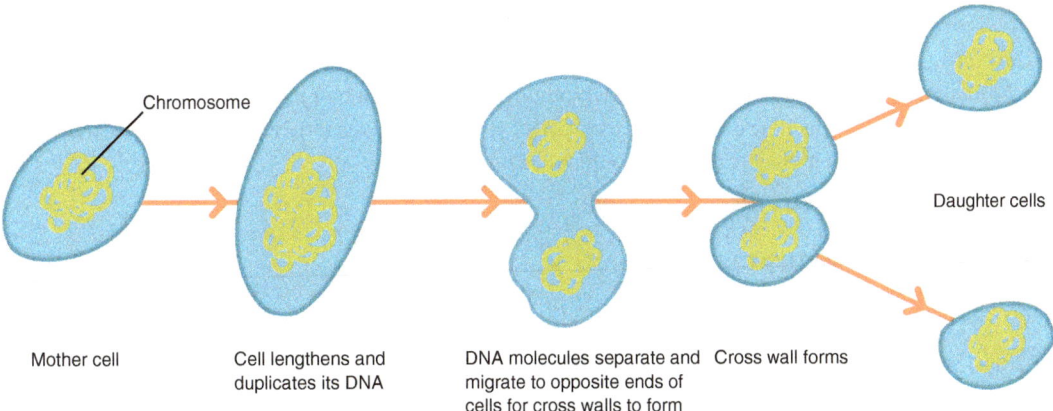

Chromosome

Daughter cells

Mother cell

Cell lengthens and duplicates its DNA

DNA molecules separate and migrate to opposite ends of cells for cross walls to form

Cross wall forms

Figure 4.5 Binary fission

Binary fission continues with the cell numbers doubling on each occasion, until the limits of the supplies of requirements for cell division are reached. Figure 4.6 illustrates the four phases of bacterial cell division:

- **Lag phase** is the period during which the bacteria are adjusting to the environment and preparing for cell division.
- **Log phase** is when the bacterial cell numbers are doubling at a constant rate and the population increases very rapidly. With pathogens, it is usually during this phase that symptoms develop.
- **Stationary phase** is when the bacterial numbers level off as cell growth and cell death are equal. This occurs as the nutrients are used up and waste products of metabolism (toxins) begin to accumulate.
- **Death phase** has a steady fall in cell numbers as there is an inadequate nutrient supply or high toxin levels.

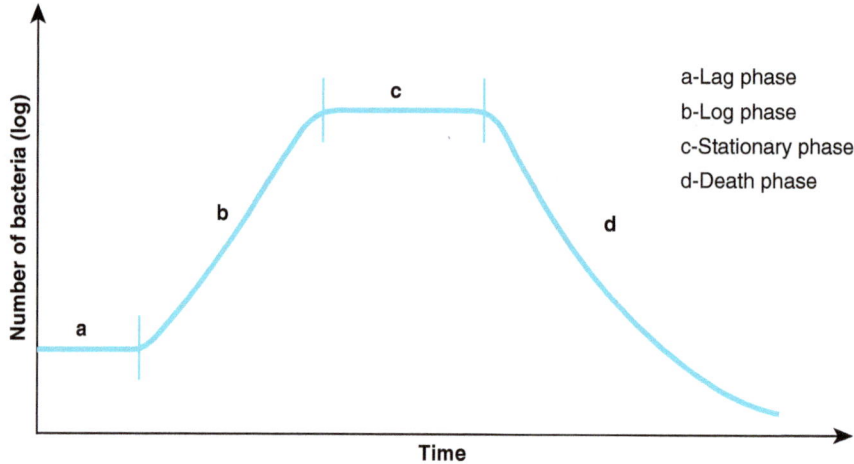

Figure 4.6 Bacterial growth

Archaea

These are similar in appearance to bacteria in that they have no membrane-bound organelles. However, some of their genes and metabolic pathways are more similar to those of eukaryotes although some other aspects of their metabolism are unique. They only reproduce asexually and do not form spores. These were initially thought to live only in extreme environments such as hot springs and salt lakes but are now known to exist within a wide range of habitats, including the human colon.

Methanogens are one type of archaea which produce methane as a by-product of metabolism in anaerobic conditions, such as in the Gastrointestinal Tract (GIT) of humans (and ruminants – contributing to greenhouse gases and global warming) (Gould and Brooker, 2008).

Neither prokaryotes nor eukaryotes

Viruses

Viruses are intracellular parasites without independent life outside the cells they infect; all life is host to one or more viruses. Whether or not these are living organisms and their classification as a separate kingdom is appropriate is an ongoing debate. Viruses vary in shape and size (Figure 4.7 illustrates two types) and, in their simplest form, consist of the genetic material of nucleic acid (either DNA or RNA) surrounded by a protein capsule. Some are further surrounded by a membrane envelope. Viruses have generally been considered to be causes of disease, although the HMP and other research is amending this view. Most of the viruses in the human microbiome are bacteriophages, that is, they infect bacteria (Gould and Brooker, 2008).

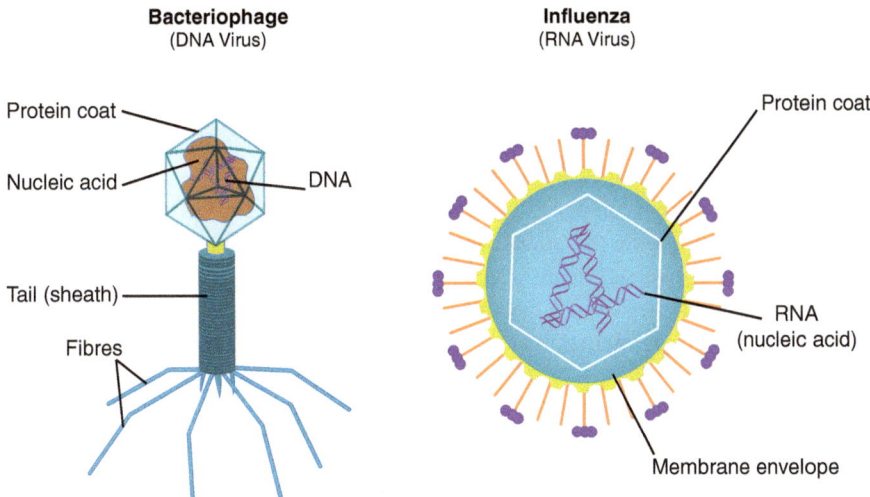

Figure 4.7 **Examples of viruses**

Viruses multiply within the cells of their host and are then released into their environment. Figure 4.8 illustrates the life-cycle.

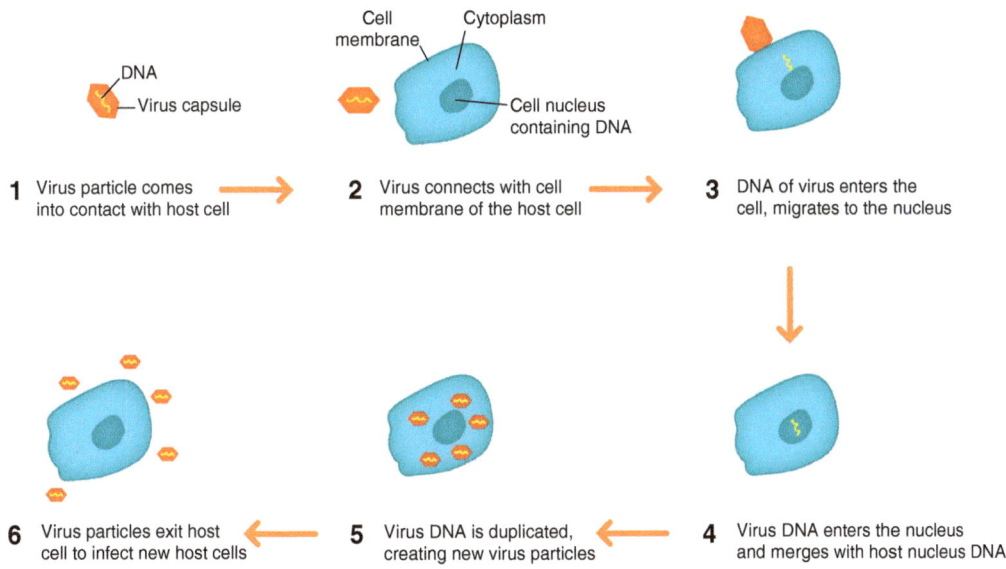

Figure 4.8 **Life-cycle of a virus**

The virome (i.e. the viruses within the microbiome) has been found to be relatively stable in an individual, although even people eating the same diet had differences in viral composition. However, alteration in diet resulted in changes in the proportion of a particular virus so that the viromes in those on similar diets were more similar than in those eating differently. Study of the human virome is continuing but there is still much to discover (Minot et al., 2011).

Eukaryotes

These all have an identifiable nucleus and other organelles surrounded by membrane.

Fungi

These are eukaryotes but are neither plants nor animals. Some have a simple structure and exist as unicellular organisms, for example yeasts. Others are more complex and exist as a mycelium, an interwoven mat of tubular filaments or hyphae. Figure 4.9 illustrates these.

Some fungi are found in healthy individuals but can cause disease if the environment becomes compromised. For example, commensal *Candida* can become pathogenic because:

> defects in the immune system, genetic predispositions, breaches in skin and mucosal barrier integrity, as well as microbial dysbiosis *(imbalance)* can all be factors predisposing to *Candida* infection and invasion. (Iliev and Underhill, 2013: 369)

The evidence of benefit to the host of resident fungi is still sparse – further research is required (Huffnagle and Noverr, 2013).

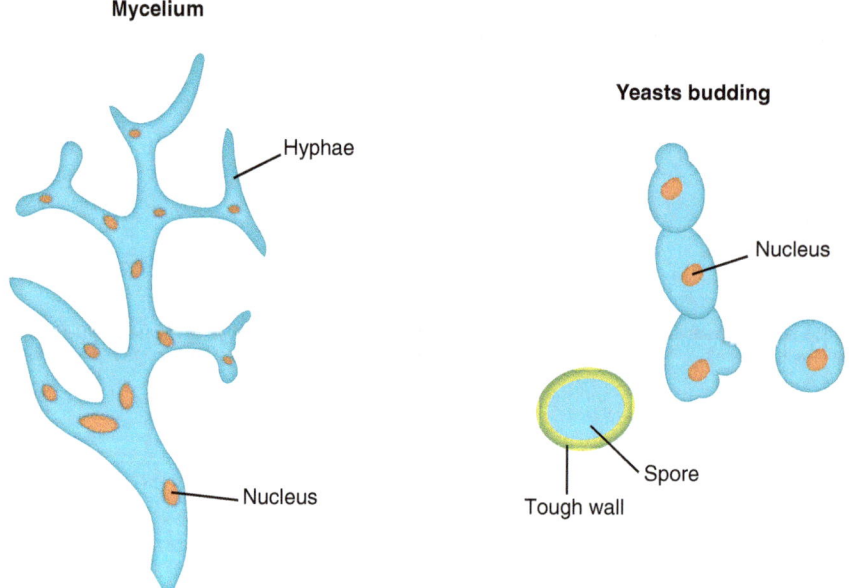

Figure 4.9   Fungi

Protozoa/protists

These are unicellular microorganisms most of which are harmless, but some are pathogens or opportunists. Examples of particularly unpleasant GIT pathogens are *Cryptosporidium* and *Giardia*. Figure 4.10 illustrates some protozoa.

Helminths (worms)

Helminths (Figure 4.11) are uncommon residents of the human gut in developed countries but more common where contamination of water and food is prevalent. While the idea of having worms resident

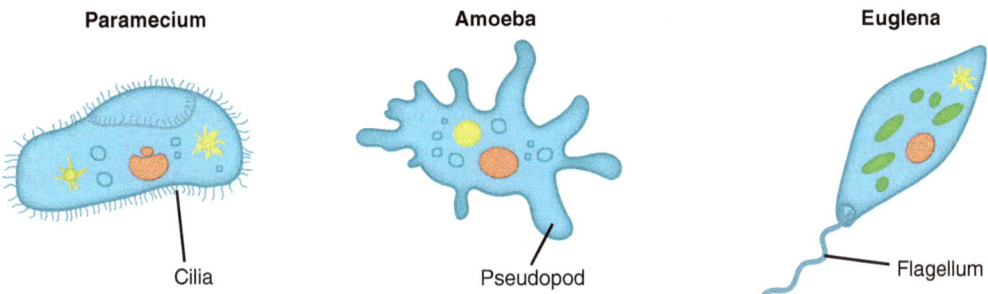

Figure 4.10 Examples of protozoa

in one's gut is unpleasant, some may have certain health benefits particularly in relation to allergic disorders (Erb, 2009). Epidemiological studies have shown a relationship between helminth infections and reduced incidence of allergic conditions, mainly in underdeveloped countries where such infections are more common than in more advanced countries (Johnston et al., 2014). However, the results from studies are varied and further research is required.

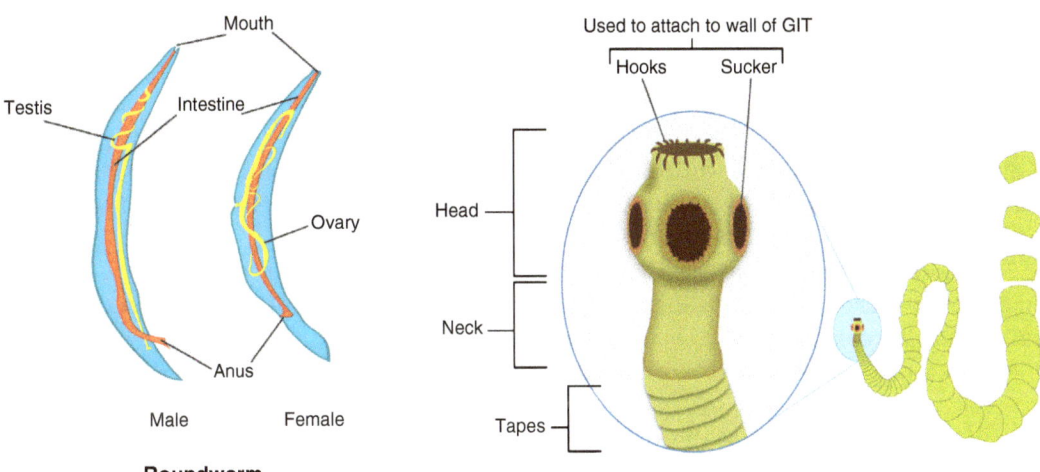

Figure 4.11 Examples of helminths

HUMAN MICROBIOME

Introduction

In-depth study of the human microbiome began with the Human Microbiome Project in 2008 using many of the methods developed within the Human Genome Project, which aimed to map the genes on the human chromosomes. The understanding of the role of the microbiome in contributing to human health is growing but it is already clear that it plays an important part in a number of host physiological activities which impinge on health status.

Pathogenic microbes in disease are discussed in detail in pathophysiology, clinical microbiology and nursing books in the context of infection control. Here we are primarily focusing on microbes in relation to the major sites of the human microbiome and health. Many different species of microbes live in and on the human body and are thought to account for 1–3% of total body mass. They exist on all surfaces exposed to the external environment particularly:

- **Gastrointestinal tract:** in essence a long tube joining the external environment at either end, it contains the majority of microbes.
- **Skin:** the largest organ of the body, which protects the internal organs from exposure to the external environment.
- **Mouth:** exposed to a wide range of microbes through food and drink.
- **Upper respiratory tract:** exposed to microbes in the air.
- **Vagina:** exposed to microbes from other individuals during sexual activity.

The makeup of the microbiota of the different sites is now being defined by genomic analysis, identifying the DNA composition of the site using similar techniques to those in the Human Genome Project. However, it is clear that there is much research still to be undertaken.

Development of the microbiome

In a review of the impact of the gut microbiota Clemente et al. (2012) discuss how the microbiota develops through life with the major changes occurring within the early years. The functions carried out by the microbiome alter according to the bacterial (and other microbial) composition.

Early development

Until recently it had been thought that the fetus was sterile and only acquired microorganisms after birth. However, there is now some evidence that a limited number and type of bacteria are present in the amniotic fluid surrounding the fetus (Jiménez et al., 2008).

The microbiota proper begins rapid development at birth as the infants are exposed to the microbes in their environment. Dominguez-Belloa et al. (2010) studied the influence of the mode of delivery of an infant on its microbial residents. A small sample of mothers and babies delivered vaginally (four) or by Caesarean Section (CS) (five mothers and six babies) had microbial smears taken as follows:

- **Mothers:** skin, oral mucosa, vagina – 1 hour before delivery.
- **Babies:**
  ○ skin, oral mucosa, nasopharyngeal aspirate – within five minutes,
  ○ meconium (earliest stool passed) – within 24 hours of delivery.

The mothers' results varied and showed that the microbial communities at the different sites examined on the body were different, but the infants' results were completely different. Each infant showed little variation across the body, but the bacteria identified were different for the two groups of babies. All the infants delivered vaginally had bacteria from all parts of their body that were similar to the vaginal microbes of the mother. Those born by CS had no maternal vaginal bacteria but had bacteria similar to those on the mother's skin. These differences may result in persistent differences and delays in the development of the adult microbiome.

In early life when the diet is primarily milk, the microbiome contains microbes with the ability to utilise lactate, and the composition of the gut microbiota begins to change to utilise plant foods even before solid food is introduced to the infant's diet. Initially, the microbes in the gut are oxygen tolerant, but are replaced by anaerobic bacteria as found in adults. The number of different types of bacteria and viruses in the newborn infant is relatively low, but with solid food the infant microbiota diversifies further. Gradually over time it becomes more like that found in adults.

APPLY

The microbiome in early life

Danielle was born naturally, and at two months, her microbiota is still similar to her mother's vaginal microbes. As she reaches six months she will be introduced to solids, which may be in the form of pureed food from her parents' diet. She is likely to develop a microbiota similar to her parents.

The makeup of the initial microbiota may influence functions associated with nutrition and immune systems. It has been reported that CS delivered infants are more liable to develop allergies and asthma than those vaginally delivered (for example: Bager et al., 2008; Tollånes et al., 2008). However, this has been disputed by Maitra et al. (2004) in a large Dutch study.

BODIE FAMILY

Later development

By the first birthday, the microbiota is beginning to resemble that of adults and normally reaches this state by two-and-a-half years of age. Usually it then remains stable during adulthood if diet, disease, environment, etc., remain constant. In older people there is greater variability in the proportions of types of microbes than in younger adults. This may be due to changes associated with ageing and the effects of medication use. The microbiota of the same site in different individuals are more similar than of different sites in the one individual, fitting the microbiological concept proposed in 1934 by Baas Becking and Beijerinck (cited by De Wit and Bouvier, 2006) of 'Everything is everywhere, but the environment selects'.

MICROBIOME, HEALTH AND DISEASE

The microbiome, the overall genetic makeup of the microbiota, plays an important role in maintaining health through interaction between the different microbes and with their host. Most microbes in and on us are commensals and some perform functions that our own cells have not evolved to undertake themselves. While we specify some microbes in this chapter, more important is understanding the functions that they perform – as many different species can play identical roles.

Gastrointestinal tract (GIT)

Within an individual the makeup of the microbiota remains relatively stable. The level of bacterial content varies through the gut with the stomach virtually sterile and the content increasing through the small intestine to the highest level in the colon (Canny and McCormick, 2008). Although there is variation between individuals, Ley et al. (2006) demonstrate that the gut microbiota is inherited from the mother and is similar in closely related individuals. However, it is suggested that 'each host has a unique biological relationship with its gut microbiota and this influences an individual's risk of disease' (Kinross et al., 2011: 14).

GO DEEPER

Gut microbiota

The gut microbiota can include several hundred species which tend to be relatively stable in any one individual although they vary between people. The main groups of bacteria identified (Eckburg et al., 2005; cited by Matsuki and Tanaka, 2014) are:

(Continued)

- two dominant bacterial groups:
  - firmicutes,
  - bacteroidetes;
- followed by:
  - actinobacteria,
  - proteobacteria,
  - verrucomicrobia,
  - and others.

It is also useful to note that the microbial diversity in the lumen varies compared to that of the surface of the intestine.

Functions of the gut microbiota

The relationship between gut microbes and host is a symbiotic one in which both sides benefit. Research with germ-free animals has helped to clarify the contribution to health of the microbiota. Functions carried out by the GIT microbiome are discussed by Matsuki and Tanaka (2014).

Energy metabolism

This involves the metabolism of carbohydrates and formation of Long Chain Fatty Acids (LCFAs). The microbiota of the GIT plays an important role in utilisation of 'resistant' starch foods. Essentially these are high-fibre carbohydrates derived from plants which are not digested effectively within the small intestine, but undergo fermentation and assimilation in the large intestine, meeting perhaps 10% of the calorie needs of the body

The metabolic activity of gut microbes creates a range of Short Chain Fatty Acids (SCFAs) with acetate, butyrate and propionate being the three most common, occurring in varying proportions in different people (Figure 4.12).

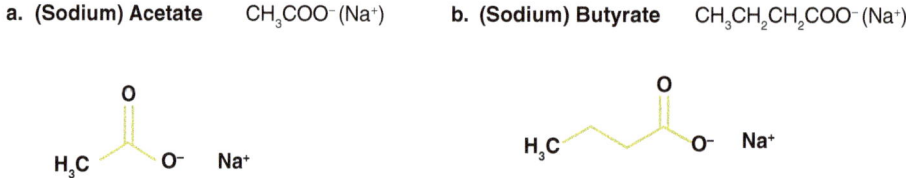

a. (Sodium) Acetate CH_3COO^- (Na^+) **b. (Sodium) Butyrate** $CH_3CH_2CH_2COO^-$ (Na^+)

c. (Sodium) Propionate $CH_3CH_2COO^-$ (Na^+)

Figure 4.12   Main short chain fatty acids (SCFAs)

GO DEEPER

Short chain fatty acids and metabolism

Acetate is produced by most anaerobic bacteria, while the bacteroides and firmicutes are the main producers of butyrate and propionate. These have an important effect on the quality of the environment of the large intestine and absorption from the gut into the internal environment. These three SCFA play important but different roles in metabolism (Russell et al., 2013) although all are utilised by the colonic mucosa:

- Acetate reaches the highest levels in the plasma and is used in lipogenesis (formation of lipids, Chapter 9) in both liver and fat cells.
- Butyrate is important as an energy source for the epithelium of the colon, but it also appears to help prevent inflammation, and possibly prevents colorectal cancer by inhibiting cellular division and promoting apoptosis (Fung et al., 2012). Decreased numbers of butyrate-producing bacteria are found in those with Crohn's disease (Sokol et al., 2008; cited by Russell et al., 2013).
- Propionate is used in gluconeogenesis (formation of new glucose, Chapter 9). It is thought to have an important role in satiety, protection against diet-induced obesity and improving insulin sensitivity (Chapter 7) (Arora et al., 2011).

Formation of vitamin K

Vitamin K is a fat soluble vitamin that plays an important role in blood clotting and a deficiency can result in low prothrombin levels and haemorrhage. While deficiency is rare in adults, it is common in newborn babies who have low levels of vitamin K, which is used up fairly quickly and leaves the baby at risk of haemorrhage. In the UK they are usually routinely administered vitamin K by injection. Microbes in the gut also manufacture vitamin K which is then absorbed into the body. However, in someone who has large amounts of antibiotics the microbes in the gut may be altered leading to a reduction in the amount of vitamin K formed, and this can result in hypoprothrombinaemia and increase the risk of bleeding.

There is also some evidence that vitamin K plays a role in bone density (Bügel, 2008) with insufficiency associated with low bone density and increased fractures. Human studies have demonstrated that vitamin K increases bone mineral density in people with osteoporosis and reduces fracture rates particularly in combination with vitamin D (Weber, 2001).

UNDERSTAND

Hypoprothrombinaemia

Hypo- = lower than normal,

prothrombin = the precursor to thrombin,

-aemia = in the blood.

Hypothrombinaemia = low levels of prothrombin in the blood

Drug metabolism

Some drugs and other xenobiotic substances (i.e. not normally found in the body) are metabolised by microbes in the gut (Kinross et al., 2011), which may reduce the efficacy of the drugs.

APPLY

Digoxin and gut microbiota

Digoxin is an example of a drug shown to be influenced by the gut microbiota. In about 10% of people, a particular bacterium in the gut was found to convert a significant amount of digoxin to an inactive substance; antibiotic treatment was also found to increase plasma digoxin levels (Lindenbaum et al., 1981; Saha et al., 1983; both cited by Matsuki and Tanaka, 2014).

Development of intestinal structure

Although more research is needed, it appears that substances produced by the gut microbiota regulate the formation of the normal structure of the intestinal epithelium (Chapter 8). In addition, these microbes stimulate the formation of Paneth cells (endocrine cells in the crypts of the epithelium), which stimulate the formation of the intestinal capillary network.

Maturation of the immune system

The microbiota is constantly in contact with the intestinal mucosa and helps to define the barrier between the external and the internal environment. The microorganisms interact with the immune system and stimulate the development of the Peyer's patches (aggregations of lymph tissue within the lower part of the ileum) (Chapters 8 and 13) and lymphoid tissue. The commensal microorganisms play an important part in the maturation of the gut and its immune system, as well as influencing the immune system more widely.

Prevention of pathogenic infection

The commensals in the intestine fill various 'ecological niches' and compete with transient pathogens for nutrients, thus inhibiting them from becoming resident. *Lactobacilli*, common in the GIT, secrete lactic acid and other molecules which also inhibit growth of competing microbes. Microbial diversity varies between the lumen of the gut and the surfaces of the intestine.

Illness and the gut microbiota

In recent years, researchers have identified relationships between the human microbiome and a variety of human diseases. Not unexpectedly, some relationships between the microbiome and some intestinal disorders, for example, Inflammatory Bowel Disease (IBD), have been identified. However, a number of other disorders have also been linked to the gut microbes (Figure 4.13).

While we all know about GIT disorders caused by a specific microbe, it has also been suggested that there is another circumstance which can cause disease. The microbiota as a community may be pathogenic in combination with other risk factors such as host genotype, behaviour and diet. An example suggested by Ley et al. (2006) is a person with ready access to food and a microbiota which is highly efficient in extracting energy: this person will be at risk of becoming obese, with the possibility of developing diabetes mellitus.

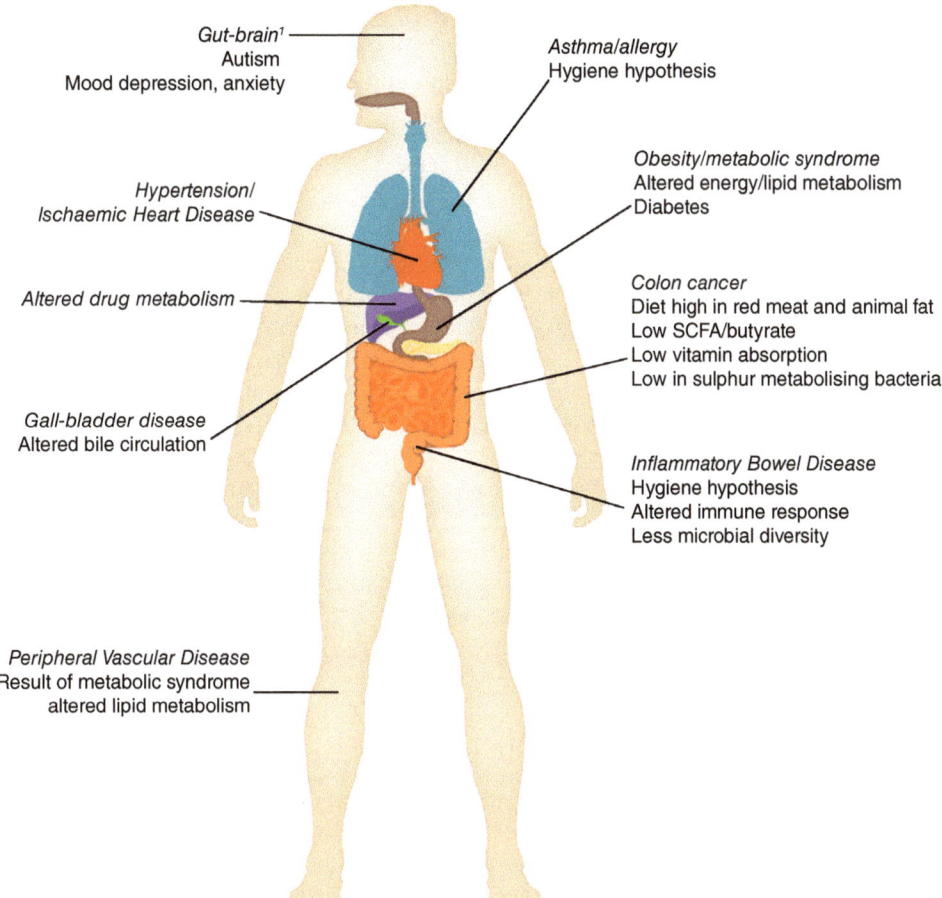

Gut-brain[1]
Autism
Mood depression, anxiety

Asthma/allergy
Hygiene hypothesis

Obesity/metabolic syndrome
Altered energy/lipid metabolism
Diabetes

Hypertension/
Ischaemic Heart Disease

Altered drug metabolism

Colon cancer
Diet high in red meat and animal fat
Low SCFA/butyrate
Low vitamin absorption
Low in sulphur metabolising bacteria

Gall-bladder disease
Altered bile circulation

Inflammatory Bowel Disease
Hygiene hypothesis
Altered immune response
Less microbial diversity

Peripheral Vascular Disease
Result of metabolic syndrome
altered lipid metabolism

Figure 4.13 **Gut microbiota and disease**

Adapted from: Kinross et al (2011: 14) by permission of Springer Nature. Gut microbiome-host interactions in health and disease, *Genome Medicine*, Kinross et al. © 2011.

[1]This refers to the gut-brain axis in which there is communication between the gastrointestinal tract and the brain, including the intestinal microbes, through neural, endocrine and immune pathways. This is important in promoting healthy brain function and behaviour.

Changes in diet can alter the microbiome remarkably fast with clear differences occurring within 24 hours when human subjects changed their diets from high-fat/low-fibre to low-fat/high-fibre (Wu et al., 2011). However, one of the difficulties in this area is relating changes in species type and abundance with change in function: the genetic plasticity of bacteria means that this is not necessarily the case.

Antibiotic use is one of the factors that has most effect on the gut microbiota in particular, tending to result in reduced diversity in the bacteria present and slow return to normal. This can allow colonisation by 'foreign' bacteria with permanent changes in the structure of the microbiota and, potentially, increased occurrence of disease. Use of broad-spectrum antibiotics has been associated with development of *Clostridium difficile* associated disease. It is important to note that antibiotics are ineffective against viral infections.

—— **GO DEEPER** ——

Antibiotics and resistance

Repeated use of antibiotics in human medicine is considered to be the main cause of the development of microbial antibiotic resistance. Antibiotics will kill many of the normal microbiota (as well as the pathogens) leaving behind those that are resistant to the antibiotic. These multiply and there is also some transfer of the antibiotic resistant genes across species.

However, in 1969 the Swann Report hypothesised that the use of antibiotics as growth factors in animal husbandry would result in the development of antibiotic resistant bacteria that could enter the human food chain. Some action was taken on this, including antibiotic use in animals being restricted to those drugs not used in humans, resulting in some reduction in antibiotic resistance. However, the similarity in structure between some antibiotics used in animals and humans was not foreseen and this use of antibiotics is still a matter of concern in the worrying growth of antibiotic resistance (Marshall and Levy, 2011).

One major example of antibiotic resistance is Methicillin-resistant *Staphylococcus aureus* (MRSA), sometimes known as a 'superbug'. This particular bacterium is resistant to a number of different antibiotics and thus causes a number of particularly difficult to treat infections. The general public are less likely to develop nosocomial infections (cross infection) than those in places where people have invasive devices, open wounds, and weakened immune systems including hospitals, prisons, and nursing homes. Careful hygiene practice is essential to minimise infection spreading.

Management of altered gut microbiota

A number of approaches involving the microbiome are being used to enhance health, including the use of probiotics, prebiotics and faecal transplants.

Advertisements for food or food supplements often mention probiotics or prebiotics in promoting health. These are defined as:

- **Probiotics:** Live microorganisms, which, when administered in adequate amounts, confer a health benefit on the host (e.g. yogurt containing live bacteria).

—— **APPLY** ——

Probiotic supplementation

The use of probiotics with premature babies is now recommended by the Cochrane Collaboration. It appears to reduce the incidence of Necrotising Enterocolitis (NEC) (affecting the bowel), which can occur in premature babies during the early weeks of life. A review by AlFaleh and Anabrees (2014) found that probiotics (supplements containing potentially beneficial bacteria or yeasts) reduced NEC and death in premature babies less than 1,500 grams in weight. However, the data on babies less than 1,000 grams in weight is not conclusive.

- **Prebiotics:** A food ingredient not digested in the small intestine (e.g. some fruits and vegetables, beans, unrefined grains) or that beneficially affects the host by selectively stimulating the growth and/or activity of one or a limited number of bacteria that can improve the host health (Reid, 2003; Gibson and Roberfroid, 1995; both cited in Rastall et al., 2005).

- Faecal transplants involve the introduction of a colony of microbes from a healthy donor to a recipient's colon. This treatment has now been tested in a number of studies and appears to be beneficial in a number of disorders including: multiple sclerosis, chronic fatigue syndrome, idiopathic thrombocytic purpura, ulcerative colitis, irritable bowel syndrome and diabetes mellitus (Vrieze et al., 2013). Recurrent diarrhoea caused by *Clostridium difficile* has been successfully treated by faecal microbiota transplantation (Orenstein et al., 2013).

ACTIVITY 4.2: UNDERSTAND

Watch the following video online to help develop your understanding of resistant starch and health.

This online video can be accessed by **scanning the QR code** with your smart phone or via https://study.sagepub.com/essentialap2e.

THE HUNGRY MICROBIOME
(4:03)

APPLY

Understanding diet for health

Hannah and Richard Jones of the Bodie family both enjoy wholemeal bread and plenty of fruit and vegetables in their diet. As one of his roles as a school teacher, Richard has been given responsibility for the introduction of free school meals for all school children in their first few years of primary school. He is aware that this is not just about ensuring that the children receive a good diet but also about their beginning to learn about the food they should be eating. He is ensuring that the menus supplied will provide good balanced diets with adequate prebiotic food every day and some probiotic foods during the week.

BODIE FAMILY

Skin

The skin is the body's largest organ (approximately 1.8–2 m²) and, as the interface with the external environment, we need to consider how it interacts with the external environment. The skin has its own ecosystem with a wide range of resident microbes including bacteria, viruses, fungi and, sometimes, mites. The rough texture of the epidermis means that many microbes can be resident in the different grooves of papillae, in hair follicles and in ducts of sweat glands (Figure 4.14). Most of these are commensals and protect against invasion by non-resident flora through action on sweat, partly by producing fatty acids which inhibit growth of fungi. Mites reside in the areas where hair covers the skin and are considered part of the normal flora, although they can sometimes cause skin disorders. The average adult has approximately 1,000 species of bacteria colonised on their skin, amounting to about 1 trillion bacteria, but the good news is that, largely, we are immune to our own skin microflora (Zulkowski, 2013). At the slightly acidic pH of skin (approximately 5.5) the normal flora of the skin is maintained, whereas an alkaline environment associated with urinary or faecal incontinence can result in skin irritation, dermatitis and excoriation (Andrews, 2012).

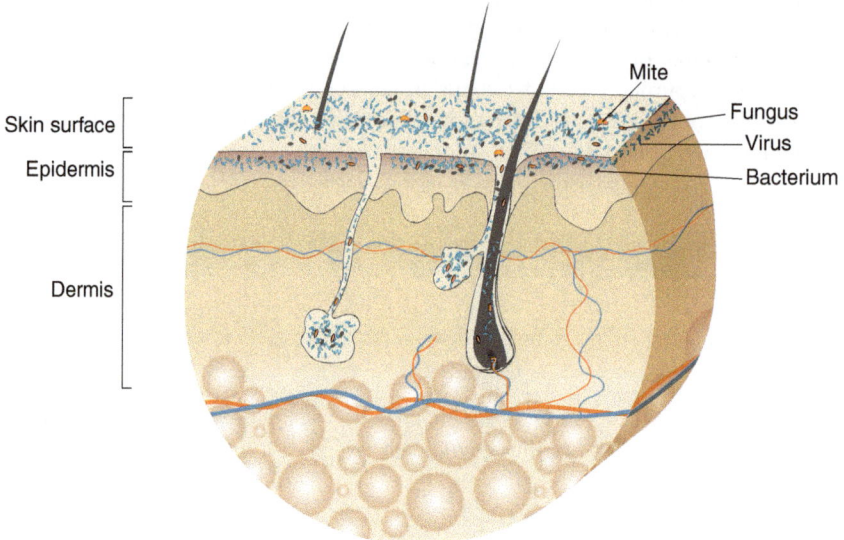

Figure 4.14   Microbiota of skin

Adapted by permission from Springer Nature: Grice, E.A. and Serge, J.A. 'The skin microbiome', *Nature Reviews Microbiology*, 9(4): 244-253, © 2011.

The microbial community varies with the environmental differences of the various sites of the body due to the range of factors identified (Grice and Serge, 2011) (Figure 4.15). A major factor in bacterial variation seems to be moisture and warmth, with such areas as the axillae and groins being more likely to have higher levels of microflora. Other sites have considerably lower microbial counts with most found in the sweat glands. An outline of the most common bacteria of the skin microbiota is given in Table 4.2. Further research is required to identify the fungi and viruses in the skin microbiota.

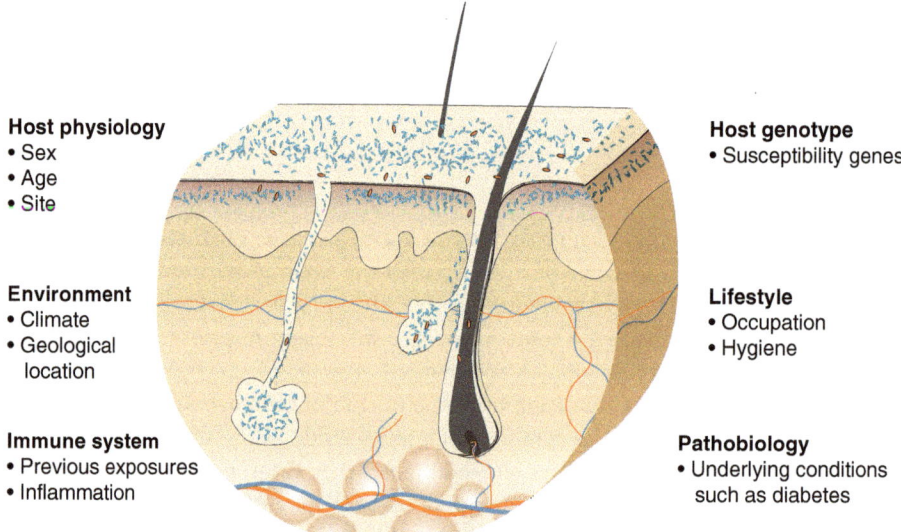

Figure 4.15   Environmental factors influencing skin microbiota

Adapted by permission from Springer Nature: Grice, E.A. and Serge, J.A. 'The skin microbiome', *Nature Reviews Microbiology*, 9(4): 244-253, © 2011.

Table 4.2 Microbes in different areas of skin

Skin environment	Skin sites	Common microbes present	Notes
Moist	Umbilicus, axilla, groin areas, soles of feet, gluteal crease (between buttocks), popliteal fossa (behind knee), antecubital fossa (inner elbow)	Most abundant organisms are: • *Staphylococcus* species • *Corynebacterium* species	The actions of these microbes in sweat results in unpleasant odour
Dry	Forearm, buttocks, parts of hand	Mixed microbiota from bacterial phyla • *Actinobacteria* • *Proteobacteria* • *Firmicutes* • *Bacteroidetes*	Greater diversity than gut or oral cavity of same individual
Sebaceous (sebum (lipid-rich substance) secreting glands)	Forehead, behind ear, the back, side of nostrils	Dominant microbes are lipophilic (lipid-liking) • *Propionibacterium* species	Diversity lowest

Adapted by permission from Macmillan Publishers Ltd: Grice, E.A. and Serge, J.A. 'The skin microbiome', *Nature Reviews Microbiology*, 9(4): 244-253, ©2011.

APPLY

Preventing self-contamination

Considering that we have bacteria residing on our skin, there is a necessity to consider how to remove such bacteria in certain situations. For example, those going for surgery are at risk of self-contamination from their skin flora getting into open wounds. Few studies have been conclusive in identifying how to effectively decolonise the skin. Darouiche et al. (2010) influenced the Department of Health (2011) when they identified that 2% chlorhexidine in 70% isopropyl alcohol is more effective than povidone-iodine in reducing the risk of post-operative wound infection.

In addition to the normal skin flora which generally do not give rise to infection or cause cross-contamination, transient microbes are those we pick up in everyday activities which can be pathogenic (Gould, 2012). Once a person becomes immunocompromised, commensals also have the potential to become pathogenic.

ACTIVITY 4.3: APPLY

Skin care and infection control

We know from different studies that people with chronic conditions such as diabetes, chronic renal failure and dermatitis have an increased likelihood of *Staphylococcus aureus* skin colonisation

(Continued)

(Young, 2011). People in hospital are also likely to have Gram-negative bacteria on their skin with an increased likelihood of transient and pathogenic bacteria available for cross contamination. When we shed dead skin cells, they also house bacteria. Some people in hospitals may be immunocompromised, and the above issues create a particular risk.

Task

Many things we do in practice can minimise or exacerbate these risks. As a person-centred nurse, you need to manage these aspects of practice without diminishing someone's sense of self-worth. To be knowledgeable and informed, take some time now to check out the impact of the following on managing skin flora and preventing cross contamination. What is the evidence telling you in relation to how effective the following are in reducing cross contamination?

- The use of paper towels to dry hands after washing them.
- Use of air hand dryers.
- The effectiveness of soap when used to cleanse skin.
- The effectiveness of skin wipes when used to cleanse skin.
- The effectiveness of hand washing by healthcare staff.
- The effectiveness of changing bed linen in care settings (hospitals, nursing homes, people's homes).

You may need to use academic literature databases such as Ovid™ or Medline™ to source the evidence, or you can also try using Google Scholar™.

Mouth

The oral microbiota is the second most complex (after the GIT) containing mainly bacteria, archaea, viruses and fungi (see Table 4.3). We have already discussed the development of the microbiome in general, but the mouth has some distinct characteristics. Toothless infants already have a diverse microbiota but once teeth erupt there are considerably more ecological niches open to colonisation by a greater range of microbes (Zaura et al., 2014).

APPLY

Oral hygiene

Some of the microbes (of a single or mixed species) in the mouth form biofilms in which bacteria stick to a surface (teeth, dentures, etc.) and secrete a slimy, glue-like substance which sticks to the surface. This protects the bacteria from harm and forms the plaque which dental hygienists clean off to prevent gum disease, gingivitis (early stage) or more severe periodontitis in which the supporting tissues of the teeth may become affected.

Good mouthcare including regular brushing of the teeth, flossing between the teeth and use of mouthwashes all help to remove the biofilms and reduce development of gum disease. If it does develop, chemical agents in the toothpaste, etc., will kill the bacteria in the biofilm.

Table 4.3 Oral microbiota

Microbe	Number and type(s)	Comments
Bacteria	Around 1000 species: about half still to be cultured	No significant differences across geographical sites
	Diverse collection of obligate aerobes, facultative and obligate anaerobes	A lot of fermentable carbohydrates results in bacteria that create acid and cause caries (dental)
	Six phyla contribute 96% of species: firmicutes, bacteroidetes, proteobacteria, actinobacteria, spirochaetes, fusobacteria	Oral microbiota inhibits pathogenic colonisation - few binding sites for pathogens
	Three distinct bacterial communities distributed as follows:	
	1. Buccal mucosa, gingivae, hard palate	Conversion of nitrate to nitrite (converted to nitric oxide)
	2. Saliva, tongue, tonsils, throat	Help keep blood vessels flexible resulting in anti-hypertensive effect
	3. Supra- and sub-gingival plaque	
Viruses	Mainly bacteriophages (i.e. attack bacteria) - fits with range of bacteria	Primarily associated with disease
Fungi	*Candida* species carried by about half the population without symptoms	
	85 fungal genera reported	
Protozoa/ protists	Mainly *Entamoeba gingivalis* and *Trichomonas tenax*. Harmless saprophytes	Nutritional link with oral disease - food debris is food source for protozoa
	Numbers raised in those with poor oral hygiene	
Archaea	Minor part of microbiome. All species are methanogens (form methane)	Numbers raised in those with periodontitis

Adapted from: Wade (2013) and Grice and Segre (2012).

APPLY

Oral microbes and illness

We can understand that the oral microbiota could be linked to disorders of the mouth, but it also has health effects unrelated to the GIT, possibly through its influence on the immune system or on inflammation. Diabetes and atherosclerosis are examples. Bacterial types and numbers in the mouth correlate with those found in atherosclerotic plaque (fat and other substances that are laid down in the lining of the artery wall) (Koren et al., 2011). They can also result in infectious endocarditis (inflammation of the lining of the heart), and brain and liver abscesses.

While airborne bacteria are the commonest cause of lung infections in those with cystic fibrosis, oral microbes have also been found in the lungs of these people (Wade, 2013).

Respiratory tract

A large number of different species of microbes inhabit the Upper Respiratory Tract (URT), some of which are shared with the mouth. *Staphylococcus epidermidis* and corynebacteria are common in the nose, and *Staphylococcus aureus* (pathogenic) is not uncommonly carried here. The throat (pharynx) microbiota usually includes a range of different streptococci and other microbes.

--- **APPLY** ---

MRSA

MRSA (Methicillin Resistant *Staphylococcus aureus*) is often carried in the nose without causing an infection. It used to be acquired in hospital settings but is now also found in some healthy individuals in the community. However, it is a major cause of cross-infection in health facilities.

It used to be considered that the Lower Respiratory Tract (LRT) was sterile. However, it has now been demonstrated that, in health, organisms similar to those in the URT, but in much smaller numbers, are present (Charlson et al., 2011). The ciliated epithelial cells of the LRT play an important part in minimising the numbers of microbes in the lungs by wafting mucus containing transient microorganisms up to the URT from which they are eliminated (Todar, 2012).

--- **APPLY** ---

Microbiota of the respiratory tract and illness

When the efficiency of pulmonary defence mechanisms is diminished, for whatever reason, inhaled or resident microorganisms can proliferate and cause disease – acute or chronic infections. Predisposing causes can include inherent defects as in cystic fibrosis. Impairment of normal airway function can occur from inhaled particles or smoke (e.g. from cigarette smoking) and, over the long term, can lead to Chronic Obstructive Airway Disease (COAD) also known as Chronic Obstructive Pulmonary Disease (COPD).

Vagina

The microbiota of the vagina changes through the different stages of the reproductive life of a woman because of variations in hormone levels, sexual activity and hygiene, and it plays an important role in minimising disease. A considerable number of different bacteria are found colonising the vagina in different combinations, although lactobacilli are the most important.

Childhood

During a normal birth a girl baby acquires lactobacilli in her vagina as bacteria are transferred from the mother's vagina. She also has some oestrogens left from her mother enabling the lactobacilli to thrive, but these diminish as the oestrogen levels fall. During childhood the pH of the vagina is slightly alkaline or neutral. A mixed colony of microbes has been described with some bacteria at fairly high levels and others lower. Lactobacilli were found more often among older girls (in 88% of those aged 11), while this was lower in younger children (in 45% of those under two years) (Hammerschlag et al. 1978; cited by Farage et al., 2010)

At puberty the rising levels of oestrogen promote division of the epithelial cells and an increase in intracellular glycogen at the mid-point of the menstrual cycle. The lactobacilli metabolise the glycogen, produce lactic acid and lower the pH of the vagina as it changes towards the microbiota of adult women.

Reproductive years

In the reproductive years, eight different *Lactobacillus* species are the commonest strains found but 92% of women studied only carried one of these. A range of other bacteria can be part of the normal flora for each woman. These bacteria maintain the pH of the vagina at 4.0–4.5 (acidic) by forming lactic acid, which maintains the health of the vaginal epithelium (Farage et al., 2010). *Lactobacillus* species stick to the epithelial cells of the vagina and thus prevent pathogens colonising this site. The levels of oestrogens vary during the menstrual cycle causing variation in the lactobacilli. During pregnancy the microbiota is more stable than at other times and it is suggested that this protects against vaginal infections during pregnancy (Romero et al., 2014).

Post-menopausal years

During the menopause, as oestrogen levels drop, the glycogen content of the vaginal cells also falls leading to diminishing lactobacilli numbers. The resulting fall in lactic acid results in a rise in vaginal pH which facilitates the growth of pathogenic bacteria. Women taking Hormone Replacement Therapy (HRT) have lower pH levels than those without. In addition, of those not taking HRT only 20% had lactobacilli in their vaginas while 83% of those on HRT did carry these microbes (Farage et al., 2010).

CONCLUSION

This chapter has introduced the importance of the microbiome in human health. This is a topic that has received little attention in most anatomy and physiology books for nurses but is now of growing importance in research and in understanding how the body maintains health. Approaches to maintaining the health of the microbiome, with particular emphasis on the potential damage that can be caused by the incorrect use of antibiotics, have been reviewed. We hope this understanding will be useful in achieving person-centred practice.

This chapter has mentioned some disorders where alteration in the microbiome is implicated. However, it has not attempted to cover all the material relevant to pathogens and disease. These will be examined later in your course.

GO DEEPER

Further reading

Burton-Shepherd, A. (2015) 'Prebiotics and probiotics as novel therapeutic agents for obesity', *Nurse Prescribing*, 13(3): 136-9.

Capone, K., Nikolovski, J., Stamatas, G., Green, M., Cox, S. et al. (2012) 'Exploration of bacteria comprising the human skin microbiome throughout the first year of life', *Journal of Obstetric, Gynecologic and Neonatal Nursing*, 41: S147-8.

Ferranti, E., Dunbar, S., Dunlop, A. and Corwin, E. (2014) 'Things you didn't know about: the human gut microbiome', *Journal of Cardiovascular Nursing*, 29(6): 479-81.

KEY POINTS

1 There are a wide range of different types of microbes that occur in the human microbiota, a number of which are similar in structure to some pathogens, and which protect the body from colonisation by pathogens. Understanding this will help you to protect people from infections.

2 The development and the intrapersonal and interpersonal variability of the human microbiome occur due to diet, hygiene practices and environmental factors, and this understanding will help you to realise person-centred practice.

3 The value of diet in enhancing the microbiota has been introduced.

4 Understanding the appropriate use of antibiotics provides the background for patient education in relation to completing courses of treatment and will be valuable in providing care.

REVISE

This chapter covers a lot of material and it is best to break your revision into components. Consider revising in the following areas:

TEST YOUR KNOWLEDGE

1 The different relationships that can occur between microbes and the human body.

2 The structure of the different types of microbes.

3 The microbiome of the different systems of the body.

4 The development of the human microbiome throughout life.

5 The microbiome in health and disease in the different systems.

6 Approaches to promoting health of the GIT.

In order to help you revise, consider the following questions, answers for which can be found by visiting **https://study.sagepub.com/essentialap2e**.

Test yourself by revising the chapter first, and then answer these questions without looking at the book. Afterwards compare your answers with the text and with the notes you made. Did you miss anything in your notes? Have you looked at the figures? Here are the questions:

1 Describe the relationship between man and microbe in the following:

- Transients.
- Commensals.
- Opportunists.
- Pathogens.

2 What are the major groups of organisms and the major differences between them?

3 What are the main environments inhabited by the human microbiome?

4 What are the types of microbes found in the different areas inhabited by the human microbiome?

5 List and write brief notes on the major functions of the microbiome in the gastro-intestinal tract (GIT).

6 Write brief notes on the implications of antibiotic use for human health.

7 What approaches can be used in the management of altered GIT microbiota?

8 Outline how the vaginal microbiota varies throughout life.

For additional revision resources visit: **https://study.sagepub.com/essentialap2e**.

REVISE

ACE YOUR ASSESSMENT

* Revise key terms relevant to this chapter with interactive flashcards.

* Test yourself with quizzes and multiple-choice questions.

* Access the glossary with audio to hear how complex terms are pronounced.

* Explore recommended websites suitable for revision.

REFERENCES

AlFaleh, K. and Anabrees, J. (2014) 'Probiotics for prevention of necrotizing enterocolitis in preterm infants', *Cochrane Database of Systematic Reviews*, Iss. 4. Art. No.: CD005496.

Andrews, H. (2012) 'The fundamentals of skin care', *British Journal of Healthcare Assistants*, 6(6): 285–90.

Arora, T., Sharma, R. and Frost, G. (2011) 'Propionate: anti-obesity and satiety enhancing factor?', *Appetite*, 56(2): 511–15.

Bager, P., Wohlfahrt, J. and Westergaard, T. (2008) 'Caesarean delivery and risk of atopy and allergic disease: meta-analyses', *Clinical and Experimental Allergy*, 38: 634–42.

Bügel, S. (2008) 'Vitamin K and bone health in adult humans', *Vitamins and Hormones*, 78: 393–416.

Canny, G.O. and McCormick, B.A. (2008) 'Bacteria in the intestine, helpful residents or enemies from within?', *Infection and Immunity*, 76(8): 3360–73.

Charlson, E.S., Bittinger, K., Haas, A.R., Fitzgerald, A.S., Frank, I. et al. (2011) 'Topographical continuity of bacterial populations in the healthy human respiratory tract', *American Journal of Respiratory and Critical Care Medicine*, 184: 957–63.

Clemente. J.C., Ursell, L.K., Parfrey, L.W. and Knight, R. (2012) 'The impact of the gut microbiota on human health: an integrative view', *Cell*, 148(6): 1258–70.

Darouiche, R.O., Wall, M.J., Itani, K.M., Otterson, M.F., Webb, A.L. et al. (2010) 'Chlorhexidine–alcohol versus povidone–iodine for surgical-site antisepsis', *New England Journal of Medicine*, 362(1): 18–26.

Department of Health (2011) *High Impact Intervention*. Care Bundle to Prevent Surgical Site Infection. London: Department of Health.

De Wit, R. and Bouvier, T. (2006) '"Everything is everywhere, but, the environment selects"; what did Baas Becking and Beijerinck really say?', *Environmental Microbiology*, 8(4): 755–8.

Dominguez-Belloa, M.G., Costello, E.K., Contreras, M., Magris, M., Hidalgo, G. et al. (2010) 'Delivery mode shapes the acquisition and structure of the initial microbiota across multiple body habitats in newborns', *Proceedings of the National Academy of Sciences*, 107(26): 11971–5.

Erb, K.J. (2009) 'Can helminths or helminth-derived products be used in humans to prevent or treat allergic diseases?', *Trends in Immunology*, 30(2): 75–82.

Farage, M.A., Miller, K.W. and Sobel, J.D. (2010) 'Dynamics of the vaginal ecosystem: hormonal influences', *Infectious Diseases: Research and Treatment*, 3: 1–15.

Fung, K.Y., Cosgrove, L., Lockett, T., Head, R. and Topping, D.L. (2012) 'A review of the potential mechanisms for the lowering of colorectal oncogenesis by butyrate', *British Journal of Nutrition*, 108(5): 820–31.

Gould, D. (2012) 'Skin flora: implications for nursing', *Nursing Standard*, 26(33): 48–56.

Gould, D. and Brooker, C. (2008) *Infection Prevention and Control: Applied Microbiology for Healthcare*, 2nd edn. Basingstoke: Palgrave Macmillan.

Grice, E.A. and Segre, J.A. (2011) 'The skin microbiome', *Nature Reviews Microbiology*, 9(4): 244–53.

Grice, E.A. and Segre, J.A. (2012) 'The human microbiome: our second genome', *Annual Review of Genomics and Human Genetics*, 13: 151–70.

Huffnagle, G.B. and Noverr, M.C. (2013) 'The emerging world of the fungal microbiome', *Trends in Microbiology*, 21(7): 334–41.

Iliev, I.D. and Underhill, D.M. (2013) 'Striking a balance: fungal commensalism versus pathogenesis', *Current Opinion in Microbiology*, 16: 366–73.

Jiménez, E., Marín, M.L., Martín, R., Odriozola, J.M., Olivares, M. et al. (2008) 'Is meconium from healthy newborns actually sterile?', *Research in Microbiology*, 159: 187–93.

Johnston, C.J.C., McSorley, H.J., Andertom, S.M., Wigmore, S.J. and Maizels, R.M. (2014) 'Helminths and immunological tolerance', Transplatation, 97(2): 127–32.

Kinross, J.M., Darzi, A.W. and Nicholson, J.K. (2011) 'Gut microbiome–host interactions in health and disease', *Genome Medicine*, 3(3): 14.

Koren, O., Spor, A., Felin, J., Fåk, F., Stombaugh, J. et al. (2011) 'Human oral, gut, and plaque microbiota in patients with atherosclerosis', *Proceedings of the National Academy of Sciences of the USA*, 108(Suppl. 1), 4592–8.

Lederberg, J. and McCray, A.T. (2001) "'Ome Sweet 'Omics: a genealogical treasury of words', *The Scientist*, 15(7): 8.

Ley, R.E., Peterson, D.A. and Gordon, J.I. (2006) 'Ecological and evolutionary forces shaping microbial diversity in the human intestine', *Cell*, 124: 837–48.

Maitra, A., Sherriff, A., Strachan, D., ALSPAC Study Team and Henderson, J. (2004) 'Mode of delivery is not associated with asthma or atopy in childhood', *Clinical and Experimental Allergy*, 34: 1349–55.

Marshall, B.M. and Levy, S.B. (2011) 'Food animals and antimicrobials: impacts on human health', *Clinical Microbiology Reviews*, 24(4): 718–33.

Matsuki, T. and Tanaka, R. (2014) 'Function of the gut microbiota', in J.R. Marchesi (ed.), *The Human Microbiota and Microbiome (Advances in Molecular and Cellular Microbiology)*. Wallingford, Oxfordshire: CABI Publishing.

Minot, S., Sinha, R., Chen, J., Li, H., Keilbaugh, S.A. et al. (2011) 'The human gut virome: inter-individual variation and dynamic response to diet', Genome Research, 21(10): 1616–25.

NIH (2012) *NIH Human Microbiome Project defines normal bacterial makeup of the body*. Released online 13 June. Available at: www.nih.gov/news/health/jun2012/nhgri-13.htm (accessed 4 May 2020).

NIH HMP Working Group (2009) 'The NIH Human Microbiome Project', *Genome Research*, 19(12): 2317–23.

O'Hara, A.M. and Shanahan, F. (2006) 'The gut flora as a forgotten organ', *EMBO Reports*, 7(7): 688–93.

Orenstein, R., Griesbach, C.L. and DiBaise, J.K. (2013) 'Moving fecal mibrobiota transplantation into the mainstream', *Nutrition in Clinical Practice*, 28(5): 589–98.

Rastall, R.A., Gibson, G.R., Gill, H.S., Guarner, F., Klaenhammer, T.R. et al. (2005) 'Modulation of the microbial ecology of the human colon by probiotics, prebiotics and synbiotics to enhance human health: an overview of enabling science and potential applications', *FEMS Microbiology Ecology*, 52: 145–52.

Romero, R., Hassan, S.S., Gajer, P., Tarca, A.L., Fadrosh, D.W. et al. (2014) 'The composition and stability of the vaginal microbiota of normal pregnant women is different from that of non-pregnant women', *Microbiome*, 2: 4.

Russell, W.R., Hoyles, L., Flint, H.J. and Dumas, M.E. (2013) 'Colonic bacterial metabolites and human health', *Current Opinion in Microbiology*, 16: 246–54.

Swann, M.M. (1969) *Use of Antibiotics in Animal Husbandry and Veterinary Medicine. UK Joint Committee Report*. London: HM Stationery Office.

Todar, K. (2012) 'The normal bacterial flora of humans', in *Todar's Online Textbook of Bacteriology*. Madison, WI: Kenneth Todar. Available at: www.textbookofbacteriology.net/ (accessed 4 May 202020).

Tollånes, M.C., Moster, D., Daltveit, A.K. and Irgens, L.M. (2008) 'Cesarean section and risk of severe childhood asthma: a population-based cohort study', *The Journal of Pediatrics*, 153(1): 112–16.

Vrieze, A., de Groot, P.F., Kootte, R.S., Knaapen, M., van Nood, E. et al. (2013) 'Fecal transplant: a safe and sustainable clinical therapy for restoring intestinal microbial balance in human disease?', *Best Practice & Research Clinical Gastroenterology*, 27: 127–37.

Wade, G.W. (2013) 'The oral microbiome in health and disease', *Pharmacological Research*, 69(1): 137–43.

Weber, P. (2001) 'Vitamin K and bone health', *Nutrition*, 17(10): 880–7 (*Erratum in Nutrition*, 17 (11–12): 1024).

Wu, G.D., Chen, J., Hoffman, C., Bittinger, K., Chen, Y.Y. et al. (2011) 'Linking long-term dietary patterns with gut microbial enterotypes', *Science*, 334: 105–8.

Young, T. (2011) 'Wound infection assessment and management', *Nursing & Residential Care*, 13(8): 384–7.

Zaura, E., Koopman, J.E., Mostajo, M.F. and Crielaard, W. (2014) 'The oral microbiome', in J.R. Marchesi (ed.), *The Human Microbiota and Microbiome (Advances in Molecular and Cellular Microbiology)*. Wallingford, Oxfordshire: CABI Publishing.

Zulkowski, K. (2013) 'Skin bacteria: implications for wound care', *Advances in Skin & Wound Care: The Journal for Prevention and Healing*, 26(5): 231–6.

PART 2

CONTROL AND COORDINATION

This section of the book focuses on control and coordination of body function through the nervous and endocrine systems. It builds on the introduction to the nervous and endocrine systems in Chapter 1 and consists of the following three chapters:

- *Chapter 5. The Nervous System: Control of Body Function*

  The nervous system is one of the two main control systems of the human body. It receives nervous impulses from the external surface of the body and the internal structures, integrates this information and sends nervous impulses to initiate activity of muscles and glands (both voluntary and involuntary). This system acts rapidly.

- *Chapter 6. Special and General Senses: Responding to the Environment*

  This chapter examines the range of different special sense organs (for vision, hearing and balance, taste and smell) and their function. It also considers the general senses – many located in the skin but some in the interior of the body. These include the senses of touch (pressure, vibration and proprioception), pain and heat.

- *Chapter 7. The Endocrine System: Control of Internal Functions*

  The endocrine system functions more slowly in the regulation of body function. In the classical model of endocrine function, hormones are secreted into the blood stream and carried round the body to the target organs where they link with receptors on the cell membrane (or in the cell nucleus), which alter cell function. More recently a number of different modes of function have been identified and are discussed. The endocrine system regulates a range of bodily functions.

THE NERVOUS SYSTEM
CONTROL OF BODY FUNCTION

5

LEARNING OUTCOMES

When you have finished studying this chapter you will be able to:

1. Describe the role of the nervous system in maintaining homeostasis
2. Describe the different kinds of nervous tissue and their roles
3. Explain how nerve cells communicate
4. Outline the structures and divisions of the nervous system, including the role of functional areas
5. Describe how the brain and spinal cord are protected and nourished
6. Discuss the interface between the nervous system, cognition and mental health

INTRODUCTION

This chapter will discuss the role and functions of the nervous system in contributing to the maintenance of homeostasis, including its relationship with and regulation of other systems in the body. You will learn about cells of the nervous system (neurons and neuroglia), how they communicate through transmission of nervous impulses and support each other, and how the nervous system is structurally and biochemically protected. This chapter will also return to the structural and functional divisions of the nervous system previously introduced in Chapter 1. Additionally, this chapter will enable readers to make links between body and mind through looking at the interface between the nervous system, cognition and mental health. This approach will help readers to place their knowledge and understanding within the perspective of the PCN Framework.

As the nervous system is a major control system in the body, this chapter contains a lot of information that you will need to learn. As a result, it is structured as follows:

1. Roles and functions of the nervous system in maintaining homeostasis.
2. Nerve cells.
3. Organisation of the nervous system:

    i. The Central Nervous System (CNS);
    ii. The Peripheral Nervous System (PNS).

4. Nutrition and protection of the nervous system.
5. Consciousness and sleep.
6. The nervous system and personhood.

The order of the content is set out in a way that will ensure you can understand information readily as you work through the chapter. For example, it is important to understand nerve cells before knowing how they communicate and are organised in the body.

Context

The nervous system is central to who we are as functioning, living, feeling people. It shapes who we are, how we experience life and how we function physically, socially and emotionally. As we go through the journey of life, the nervous system develops, and continues to develop, coordinating our development, influencing our life choices and cataloguing our experiences.

This is evident in all members of the Bodie family. Danielle is in the early stages of development – her nervous system is yet to mature. She is learning fast how to coordinate her movements, develop her speech and respond emotionally. Her experiences at this age will influence her mental health, her sense of security and how much she is exposed to learning and development experiences. The young adults of the family (Thomas (30), Derek (29), Michelle and Margaret (27), Kwame (28), Jack (28)), will largely have formed their own personalities and identities. Their development will have been influenced by environment and genetics, and will have guided them into their chosen careers, as it did the mature members of the family. For example, George developed fine motor skills for his job as an engineer. He also needed the ability to analyse and problem solve, all components largely undertaken by the frontal lobe.

Maud needed the ability to communicate well in order to be successful in working with her colleagues as a team when she was a dinner lady. She had to be able to multitask, have good psychomotor skills and to be skilled in calculations and observation. Both George and Maud will have very developed central nervous systems. Its optimal functioning is as relevant now as it was throughout their lives, enabling them to enjoy walking their dog, experience the joy of grandchildren and great-grandchildren, experience

BODIE FAMILY

their surroundings and remember key moments in their lives. Almost all of the members of the Bodie family have established meaningful relationships with someone, illustrating they have developed to some degree emotionally, a vital function of the nervous system that is central to quality of life. These are some aspects of the influence the nervous system has on personhood. As you work through this chapter, you will learn to apply this system more and consider influences on your practice as a person-centred nurse.

ROLE AND FUNCTIONS OF THE NERVOUS SYSTEM IN MAINTAINING HOMEOSTASIS

The major functions of the nervous system are to maintain homeostasis by:

- receiving information and transmitting it to the central nervous system for processing,
- integrating and analysing the different sources of information,
- making decisions,
- sending instructions to the muscles and glands of the body to:
  - carry out voluntary movement
  - influence endocrine function (Chapter 7)
  - regulate unconscious activities through the autonomic nervous system.

So how does the nervous system contribute to homeostasis? Largely, the nervous system responds to the internal and external environment. It does this by being one of the major coordinating systems of the body, with the brain acting as the coordinating centre (Figure 5.1). The brain receives information from both outside and inside the body, integrates the information received and coordinates the response, which is carried out through the actions of muscles and glands. Information to and from the brain is carried by nerve fibres that transmit impulses very quickly and result in an extremely rapid response.

The nervous system does not work alone in coordinating activities within the body: it is integrated very closely with the endocrine system whereby the hypothalamus acts as the control centre while also being part of the nervous system. This is discussed further in Chapter 7.

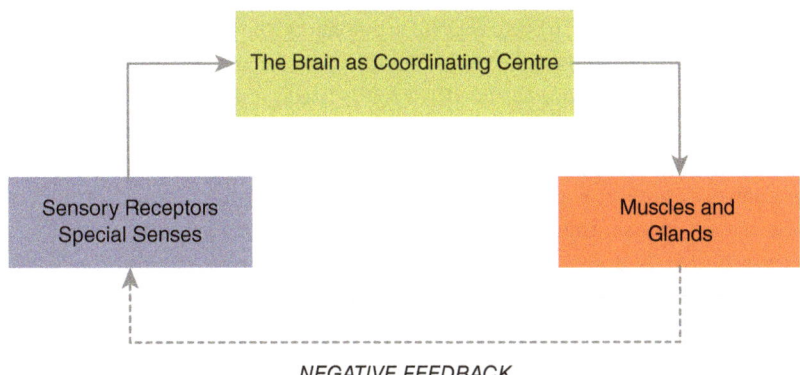

Figure 5.1 The nervous system as a control system

NERVE CELLS

In order to understand how the nervous system operates as a control system, it is necessary to understand how the different cells within the brain are structured and how they work. The nervous system is composed of two major types of nerve cells, neurons and neuroglia (Table 5.1).

Table 5.1   Types of cells in the nervous system

Nervous system cell	Location	Subtype	Function
Neurons	Central and peripheral nervous system	Unipolar, bipolar or multipolar	To generate action potentials for sending and receiving information
Neuroglia	Central nervous system	Astrocytes	Secure neurons to their blood supply. Regulate external chemical environment of neurons by removing excess ions (such as potassium) and promote re-uptake of neurotransmitters released during synaptic transmission. Form blood-brain barrier
		Microglia	Specialised macrophages capable of phagocytosis. Thus, they protect neurons from pathogens
		Ependymal cells	Thought to be stem cells in the nervous system. Create and secrete cerebrospinal fluid (CSF) and circulate it by cilia activity. A role in reabsorption of CSF
		Oligodendrocytes	Produce myelin to coat axons of neurons
	Peripheral nervous system	Schwann cells	Produce myelin to coat axons of neurons. Protective role - phagocytotic and remove debris to allow growth and regrowth of neurons
		Satellite cells	Regulate external chemical environment of neurons, particularly calcium ions. Thought to play role in chronic pain as they are sensitive to injury and inflammation

Neurons are invaluable cells within the nervous system. They are often referred to as nerve cells and their function is to rapidly transmit information to, from and within the brain and spinal cord. The number of neurons in the central nervous system is estimated as being 100 billion, supported by one trillion neuroglial cells (Woolsey et al., 2008). When neurons connect together, they form neural networks.

Neurons use the most energy in the brain, particularly for synaptic transmissions whereby one neuron triggers another (Harris et al., 2012). Glucose is the primary source of energy alongside lactate oxidation (Wyss et al., 2011) with 93% of ATP generated by neuronal mitochondria and the rest from glycolysis (Harris et al., 2012).

Neuroglial cells are very different to neurons in both structure and function. They are smaller than neurons and almost ten times more numerous. Their role is largely a supportive one. They provide oxygen and nutrients to neurons, create a barrier between one neuron and another, provide structural support to neurons and protect neurons by destroying pathogens. They also remove dead neurons. Additionally, they are now known to support the process of a neuron's nervous impulse (action potential), regulating the uptake of neurotransmitters back into the neuron from the synaptic cleft (gap between two communicating neurons), and also playing an important role in the growth and repair of neurons. A neurotransmitter is a chemical released by a neuron to transmit an impulse across the synaptic cleft (gap between two communicating neurons) to another neuron.

Neurons communicate by sending nervous impulses, known as action potentials, through the cell. This passes from one cell to another by either an electrical or chemical transmission across the gap between two neurons enabled by the structure of the neuron.

Structure of the neuron

The neurons' structure allows them to communicate efficiently and effectively throughout the body. Figure 5.2 shows typical neurons in the peripheral nervous system. The nucleus is large and directs the metabolic activities of the cell, including protein synthesis. The nucleus is located in the cell body, which has projections from it known as dendrites that receive information from the connecting neurons. The cell body also contains neurofibrils (neurofilaments), exclusively found in neurons, which contribute to the transport of cellular material and facilitate axon movement and growth. The axon distinguishes the neuron and is a long projection which carries action potentials. It is surrounded by Schwann cells (peripheral nervous system) or oligodendrocytes (central nervous system) forming the myelin sheath insulating the axon and preventing passive movement of ions across the cell membrane. The Nodes of Ranvier are gaps between areas of myelination on the axon where ions can easily flow into the Extracellular Fluid (ECF). The end of the axon (axon terminal) contains synaptic vesicles holding neurotransmitters.

How is information transmitted?

Action potential

Nervous impulses are transmitted by action potentials along the axon. At rest all cells have a potential difference, i.e. a differing electrical charge, across the cell membrane due to the distribution of the electrolytes in the Intracellular Fluid (ICF) and ECF with the ICF at a lower charge than the ECF (indicated by the negative charge). Neurons have a potential difference of –70 mV (millivolts). Neurons are specialised to be able to change this electrical charge difference, enabling them to produce an action potential, and then return it to its original state (Figure 5.3). This is action on the potential difference and explains why a nervous impulse is called an action potential.

An action potential has three stages:

1. **Depolarisation.** A stimulus makes the cell membrane more permeable to sodium which, attracted by the negative charge in the cell, allows a small amount of sodium into the cell. If depolarisation is large enough, the sodium channels open, allowing considerably more sodium to pass down the concentration gradient through the cell membrane into the neuron. The potential difference moves towards a positive value: the maximum positive value that can be reached is the action potential and for most neurons is +40 mV. Depolarisation triggers adjacent sodium channels to open, and then the action potential progresses along the axon. When the potential difference reaches +30 mV, sodium channels are inactivated, stopping more sodium moving into the neuron.
2. **Repolarisation.** As a result of inactivation, sodium channels close. Potassium channels open, causing a rush of potassium out of the neuron to restore the potential difference to negative (–70 mV).
3. **Refractory period.** After repolarisation, the potential difference falls to below –70 mV and the neuron will not respond to another stimulus for a brief period, the Refractory Period. This is as a result of **hyperpolarisation**, which occurs as a result of movement of extra potassium out of the neuron for a brief period. This inhibits an action potential as it increases the stimulus threshold for one to occur. In order to restore the conditions for another action potential to occur, the sodium–potassium pump moves sodium out of and potassium into the neuron to restore the original resting potential and ionic concentrations.

Action potentials pass between the Nodes of Ranvier by saltatory conduction, when electrical activity jumps between the gaps and thus moves more rapidly than along unmyelinated nerve fibres. Synaptic transmission is by neurotransmitters released from one neuron, passing across the synaptic cleft and attaching to receptors on the next neuron. It is important to note that an action potential is an 'all-or-nothing' phenomenon in which the change in potential difference must reach a certain level/threshold for the action potential to occur. A stimulus below the threshold will not result in any action potential, and a stimulus above the threshold will produce a full action potential.

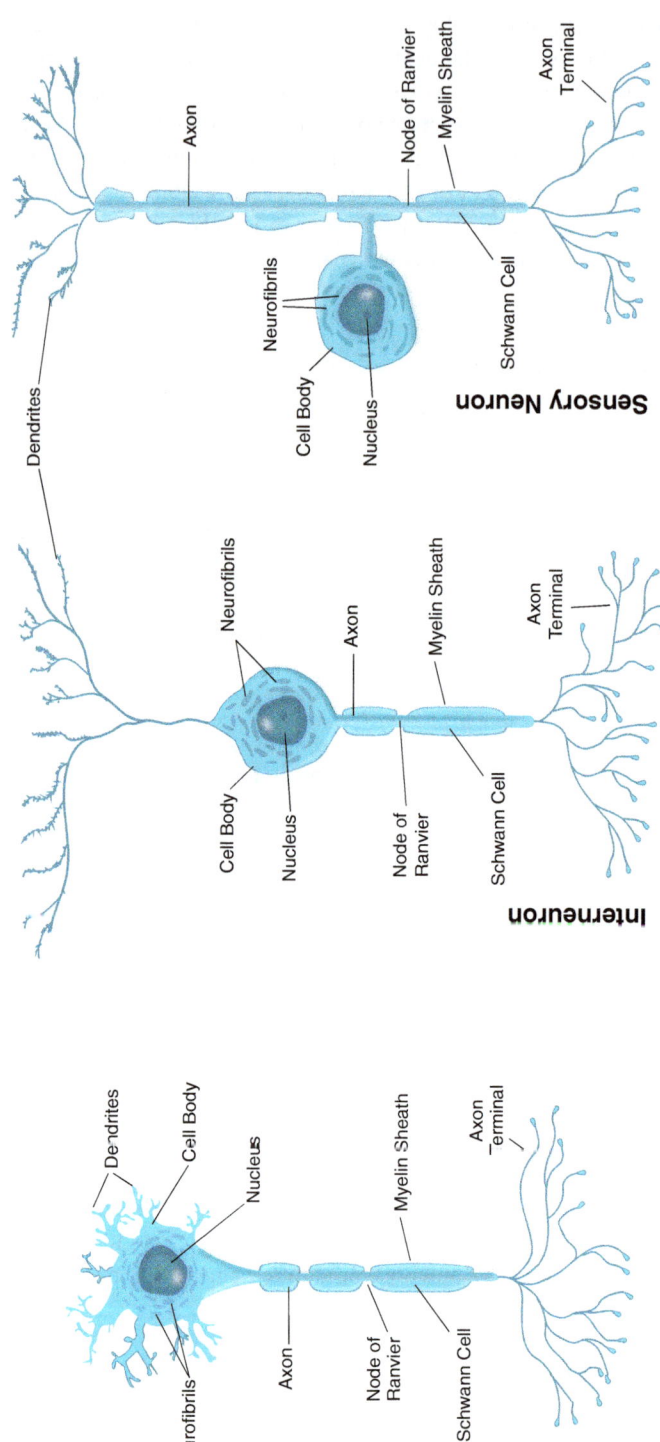

Figure 5.2   Structure of a neuron

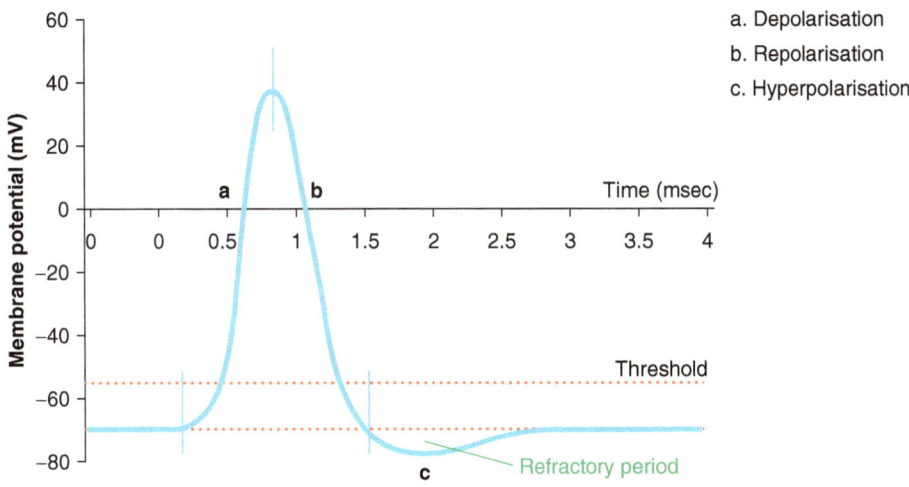

a. Depolarisation
b. Repolarisation
c. Hyperpolarisation

Figure 5.3 Action potential

ACTIVITY 5.1: UNDERSTAND

To see saltatory conduction in action, watch the following short video clips.

These online videos can be accessed by **scanning the QR codes** with your smart phone or via https://study.sagepub.com/essentialap2e.

SALTATORY CONDUCTION (3:11)   THE NERVE IMPULSE (5:41)

Synaptic transmission

Once the action potential reaches axon terminals, it crosses the synaptic cleft for the signal to continue in another (the post-synaptic) neuron. Synaptic transmission starts in the synaptic knob and ends in the dendrite of the receiving neuron. There are two types of synaptic transmission, electrical and chemical.

Electrical: In electrical synaptic transmission, the gap between the axon terminal/synaptic knob and the dendrite is much smaller than in a chemical synaptic transmission. This is known as a gap junction, which contains gap junction channels that cross the membranes of both neurons (sending and receiving) (Figure 5.4). These permit ions and small molecules to flow from one neuron to the other, enabling depolarisation.

Chemical: Chemical synaptic transmission is the more usual form of transmission. It is slower and more complex (Figure 5.5):

1. Depolarisation triggers synaptic vesicles filled with neurotransmitters to merge with the presynaptic membrane.
2. Calcium channels open in the pre-synaptic membrane, calcium moves into the synaptic bulb and triggers the synaptic vesicles to release the neurotransmitter into the synaptic cleft.

3. Neurotransmitters diffuse across the synaptic cleft and interact with receptors in the post-synaptic membrane, the stimulus for triggering an action potential in the receiving neuron.

4. Neurotransmitters are either broken down in the synaptic cleft or reabsorbed from the ECF by astrocytes, which return them to the neuron for reuse, or are reabsorbed directly back into the synaptic knob.

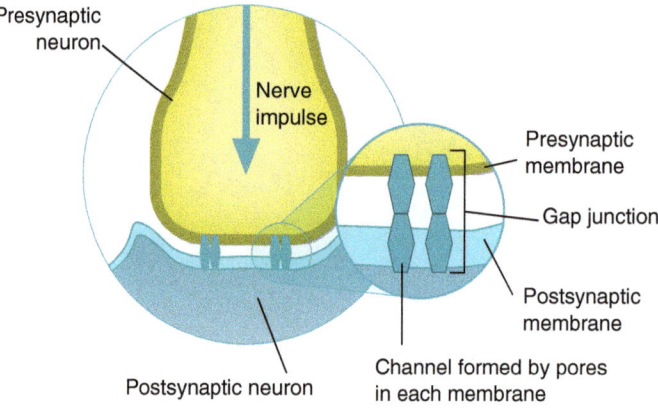

Figure 5.4 Gap junction

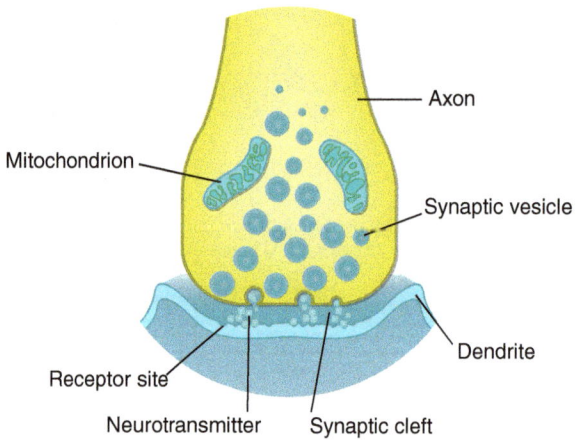

Figure 5.5 Chemical synapse

Neurotransmitters

Neurotransmitters are the chemical messengers in the nervous system. There are four main types (see Table 5.2), and each neurotransmitter has a specific action.

Table 5.2 Neurotransmitters

Class of neurotransmitter	Role
Acetylcholine	Normally an excitatory function (sympathetic nervous system)
	Parasympathetic effect on some organs (e.g. reduces heart rate)
	Sometimes classed as a monoamine
Monoamines (biogenic amines)	Synthesised from amino acids
	Include catecholamines such as adrenaline (epinephrine) and noradrenaline (norepinephrine)
	Thought to mediate emotion, arousal and cognition
Amino acids	Inhibitory Amino Acids (IAA) or Excitatory Amino Acids (EAA)
	Glycine and glutamate are two amino acid neurotransmitters
	Glycine is inhibitory in the spinal cord, brainstem and retina
	Glutamate is excitatory and involved in learning and memory
Neuropeptides	Neuropeptides work at lower concentrations and are not recycled
	Effects last longer than other neurotransmitters
	Associated with analgesia, metabolism, reproduction, social behaviours, learning and memory

ORGANISATION OF THE NERVOUS SYSTEM

Neurons and neuroglia make up the nervous system. Neuronal cell bodies are organised in identified groups (ganglia) or layers as in the cerebral cortex. The axons form pathways between the different parts of the nervous system (nerve tracts) or between the nervous system and other parts of the body (nerves). The nervous system is divided into the CNS and PNS (Figure 5.6).

Nerve fibres are grouped in tracts or pathways. Two major tracts are:

- the corpus callosum, which joins the two sides of the brain
- the internal capsule, which consists of both sensory and motor fibres that carry information to and from the cerebral cortex.

Grey matter

The grey matter (nerve cell bodies) is distributed mainly around the outside of the brain with a convoluted surface to increase the number of brain cells that can fit within the surface area of the brain. This part of the brain is concerned with higher functions and specific parts have different functions. There are also clumps of grey matter deep within the brain with specific functions, mainly those not under conscious control. In evolutionary terms, these are the older parts of the brain.

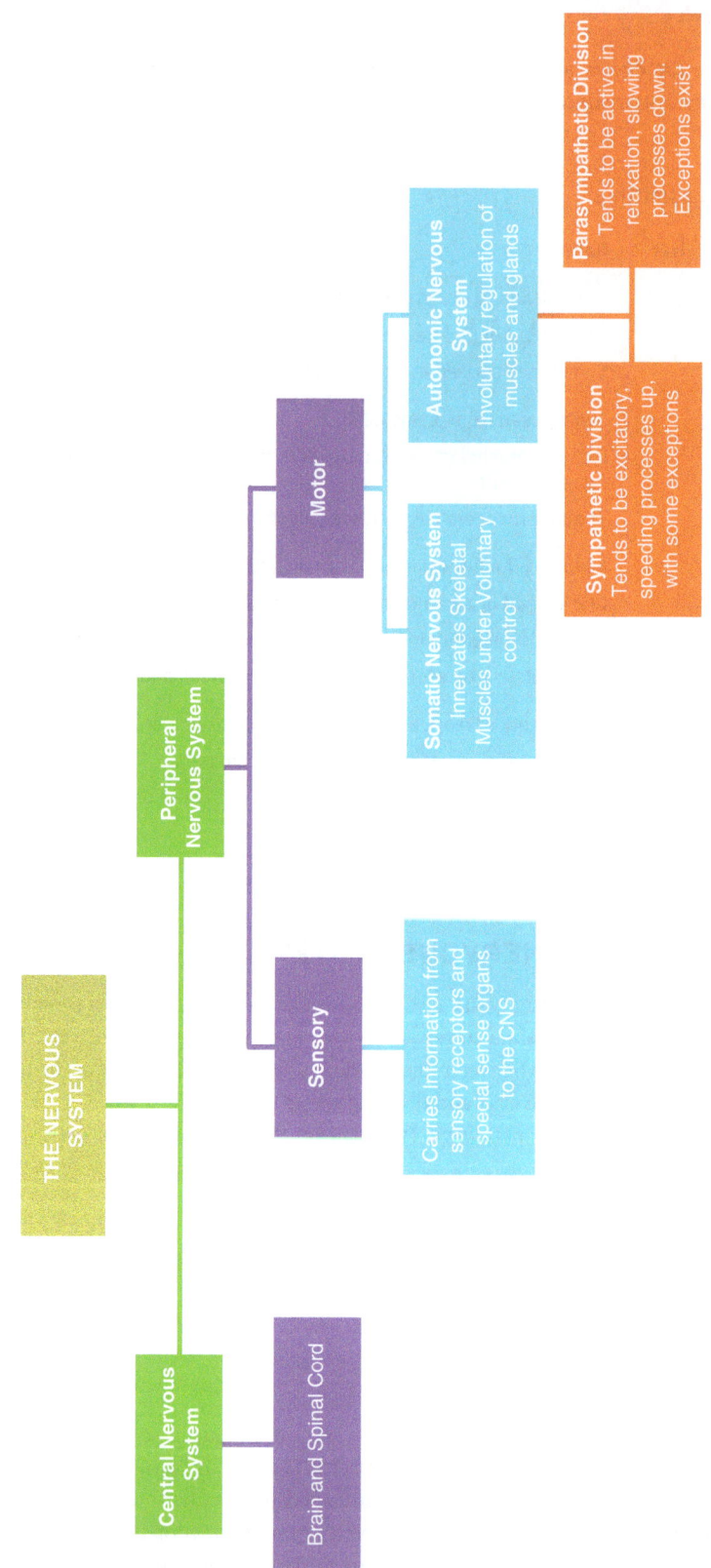

Figure 5.6   Organisation of the nervous system

White matter

The white matter (nerve cell axons) connects different parts of the brain, carrying entering information to the appropriate parts of the brain, and carrying instructions out of the brain. Much of the white matter runs in clearly defined tracts.

The spinal cord connects to the base of the brain and runs down within the spinal column. The distribution of grey and white matter is different from within the brain. Grey matter is within the centre of the spinal cord with nerve axons running in tracts around the outside.

THE CENTRAL NERVOUS SYSTEM

The structure of the brain

In the textbooks, the structure of the brain is described in two main ways (Figure 5.7). In this book, we are using Structure 1. Figure 5.8 shows the parts of the brain.

The forebrain

The forebrain is the largest section of the brain (Figure 5.9). It is associated with control of body temperature, reproduction, eating, sleeping, cognition and emotional responses. It consists of the diencephalon and the cerebrum.

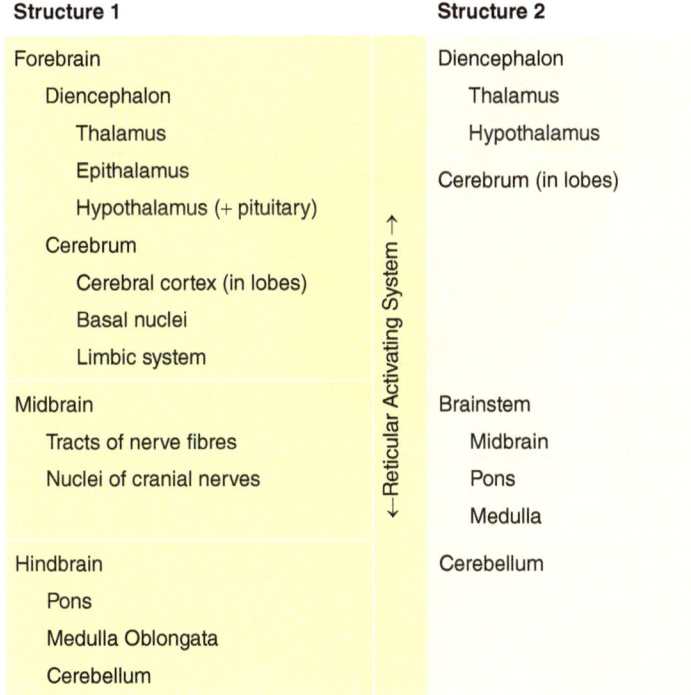

Structure 1

Forebrain
    Diencephalon
        Thalamus
        Epithalamus
        Hypothalamus (+ pituitary)
    Cerebrum
        Cerebral cortex (in lobes)
        Basal nuclei
        Limbic system

Midbrain
    Tracts of nerve fibres
    Nuclei of cranial nerves

Hindbrain
    Pons
    Medulla Oblongata
    Cerebellum

←—Reticular Activating System →

Structure 2

Diencephalon
    Thalamus
    Hypothalamus
Cerebrum (in lobes)

Brainstem
    Midbrain
    Pons
    Medulla

Cerebellum

Figure 5.7 Structure of the brain

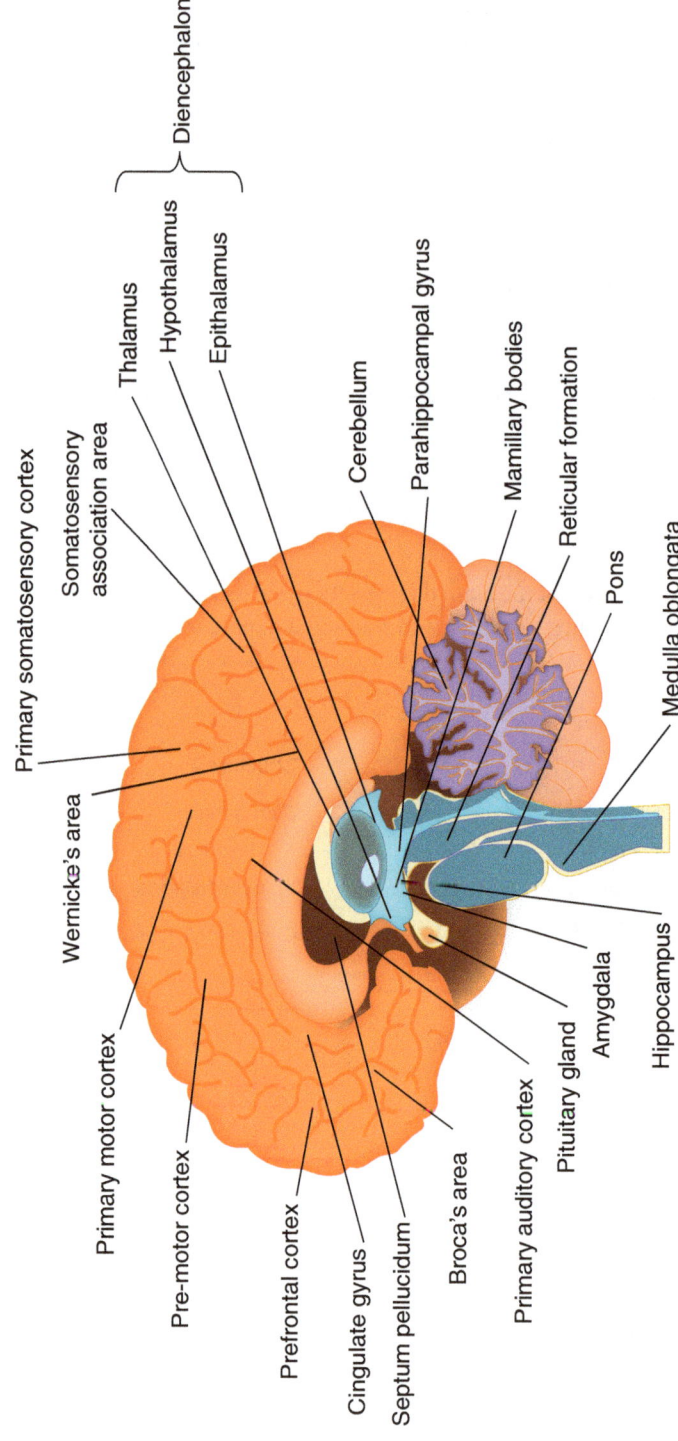

Figure 5.8 Parts of the brain

Primary somatosensory cortex

Somatosensory association area

Thalamus

Hypothalamus

Epithalamus

Diencephalon

Cerebellum

Parahippocampal gyrus

Mamillary bodies

Reticular formation

Pons

Medulla oblongata

Wernicke's area

Primary motor cortex

Pre-motor cortex

Prefrontal cortex

Cingulate gyrus

Septum pellucidum

Broca's area

Primary auditory cortex

Pituitary gland

Amygdala

Hippocampus

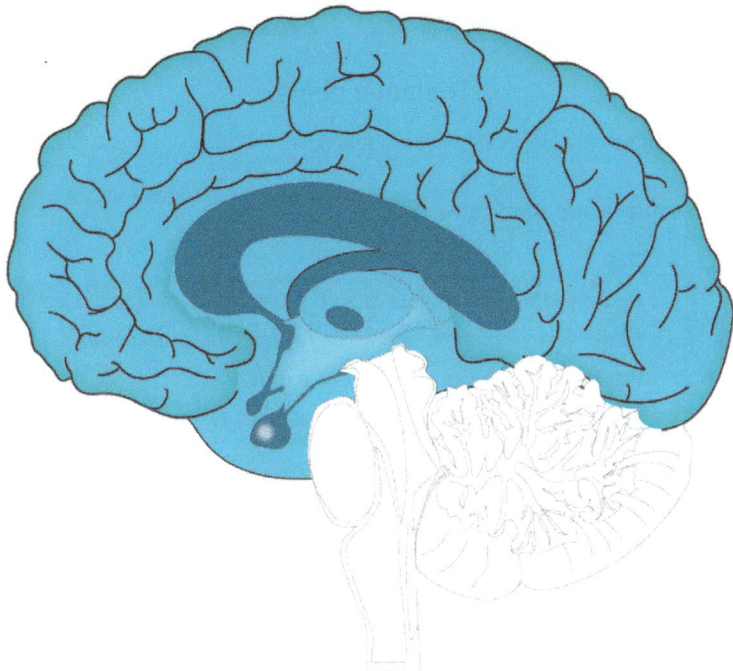

Figure 5.9 The forebrain

The diencephalon

The diencephalon is composed of the thalamus, hypothalamus and epithalamus:

- **The thalamus:** is a relay centre for nervous impulses moving to and from the cerebrum. It is a large mass of grey matter that is positioned deep within the brain on either side. Except for olfaction, all sensory pathways go via the thalamus to be processed and relayed to the appropriate area of the cerebral cortex. It also has a role in processing painful stimuli, temperature and attention in collaboration with the reticular activating system.
- **The hypothalamus:** lies below the thalamus and includes several nuclei (groups of neuronal cell bodies) and tracts of axons. There are three groups of hypothalamic functions:

  - Control of the autonomic nervous system.
  - Control of the neuro-endocrine system.
  - Control of the limbic system.

- The pituitary gland is a small pea-sized extension below the hypothalamus. It is considered to be part of the endocrine system and is the key link between the nervous and endocrine systems (Chapter 7).
- **The epithalamus:** is composed primarily of the pineal gland and the habenula. The habenula is a relay from the limbic system, and deals with sleep, stress, pain and reinforcement processing (Lawson et al., 2013). The pineal gland secretes serotonin during the day and melatonin at night in regulating the sleep–wake cycle. The large pineal gland in childhood is thought to inhibit the onset of puberty by secreting melatonin in high amounts before shrinking to a small size in adulthood.

— **GO DEEPER** —

Thalamic nuclei

Some nuclei in the thalamus have particular roles (Table 5.3).

Table 5.3 Thalamic nuclei and their roles

Thalamic nuclei	Role
Medial geniculate	Relays auditory information from the midbrain to the primary auditory cortex
Lateral geniculate	Relays visual information from the retina to the visual cortex
Lateral dorsal	Works with the limbic system to form memories
Mediodorsal	Works with the limbic system and prefrontal cortex to manage cognitive processes such as reasoning, thoughts and mood
Ventrolateral	Relays signals from the cerebellum to the primary motor cortex in coordinating movement

— **GO DEEPER** —

Hypothalamic nuclei

The hypothalamus is a major control system in the body occurring through 11 groups of nuclei regulating numerous homeostatic functions (Table 5.4).

Table 5.4 Hypothalamic nuclei and their functions

Hypothalamic nuclei	Functions
Paraventricular	Fluid regulation, anterior pituitary control, production of oxytocin
Preoptic	Thermoregulation, sexual arousal
Anterior	Thermoregulation, sexual arousal
Suprachiasmatic	Circadian rhythms such as the sleep/wake cycle
Supraoptic	Fluid regulation, anterior pituitary gland control, production of oxytocin
Dorsomedial	Emotional expression
Ventromedial	Appetite/hunger, satiety, fear and aggression
Arcuate	Growth, dopamine release
Posterior	Thermoregulation
Mamillary	Emotion (via the limbic system) and recognition memory
Lateral	Hunger

The cerebrum

The cerebrum consists of the cerebral cortex (grey matter) and underlying white matter, the basal nuclei and the limbic system. The cerebral cortex and some of the underlying white matter are folded in gyri to maximise surface area. A deep fissure divides the cerebrum into two halves, the cerebral hemispheres. It is further divided by folds (sulci) into five functional lobes (Figures 5.10a and 5.10b).

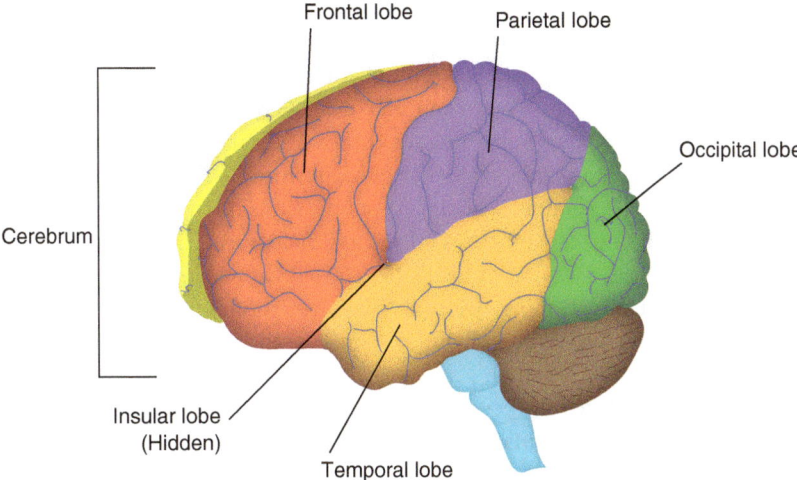

Figure 5.10a Lobes of the brain

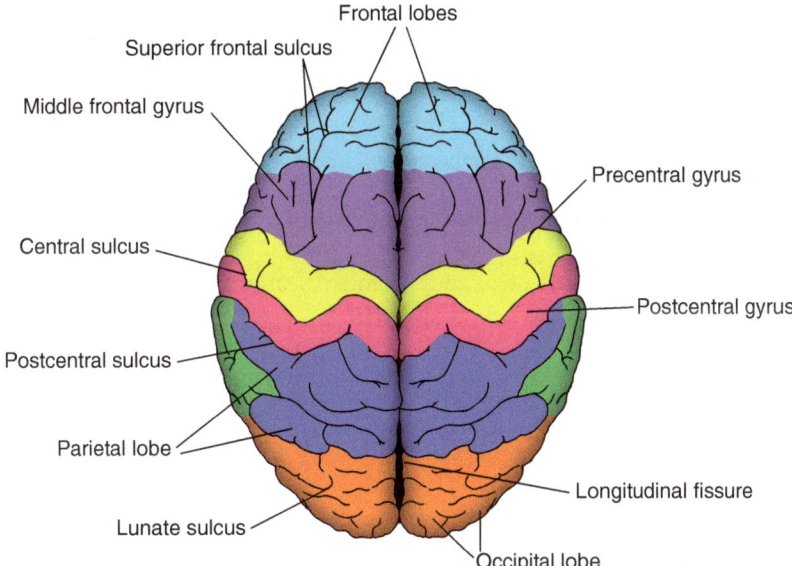

Figure 5.10b Divisions of the brain

Cerebral cortex

1. **The frontal lobe.** This regulates voluntary control of movement, including learned complex movement. Planning and sequencing of movement occurs in the pre-motor cortex. The primary motor cortex enables muscles to work in synergy for coordinated movement. The neuronal axons from the primary motor cortex extend to the spinal cord and down to connect with the spinal

nerves. Broca's area, associated with motor actions of speech such as word formation including articulation, pronunciation and expression, is also in the frontal lobe.

The prefrontal cortex is primarily involved with higher cognitive function (reasoning, understanding, foresight and thinking). It has extensive connections with the other lobes of the cerebral cortex, and is thus associated with personality.

2. **The parietal lobe.** This area processes sensory information. The primary somatosensory cortex processes sensations of touch, vibration and proprioception (awareness of position) and is conveniently located beside the primary motor cortex. This enables the brain to respond rapidly. The somatosensory association area processes stereognosis – the recognition of an object by feeling shape, texture and temperature. Spatial perception, that is interpreting the nature of objects in the environment, including recognition of images and shapes, is processed in the parieto-occipital association areas, largely in the right parietal lobe. The left parietal lobe contributes to calculations, writing and reading. Finally, the optic nerve pathways pass through the parietal lobe.

3. **The occipital lobe.** This deals with visual stimuli, recognising, interpreting and finally memorising objects.

4. **The temporal lobe.** This focuses on processing the special senses, particularly taste (gustation), smell (olfaction) and hearing (audition). Additionally, it has roles in learning, memory, visual recognition and emotional actions. The primary auditory cortex interprets volume and tone of sound, and links with Wernicke's area to interpret spoken language. Wernicke's area is also associated with the production of spoken language. The medial areas of the temporal lobe integrate into the limbic system in relation to learning, memory and emotion.

5. **Insular lobe.** This is often referred to as the fifth lobe. It is a small area of dense tissue positioned under the temporal, frontal and parietal lobes and located under arteries and veins, making it difficult to isolate and access. Its limited accessibility has resulted in little understanding of its functions. They are thought to include (Stephani et al., 2011):

   - Thermosensation (perception of temperature).
   - Nociception (perception of pain).
   - Somatosensation (sensations originating mainly in the skin including proprioception, touch and temperature).
   - Viscerosensation (sensations originating in internal organs including pain, palpitations and spasms).
   - Gustation (taste).
   - It is also thought to have a role in human emotions and associated behaviours.

Basal nuclei

The basal nuclei (or basal ganglia) refer to nuclei of grey matter buried within the white matter. These include the corpus striatum (caudate nucleus, putamen and globus pallidus), the substantia nigra and the subthalamic nucleus. The basal ganglia have an important role in motor and thought control and generally inhibit muscle tone.

The limbic system

The limbic system is a system, or set of systems, within the brain that regulates emotions, motivation, learning and memory through a variety of interconnected structures that includes nuclei, tracts and cortical areas. The system borders the brainstem and focuses on emotional state, instinct and motivation/ drive, which can include aspects of social behaviour and feelings of pleasure. It has links into the olfactory system, the hypothalamus, the midbrain and the reticular formation.

GO DEEPER

Structures of the limbic system

The structures of the limbic system include:

1. **The amygdala.** This contributes to storage of emotional experiences as memories and regulates emotional learning. It is associated with responding to stimuli that are potentially harmful/threatening, processing distress, and recognising and evaluating other emotive stimuli such as facial expressions (Sergerie et al., 2008).
2. **The hippocampus.** The hippocampus is primarily associated with formation of memory, including short-term (working) and long-term memory. Short-term memory works on quick recall over seconds and minutes, whereas long-term memory involves memory consolidation. Long-term memory can be subdivided into declarative memory (recalling personal experiences and language) and procedural memory (related to learning and recalling how to complete a task). The hippocampus facilitates declarative memory, whereas the basal nuclei facilitate procedural memory. The hippocampus also organises sensory and cognitive experiences for storage in the frontal and temporal lobes of the cerebral cortex.
3. The **septum pellucidum.** Situated between the lateral ventricles at the midline of the brain, this contains no grey matter but comprises myelinated axons connecting parts of the limbic system (Raybaud, 2010). Its role is therefore unclear but when it is damaged or malformed people experience disorders relating to rage, pleasure and mood as it links parts of the limbic system that control these.
4. The **cingulate gyrus.** The cingulate gyrus is thought to integrate emotion and sensory experiences, for example the sensation of pain and the emotional response (e.g. distress, anger).
5. The **insula.** The insula (or insular lobe) is discussed previously under the lobes of the brain.
6. The **parahippocampal gyrus.** This structure of the limbic system works with the hippocampus in processing declarative memory.
7. **Mamillary bodies.** Although not fully understood, the mamillary bodies (sometimes spelt mammillary) are thought to be a relay centre with a distinct role in memory operations (Tagliamonte et al., 2015).

Midbrain

The midbrain joins the part of the brain containing the thalamus and hypothalamus with the pons, which links to the lowest part of the brain, the medulla oblongata (Figure 5.11). These parts of the brain contain the nuclei for the cranial nerves described later. The midbrain consists of:

- Tracts of nerve fibres.
- A number of nuclei of cranial nerves.

An important component is the substantia nigra, which works with the basal nuclei to regulate movement through inhibiting the neurotransmitter/hormone dopamine produced in the substantia nigra. The cranial nerves responsible for eye movement originate in the midbrain, as do nuclei for the response to auditory and visual stimuli, such as turning your head to respond to a sound or visual stimulus. Corticospinal tracts pass through the midbrain on their way to the medulla, and reticulospinal tracts associated with the experience of pain are present.

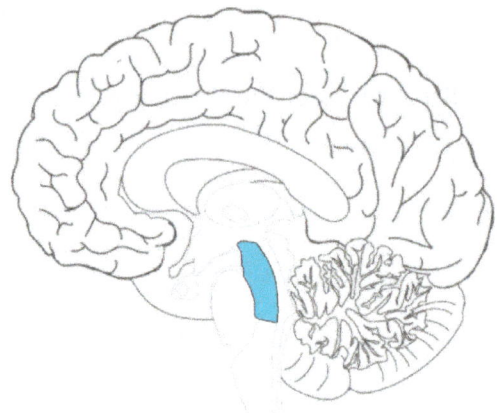

Figure 5.11 The midbrain

The hindbrain

The hindbrain (Figure 5.12) is composed of:

- the pons
- the medulla oblongata (or medulla)
- the cerebellum.

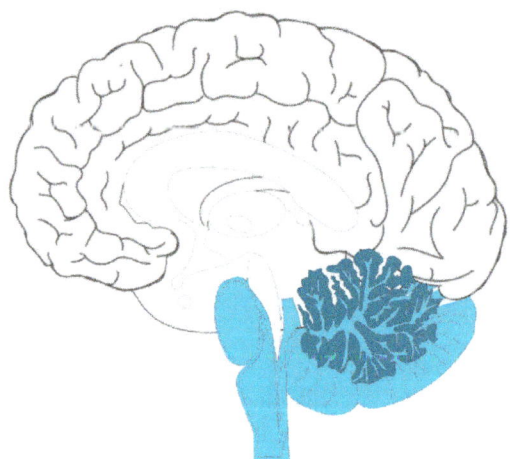

Figure 5.12 The hindbrain

Pons

The pons is in front of the cerebellum, below the midbrain and above the medulla. It consists mainly of nerve fibres, which connect the two hemispheres of the cerebellum and fibres passing between the higher levels of the brain and the spinal cord. There are nuclei within the pons, which act as relay stations and some are associated with cranial nerves. The pons also has a role in regulating breathing through the pneumotaxic and apneustic centres (Chapter 10).

Medulla oblongata

The medulla links the brain with the spinal cord and fits just within the cranium. As with the spinal cord, the grey matter lies centrally with the white matter surrounding it. Some nuclei act as relay stations for sensory information coming from the spinal cord. An area in the medulla, known as the pyramids due to its shape, is where the corticospinal and some sensory pathways cross over from left to right and vice versa. This is termed decussation and is why the right cerebral cortex regulates movement in the left-hand side of the body and vice versa.

Also within the medulla are a number of vital centres that are concerned with coordination of the autonomic nervous system. These are the:

- Cardiac centre, regulating heart rate and force of contraction.
- Respiratory centre, or rhythmicity centre, regulating the pattern of breathing.
- Vasomotor centre, regulating the diameter of blood vessels.
- Reflex centres of vomiting, coughing, sneezing and swallowing.

--- **APPLY** ---

The medulla and monitoring health status

The medulla could be thought of as the nurses' friend. Its role in maintaining homeostasis means that it can tell us when something is wrong in the body. For example, if a person is in pain, the cardiac centre may increase heart rate and the respiratory centre the rate of breathing. This could happen if the person is in physiological shock from bleeding, for example. Vomiting may be a sign of gastrointestinal infection. The body automatically tries to compensate for illness, and these compensations are vital sources of data for the nurse to identify and interpret.

Cerebellum

The cerebellum coordinates the muscles of the body and regulates muscle tone and posture. In effect, it contributes to the fine tuning of motor commands and sensorimotor adjustment needed for motor learning (Brooks and Cullen, 2013). It achieves this through information about muscle stretch, position of parts of the body and other coordinating information from the pons and cerebral cortex. It relays back to the thalamus and cerebral cortex to respond appropriately. A role in cognition has been confirmed, with the cerebellum now known to both send and receive information to non-motor regions of the cerebral cortex. This includes prefrontal areas that regulate higher cognitive functions (Buckner, 2013).

ACTIVITY 5.3: APPLY

Neurological control and occupation

Think about George Bodie. Why would an effectively functioning cerebellum be important in his career and in his pursuit of leisure activities now?

BODIE FAMILY

Reticular activating system

The Reticular Activating System (RAS) is composed of a number of nuclei that connect throughout the forebrain, midbrain and hindbrain. It controls arousal mechanisms used in maintaining consciousness and awake states essential for selective attention and purposeful responses. Arousal regulation is not confined to the RAS; neural pathways descend to the spinal cord and ascend to the cerebral cortex, referred through the thalamus and the suprachiasmatic nuclei (pair of neuron clusters in the hypothalamus situated directly above the optic chiasma[1] of the hypothalamus). Through these processes the sleep–wake cycle is regulated.

The RAS contains the reticular formation, a core of nerve cell bodies which extend from the spinal cord up through the medulla, pons and midbrain to the hypothalamus and thalamus and with connections to the cerebral cortex. Functions of the reticular formation include:

- Skeletal muscle tone.
- Autonomic control as it forms part of the cardiovascular and respiratory centres in the pons and medulla.
- Somatic[2] and visceral[3] sensations, such as pain.

THE SPINE AND SPINAL CORD

When we refer to the spine, we are referring to the spinal (or vertebral) column and the mechanical structures that it is composed of, such as the vertebrae and the ligaments and tendons that connect them together (Chapter 15). The spinal cord refers to the neural tissue encased within the spine that runs from the medulla, where it joins the brain at the level of the foramen magnum (the opening on the underside of the skull), to the level of the first or second lumbar vertebrae.

Spinal cord

The spinal cord runs down through the space in the spinal vertebrae within three groups of neurons:

- **Ascending (afferent) neurons** carrying information up the spinal cord to the brain for processing.
- **Descending (efferent) neurons** carrying instructions down the spinal cord from the brain to the muscles and glands of the body.
- **Interneurons (association neurons)** acting as connections between descending and ascending neurons.

The spinal cord is an extension of the brain from the medulla with two main enlargements. The cervical enlargement (through the brachial plexus) provides neurons for the arms, and the lumbar enlargement (through the lumbosacral plexus) provides neurons for the legs. Below this, the cord narrows into a conical shape, the conus medullaris. Nerves resembling a horse's tail (called the cauda equina) extend further down to the sacrum.

---

[1]Point at the base of the brain where the optic nerves cross over each other.

[2]Sensations from the body surface (skin) or musculoskeletal tissues.

[3]Sensations from the internal organs.

Spinal grey matter

The grey matter of the spinal cord is set out in an H shape and is divided into 'horns' (Figure 5.13):

1. **Anterior (ventral) horn.** This contains cell bodies of the somatic motor nerves that stimulate skeletal muscle. These run the length of the spinal cord.
2. **Posterior (dorsal) horn.** This contains cell bodies of the somatic and autonomic sensory (afferent) neurons running the length of the spinal cord.
3. **Lateral horns.** This grey matter contains cell bodies of the autonomic motor nerves that stimulate smooth muscle, cardiac muscle and glands. They run from T2 to L1 and link with the thoracolumbar ganglia running alongside the spinal cord and supplying the sympathetic nervous system.

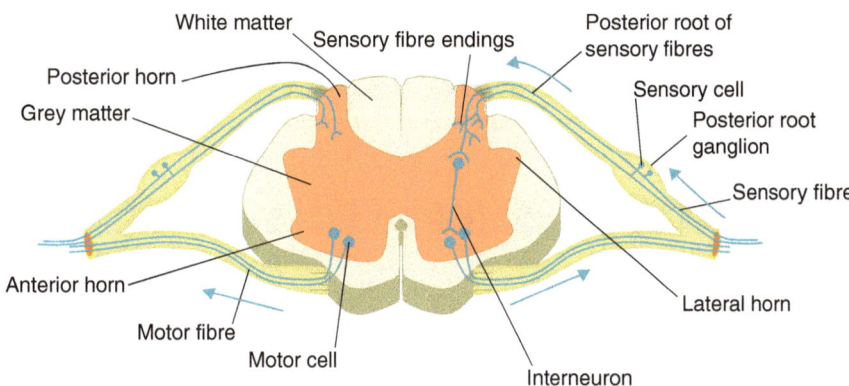

Figure 5.13 Grey matter of the spinal cord (horns)

Spinal white matter

White matter surrounds the grey matter and is structured into three columns: the anterior, posterior and lateral columns. They are then subdivided into tracts – distinct sets of fibres going to or from the same place. The name of the tract indicates its path; for example, the corticospinal tract carries information from the cortex to the spinal cord.

THE PERIPHERAL NERVOUS SYSTEM (PNS)

The PNS refers to the nerves and ganglia outside the CNS, largely the cranial and spinal nerves. The PNS carries information to and from the CNS where information is processed. Structurally, the PNS is divided into two divisions: the sensory division/system and the motor division/system.

Sensory division/system

This division of the PNS carries sensory information through afferent neurons to the CNS for processing. This includes information from viscera of the thoracic and abdominal cavities (visceral sensations) and the skin, muscles, bones and joints (somatic sensations). Sensory input about the external and internal environments is received in the CNS. Some sensory input is consciously perceived, although considerable information for maintaining homeostasis does not need this level of awareness. Some sensory input is carried directly to the brain by the cranial nerves. Other input enters via the spinal nerves and is carried to the brain in various ascending tracts.

Sensory receptors are organs that convert energy from one form (e.g. heat, light, pressure) into electrical energy, which is transmitted through the sensory nerve fibres. The sensitivity to a particular modality (i.e. touch, temperature, etc.) varies with:

- The density of distribution of the specific receptors for that modality/sensation.
- The degree of overlap of receptor fields.
- The size of the receptor fields.

Receptors vary considerably in structure including free nerve endings, specialised cells or within specialist organs (such as in the eye) (Chapter 6).

Many receptors demonstrate decreasing sensitivity to a continued stimulus thus permitting adaptation to, for example, continued pressure. Some receptors respond rapidly, others much more slowly. Electrical activity carrying sensory information is transmitted via sensory neurons to the CNS with most eventually arriving in the brain. Most sensory information is relayed through the thalamus.

ACTIVITY 5.4: APPLY

Sensory fatigue

Decreasing sensitivity in sensory receptors can be easily demonstrated. Have you ever been in a shop that sells scented candles? You can smell the first few very easily but after smelling a lot, you stop being able to differentiate the smells. Test this out the next time you are in a candle shop, or try with common smells at home, like lemons, coffee, etc.

Receptors are classified in a number of different ways depending on the stimulus detected or the environment monitored. They may be classified as:

- **Enteroreceptors:** the internal environment.
- **Exteroreceptors:** the external environment located in the skin.
- **Proprioceptors:** relative position of parts of the body.

Sensory information is normally carried to the brain through three sets of neurons: first-, second- and third-order neurons (Figure 5.14).

Sensory pathways

Sensory pathways are ascending and organised in three distinct groups:

1. **Anterolateral pathways.** Temperature, pain and coarse touch sensations to the brain.
2. **Medial lemniscal pathways.** Discriminative touch, vibration and proprioception sensations to the brain.
3. **Spinocerebellar tracts** (anterior and posterior). Sensations for muscles and tendons (stretch) to the cerebellum to coordinate skeletal muscle movement.

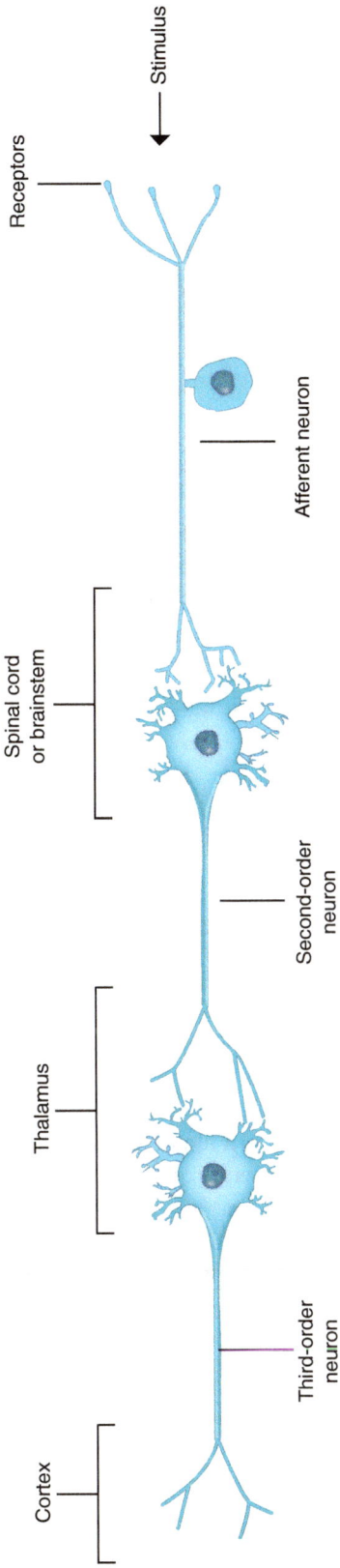

Figure 5.14 Sensory order neurons

The distribution of cells within the primary somatosensory cortex receive information from the different parts of the body as represented pictorially by the sensory homunculus. Those very sensitive body areas with many sensory receptors (e.g. lips, tongue) appear very large in the homunculus.

ACTIVITY 5.5: UNDERSTAND

Using a web-based search engine, look up an image of a sensory homunculus. This will show you how parts of the body that are very sensitive have a greater representation in the brain.

Posture and movement

For the brain to send the correct instructions to the skeletal muscles, it needs first to receive information about tension and stretch from the various receptors. Sensory information from muscles, tendons and joints passes to the sensory cortex and the cerebellum which provide feedback about the effects of motor activity.

Motor division/system

The motor division/system refers to nervous impulses sent from the CNS to cells/organs/muscles to initiate responses. It is subdivided into two subdivisions – somatic (motor) and visceral (autonomic nervous system).

The somatic nervous system (motor)

The somatic nervous system is responsible for voluntary (conscious) control of body movements through stimulating contraction of skeletal muscles. The motor cortex controls movement of the opposite side of the body and gives instructions to move different parts of the body. Efferent neurons carry impulses to the muscles of the skeleton to stimulate muscle contraction and regulate movement. The area of the cortex concerned with movements of different parts of the body for a particular activity is related to the importance of that function. A motor homunculus represents the parts of the body in proportion to the complexity of the movements controlled.

ACTIVITY 5.6: UNDERSTAND

Using a web-based search engine, look up an image of a motor homunculus. This will show you how parts of the body that require complex coordination have greater representation in the brain.

Movement is controlled at two levels through three motor pathways. The parallel direct and indirect pathways are designated as upper motor neurons. Both pathways end on the same motor neuron cell body in the spinal cord and travel to the muscle through the same nerve axon, the lower motor neuron.

1. **Direct (formerly pyramidal) pathway** (Figure 5.15). These upper motor neurons from the cerebral cortex pass down to the spinal cord through the direct pathway, i.e. without any synapses. These impulses travel directly from the brain to the level of the spinal cord where it synapses with the lower motor neuron. They are therefore called corticospinal tracts and carry impulses needed for fast or skilled movement.

2. **Indirect (formerly extrapyramidal) pathway** (Figure 5.16). The indirect pathway consists of all the other nervous input to motor function and also involves upper motor neurons. It is much more complex than the direct pathway. Information travelling through the indirect pathway passes through a series of synapses between some of the basal nuclei, the cerebellum and the motor cortex. It is, thus, slower in carrying information to the lower motor neuron.

The basal nuclei and cerebellum are concerned with regulation and coordination of movement through moderating the activity of the motor areas of the cerebral cortex and brainstem. The basal nuclei carry out subconscious patterns of previously learnt movement. The cerebellum regulates fine movements through receiving information from receptors concerned with position and movement of the body (eyes, ears – balance, joint and muscle receptors). It also receives information from the motor cortex and compares intended with actual movement and makes adjustments to ensure smooth movement. Limb movement is largely determined through the reticulospinal tracts and balance and posture through the vestibulospinal tracts.

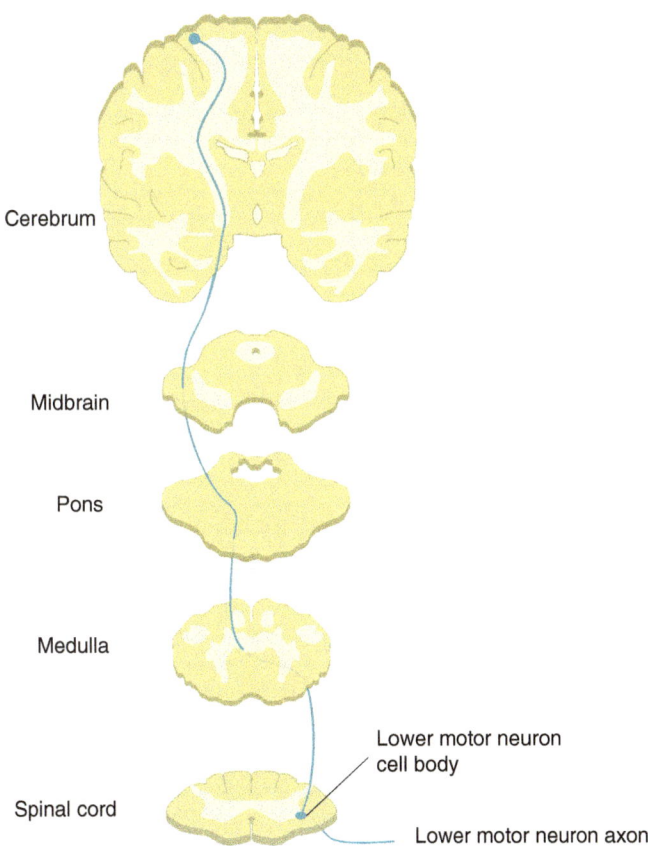

Cerebrum

Midbrain

Pons

Medulla

Lower motor neuron
cell body

Spinal cord

Lower motor neuron axon

Figure 5.15 Direct motor pathways

3. **The spinal cord (lower motor neuron)** (Figure 5.17). The spinal cord plays an important part in control of movement. Since both upper pathways (i.e. direct and indirect pathways) end on the same neuron, the motor part of the spinal nerve or lower motor neuron is also known as the final common pathway. The cell body of the lower motor neuron receives and integrates impulses from:

- The direct and indirect pathways.
- The sensory nerves carrying information from the tissues at that spinal level.
- Association neurons carrying information from higher and lower segments of the spinal cord.

Motor instructions are then sent to the muscles supplied by the lower motor neuron.

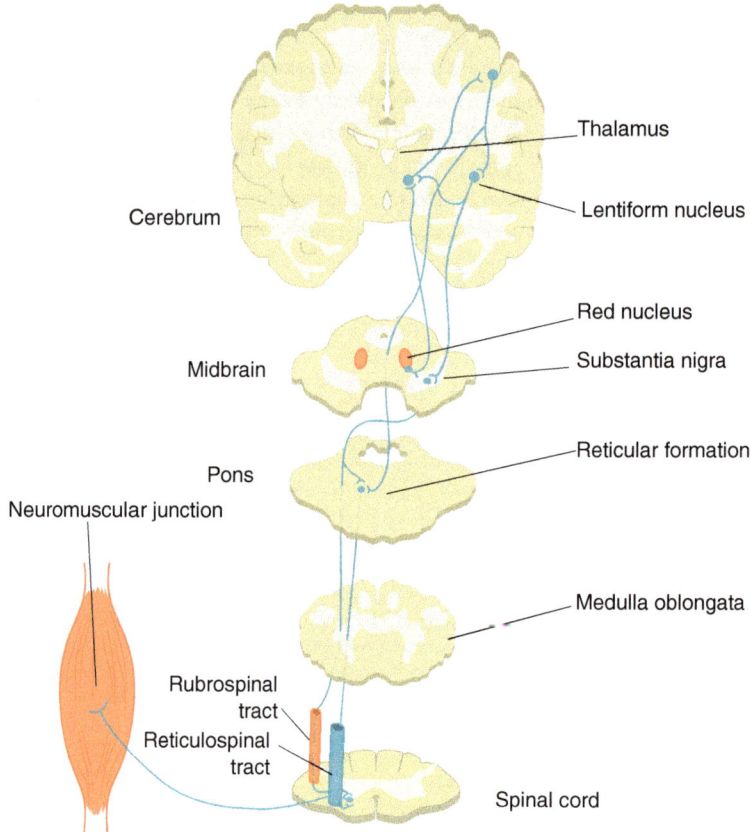

Figure 5.16   Indirect motor pathways

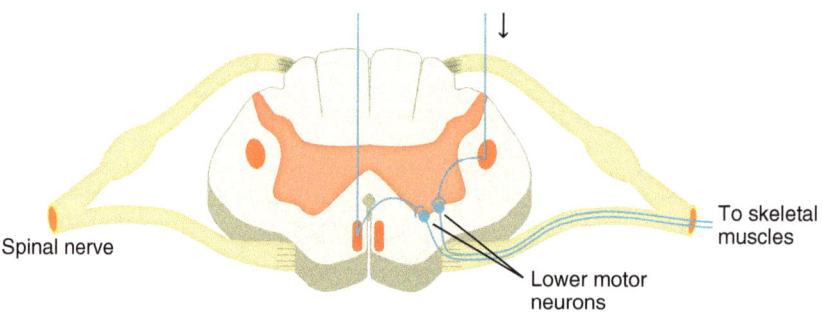

Figure 5.17   Lower motor neuron

Reflexes

The spinal reflexes are important in coordination of movement at a subconscious level. More complex reflexes involve spinal segments and the associated nerves higher and lower in the spinal cord and/or on the opposite side of the spinal cord ensure correct coordination. The spinal cord also contains neural circuits that create walking movements.

A reflex arc exemplifies integration and coordination at the level of the spinal cord. It enables a very rapid response to noxious stimuli (Figure 5.18). More complex reflexes involve spinal segments and associated nerves higher and lower in the spinal cord and/or on the opposite side of the spinal cord. This ensures that actions are coordinated correctly. For example, if you stand on a drawing pin, the simple reflex will withdraw your foot very rapidly and would result in you falling over. The complex reflexes associated with this simple reflex ensure that your other leg and body adjust their position so that you keep your balance and do not fall over.

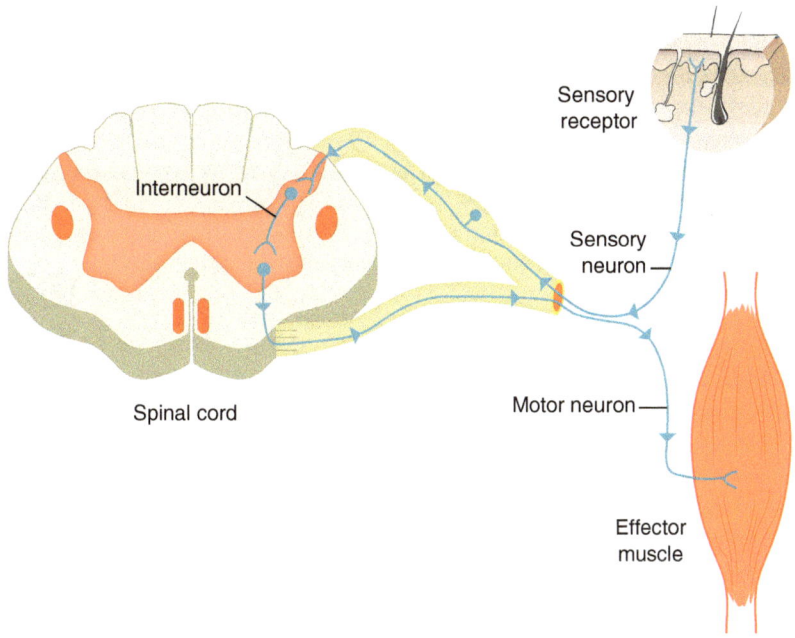

Figure 5.18 Reflex arc

APPLY

Reflex arc

A reflex is an involuntary reaction that occurs to protect us from danger. Consider Figure 5.18 and the following example:

> When you touch a hot substance (e.g. the hot handle of a frying pan or the surface of a hot clothes iron), this is detected by the sensory receptors and relayed to the spinal cord by a sensory neuron. An interneuron, or relay neuron, then picks up this message from the sensory neuron and links directly to a motor neuron to activate it, resulting in the stimulation of effector muscles to pull your hand away from the heat source. This occurs rapidly and without conscious thought as it does not involve processing in the brain.

The autonomic (visceral motor) nervous system

The autonomic nervous system is a subconscious control system for visceral organs including those of the circulatory, digestive and respiratory systems. Most of the pathways in the autonomic nervous system are motor pathways, but include some sensory pathways, indicating that the autonomic nervous system works in synergy with the sensory system in places. While technically part of the peripheral nervous system, the autonomic nervous system has elements in the CNS. Cranial and spinal nerves innervate the visceral organs, but the hypothalamus and medulla represent the control centres for the autonomic nervous system.

The autonomic nervous system has two divisions, the sympathetic and the parasympathetic nervous systems. Although anatomically and functionally different, both divisions normally stimulate the same organs. The sympathetic and parasympathetic nervous system work in harmony to maintain homeostasis. For example, when you are exercising, the sympathetic nervous system raises your heart rate. Following exercise, it is slowed by the parasympathetic nervous system. Of course, both are working in synergy to ensure the heart rate continues to meet the body's needs and is neither too fast nor too slow.

ACTIVITY 5.7: UNDERSTAND

In order to further your understanding of the autonomic nervous system, watch this online video clip.

This online video can be accessed by **scanning the QR code** with your smart phone or via https://study.sagepub.com/essentialap2e.

AUTONOMIC NERVOUS SYSTEM
(11:53)

Parasympathetic nervous system

The parasympathetic nervous system mainly slows activities (the 'rest and digest' system) including:

- Slowing and pacing the heart rate.
- Constricting coronary arteries.
- Constricting bronchioles in the respiratory system.
- Reducing respiratory rate.
- Increasing motility of the digestive system and release of enzymes for digestion.
- Promoting conversion of glucose to glycogen to create an energy store.
- Relaxing gastrointestinal sphincter muscles to facilitate movement of gastrointestinal contents.
- Relaxing the internal urethral sphincter muscle and contraction of the bladder to permit urination/micturition.
- Constricting pupils and bulging the lens of the eye to permit close vision.

The vagus nerve plays a particularly important role in the parasympathetic system.

Sympathetic nervous system (SNS)

The sympathetic nervous system speeds activities through the fright, flight or fight response. Noradrenaline (norepinephrine) is the principal neurotransmitter in the SNS. Some activities include:

- Increasing rate and force of myocardial contraction.
- Vasodilating the coronary arteries.
- Dilating the respiratory bronchioles.
- Increasing respiratory rate.
- Reducing the motility of the digestive system and release of digestive enzymes.
- Constricting gastrointestinal sphincter muscles to reduce gastrointestinal activity.
- Promoting conversion of glycogen to glucose for energy for increased activity.
- Constricting the internal urethral sphincter muscle and relaxing the bladder to reduce urination.
- Producing renin for raising blood pressure (Chapter 12).
- Dilating pupils to increase light entering the eye for visual accuracy.

Structure of the autonomic nervous system

Through the sympathetic ganglion chain running either side of the spinal cord, the sympathetic nervous system innervates organs through thoracic and lumbar spinal nerves, whereas the parasympathetic nervous system uses cranial nerves and sacral spinal nerves (Figure 5.19).

APPLY

Autonomic imbalance

The structure of the autonomic nervous system is important when someone has a spinal cord injury. The level of the injury can mean that some organs may have parasympathetic stimulation but not sympathetic. For example, in a cervical cord injury, the parasympathetic stimulus for the heart from the cranial nerves (the vagus nerve) may lead to a slow heart rate and low blood pressure. The sympathetic stimulus from thoracic spinal nerves, below the level of injury, cannot counteract this.

CRANIAL AND SPINAL NERVES

Cranial nerves

Cranial nerves emerge directly from the brain and brainstem whereas spinal nerves emerge from the spinal cord. There are 12 cranial nerves: some all motor, some all sensory, some mixed. The nuclei for the cranial nerves are within different parts of the brain. The nuclei for the first (olfactory) and second (optic) cranial nerves connect with parts of the forebrain concerned with smell and vision. The other cranial nerve nuclei are in the brainstem. However, information entering the brain is sent to various parts of it. The cranial nerves mainly supply muscles and sensory receptors of the face, head and neck. Table 5.5 identifies the function of each cranial nerve, although some work together to control particular actions and interpret different sensations.

The following mnemonic may help you in remembering these nerves: **O**n **O**ld **O**lympus' **T**owering **T**ops **A** **F**inn **A**nd **G**erman **V**iewed **S**ome **H**ops.

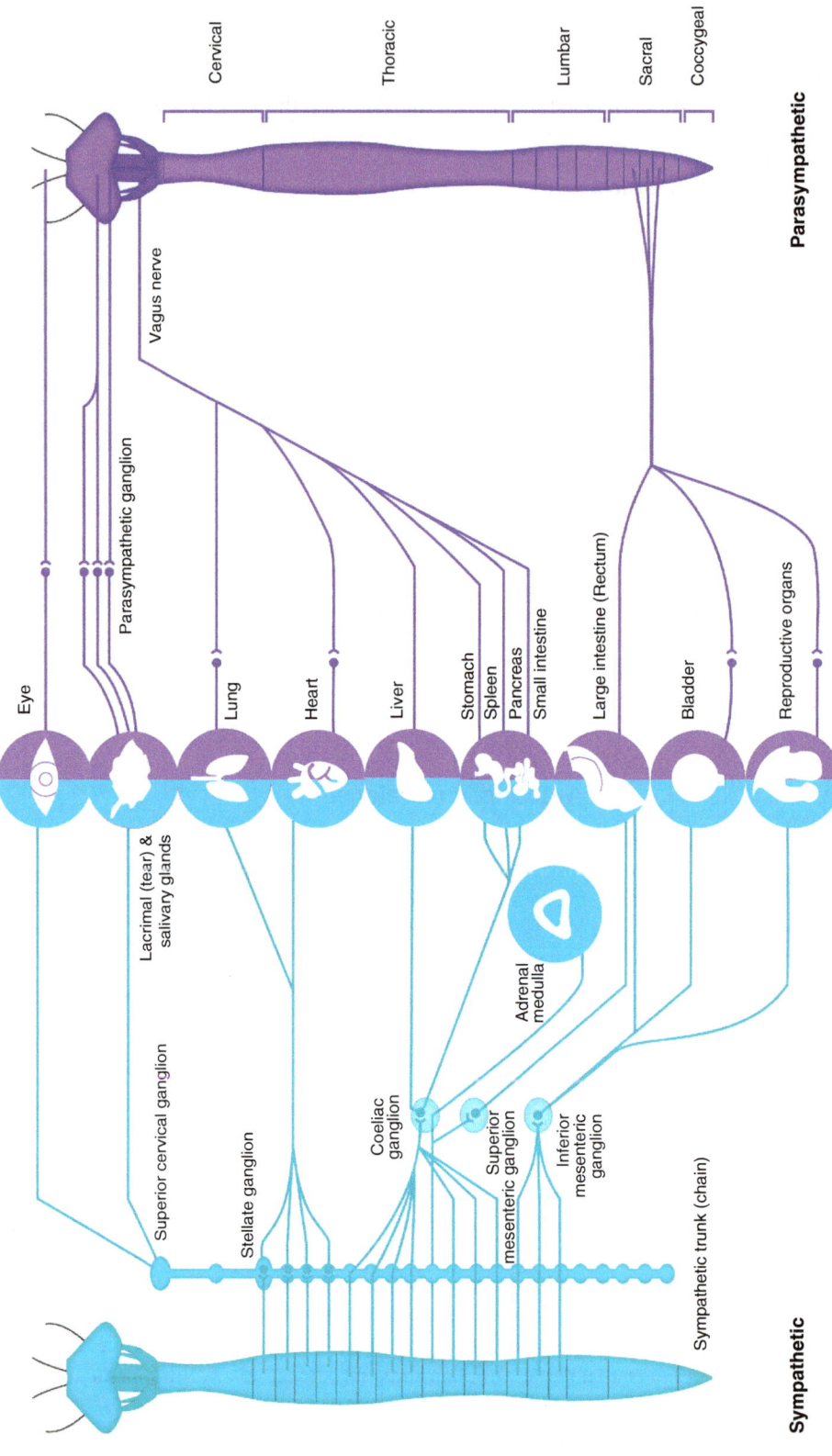

Sympathetic

Parasympathetic

Cervical

Thoracic

Lumbar

Sacral

Coccygeal

Vagus nerve

Parasympathetic ganglion

Eye

Lung

Heart

Liver

Stomach

Spleen

Pancreas

Small intestine

Large intestine (Rectum)

Bladder

Reproductive organs

Lacrimal (tear) & salivary glands

Superior cervical ganglion

Stellate ganglion

Coeliac ganglion

Superior mesenteric ganglion

Inferior mesenteric ganglion

Adrenal medulla

Sympathetic trunk (chain)

Figure 5.19   The autonomic nervous system

Table 5.5 The cranial nerves

Cranial nerve		Type	Function
I	Olfactory	(O) Sensory	Olfaction (smell): originates in olfactory mucosa in nasal cavity, terminates in olfactory bulb beneath the frontal lobe
II	Optic	(O) Sensory	Vision: originates in retina, terminates in thalamus from where information is relayed to the visual cortex
III	Oculomotor	(O) Motor	Eye movement: opening eyelids, constricting pupil, focusing sight, proprioception. Originates in midbrain, terminates in muscles moving eye, iris and lens (ciliary muscles)
IV	Trochlear	(T) Motor	Eye movement and proprioception. Originates in midbrain, terminates in oblique muscles of the eye
V	Trigeminal	(T) Motor and sensory	Three sensory divisions, originate in face and terminate in pons:
			• Ophthalmic division: Main sensory nerve of upper face for touch, temperature and pain
			• Maxillary division: Main sensory nerve of middle face for touch, temperature and pain
			• Mandibular division: Main sensory nerve of lower face for touch, temperature and pain
			Motor function: stimulates the muscles for chewing (mastication)
VI	Abducens	(A) Motor	Eye movement. Originates in pons, terminates in lateral rectus muscles of the eye
VII	Facial	(F) Sensory and motor	Motor for muscles of facial expression, control of tear, nasal, palatine, and salivary glands, originates in pons, terminates in facial muscles and glands, and muscles of middle ear
			Sensory for taste on anterior two thirds of tongue, originates in taste buds, terminates in thalamus
VIII	Vestibulocochlear (Auditory)	(A) Sensory, but some motor	Hearing and balance (equilibrium). Sensory originates in inner ear, terminates in pons and medulla. Motor fibres originate in pons, terminate in outer hair cells in cochlea
IX	Glossopharyngeal	(G) Motor and sensory	Sensory: tongue and pharyngeal sensations (touch, pressure, taste and pain), outer ear sensations (touch, pain and temperature)
			Swallowing, production of saliva, gagging, regulation of blood pressure and breathing
			Motor fibres originate in medulla and terminate in parotid salivary gland, glands of posterior tongue and stylopharyngeal muscle
			Sensory fibres originate in pharynx, middle and outer ear, posterior of tongue and internal carotid arteries
X	Vagus	(V) Motor and sensory	Swallowing, taste, speech, respiratory, cardiovascular and gastrointestinal regulation, hunger, satiety and intestinal discomfort
			Motor fibres originate in medulla and terminate in tongue, palate, pharynx, larynx, thorax and abdomen
			Sensory fibres originate in thorax, abdomen, root of the tongue, epiglottis, pharynx, larynx, outer ear and dura mater, terminate in medulla
XI	Spinal accessory	(S) Motor	Swallowing, head neck and shoulder movement. Originates in medulla and C1 to C5 or C6 of spinal cord, terminates in palate, pharynx and muscles of shoulders, head and neck
XII	Hypoglossal	(H) Motor	Movements of the tongue for speech, moving food and swallowing
			Originates in medulla, terminates in muscles of the tongue

ACTIVITY 5.8: GO DEEPER

Cranial nerve assessment

Now you know the function of each cranial nerve, how could you assess each one? You will need to do some further reading and consider that some cranial nerves are assessed together.

Spinal nerves

There are 31 pairs of spinal nerves emanating from the spinal cord, each distributed to a specific part of the body. During fetal development, the nerve fibres get stretched out as the body develops, resulting in a distribution of dermatomes over the body, showing the area supplied by each spinal nerve. Dermatomes are therefore areas of skin normally supplied by a single spinal nerve.

Some of these nerve fibres arise from the spinal cord (T2 to T12) while the others originate from the four plexuses or nerve networks:

- cervical
- brachial
- lumbar
- sacral.

Each plexus originates from nerve fibres from several segments of the spinal cord and forms several major nerves. Trauma to a nerve plexus can cause major loss of function to the part supplied.

ACTIVITY 5.9: APPLY

Dermatomes

Using a web-based search engine, look up an image of the body's dermatomes. Note how you can trace an area of the skin back to each spinal nerve.

NUTRITION AND PROTECTION OF THE NERVOUS SYSTEM

As the nervous system is home to the major control systems in the body, its nourishment and protection are vital for survival.

Nutrition

Cells of the nervous system must receive adequate nutrition and oxygen; this occurs through the cerebral and spinal circulation which distributes blood to the various parts of the central nervous system. Cerebrospinal Fluid (CSF) also provides nutrition to parts of the central nervous system (discussed later).

Cerebral circulation

The main arteries supplying the brain are the two internal carotid arteries and the basilar artery (arising from the two vertebral arteries) (Figure 5.20). The Circle of Willis is a circle of blood vessels that supplies blood to the different parts of the brain. It is formed from the main arteries supplying the brain joined by the connecting arteries. It enables regulation of blood flow and blood pressure within the brain and facilitates collateral circulation if arteries in the brain are narrowed or blocked. The blood from the brain drains into sinuses (wide vessels with no muscle in the walls), which return it to the main circulation at the internal jugular vein in the neck.

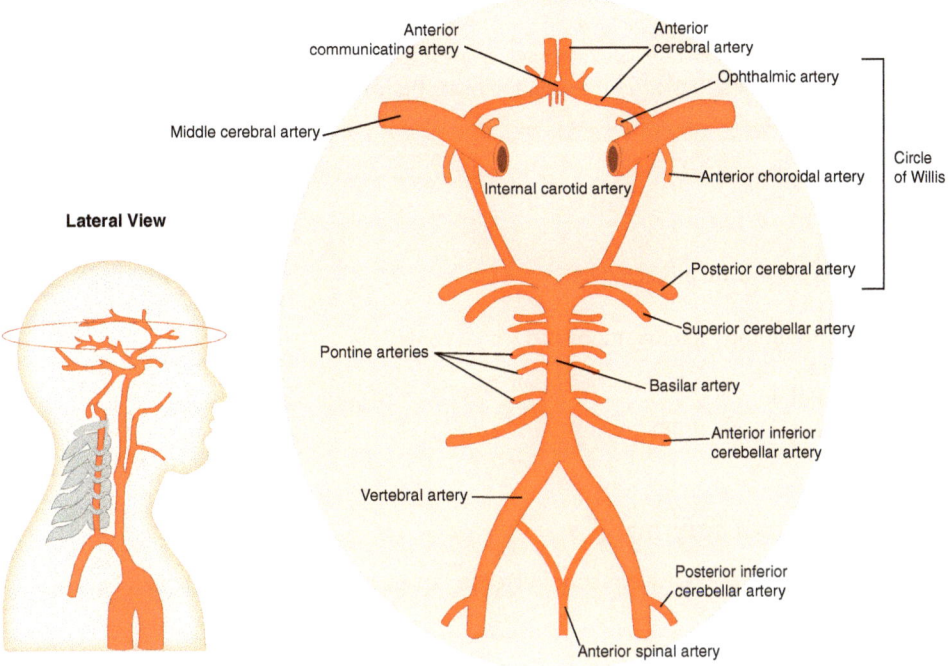

Figure 5.20 Cerebral circulation

Presentation of neurological impairment

Earlier in this chapter you learned about the different parts of the brain and their functions. If circulation was interrupted or occluded (blocked) to the following parts of the brain, what signs and symptoms could a person have?

1. The temporal lobe.
2. The frontal lobe.
3. The occipital lobe.

Spinal circulation

The spinal cord is nourished by three main arteries and an arterial network although the CSF supplies the nutrients and oxygen to the tissues of the brain and spinal cord.

GO DEEPER

Spinal circulation

The anterior spinal arteries supply the anterior two thirds of the spinal cord and the posterior spinal arteries supply the posterior surface. Additionally, the posterior spinal arteries give rise to the vasocorona, which supplies blood to the superficial layer of the anterolateral surface of the spinal cord. Posterior spinal veins and anterior spinal veins drain into the radicular veins to return venous blood to the vertebral venous plexus. This drains blood into the larger veins in the neck, thorax and abdomen.

Protection

The brain and spinal cord are protected by:

- The bone surrounding them (cranium and spinal column).
- The meninges.
- Cerebrospinal fluid.
- The Blood–Brain Barrier (BBB).

Cranium and spinal column

These bony structures all provide a structural defence for the brain and spinal cord and are explored in Chapter 15.

The meninges

The meninges are the three layers of connective tissue that surround the brain and spinal cord (Figure 5.21):

The dura mater: This is the outermost layer of the brain and spinal cord composed of the outer periosteal layer (absent from the spinal cord) and inner meningeal layer. These are largely closely joined except where they separate in the cranium to form venous sinuses that assist in returning blood to the heart. The dura is tough to prevent friction against the skull and is a lining that retains CSF within the CNS. It also suspends the brain in the cranial cavity.

The arachnoid mater: The arachnoid mater is next, separated from the dura mater by the dural space and from the pia mater below by the subarachnoid space. Web-like extensions connect the arachnoid and pia mater and the subarachnoid space is filled with CSF. The arachnoid mater contains large blood vessels and the arachnoid villi from these reabsorb CSF as it circulates through the CNS.

The pia mater: The pia mater is a thin, impermeable fibrous membrane adjacent to the brain and spinal cord ensuring that CSF remains within the subarachnoid space. Blood vessels pass through the pia mater to the brain and spinal cord.

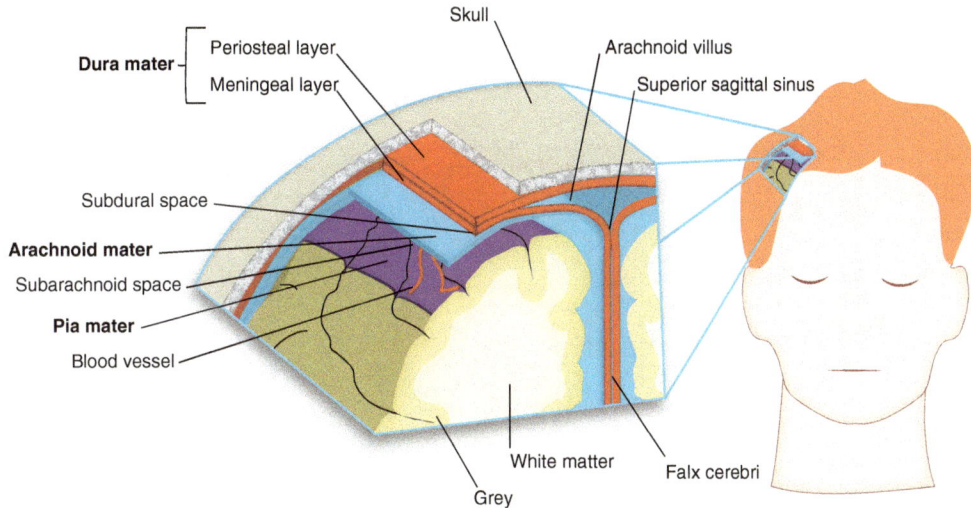

Figure 5.21 The meninges

Cerebrospinal fluid (CSF)

CSF protects and nourishes the CNS. About 400–500 ml of CSF daily is formed by the choroid plexus in the sides and floor of the lateral ventricles and the roof of the third and fourth ventricles. Only about 150–180 ml of this is in circulation as it is readily absorbed by the arachnoid villi back into the blood. The flow of CSF is as follows (see Figure 5.22):

1. CSF is created in the choroid plexus.
2. It moves through the foramen of Monro to the third ventricle.
3. It passes through the cerebral aqueduct to the fourth ventricle.
4. It then passes through three apertures (two lateral and one median) to the cerebellomedullary cistern.
5. From here it circulates over the spinal cord and enters the subarachnoid space where it is reabsorbed.

CSF is normally clear and odourless containing glucose, protein and white cells (largely lymphocytes, monocytes and macrophages) but no red blood cells. As CSF flows in one direction it effectively washes out waste products and metabolites. With the meninges it provides protection by acting as a cushion for impact and lubricates the meninges for frictionless movement of the brain.

Blood-brain barrier

As a structural and chemical barrier, the BBB strictly regulates substances passing from the circulatory system into the nervous system.

Structural: A mechanical barrier occurs as capillary endothelial cells in the brain are packed very tightly together with no gaps (fenestrations) in the cell walls. The capillary lumen also has a reduced surface area.

Chemical: Capillary endothelial cells in the brain have large numbers of mitochondria for high production of energy. Additionally, specific proteins inhibit lipid-soluble drugs crossing from the blood into the brain by actively transporting them into the blood stream. Should substances pass, these cells contain drug-metabolising enzymes to eliminate or deactivate them before reaching CNS tissues.

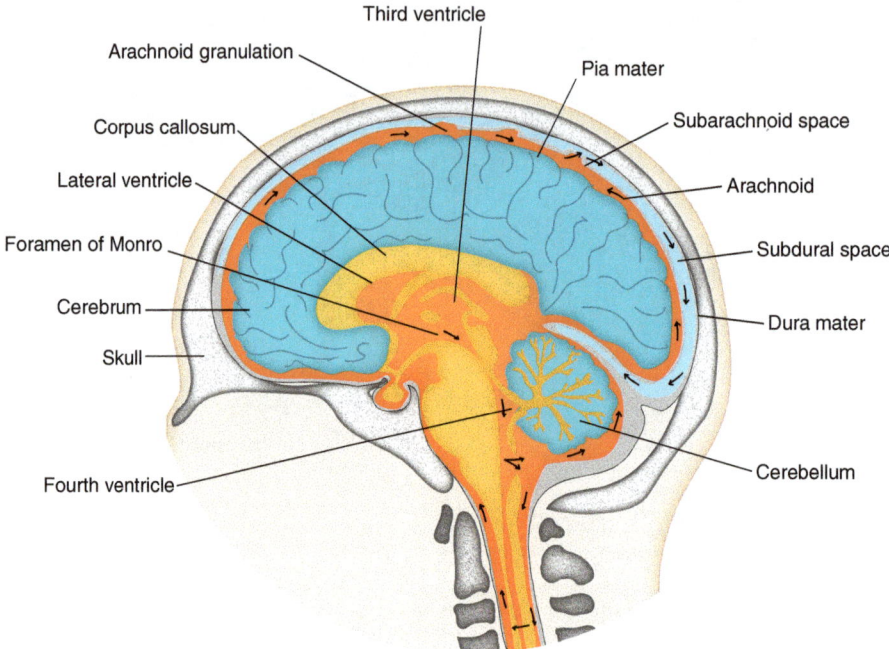

Figure 5.22 Cerebrospinal fluid (CSF) circulation and ventricles of the brain

Routes for substances to cross the BBB are:

1. Transport mechanisms control movement of water-soluble substances through ion channels, transport proteins or, for larger items, transcytosis. Transcytosis moves large molecules by capturing them in vesicles, carrying them across the cell, and exporting them out the other side.
2. Fats move across by diffusion.

APPLY

BBB and drug administration

The BBB is useful in preventing substances entering the brain. However, sometimes the brain can become infected by bacteria and viruses. In these circumstances, drugs are often required to treat the infection. When these drugs enter the body outside of the nervous system (i.e. orally, intravenously or by intramuscular injection), the therapeutic dose may not be achieved within the central nervous system as the BBB may impede the drugs from entering. In such situations, certain drug preparations can be given directly into the ventricles of the brain (intraventricular route) or the subarachnoid space (intrathecal route). This ensures that the therapeutic level of the drug can circulate and act within the central nervous system. Such drugs are given under strict conditions and with specialist education and training.

Caffeine and alcohol cross the BBB freely, which is why they both can have a relatively acute effect on the nervous system.

— **GO DEEPER** —

Neurogenesis (growth of new neuronal tissue)

The brain grows and develops from conception into maturity through stem cells in the walls of the ventricles. For many years it was thought that mature adult brain cells do not repair or regenerate. Over the past 20 years we have found this not to be true and that new nerve cells are constantly being produced in the brain to maintain homeostasis. For example, Spalding et al. (2013) determined that the hippocampus generates 700 new neurons daily, about 1.75% regeneration. Their study concluded that neurons are generated throughout adulthood, albeit with some decline as a result of ageing.

Neurogenesis, or the growth of new neuronal tissue, appears to be limited to the hippocampus and the olfactory bulb where there are active stem cells. Stem cells in other regions of the brain do not appear to be active in producing new nerve cells. When neurons are damaged in the CNS, they generally do not repair well. In contrast, neurons of the PNS are much better at adapting to injury by growing new axons and making new synaptic connections (Purves et al., 2011). The difference is down to genetic programming – adult CNS neurons tend to no longer express the gene that causes axonal growth, whereas the same genes can be reactivated in adult PNS neurons. The reactivation occurs in the PNS as macrophages efficiently remove cell debris in damaged neurons, and Schwann cells provide the optimal conditions to promote regeneration. This is through enhancing cell adhesion molecules, the extracellular matrix, and a variety of growth factors called neurotrophins. That same efficiency is not present in the CNS and the debris from damaged neurons can remain, impeding repair.

Additionally, oligodendrocytes produce a substance called Nogo, which inhibits axons extending to form new connections (Purves et al., 2011). Astrocytes compound this by releasing more factors that also inhibit axons extending in the presence of injured cells. Even if CNS neurons do reactivate the gene for repair, the surrounding environment inhibits it. As a result, there is limited repair. However, some neurons in the CNS do manage to repair and reconnect, and so some level of recovery can occur. Research continues into how to promote this process for those with neuronal damage in the CNS.

While we are born with most central neurons created, it is through synaptogenesis (formation of synapses between neurons), pruning (removal of unnecessary neuronal connections and strengthening important ones) and apoptosis that we modify and refine neuronal function and efficiency over time and as we develop. This is why infants develop according to milestones, gaining skills sequentially, such as learning to control the head, sit and stand before walking.

CONSCIOUSNESS AND SLEEP

Consciousness

Consciousness is a state of explicit awareness dependent on both biological arousal in the brain, mainly by the reticular activating system (RAS), and the processing of experiences (perception). A person can process, interact and experience their surroundings, and is influenced by their state of mind. Consciousness is crucial not only to survival but also to a person's experience of their reality, and therefore is an integral component of personhood. Only through consciousness can the brain process information, analyse it and make decisions that influence the present while planning for the future.

There is significant debate on how consciousness is achieved. However, there is a general consensus that both cerebral hemispheres are functioning simultaneously and both the central and peripheral nervous systems are involved. The reticular formation receives signals from various sources in the nervous system and directs them to the thalamus. The reticular formation contains the reticular activating system and has a role in regulating signals that go to the thalamus. The thalamus then directs signals out across the cerebral hemispheres to the cerebral cortex. The thalamus also communicates with the hypothalamus in this regard to regulate the sleep–wake cycle. Being awake is a vital component of consciousness. The prefrontal cortex is stimulated by the thalamus and is central to regulating perception and experience of the environment to produce a response. The limbic system integrates this to provide an emotional element to that response.

Sleep

Sleep comprises two out of three states of human existence:

- Wakefulness.
- Non-Rapid Eye Movement (NREM) sleep.
- Rapid Eye Movement (REM) sleep.

The electrical activity present in the brain can tell us a lot about how alert or sleepy a person is. This activity can be viewed as brain waves. Figure 5.23 describes these briefly. We spend around a third of our life asleep and sleep is considered central to homeostasis and health. More and more studies now highlight that our overall health is dependent on the duration and quality of sleep. Sleeping more or less than

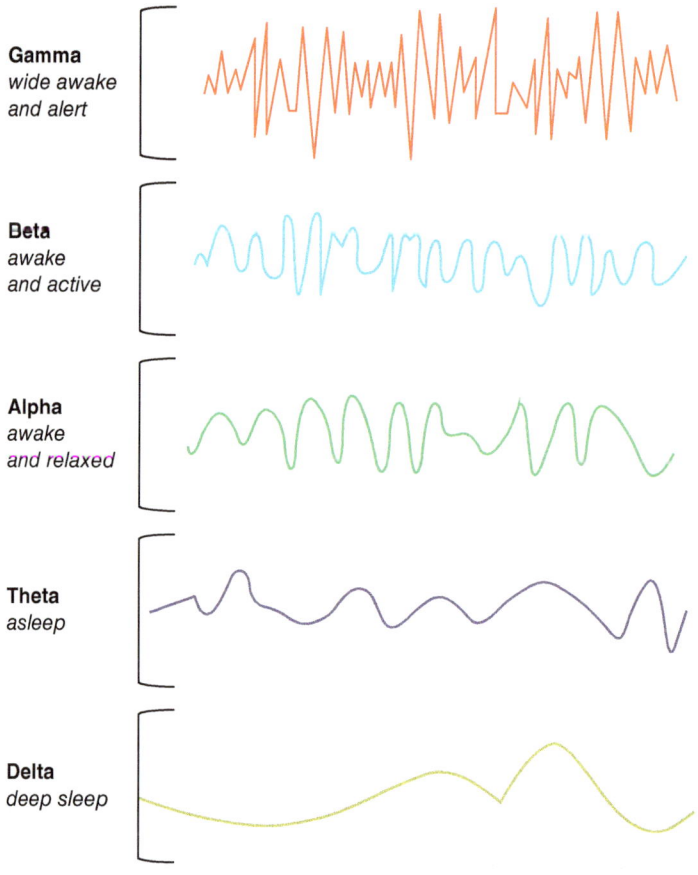

Gamma
*wide awake
and alert*

Beta
*awake
and active*

Alpha
*awake
and relaxed*

Theta
asleep

Delta
deep sleep

Figure 5.23   Brain waves

seven hours at night has been shown to increase mortality and morbidity (Ferrie et al., 2007; Cappuccio et al., 2011). However, we still do not fully understand all the components of sleep and its entire role in homeostasis. Figure 5.24 shows the structure and functions of sleep.

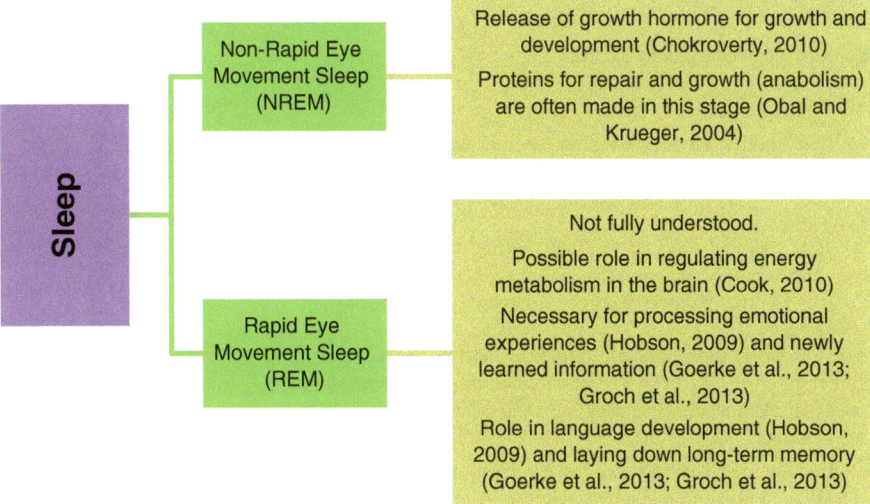

Figure 5.24 Structure of sleep

Sleep operates in cycles largely based around a 24–25-hour day. There are five stages within each cycle lasting about 90–100 minutes. Four stages are NREM stages and the fifth is REM. Figure 5.25 illustrates the structure of a cycle of sleep. A healthy night's sleep normally consists of four to five cycles. However,

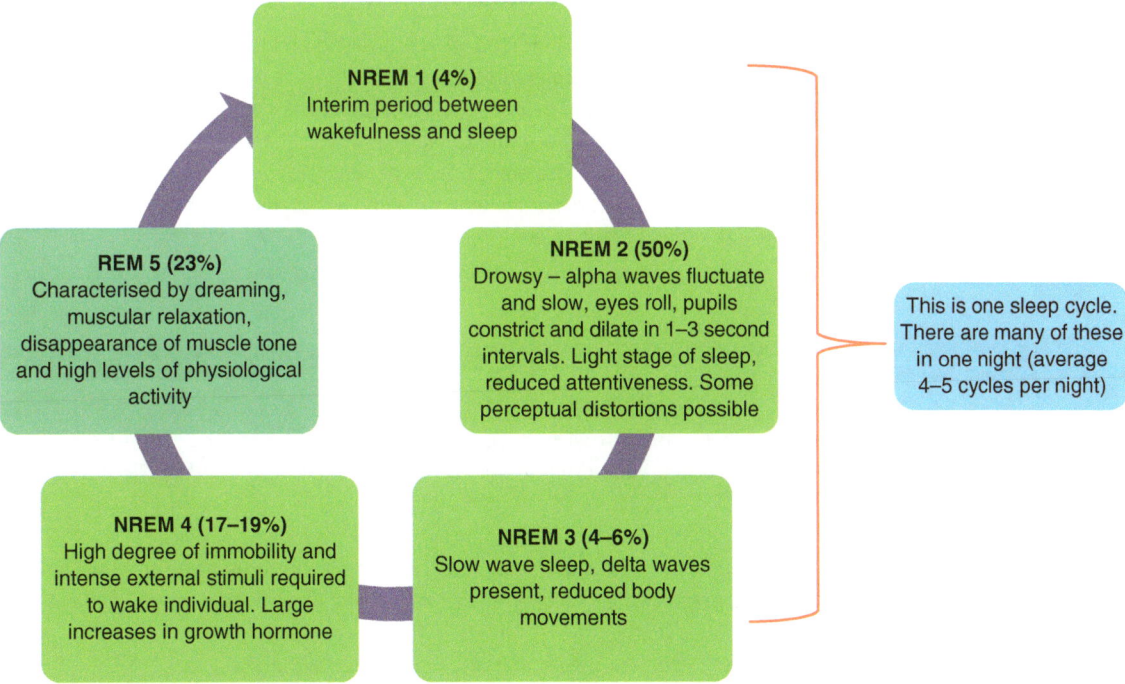

Figure 5.25 Stages of sleep

there are two natural periods of sleepiness during the day. The obvious one is at night, and the second is in the early afternoon, roughly between 3 pm and 5 pm. People have more periods of sleepiness in early life as they are growing and learning rapidly. Older adults also have more, usually shorter, periods of sleep over the 24 hours. These cycles are regulated by various biological and social factors (Table 5.6).

Table 5.6 Factors regulating sleep

Biological factors	Social factors

• The hypothalamus is the body's biological clock, determining the **circadian rhythm**. It regulates sleep in the nervous system, influenced by social factors regarding timing; triggered by light and melatonin from the pineal gland
• The Ascending Reticular Activating System (ARAS) sends signals to the forebrain and hypothalamus to regulate sleep; signals to the thalamus are forwarded to the cerebral cortex - pathways similar to those of consciousness
• The forebrain works with the RAS to regulate NREM sleep
• The pons regulates REM sleep and reactivates the brain from sleep
• Three neurotransmitters/hormones from the pons are central to controlling sleep:
  o Noradrenaline arouses the cerebral cortex to achieve wakefulness; low levels are present in sleep states
  o Acetylcholine achieves wakefulness and regulates REM sleep; low levels are present in sleep states
  o Serotonin, involved in regulating REM and NREM sleep
	• Exposure to light
	• Social activities such as:
	  o Regular meal times
	  o Regular times of going to bed and getting up
	  o Levels of stress
	  o Emotional state
	  o Undertaking exercise

APPLY

Sleep and development

BODIE FAMILY

Think of baby Danielle Zuma. She is two months old and will be growing and learning faster than at any other stage in her life. Danielle therefore needs plenty of sleep. This allows release of growth hormone in stage 4 of sleep, helping to achieve the expected growth and development. As she grows through her first few years, REM sleep will be crucial in enabling her brain to learn and recall new information and develop her language skills. It will also allow her to process emotional experiences and is therefore central to her emotional development. Having a pattern to the day, including settling her to rest in relation to feeding and emotional state, will be central to regulating her sleep.

ACTIVITY 5.11: APPLY

As you will have learned (and probably experienced), you have two natural periods of sleepiness in the day. The obvious one is overnight, the other in the early afternoon.

• Does society enable this natural process to occur?
• Think about the care provided on hospital wards. What normally happens in the afternoon? How could this impact on the person who is recovering or dealing with illness? How could practice be more person-centred in this regard?
• How do hospital activities impact on the regulation of circadian rhythms?

THE NERVOUS SYSTEM AND PERSONHOOD

Personhood refers to the quality of being an individual person. While there are many components of personhood, which will not be debated here, the biological components are largely based within the nervous system. The reticular formation and limbic system are thought to play key roles in personhood; the reticular formation being key in regulating arousal and consciousness. Both are necessary for awareness. You have already learned how the limbic system regulates emotion, behaviour, motivation and long-term memory. Combined, a person's ability to interact and respond emotionally is regulated biologically. Central to this will be a person's life experiences, analysed and laid down in memory through the limbic system linking experiences with emotion.

The nervous system and mental health

As we learn to understand the nervous system better, we are discovering more and more how it is central to maintaining a person's mental health. We know that there are areas of the brain involved in maintaining mental health and that some hormones are also important.

There are four key areas of the brain that contribute to a person's mental health status: the amygdala, hippocampus, anterior cingulate cortex and prefrontal cortex. Some of these we have already looked at earlier in this chapter, particularly as three are core areas of the limbic system. Figure 5.26 looks at these in more detail in relation to mental health. A number of neurotransmitters, or hormones, in the central nervous system are involved in regulating mental health. Table 5.7 identifies three of these and their roles.

Amygdala

Regulates response to fear through fright, flight, fight response. Part of this is recalling previous experiences that caused fear and using those to regulate the response being experienced at that moment. The amygdala is one component involved with phobic experiences as it links fear with a previous unpleasant experience. However, that response can be relearnt if the person is exposed to the experience again and it is positive, part of the management of phobias and post-traumatic stress disorder.

Anterior Cingulate Cortex

Role in regulating emotional responses. Influences feelings of motivation, focus and realisation. For example, realising when we did something well, or need to improve something. As a result, when the anterior cingulate cortex operates ineffectively, people can experience depression, lack of motivation or drive. Dysfunction in this area of the brain is linked with behaviour disorders and schizophrenia.

Key Functional Areas in the Brain Regulating Mental Health

The Hippocampus

This area of the limbic system is associated with creating memories and working with parts of the cerebral cortex to lay them down in long-term memory. Memories are very important to a person's health-related quality of life, from being able to reminisce to being able to learn new skills to be independent and self-caring. It has a key role in regulating mood.

The Prefrontal Cortex

Within this area of the cerebral cortex, the brain coordinates cognitive functioning. This includes executive functioning, the ability to take a problem, solve it and make decisions, all underpinned by making judgements. To achieve this, the prefrontal cortex accesses short- and long-term memory and regulates the amygdala in times of heightened stress.

Figure 5.26 Functional areas in the brain regulating mental health

Table 5.7 Hormones regulating mental health

Serotonin	Dopamine	Glutamate
Role in regulating numerous functions within the body: regulating mood, regulating sleep; regulating the mood of a person Depression is linked with low levels of serotonin Some types of antidepressants promote the presence of serotonin within the synaptic cleft so that the receiving neuron works better in regulating mood. This is achieved by drugs that stop serotonin being recycled or reabsorbed back into the sending neuron	Dopamine commonly known for role in controlling movement It is also involved in regulating reward-motivated behaviour Rewards increase the level of dopamine and a lack of dopamine is linked with disorders such as schizophrenia and attention deficit hyperactivity disorder	This very common neurotransmitter attributed to learning and memory, specifically in the hippocampus. It is very important in brain development and increases the excitability of neurons, promoting sending of nervous impulses

ACTIVITY 5.12: APPLY

Perception and personhood

Think about the experiences people have during their lives and how these can vary from person to person so vastly. How can environment influence personhood and how is this facilitated by the central nervous system?

People's past experiences will influence how they react to a number of things, whether it be psychological triggers that have resulted in a mental health problem, or how they learn to manage an aspect of their health. In order to be person-centred, we must try to empathise with people and understand what has influenced their view of the world. Your view of the world will influence your outlook, and so you may not fully appreciate the outlook of those in your care. How can you overcome this to be person-centred? Consider the Person-Centred Nursing Framework in your reflection. Finally, think about your role as a nurse and how you respond to different personalities.

CONCLUSION

By now you should have a good understanding of how the nervous system works and plays a key role in regulating homeostasis. You should also appreciate how the nervous system plays an essential role in a person's experiences of life, being the link between body and mind through regulating emotions and experiences, linking these with memory. An understanding of the nervous system is therefore central to understanding mental health. Furthermore, we have seen that the adaptive responses by the nervous system to maintain homeostasis provide nurses with key information about what is going on in the body. This is a rich form of data that is vitally important in caring for people.

However, the nervous system cannot coordinate homeostasis alone. It integrates with the endocrine system to achieve this. You will learn more about this second control system in Chapter 7. Next, however, we will go into the nervous system a little further in Chapter 6 and look at the special senses.

GO DEEPER

Further reading

Braine, M.E. (2009) 'The role of the hypothalamus, part 3: a key to the control of emotions and behaviour', *British Journal of Neuroscience Nursing*, 5(4): 157–64.

Khan, E. (2006) 'The blood-brain barrier: its implications in neurological disease and treatment', *British Journal of Neuroscience Nursing*, 2(1): 18–25.

Mayer, J. and Garner, A. (2009) 'Action potentials: understanding generation, propagation and their clinical relevance', *British Journal of Neuroscience Nursing*, 5(8): 367–71.

- The nervous system is one of two control systems in the human body, regulating all biological activities in the body along with the endocrine system. Without the nervous system, we would not be able to respond to the internal or external environment.

- There are two types of nervous tissue – neurons, which send electrochemical impulses in response to stimuli received and to effect an action as a result of stimuli, and neuro-glia, which support and protect neurons.

- Neurons communicate by sending electrochemical impulses down the axon of the cell (called an action potential). These impulses move from neuron to neuron crossing gaps between neurons, called synapses, either chemically or electrically.

- Nervous tissue is organised into two main divisions of the nervous system – the CNS and the PNS. There are further subdivisions of both systems that have different roles in maintaining homeostasis.

- Different lobes of the brain have dedicated functions. However, they often interlink throughout the nervous system.

- Consciousness is controlled by the nervous system. Sleep is a state of consciousness and is fundamental to learning and development as well as for reparative processes in the body.

- A person's life experiences are recorded and associated with emotion by the nervous system. These can be recalled from memory and influence how a person reacts to a variety of situations.

REVISE

TEST YOUR KNOWLEDGE

Revision of this chapter will require you to be structured. It is important you revise it in the order it is laid out. This means there are six key areas to revise:

1 Roles and functions of the nervous system in maintaining homeostasis.

2 Nerve cells.

3 Organisation of the nervous system:
- the central nervous system
- the peripheral nervous system.

4 Protection and nutrition of the nervous system.

5 Consciousness and sleep.

6 The nervous system and personhood.

In order to help you revise, consider the following questions, answers for which can be found by visiting **https://study.sagepub.com/essentialap2e**.

Test yourself by revising the chapter first, and then answering these questions without looking at the book. Afterwards compare your answers with the text and with the notes you made. Did you miss anything in your notes? Here are the questions:

1   What are the major functions of the nervous system?

2   Describe the divisions of the nervous system and their functions. Support this with a diagram of the structural divisions of the nervous system.

3   Describe how nerve impulses are transmitted.

4   Describe in outline the major structural elements of the CNS and their functions.

5   How many cranial nerves are there? What is the 5th cranial nerve called and what are its characteristics? Which cranial nerve innervates the organs of the thorax and abdomen?

6   What happens to motor nerves as they pass through the medulla oblongata? Where does this happen for sensory nerves? What is the significance of this?

7   What is a reflex arc? Illustrate this in a diagram.

8   Briefly explain how information is transmitted to the central nervous system and represented in the cerebral cortex.

9   What activities are the autonomic nervous system concerned with? What are the effector organs for this system? How is the autonomic nervous system structured, and what are the functions of its divisions?

10   Where is CSF formed and reabsorbed in the brain? How many fluid-filled spaces are there filled with CSF? Why do we have CSF?

11   What are the functions of the meninges surrounding the central nervous system? Describe the three layers of the meninges.

12   Briefly describe how blood is supplied to the brain.

REFERENCES

Brooks, J.X. and Cullen, K.E. (2013) 'The primate cerebellum selectively encodes unexpected self-motion', *Current Biology*, 23(11): 947–55.

Buckner, R.L. (2013) 'The cerebellum and cognitive function: 25 years of insight from anatomy and neuroimaging', *Neuron* 80(3): 807–15.

Cappuccio, F.P., Cooper, D., D'Elia, L., Strazzullo, P. and Miller, M.A. (2011) 'Sleep duration predicts cardiovascular outcomes: a systematic review and meta-analysis of prospective studies', *European Heart Journal*, 32: 1484–92.

Chokroverty, S. (2010) 'Overview of sleep and sleep disorders', *Indian Journal of Medical Research*, 131: 126–40.

Cook, N.F. (2010) 'The physiology of sleep: homeostasis and health', *British Journal of Wellbeing*, 1(8):16–20.

Ferrie, J.E., Shipley, M.J., Cappuccio, F.P., Brunner, E., Miller et al. (2007) 'A prospective study of change in sleep duration: associations with mortality in the Whitehall II Cohort', *Sleep*, 30(12): 1659–66.

Goerke, M., Cohrs, S., Rodenbeck, A. Grittner, U., Sommer, W. et al. (2013) 'Declarative memory consolidation during the first night in a sleep lab: the role of REM sleep and cortisol', *Psychoneuroendocrinology*, 38(7): 1102–11.

Groch, S., Wilhelm, I., Diekelmann, S. and Born, J. (2013) 'The role of REM sleep in the processing of emotional memories: evidence from behavior and event-related potentials', *Neurobiology of Learning and Memory*, 99: 1–9.

Harris, J.J., Jolivet, R. and Attwell, D. (2012) 'Synaptic energy use and supply', *Neuron*, 75(5): 762–77.

Hobson, J.A. (2009) 'REM sleep and dreaming: towards a theory of protoconsciousness', *Nature Reviews Neurosciences*, 10: 803–14.

Lawson, R.P., Drevets, W.C. and Roiser, J.P. (2013) 'Defining the habenula in human neuroimaging studies', *Neuroimage*, 64: 722–7.

Obal, F. Jr and Krueger, J.M. (2004) 'GHRH and sleep', *Sleep Medicine Reviews*, 8(5): 367–77.

Purves, D., Augustine, G.J., Fitzpatrick, D., Hall, W.C., LaMantia, A.S., McNamara, J.O. and White, L.E. (eds) (2011) *Neuroscience*, 5th edn. Sunderland: Sinauer Associates.

Raybaud, C. (2010) 'The corpus callosum, the other great forebrain commissures, and the septum pellucidum: anatomy, development, and malformation', *Neuroradiology*, 52(6): 447–77.

Sergerie, K., Chochol, C. and Armony, J.L. (2008) 'The role of the amygdala in emotional processing: a quantitative meta-analysis of functional neuroimaging studies', *Neuroscience & Biobehavioral Reviews*, 32(4): 811–30.

Spalding, K.L., Bergmann, O., Alkass, K., Bernard, S., Salehpour, M. et al. (2013) 'Dynamics of hippocampal neurogenesis in adult humans', *Cell*, 153(6): 1219–27.

Stephani, C., Fernandez-Baca Vaca, G., Maciunas, R., Koubeissi, M. and Lüders, H.O. (2011) 'Functional neuroanatomy of the insular lobe', *Brain Structure and Function*, 216(2): 137–49.

Tagliamonte, M., Sestieri, C., Romani, G.L., Gallucci, M. and Caulo, M. (2015) 'MRI anatomical variants of mammillary bodies', *Brain Structure and Function*, 220(1): 85–90.

Woolsey, T.A., Hanaway, J. and Gado, M.H. (2008) *The Brain Atlas: A Visual Guide to the Human Central Nervous System*, 3rd edn. Hoboken: John Wiley and Sons.

Wyss, M.T., Jolivet, R., Buck, A., Magistretti, P.J. and Weber, B. (2011) 'In vivo evidence for lactate as a neuronal energy source', *The Journal of Neuroscience*, 31(20): 7477–85.

SPECIAL AND GENERAL SENSES
RESPONDING TO THE ENVIRONMENT

UNDERSTAND: CHAPTER VIDEO

Before reading this chapter, you might find it useful to remind yourself of the special and general senses by watching the following video.

The video can be accessed by **scanning the QR code** with your smart phone camera or via https://study.sagepub.com/essentialap2e.

THE SENSORY SYSTEM (10.31)

LEARNING OUTCOMES

When you have finished studying this chapter you will be able to:

1. Identify the five special senses and three general senses
2. Explain the structure of the special sense organs in relation to their function and describe the nervous pathways by which information is transmitted to the central nervous system
3. Relate types of sensory receptors to their function and the nervous pathways they take to the central nervous system

INTRODUCTION

The sensory system is fundamental to a person's relationship with their environment. Additionally, the senses are fundamental to health-related quality of life in that they enable people to derive pleasure through the experience of their environment. This chapter will enable you to understand how people react to environmental stimuli through the special senses of vision, hearing and balance, taste and smell and the organs responsible. Additionally, this chapter will address the general senses of touch (pressure, vibration and proprioception), pain and heat. An understanding of these structures and their functions will equip you to understand the relationship between the person and their experience of their environment, an important consideration in person-centred nursing. This chapter builds on concepts and principles discussed in Chapter 5, The Nervous System, which you need to study before this one, particularly the sensory system.

Context

You have already discovered how the nervous system is central to who we are as functioning, living, feeling people. Special and general senses are key to this as they are the way we sense life to experience it. These senses feed into our experiences of life as moments of pleasure, pain, wonderment and awe. We learn to respond to senses across the lifespan and can lose the acuity of these as we age.

BODIE FAMILY

Thinking about the Bodie family, we can apply this. Danielle is in the early stages of development still. Her visual acuity is not yet at its prime, she is learning about shapes, colours, her 3D vision – all factors that influence her ability to interact safely in the environment as well as to experience pleasures in life. Her balance will be central to her ability to walk, her sight key to responding to danger and in learning psychomotor skills. At the other end of the spectrum, Maud and George will have had these senses mature and influence their development also. However, at this stage of their lives, visual acuity may not be as accurate. The wear and tear of loud sounds and strong vibrations over a lifetime can damage the stereocilia (sensory hairs) of the ear, affecting hearing. Balance can similarly be affected. These factors may compromise safety by a person not hearing or seeing a danger approach, increasing the risk of falling, and reducing the enjoyment and pleasure derived from life.

For the remaining adults in the Bodie family, the special senses are central to their ability to succeed in their chosen professions – Michelle relies on her hearing as part of her role as a translator, Thomas relies on sight to be a safe and effective pilot, as does Matthew to use the fine motor skills needed to be an electrician. Our sense of smell is closely linked to memory and emotion, and all of the Bodie family members will have memories triggered by experiencing smells that they associated with happy or challenging moments in their lives.

As you work through this chapter, you will learn to apply this system more and consider influences on your practice as a person-centred nurse.

SENSATION

Before we begin to look at the different types of sensations, we need to consider how sensations are picked up and translated to the nervous system. Sensations all start with a stimulus which has to be received and transmitted through the nervous system. Receptors detect the stimulus and convert it to electrochemical energy – the action potential (Chapter 5). Receptors may be within specialist sense organs, or widely distributed, for example in the skin. The nervous system needs to know three things about the stimulus:

1. **Modality:** i.e. what type of stimulus it is, e.g. painful or visual.
2. **Location:** the receptive field that the stimulus has come from and the use by the CNS in response.

3. **Intensity:** determined in four ways:

   i. Are only the most sensitive nerve fibres being triggered? This happens with a weak stimulus.
   ii. How often is the stimulus triggering a nervous reaction (action potential)? If it is frequent and fast, then the stimulus is intense.
   iii. Is the stimulus triggering numerous sensory nerve fibres to react? This also indicates a strong stimulus.
   iv. What is the duration? Duration indicates intensity, but sensory nerves will fatigue with continuous stimuli.

Chapter 5 has already discussed some of the types of receptors in the body and you may need to reread that. Sensations, though, can be categorised as follows:

Vision: the conversion of electromagnetic radiation to electrochemical energy in order to produce images of our surroundings.

Chemical senses: the sensations of smell (olfaction) and taste (gustation).

Hearing and equilibrium: the ability to interpret sound from vibration and the senses of balance and motion.

General senses: the sensations of pain, temperature and touch (pressure, vibration and proprioception).

The special senses have dedicated sense organs, and the general senses are largely associated with sensory receptors in the skin (integumentary system). We will start by looking at the special senses.

VISION

Vision, or sight, is the perception of objects in our surroundings. The image of these objects is created by interpreting electromagnetic radiation. Light creates a photochemical reaction within the eye, which contains more than half the sensory receptors of the body. To understand vision, we must look at the structure and function of the eye and its accessory structures.

ACTIVITY 6.1: APPLY

Vision and quality of life

Think of the Bodie family members. Take ten minutes to reflect on how and why vision is important in their lives. Can you think of how a visual impairment could impact upon health from a holistic perspective?

BODIE FAMILY

Accessory structures of the eye

The accessory structures of the eye are in and around the orbit of the eye and refer to the:

- Eyelids.
- Eyebrows.
- Lacrimal apparatus.

- Conjunctiva.
- Extrinsic eye muscles.

Each of these has a specific role that enables vision by supporting the eye's functions.

Eyelids

The eyelids are also known as palpebrae and they shade the eyes during sleep, preventing interruption by visual stimuli. They also protect against excessive light and foreign bodies. The tarsal glands in the eyelid produce a protective lubricant, an oily substance that helps prevent water/tear evaporation from the eye. The eyelashes are protective structures that shield the eye from environmental debris and produce the blink reflex to close the eyelids to prevent anything touching the eyes. The gap between open eyelids is known as the palpebral fissure.

Eyebrows

While largely a feature that enhances facial expression, the eyebrows help to reduce glare to the eyes and prevent perspiration and other liquids running into the eye.

Lacrimal apparatus

The lacrimal, or tear, apparatus is composed of lacrimal glands and ducts that manage the flow of tears. The lacrimal fluid (tears) contains water, salts, mucus and lysozyme which lubricate the eye. Lysozyme is a protective bactericidal enzyme. The lacrimal ducts carry tears from the lacrimal gland to the surface of the eye. Tears pass over the anterior surface of the eyeball to the lacrimal puncta (small holes that enable drainage of tears) and are then carried away from the eye through the lacrimal canals. From there they drain into the nasolacrimal duct and the inner nose. Lacrimal glands are innervated by the facial nerve, part of the parasympathetic nervous system.

Conjunctiva

The conjunctiva is a thin, transparent protective membrane that coats the anterior surface of the eye (except for the cornea) and the inner surface of the eyelid. It is highly vascular and sensitive and secretes mucus to keep the eye hydrated. Because it is so vascular it heals readily.

Extrinsic eye muscles

These six muscles, as the name suggests, are outside the eye. They are also referred to as the extraocular muscles. They connect the walls of the orbit, where the eye sits, to the eye itself. Together they can move each eye in almost any direction. These muscles, and their association with eye movement, are identified in Table 6.1.

The lateral and medial rectus muscles, superior and inferior rectus muscles, and superior and inferior oblique muscles work antagonistically together to control eye movement (Purves et al., 2011). Most of these muscles are supplied by the oculomotor nerve. However, the superior oblique is innervated by the trochlear nerve and lateral rectus by the abducens. You will remember from Chapter 5 that these are cranial nerves.

Table 6.1 Extraocular muscles and associated eye movements

Muscle	Movement
Superior rectus	Vertical elevation
Inferior rectus	Vertical depression
Lateral rectus	Horizontal abduction
Medial rectus	Horizontal adduction
Superior oblique	Vertical depression, torsion
Inferior oblique	Vertical elevation, torsion

The eye

We will now examine the structure and function of the eye itself. Figure 6.1 details the components of the eye. The eye is a sphere about 2.5 cm in diameter and five sixths of it is concealed within the orbit, with one sixth visible. It has three principle components:

1. The three-layered wall.
2. The optical components that focus light and regulate its entry to the eye.
3. Neurological components that convert light to electrochemical energy to generate images.

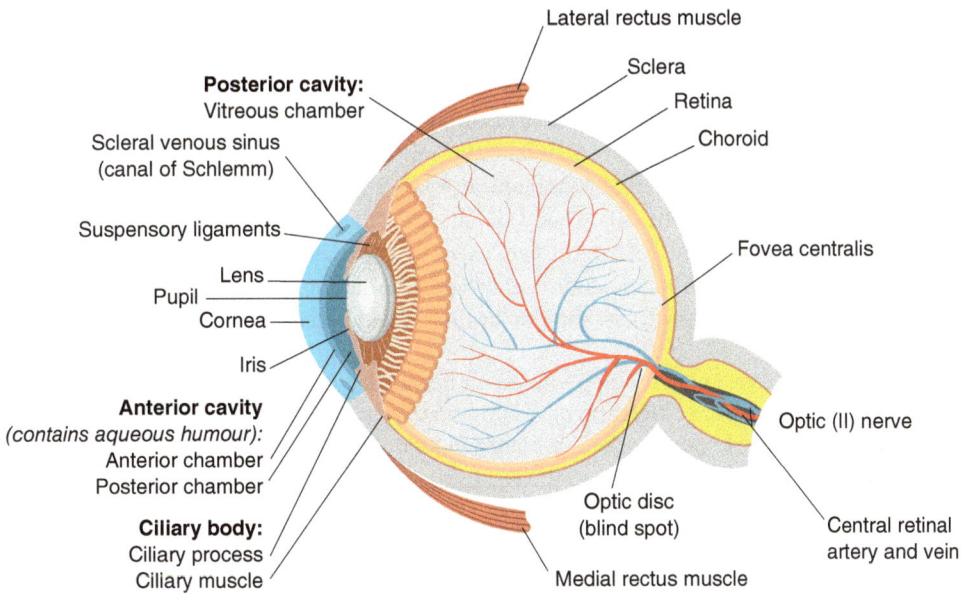

Figure 6.1 Components of the eye

The wall of the eye

The wall of the eye has three layers: the fibrous tunic (tunica fibrosa), the vascular tunic (tunica vasculosa) and the neural tunic (tunica interna).

Fibrous tunic (tunica fibrosa)

The fibrous tunic has two main areas, the sclera and the cornea. The sclera is the white of the eye and is a tough layer of connective collagen fibres that contains blood vessels and nerves. It provides shape to the eyeball, makes it more rigid, protects its inner parts and serves as a site for attachment of the extrinsic eye muscles. The sclera also contains a sinus, known as the scleral venous sinus, that drains the aqueous humour. The cornea is continuous with the fibrous tunica but is transparent to allow light through into the eye.

Vascular tunic (tunica vasculosa)

The vascular tunic is divided into three sections – the choroid, ciliary body and iris. The choroid sits behind the retina and is heavily pigmented stopping stray light scattering and reflecting around the eye. It has a rich vascular supply which nourishes the retina.

The ciliary body is a continuation of the choroid. It contains the ciliary muscle and ciliary processes. The ciliary muscle encircles the lens of the eye and changes the shape of the lens to focus the light for vision. The aqueous humour is secreted from the ciliary processes.

The iris is the coloured part of the eye around the pupil. It controls the amount of light entering the eye through the pupil regulated by the autonomic nervous system in response to the levels of light. Bright light stimulates the oculomotor nerve to contract the iris to reduce the pupil size; low levels of light stimulate the iris to dilate the pupils. The colour of the iris is determined by the amount of melanin present – brown eyes have a high concentration of melanin in the iris, blue eyes a low concentration, and green eyes a moderate concentration. The pupil is effectively a gap in the iris and it appears dark as you are seeing through the eye back to the heavily pigmented choroid layer and retina.

Neural tunic (tunica interna)

The neural layer (inner layer) contains the retina and optic nerve. The retina is functionally part of the brain, being an extension of the diencephalon, and the only part you can see without undertaking a dissection. The retina is a thin transparent membrane attached within the eye at the optic disc and at the anterior margin (where it meets the ciliary body and muscle). The vitreous humour fills most of the eye, holding the retina flat against the back of the eye. There are two main layers to the retina. The posterior pigmented layer contains melanin that absorbs stray light. The anterior neural, or sensory layer, is rich in sensory receptors and neural tissue. This anterior layer has three further layers:

- **Photoreceptor layer:** containing photoreceptor cells that absorb light and convert it to electrochemical energy. There are different types of photoreceptors. Only two are functional and are called rods and cones. Rods are responsible for night vision, only seeing black, shades of grey and white. The three types of cones are sensitive to blue, green and red light and allow colour vision and are important in day vision.
- **Bipolar cell layer:** these cells receive information from rods and cones and are the first-order neurons of the visual pathway.
- **Ganglion cell layer:** these cells are the second-order neurons in the visual pathway. They receive information from the bipolar cells and make up the optic nerve. Some ganglion cells absorb light directly and communicate with the brainstem to regulate pupil size and sleep cycles. These particular ganglion cells do not, however, contribute to vision.

At the optic disc of the retina the ganglion cells converge to form the optic nerve which is a blind spot with no rods or cones. The fovea centralis, with a high density of cones, sits about 3 mm from the optic disc and is responsible for detailed images.

Optical components

The optical components permit entry of light into the eye and focus the rays onto the retina. These include the:

- Cornea.
- Aqueous humour.
- Lens.
- Vitreous humour.

There are essentially two cavities in the eye. The largest (posterior cavity) contains the vitreous humour at the posterior of the eye. The smallest (anterior cavity) is to the front of the eye and contains the aqueous humour and lens (supported by the ciliary body). The aqueous humour is a clear fluid secreted by the ciliary processes (in the ciliary body) into a chamber behind the iris and lens. However, it flows forward through the pupil to the anterior chamber and nourishes the lens and cornea. The scleral venous sinus drains this fluid at the same rate that it is produced. The lens of the eye is supported by suspensory ligaments to the ciliary body within this fluid. Tension on these ligaments changes the shape of the lens to focus light appropriately. The vitreous humour keeps the retina flush against the choroid for a smooth surface to generate clear images transmitted through the neural layer.

ACTIVITY 6.2: UNDERSTAND

Watch this useful online video clip to further your understanding of the anatomy of the eye.

This online video can be accessed by **scanning the QR code** with your smart phone or via https://study.sagepub.com/essentialap2e.

ANATOMY OF THE EYE (11.25)

Image formation

Now that you understand the components and functions of the eye and its accessory structures, how do they work together to generate an image? There are three processes in forming an image on the retina through:

- **Refraction:** bending of light by the lens and cornea.
- **Accommodation:** change in shape of the lens.
- **Constriction:** narrowing of the pupil.

Refraction

As light enters the eye, it is bent by the fixed surface of the cornea and then by the lens (both on entering and exiting the lens). Some 75% of refraction occurs in the cornea and is unchanging while 25% of refraction in the lens is adjustable (Figure 6.2).

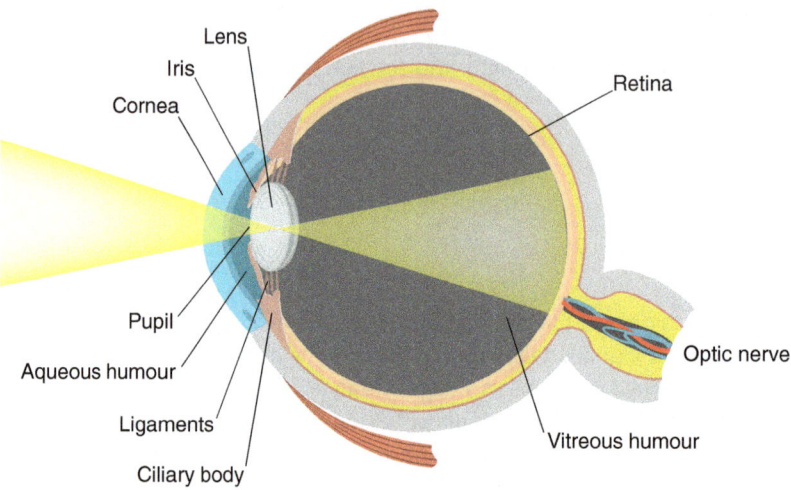

Figure 6.2 Light refraction in the eye

Accommodation

Accommodation refers to changing the shape of the lens to focus light for near or distant object interpretation. Near vision is achieved by contraction of the ciliary muscle, which relaxes the suspensory ligaments and increases the curvature of the lens thus focusing (or increasing the bend of) the light rays and bringing forward the image. The lesser the curvature with relaxed suspensory ligaments, the better the distance vision. When relaxed, the eye is prepared for distant vision. It has to actively accommodate for near vision.

Constriction

This is an autonomic reflex that occurs simultaneously with accommodation and prevents light rays entering the eye through the periphery of the lens. Peripheral light would not be brought to a focus on the retina resulting in blurred vision. How this occurs is described above under the vascular tunic.

Neural response to light

Once light reaches the retina after refraction, accommodation and entry through the pupil, it is absorbed by the cones and rods – the receptors. This creates a receptor potential, which generates the action potential along the axons of the nerve cells in the bipolar and ganglion cells. As previously described, the ganglion cells converge at the optic disc to form the optic nerve. The optic nerve takes the nervous impulse to the brain. At the optic chiasm the nerve fibres from the inner side of each retina cross over (Figure 6.3). The optic tracts carry information of vision from one side of the body to the lateral geniculate body on the same side and the visual information then travels to the occipital (primary visual) cortex of the cerebral hemisphere on the same side. Here, the brain interprets the visual image.

ACTIVITY 6.3: APPLY

Adapting to visual impairment

It is important to understand the visual pathways for vision so that you can appreciate the impact of damage to this process. This can occur frequently when a person has a stroke (around 10% of people with a stroke are affected). A stroke in the left hemisphere of the brain can inhibit the ability to see the right visual field of each eye, whereas the left visual field of each eye is impaired in a stroke in the right hemisphere. This is called a hemianopia, or blindness in one half of the visual field. The most common form of this is a homonymous hemianopia, which means that the vision loss is on the same side of each eye.

As a nurse, what would you need to consider for a person with a homonymous hemianopia? How could it impact on their independence?

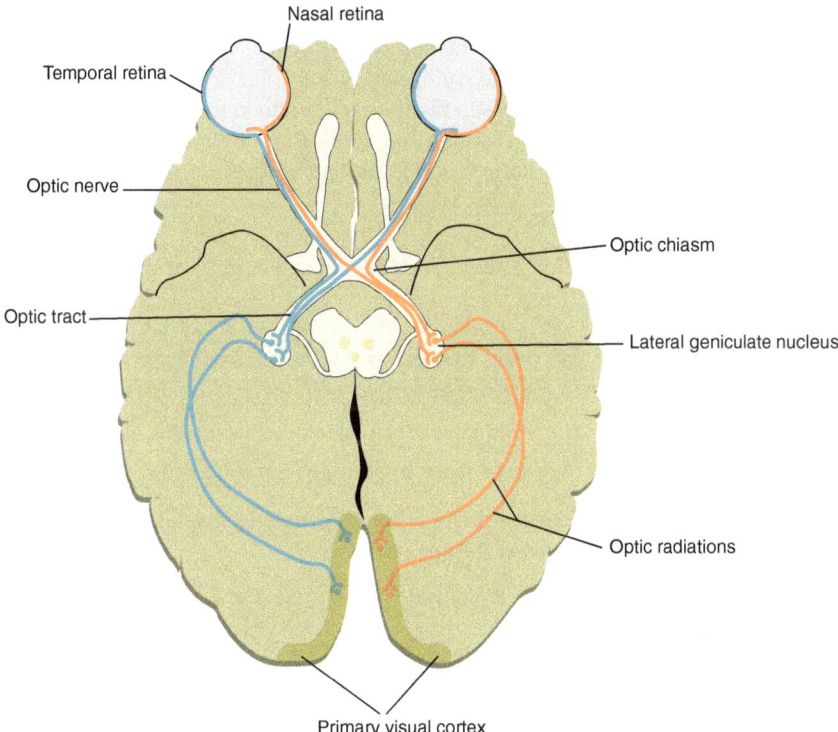

Figure 6.3 Visual pathways in the brain

CHEMICAL SENSES

The chemical senses refer to taste (gustation) and smell (olfaction). Both of these involve receptors stimulated by environmental chemicals. There is some debate as to whether olfaction and gustation influence each other, i.e. whether how good something tastes can be affected by how it smells. While olfactory and gustatory nerve fibres are not located together peripherally (at the point of stimulus), incoming stimuli from both come together in the nervous system at the orbitofrontal cortex, along with sensory information through the trigeminal nerve (Landis et al., 2010). As a result, all three sensory modalities interact frequently and are thought to influence the perception of each other. Stinton et al. (2010) examined this and found that the loss of olfaction had no meaningful influence on gustation. In contrast, Landis et al. (2010) also examined this relationship and found that when olfaction is weakened, gustation also appears to be weakened. Further supporting the case for the interlink between the perceptions of taste and smell, Fortis-Santiago et al. (2010) also determined that activation of the gustatory and olfactory cortices in the brain were co-dependent in terms of one influencing the other, and that from both together the perception of food is formulated. It therefore appears that the evidence is growing that olfaction affects gustation and vice versa.

Gustation (taste)

Gustation occurs from chemicals reacting with taste buds (of which there are about 10,000). These are primarily located on the tongue, but also on the inside surface of the cheeks and the soft palate, pharynx and epiglottis. These taste buds can distinguish five primary tastes (Figure 6.4 shows the location of taste zones on the tongue):

- **Sweet:** Mostly stimulated by sugary carbohydrates.
- **Sour:** Stimulated mainly by acids such as fruit acids.
- **Bitter:** Primarily stimulated by alkaloids in plant leaves and food that has spoiled.
- **Salty:** triggered by metal ions, such as sodium and potassium.
- **Umami:** This is the meaty/savoury taste found in protein foods such as meat and fish (no specific zone on tongue).

The tongue is the major sense organ for taste. Its surface contains bumps, or the four different types of lingual papillae (Figure 6.4):

Filiform: These filaments are small spikes that lack any taste buds, but are involved with the experience of food texture.

Foliate: These are on the sides of the tongue around two thirds of the way back. In youth we have some taste buds active here, but these degenerate by the age of around three years.

Fungiform: These mushroom-shaped structures contain three to five taste buds and are scattered across the tongue but are concentrated at the apex (tip).

Circumvallate: These large papillae containing about 100–300 taste buds are arranged in an inverted V shape at the back of the tongue.

Taste buds

The taste buds are onion/lemon-shaped and consist of three types of cells (see Figure 6.5):

Taste (gustatory) cells: Taste cells contain taste hairs that lead into a taste pore with receptors that synapse with a neuron, but are not neurons themselves. These cells have a short lifespan of up to ten days, before replacement.

Supporting cells: Supporting cells lie between taste cells but do not contain taste receptors.

Basal cells: Basal cells are the stem cells located at the periphery of the taste bud which mature into taste cells to replace those that have died.

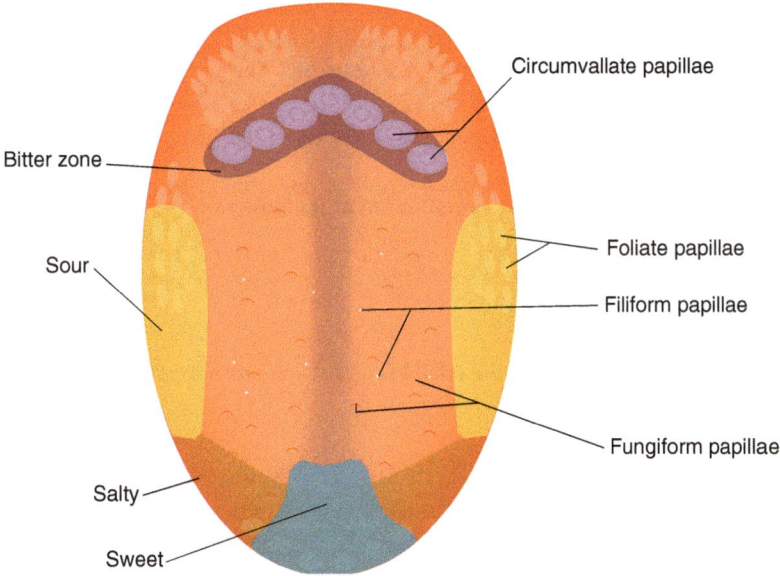

Circumvallate papillae

Bitter zone

Sour

Foliate papillae

Filiform papillae

Fungiform papillae

Salty

Sweet

Figure 6.4 Taste zones and papillae of the tongue

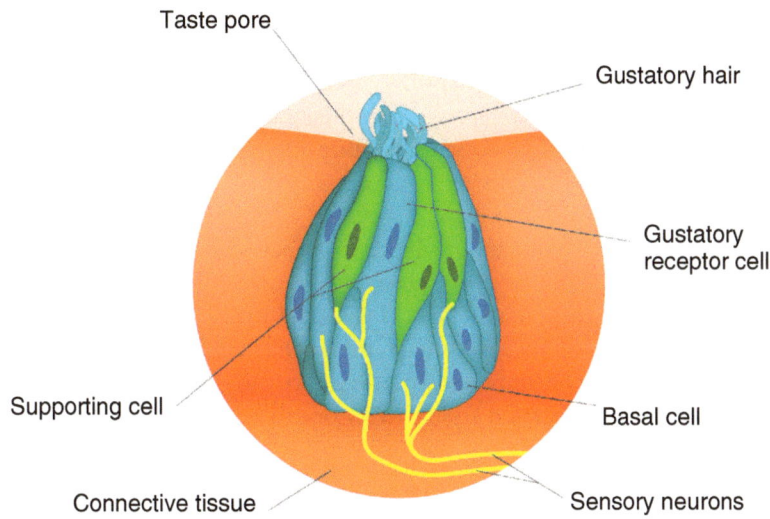

Taste pore

Gustatory hair

Gustatory
receptor cell

Supporting cell

Basal cell

Connective tissue

Sensory neurons

Figure 6.5 Taste bud

ACTIVITY 6.4: APPLY

Taste and health

Taste plays an important role in maintaining homeostasis. It is part of the stimulus for digestion, increases the pleasure from food, and is closely associated with memory and emotions. It also protects us from eating food that has spoiled by alerting us. Think about these points and make a list of potential problems a person may experience due to a loss of taste. How could this compromise their health?

How does taste occur?

Taste occurs from a tastant (chemical) stimulating the taste cells. The tastant is dissolved in saliva to make contact with the plasma membrane of the taste cell hairs. Once the receptors are stimulated, they trigger cranial nerves – which one is stimulated depends on the location of the receptor:

Glossopharyngeal: stimulated for taste sensations at the posterior one third of the tongue.

Facial: stimulated for taste sensations in the anterior two thirds of the tongue.

Vagus: stimulated for taste sensations in the palate, pharynx and epiglottis.

Regardless of which cranial nerve is stimulated, all project to the solitary nucleus in the medulla. From here the signal is relayed to two areas of the brain:

1. Nuclei in the hypothalamus and amygdala that trigger salivation, vomiting and gagging.
2. The thalamus, which relays signals to the insula and gyrus for the conscious experience of taste. The thalamus also relays signals to the gustatory centre in the parietal lobe.

Olfaction

Olfaction, or smell, is the second chemical sense with the primary sense organ in the olfactory mucosa in the superior nasal cavity. Olfaction is highly sensitive, necessary in such a poorly ventilated space. People can detect up to 4,000 different smells on average, some being able to smell up to 10,000 from the 10–20 million olfactory cells in the olfactory mucosa. Supporting cells provide electrical insulation, protection and nourishment to the olfactory neurons while also detoxifying chemicals, while the basal cells divide and differentiate into new olfactory neurons.

Olfactory cells are different to gustation cells in that they are neurons. The dendrites of these neurons are exposed and contain fixed cilia (known as olfactory hairs) where odour molecules bind. Above the olfactory mucosa is the ethmoid bone of the skull, with pores (called cribriform foramina) that allow the axon of the neurons to pass through to the olfactory bulb. These axons merge to form the olfactory nerve. The olfactory neurons are exposed directly to the environment making them vulnerable to damage, but they are replaced around every one to two months (Bermingham-McDonogh et al., 2012).

ACTIVITY 6.5: APPLY

Olfaction and complementary therapies

Olfaction is the basis of some complementary therapies such as aromatherapy. Essential oils are the basis for aromatherapy practice.

Read the following article to learn more about the physiology underpinning this therapy:

Cook, N.F. and Lynch, J. (2008) 'Aromatherapy: reviewing evidence for its mechanisms of action and CNS effect', *British Journal of Neuroscience Nursing*, 4(12): 595-601.

How does olfaction occur?

Olfactory glands are present in the mucosa and produce mucus that lubricates the surface of the olfactory epithelium (Figure 6.6). Odorants dissolve before binding with the olfactory hairs. Olfactory glands and supporting cells are innervated by autonomic neurons from the facial nerve. This autonomic stimulant can result in production of tears and nasal mucus following stimulation from odorants.

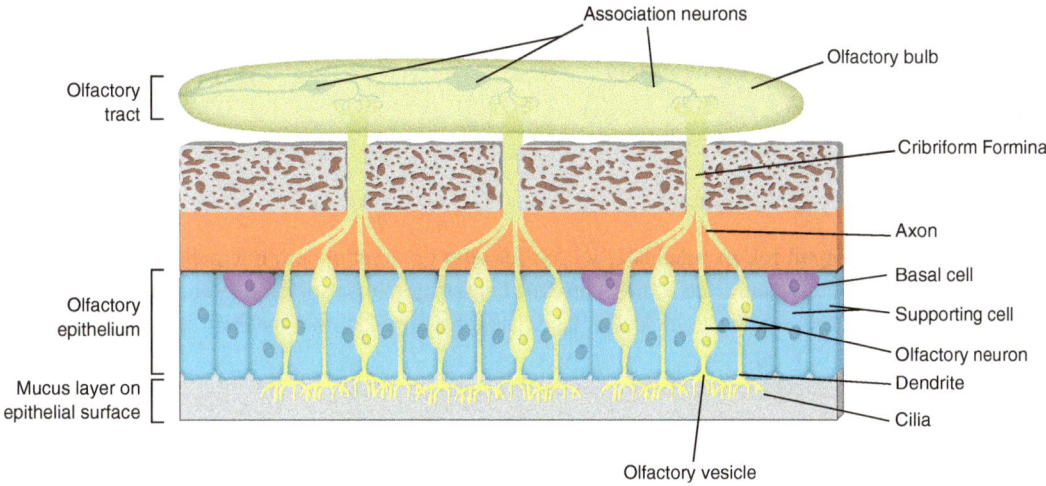

Figure 6.6   Olfaction

Once olfactory receptors are stimulated, nerve stimuli pass along the axon of the neurons to the olfactory bulb – a mass of grey matter in the brain. They are relayed through a complex network of neurons to the temporal lobes of the brain. This includes signals going to:

- The hypothalamus where they are further relayed to the reticular formation. Smells can cause a visceral reaction here.
- Directly to the olfactory cortex, resulting in conscious perception of smell, before being directed to:
  - The thalamus and from there to the orbitofrontal cortex where conscious smell perception occurs.
  - The thalamus and on to the amygdala, where the emotional response to smells occurs, and from there to the hypothalamus.
  - The hippocampus where the sensation of smell is linked with memory. This is why a smell can trigger a memory, whether it is pleasant or unpleasant. Some nerve pathways bring signals directly to the hippocampus.

HEARING AND EQUILIBRIUM

Hearing

Sound provides so much pleasure in life, whether it is the reassurance of someone's voice, the relaxation of hearing nature around us, or the emotional pleasure derived from music. It also warns of impending danger, helps us with orientation and is central to communication with others. Sound is audible vibration of molecules and can travel through liquids, solids and gases, but not vacuum. The human body has developed to maximise hearing by funnelling sounds into a system where it can be sensed and interpreted. This starts with the ear.

ACTIVITY 6.6: APPLY

Hearing and person-centred nursing

Refresh yourself on the components of the Person-Centred Nursing Framework. Identify the different elements and make some notes on where hearing is fundamental to the success of those elements.

The ear

The ear has the important role of funnelling sounds for sensory reception and interpretation. It also has a role in equilibrium. Equilibrium refers to the sense that helps maintain balance and awareness of orientation in space (proprioception), which we will look at later. The ear is divided into three main regions (Figure 6.7):

External ear: to collect sound waves and channel them inwards.

Middle ear: to convey sound vibrations to the oval window and inner ear.

Internal (inner) ear: to house receptors for hearing and equilibrium.

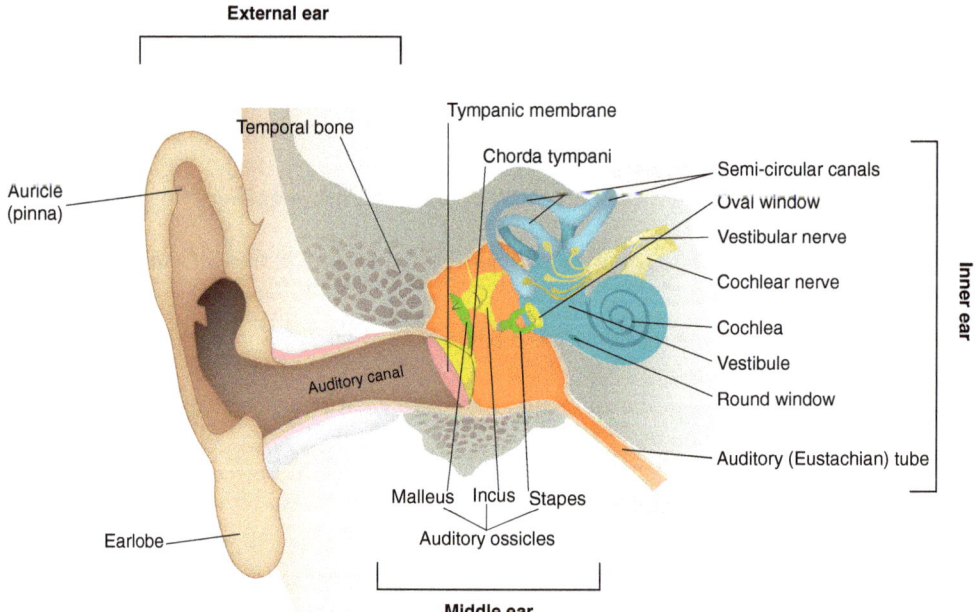

Figure 6.7 Anatomy of the ear

External ear

The external ear is largely composed of the auricle, or pinna, the fleshy part of the ear. The upper rim of the auricle is the helix, and the ear has a range of recesses that function as funnels to direct sound into the auditory canal. The auditory canal passes through the temporal bone, which provides structural support, and is lined with sebaceous and ceruminous glands that secrete oil. This oil combines with dead skin cells to form cerumen (earwax) which prevents dust and foreign objects entering the ear. It also protects against insects and water moving into the ear. The auditory canal terminates at the eardrum, or tympanic membrane.

Middle ear

The middle ear commences with the tympanic membrane, a thin semi-transparent partition between the auditory canal and the middle ear. This membrane, approximately 1 cm in diameter, vibrates freely to sound and is innervated by the vagus and trigeminal nerves. It is highly sensitive to pain.

The tympanic cavity lies behind the eardrum. It is an air-filled cavity that connects to the nasopharynx by the eustachian tube (or auditory tube). This tube is normally closed but swallowing or yawning opens it, allowing air to move in to equalise air pressure on either side of the tympanic membrane. Within the tympanic cavity are three small bones, the auditory ossicles (the smallest bones in the human body). These connect the eardrum, transferring and magnifying sound vibration from the auditory canal to the inner ear. These bones are the:

Malleus: (the hammer): connects to the eardrum with a small process that connects to the next auditory ossicle, transferring the vibration.

Incus (the anvil): the middle of the three ossicles and transfers sound vibration from the malleus to the stapes.

Stapes (the stirrup): continues the sound vibration to the oval window.

Two small skeletal muscles attach to the ossicles. These are the:

Tensor tympani muscle: Supplied by the mandibular branch of the trigeminal nerve, it limits movement of the ossicles and increases tension on the eardrum to prevent damage to the inner ear from loud noises.

Stapedius muscle: Supplied by the facial nerve, it dampens large vibrations of the stapes that can occur from loud noises. This protects the oval window and decreases sensitivity of hearing.

These muscles take a fraction of a second to contract and so can protect the inner ear from prolonged noises, but not from short, sharp loud noises.

Inner ear

The inner ear is a bony labyrinth lined by a fleshy membrane known as the membranous labyrinth. It is a series of cavities within the temporal bone that contain the main organs of balance and hearing. It is further subdivided into three parts: the vestibule and semi-circular canals are the organs of balance, while the cochlea deals with hearing.

The vestibule consists of a pair of membranous sacs, saccule and utricle, which house the receptors for sensations of gravity and linear acceleration. The three semi-circular canals are in the three spatial dimensions and contain receptors that are stimulated by movement of the head.

The cochlea is a spiral-shaped bony chamber that contains the three cochlear ducts of the membranous labyrinth which spiral together for 2½ turns (Figure 6.8):

- **The scala vestibuli** (vestibular duct) is filled with perilymph and connects with the oval window and subsequently the tympanic duct.
- **The scala tympani** (tympanic duct) is filled with perilymph and connects with the vestibular duct and subsequently the round window. The round window is covered by a membrane known as the secondary tympanic membrane.
- **The scala media** (cochlear duct) is filled with endolymph. It is separated from the vestibular duct by the thin vestibular membrane and from the tympanic duct by the thicker basilar membrane that contains the organ of Corti, which converts sound vibrations into nervous impulses through long, stiff hairs called stereocilia. On top of these hairs lies the tectorial membrane and the tips of each stereocilium contain an ion channel. When a stereocilium moves, the ion channel opens, potassium moves in and triggers a nervous impulse (action potential) to the cochlear nerve.

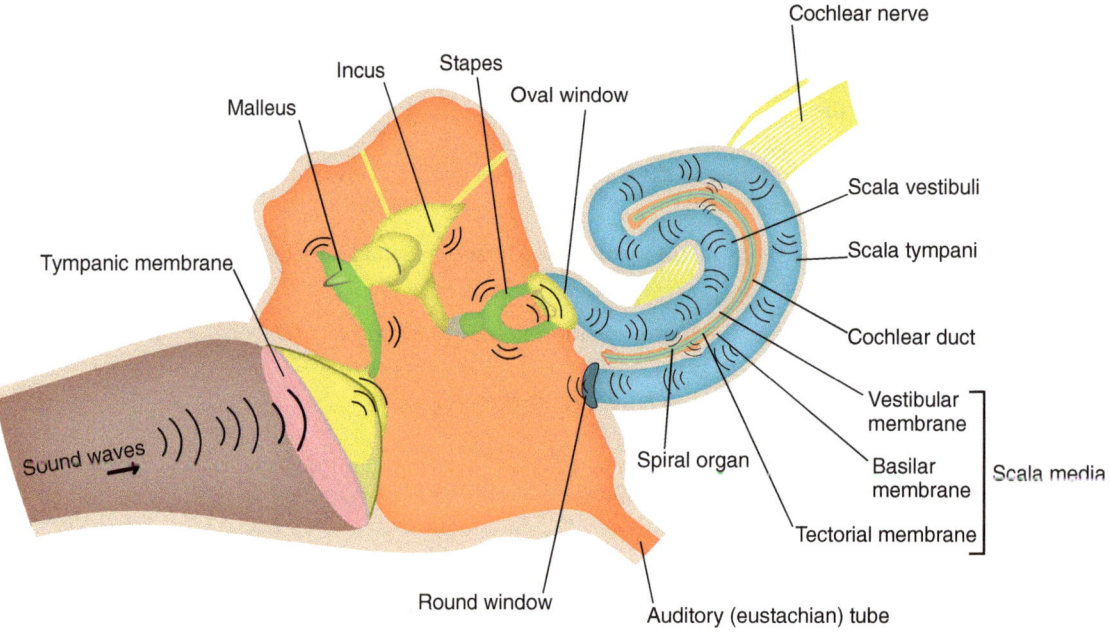

Figure 6.8 The cochlea

ACTIVITY 6.7: UNDERSTAND

Watch this online video clip to further your understanding of the anatomy of the ear.

The external video link can be accessed by **scanning the QR code** with your smart phone or via https://study.sagepub.com/essentialap2e.

ANATOMY OF THE EAR (5.18)

How does hearing occur?

The way molecules vibrate through the ear to create sound can be condensed into a number of steps shown in Figure 6.9.

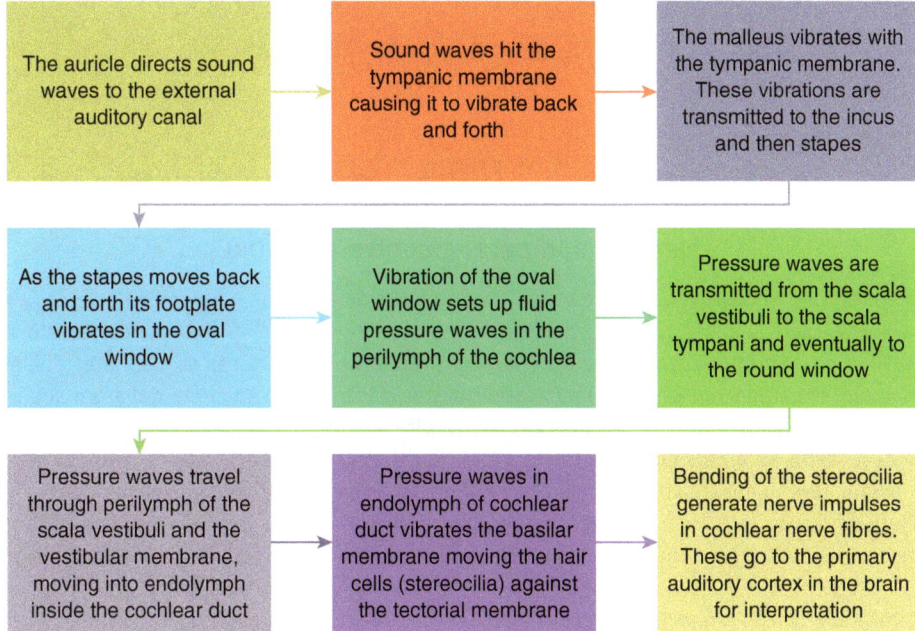

Figure 6.9 How hearing happens

Sensory impulses are carried through the eighth cranial nerve (the vestibulocochlear or auditory nerve) to the cochlear nuclei in the brainstem. Impulses then pass to the thalamus, which acts as a relay station and transmits impulses to the auditory area of the cortex within the temporal lobe. Information is also carried to the reticular formation involved in sleeping and waking. Loudness is determined by the intensity of the sound waves. High-intensity waves cause larger vibrations of the basilar membrane, and a higher frequency of nerve impulses reaching the brain. Louder sounds may also stimulate a larger number of stereocilia. However, the stereocilia are vulnerable to damage from loud sounds, infection, and certain drugs such as antibiotics and chemotherapy (Bermingham-McDonogh et al., 2012). Damage to these hairs cannot be repaired and hair cells are never reproduced after birth: we must therefore protect them.

ACTIVITY 6.8: APPLY

Hearing and quality of life

Reflect on the impact the loss of hearing would have on your life. How might that feel? How might it affect your life and health?

(Continued)

BODIE FAMILY

Now think about the Bodie family. Which members may not have optimal hearing and why? What implications could this have for them? Try not to just think of the obvious things, dig deep and think about all aspects of their lives.

What can you do to adapt your practice for those whose hearing is impaired? Think across the lifespan.

APPLY

Hearing and person-centred nursing

Impairment or loss of hearing can have a significant impact on a person. They may not be as aware of dangers in their surroundings normally made more obvious by sound (e.g. the sound of traffic coming when walking on a country road). They also may not hear music and voices as clearly or distinctly as they once did, impairing not only communication but also the enjoyment they get from sounds with which they may have an emotional connection. Those with a hearing loss are also at risk of social isolation (McKee, 2013). As a person-centred nurse it is vital that you can optimally communicate with the people in your care to ensure your therapeutic relationship is collaborative and considerate of the need to facilitate empowerment. Hearing loss is often associated with ageing, but we must also consider that it can be across the lifespan. The following should be considered in your interactions and plans of care:

- Do you know how to screen for a hearing impairment in your practice as a nurse? For example, can you test for air and bone conduction of sound?
- Does the person have a hearing impairment or is their interpretation or processing of speech the primary issue? For example, could the person have a receptive dysphasia whereby they can hear without impairment but cannot interpret what is being said?
- Are you speaking in the person's native language? Could there be a language barrier rather than a hearing impairment? Is there an interpreter service available or could the family provide information/ support on how best to communicate with their relative?
- Does the person have a hearing aid and do you know how it works? Does it work for them? The three primary reasons for not wearing a hearing aid are (Hartley et al., 2010):
  o It is not helpful.
  o It is too noisy.
  o It is uncomfortable.

The hearing aid must work optimally for the person wearing it for it to be effective. Most contemporary hearing aids are digital hearing aids, whereas older types are analogue. Analogue hearing aids detect sound, convert it to an electrical signal and amplify it before transmitting it into the ear for detection. If they have automatic gain control, they can also determine the volume of sounds and decide which ones need to be amplified and which ones are already loud enough. Digital hearing aids undertake a more sophisticated assessment of sound, assessing its volume and quality. They convert sound to be clearer and more audible, filter out unwanted sound and also determine the volume for sound. They are more adaptable as they can be set for different situations depending on the extent of background noise. However, this requires the user to know when to change the setting and to have the ability to change it.

- Has the use of Assistive Listening Devices (ALDs) been considered? These devices minimise the distance between the source of sound and the listener, increasing the signal-to-noise ratio (Aberdeen and Fereiro, 2014). They can therefore make speech easier to understand and also reduce background noise, improving the function of hearing aids.
- Do you have a documented plan of care in place that considers the person's individual needs?
- Ensure that you follow some fundamental communication principles (McKee, 2013):
  - Face the person, making eye contact.
  - Speak clearly without slowing your pace (unless you usually speak very fast).

These principles can help those who lip read.

- Have you considered social inclusion strategies in your plan of care?
- Is the use of visual cues or sign-language/Makaton relevant/appropriate?
- Could a word/letter board and/or use of pen and paper enhance communication? Many people can also use advanced technologies such as text-to-voice applications on smart devices (tablets and phones).

This list is not exhaustive and many other options are available. Charitable and government agencies often have highly useful resources and advice that can complement your practice and also be a source of support for the person and their family.

For examples of these resources, go to http://study.sagepub.com/essentialap2e to access the online readings **Action on Hearing Loss** and **Deafness and Hearing Loss**.

Equilibrium

Balance, or equilibrium, is central to our ability to move purposively and is therefore essential for remaining safe and being able to function independently. There are two types:

- **Static equilibrium** is maintenance of the body position relative to the force of gravity. This is perceived through movements such as tilting of the head, and linear acceleration and deceleration (such as a car speeding up or slowing down).
- **Dynamic equilibrium** is maintenance of the body position in response to sudden movements and is perceived through rotational movement.

The vestibular apparatus is the organ of balance. It consists of the vestibule and the semi-circular canals described earlier (Figure 6.10).

Vestibule

The vestibule consists of a pair of membranous sacs, the saccule and utricle. Both of these contain maculae essential for maintaining appropriate posture and balance as they provide sensory information on the position of the head in space. Maculae consist of two kinds of cells: hair cells, which are the sensory receptors, and support cells, which secrete a thick gelatinous glycoprotein layer known as the otolithic membrane.

The otolithic membrane sits on top of the macula. When the head tilts down, the otolithic membrane is pulled by gravity. If the head jerks forward from an upright position, the otolithic membrane lags behind and pulls hair bundles and bends them in the opposite direction. This results in depolarisation in the hair cells, generating a stimulus for first-order neurons in the vestibular branch of the vestibulocochlear nerve.

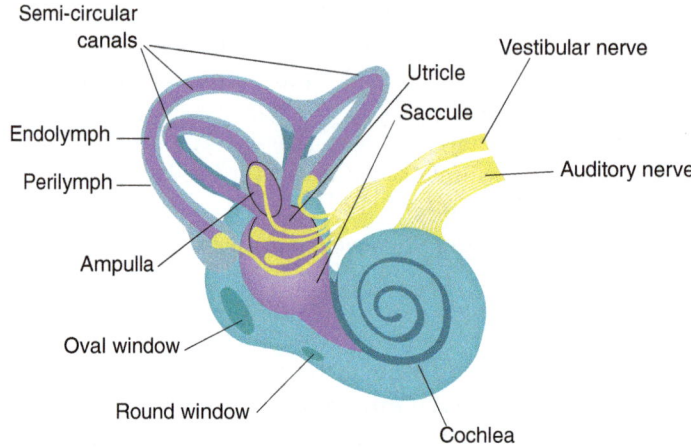

Figure 6.10   Vestibular apparatus

Semi-circular canals

The three semi-circular canals that function in dynamic equilibrium are:

* Anterior.
* Posterior.
* Lateral.

The anterior and posterior canals lie vertically while the lateral duct lies horizontally. This positioning permits the detection of rotational acceleration/deceleration as the endolymph in the canals moves with head rotation. The ducts open into the utricle, with a sac called an ampulla at one end. This sac contains hair cells that are embedded in a gelatinous membrane, the cupula. When the head moves, the semi-circular canals and hairs move with it but the endolymph lags behind. The moving hairs drag along the stationary endolymph, bending the hair bundles which create receptor potentials. This generates nerve impulses along the vestibular branch of the vestibulocochlear nerve, sending information on movement to the brain.

Together, the vestibule and semi-circular canals stimulate the sensory fibres of the vestibular nerve, which combine with the cochlear nerve to form the vestibulocochlear nerve. These impulses are relayed to:

* The vestibular nucleus in the pons and then to the spinal cord. From there they project to the brainstem nuclei of the cranial nerves controlling eye movement and to the accessory nerve, which controls head and neck movement. The brain combines these sensory inputs from both ears with sensory input from the eyes and stretch receptors in the neck to determine head orientation.
* The cerebellum, where muscles of the body are coordinated to regulate muscle tone and posture.

It is through these processes that we achieve equilibrium.

ACTIVITY 6.9: APPLY

Impaired balance and risk of falls

Balance difficulties are a primary cause of falls in older people today. As a result, nurses often use fall risk assessment tools to identify those at risk of a fall. These tools are very useful in identifying what factors may have impaired the person's ability to maintain balance. When nurses look at risk factors, there are two broad categories:

- **Intrinsic factors:** within the individual.
- **Extrinsic factors:** external to the person.

Often risk factors can be eliminated or reduced, and can form part of the care we provide to people.

Through a literature search, determine the key intrinsic and extrinsic factors which can affect balance. Can you identify what interventions can eliminate or minimise these factors? Keep focused on using evidence-based literature to inform those nursing interventions.

THE GENERAL SENSES

The general senses refer to the sensations of pain, temperature and touch (pressure, vibration and proprioception). Their receptors' structure and physiology are largely quite simple and generally composed of a sensory nerve fibre and a small amount of connective tissue. Sensory receptors for the general senses are classified as either encapsulated or unencapsulated (Table 6.2).

Table 6.2 Types of general sensory receptors

Unencapsulated		Encapsulated	
Type	**Function**	**Type**	**Function**
Free nerve endings	Respond to temperature changes of heat and cold, as well as pain. Widespread in epithelial and connective tissues	Tactile corpuscles	Found in the fingertips, palms, eyelids, the mouth (lips and tongue), nipples and genitalia, detect light touch and texture
		Ruffini corpuscles	Located in the dermis, subcutaneous tissues and joint capsules, detect heavy touch, pressure, and stretch
Tactile discs and Merkel cells	Located in the stratum basale of the epidermis, these flattened receptors detect the tactile sensations of light touch, texture, shape and edges	Krause end bulbs	Found in the mucous membranes, similar to tactile corpuscles
		Muscle spindles	Detect muscle stretch for proprioception, found in skeletal muscle near tendons
Hair receptors (root hair plexus)	Found around hair follicles, these interpret touch by movement of the hair	Lamellated corpuscles	Detect deep pressure, stretch, vibration and tickle sensations. Found in dermis, joints and parts of the skin
		Golgi tendon organs	Found in tendons and detect tension as part of proprioception

Nerve fibres for touch, pressure and proprioception are myelinated, resulting in fast signals. Those for heat and cold are slower due to the absence of myelin. Examples of some general sensory receptors in the skin are illustrated in Figure 6.11.

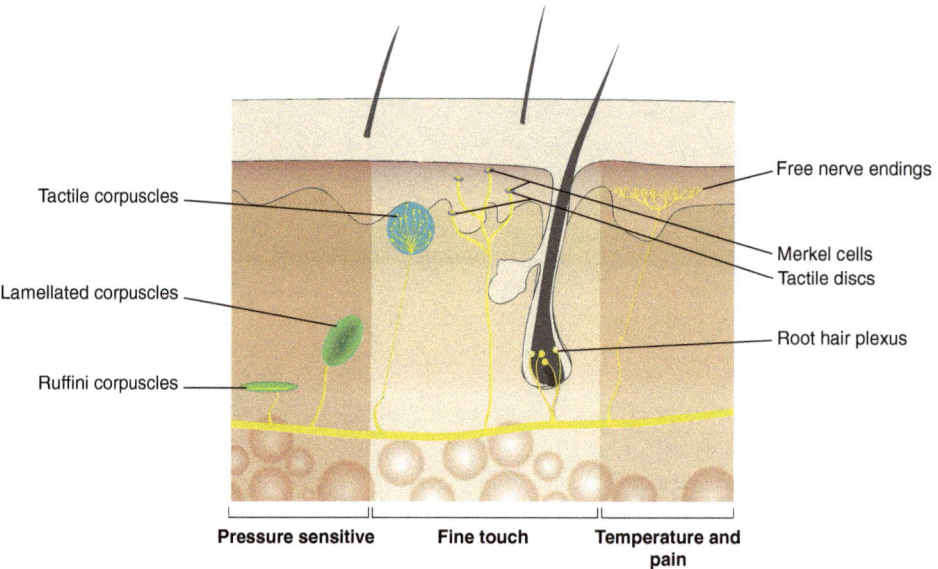

Figure 6.11   General sensory receptors in the skin

Sensory pathways

Sensory stimuli from the head travel through the cranial nerves to the pons and medulla and are relayed to the thalamus and finally to the cerebrum. Those below the head are relayed by first-order neurons to the dorsal horn of the spinal cord. From there they are transmitted through the following pathways:

1.   **Anterolateral pathways** transmit temperature, pain and coarse touch sensations.
2.   **Medial lemniscal pathways** transmit discriminative touch, vibration and proprioception sensations.
3.   **Spinocerebellar tracts (anterior and posterior)** transmit sensations for muscles and tendons (stretch) to the cerebellum to coordinate skeletal muscle movement.

Pain

Pain is a complex physiological and psychological response that can occur in relation to physical injury and the unique perceptual experience of a person to a variety of stimuli. From a purely physiological perspective, nociception is the process by which noxious stimuli are transmitted to the brain for processing. Noxious stimuli are those that cause pain, such as mechanical (pressure, shearing), chemical (acid, certain biological substances) and thermal (heat, UV radiation). Mechanical and thermal stimuli are usually the causes of most pain, but the damage to tissue causes chemical mediators to be released that also result in pain or increase the sensitivity to pain. It must be remembered that nociception does not take into account the experience of pain, making pain a very personally unique experience for every individual (Steeds, 2013). Often, physical injury may be absent and yet the person can be experiencing pain. This may be as a result of microscopic processes occurring whereby nociceptors are being triggered, or due to a psychological stimulus. It is for these reasons that being person-centred requires the nurse to acknowledge, appreciate and respond to reports of pain in a compassionate manner. Understanding how pain works biologically will enable you, as a nurse, to be knowledgeable alongside that compassion.

Somatic pain refers to nociception that originates in the skin, muscles, connective tissues, bones and joints. Visceral pain refers to nociception that originates in the visceral organs – stomach, kidney, gallbladder, intestines and bladder (Godfrey, 2005; Wood, 2008). Nociceptive pain occurs in five phases (Figure 6.12).

Figure 6.12 Five phases of nociception

Transduction

Free nerve endings (which are unencapsulated) are the sensory receptors for nociception (nociceptors). It is at these free nerve endings that nociception starts. These sensory receptors are equipped to detect noxious stimuli, converting them to electrochemical signals that can be relayed to the peripheral and central nervous systems through afferent pathways. There are two main types of nociceptors (Steeds, 2013):

High Threshold Mechanoreceptors (HTM): These detect mechanical injury.

Polymodal Nociceptors (PMN): These detect chemicals released when tissue is damaged.

— **GO DEEPER** —

Table 6.3 Chemical mediators that stimulate polymodal nociceptors

Chemical mediator	Role
Ionic disturbance (hydrogen, potassium)	Sensitise receptors, making them more sensitive to low-intensity stimuli
Serotonin	Directly stimulate receptors to cause pain (Steeds, 2013)
Cytokines	Activate central glia in the spinal cord and brain, promoting central neuroinflammation that results in chronic pain (Shubayev et al., 2010)
Bradykinin (most potent)	Sensitise receptors, making them more sensitive to low-intensity stimuli (Steeds, 2013)
Histamine (released by mast cells)	Directly stimulate receptors to cause pain (Steeds, 2013)
Prostaglandins (lipid mediators released from damaged cells)	Sensitise receptors, making them more sensitive to low-intensity stimuli (Steeds, 2013)
Leukotrienes (chemicals released by leucocytes)	Sensitise receptors, making them more sensitive to low-intensity stimuli (Noguchi and Okubo, 2011)
Substance P	Neurotransmitter that stimulates second-order neurons

When tissue is damaged, a variety of chemical mediators are released to induce inflammation for a combined healing and immune response. The chemical mediators that trigger PMNs are identified in Table 6.3. Nociceptors contain specialised protein receptors within their membrane. Once stimulated by noxious stimuli, these protein receptors reshape to create ion channels causing ions to flow through, stimulating an action potential in nociceptive neurons (Godfrey, 2005). The amount of ions moving through these channels will influence whether the action potential is generated – a threshold level must be reached in order for it to occur. The greater the stimulus, the greater the voltage change and the greater the number of action potentials generated (Godfrey, 2005).

APPLY

Pain and Non-Steroidal Anti-Inflammatory Drugs (NSAIDs)

As inflammatory processes have a significant role in releasing chemical mediators of noxious stimuli, NSAIDs work to block such chemical mediators. NSAIDs inhibit the enzymes Cyclo-Oxygenase 1 (COX-1) and Cyclo-Oxygenase 2 (COX-2). COX-1 is involved in the production of prostaglandins – potent chemical mediators for pain (Waterfield, 2008). COX-1 inhibitors therefore work locally at the site of sensory receptors for pain (nociceptors) to block prostaglandins stimulating a pain response. COX-2 is thought to inhibit the inflammatory response to tissue injury. This makes NSAIDs ideal for tissue injury pain relief (analgesia).

Conduction

Once an action potential is generated, it has to be conducted along nerve fibres. The speed of this conduction is related to the diameter of the fibre. There are three main types of conducting nerve fibres:

A-delta fibres: Fast-conducting myelinated fibres that are connected to HTMs. Stimuli are identified as being localised, sharp, stinging and pricking through these fibres (Godfrey, 2005; Wood, 2008).

C fibres: Slower-conducting fibres that are connected to PMNs. Stimuli are identified as being diffuse, dull and aching through these fibres (Godfrey, 2005; Wood, 2008).

A-beta fibres: Fast-conducting myelinated fibres that are connected to mechanoreceptors that detect touch, pressure and temperature. These do not carry nociceptive signals, but are involved in the modulation of pain (discussed later).

Transmission

All first-order neuron pain nerve fibres terminate in the dorsal horn of the spinal cord (Steeds, 2013). Most (70%) enter at the ventrolateral bundle of the dorsal root, with the remaining 30% entering though the ventral root. First-order neurons synapse with second-order nociceptive neurons in the dorsal horn and are then directed to the brain, being relayed out to various parts of the brain through third-order neurons (Godfrey, 2005; Steeds, 2013). It is only through ascending pathways that pain is processed and experienced.

Within the dorsal horn there are three types of second-order neurons, but only two of these respond to harmful stimuli (Godfrey, 2005; Steeds, 2013):

Nociceptive Specific (NS): respond to high-threshold noxious stimuli.

Wide Dynamic Range (WDR): respond to a variety of low-threshold stimuli.

Low-threshold: respond only to non-harmful stimuli.

Depending on how the nociceptive signals are modulated, ascending signals travel through either the spinothalamic or spinoreticular pathway (Steeds, 2013). Many first-order neurons can converge onto the same second-order neuron, which can make it difficult to determine where the exact origin of the pain is. This can result in referred pain. A good example of this is when visceral pain from the heart (experienced during a myocardial infarction, or heart attack) is felt as somatic pain that radiates into the left arm and shoulder.

Modulation

The signalling of pain from the peripheral nervous system to the central nervous system is modulated by a number of mechanisms. Such modulation can block pain signals ascending to higher centres within the central nervous system. The site of modulation is where primary afferent neurons synapse with second-order neurons in the dorsal horn. Here, descending inhibitory tracts from the brain can prevent pain signals from ascending to be processed and therefore they are not experienced by the person. This process is called 'spinal gating', whereby the gate can be opened to permit pain signals to ascend to the brain for processing, or it can be closed to prevent this occurring. This theory of pain modulation is referred to as the gate control theory (Melzack and Wall, 1965).

Once pain signals pass through the dorsal root and synapse with ascending second-order neurons to ascend up to the brain, the gate is said to be open. Where the first-order and second-order neurons synapse are inhibitory interneurons with the ability to block ascension through two primary processes:

1. Descending reticulospinal fibres from the reticular formation synapse with inhibitory neurons that meet at the junction of the first- and second-order neurons. The descending fibres excite the inhibitory neuron which then releases endorphins, enkephalins and dynorphins. These inhibit the first-order neurons from secreting their neurotransmitters (substance P), preventing an action potential from occurring in the second-order neuron. This is known as endogenous opioid analgesia (Figure 6.13).

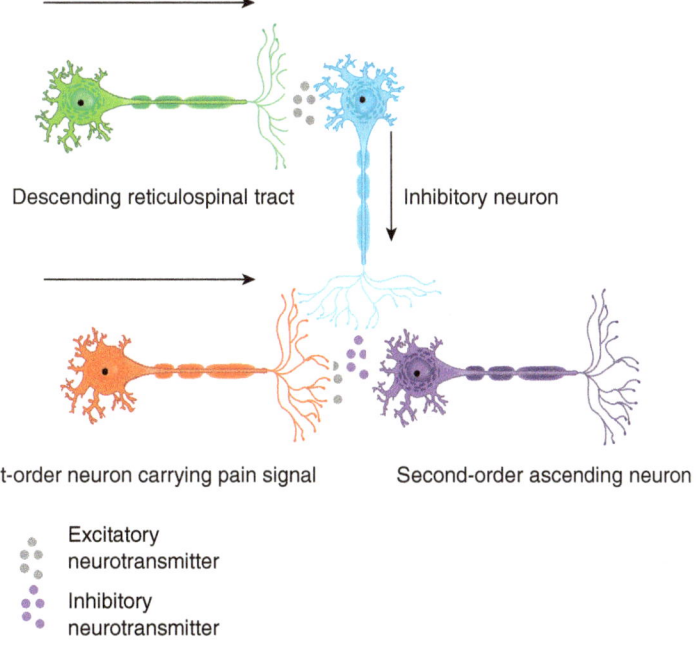

Descending reticulospinal tract · Inhibitory neuron

First-order neuron carrying pain signal · Second-order ascending neuron

Excitatory neurotransmitter

Inhibitory neurotransmitter

Figure 6.13 **Descending neuromodulation of pain**

Adapted and redrawn from: Godfrey, H. 'Understanding pain, part 1: physiology of pain', *British Journal of Nursing*, 14 (16): 846–52; Copyright ©2005 MA Healthcare Ltd. Reproduced by permission of MA Healthcare Ltd.

2. Sensory neurons (A-beta fibres) that respond to touch, pressure and temperature stimuli (non-nociceptive), can also stimulate the inhibitory interneurons, relaying their signals onto the second-order neurons and also stimulating the inhibitory interneurons, preventing the A-delta fibres from transmitting their pain/nociceptive signals from ascension to higher centres in the brain. It is through this process that rubbing a painful area and applying pressure can often decrease pain. It is also the principle behind a number of interventions used to treat pain, such as Transcutaneous Electrical Nerve Stimulation (TENS).

Of course, both these mechanisms may work at the same time (Figure 6.14), complementing each other.

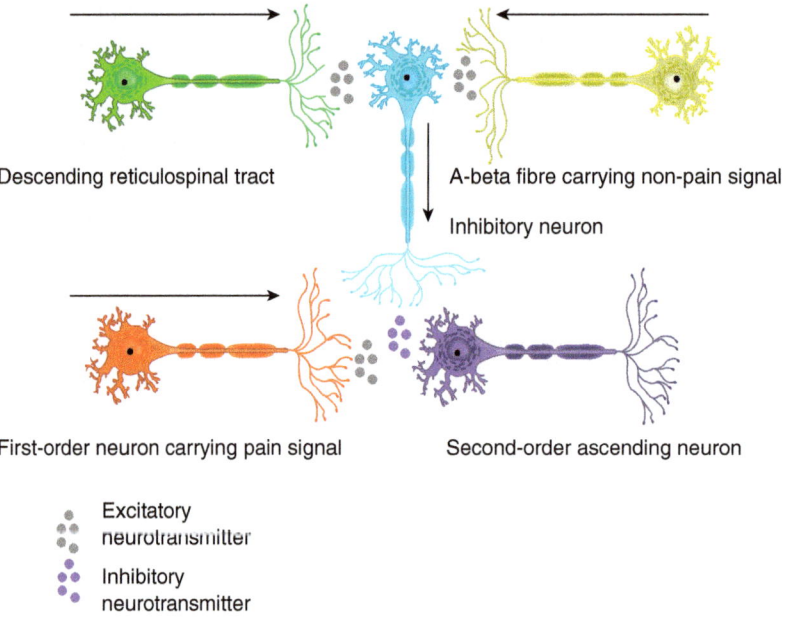

Descending reticulospinal tract

A-beta fibre carrying non-pain signal

Inhibitory neuron

First-order neuron carrying pain signal

Second-order ascending neuron

Excitatory neurotransmitter

Inhibitory neurotransmitter

Figure 6.14 Combined descending and co-stimulus modulation of pain

Adapted and redrawn from: Godfrey, H. 'Understanding pain, part 1: physiology of pain', *British Journal of Nursing*, 14 (16): 846–52; Copyright ©2005 MA Healthcare Ltd. Reproduced by permission of MA Healthcare Ltd.

--------------------- **APPLY** ---------------------

Opioids

You have learned in this section that endogenous opioids are used by inhibitory neurons to block pain signals from being passed to second-order neurons. This occurs by opioids binding to opioid receptors on the first-order neurons to prevent the release of substance P. They are also thought to act on the second-order neuron. It is by this principle that opioid analgesia works, whereby we provide opioids (natural from opium poppies or (at least partly) synthetic) to achieve the same effect. An example is the use of morphine in acute pain. Opioids are therefore considered to be agonists – drugs that act at a cell receptor site to produce the same or similar effect to that of a substance produced by the body (Hall, 2009). An agonist causes an action, while an antagonist blocks the action of an agonist. Examples of competitive antagonists to opioids are naloxone and naltrexone which have higher affinity to the opioid receptors but do not activate the receptors. Thus, the action of the opioids (and endorphins) is blocked.

Perception

The final phase of nociception is perception, the conscious experience of pain. It is generally accepted that there are two interrelated parts to the perception of pain: sensory-discriminative and affective-motivational (Godfrey, 2005). Sensory-discriminative components travel through spinothalamic tracts and affective-motivational components travel through spinoreticular tracts:

Spinothalamic: A-delta and C fibres synapse with second-order neurons in the dorsal horn and decussate (cross-over) to ascend this tract to the lateral and medial nuclei of the thalamus. There they synapse with third-order neurons to project out to the frontal and somatosensory cortex. The latter enables the location of pain to be determined. Processing of pain signals through this route enables the location, modality and intensity of the stimulus/stimuli to be determined and consciously experienced.

Spinoreticular: Pain fibres in this tract ascend to the reticular formation (brainstem) from the dorsal horn, synapsing with third-order neurons to project out to the primary and secondary somatosensory cortices, insula, anterior cingulate gyrus and prefrontal cortex. Processing of pain signals through this route enables the emotional responses to pain to be generated, such as anxiety and distress, and also results in the actions to respond to the pain (i.e. removing the stimulus, rubbing the area, reacting with tears or crying out). As the reticular formation is involved in this pathway, the autonomic responses to pain (alterations in heart and respiratory rate and in blood pressure) are thought to originate here.

It is the combination of both interrelated parts of the perception of nociceptive stimuli that results in the full experience of what pain is. The intensity, location and modality of the stimulus coupled with the emotional state of the person at that time, their past experiences of pain and degree of vulnerability means that every person's perception of pain is vastly different. Furthermore, the degree of tissue damage or imbalance of chemical mediators of pain can result in continuing stimulation of C fibres, resulting in ongoing pain after A-delta fibres are no longer stimulated. Godfrey (2005) further advocates that the combination of cultural factors, sensory stimuli, emotions, memories and endogenous opioids, endocrine, immune and neurological processes all influence the full pain experience. This is known as the neuromatrix theory, which complements the gate control theory alongside recognising how perception is processed while feeding into the experience of pain itself.

ACTIVITY 6.10: UNDERSTAND

Watch this online video clip to further your understanding of how pain works.

The external weblink for this video clip can be accessed by **scanning the QR code** with your smart phone or via https://study.sagepub.com/essentialap2e.

INTRODUCTION TO PAIN (12:16)

APPLY

Neuropathic pain

While nociceptive pain is largely considered to be visceral or somatic, neuropathic pain is a particular type of experience that is described as pain or an abnormal sensation secondary to neurological dysfunction. It is not considered to be nociceptive as it does not have a purpose (Taverner, 2014), whereas nociceptive pain is intended to signal injury or disease to initiate a response. Often there is no overt stimulus and the nociceptive sensation can occur suddenly or be constant (Steeds, 2013). However, people with neuropathic pain may also have nociceptive pain, and while they are treated differently, both need to be attended to.

Neuropathic pain can be peripheral or central:

- **Peripheral:** associated with damage to peripheral nerves, plexus, dorsal root ganglia or roots.
- **Central:** associated with neurological dysfunction within the brain or spinal cord.

In neuropathic pain, A-delta fibres or C fibres are sheared or damaged. Neuroplasticity occurs in an attempt to restore the nerve fibres to their original state, resulting in axonal sprouting and/or neuroma formation (Bass, 2010; Steeds, 2013). These can result in them being unstable in terms of their excitatory properties, resulting in ectopic stimulation of second-order neurons in the dorsal root of the spinal cord (Steeds, 2013). How this is experienced by people is varied and complex. Table 6.4 identifies how neuropathic pain can be evoked (Steeds, 2013; Taverner, 2014).

Table 6.4 Evoked experiences of neuropathic pain

Evoked sensation	Description
Allodynia	Pain that results from an innocuous stimulus, i.e. non-nociceptive stimulus like light touch
Hyperalgesia	Heightened pain response to a noxious (nociceptive) stimulus
Hyperaesthesia	Increased sensitivity to a stimulus (e.g. tactile)
Dysaesthesia	Evoked or spontaneous unpleasant sensation. Often described as discomfort rather than pain
Hyperpathia	Sustained, exaggerated and volatile response to a stimulus, often delayed in onset
Hypoalgesia	Reduced response to a noxious stimulus
Hypoaesthesia	Reduced sensitivity to a stimulus
Paraesthesia	Abnormal sensation, such as tingling or prickling

Neuropathic pain can be complex to treat and less responsive to traditional forms of analgesia. Any treatment should be preceded by a comprehensive assessment using effective tools such as the Leeds Assessment of Neuropathic Symptoms and Signs (S-LANSS) (Bennett et al., 2005) alongside the person's subjective history that can yield factors associated with the pain experience. Evidence-informed pharmacological and non-pharmacological treatments should both be considered following assessment, with a treatment plan that is focused on improving health-related quality of life. Pharmacological treatments for the most part do not include opioids and NSAIDs but rather anti-epileptics, antidepressants and capsaicin preparations (NICE, 2013). These drugs largely work by reducing neuroexcitability. Treatment plans should be holistic, and consider the impact that pain has on the person's life and what support is needed to ensure quality of life is maximised in all areas, including their family life, vocational pursuits and physical independence.

ACTIVITY 6.11: APPLY

Read through the NICE (2013) (updated July 2019) guidelines on **managing neuropathic pain**.

The weblink to the guidelines can be accessed via http://study.sagepub.com/essentialap2e.

CONCLUSION

Now that you have read through this chapter and undertaken the activities, you should have a good understanding of how the special senses and general senses function. These are fundamental to a person experiencing the pleasures of life as well as protecting us from harm. As a result, the senses are central to health-related quality of life and protecting people from harm.

GO DEEPER

Further reading

Oliver, D. (2010) 'The assessment and management of pain in neurological disease', *British Journal of Neuroscience Nursing*, 6(2): 70-2.

Taverner, T. (2014) 'Neuropathic pain: an overview', *British Journal of Neuroscience Nursing*, 10(3): 116-23.

Winter, G. (2008) 'Eyesight', *British Journal of Neuroscience Nursing*, 4(11): 556.

KEY POINTS

- The special senses are vision, olfaction, gustation, hearing and equilibrium.

- The general senses are pain, temperature and touch (pressure, vibration and proprioception).

- The senses are central to a person's experience of the pleasures and physical pains of life, while also enabling people to detect and respond to dangers.

- The special senses involve sense organs that have unique structures to enable the particular sense to be received.

- The general senses are more widely distributed and simpler in structure.

REVISE

TEST YOUR KNOWLEDGE

Revision of this chapter builds on Chapter 5 on the nervous system which you need to revise first. Then start with the special senses. When revising these, try to remember the diagrams included. Can you label them? Do you understand the functions of each component? Can you relate these components to the way the nervous system receives the stimulus and brings it to the central nervous system for interpretation?

Then review the general senses. Which types of receptor detect the different general senses and where are they located? Can you explain the sensory pathways to the central nervous system for these, depending on where they occur in the body?

In order to help you revise, consider the following questions, answers for which can be found by visiting **https://study.sagepub.com/essentialap2e**.

Test yourself by revising the chapter first, and then answer these questions without looking at the book. Afterwards compare your answers with the text and with the notes you made. Did you miss anything in your notes? Here are the questions:

1 What is the stimulus for sight?

2 What is the stimulus for equilibrium?

3 What is the receptor organ for hearing?

4 How are sounds converted to mechanical energy for hearing?

5 In terms of sight, what is a direct reaction?

6 What are the different types of sensory organ and the types of sensation that they transmit?

7 What protects the eye?

8 What are the three layers of the wall of the eye?

9 What is the retina made up of? How does it communicate to the brain?

10 What prevents the eye from collapsing?

11   What is the function of the middle ear?

12   What are the two functions of the inner ear?

13   What is proprioception?

14   Where are proprioceptors located?

15   What are proprioceptive stimuli needed for?

16   Draw and label the parts of the eye. Identify the function of each part. Once you have done this, compare your answer to the content of this chapter.

17   Draw and label the parts of the ear. Identify the function of each part. Once you have done this, compare your answer to the content of this chapter.

For additional revision resources visit: https://study.sagepub.com/essentialap2e.

REVISE

ACE YOUR ASSESSMENT

- Revise key terms with interactive flashcards.

- Test yourself with multiple-choice questions.

- Access the glossary with audio to hear how complex terms are pronounced.

- Explore recommended websites suitable for revision.

REFERENCES

Aberdeen, L. and Fereiro, D. (2014) 'Communicating with assistive listening devices and age-related hearing loss: perceptions of older Australians', *Contemporary Nurse*, 47(1–2): 119–31.

Bass, M. (2010) 'Anatomy and physiology of pain and the management of breakthrough pain in palliative care', *International Journal of Palliative Nursing*, 16(10): 486–92.

Bennett, M.I., Smith, B.H., Torrance, N. and Potter, J. (2005) 'The S-LANSS score for identifying pain of predominantly neuropathic origin: validation for use in clinical and postal research', *The Journal of Pain*, 6(3): 149–58

Bermingham-McDonogh, O., Corwin, J.T., Hauswirth, W.W., Heller, S., Reed, R. et al. (2012) 'Regenerative medicine for the special senses: restoring the inputs', *The Journal of Neuroscience*, 32(41): 14053–7.

Fortis-Santiago, Y., Rodwin, B.A., Neseliler, S., Piette, C.E. and Katz, D.B. (2010) 'State dependence of olfactory perception as a function of taste cortical inactivation', *Nature Neuroscience*, 13: 158–9.

Godfrey, H. (2005) 'Understanding pain, part 1: physiology of pain', *British Journal of Nursing*, 14(16): 846–52.

Hall, S. (2009) 'Opioids: prescribing rationale and uses in pain management', *Nurse Prescribing*, 7(5): 212–18.

Hartley, D., Rochtchina, E., Newall, P., Golding, M. and Mitchell, P. (2010) 'Use of hearing aids and assistive listening devices in an older Australian population', *American Academy of Audiology*, 21: 642–53.

Landis, B.N., Scheibe, M., Weber, C., Berger, R., Bramerson, A. et al. (2010) 'Chemosensory interaction: acquired olfactory impairment is associated with decreased taste function', *Journal of Neurology*, 8: 1303–8.

McKee, M.M. (2013) 'Caring for older patients who have significant hearing loss', *American Family Physician*, 87(5): 360–6.

Melzack, R. and Wall, P.D. (1965) 'Pain mechanisms: a new theory', *Science*, 150(699): 971–9.

NICE (National Institute for Health and Care Excellence) (2013) *Neuropathic Pain – Pharmacological Management: The Pharmacological Management of Neuropathic Pain in Adults in Non-Specialist Settings.* CG173. London: NICE.

Noguchi, K. and Okubo, M. (2011) 'Leukotrienes in nociceptive pathway and neuropathic/inflammatory pain', *Biological and Pharmaceutical Bulletin*, 34(8): 1163–9.

Purves, D., Augustine, G.J., Fitzpatrick, D., Hall, W.C., LaMantia, A.S. and White, L.E. (eds) (2011) *Neuroscience*, 5th edn. Sunderland, MA: Sinauer Associates.

Shubayev, V.I., Kato, K. and Myers, R.R. (2010) 'Cytokines in pain', in L. Kruger and A.R. Light (eds), *Translational Pain Research: From Mouse to Man.* Boca Raton: CRC Press. pp. 187–214.

Steeds, C.E. (2013) 'The anatomy and physiology of pain', *Surgery*, 31(2): 49–53.

Stinton, N., Atif, M.A., Barkat, N. and Doty, R.L. (2010) 'Influence of smell loss on taste function', *Behavioral Neuroscience*, 124(2): 256–64.

Taverner, T. (2014) 'Neuropathic pain: an overview', *British Journal of Neuroscience Nursing*, 10(3): 116–23.

Waterfield, J. (2008) 'Non-opioid analgesics: prescribing rationale and uses', *Nurse Prescribing*, 6(11): 496–501.

Wood, S. (2008) 'Anatomy and physiology of pain', *Nursing Times*. Available at: www.nursingtimes.net/nursing-practice/specialisms/pain-management/anatomy-and-physiology-of-pain/1860931.article (accessed 4 May 2020).

THE ENDOCRINE SYSTEM
CONTROL OF INTERNAL FUNCTIONS

7

UNDERSTAND: CHAPTER VIDEO

Before reading this chapter, you might find it useful to familiarise yourself with the endocrine system. Watch the following video clip to enhance your understanding.

This video can be accessed by **scanning the QR code** with your smart phone camera or via https://study.sagepub.com/essentialap2e.

THE ENDOCRINE
SYSTEM (3:47)

LEARNING OUTCOMES

When you have finished studying this chapter you will be able to:

1. Identify the different modes of action and types of hormones
2. Describe the organisation of the endocrine system and how hormone secretion is controlled
3. Identify the range of functions of the endocrine system
4. Describe how each of the specified functions is managed by the endocrine system including:

   i. Reproduction and sexual differentiation (Chapter 16)
   ii. Development and growth (Chapter 17)
   iii. Maintenance of the internal environment
   iv. Regulation of metabolism
   v. Nutrient supply
   vi. Stress response

INTRODUCTION

This chapter will discuss the role and functions of the endocrine system in contributing to the maintenance of homeostasis, including its relationship with other systems in the body, particularly the nervous system. Within this chapter you will learn how hormones influence the function of glands to regulate the internal environment, including metabolism, growth and development, tissue function, sleep and mood. This will enable you to understand how the endocrine system operates in supporting health and well-being through its impact on physical and mental health. This approach facilitates readers in placing their knowledge and understanding within the perspective of the Person-Centred Nursing Framework.

Context

The endocrine system is of major importance at all stages throughout life. In Danielle, her rapid coordinated growth and development during early life is largely dependent on her endocrine system functioning normally. She also needs the best possible nutrition, which is obtained through breast-feeding and dependent on the hormones prolactin and oxytocin in her mother (Michelle).

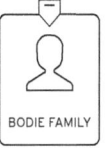

BODIE FAMILY

Hannah is going through the menopause and having some of the symptoms discussed in Chapter 17. Her general practitioner has prescribed Hormone Replacement Therapy (HRT) to enhance her quality of life while passing through this difficult period of change.

Maude (77 years) has recently been diagnosed with subclinical hypothyroidism (i.e. low levels of thyroid hormones without overt signs and symptoms), a not uncommon condition in older people, particularly women (Bensenor et al., 2012). The symptoms are similar to those of some other conditions, including anaemia, and a correct diagnosis is achieved by measuring blood levels of Thyroid Stimulating Hormone (TSH) and thyroid hormones.

HORMONES AND THE ENDOCRINE SYSTEM

The endocrine (endo – within, crine – to secrete) system works alongside the nervous system to coordinate the functions of all body systems. The nervous system (Chapter 5) tends to work rapidly using nerve impulses relayed along the neuron (action potentials) and the release of chemicals (neurotransmitters) to control functions, whereas the endocrine system works more slowly and is often responsible for the regulation of longer-term processes. It does this by releasing chemical mediators known as hormones. Hormones are chemical messengers secreted into the blood or extracellular fluid by one cell and have an effect on the functioning of other cells in other parts of the body. As hormones circulate in the blood, they come into contact with virtually every cell in the body, but they only exert their specific effect on those cells that have receptors for that hormone (target cells). Hormone receptors are either found within the target cell or on its surface: the site of the receptor is dependent on the type of hormone (discussed later). Some mediators can act as both neuro-transmitters and hormones. For example, noradrenaline/norepinephrine, when released by postganglionic neurons (Chapter 5), acts as a neurotransmitter but when released from the adrenal medulla acts as a hormone.

The endocrine system is comprised of endocrine glands that include the pituitary, thyroid, parathyroid, adrenal and pineal glands and several organs and tissues that contain cells that can secrete hormones, including the hypothalamus, thymus, pancreas, ovaries, testes, kidneys, liver, stomach, small intestine, skin and heart. These will be discussed throughout this chapter.

Modes of hormone action

Chapter 1 discusses how hormones link with receptors to modulate their function in the different modes of hormone action shown in Figure 7.1:

- Classical endocrine as indicated in the first paragraph in this section.
- Paracrine in which hormones secreted act on other cells in the neighbourhood.
- Juxtacrine in which hormones from one cell act on receptors on their immediate neighbours.
- Autocrine in which hormones act on the cells which produce them.
- Intracrine in which activation of the hormones occurs within the cell where they are created and they bind with nuclear receptors to modify the function of that cell.

There are three issues to understand in relation to hormones: types of hormones; control of secretion of hormones; and bodily organisation of hormone secretion. We will then go on to consider the major functions of the endocrine system.

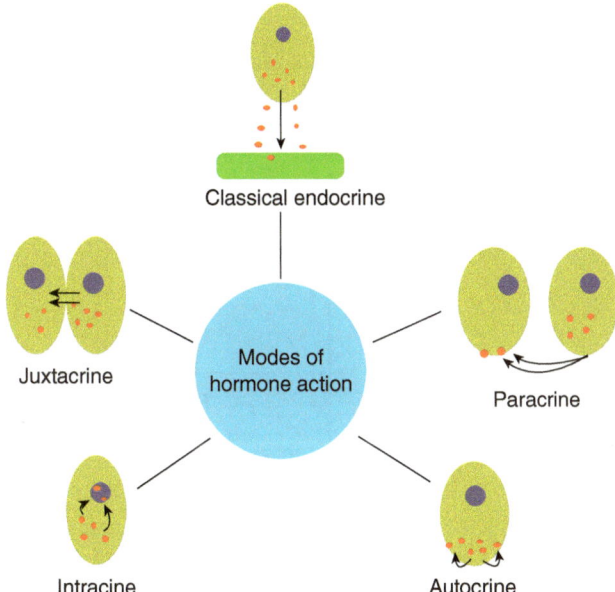

Figure 7.1 Modes of hormone action

Adapted from: Zouboulis, C.C. (2009) 'The skin as an endocrine organ', *Dermatoendocrinology*, 1(5): 250-252, Taylor & Francis Group.

Types of hormones

The hormones produced by the body each have one of three types of structure – steroid, protein/polypeptide or modified amino acids – and are either lipid soluble (e.g. steroid, thyroid hormones) or water soluble (e.g. adrenaline, noradrenaline, insulin, human growth hormone). Lipid-soluble and water-soluble hormones exert their effects differently. Water-soluble hormones attach to receptors on the plasma protein on the outside of the target cell as they are unable to diffuse through the lipid bilayer of the cell membrane – therefore, they must pass on their message to a receptor located at the surface of the cell. They do not directly affect the transcription of target genes, but instead initiate a signalling cascade that is carried out by a molecule called a second messenger (e.g. cyclic adenosine monophosphate (cAMP)). Lipid-soluble hormones attach to receptors inside the target cell, thereby altering the genes within the cell nucleus, which in turn alters the function of that cell to exert the desired response.

GO DEEPER

Figure 7.2 shows the chemical structure of cholesterol and steroid hormones.

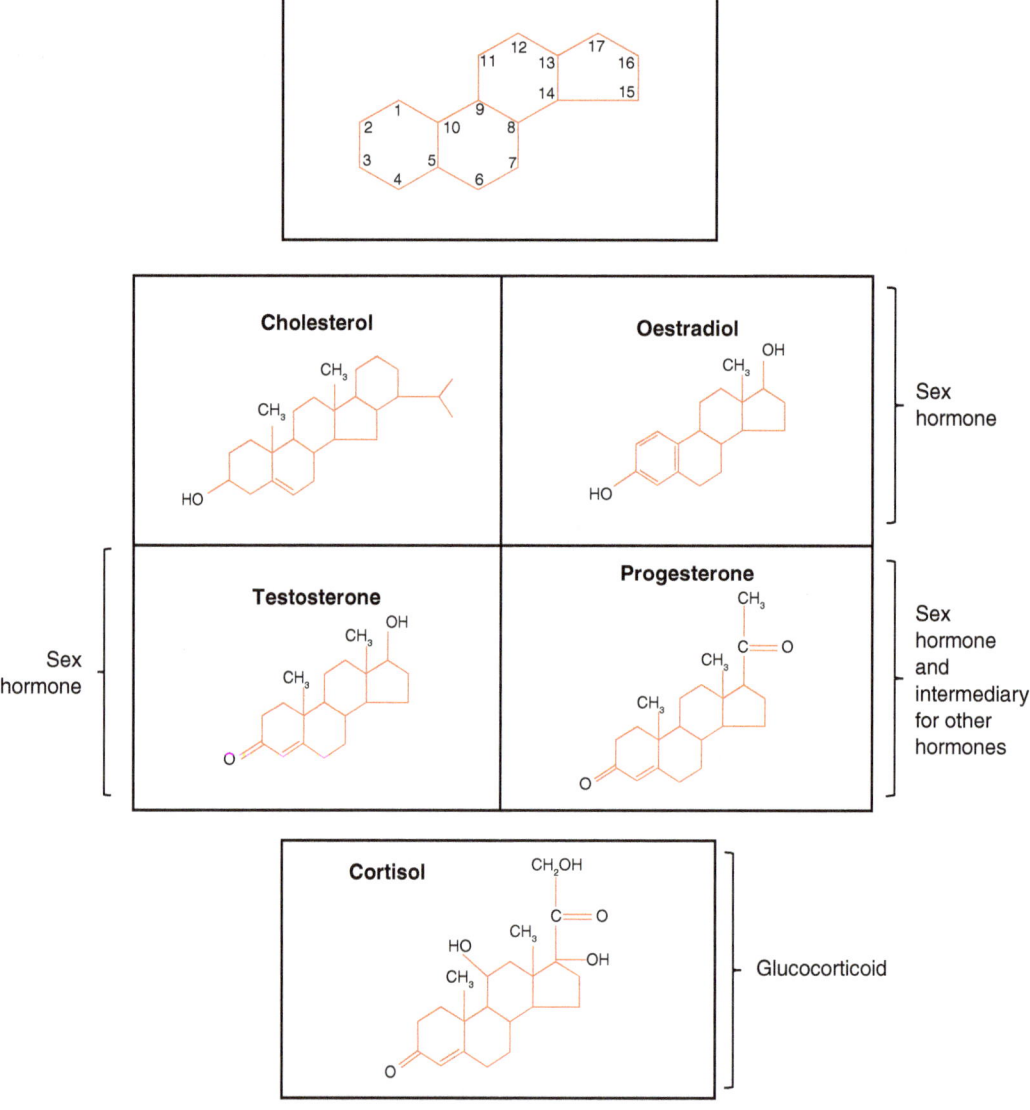

Figure 7.2 Cholesterol and steroid hormones

Steroid hormones

These are formed from cholesterol, manufactured in the liver and acquired from the diet. Figure 7.2 shows cholesterol and some examples of steroid hormones. The adrenal cortex, ovaries and testes create steroid hormones. In addition, vitamin D made by the skin or taken in the diet and activated in the liver is also formed from cholesterol.

Steroid hormones are mainly transferred in the blood in combination with carrier proteins in equilibrium with a small amount free in the blood. It is the free hormone that can pass through cell membranes, combine with receptors inside the cell and modulate the cell function. This is largely through regulating the transcription of specific genes in the DNA and the formation of specific messenger RNA and/or protein molecules. The onset of steroid hormone function may take several hours due to the time taken in transcription, and the half-life[1] is longer than that of other hormones, also lasting hours. Steroid hormones are eventually metabolised and excreted via the liver.

Protein/polypeptide hormones

These hormones vary in size from three amino acids (e.g. Thyroid Stimulating Hormone – TSH) to much larger chains with a sub-structure (e.g. Luteinising Hormone – LH). These molecules are formed (often as pro-hormones) from the genes coding for them within the endocrine cell, processed into the hormone itself, and then stored within vesicles or secretory granules until needed. Secretion is stimulated by environmental triggers (such as calcium ions (Ca^{2+})) and, because the hormone is stored in the cell, increased secretion can be initiated rapidly.

The half-life of these hormones is usually short (minutes) and mostly they are not carried in combination with transfer proteins. They combine with protein or glycoprotein receptors on the cell membrane but interact in both the intracellular or extracellular environments and may initiate a series of interactions leading to a number of changes in the cell, which may include: initiating secretion of some substance from the cell; increasing uptake of a substance into the cell; stimulating cell division (mitosis).

Modified amino-acid hormones

This is a small group of hormones formed mostly by modification of the amino acids tyrosine or tryptophan and include:

- From tyrosine:
  - **Thyroid hormones:** thyroxine (T_4) and triiodothyronine (T_3). These are manufactured and stored in the thyroid gland as part of the large molecule thyroglobulin. Thyroglobulin is taken up into the cells of the thyroid gland and the thyroid hormones are released and secreted from the cells.
  - **Catecholamines:** adrenaline (epinephrine) and noradrenaline (norepinephrine). The tyrosine is modified by various chemical reactions and secreted from the adrenal medulla into the blood stream as free hormone. They act via receptors on the cell surface and have very short half-lives (< five minutes).
- From tryptophan:
  - **Serotonin:** a neurotransmitter which is then converted to the hormone melatonin in the pineal gland.

---

[1]The time required for any specified property, e.g. the concentration of a hormone or drug in the body, to fall to half of its original value.

Control of secretion of hormones

There are three main ways by which endocrine secretion is stimulated or inhibited:

- Via the hypothalamus and pituitary gland.
- Through the influence of factors in the blood stream.
- By nervous stimulation.

Hypothalamus and pituitary gland

The hypothalamus with the pituitary gland plays an important role in regulating endocrine secretion. Nuclei in the hypothalamus of the brain receive information from elsewhere in the body and send information to the pituitary gland:

- To the anterior and intermediate pituitary by hormones secreted into the blood stream and delivered to the gland where secretion of hormones is enhanced or inhibited by the hypothalamic hormones.
- To the posterior pituitary gland along nerve axons in which the hormones are stored until their release is initiated.

Factors in the blood stream

Some hormones are secreted in direct response to the level of the substance that they control. For example, the hormones insulin and glucagon from the pancreas are secreted in response to the level of glucose in the blood. Insulin lowers the blood glucose and glucagon raises it. Parathyroid Hormone (parathormone) (PTH) and calcitonin are secreted in response to serum Ca^{2+} levels, which they maintain within normal limits.

Nervous stimulation

Some hormones are secreted in response to nervous stimulation. The adrenal medulla functions as part of the sympathetic nervous system and secretes adrenaline (epinephrine) in response to sympathetic nervous stimulation.

ORGANISATION OF HORMONE SECRETION

Major endocrine organs

The major organs of the endocrine system are shown in Figure 7.3. While these organs may perform some other activities, secretion of hormones is a major function.

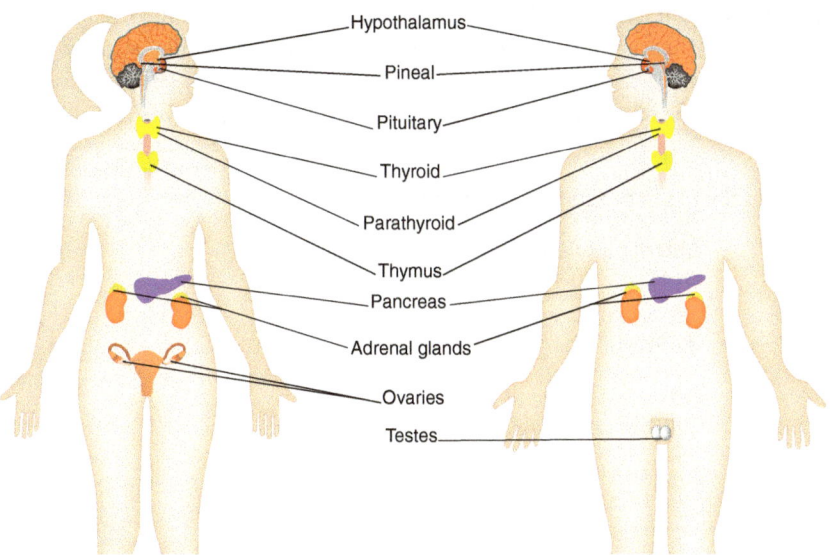

Figure 7.3 Major endocrine organs

Hypothalamus, pituitary gland and pineal gland

Hypothalamus and pituitary gland

The hypothalamus at the base of the brain is linked by the pituitary stalk to the pituitary gland, protected by a small depression in the skull. It produces a number of hormones that regulate the secretion of hormones from the pituitary gland. The pituitary gland has three parts – anterior, posterior and

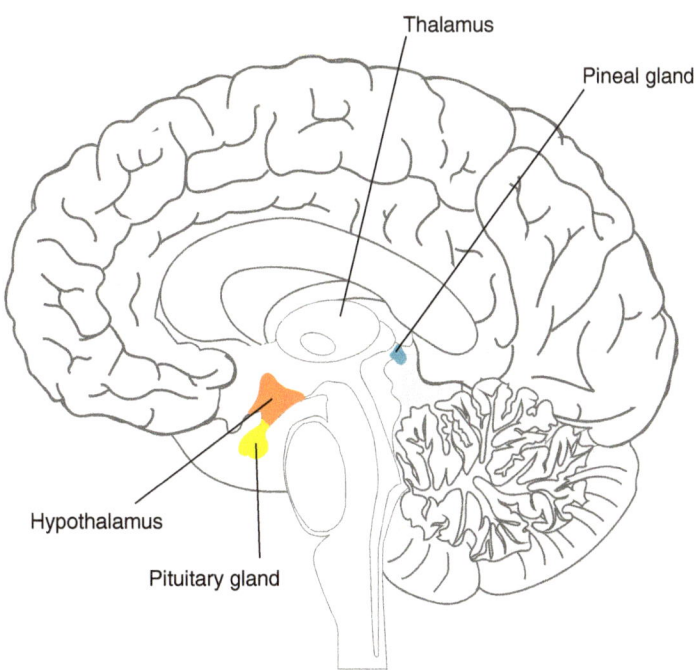

Figure 7.4 The hypothalamus, pituitary gland and pineal gland

intermediate (although often included in the anterior gland in humans) – and the hypothalamus regulates the function of all three sections. In addition, within the brain is the pineal gland (Figure 7.4). The activity of the hypothalamus on the pituitary gland is outlined in Table 7.1. The pituitary gland produces a range of hormones involved in a wide range of endocrine functions given in Table 7.2.

Table 7.1 Hypothalamus in regulation of pituitary gland

Hypothalamic ⇒ hormones	Anterior pituitary gland	Function
Thyrotrophin RH	Thyroid Stimulating Hormone (TSH)	Stimulates thyroid gland to secrete thyroid hormones (thyroxine (T_4) and T_3) Regulates metabolic rate
Corticotrophin RH	Adrenocorticotrophic Hormone (ACTH)	Stimulates adrenal cortex to secrete glucocorticoid hormones Response to stress
Gonadotrophin RH	Luteinising Hormone (LH) Follicle Stimulating Hormone (FSH)	Stimulates gonads (see Figure 7.9) male (♂) and female (♀) LH: causes ovulation (♀) and release of testosterone (♂) FSH: regulates oogenesis (♀) and spermatogenesis (♂)
Prolactin RH Prolactin IH	Prolactin	Stimulates milk production from breasts
Growth hormone RH Growth hormone IH (somatostatin)	Human growth hormone (hGH)	Stimulates growth of bones, muscles, cells in general
Hypothalamic ⇒ nuclei	**Posterior pituitary gland**	**Function**
Supraoptic and paraventricular nuclei nerve impulses	Oxytocin	Mainly neuromodulator in brain. 'Bonding hormone' (orgasm, social recognition, pair bonding, anxiety, etc.). Role in childbirth and breast-feeding
Suprachiasmatic nucleus nerve impulses	Antidiuretic Hormone (ADH) (vasopressin)	Water retention, vasoconstriction → raised blood pressure
	Intermediate pituitary gland	
Regulated by a number of different factors	Melanocyte Stimulating Hormones (MSH)	Production and release of melanin from melanocytes in skin and hair. Helps activation of regulatory T-cells in immune system (Namba et al., 2002)

Male = ♂   Female = ♀   RH = Releasing Hormone   IH = Inhibiting Hormone

Pineal gland

This small gland develops from the roof of the third ventricle of the brain under the posterior end of the corpus callosum. Although part of the brain, it is not isolated from the rest of the body by the blood–brain barrier and it has a very considerable blood flow, implying an important role in physiology. This gland secretes the hormone melatonin derived from serotonin: at higher levels in the dark and lower in daylight. It is thought that the cycle of melatonin secretion controls the circadian (24-hour) rhythm of endocrine secretion and other bodily functions. It also plays an important role in enhancing sleep patterns.

Table 7.2 Pituitary hormones

Hormone type	Functions
Anterior pituitary	
Somatotrophins	Human growth hormone (hGH), or growth hormone (GH), belongs to this group of hormones. hGH is also known as somatotrophin. It stimulates growth and is released in the presence of hypothalamic growth hormone-releasing hormone (GHRH). The hypothalamus also produces somatostatin to inhibit its release
Thyrotrophins	Thyroid Stimulating Hormone (TSH) belongs in this group. TSH stimulates the thyroid gland to produce thyroxine (T_4), and then triiodothyronine (T_3), which stimulates metabolism. Thyrotrophin-Releasing Hormone (TRH) from the hypothalamus promotes its release and somatostatin inhibits it
Corticotrophins	The hypothalamus produces Corticotrophin-Releasing Hormone (CRH), which stimulates the release of adrenocorticotrophic hormone (ACTH), and beta-endorphin from the pituitary. ACTH increases the production and release of corticosteroids. Beta-endorphin is an opioid receptor agonist
Lactotrophins	Prolactin (PRL), also known as 'Luteotrophic' Hormone (LTH) is released under the influence of several hormones. It is primarily associated with the production of breast milk but is also essential in metabolism, immunity and pancreatic development
Gonadotrophins	Luteinising hormone (also referred to as 'Lutrophin' or 'LH') and Follicle stimulating hormone (FSH), both of which are released under influence of Gonadotrophin-Releasing Hormone (GnRH). LH stimulates ovulation in females and testosterone production in men
Intermediate pituitary	
Melanocyte-stimulating hormone (MSH)	MSH stimulates the production and release of melanin (melanogenesis) by melanocytes in the skin and hair. MSH also has a role in appetite and sexual arousal
Posterior pituitary	
Antidiuretic hormone (ADH), (or vasopressin and arginine vasopressin AVP)	Results in the reabsorption of water in the renal tubule
Oxytocin (also released by the hypothalamus)	Increases uterine contractions Also influences relationships

Thyroid and parathyroid glands

Thyroid gland

This is situated at the front of the neck and has four parathyroid glands embedded in it (Figure 7.5).

The thyroid gland produces the hormones thyroxine (T_4) and triiodothyronine (T_3) under the influence of TSH from the anterior pituitary gland controlled by thyrotrophin releasing hormone (TRH) from the hypothalamus. The thyroid hormones increase the basal metabolic rate of the body, regulate metabolism of all nutrients and are essential for normal growth and development of all body cells. They also stimulate heat generation. The parafollicular cells, also in the thyroid gland, secrete the hormone calcitonin which, with parathyroid hormone (PTH), is involved in calcium metabolism.

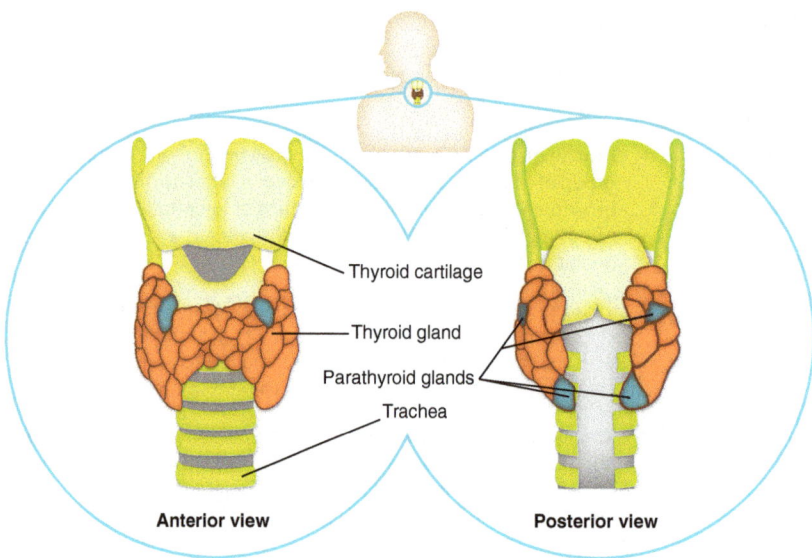

Figure 7.5 The thyroid and parathyroid glands

GO DEEPER

Formation of thyroid hormones

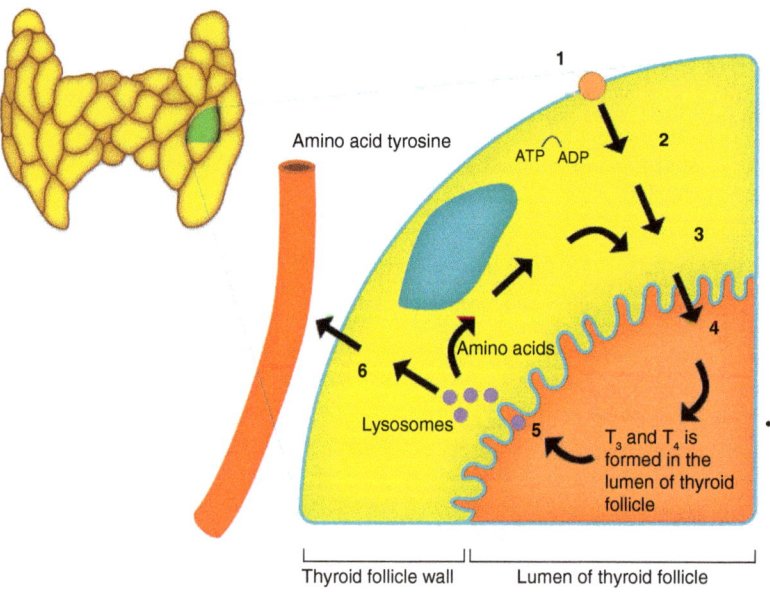

Figure 7.6 Formation of thyroid hormones

Thyroid gland tissue is composed of follicles containing the colloid thyroglobulin (a gelatinous glycoprotein containing suspended particles) surrounded by the follicular cells which make it. Figure 7.6 illustrates the process of formation of the thyroid hormones. It starts with two parallel pathways:

- Iodine is trapped and absorbed into the follicular cells as active iodine and passed into the follicle.
- Thyroglobulin is formed in the follicular cells and also passes into the follicle.

In the follicle, the iodine becomes attached to the amino acid tyrosine in the thyroglobulin to form mono-iodotyrosine (MIT) and diiodotyrosine (DIT). Two DITs combine to form thyroxine (T_4) while one MIT with one DIT forms triiodothyronine (T_3). Small amounts of the thyroglobulin are taken back into the follicular cells and digested to release T_3 and T_4 which are then transported into the blood stream.

Parathyroid glands

The four parathyroid glands embedded in the posterior of the four lobes of the thyroid gland (Figure 7.5) secrete PTH. Blood calcium levels are raised by PTH and lowered by calcitonin by acting on bone, intestine and kidneys (Table 7.3).

Table 7.3 Major hormonal actions on calcium and phosphate metabolism

| Tissue | Calcium levels | | Phosphate levels | |
	Parathormone action	Calcitonin	Parathormone action	Calcitonin
Bone	Enhances calcium release from bones into blood stream by increasing bone resorption (the normal destruction of bone by osteoclasts)	Inhibits osteoclast activity and stimulates osteoblasts, reducing Ca^{2+} released into blood	Enhances uptake of phosphate from bones	
Kidney	Enhances reabsorption of calcium and magnesium from distal tubules and thick ascending limb	Inhibits renal tubular cell reabsorption of Ca^{2+}	Reduces reabsorption of phosphate from proximal tubule of kidney	Inhibits phosphate reabsorption by kidney tubules
Intestine	Enhances absorption of calcium in intestine by increasing vitamin D activation in the kidney	Inhibits Ca^{2+} absorption by the intestines	Enhances uptake of phosphate from intestine into blood	
Summary	Overall raises serum Ca^{2+}	Overall lowers serum Ca^{2+}	Overall small net drop in serum phosphate concentration	

Thymus gland

The thymus gland is positioned in the anterior mediastinum in front of the great vessels. It grows throughout childhood reaching its largest at puberty and decreasing in size up to about 50 years of age. It produces a number of polypeptide hormones, which contribute to the function of the immune system including differentiation of T-lymphocytes (Marchevsky and Wick, 2014) (Chapter 13).

ACTIVITY 7.2: UNDERSTAND

Watch this online video to further your understanding of the thyroid gland and hormones.

This external video can be accessed by **scanning the QR code** with your smart phone or via https://study.sagepub.com/essentialap2e.

THYROID GLAND AND
HORMONES (12: 46)

Adrenal glands

The two adrenal glands are positioned on the top of the kidneys and each consists of two separate parts – the cortex and the medulla – both of which secrete hormones (Figure 7.7).

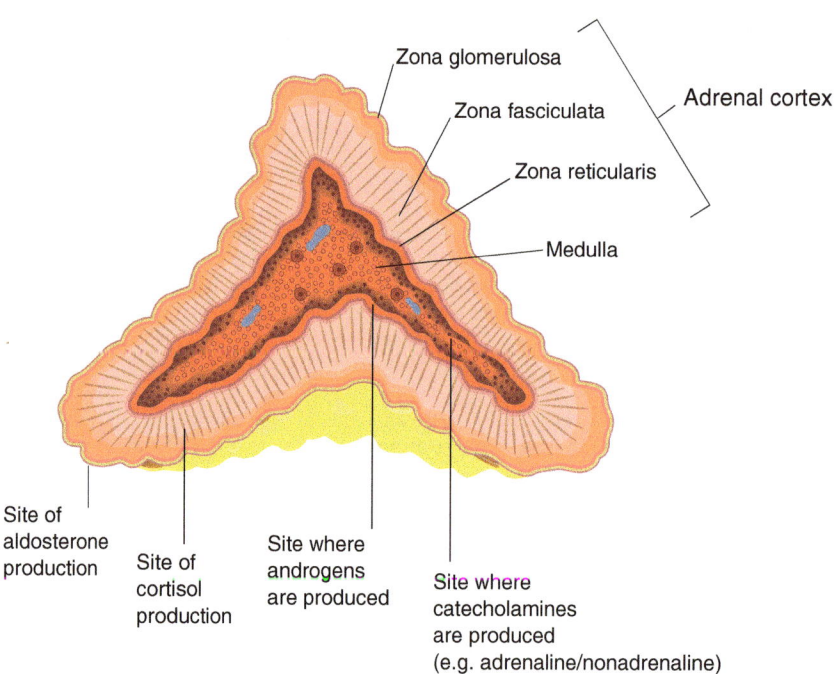

Figure 7.7 Adrenal gland

Adrenal cortex

This is the outer layer of the adrenal gland and consists of three layers, each of which secrete different steroid hormones:

- Zona glomerulosa: secretes aldosterone, the main mineralocorticoid hormone, involved in regulation of fluid and electrolyte balance and blood pressure.
- Zona fasciculata: secretes glucocorticoid hormones, including cortisol, which play an important role in the response to stress.
- Zona reticularis: produces androgens (the male sex hormones).

While each of these groups of hormones have the specific main areas of activity described, they also have some activity in the other two zones. The adrenal cortex secretions are regulated by the hormones from the pituitary gland controlled by the hypothalamus, and by the Renin Angiotensin Aldosterone System (RAAS).

Adrenal medulla

The adrenal medulla is functionally part of the sympathetic nervous system with the secretion of catecholamine hormones (80% adrenaline and 20% noradrenaline) formed from tyrosine. These hormones play an important role in the flight or fight response in stress. Their secretion is mainly controlled by the sympathetic nervous system from the preganglionic fibres from the thoracic spinal cord and, in essence, the adrenal medulla performs as a specialised sympathetic ganglion, although its hormones are released directly into the blood stream.

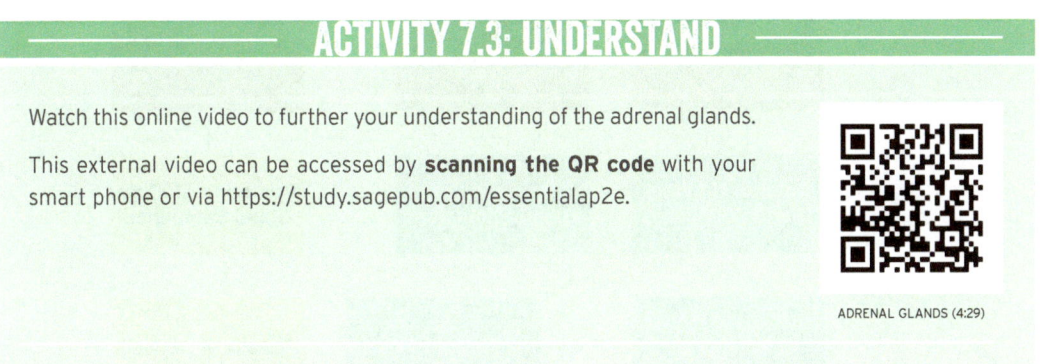

ACTIVITY 7.3: UNDERSTAND

Watch this online video to further your understanding of the adrenal glands.

This external video can be accessed by **scanning the QR code** with your smart phone or via https://study.sagepub.com/essentialap2e.

ADRENAL GLANDS (4:29)

Pancreas (islets of Langerhans)

The pancreas lies in the abdomen and plays an important role in digestion as an exocrine gland secreting its enzymes into the duodenum for digestion of carbohydrates, lipids and proteins. Its endocrine function is concerned with glucose metabolism through the secretion of hormones from the specialised cells of the islets of Langerhans (Figures 7.8 and 7.12) (Chapter 9).

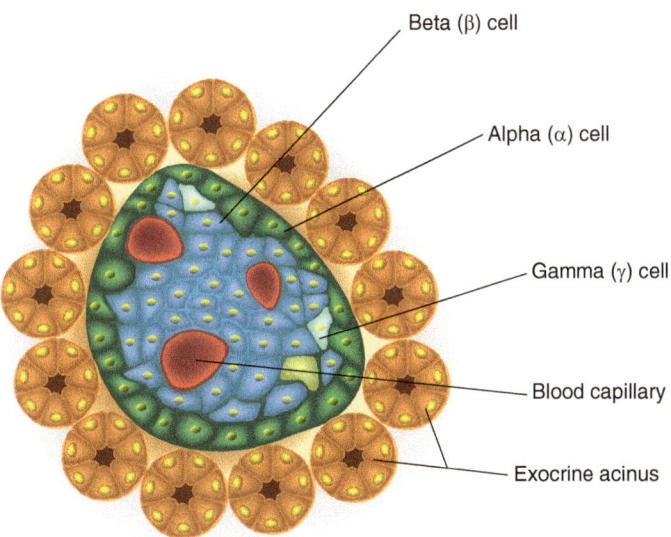

Beta (β) cell
Alpha (α) cell
Gamma (γ) cell
Blood capillary
Exocrine acinus

Figure 7.8 Islet of Langerhans – endocrine part of pancreas

Ovaries and testes

The endocrine functions of these organs are concerned with the overall function of reproduction and sexual differentiation and are discussed in Chapter 16, The Reproductive Systems, along with the hypothalamus and pituitary glands in the overall regulation of the reproductive systems. Here the hormones involved are summarised in Figure 7.9.

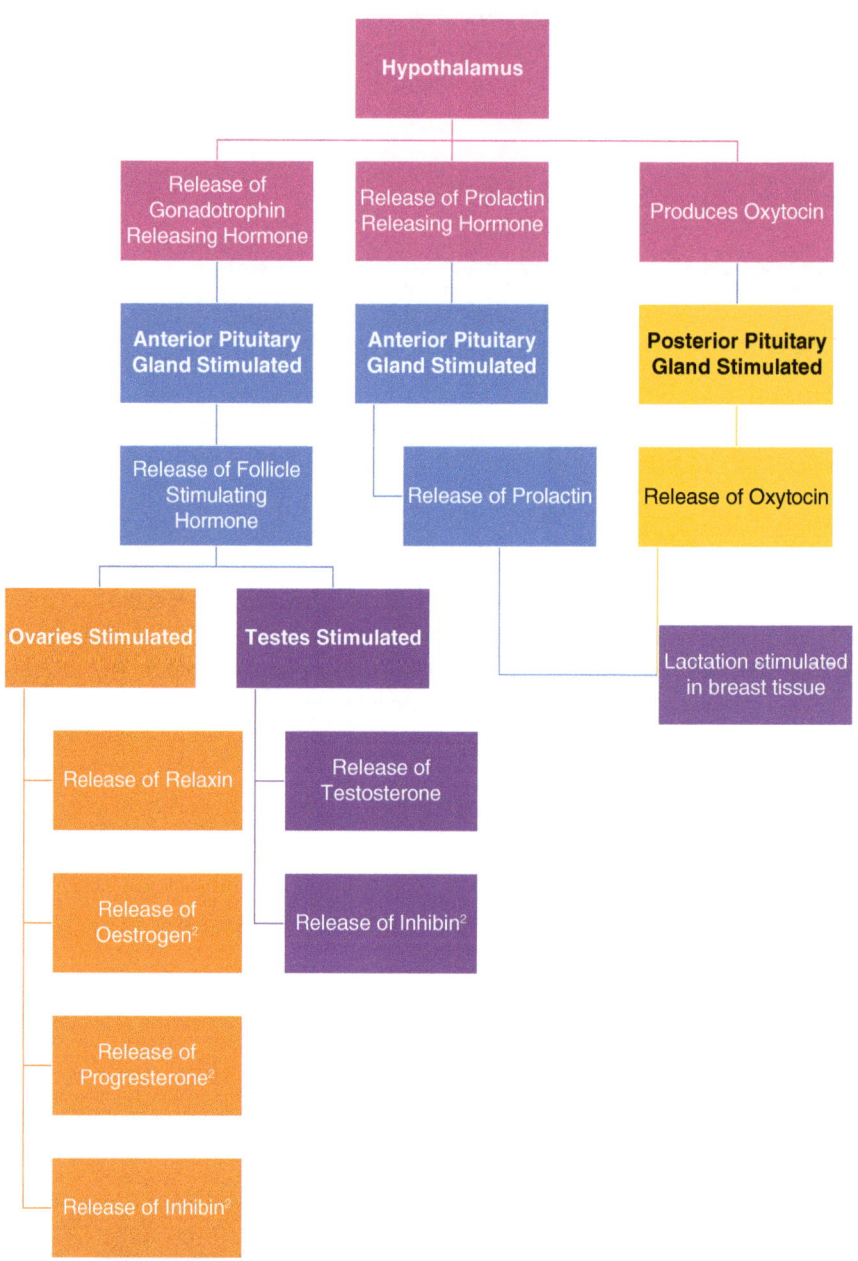

[2] These hormones reduce/stop the release of Follicle Stimulating Hormone from the anterior pituitary gland

Figure 7.9   Hormones involved in reproduction

Other organs of the body with endocrine function

Some other organs/tissues in the body with important non-endocrine functions also secrete certain hormones. These include the heart, kidneys, adipose tissue, skin and gastrointestinal tract (as well as some of those identified above).

Heart

Specialised cardiac cells in the atria of the heart respond to increased blood volume by releasing the hormone Atrial Natriuretic Peptide (ANP) (Chapter 11). This acts on the kidneys to reduce sodium (Na^+) reabsorption and thus increase excretion of Na^+ with water from the kidneys and so reduce blood volume. It also acts to alter secretion of other hormones: renin (from the kidneys) and aldosterone (from the adrenal cortex).

Kidneys

There are two main sets of hormones secreted by the kidneys, involving regulation of fluid and electrolyte balance, and red blood cell count.

Renin angiotensin aldosterone system (RAAS)

This is an important system in relation to fluid and electrolyte balance and blood pressure (Figure 7.10). It also demonstrates the interaction between different organs in endocrine control. The liver produces the protein angiotensinogen, which is converted to an active hormone (angiotensin I) by an enzyme secreted by the kidney (renin). Angiotensin Converting Enzyme (ACE), partly secreted by the lungs, converts angiotensin I to angiotensin II. ACE inhibitors are a type of drug used in treating hypertension (raised blood pressure).

Angiotensin II has five main actions:

- It causes vasoconstriction and thus raises blood pressure.
- It stimulates the release of aldosterone from another endocrine organ (the adrenal cortex), which adjusts renal function to increase the blood volume, also raising the blood pressure. This is achieved through stimulating the reabsorption of sodium in the nephron, drawing with it water and thus raising blood volume and blood pressure.
- It stimulates the release of ADH (Antidiuretic Hormone) from the pituitary gland, which acts on the kidney tubules to increase reabsorption of water into the blood, also raising blood volume and blood pressure.
- It stimulates thirst to increase water intake. The sensation of thirst is often diminished in older people, leading to an increased risk of dehydration.
- It will result in cardiac and vascular hypertrophy which, over time, will increase blood pressure by reducing the elasticity of the heart and arterial walls.

Renin secretion is the key element in this system and is regulated through the Juxtaglomerular Apparatus (JGA) (or granular cells) positioned as shown in Figure 7.11. The macula densa is composed of specialised columnar epithelium cells, which sense the concentration of sodium in the distal part of the renal tubule. If this rises two changes are initiated:

- Glomerular filtration increases back to normal.
- Renin release from the JGA cells is also increased, which then initiates the renin angiotensin aldosterone system (Chapter 12).

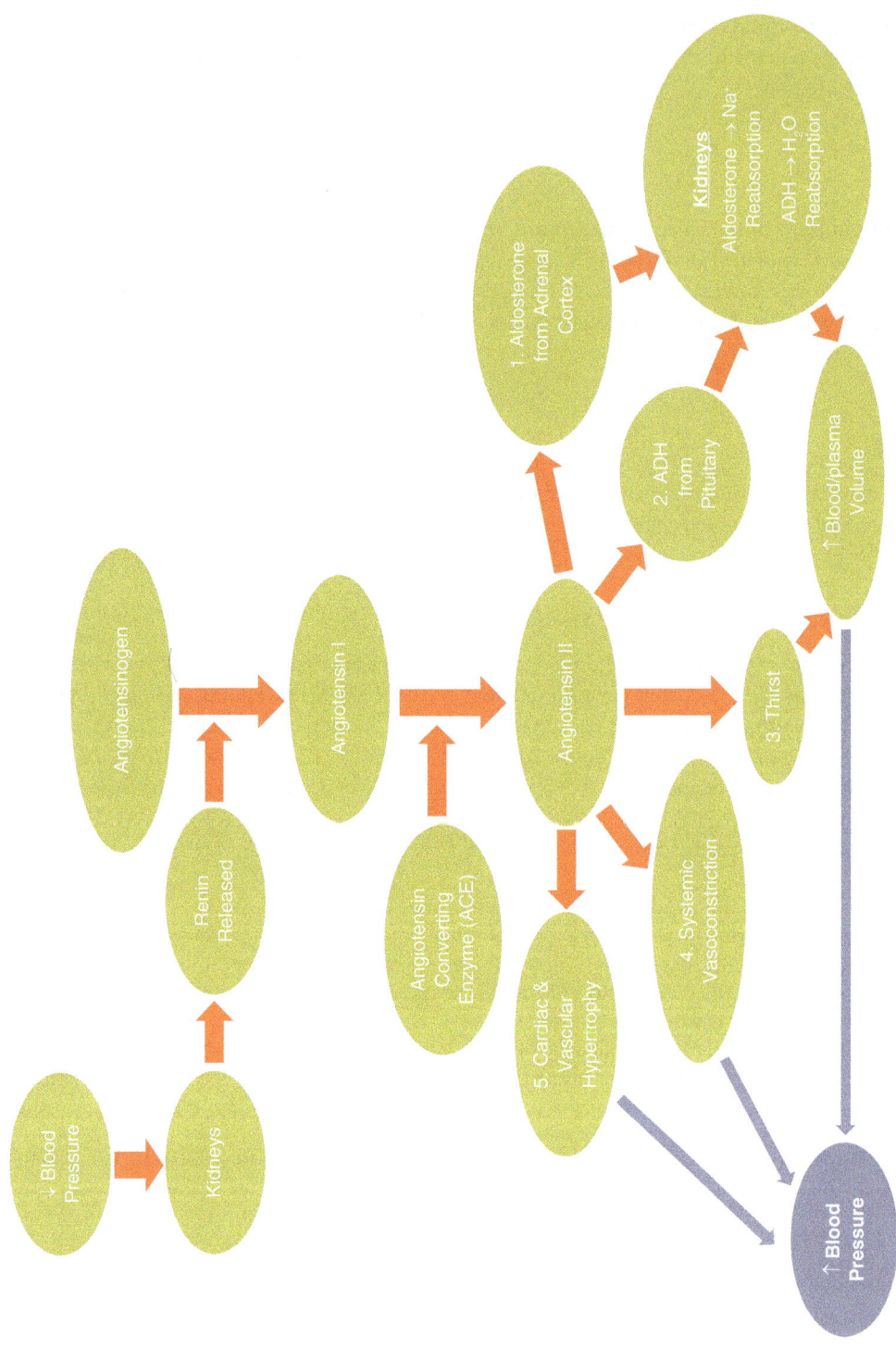

Figure 7.10 The renin angiotensin aldosterone system

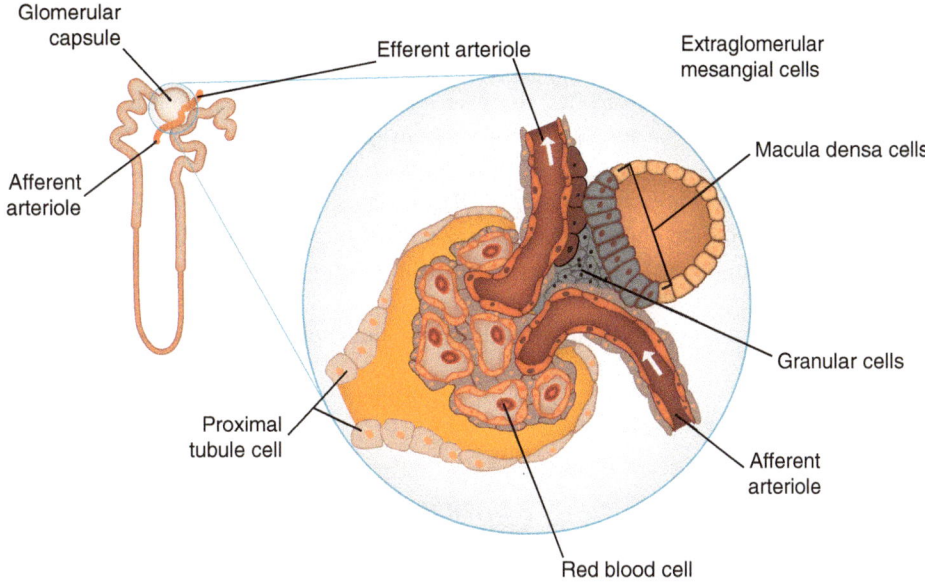

Figure 7.11 The juxtaglomerular apparatus

Erythropoietin (EPO)

This is a protein hormone released from the kidney in response to low blood oxygen levels which stimulates formation of red blood cells from the bone marrow.

APPLY

EPO, performance and anaemia

EPO increases the red blood cell (erythrocyte) count and thus increases the oxygen available to the organs. Although this is not permitted, it has been used by some athletes in order to artificially improve their performance. Because increased red blood cells increase the viscosity (thickness or how resistant a liquid is to flowing) of the blood it does have health risks associated with it. EPO is also given to people in renal failure as their kidneys no longer produce it in sufficient amounts. This is essential in preventing anaemia.

Adipose tissue

Adipose tissue is one of the connective tissues studied in Chapter 2, distributed widely in the body, which produces the hormone leptin. This is the 'satiety hormone', which regulates how much fat is stored in the body by adjusting the sensation of hunger and the amount of energy expended. Leptin is released and circulated when fat stores reach a certain level and activates the leptin receptors in the hypothalamus, inhibiting hunger. Its effect is opposite to that of the hormone ghrelin, the 'hunger hormone'.

Skin

The endocrine function of the skin has only been recognised fairly recently. It contains hormone receptors and responds to a number of hormones. However, it also produces hormones which act throughout the body (Zouboulis, 2009). It creates sex hormones from precursors and is particularly important in post-menopausal women.

However, the skin is the only organ which synthesises cholecalciferol (a form of vitamin D_3) and vitamin D_2 (ergocalciferol) from cholesterol in the presence of sunlight (Chapter 14). While called vitamins and available within the diet, they fulfil the criteria of a hormone. Vitamin D is carried to the liver and metabolised to specific modified vitamin D molecules, some of which are converted to the biologically active form of vitamin D in the kidney. With PTH and calcitonin, it contributes to the regulation of calcium and phosphate.

Vitamin D deficiency

Vitamin D plays an important role in health. However, it is estimated that many people are vitamin D deficient, particularly the middle aged and elderly, in countries at high latitude such as the UK and Canada (Chapters 14 and 17). There are several risk factors including:

- **Old age:** due to reduced synthesis of vitamin D and reduced exposure to sunlight.
- **Dark skin:** the larger amounts of melanin in the skin limit ultraviolet B (UVB) light necessary for vitamin D production.
- **Season, latitude and time of day:** influence the amount of UVB reaching the skin.
- **Sunscreen use:** reduces vitamin D formation (Zhang and Naughton, 2010).

A study at age 45 years of a cohort of British adults born in 1958 found that 90% had hypovitaminosis D in winter and spring, with 60% having lowered levels all year. It appears that north of London, no vitamin D is formed in the skin in December and January with hormone formation being much reduced by cloud in other parts of the year (Hyppönen and Power, 2007).

The UK Chief Medical Officers (2012) have issued advice on vitamin D supplements.

Gastrointestinal tract

The gastrointestinal tract (GIT) is discussed in Chapter 8 including an introduction to its regulation. However, endocrine regulation of the GIT is highly complex, including a large number of different hormones formed in cells in the mucosa distributed throughout the GIT (rather than in defined glands). They exert autocrine and paracrine effects which help to integrate the function of the whole GIT. Many also act as neurotransmitters in the central and peripheral nervous system. The main hormones are given in Table 7.4.

Table 7.4 Major hormones of the gastrointestinal tract

Hormone	Source	Secretion stimulated by	Secretion inhibited by	Action
Gastrin	From G-cells in antrum of stomach	• Distension of stomach • Products of protein digestion in stomach • Vagus nerve activity • Adrenaline and Ca^{2+}	• Acid in stomach • Somatostatin • Secretin • Glucagon	Stimulates: • Acid and pepsin secretion • Gastric motility • Insulin secretion after protein meal • Growth of GI mucosa
Cholecystokinin	Mucosal I-cells of upper small intestine	• Products of digestion (particularly proteins) acting on intestinal wall • Fatty acids in duodenum	• Digestive products moving to lower portions of GI tract	Causes: • Contraction of gallbladder to release alkaline bile • Secretion of pancreatic juices • Inhibits gastric emptying
Secretin	Mucosal S-cells in glands of upper small intestine	Products of protein digestion and acid in upper small intestine	• By neutralisation of acid entering small intestine	Causes: • Secretion of bicarbonate from pancreas • Decrease in gastric acid secretion • Contraction of pyloric sphincter
Somatostatin	D-cells in GI mucosa (as well as in pancreas)	• Acid in lumen		Inhibits: • Secretion of gastrin, secretin and others • Pancreatic exocrine secretion • Gastric acid secretion and motility • Gallbladder contraction
Ghrelin	Stomach cells	• Fasting	• Eating food	Functions to increase food intake, in opposite way to leptin (see Adipose Tissue above)

HORMONE FUNCTIONS

Hormones play a number of different roles in bodily function and maintaining homeostasis, which can be grouped into several categories. Each of these will be affected by several hormones, and each hormone may well influence more than one of these functions. The main groups of functions are:

- Reproduction and sexual differentiation (Chapter 16).
- Development and growth.
- Maintenance of the internal environment.
- Regulation of metabolism.
- Nutrient supply.
- Stress response.

Development and growth

In Chapter 17 you will consider the pattern of physical growth throughout life. In this chapter we are going to consider the role of a number of hormones in achieving normal growth and development. This depends on genetic makeup, but is also influenced by environmental factors and various aspects regulated by a number of different hormones. The major hormones involved are identified below; in addition, those hormones involved in regulation of these hormones are equally important.

- **Human growth hormone (hGH):** As its name suggests, hGH has a major role in regulating growth.
- **Insulin-Like Growth Factor-1 (IGF-1) and insulin:** IGF-1 is similar in structure to insulin. IGF-1 is a primary mediator of the effects of growth hormone and plays an important role in growth throughout childhood while continuing to have anabolic effects during adulthood. Children with inadequate insulin secretion may be short in stature (Laron, 2008).
- **Thyroid hormones:** These are necessary for normal growth although they do not initiate growth without hGH. They are particularly important for development of the bones and nervous system.
- **Calcitonin, PTH, vitamin D:** These are essential for normal calcium metabolism and, thus, for normal growth and calcification of bones.
- **Reproductive hormones:** These are necessary for normal development of reproductive organs and secondary sex characteristics of the adult body (Chapter 16).

Disorders of growth and development

Disturbances in secretion of any of these hormones can result in abnormalities of growth and development.

Restricted growth (sometimes referred to as dwarfism, a term now considered insensitive) is a condition characterised by short stature. In American literature it often refers to an adult who is 4 foot, 10 inches (147 cm) or less, although the average height of those affected is around 4 feet (122 cm), due to genetic or medical conditions (Mayo Clinic, 2018). There are many different causes, but they fall into two main groups:

- **Disproportionate:** with some body parts relatively larger or smaller than the rest of the body. An example is the genetic disorder, achondroplasia, in which the limbs are abnormally short and the head tends to be larger than average.
- **Proportionate:** with a normally proportioned, but small, body, such as pituitary dwarfism due to lack of hGH.

Excessive hGH secretion is usually due to a pituitary tumour:

- A child with excessive hGH secretion grows to a height between 7 feet (2.13 m) and 9 feet (2.74 m): this is known as gigantism.

An adult can develop acromegaly – swelling of soft tissue with enlargement of hands, feet, nose, lips and ears, and thickening of the skin. Parts of the face, jaw and skull also become enlarged. When this is caused by a tumour of the pituitary gland (pituitary adenoma), the tumour can be removed by neurosurgery, accessing the skull via the nasal cavity (transsphenoidal hypophysectomy).

MAINTENANCE OF THE INTERNAL ENVIRONMENT

In Chapter 1 the importance of homeostasis in maintaining the constancy of the internal environment is indicated and Chapter 2 considers the importance of the plasma membrane in regulating the movement of water and electrolytes between the ECF and ICF. Important hormones in this area are aldosterone and Antidiuretic Hormone (ADH) (see Figure 7.10) discussed mainly in Chapters 11 and 12.

Regulation of metabolism

In Chapter 9 we examine metabolism, i.e. anabolism and catabolism, of the different nutrients required. Earlier in this chapter the hormones involved in glucose metabolism are discussed and an outline of the main effects is shown in Figure 7.12. Blood glucose regulation is also discussed in detail in Chapter 9.

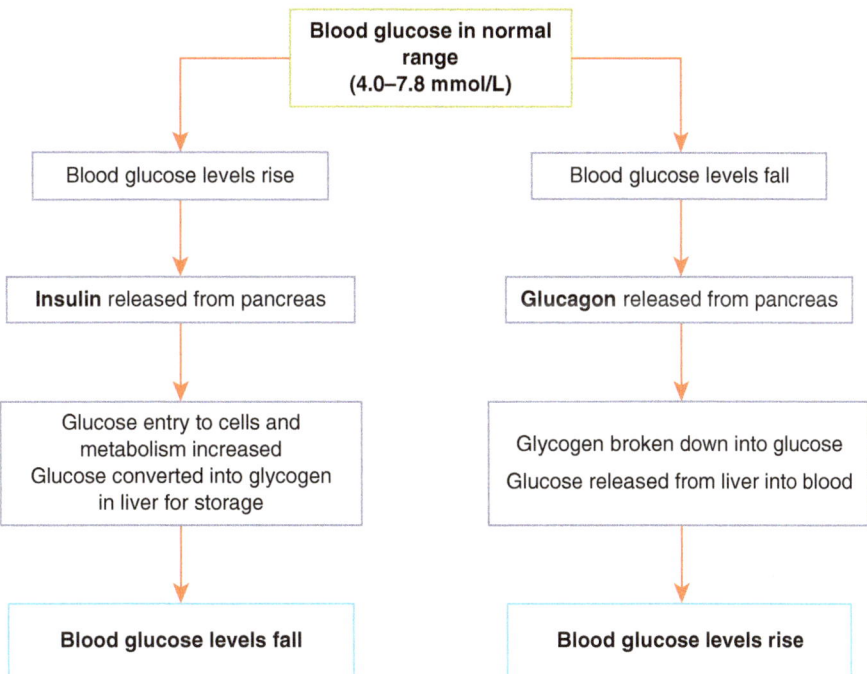

Figure 7.12   Regulation of blood glucose

Metabolism is increased by adrenaline, which initiates physiological changes, including increased heart rate, necessary for the flight or fight response. The thyroid hormones interact with catecholamines and enhance the response of the heart to catecholamines. In addition, the thyroid hormones have a major influence in regulating the Basal Metabolic Rate (BMR), i.e. the amount of calories needed by the body at rest. They affect most of the cells of the body in various ways including developmental (see above) and metabolism. The increase in metabolic rate increases oxygen consumption and heat produced by metabolically active tissues.

APPLY

Altered thyroid function

As with other endocrine glands, thyroid function can be increased above or decreased below normal. In adults with hypothyroidism (myxoedema - severe hypothyroidism) the metabolic rate is reduced and the person concerned feels the cold, is slower in physical and mental functioning and puts on weight, the hair is coarse and the voice husky. Babies who are hypothyroid before birth, if not treated appropriately, do not grow to expected milestones and can become intellectually disabled (originally known as cretinism, a term not now used).

The major change in hyperthyroidism is increased metabolic rate resulting in weight loss, intolerance of heat, nervousness, sweating and raised heart rate. Graves' disease is the commonest type of hyperthyroidism and is an autoimmune disorder. Often the person develops exophthalmos in which the eyeballs protrude (Ganong, 2005).

Nutrient supply

Table 7.4 earlier in this chapter identified the main hormones involved in digestion and absorption of nutrients (Chapter 8). However, the other area concerned in nutrient supply is appetite, which is also regulated by hormones. The two main hormones involved are:

- **Leptin:** mentioned earlier, this satiety hormone inhibits the sensation of hunger when the amount of fat stored in the body reaches the appropriate level. Leptin is secreted by the adipose cells and circulated in the blood. It reaches and activates the leptin receptors in the hypothalamus of the brain and reduces hunger.
- **Ghrelin:** also known as the 'hunger hormone', increases the sensation of hunger and increases the amount of food eaten. It is secreted from specialist cells mainly in the stomach and duodenum but also in various other organs in the body. The blood–brain barrier is not a barrier to ghrelin and it stimulates food intake by acting on 'hunger' centres, including in the hypothalamus (Diz-Chaves, 2011).

APPLY

Leptin and managing obesity

In the future, it is possible that the administration of leptin will be functional in the management of obesity by reducing appetite and the amount of food eaten. This could be a valuable tool in promoting public health across the world. Currently, the use of leptin has not been successful, as people who are overweight often have leptin resistance.

Response to stress

The body responds to stress in two main ways. The factors which can cause stress are known as stressors and can be a range of different internal or external factors that disrupt homeostasis. Examples can include:

- environmental stressors (noise, pollution, overcrowding);
- events in daily living (commuting to work, lost keys);
- life events (divorce, bereavement);
- stressors at work (high demand with low control in the job, high levels of physical exertion);
- chemical stressors (e.g. tobacco, alcohol, drugs);
- social and psychological events (family demands, anxiety, health worries).

Acute stress response

The acute stress response is also known as the flight or fight response and is regulated by the sympathetic nervous system, which stimulates the secretions of the adrenal medulla gland: adrenaline and noradrenaline (catecholamines) (Figure 7.13). These hormones act rapidly in providing energy to deal with the physical challenges imposed. The functions that enable a strong and rapid response are increased, such as the heart rate, respiratory rate and depth, the blood supply to muscles, and release of glucose from glycogen in the liver and muscles. Blood supply to organs that do not contribute to this response, such as the gastrointestinal tract, the skin and the kidneys, is reduced.

The plasma half-life of adrenaline is two to three minutes, so the physiological changes only last a short time.

Long-term stress response

Some stressors, such as illness or anxiety about one's children, can last for a longer period of time than the energy released by adrenaline in the acute stress response will last. Thus, a mechanism is needed that will ensure an adequate energy source for this longer period. Glycogen stores will last for several hours but after that, other energy sources are necessary for metabolism.

The hypothalamic hormone corticotrophin releasing hormone (CRH) stimulates the release of adrenocorticotrophic hormone (ACTH) from the anterior pituitary gland; ACTH initiates the release of corticosteroid hormones, mainly glucocorticosteroids, from the adrenal cortex (Figure 7.13). These stimulate gluconeogenesis, mainly from protein, for neural cells which require glucose. Fat metabolism provides energy for most other tissues of the body in long-term stress. The mineralocorticoid actions of glucocorticoids and aldosterone both cause sodium and water retention resulting in increased blood volume and raised blood pressure.

Glucocorticoids also inhibit the function of the immune system and act as anti-inflammatory agents. While this can be helpful in limiting inflammation due to disease or after injury, it can also be harmful if prolonged due to its immune suppression action and can permit spread of infection.

Altered corticosteroid levels

As with other endocrine organs, hypo- and hypersecretion of corticosteroids can occur. In Cushing's syndrome (hypersecretion), the fat storage areas shift, leading to a puffy face and neck. It also results in hyperglycaemia (raised blood glucose). Addison's disease results from hyposecretion, which gives rise to bronzing of the skin, reduced electrolyte levels and hypoglycaemia.

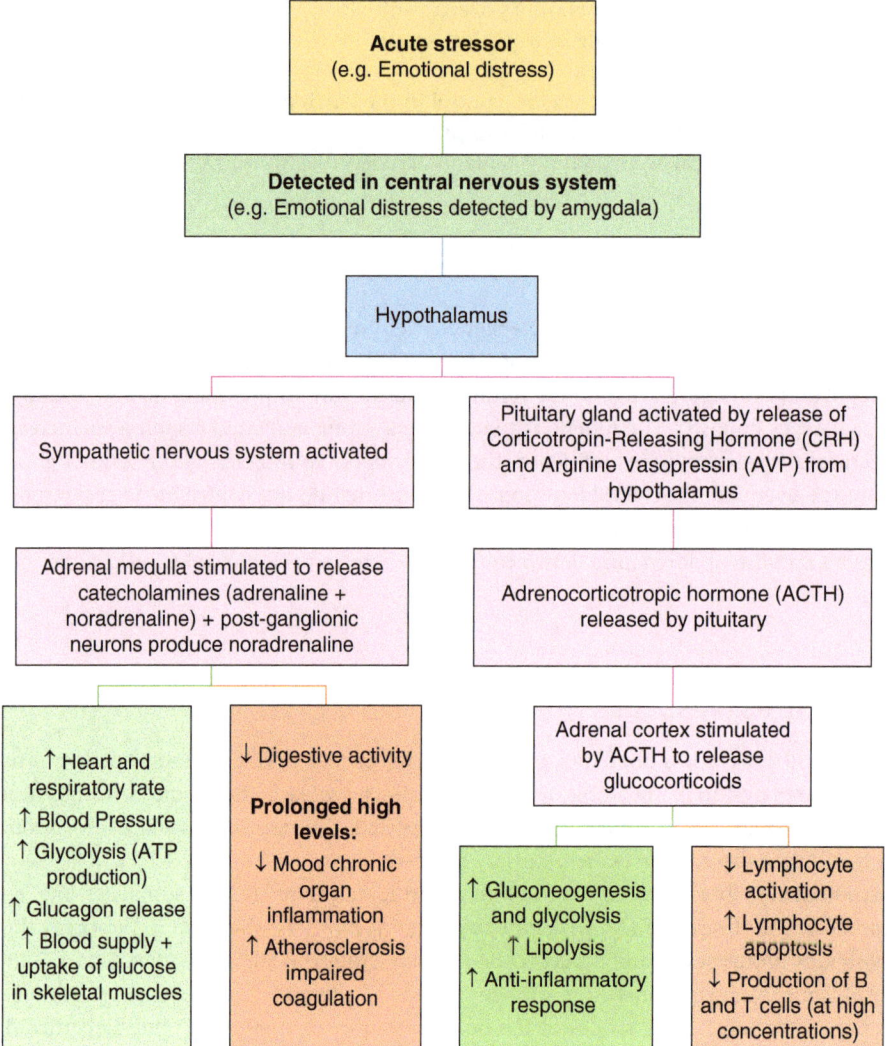

Figure 7.13   Acute stress response

APPLY

Stress and person-centred nursing

In person-centred nursing, the concept of stress is really important. People in our care and their families may have a wide range of stressors that last for a long time. Working with people to minimise stress can improve their physical and psychological state and have significant positive effects on their well-being (Rice, 2000). One example is that pre-operative psychological preparation reduces the incidence of post-operative infection (Boore, 1978). However, this is dependent on the information being delivered in a therapeutic and individualised way, tailored to the person's need, to reduce their anxiety.

CONCLUSION

This chapter has examined the endocrine system and how it works in partnership with the nervous system in regulating body function. In addition, there are numerous paracrine, autocrine, juxtacrine and intracrine substances that are also involved in regulating cell function, although we have not examined these in any detail. The ways in which hormones are regulated and the main organs that secrete hormones have been discussed, and the main functions of the body in which endocrine regulation is important are identified and considered, although the reproductive systems are considered in Chapter 16. These functions include: development and growth; maintenance of the internal environment; regulation of metabolism, nutrient supply, and the stress response. The understanding gained will enable the reader to understand how the endocrine system operates in supporting health and well-being through its impact on physical and mental health. This approach facilitates nurses in placing their knowledge and understanding within the perspective of the PCN Framework.

GO DEEPER

Further reading

Braine, M.E. (2009) 'The role of the hypothalamus, part 1: the regulation of temperature and hunger', *British Journal of Neuroscience Nursing*, 5(2): 66-72.

Braine, M.E. (2009) 'The role of the hypothalamus, part 2: regulation of sleep-wake cycles and effect on pain', *British Journal of Neuroscience Nursing*, 5(3): 113-20.

Hendry, C., Farley, A., McLafferty, E. and Johnstone, C. (2014) 'Endocrine system: part 2', *Nursing Standard*, 28(39): 43.

Johnstone, C., Hendry, C., Farley, A. and McLafferty, E. (2014) 'Endocrine system: part 1', *Nursing Standard*, 28(38): 42.

- Development and growth: growth and development are dependent on a range of factors, including nutrient intake, and are discussed in Chapter 17. A range of hormones are involved in different aspects of growth and development, with growth hormone and thyroid hormones playing the most central roles.

- Maintenance of the internal environment: regulation of the fluid and electrolyte balance is essential for normal cell function. Two major hormones involved are aldosterone and ADH (vasopressin) and their joint function significantly influences blood volume and blood pressure (Chapters 11 and 12).

- Regulation of metabolism through the processes of anabolism and catabolism are discussed in Chapter 9. The hormones of the pancreas, particularly insulin and glucagon, play an important role in regulating glucose metabolism. However, thyroid hormones regulate the basal metabolic rate thus modifying the rate at which nutrients are used at rest: body weight, heart and respiratory rates are indicative of the intensity with which the body is functioning.

- Nutrient supply: an adequate but not excessive intake of nutrients is necessary for health. Two hormones play an important role in regulating appetite:

  o Leptin (from fat cells) reduces hunger.

  o Ghrelin (secreted by the stomach) increases hunger.

  It is possible that future research developments will contribute to the management of conditions such as obesity and anorexia nervosa. The hormones which regulate gastrointestinal activity are discussed in Chapter 8.

- Stress response: knowledge and understanding of the stress response is particularly important for nurses. Many (possibly most) people will have stressors influencing their physiological function, as well as their behavioural response. Family members may also have major concerns which act as stressors. Nursing approaches to minimise stress can have important implications for promoting health by reducing infections, lowering blood pressure, and generally improving the quality of life.

REVISE

TEST YOUR KNOWLEDGE

Revision of this chapter is important as the endocrine system is one of the key elements in controlling body function. There are three key areas to revise:

1 The ways in which hormones function, the types of hormone, and the ways in which their secretion is controlled.

2 The organisation of hormone secretion.

3 The functions of the hormones.

In order to help you revise, consider the following questions, answers for which can be found by visiting **https://study.sagepub.com/essentialap2e**.

Test yourself by revising the chapter first, and then answering these questions without looking at the book. Some answers may need knowledge from different parts of the chapter. Afterwards compare your answers with the text, the figure and the notes you made. Did you miss anything? Here are the questions:

1   Define hormones and identify the different types within the human body, giving an example of each.

2   Identify the ways in which endocrine secretion is controlled and give examples of each.

3   Outline the major functions of the endocrine system and identify how these differ from the nervous system.

4   Describe the role of the endocrine system in the maintenance of homeostasis.

5   Label the names of the endocrine glands shown in Figure 7.14.

6   Identify five organs that all have a range of other functions but which also have endocrine function and identify at least one hormone secreted by each.

7   Outline how blood glucose levels are regulated.

8   Identify the different hormones and their roles in control of growth and development.

9   Outline the stress response and the hormones involved. Indicate the importance of understanding this for person-centred nursing.

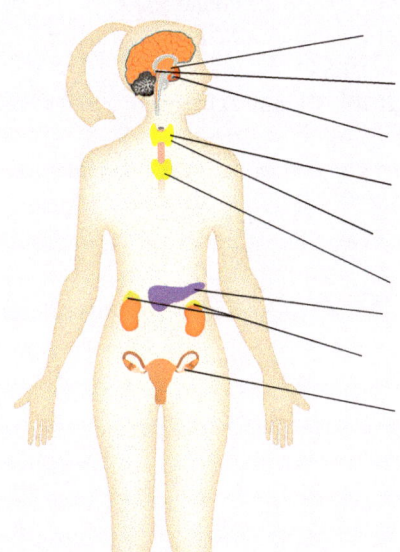

REFERENCES

Bensenor, I.M., Olmos, R.D. and Lotufo, P.A. (2012) 'Hypothyroidism in the elderly: diagnosis and management', *Clinical Interventions in Aging*, 7: 97–111.

Boore, J.R.P. (1978) *Prescription for Recovery*. London: Royal College of Nursing.

Diz-Chaves, Y. (2011) 'Ghrelin, appetite regulation, and food reward: interaction with chronic stress', *International Journal of Peptides*, Article ID 898450.

Ganong, W.F. (2019) *Review of Medical Physiology*, 26th edn. New York: Lange Medical Books/ McGraw-Hill.

Hyppönen, E. and Power, C. (2007) 'Hypovitaminosis D in British adults at age 45 y: nationwide cohort study of dietary and lifestyle predictors', *The American Journal of Clinical Nutrition*, 85(3): 860–8.

Laron, Z. (2008) 'Insulin: a growth hormone', *Archives of Physiology and Biochemistry*, 114(1): 11–16.

Marchevsky, A.M. and Wick, M.R. (2014) 'The thymus gland', in *Pathology of the Mediastinum*. Cambridge: Cambridge University Press, pp. 37–50.

Mayo Clinic (2018) *Dwarfism*. Available at: www.mayoclinic.org/diseases-conditions/dwarfism/basics/ definition/CON-20032297?p=1 (accessed 4 May 2020).

Namba, K., Kitaichi, N., Nishida, T. and Taylor, A.W. (2002) 'Induction of regulatory T cells by the immunomodulating cytokines α-melanocyte-stimulating hormone and transforming growth factor-β2', *Journal of Leukocyte Biology*, 72: 946s–952.

Rice, V.H. (2000) *Handbook of Stress, Coping and Health: Implications for Nursing Research, Theory and Practice*, 9th edn. Thousand Oaks, CA: Sage Publications.

UK Chief Medical Officers (2012) *Vitamin D: Advice on Supplements for at Risk Groups. Letter from UK Chief Medical Officers*. London: DHSS.

Zhang, R. and Naughton, D.P. (2010) 'Vitamin D in health and disease: current perspectives', *Nutrition Journal*, 9: 65.

Zouboulis, C.C. (2009) 'The skin as an endocrine organ', *Dermatoendocrinology*, 1(5): 250–2.

PART 3

PRESERVATION OF THE INTERNAL ENVIRONMENT

The chapters in this section are all concerned with the maintenance of homeostasis – that is the stability of the internal environment for the cells of the body. It consists of the following five chapters:

- *Chapter 8. The Digestive System: Nutrient Supply and Waste Elimination*

  This system is composed of the gastrointestinal tract (GIT) and accessory organs. The GIT is essentially still part of the external environment passing from the mouth to the anus, where waste products are eliminated. This chapter describes how foodstuffs are taken into the body, broken down into small molecules of nutrients and absorbed into the internal environment of the body.

- *Chapter 9. Metabolism and Liver Function*

  The key requirement for bodily function is energy and the previous chapter has discussed how nutrients – the source for energy – enter the body. In this chapter, we consider how they are metabolised: broken down to provide energy (catabolism) and built up into new structures (anabolism). The liver plays an important role in metabolism but also has a number of other functions, which are discussed here.

- *Chapter 10. The Respiratory System: Gaseous Exchange*

  The other essential requirement for effective energy production is oxygen, used in the formation of ATP from the breakdown, primarily, of glucose. The respiratory system is responsible for getting oxygen into the body and removing carbon dioxide.

- *Chapter 11. The Renal System: Fluid, Electrolyte and Acid–Base Balance*

  One of the essential functions in physiology is to maintain the stability of the constituents of the fluid elements of the body. Water makes up a very large proportion of the body: by weight an average human male is approximately 65% water. The proportion varies with age, amount of fat, and gender: 73% of a baby down to 45% in obese people. The balance of electrolytes and the acid–base balance of the fluid in the different compartments of the body is essential for normal cell function. The renal system plays a major role in regulating these functions.

- *Chapter 12. The Cardiovascular and Lymphatic Systems: Internal Transport*

  While all these areas are important, the movement of these substances around the body is also essential and is carried out by these two connected systems. In the cardiovascular system, substances are carried in the blood, through a system of tubes (the circulation), pumped by the heart. The lymphatic system gathers up excess fluid from the tissues and returns it to the circulation.

THE DIGESTIVE SYSTEM

NUTRIENT SUPPLY AND WASTE ELIMINATION

UNDERSTAND: CHAPTER VIDEO

Before reading this chapter, you might find it useful to have a quick overview of the digestive system. Watch the following online video to enhance your understanding.

This video can be accessed by **scanning the QR code** with your smart phone camera or via https://study.sagepub.com/essentialap2e.

THE DIGESTIVE SYSTEM (14:43)

LEARNING OUTCOMES

When you have finished studying this chapter you will be able to:

1. Identify the macro- and micronutrients, specifying their roles in maintaining homeostasis
2. Identify the structure and function of the different parts of the digestive system
3. Discuss how the digestive system plays a role in health and well-being

INTRODUCTION

The digestive system represents a vital link between the external and internal environments. Its contribution to homeostasis is through the provision of nutrients to the internal environment and the removal of waste products. This chapter examines the variety of nutrients in the body and their key roles. We then study the structure of the digestive system, relating the functions of the different parts of the tract to undertake these functions. In Chapter 4 we considered the importance of the microbes in the gut in promoting immune function and health of the digestive system. In this chapter, issues related to the digestive system and individual well-being will be considered, including differing requirements at stages of life and the importance of fibre in maintaining function.

Nutrients provide the resources needed for life – the building blocks of the human body. They also influence our experience of life, the pleasures of food, the social nature of eating with family and friends, and the enhancement that food can provide to our mood. The digestive system is central to obtaining nutrients within this social framework: we organise our day around our appetite and social interactions, we seek privacy for elimination, and we feel up to life and its challenges when we feel nourished.

In this chapter, the need for a balanced intake of nutrients will be explored, including the roles of nutrients. However, you will also be challenged to think of the variety of factors that make a healthy and nutritious diet possible. As a person-centred nurse, understanding these matters is crucial to understanding how homeostasis is maintained. This knowledge underpins your activities in health education and promotion, as well as ensuring that the care you provide supports people in having a healthy nutritional intake and optimally functioning digestive system.

Context

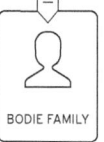

BODIE FAMILY

Think of the Bodie family. Danielle is growing at a rapid pace, requiring nutrients for energy, as well as to build new cells for growth and development. She is reliant on her parents to provide these to her and they must bear in mind that her digestive system is not fully developed. At the age of two months she is currently breast-fed and her needs are fully met through breast milk. As she grows older, her parents will have to modify her diet to meet her growth and development needs. Additionally, Danielle has not yet developed continence, requiring her parents to assist with keeping her skin clean and healthy after elimination.

At the other end of the age spectrum, George and Maud have well-established patterns of eating, although their appetites may be reduced due to a reduction in their activity. While they are still active and well now, in years to come they may face challenges with the integrity of their teeth for eating, their ability to access shops that supply healthy foods, and their dexterity to prepare food in a healthy way. Their children and grandchildren lead hectic lives. This can result in them skipping meals or eating on-the-go with convenience foods. Neither of these will provide optimal nutrition, nor will they provide the necessary social exchanges that add to health-related quality of life.

NUTRIENTS

Nutrients are substances that are ingested, digested, absorbed and metabolised to maintain homeostasis through their roles in structure and function within the body. There are many nutrients needed, all with specific purposes. This variety means that people require a mixed and balanced diet in order to support the maintenance of homeostasis, and therefore health. Insufficient nutrient intake will lead to malnutrition and ill health, and excess intake of nutrients can result in toxicity and obesity, also leading to a variety of health problems. We need nutrients to physically survive, but we also need them to enable us to live happy and productive lives. By preserving optimal health and homeostasis, a balanced and healthy nutritional intake enables people to feel good and preserve their independence, providing them with the energy to engage in activities that are central to their social and psychological well-being.

APPLY

Healthy eating in the first 1,000 days of a child's life

In recent years a campaign has arisen to focus on healthy eating in the first 1,000 days of a child's life, from conception through the child's second birthday. This period is considered to be the peak phase of vulnerability from a nutritional and developmental perspective (Ruel and Alderman, 2013). During this time, key phases of brain growth and development occur, and nutritional deficits during this period are associated with increased mortality and morbidity (Ruel and Alderman, 2013).

ACTIVITY 8.1: APPLY

Visit the website **Thousand Days** which emphasises the importance of healthy eating in the first 1,000 days of a child's life.

A link to the website can be accessed via http://study.sagepub.com/essentialap2e.

Getting the best from nutrients – metabolism

In Chapter 9 you will learn in detail about metabolism. Metabolism, briefly, refers to all of the organic and chemical reactions in the body, of which there are two main types:

- **Anabolism:** the building up or synthesising of large and complex molecules. This process requires energy.
- **Catabolism:** the breaking down of substances to provide energy and raw materials for anabolism.

In essence, metabolism uses all of the nutrients that the body takes in to create molecules for structure, chemical reactions, and for energy. This chapter will focus more on the nutrients themselves and their roles.

There are two main groupings of nutrients: macronutrients (macro means large) and micronutrients (Table 8.1).

Table 8.1 Nutrient groups

Macronutrients	Micronutrients
• Water	• Vitamins
• Carbohydrates	• Minerals
• Proteins	• Trace elements
• Fats	

Both macro- and micronutrients are essential in maintaining homeostasis. We will now look at these individually.

Macronutrients

Carbohydrates

Carbohydrates contain carbon (C), oxygen (O) and hydrogen (H) (see Chapter 9 for structures). Carbohydrates are usually the key source of energy for the body, with 40–45% of calories in the diet coming from the different types of carbohydrates (Goodman, 2010). They are also key components of other molecules in the body such as nucleic acids (e.g. DNA), glycoproteins and glycolipids and can be converted into amino acids and fats for storage. Most of the body uses carbohydrates as a source of energy alongside fats; however, neurons (nerve cells) rely primarily, and erythrocytes (red blood cells) solely, on carbohydrates for energy. The term saccharide, meaning sugar, is used to describe carbohydrate groups, with monosaccharides being the simplest form. As more saccharides link together, they become more complex and have higher molecular weights, taking longer to digest. There are four main categories of carbohydrates described in the following ways:

- **Monosaccharides:** These are often referred to as simple carbohydrates and include glucose, galactose and fructose. They are easily digested and absorbed in the body.
- **Disaccharides:** These are the most common types of carbohydrates consumed in the diet (30–40%) (Goodman, 2010) and include sucrose and lactose. They take longer to break down and are composed of two monosaccharides. Both monosaccharides and disaccharides are commonly referred to as sugars or simple carbohydrates
- **Oligosaccharides:** These carbohydrates are composed of approximately three to nine monosaccharides. These are used in the body to make glycoproteins and glycolipids and one of their main functions is in cell recognition, for example determining blood group.
- **Polysaccharides:** These carbohydrates are made up of many monosaccharides, usually more than ten, and are often used as a form of energy storage, such as glycogen in the liver and muscle cells.

Carbohydrates are stored in the body as glycogen. Glycogen stores are found in the liver, skeletal muscles and also in cells of the brain, primarily astrocytes.

Dietary fibre (non-starch polysaccharide, or cellulose) is largely composed of oligosaccharides and polysaccharides. It is mainly found in fruit and vegetables and is relatively indigestible. Fibre is classified as being either soluble in water or insoluble, with soluble fibre having the greatest health benefits (Table 8.2).

Table 8.2 Fibre classifications

Soluble fibre	Insoluble fibre
<ul><li>As it dissolves in water, it ferments and produces gaseous by-products</li><li>Forms viscous solutions that prolong gastric emptying. This prevents the movement of glucose, triglycerides and cholesterol across the intestinal wall into the blood stream (Kaczmarczyk et al., 2012)</li><li>Enhances glucose tolerance and delays gastric emptying (Kaczmarczyk et al., 2012)</li><li>Provides bulk to faeces, helping to improve intestinal motility (Suares and Ford, 2011)</li></ul>	<ul><li>Adds bulk to faeces and promotes elimination of toxins and waste by stimulating peristalsis</li><li>Insoluble fibre from cereal and vegetable sources reduces the risk of coronary heart disease and cardiovascular disease (Threapleton et al., 2013)</li></ul>

Both types of dietary fibre play an important role in the digestive system. Table 8.2 indicates that fibre provides bulk that enhances intestinal motility, enabling food and waste to move through the digestive tract. It also helps to reduce sugar, fat and cholesterol absorption, contributing to a healthier diet.

As soluble fibre is fermented by the microbes in the gut, it produces energy. Fibre also can prevent over-eating as it delays gastric emptying and provides a sense of satiety due to its bulk (Smith, 2012).

Proteins

Proteins contain carbon (C), oxygen (O), hydrogen (H) and nitrogen (N) (see Chapter 9 for structures). Some also contain minerals such as sulphur, iron, copper, zinc and phosphate. Adults require, on average, 70–100 g of dietary protein daily (Goodman, 2010). Proteins are made up of amino acids. Globular amino acids are needed in the body to make enzymes and hormones; fibrous amino acids are used in structures such as cell membranes, collagen, and keratin in the skin and hair. There are 20 amino acids in the human body, nine of which are essential, meaning they cannot be created/synthesised in the body (Table 8.3). A further six are conditionally essential amino acids, meaning their production can be inhibited by certain pathophysiological conditions. Two of these can only be created from essential amino acids. The remaining five are non-essential, meaning they can be synthesised in the body. Naturally, this means that all the essential amino acids must be in the diet. A complete protein contains all 20 amino acids, an incomplete protein is missing one or more. The body cannot store proteins or amino acids. This means that they must be readily available when needed, making them an essential part of the daily diet.

Table 8.3   Essential and non-essential amino acids

Essential amino acids	Conditionally essential amino acids	Non-essential amino acids
• Isoleucine	• Arginine	• Alanine
• Leucine	• Cysteine (from methionine)	• Asparagine
• Lysine	• Glycine	• Aspartate (aspartic acid)
• Methionine	• Glutamine	• Glutamate (glutamic acid)
• Phenylalanine	• Proline	• Serine
• Threonine	• Tyrosine (from phenylalanine)	
• Tryptophan		
• Valine		
• Histidine (infants)		

Structurally, amino acids combine together to form peptide bonds (Chapter 9). Two amino acids combine to form dipeptides, three for tripeptides, and polypeptides contain many amino acids. These peptides are termed proteins when polypeptides combine. The order of how amino acids combine in proteins is determined by DNA (Chapter 3).

Proteins are an important source of nitrogen in the body, with urea containing nitrogen being a by-product of the breakdown of amino acids. Nitrogen is used in the body if needed to form non-essential amino acids, or excreted. In children, their rate of growth and development requires more nitrogen than in adults, resulting in children having a positive nitrogen balance (i.e. more nitrogen entering the body than leaving it). Proteins, of which there are over 140,000 different types, are also a source of energy for the body. However, they are only used by the body when there is no carbohydrate or fat available for energy (Chapter 9). Protein is largely sourced from meat, eggs and dairy products.

Fats (lipids)

Fats also contain carbon (C), oxygen (O) and hydrogen (H) (see Chapter 9 for structures). However, their chemical structure is different from carbohydrates. Fats represent the largest store of energy in the body, being stored as adipose tissue. They are an important source of energy in the body as they provide over twice as much energy as carbohydrates. Fats are considered to be glucose and protein sparing in that they can be used as the source of energy, preserving proteins from being used for energy and making glucose available for neurons and erythrocytes to use. Fats are also a vital source of fat-soluble vitamins, A, D, E and K, dissolved in fats for absorption across the intestinal wall into the body. Fats also have a structural role as an essential component of the cell membrane and the myelin sheath, an insulating layer around neurons.

Most fatty acids can be synthesised within the body, but some are essential fatty acids and are required from the diet as they are not produced in sufficient amounts in the body. Saturated fats are those that are solid at room temperature, unsaturated (i.e. with some double bonds between carbon atoms) are normally oils at room temperature. The former mainly come from animal sources, the latter from plants and vegetables. Fats are largely stored in the body in the form of triglycerides in adipocytes in adipose tissue.

There are three main types of fats:

- **Triglycerides:** These are composed of one molecule of glycerol joined with three fatty acid molecules, hence the name. They are important in providing insulation and energy storage within the body.
- **Phospholipids:** These fats are the structural components of cell membranes. They are made up of one glycerol molecule, two fatty acids, and a phosphate group.
- **Lipoproteins:** Lipoproteins are triglycerides and cholesterol units combined with proteins and phospholipids. This enables them to be recognised within the body and to remain in suspension in the blood. The most well-known lipoprotein is a cholesterol derivative. There are four classifications of lipoproteins:
  - **Chylomicrons:** Easily absorbed substances created in the small intestine to aid absorption. These are converted to triglycerides in adipocytes (fat cells).
  - **High Density Lipoproteins (HDLs):** HDLs mop up cholesterol in the blood, bringing it to the liver to be eliminated in bile.
  - **Low Density Lipoproteins (LDLs):** LDLs are largely composed of cholesterol and are absorbed by cells when they need cholesterol. The cells use enzymes to break down the lipoprotein to release the cholesterol.
  - **Very Low Density Lipoproteins (VLDLs):** Precursors to LDLs, these lipoproteins have their triglycerides removed in adipocytes and become LDLs.

ACTIVITY 8.2: UNDERSTAND

Watch the following video clip online to further your understanding of lipoproteins. You may need to take your time with this video and perhaps go back over some parts of it.

This online video can be accessed by **scanning the QR code** with your smart phone or via https://study.sagepub.com/essentialap2e.

LIPOPROTEINS (11.04)

APPLY

LDLs and HDLs and cardiovascular disease

Lipoproteins carry cholesterol, and so their levels are associated with cardiovascular disease. LDLs contain high amounts of cholesterol, and HDLs low amounts. LDLs bring cholesterol to cells for hormone or cell membrane synthesis. However, when present in high amounts, LDLs can also deposit cholesterol in the endothelium of arterial walls, leading to atheromatous plaque formation. Largely, it is considered healthy to have a high amount of HDLs (>3.3 mmol/L) to mop up excess cholesterol, and to have low amounts of LDL to prevent deposits in the arterial walls. It is considered optimal to have LDL levels less than 3 mmol/L and to have HDL levels greater than 1 mmol/L (Bond, 2014). Studies are ongoing, as looking at the LDL levels alone may not be an appropriate indicator for cardiovascular disease risk, rather a percentage reduction may be a more appropriate target (Cowan, 2014). However, it is important to remember that cholesterol is essential in the body and so lowering LDL levels too much can leave insufficient amounts of cholesterol available for use by cells.

Statins are drugs that can be given to reduce cholesterol levels. These drugs inhibit an enzyme responsible for cholesterol formation, known as HMG-CoA reductase, and subsequently causes an increase in synthesis of LDL receptors. LDL clearance increases, lowering LDL levels and thus cholesterol mobilisation (Cowan, 2014). There are many natural ways to lower cholesterol (Bond, 2014):

- Reduce saturated fat and favour non-saturated fat in the diet.
- Increase soluble fibre in the diet to lower cholesterol absorption.
- Increase oat content in the diet – beta-glucan in oats reduces LDL.
- Choose nuts as a snack – nuts with skins (pecans, walnuts, almonds, peanuts and pistachios) are high in soluble fibre and unsaturated fats.
- Choose natural foods with natural plant stanols and sterols to lower LDL. Stanols and sterols are chemically different in structure, but both are effective in lowering LDLs.
- Exercise and stop smoking to raise HDL levels.

As a person-centred nurse, you need to consider all these factors alongside culture, access to education and people having the finances to afford these healthy choices.

Micronutrients

Vitamins

Vitamins are substances required in small quantities and necessary for metabolism. They largely cannot be synthesised in the body, except for vitamin D (Chapter 14), which goes through a variety of processes to be created in the body and is also considered a hormone as well as a vitamin. There are two main groups of vitamins:

- **Water-soluble vitamins:** These are readily absorbed into the body and any excess excreted, largely in urine.
- **Fat-soluble vitamins:** The absorption of these vitamins, A, D, E and K, is dependent on fat absorption and they are stored in the body. In excess they can be harmful.

GO DEEPER

Vitamin A toxicity

Polar bear liver contains very high amounts of vitamin A, considerably greater than in humans and within a toxic level for humans. If eaten by humans, it can cause death.

APPLY

Liver food products and pregnancy

Pregnant women are advised to avoid eating liver products as they contain high amounts of vitamin A, which can lead to fetal toxicity (the maximum amount should be 10,000 IU daily). Liver is effectively the only dietary source of vitamin A likely to lead to such an occurrence (WHOFAOUN, 2004).

APPLY

Vitamin D deficiency and sunscreen

In recent years it has become apparent that sunscreen can lead to vitamin D deficiency, particularly in countries like New Zealand and Australia where there is a high usage. Recommended sunscreens mostly create a layer on the skin that is thicker than 2 mg/cm² (Faurschou et al., 2012). Above this thickness vitamin D production may be prevented (Faurschou et al., 2012). This has led to use-of-sunscreen guidelines coming under scrutiny in order to ensure sufficient exposure to the sun to allow adequate vitamin D to be synthesised.

There are a number of vitamins utilised in the body, all with a variety of functions (Table 8.4).

Table 8.4 Vitamins and their roles

	Vitamin	Role
Fat soluble	A (retinol)	Necessary for cell division and is central to healthy skin and hair. Also has a role in vision in low light. Vitamin A has a role in promoting gastrointestinal health, particularly in children, and also in promoting an optimal immune system. In addition, it is involved in the maintenance of epithelial cellular integrity (WHOFAOUN, 2004)
	D	Promotes absorption and use of calcium and phosphate for healthy bones and teeth. Role in controlling the activity of monocytes, macrophages, lymphocytes, and epithelial cells, contributing to protecting the body from pathogens (Herr et al., 2011). Also influences expression of genes that have roles in preventing neurological and mental health disorders, most notably multiple sclerosis (DeLuca et al., 2013)
	E	An antioxidant that protects cell membranes. Also has been identified as neuroprotective in recent years, preventing neurotoxicity (Pace et al., 2010)
	K	Required for blood coagulation and synthesis of proteins found in plasma, bone, and the kidneys
Water soluble	C (ascorbic acid)	An antioxidant used in the synthesis of collagen and protects cells from the damage of oxygen free radicals. Supports vitamin D in its role in immunity
	B_1 (thiamine)	Needed for the release of energy from carbohydrate metabolism. Supports neural function, particularly as neurons use carbohydrate as a source of energy
	B_2 (riboflavin)	Necessary for carbohydrate and protein metabolism. Also has a role in tissue homeostasis, particularly in the skin and eyes
	B_3 (niacin)	Used in releasing energy from carbohydrates during metabolism. Inhibits cholesterol production and aids fat breakdown. Currently being investigated as an up-and-coming form of cholesterol-lowering medication with early indications showing promise (Gouni-Berthold and Berthold, 2013)

Vitamin	Role
B$_6$ (pyridoxine)	Amino acid metabolism and synthesis of non-essential amino acids. Also has a protective role in reducing the damage caused to cells by oxidative stress
B$_{12}$ (cobalamin)	Essential in nucleic acid synthesis, the formation and maintenance of myelin (the protective sheath that insulates neurons) (Stabler, 2013)
Folate (folic acid)	Essential in nucleic acid synthesis and cell division. Has a distinct role in neural tube development in the embryo

Minerals and trace elements

Minerals are chemical substances needed for a variety of reasons in the body, largely to do with structure, fluid balance, nervous and muscular activity and blood clotting. Table 8.5 outlines the main minerals and their roles.

Table 8.5 Minerals and their roles

Mineral	Role
Calcium	Essential for neural functioning, including the release and generation of neurotransmitters necessary to carry nervous impulses from one nerve to another. Also essential in bone structure and density. Necessary for erythrocyte production (Gupta et al., 2014)
Phosphorus	As for calcium, essential for bone structure and density. Also essential in the nervous system as it has a key role in nervous impulses. Used in balancing the pH of the blood, assisting in alkalising blood when necessary (Gupta et al., 2014)
Sodium	Essential for muscular contraction and transmission of nervous impulses. Helps to maintain fluid balance
Potassium	Involved in the synthesis of proteins, intestinal motility, and necessary for generating nervous impulses. Necessary for erythrocyte production (Gupta et al. 2014). Essential for muscular contraction and transmission of nervous impulses. Helps to maintain fluid balance
Magnesium	Necessary for erythrocyte production (Gupta et al., 2014)
Iron	Necessary for the transport of oxygen in the blood. Also needed for nucleic acid synthesis. Has a role in activating enzymes modulating neurotransmitters (Gupta et al., 2014)
Iodine	Involved in the synthesis of thyroxine and triiodothyronine. These hormones are necessary for metabolism, growth and development. Essential for the development and function of the nervous system (Zimmermann, 2011)
Zinc	Necessary for cell division, replication and enzyme activity. Has a central role in cells of the immune system, promoting optimal function

ACTIVITY 8.3: APPLY

Nutrients, mental health and development

Think about the nutrients you have just explored. Can you make any links between nutrients and a person's mental health? Can you think which nutrients may be more important in a child and why?

Exploring the web will help you to make these links.

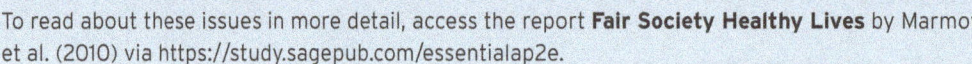

ACTIVITY 8.4: APPLY

Accessing nutrition and health

Now that you have read about the nutrients needed in the body, how easy is it to obtain these nutrients in healthy quantities? There are various government recommendations about a balanced, healthy diet, but there are a variety of social and physical challenges that can result in a less than favourable diet. These include: low income; difficulty accessing affordable, healthy foods; and a lack of education on what to eat and how to prepare foods healthily. Children will learn from their parents and from what is socially promoted and accessible. Children can then take these habits and learned behaviours with them into adulthood. Unhealthy food is often cheaper than healthy food. Being able to access food can be difficult for those with no transport, people with mobility problems and those who live in isolated settings.

To read about these issues in more detail, access the report **Fair Society Healthy Lives** by Marmot et al. (2010) via https://study.sagepub.com/essentialap2e.

THE DIGESTIVE SYSTEM AND HOMEOSTASIS

The digestive system (GIT and accessory organs) is a key contributor to homeostasis as it ensures that food is digested and absorbed so that an adequate nutrient supply is available for the cells of the body. Without these, the body does not have the resources to grow and repair or to undertake the necessary chemical reactions. The digestive system's primary role is to convert nutrients ingested in the diet into a form where they can be readily absorbed into the body. The structure of the digestive system is adapted at various points to achieve this, and we will explore this as we work through the different components of the system. In addition, the digestive system eliminates waste composed of nutrients that have not been digested and absorbed from the digestive tract. The final contribution to homeostasis is protection. The lymphoid tissue contained within the digestive tract comprises 70% of the body's immune system (Chapter 13).

There are five key activities of the digestive system:

- **Ingestion:** the taking in of food into the system.
- **Propulsion:** moving food along the full length of the GIT.
- **Digestion:** breaking down food into individual nutrients both mechanically and chemically.
- **Absorption:** once nutrients are broken down to a state where they can be absorbed, they cross the intestinal wall into the blood and lymphatic circulation.
- **Elimination:** the removal of undigested and unabsorbed food from the digestive tract in the form of faeces.

Structure and function of the digestive system

Technically, the digestive tract is outside the body and therefore part of the external environment: it is essentially a tube through the body with the purpose of moving nutrients from the external environment into the internal environment. The digestive tract runs through the body from the mouth to the anus, supported in its role by a number of accessory organs. The tract itself is modified along its length for specific functions. The components are identified in Figure 8.1 and we will explore each in the order of the digestive journey of food through the system.

Tubular structures are muscular, enabling them to move contents within and through them, and the digestive tract uses this movement to mechanically digest (i.e. break down into smaller particles) and propel contents along its length. Alongside this, enzymes are secreted to chemically digest the nutrients. Digestion is supported by the actions of accessory organs, such as the liver (Chapter 9), gallbladder and pancreas (Figure 8.1). The digestive tract is also highly vascular, enabling the absorption of nutrients. The complex coordination of the digestive system is achieved by both the nervous and endocrine systems.

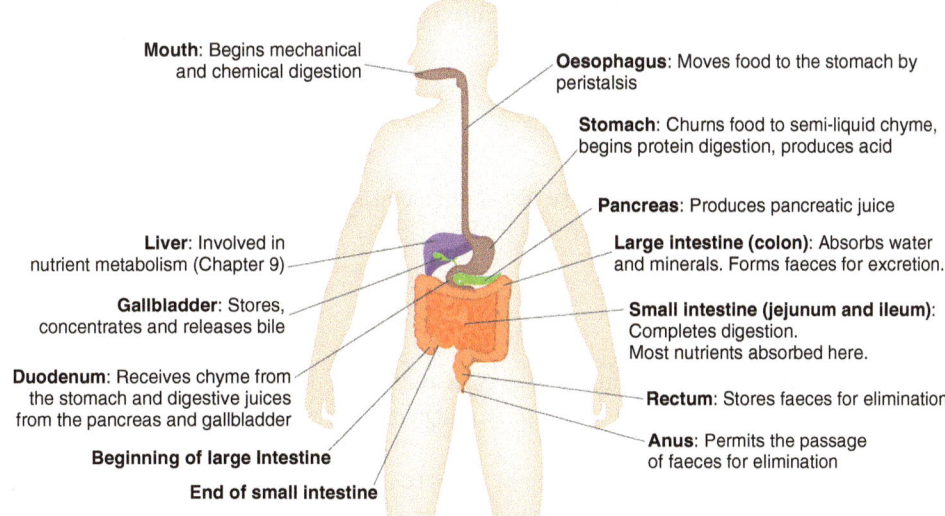

Figure 8.1 Components of the digestive system

Tubular structure

The tubular structure of the digestive tract consists of four layers (Figure 8.2).

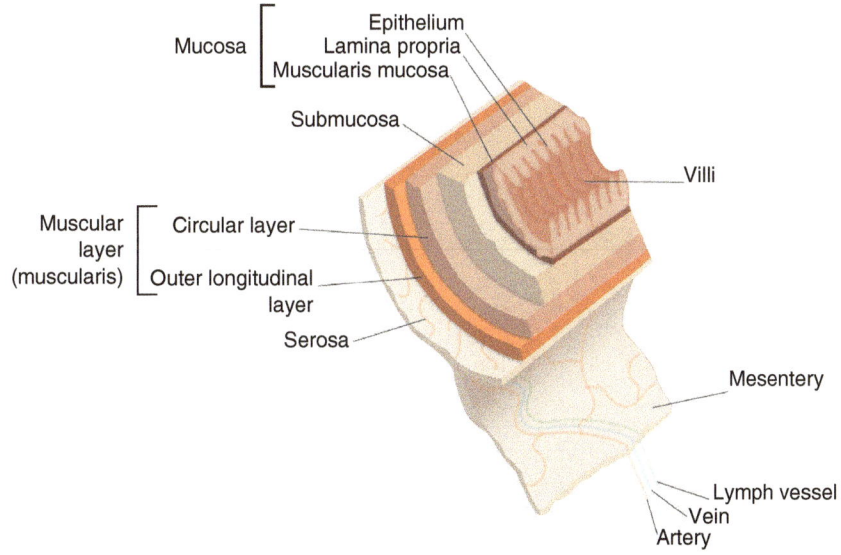

Figure 8.2 Tubular structure of the digestive tract

Mucosa

This innermost layer is involved in secreting enzymes for digestion in parts of the digestive tract and in absorbing nutrients. In the small intestine, absorption is maximised by structuring the mucosa's finger-like projections, known as villi. These increase the surface area, thus enhancing absorption. When the stomach is empty, the mucosa falls into folds called rugae which are stretched flat when the stomach is full.

The mucosa can be divided into three sublayers: epithelium, lamina propria (connective tissue), and the muscularis mucosae (muscular layer that maintains the folds and enables secretion of enzymes through movement).

Submucosa

This layer is structural, containing connective tissue with blood and lymphatic vessels and nerves passing through it.

Muscular

This layer contains smooth muscle arranged in circular and longitudinal layers. This muscle is essential for mixing the contents and for propulsion within the digestive tract. The two types of movement in the gut are peristalsis and segmentation (Figure 8.3).

Peristalsis describes ripple-like waves created by the relaxation and contraction of muscle along the entire digestive tract. This results in the movement of material along the digestive tract. In general, the parasympathetic nervous system stimulates peristalsis, whereas the sympathetic nervous system inhibits it. This is achieved through the vagus nerve and also the enteric nervous system. Segmentation mixes the contents of the gut, i.e. food and digestive enzymes, to promote digestion.

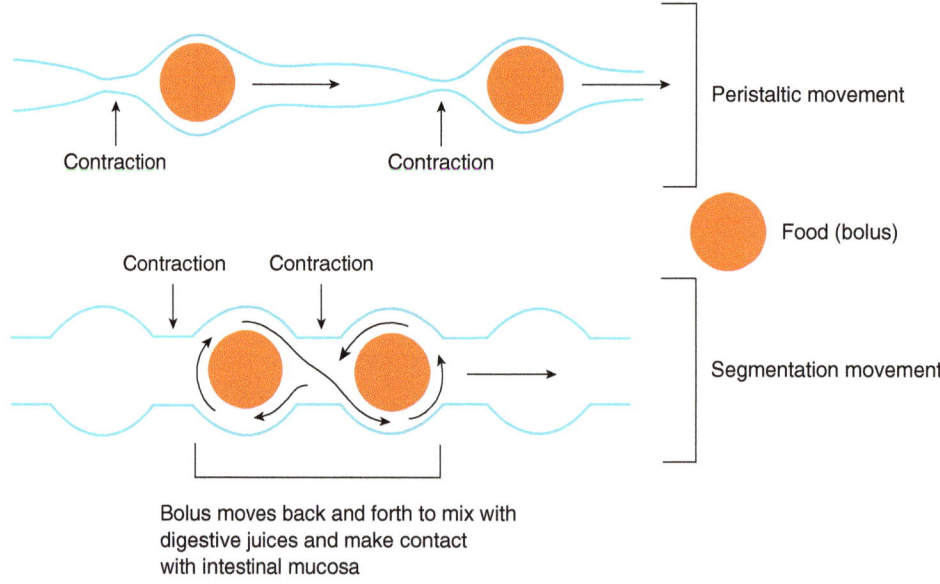

Figure 8.3 Movements of the gut

Serosa

This outermost layer is fibrous and elastic. Over the small and large intestine, this layer is the peritoneum and is composed of two serous membranes – the parietal, lining the walls of the abdomen, and visceral, covering the abdominal organs. This layer is not present over the oral cavity, pharynx, oesophagus and rectum.

ACTIVITY 8.5: UNDERSTAND

Watch the following video clip to help develop your understanding of peristalsis and segmentation.

This external video can be accessed by **scanning the QR code** with your smart phone or via https://study.sagepub.com/essentialap2e.

PERISTALSIS AND
SEGMENTATION (0:45)

The mouth

The digestive journey begins in the mouth. The mouth is formed by the hard and soft palates at the roof, the tongue at the floor and the rest covered by the oral mucosa as part of the tubular structure of the digestive system. Here, both mechanical and chemical digestion of nutrients begins. The muscular tongue manipulates food into position for mastication (chewing) to mechanically break down food. This increases the available surface area for enzymes to make contact with nutrients to begin chemical digestion. The tongue also assists by moving the bolus of food back into the oropharynx for swallowing. In addition, the tongue has a role in speech and is the organ of taste (Chapter 6).

Teeth

Teeth are designed for biting, chewing (mastication) and grinding food when the upper and lower teeth are moved against each other by the muscular movement of the mandible (jaw). The muscles used in mastication are the masseter, temporalis and medial and lateral pterygoid muscles. Teeth are composed of four primary layers (Figure 8.4):

- **Enamel:** This is a hard, bone-like structure that protects the tooth. Above the gum line it forms the crown, while below it is known as cementum.
- **Dentine:** Dentine is hard but porous, requiring enamel to protect it. It makes up the bulk of the tooth.
- **Cementum:** is also a hard, bone-like tissue but is softer than enamel and dentine. It is a thin layer that covers the root of the tooth and helps to anchor the tooth within bone.
- **Pulp:** This is the core layer of the tooth, containing a rich supply of blood and nerves entering from the root of the tooth through a gap called a foramen (root end opening). The pulp is necessary for the growth of a tooth and for sensation. Once a tooth is fully developed, the tooth can survive without the pulp (Nazarko, 2008).

People have two sets of teeth. The first 20 (deciduous or primary teeth) appear in the first two years of life. The 32 permanent teeth emerge and begin replacing primary teeth from approximately the age of six, taking up to the age of 25 to complete. There are four types of teeth (Figure 8.5):

- **Incisors:** These are the sharpest teeth, for biting into food.
- **Canines:** Located beside the incisors, these teeth have large, cone shaped crowns to hold, grab and tear food.
- **Premolars (bicuspids):** Only present in permanent teeth, these have rounded crowns and are structured for holding, crushing and grinding food.
- **Molars:** The largest of the teeth, molars have large, wide crowns and grind food. The third molars are also known as 'wisdom teeth' because they usually erupt between the ages of 17 and 25. They often cause problems through impaction because, in many people, the jaws are not large enough for the teeth to grow properly.

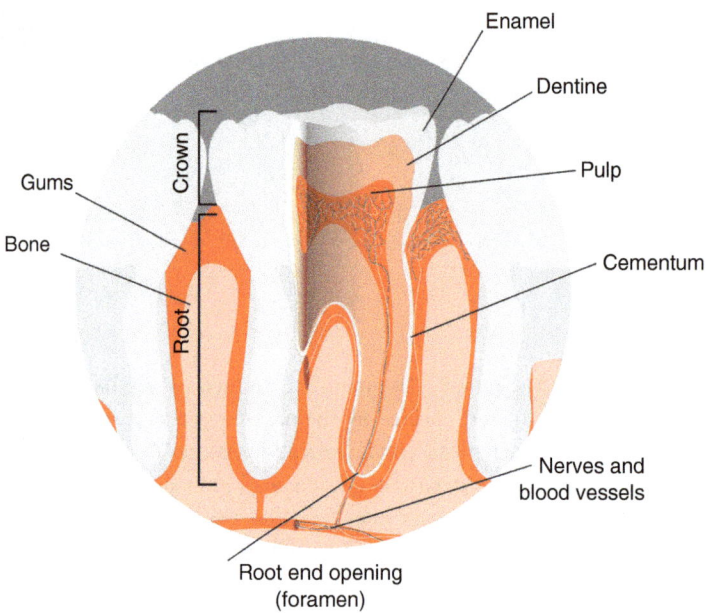

Figure 8.4 Tooth anatomy

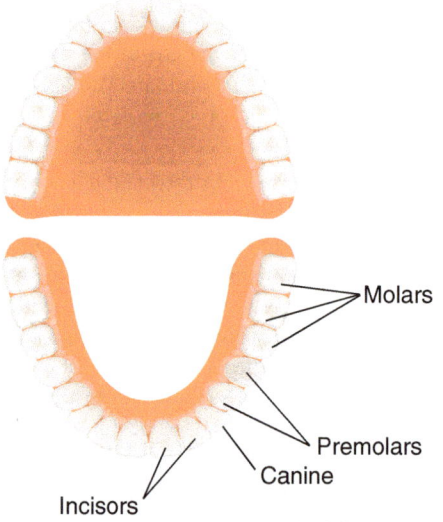

Figure 8.5 Types and location of teeth. From centre: 2 incisors, 1 canine, 2 premolars, 3 molars

Saliva

Opening into the mouth are three pairs of salivary glands shown in Figure 8.6: parotid, submandibular and sublingual. These produce and secrete saliva, a slightly acidic liquid, into the mouth through ducts. The pH range of saliva is between 6.2 and 7.6 (average of 6.7) (Baliga et al., 2013). Some sources identify it as being as low as 5.5 and this variation is a result of how the pH is tested, creating a diversity of opinion on the true value.

Saliva is produced in response to sight, smell, thought and presence of food. Whilst mostly composed of water, saliva contains a variety of enzymes, hormones, antibodies and antimicrobial substances that fulfil the following roles (Spielmann and Wong, 2011):

- Carbohydrate (polysaccharide) digestion by the enzyme salivary amylase, producing disaccharide maltose.
- Lipase for fat digestion, activated once in the stomach by hydrochloric acid.
- Enables food swallowing and tasting through lubrication.
- Lubricates the tissues of the oral cavity.
- Maintenance of tooth integrity and oral hygiene by the antibacterial substances, lysozyme and immunoglobulins.
- Antibacterial and antiviral protection.

Swallowing

Once the mouth has begun the mechanical and chemical digestion of food, the lubricated food is formed into a bolus and moved by the tongue into the oropharynx for swallowing. Once inside the pharynx, swallowing occurs involuntarily. The three stages of swallowing are:

1. **Oral:** The food bolus is moved into the oropharynx by the tongue and cheeks.
2. **Pharyngeal:** Involuntary contraction of the muscles in the oropharynx moves the food bolus into the oesophagus.
3. **Oesophageal:** The food bolus is moved by peristalsis, the cardiac sphincter relaxes and food enters the stomach.

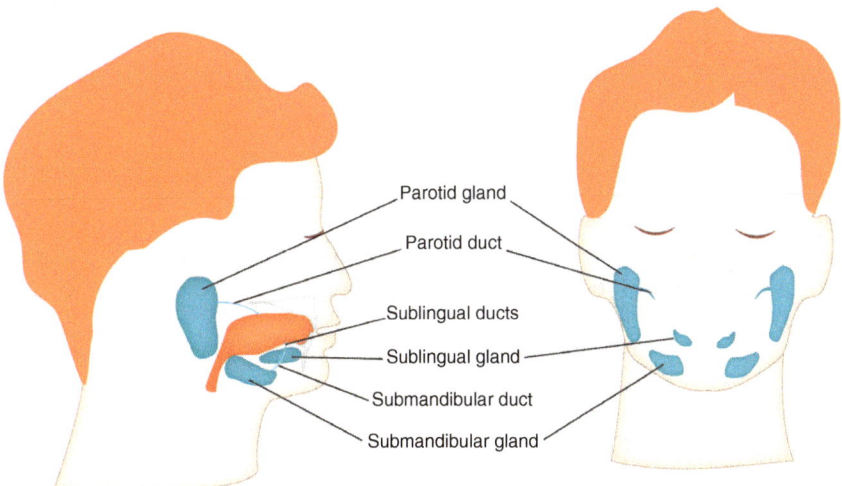

Parotid gland

Parotid duct

Sublingual ducts

Sublingual gland

Submandibular duct

Submandibular gland

Figure 8.6 Salivary glands

Sensory fibres in the oropharynx respond to temperature, pressure or both together to initiate the swallowing reflex. It is thought that taste is also involved in coordinating swallowing. This sensory input travels along five pairs of cranial nerves (V, VII, IX, X and XI), relaying signals to nuclei in the brainstem. These signals are interpreted and generate a motor response through six pairs of cranial nerves (V, VII, IX, X, XI and XII) to the oropharyngeal muscles, contracting them to move the food bolus into the oesophagus (Malandraki et al., 2011). The soft palate rises during this process and the epiglottis prevents food entering the trachea. Food is now in the oesophagus.

Oesophagus

The oesophagus is a muscular tube that connects the oropharynx to the stomach, passing through the diaphragm. It is approximately one foot (18–25 cm) in length and lined with mucous membrane to lubricate the passage of food by secretion of mucus. Innervated by the vagus nerve, the oesophagus moves food through to the stomach by peristalsis. The cardiac sphincter muscle is located at the entrance to the stomach (Figure 8.7). It permits the bolus to move into the stomach but also prevents the contents from refluxing back into the oesophagus.

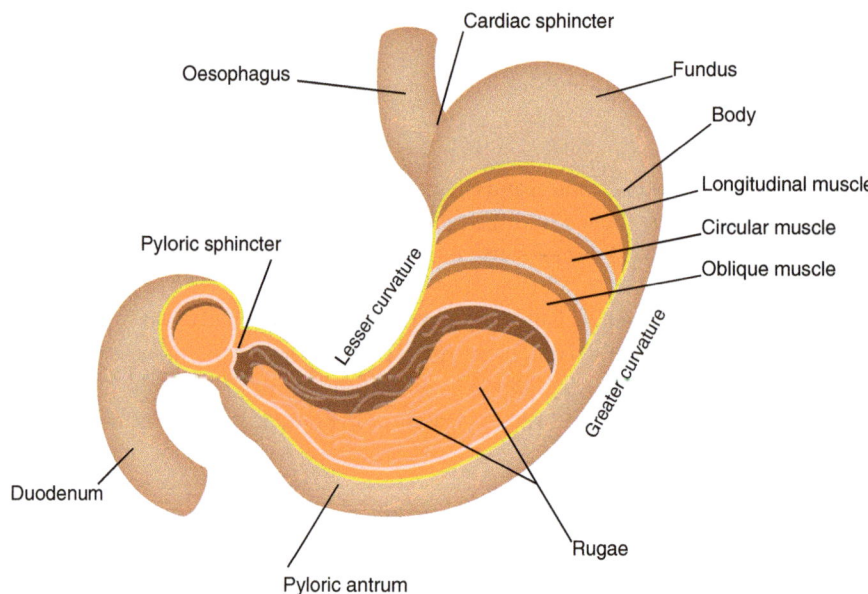

Figure 8.7   Parts of the stomach

The stomach

The stomach is a bean-shaped sac designed to house food for digestion (Figure 8.7). As the food bolus moves into the stomach, the pyloric sphincter (at the exit from the stomach) closes to retain the food for digestion. Chemical and mechanical digestion both occur in the stomach.

Mechanical digestion

While the stomach has the same overall tubular structure as the rest of the GIT, the muscular layer has three strata of muscle compared to two in the rest of the tract. The additional oblique muscles are present to enable the stomach to churn the contents, maximising digestion by breaking down the food bolus and mixing it with the gastric juices.

Chemical digestion

Chemical digestion occurs from gastric juices secreted by the mucosal layer of the stomach. Parietal cells produce hydrochloric acid and intrinsic factor (required for the absorption of vitamin B_{12} from the lower part of the small intestine), and chief cells secrete pepsinogen, an inactive enzyme, and rennin. Gastric juices also contain water, salts and mucus. All of these individual constituents have a role in digestion (Table 8.6).

Table 8.6 Components and functions of gastric juice

Gastric juice component	Function
Water	A liquid medium for chemical reactions and acts as a lubricant
Hydrochloric acid	Converts pepsinogen to its active form, pepsin. Lowers pH to inactivate amylase and promote pepsin function
	Kills ingested pathogens
Pepsin (from pepsinogen)	Begins digestion of proteins
Rennin	Digests milk proteins in infants
Intrinsic factor	Enables absorption of vitamin B_{12} in the ileum
Mucus	Protects the mucosal layer from chemical damage by hydrochloric acid and digestive enzymes
	Lubricates food

In addition, the lipase from the saliva is activated to work in the acidic conditions of the stomach, and digests 10% of fats there, breaking triglycerides into two fatty acids and one monoglyceride. In infants, the stomach can also produce a small amount of lipase.

Production of gastric acid

The control of secretion of gastric juices occurs in three stages:

Cephalic phase: the sight, smell and taste of food are the stimuli for action of the vagus nerve on the stomach before food ever reaches the stomach.

Gastric phase: once food enters the stomach, its presence stimulates the production of gastrin, a hormone produced in the pylorus of the stomach (and in the duodenum). This hormone passes into the blood stream, is circulated back to the stomach and enhances gastric juice secretion until the pH in the stomach drops to about 1.5.

Intestinal phase: it is not until the partially digested food, now known as chyme, moves into the small intestine that the production of gastric juice stops. The acidic, fatty chyme in the small intestine stimulates release of the hormones secretin and cholecystokinin from the duodenum into the blood stream. Once they reach the stomach, they inhibit gastric juice production and reduce gastric contractions to slow stomach emptying.

Movement of food through the stomach

Once food enters the stomach, the proximal stomach relaxes and functions as a reservoir, enabling a large amount of food to enter without pressure. Distally, food is mixed and ground by contractions, followed by peristalsis creating a controlled flow of chyme into the duodenum via the pyloric sphincter (Janssen et al., 2011).

The speed at which the stomach empties depends on the volume of its contents. One to three percent of the contents move into the duodenum per minute with carbohydrates emptying faster than fats (Yasmin et al., 2009). Liquids exit even quicker, with clear fluids (water, juice, black tea/ coffee) mostly within one hour (90%) (Yasmin et al., 2009). In a study by Hellmig et al. (2006) the mean length of time for half the volume of fluid to leave the stomach was 80.5 minutes and for solids 143.6 minutes.

APPLY

Gastric emptying and nursing practice

The speed of gastric emptying is significant for nurses to understand, particularly when preparing people for a General Anaesthetic (GA) or for investigations of the gastrointestinal tract. While under a GA, there is the potential for stomach contents to be regurgitated back into the oesophagus and further into the respiratory tract. This is clearly not desirable as it can result in respiratory complications and also irritation of the oesophagus. As a result, people are 'fasted' (i.e. they do not have anything to eat or drink) for certain periods before the administration of a GA. Traditionally this has not been optimally managed, with people fasting for excessive periods. Considering the physiology of gastric emptying times, the following are largely agreed to be the optimal fasting times (Smith et al., 2011):

- Adults and children can take clear fluids up to two hours before elective procedures. There is debate whether fluids with milk can take longer to leave the stomach but there is little consensus on this issue.
- Adults and children can take solids up to six hours before elective procedures. Chewing gum or sucking on boiled sweets are largely not considered to be taking solids.
- Babies and infants can be fed (and should be fed) up to six hours before elective procedures – breast milk is largely acceptable to be taken up to four hours before and other milk up to six hours before. Fluids are the same as for adults and children.

The small intestine

The small intestine receives partially digested food with few nutrients yet absorbed into the body. It completes the digestive process, preparing nutrients for absorption. Chyme spends about three to six hours in the small intestine, moving along its approximately 6-metre length, of which there are three connecting parts between the stomach and the large intestine (Figure 8.1):

- **The duodenum:** about 25 cm long.
- **The jejunum:** the middle section about 2.5 m long.
- **The ileum:** the terminal section, approximately 3.5 m in length.

Digestion in the small intestine

The duodenum receives digestive enzymes from the pancreas and the gallbladder, and it is here that the gastric juices are neutralised. It is at this stage of digestion that the pancreas and gallbladder become accessory organs for digestion.

Pancreas   The pancreas contains both endocrine and exocrine tissue; islet cells secrete the hormones insulin and glucagon (Chapter 7), acinar cells secrete up to 1.5 litres of pancreatic juice daily. Pancreatic juice is secreted into ducts that merge to become the pancreatic duct. This duct eventually merges with the common bile duct and enters the duodenum at the hepatopancreatic ampulla where release of bile and pancreatic juice is controlled by the sphincter of Oddi (a ring of muscle around the opening).

Gallbladder   The gallbladder is a small sac located at the underside of the liver. It stores and concentrates bile, produced by the liver, ready for release to the duodenum through the bile duct and subsequently the hepatopancreatic ampulla. The mucosa of the gallbladder absorbs the water and salts from the bile, concentrating it.

Bile and pancreatic juices   In the intestinal phase of gastric function, you read that cholecystokinin was released once acidic, fatty chyme enters the duodenum. While cholecystokinin decreases gastric juice secretion, it also has three other actions:

1.   It triggers contraction of the gallbladder, releasing bile.
2.   It triggers the secretion of pancreatic juices.
3.   It triggers the hepatopancreatic sphincter to relax, releasing bile and pancreatic juices into the duodenum.

Gastrin (from the stomach) also triggers the latter two of these actions, but not as potently.

Pancreatic juice is alkaline due to sodium bicarbonate and contains three inactive proteolytic (i.e. protein digesting) enzymes (so as not to digest the pancreas itself), and a number of active enzymes. These, and their functions, are outlined in Table 8.7.

Table 8.7   Components of pancreatic juice and associated functions

Pancreatic juice component	Function
Trypsinogen	Converted to active trypsin by enterokinase (secreted by duodenal wall). Trypsin digests polypeptides (proteins) into tripeptides, dipeptides and amino acids (proteolysis), and activates chymotrypsinogen and procarboxypeptidase
Chymotrypsinogen	Converted to active chymotrypsin by trypsin. Chymotrypsin digests polypeptides (proteins) into tripeptides, dipeptides and amino acids
Procarboxypeptidase	Converted to active carboxypeptidase by trypsin. Carboxypeptidase digests polypeptides (proteins) into tripeptides, dipeptides and amino acids
Pancreatic amylase	Digests polysaccharides into disaccharides
Pancreatic lipase	Breaks triglycerides (fat) into two fatty acids and a monoglyceride
Ribonuclease	Digests RNA into nucleotides
Deoxyribonuclease	Digests DNA into nucleotides
Sodium bicarbonate	Neutralises pH of chyme, enabling amylase and lipase to work optimally

Bile is secreted by the liver and contains water, salts, mucus, cholesterol, lecithin and bilirubin. Their functions are detailed in Table 8.8.

Table 8.8   Components of bile and associated functions

Bile component	Function
Bilirubin	Converted to urobilinogen by bacteria in the large intestine. It gives faeces its brown colour
Cholesterol	Secretion of cholesterol in bile enables excretion of cholesterol from the body
Lecithin and bile acids	Emulsifies fats and cholesterol

The bile and pancreatic juices carry out further digestion of carbohydrates, proteins and fats in the duodenum. The small intestinal structure is modified to maximise absorption, with the mucosal wall structured into villi, and microvilli (also known as the brush border), increasing the surface area (Figure 8.8) for absorption and also producing some enzymes. It is at the surface of microvilli that final digestion occurs through the action of brush border enzymes.

Final carbohydrate digestion: Dextrinase and glucoamylase break down oligosaccharides. Maltase breaks maltose down to glucose. Finally, sucrase and lactase digest sucrose and lactose respectively. Carbohydrates are then ready for absorption through the walls of the small intestine. The intestine also contains two main groups of probiotic bacteria (*Bifidobacterium* and *Lactobacillus* species), which are saccharolytic in that they digest carbohydrates anaerobically into short chain fatty acids (Szmulowicz and Hull, 2014) (Chapter 4).

Final protein digestion: Carboxypeptidase, aminopeptidase and dipeptidase are also brush border enzymes, and they complete protein digestion, releasing amino acids from the beginning, end and middle of peptide chains.

Final fat digestion: Fats are digested by lipase. Bile enables this by emulsifying fats, breaking them into smaller globules for lipase to be effective. Bile coats free fatty acids, monoglycerides and other fats with bile acids to form micelles (lipid molecules composed of monoglycerides, fatty acids and bile salts).

Final nucleic acid digestion: Nucleic acids (DNA, RNA), now in the form of nucleotides after the actions of pancreatic juice, are finally digested to phosphate ions, ribose and deoxyribose by the brush border enzymes, nucleosidases and phosphatases.

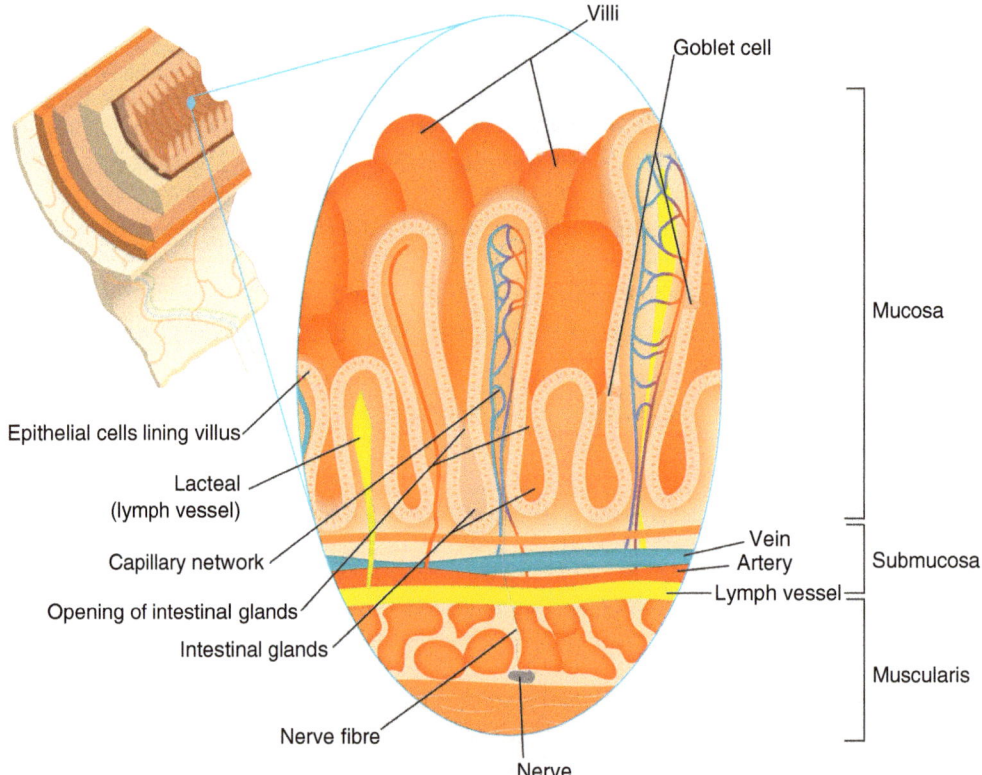

Figure 8.8   Villi of the small intestine

Absorption in the small intestine

The villi and microvilli of the small intestine are highly vascular, containing blood vessels (carrying all nutrients except lipids to the liver) and lacteals (draining lipids into lymph vessels) (Chapters 9 and 12).

Carbohydrates, as monosaccharides, are transported across the intestinal mucosal wall and into the extracellular fluid by transport proteins: they then pass into capillaries by facilitated diffusion or by following water moving by osmosis (solvent drag).

Amino acids are absorbed in a similar fashion, some of which are sodium-dependent in terms of transport. Dipeptides and tripeptides may move across the plasma membrane but are broken down to amino acids within mucosal cells before reaching the circulation.

Micelles release lipids (free fatty acids, monoglycerides, cholesterol and other fatty acids) at the microvilli of the brush border. These pass freely into the cells of the mucosal wall. Active transport proteins also transport cholesterol and fatty acids. Once in the cells, the smooth endoplasmic reticulum recreates the lipid components back into triglycerides. The Golgi complex then converts these triglycerides to chylomicrons (a type of lipoprotein created in the small intestine to aid absorption) by adding cholesterol and coating them in phospholipids and proteins. These are transported out of the cell and are absorbed into the lacteals. This collection of chylomicrons is called chyle and drains into lymph vessels. Some free fatty acids are directly absorbed into the blood capillaries.

Phosphate ions, ribose and deoxyribose are transported by membrane carriers across the plasma membrane and into blood capillaries. Vitamins are absorbed either with fats (vitamins A, D, E and K) or by diffusion (water-soluble vitamins). The exception is B_{12}, which binds with the intrinsic factor from gastric juice and is taken up by receptor-mediated endocytosis in the terminal ileum. Minerals are absorbed as described in Table 8.9.

Table 8.9 Absorption of minerals in the small intestine

Mineral	Absorption method
Sodium	Co-transported with sugars and amino acids
Chloride	Transported by an ion exchange pump in exchange for bicarbonate ions
Potassium	Diffusion
Iron (ferrous ions)	Active transport
Calcium	Absorbed from the duodenum into epithelial cells by diffusion and pumped out for absorption into the circulation by calcium-ATPase. Calbindin, a carrier protein that is created under the influence of vitamin D, enhances this process

The large intestine

The roles of the large intestine are to absorb water and minerals and to form faeces for excretion. As chyme transits the small intestine, it reaches the terminal of the ileum, which connects to the caecum, the first part of the large intestine (Figure 8.9). The small intestine is separated from the large intestine by the ileocaecal valve which opens following contraction of the small intestine stimulated by the presence of food in the stomach – the gastroileal/gastrocolic reflex. This reflex is stimulated by the sympathetic nervous system. Attached to the caecum is the vermiform appendix, a blind tube of about 9 cm in length with a large amount of lymphoid tissue.

The large intestine is then laid out in a clockwise manner, with the ascending colon on the right-hand side of the body. This becomes the transverse colon, going from right to left across the body. This becomes the descending colon on the left-hand side which then extends centrally into the sigmoid colon. At the terminal end of the large intestine, and indeed the GIT, is the rectum which stores faeces until elimination through the anus. The rectum joins the anus at the internal sphincter, a thickened layer of muscle that covers the first three quarters of the anal canal. The external sphincter surrounds the whole of the anal canal, and, together with the internal sphincter, regulates defaecation.

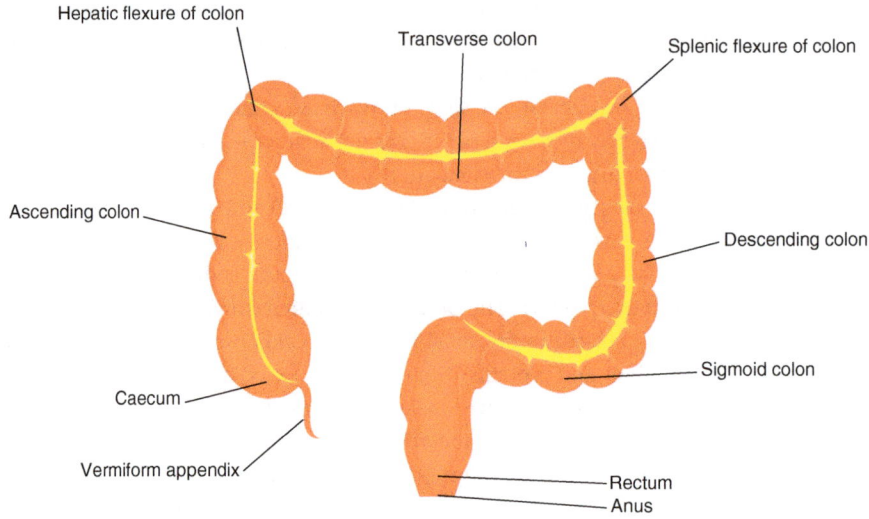

Figure 8.9 Components of the large intestine

Chyme entering the large intestine has little nutritional content as this has largely been digested and absorbed in the small intestine; it is largely composed of indigestible cellulose (fibre), water and minerals. There is a large amount of bacteria present, as the small intestine provides an environment that is conducive to bacterial growth. The colon absorbs a significant amount of water and minerals, largely by osmosis due to epithelial cells of the mucosa having a higher sodium concentration (Szmulowicz and Hull, 2014). A significant amount of bacteria die here as a result of the few available nutrients. The result is faeces, largely composed of dead bacteria and fibre (cellulose). It also contains short chain fatty acids, which are fermented in the colon by probiotic bacteria (*Bifidobacterium* and *Lactobacillus* species). These bacteria also contribute to replenishing the colonic mucosa (Szmulowicz and Hull, 2014).

The colon has small pouches (or haustrations) that give it a segmented appearance and maximise the surface area in the colon for absorption of water and minerals. As they fill with chyme, they distend causing them to contract approximately every 25 minutes. This is known as haustral churning, which propels colonic contents from one haustration to another. However, this action actually retains faecal content to maximise water and mineral absorption. High- and low-amplitude peristaltic contractions are necessary to move content along the colon towards the rectum and anus. High-amplitude peristaltic contractions occur 2–14 times daily, achieving mass movement of faecal content. These are stimulated by colonic and gastric distention, eating, short chain fatty acids and some medications (cholinergic drugs and laxatives) (Szmulowicz and Hull, 2014). Low-amplitude peristaltic contractions, referred to as long spike bursts, are less frequent and contribute to the urge to defaecate. The gastrocolic reflex is the most likely stimulant, occurring as a result of gastric distention.

 APPLY

Constipation can cause many people considerable distress. People can help to prevent constipation by responding to the urge to defaecate, anticipating it about 20-30 minutes after eating. The gastrocolic reflex is most effective after breakfast and can also be enhanced by walking and taking warm drinks. Person-centred practice should take into account cultural issues, levels of mobility and social determinants such as shift patterns, privacy and other considerations that can promote or inhibit attending to such factors.

Defaecation involves the entire colon. Faeces is collected and stored in the rectum, ready for defaecation. Once the rectum is full, mechanoreceptors sense the distention, stimulating a reflex contraction of the muscles in the rectum and along the colonic walls (Palit et al., 2012). In addition, the internal sphincter is relaxed, permitting movement of the faeces into the anal canal. The urge to defaecate becomes conscious. If this is not consciously responded to, reverse peristalsis occurs and faecal content returns to the colon. More water is absorbed in this instance and may contribute to constipation. If a person makes the voluntary choice to defaecate, intra-rectal pressure increases, anal canal muscles relax, the external sphincter opens, the pelvic floor relaxes and intra-abdominal pressure increases (often through contraction of the abdominal muscles) (Palit et al., 2012). Combined, these result in faeces being expelled from the digestive tract.

ACTIVITY 8.6: UNDERSTAND

Watch the following online video clip to help develop your understanding of the physiology of the digestive system.

(Please note that cholecystokinin is a hormone as well as an enzyme. It is pronounced in this video clip as 'cholcytokinin'.)

This online video can be accessed by **scanning the QR code** with your smart phone or via https://study.sagepub.com/essentialap2e.

DIGESTIVE SYSTEM OVERVIEW
(10.31)

REGULATION OF APPETITE

While there is a homeostatic response in the body that drives a person to feel hungry, it is also widely recognised that that there is a psychological component to appetite. The full regulation of appetite is still being researched to gain a complete understanding. It is a complex process that integrates a neural and endocrine response.

Endocrine control of appetite

There are key hormones involved in peripheral appetite regulation (i.e. they signal to the central nervous system from outside of it). These are identified in Table 8.10. You will notice that there are four primary hormones that increase the sensation of hunger, and five that increase the sensation of satiety – the opposite of hunger.

Table 8.10 Hormones regulating appetite

Hormone	Increase appetite/hunger	Hormone	Decrease appetite/hunger
	Action		**Action**
Ghrelin	Produced primarily by endocrine cells in the stomach and hypothalamus	Leptin	Released by adipocytes (fat cells) when fat storage reaches a particular level
	Stimulates the hypothalamus (opposite effect to leptin) increasing appetite		Activates leptin receptors in the hypothalamus, causing increased energy expenditure and reduction in appetite (Crispima et al., 2011)
	Increases blood glucose and inhibits insulin secretion (Crispima et al., 2011)		Inhibits appetite-stimulating neurons and enhances effects of cholecystokinin (Harrold et al., 2012)
	Highest level before a meal		
	Stimulates neuropeptide Y, potent appetite /hunger stimulators		
Orexins	The orexins are hormones released when blood glucose levels fall, ceasing when food is in the stomach Their action occurs within the hypothalamus, where they are potent in creating a hyperphagic response (making you hungry) (Harrold et al., 2012)	Cholecystokinin	Released in the intestinal phase of gastric juice secretion, cholecystokinin stimulates the brainstem to signal meal termination and early satiety (Harrold et al., 2012)
			Cholecystokinin released about 25 minutes after starting a meal and levels do not fall for approximately three hours
Neuropeptide Y	Produced in the hypothalamus, this hormone promotes meal initiation and delays the onset of satiety, resulting in hyperphagia (Harrold et al., 2012)	Glucagon-like peptide-1 and peptide YY 3-36	Secreted by distal cells of the small intestine in response to carbohydrate and fat digestion, these hormones release insulin, inhibit glucagon, delay gastric emptying, and reduce appetite via vagal nerve stimulation (Harrold et al., 2012)
Melanin-concentrating hormone	Produced in the hypothalamus and acts on the limbic system, this increases food intake and results in hyperphagia	Amylin	Once blood glucose levels rise in the blood, the beta cells of the pancreas release amylin, which circulates in the blood to signal to the brainstem, slowing gastric emptying and contributing to the sensation of satiety (Harrold et al., 2012). Amylin is known to inhibit neuropeptide Y
		Serotonin	Serotonin is known to stimulate satiety during and after meals by stimulating the hypothalamus. This occurs in response to fat digestion and carbohydrate intake (Harrold et al., 2012)

Neural control of appetite

The sight and smell of food are often the first triggers of appetite. Once we start to eat, taste receptors (chemoreceptors) enhance this feeling of hunger by relaying through the cranial nerves to the brainstem (Harrold et al., 2012). Once food enters the stomach and it becomes distended, the stretching of the stomach wall is detected by mechanoreceptors. Chemoreceptors also detect the presence of nutrients and both of these sensors in the stomach relay this to the central nervous system through the vagus nerve (Harrold et al., 2012). Once nutrients are absorbed their levels are detected by chemoreceptors both peripherally and in the central nervous system. These two mechanisms achieve the sense of satiety, and appetite decreases. Mechanoreceptors in the intestinal wall detect distention, chemoreceptors detect nutrients in the intestinal tract and blood glucose levels also stimulate the brainstem via the vagus nerve (Harrold et al., 2012).

The limbic system plays a key role in appetite regulation, processing the reward/pleasure experience of food and how that influences appetite (Harrold et al., 2012). In fact, the taste and pleasure derived from food are considered to be the key motivators of appetite.

CONCLUSION

This chapter has introduced you to nutrients and how the digestive system converts them from a form available in the external environment to a form that can be absorbed into the internal environment. Here, you have discovered how they play a vital role in supporting homeostasis through providing structural resources and energy for growth, development and repair. This is fundamental to not only physical survival, but health-related quality of life. The regulation of appetite is social and physically complex. The person-centred nurse needs to have a wider appreciation of the complexity of challenges that face people in being able to obtain a healthy and nutritious diet. An understanding of the digestive system and its interplay with nutrients is fundamental to these activities.

GO DEEPER

Further reading

Davis, G. (2006) 'Hormonal control and the endocrine system: achieving homeostasis', *Nurse Prescribing*, 4(11): 446-53.

Hendry, C., Farley, A., McLafferty, E. and Johnstone, C. (2014) 'The digestive system: part 2', *Nursing Standard*, 28(25): 37-44.

Johnstone, C., Hendry, C., Farley, A. and McLafferty, E. (2014) 'The digestive system: part 1', *Nursing Standard*, 28(24): 37-45.

- Macronutrients, namely carbohydrates, proteins and fats, are fundamental sources of energy and structural resources in the body.

- Micronutrients, namely vitamins, minerals and trace elements have specific functions that support homeostatic functions within the body.

- A balanced nutritional diet is fundamental to health-related quality of life, providing energy for activities and role functions, creating social interaction focal points, and providing a sense of well-being through the pleasures of food and being nourished.

- The digestive system uses a complex process of mechanical and chemical digestion to convert macronutrients into a form to be absorbed into the internal environment.

- The various sections of the digestive tract are adapted to provide enzymes and optimal environmental conditions that enable a staged process to digestion.

- Appetite and satiety are balanced against each other through a complex regulation achieved by both the endocrine and neurological system.

REVISE

TEST YOUR KNOWLEDGE

This chapter leads into the next where you will learn to build upon your knowledge of energy and the nutritional support of homeostasis through metabolism and the functions of the liver. This chapter requires you to understand each stage of digestion in phases. Revise each section individually and check your understanding of each section before you move to the next. Do you understand which nutrients are digested by which enzymes and where? Do you know how this is regulated and how these nutrients are absorbed? Challenge your knowledge. You also need to build on the content of this chapter by doing some further reading about healthy eating and diet (for example, looking at the **CalorieKing** database). This is addressed through this Revise section.

In order to help you revise, consider the following questions, answers for which can be found by visiting **https://study.sagepub.com/essentialap2e**.

Test yourself by revising the chapter first, and then answer these questions without looking at the book. Afterwards compare your answers with the text and with the notes you made. Did you miss anything in your notes? Here are the questions:

1   Plan a balanced diet for an average 70 kg adult for one day, identifying the main nutrient component of each item (e.g. fat, carbohydrate, protein, vitamins) and number of calories. Please go to the CalorieKing website to help you identify the values of each food item.

2   Explain why each component is needed for health and well-being. What are the main sources of each nutrient?

3   Explain how each component is digested, metabolised, stored and eliminated.

4   What are the factors that influence a person's ability to maintain a healthy diet and digestive system in society?

5   What are nutrients? What are the main nutrients in the body?

6   Describe the basic functions of the digestive system.

7 Describe the tubular structure of the digestive tract.

8 What are the three phases of gastric juice secretion?

9 Briefly describe how carbohydrates are stored in the body.

For additional revision resources visit: **https://study.sagepub.com/essentialap2e**.

REVISE

ACE YOUR
ASSESSMENT

- Revise key terms with interactive flashcards.

- Test yourself with multiple-choice questions.

- Access the glossary with audio to hear how complex terms are pronounced.

- Explore recommended websites suitable for revision.

REFERENCES

Baliga, S., Muglikar, S. and Kale, R. (2013) 'Salivary pH: a diagnostic biomarker', *Journal of Indian Society of Periodontology*, 17(4): 461–5.

Bond, H. (2014) 'Natural alternatives to tackling high cholesterol: guidance for community nurses', *British Journal of Community Nursing*, 19(8): 375–81.

Cowan, H. (2014) 'Can cholesterol be defined as just "good" or "bad"?', *British Journal of Cardiac Nursing*, 9(10): 515–16.

Crispima, C.A., Waterhouse, J., Dâmaso, A.R., Zimberg, I.Z., Padilha, H.G. et al. (2011) 'Hormonal appetite control is altered by shift work: a preliminary study', *Metabolism*, 60(12): 1726–35.

DeLuca, G.C., Kimball, S.M., Kolasinski, J., Ramagopalan, S.V. and Ebers, G.C. (2013) 'Review: the role of vitamin D in nervous system health and disease', *Neuropathology and Applied Neurobiology*, 39(5): 458–84.

Faurschou, A., Beyer, D.M., Schmedes, A., Bogh, M.K., Philipsen, P.A. et al. (2012) 'The relation between sunscreen layer thickness and vitamin D production after ultraviolet B exposure: a randomized clinical trial', *British Journal of Dermatology*, 167: 391–5.

Goodman, B.E. (2010) 'Insights into digestion and absorption of major nutrients in humans', *Advances in Physiology Education*, 34: 44–53.

Gouni-Berthold, I. and Berthold, H.K. (2013) 'The role of niacin in lipid-lowering treatment: are we aiming too high?' *Current Pharmaceutical Design*, 19(17): 3094–106.

Gupta, S., Prasad, K. and Bisht, G. (2014) 'Macro and micro minerals content in some important Indian medicinal plants', *Research Journal of Phytochemistry*, 8: 168–71.

Harrold, J.A., Dovey, T.M., Blundell, J.E. and Halford, J.C.G. (2012) 'CNS regulation of appetite', *Neuropharmacology*, 63(1): 3–17.

Hellmig, S., Von Schöning, F., Gadow, C., Katsoulis, S., Hedderich, J. et al. (2006) 'Gastric emptying time of fluids and solids in healthy subjects determined by ^{13}C breath tests: influence of age, sex and body mass index', *Journal of Gastroenterology and Hepatology*, 21: 1832–8.

Herr, C., Greulich, T., Koczulla, R.A., Meyer, S., Zakharkina, T. et al. (2011) 'The role of vitamin D in pulmonary disease: COPD, asthma, infection, and cancer', *Respiratory Research*, 12: 31.

Janssen, P., Vanden Berghe, P., Verschueren, S., Lehmann, A., Depoortere, I. et al. (2011) 'The role of gastric motility in the control of food intake', *Alimentary Pharmacology and Therapeutics*, 33: 880–94.

Kaczmarczyk, M.M., Millera, M.J. and Freund, G.G. (2012) 'The health benefits of dietary fiber: beyond the usual suspects of type 2 diabetes mellitus, cardiovascular disease and colon cancer', *Metabolism*, 61(8): 1058–66.

Malandraki, G.A., Johnson, S. and Robbins, J.A. (2011) 'Functional MRI of swallowing: from neurophysiology to neuroplasticity', *Head & Neck*, 33(S1): S14–20.

Marmot, M., Allen, J., Goldblatt, P., Boyce, T., McNeish, D., Grady, M. and Geddes, I. (2010) *Fair Society, Healthy Lives – The Marmot Review*. London: The Marmot Review.

Nazarko, L. (2008) 'Oral health: the anatomy and physiology of the mouth', *Journal of Healthcare Assistants*, 2(4): 171–3.

Pace, A., Giannarelli, D., Galiè, E., Savarese, A., Carpano, S. et al. (2010) 'Vitamin E neuroprotection for cisplatin neuropathy – a randomized, placebo-controlled trial', *Neurology*, 74(9): 762–6.

Palit, S., Lunniss, P.J. and Scott, S.M. (2012) 'The physiology of human defaecation', *Digestive Diseases and Sciences*, 57(6): 1445–64.

Ruel, M.T. and Alderman, H. (2013) 'Nutrition-sensitive interventions and programmes: how can they help to accelerate progress in improving maternal and child nutrition?', *The Lancet*, 382(9891): 536–51.

Smith, G.D. (2012) 'Dietary fibre in the management of gastrointestinal symptoms', *Gastrointestinal Nursing*, 10(2): 17–18.

Smith, I., Kranke, P., Murat, I., Smith, A., O'Sullivan, G. et al. (2011) 'Perioperative fasting in adults and children: guidelines from the European Society of Anaesthesiology', *European Journal of Anaesthesiology*, 28(8): 556–69.

Spielmann, N. and Wong, D.T. (2011) 'Saliva: diagnostics and therapeutic perspectives', *Oral Diseases*, 17(4): 345–54.

Stabler, S.P. (2013) 'Vitamin B_{12} deficiency', *New England Journal of Medicine*, 368: 149–60.

Suares, N.C. and Ford, A.C. (2011) 'Systematic review: the effects of fibre in the management of chronic idiopathic constipation', *Alimentary Pharmacology & Therapeutics*, 33(8): 895–901.

Szmulowicz, U.M. and Hull, T.L. (2014) 'Colonic physiology', in D.E. Beck, S.D. Wexner, T.L. Hull, P.L. Roberts, T.J. Saclarides, A.J. Senagore, M.J. Stamos and S.R. Steele (eds), *The ASCRS Manual of Colon and Rectal Surgery*, 2nd edn. New York: Springer. pp. 23–39.

Threapleton, D.E., Greenwood, D.C., Evans, C.E., Cleghorn, C.L., Nykjaer, C. et al. (2013) 'Dietary fibre intake and risk of cardiovascular disease: systematic review and meta-analysis', *British Medical Journal*, 347: f6879.

WHOFAOUN (World Health Organization and Food and Agriculture Organization of the United Nations) (2004) *Vitamin and Mineral Requirements in Human Nutrition*, 2nd edn. Hong Kong: WHOFAOUN.

Yasmin, R., Khan, S.A., Sarker. P.C. and Khan, Z.H. (2009) 'Pre-operative fasting guidelines: an update', *Journal of the Bangladesh Society of Anaesthesiologists*, 22(1): 32–4.

Zimmermann, M.B. (2011) 'The role of iodine in human growth and development', *Seminars in Cell & Developmental Biology*, 22(6): 645–52.

METABOLISM AND LIVER FUNCTION

9

UNDERSTAND: CHAPTER VIDEO

Before you start reading this chapter, you might find it useful to watch a short video clip on the basics of metabolism and how it works.

This external video can be accessed by **scanning the QR codes** with your smart phone camera or via https://study.sagepub.com/essentialap2e.

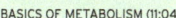

BASICS OF METABOLISM (11:04)

LEARNING OUTCOMES

When you have finished studying this chapter you will be able to:

1. Identify and describe the functions of the liver
    i. in relation to nutrient metabolism - carbohydrates, fats and proteins
    ii. immune function (Chapter 13)
    iii. drugs, toxins and hormone deactivation
    iv. bile secretion
    v. storage

2. Understand the principles of metabolism and the major metabolic pathways
3. Discuss the transport, metabolism, storage and function of the major nutrients
4. Describe energy extraction from nutrients, storage and release as necessary
5. Understand how proteins are formed from amino acids

INTRODUCTION

This chapter follows on from the previous one which looked at how nutrients entered the body. We will now examine what happens to these nutrients when they get into the body and are metabolised by cells to create and store energy and form new structures. Metabolism is a series of chemical reactions that occur in the body to maintain life. The liver plays an important role in metabolism and thus fits into this chapter; however, it also undertakes a number of additional functions which are considered here as well.

Context

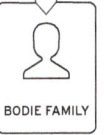

BODIE FAMILY

Nutrient requirements vary throughout life. As Danielle is breast-fed she will obtain all the nutrients and antibodies she needs in her feeds. Human breast milk is designed specifically for the requirements of human babies and has been reported as having many benefits, although there is not total unanimity in the findings. During the early stages of life as Danielle is growing fast, anabolism is greater than catabolism while body mass is increasing. After adulthood is reached, normally there is a rough balance between the two aspects of metabolism. George and Maud, in older age, will have catabolism beginning to outstrip anabolism and their body tissues will be diminishing.

An imbalance between anabolism and catabolism, as in disorders such as anorexia or in overeating, results in a decrease or increase in body stores of nutrients. Throughout life, the ideal is to ingest a balanced diet including the necessary major (carbohydrates, proteins and fats) and minor nutrients (vitamins and minerals). For all the Bodie family, a good balanced diet is really important.

APPLY

Breast-feeding and development

Findings from The United Kingdom Millennium Cohort Study (Quigley et al., 2012) corroborated the findings of a meta-analysis (Anderson et al., 1999), which identified a significant difference in cognitive ability between breast- and formula-fed infants. In the later (2012) study a large sample of infants born in the UK at the beginning of this millennium were followed up until they were five. Even allowing for confounding factors, those children breast-fed for prolonged periods were some months ahead of formula-fed children. Pre-term babies showed an even stronger effect. Further work is needed to clarify the causal relationships between type of nutrition and childhood cognitive development (Boutwell et al., 2012).

Other studies have indicated a positive relationship between breast-feeding and other outcomes including: short-term reduced morbidity and mortality in pre-term babies and those from poor conditions; long-term effects on reduced risk of obesity, type 2 diabetes and high blood pressure (Gruszfeld and Socha, 2013). Healthier lung function was demonstrated in a cohort study up to 18 years (Soto-Ramirez et al., 2012).

ACTIVITY 9.1: APPLY

Explore the link **Health Benefits of Breastfeeding** via https://study.sagepub.com/essentialap2e to read more about the benefits of breast-feeding for baby and mother.

THE LIVER

The liver contributes to the function of the body in a number of different ways. Most of this chapter is about nutrient metabolism in which the liver is important and we are starting with the liver as a foundation for the later part of the chapter.

The liver is the largest glandular organ in the body and normally weighs between 1.4 and 1.6 kg in the healthy adult. It is in the right upper quadrant of the abdomen (right hypochondrium) under the diaphragm, behind and protected by the lower ribs. As you read through the chapter you will see terms related to the liver such as hepatic or hepato from the Greek word for liver, hĕpar.

Gross anatomy divides the liver into four lobes based on surface features (Figure 9.1). When you look at the liver from the front (anterior surface) you will see the falciform ligament; this divides the liver into the left lobe and the right lobe (larger of the two). If the liver is flipped over to look at it from behind (posterior surface) you will see two more lobes between the right and left lobes; these are the caudate lobe (superior of the two) and the quadrate lobe. Also on the posterior surface is the portal fissure, which allows for entry and exit of blood vessels, lymph vessels, nerves and bile ducts.

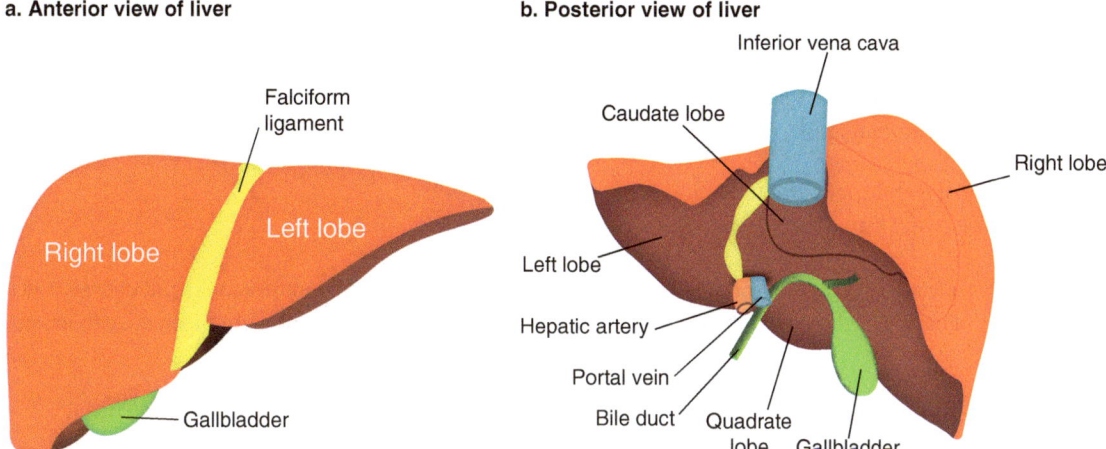

a. Anterior view of liver

b. Posterior view of liver

Figure 9.1 Anterior and posterior views of the liver

Each lobe of the liver is made up of lobules (subdivision of a lobe) (Figure 9.2). Each lobule (hexagonal in shape) consists of sheets of hepatocytes (functional cells of the liver). These hepatocytes are unipotential stem cells, i.e. they can divide into two daughter cells meaning that the liver has the ability to renew its cells (regeneration).

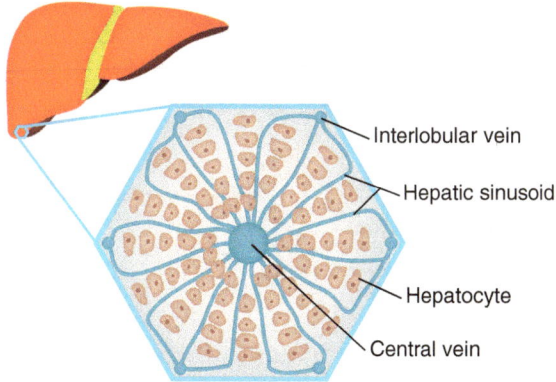

Figure 9.2 Liver lobule

Each lobule is organised around a core cluster of vessels (portal triad), i.e. a branch of the hepatic artery, a branch of the portal vein and a small bile duct. The cells are arranged in pairs of columns. Between two pairs of columns of cells are sinusoids (blood vessels whose walls are incomplete). Sinusoids are lined with highly permeable endothelium which enhances transportation of nutrients into the hepatocytes. The sinusoids also contain Kupffer cells (discussed later in the chapter).

GO DEEPER

Stem cells

There is evidence to suggest that bipotential stem cells exist in the canals of Hering (intrahepatic bile ductiles) (Ross et al., 2003). These bipotential stem cells have the ability to differentiate into hepatocytes (liver cells) or cholangiocytes (cells that line bile ducts).

Blood supply to the liver

The liver is highly vascular (Figure 9.3) receiving approximately 1,600 ml (25% of cardiac output) of blood every minute, from two vessels:

- The hepatic artery (25%).
- The portal vein (75%).

The hepatic artery branches from the coeliac artery (first branch of the abdominal aorta) and delivers a rich supply of oxygenated blood to the liver cells to ensure it is well perfused to carry out its metabolic functions.

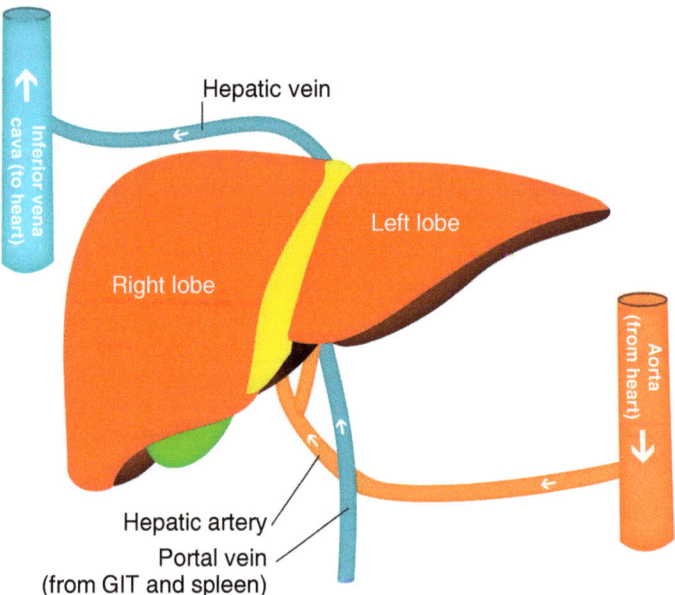

Figure 9.3 Blood supply to the liver

The portal vein delivers nutrient-rich deoxygenated blood from the stomach, pancreas, spleen and large and small intestines to the liver sinusoids so that the cells can carry out their metabolic function. Blood leaves the liver via the hepatic vein, draining into the inferior vena cava.

Following digestion most nutrients, the end products of digested food, are absorbed from the small intestine into the blood stream and transported by the **portal system** (Figure 9.4) to the liver where these nutrients are processed, stored and released as needed into the general circulation. It also delivers used red blood cells from the spleen.

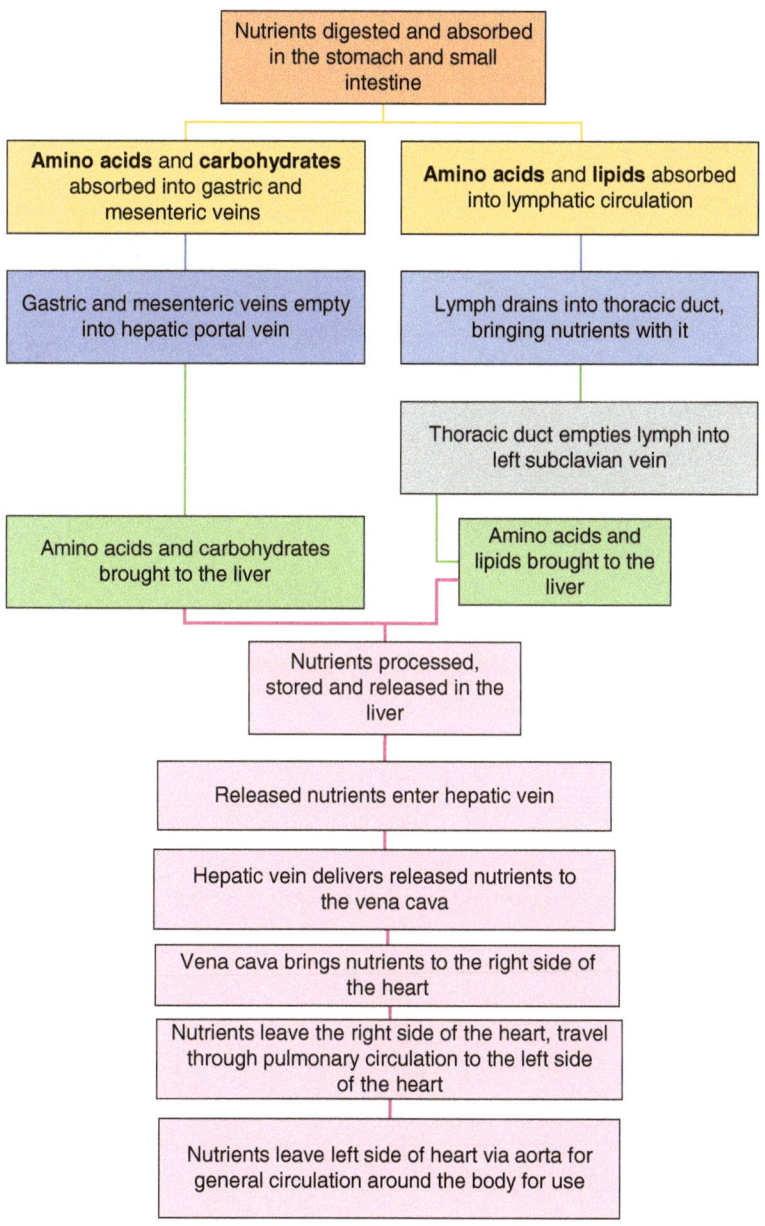

Figure 9.4 Transport of nutrients for metabolism

APPLY

Liver regeneration and scarring

The liver has the ability to regenerate, which is particularly useful when it becomes damaged. However, it can only renew itself to a certain point beyond which the liver is no longer able to regenerate. In chronic liver disease, such as cirrhosis, hepatocytes are destroyed faster than they can be regenerated (irreversible). Liver tissue is replaced by fibrotic scar tissue as well as regenerative nodules, leading to progressive loss of liver function. The formation of macro and micro nodules affects the structure of the liver and its blood supply, impeding blood flow through the liver. The resulting portal hypertension (high pressure in the portal vein) can lead to numerous complications for the person. Jaundice (yellow tint of the skin and sclera of the eye) can develop because the yellow bile is not metabolised and excreted in the bile into the small intestine, but accumulates in the blood.

ACTIVITY 9.2: UNDERSTAND

Watch this short video clip explaining the structure of the liver.

This external video can be accessed by **scanning the QR code** with your smart phone camera or via https://study.sagepub.com/essentialap2e.

LIVER ANATOMY AND BLOOD
SUPPLY (8.11)

FUNCTIONS OF THE LIVER

The liver carries out a diverse range of functions: metabolic, storage and immunological.

Metabolic functions

The liver carries out many metabolic activities which are considered here. Nutrient metabolism is discussed in some detail later in the chapter. The heat produced through metabolic activities of the liver makes a major contribution to body temperature.

Nutrient metabolism

The liver plays an important role in major nutrient metabolism, introduced here but also considered later in this chapter.

Carbohydrate metabolism

The liver carries out three major activities in relation to carbohydrates under control of the hormones insulin and glucagon (Chapter 7):

- **Glycogenesis:** the formation and storage of glycogen from glucose under the influence of the hormone insulin.
- **Glycogenolysis:** when glycogen is broken down into glucose; the hormone glucagon stimulates the release of glucose from glycogen as required.
- **Gluconeogenesis:** when there is inadequate glucose available then new glucose is formed from protein and glycerol (from fat).

Protein metabolism

The liver also performs three major roles in protein metabolism:

- **Deamination and urea formation:** when ingested amino acids are not required then the nitrogen is removed and converted into urea for excretion in urine from the kidney.
- **Transamination:** a chemical process whereby an amino group is transferred to a ketoacid to form new amino acids. The nitrogen element can be transferred from one molecule to another to form an amino acid that the body needs.
- **Protein formation:** the plasma proteins that are essential components of the plasma are mainly formed in the liver. These include:
  - **Albumin:** for the osmotic pressure of the blood (Chapter 12), and as carrier proteins.
  - **Globulins:** for the immune response, some are also formed elsewhere.
  - **Blood-clotting factors:** prothrombin, fibrinogen, factors V, VII, IX, X and XIII.

Lipid metabolism

The liver is involved in anabolism and catabolism of lipids:

- **Anabolism** (building up molecules):
  - produces triglycerides (from fatty acids and glycerol);
  - forms cholesterol for steroid and hormone synthesis;
  - synthesises lipoproteins (containing lipids and proteins) such as chylomicrons, very-low-density, low-density and high-density lipoproteins. The main function of lipoproteins is to transport triglycerides and cholesterol (Table 9.1).
- **Catabolism** (breaking down molecules):
  - triglycerides are converted to fatty acids and glycerol when an energy source is needed;
  - cholesterol is removed by hepatocytes for excretion as bile salts and sterols in bile.

Table 9.1 Lipoproteins and their functions

Lipoprotein	Function
Chylomicrons	Transport triglycerides from intestine to tissues
Very-Low-Density Lipoproteins (VLDLs)	Transport triglycerides from liver to adipose tissue and muscle
Low-Density Lipoproteins (LDLs)	Transport cholesterol from liver to tissues
High-Density Lipoproteins (HDLs)	Transport excess cholesterol from tissues back to liver

ACTIVITY 9.3: APPLY

Monitoring health through lipoproteins

Find out what the levels of lipoproteins should be in the blood and what effects changes in these might have on the person's health.

Bilirubin metabolism and bile production

As mentioned earlier, used red blood cells from the spleen are delivered to the liver via the portal vein. The Kupffer cells that line the sinusoids degrade the haemoglobin into haem and globin molecules. The haem molecule is then converted to biliverdin with iron being released; enzymes then convert biliverdin to bilirubin. Bilirubin is yellow and gives the yellow colour to bruises and the brown colour of faeces. If there is too much bilirubin in the plasma it causes discolouration of the skin and sclera (jaundice). Bilirubin is lipid soluble and cannot be excreted in bile. To make it water soluble and thus able to be excreted it combines (conjugates) with glucuronic acid (similar in structure to glucose) (or other suitable molecules) in the hepatocytes and is eliminated in the bile through the intestine. The liver is responsible for the excretion of bile and thus contributes not only to aiding digestion but acts as a means of excreting metabolites produced by the liver. The liver secretes approximately 800–1,000 ml of bile every 24 hours in the healthy adult. Bile consists of: water, mineral salts (sodium, potassium, calcium, chloride and bicarbonate), bile salts, bilirubin, cholesterol and phospholipids. The bile is stored in the gallbladder with water reabsorption leading to about 500 ml per day released into the intestine. This release is stimulated by the hormone cholecystokinin secreted from the duodenum under the stimulation of stomach contents (chyme) entering. The main function of bile is aiding the digestion and absorption of fat. As fatty acids are insoluble in water it makes them difficult to absorb through the intestinal wall. Bile emulsifies fat in the small intestine, breaking it down from large fat globules into smaller particles, thereby making them more water soluble and more easily absorbed. Bile also facilitates the absorption of fat-soluble vitamins (A, D, E and K). The bile pigments are reabsorbed and re-excreted in the entero-hepatic (intestine–liver) circulation, thus maintaining the bile necessary for fat absorption.

APPLY

Jaundice

Jaundice is characterised by elevated levels of bilirubin in the plasma. There are three main types of jaundice named in relation to the site of the disorder:

- **Pre-hepatic**: when increased red cell breakdown occurs due to disease (e.g. malaria) and it cannot all be conjugated.
- **Hepatic:** inadequate ability of hepatocytes to conjugate bilirubin due to disease (e.g. cirrhosis) or immature liver (e.g. in the newborn) leading to a build-up of unconjugated bilirubin in the blood.
- **Post-hepatic**: interruption of drainage of bile through the biliary system (e.g. due to gallstones).

Unconjugated bilirubin is not water soluble so cannot be excreted in urine. Conjugated bilirubin is water soluble and if it cannot be excreted through the biliary system it will be partly excreted through the kidneys.

ACTIVITY 9.4: UNDERSTAND

Some people, when they develop obstructive jaundice due to gallstones, have very dark yellow urine and very pale faeces. Can you explain why this might happen?

Drug metabolism

The liver metabolises the majority of drugs (and other toxins) including alcohol. Most drugs taken orally are absorbed by the small intestine and delivered to the liver via the portal vein. There are a number of different outcomes that can occur when drugs are metabolised by the liver. These are:

- Metabolites are less pharmacologically active.
- The metabolites that are produced are stronger than the drug that was initially taken.
- Inactive pro-drugs are converted to active drugs.
- Production of metabolites that are toxic.
- Elimination of metabolites which are inactive.

From the liver they are distributed through the body and act where they are needed.

 APPLY

First-pass metabolism

First-pass metabolism takes place when a drug is metabolised between its site of administration and its target tissue. The liver is usually assumed to be the major site of first-pass metabolism of a drug administered orally as it leaves the gastrointestinal tract via the portal vein. In the liver several things can happen: the drug may be unchanged; all or some of the drug may be converted to an inactive metabolite; or it may be metabolised to an active metabolite that exerts the therapeutic effect. This has implications for practice as first-pass metabolism will determine drug availability. Some drugs are completely metabolised and inactivated by the liver; this means that certain types of drugs cannot be given orally (e.g. glyceryl trinitrate taken for chest pain), therefore given under the tongue or taken as a spray. Some other drugs such as insulin (a protein) are metabolised by first-pass metabolism in the gut and also cannot be administered orally.

If drugs are partially inactivated by the liver it means that a larger dose may be required when administered orally than if administered intramuscularly, subcutaneously or intravenously for example.

Hormone synthesis and deactivation

The liver forms cholesterol molecules used in various endocrine glands to form a number of hormones, e.g. sex hormones (in the gonads), glucocorticosteroids and mineralocorticoids (in the adrenal cortex) (Chapter 7). It also contributes to the conversion of inactive vitamin D to its active form.

The liver is also responsible for deactivating most hormones through similar mechanisms to those for drug metabolism. The rate at which the liver deactivates hormones varies with the particular hormone; for example, insulin is cleared in minutes whereas it takes months to clear thyroxine.

Storage functions

The liver is responsible for storing a number of substances. These include:

- Glycogen (i.e. glucose).
- Fat-soluble vitamins A, D, E and K.
- Some water-soluble vitamins, e.g. B$_{12}$.
- Trace elements, e.g. iron (bound to ferritin in hepatocytes) and copper.

Immunological functions

The liver plays an important part in protecting the body through a number of different mechanisms including: production of acute phase proteins; Kupffer cells; tissue repair.

- Acute phase proteins are secreted by hepatocytes as an inflammatory response; synthesis depends on the presence of pro-inflammatory cytokines (small proteins involved in cell signalling). They may either increase (positive acute phase reactants) or decrease (negative acute phase reactants) in concentration. An increase in concentration (e.g. of C-reactive protein) helps to destroy or inhibit microbe growth.
- Kupffer cells are macrophages that provide a barrier preventing antigens and bacteria absorbed in the gut returning into the systemic circulation. They provide this protection by the process of phagocytosis (Chapter 13).
- Tissue repair is also facilitated by the liver. In response to the endotoxins (a toxic substance bound to the wall of bacteria) Kupffer cells release cytokines that stimulate the release of pro-inflammatory cytokines from the hepatocytes which in turn initiates an inflammatory response (Chapter 13).

--- **APPLY** ---

Plasma proteins and inflammation

As you already have learned, the liver produces plasma proteins. Acute Phase Proteins (APP) are plasma proteins whose concentrations increase (positive APP) by at least 25% during inflammatory states. APPs are released from the liver having been stimulated by interleukin-1 (IL-1) and interleukin-6 (IL-6). One of the most commonly measured APPs is C-Reactive Protein (CRP). CRP is thought to be protective, in that it binds to the surface of invading microorganisms and targets their destruction by complement and phago-cytosis (De Vita et al., 2011). APPs are detectable within hours (24-48) after the onset of the inflammatory response known as the Acute Phase Response (APR). The APR is a pathophysiological phenomenon, in which the normal homeostatic mechanisms are replaced by new values that may lead to the defensive or adaptive capabilities (Samols et al., 2002). Fever occurs through resetting the hypothalamic thermostat at a higher level. IL-1, IL-6 and Tumour Necrosis Factor (TNF) release prostaglandins in the hypothalamus, which act directly on the thermostat. An increase in the white cell count results from colony stimulating factors, stimulating the bone marrow to release leucocytes. It is important to note that the APR is not limited to acute inflammatory states but may also accompany chronic inflammation, e.g. arthritis, some carcinomas.

Clinical application

Serum concentrations of APPs are a useful indicator of the presence and intensity of inflammatory pro-cesses. Although they may not be able to diagnose specific conditions, they are helpful in differentiating between inflammatory and non-inflammatory conditions (Samols et al., 2002). They are also used to guide management during the course of inflammatory diseases. For example, if George Bodie was admitted to hospital with a chest infection, a number of bloods including CRP and full blood count (to ascertain level of white cells) may be taken and sent to the laboratory for measurement. Analysis of the blood may indicate that George's CRP is raised above the normal range (<10 mg/L) and white cell count may be beginning to elevate. These measurements would alert staff that there is an infective process and may inform the necessary treatment.

BODIE FAMILY

ACTIVITY 9.5: UNDERSTAND

Watch this video clip giving an overview of the physiology of the liver.

This online video can be accessed by **scanning the QR code** with your smart phone or via https://study.sagepub.com/essentialap2e.

LIVER PHYSIOLOGY (14:55)

APPLY

Liver function tests

In clinical practice, liver function tests are undertaken to determine if a range of biochemical markers are within their homeostatic range. Altered levels can indicate some particular illnesses that then have to be ruled out through further testing, examination and history taking. Table 9.2 below provides a summary of common liver function tests and their homeostatic ranges, including what altered levels may indicate. This is not exhaustive and ranges can vary between countries and organisations. To read about liver disease, access Chapters 12 and 13 in Cook et al.'s (2019) *Essentials of Pathophysiology for Nursing Practice*.

Table 9.2 Liver function tests

Item	Homeostatic Range (*varies by country and organisation*)	Possible Cause of Altered Levels
Bilirubin	2-17 µmol/L	**Raised:** • ↑ unconjugated levels – Gilbert's syndrome (inherited metabolic disorder) • ↑ conjugated levels – hepatobiliary obstruction, parenchymal liver disease (e.g. hepatitis), advanced cirrhosis • Excessive haemolysis

(Continued)

Table 9.2 (Continued)

Item	Homeostatic Range (varies by country and organisation)	Possible Cause of Altered Levels
Albumin	40-60 g/L	**Low:** • Malnutrition • Sepsis • Systemic inflammatory disorders • Protein malabsorption • Nephrotic syndrome (when kidney excretes too much protein in the urine)
Alkaline phosphatase (ALP)	30-120 IU/L	**Raised:** • Normal in bone growth (childhood and pregnancy) • Common bile duct obstruction • Intrahepatic duct obstruction • Metastatic bone disease • Vitamin D deficiency • Paget's disease (metabolic bone disease) • Bone fractures • Biliary cirrhosis
Aspartate aminotransferase (AST)	0-35 IU/L	**Raised:** • Hepatocyte injury or death • Pathogenic infection • Haemolytic anaemia
Alanine aminotransferase (ALT)	0-45 IU/L	**Raised:** • Hepatocyte injury or death • Pathogenic infection
Gamma-glutamyltransferase (GGT or Gamma GT)	0-30 IU/L	**Raised:** • Cirrhosis • Hepatitis • Haemochromatosis • Non-liver related (e.g. obesity, excess alcohol intake) Some medications can alter results, making them unreliable (e.g. phenytoin, carbamazepine and alcohol).
Prothrombin time (PT)	10.9-12.5 sec	**Prolonged:** • Acute or chronic liver dysfunction • Vitamin K deficiency • Chronic cholestasis (decreased bile flow or production) • Anticoagulant use
International normalised ratio (INR)	- 1.2 (not on anticoagulants) - 3.0 (on anticoagulants)	**Raised:** • Acute or chronic liver dysfunction • Vitamin K deficiency • Chronic cholestasis (decreased bile flow or production) • Anticoagulant use

METABOLISM FOR ENERGY

Building on what you now know about the liver, we are next going to look at metabolism.

Adenosine triphosphate (ATP) is the key energy currency in living beings, and the formation and utilisation of ATP are generally in balance. Metabolism consists of the two sets of opposing reactions which achieve this balance:

- Catabolism is the breakdown of nutrients from ingested food to provide energy stored in energy-rich compounds, mainly ATP, for the activities of the body, including anabolism. In addition, heat is produced.
- Anabolism is the building up of larger molecules for replacement structures and for storage of nutrients in larger molecules. It requires the energy from ATP.

The amount of energy released in catabolism is the same as would be produced by burning the nutrients in oxygen outside the body.

- 1 gram of carbohydrate releases 4.1 kcal (17.2 kJ)[1] of energy.
- 1 gram of protein releases 4.1 kcal (17.2 kJ) of energy.
- 1 gram of fat releases 9.3 kcal (38.9 kJ) of energy.

The energy output is used for a range of purposes (Figure 9.5).

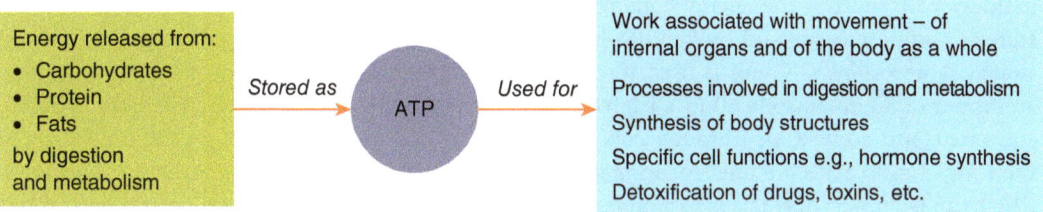

Figure 9.5 Production and use of energy

Energy requirements vary from one person to the next, depending on factors such as age, sex, body composition and physical activity level. Energy usage is the sum of the basal metabolic rate (the amount of energy expended while at complete rest which is controlled by thyroid hormones – Chapter 7), the Thermic Effect of Food (TEF, the energy required to digest and absorb food) and the energy expended in physical activity.

Major nutrients and energy metabolism

We are now going to examine what happens to these major nutrients in the body.

---

[1]Energy intake is often measured in Joules (J) (or kiloJoules (kJ)) but many people are more familiar with calories (kcal).

- 1 kiloJoule (kJ) = 1,000 Joules
- 1 megaJoule (MJ) = 1,000,000 Joules
- 1 kilocalorie (kcal) = 1,000 calories, or 1 large calorie

To convert from one unit to another:

1 kcal = 4.184 kJ, so a 1,000 kcal diet provides 4.184 MJ or 4184 kJ
1 MJ = 239 kcal.

Major pathways for energy formation

Here we are going to examine the main chemical pathways by which the major nutrients are metabolised in relation to energy management, and the relationships between them. It is important to remember that (under normal circumstances) the brain can only metabolise glucose for energy while other tissues can also use fatty acids through breakdown to acetyl coenzyme A (acetyl CoA). Thus, the continued supply of glucose for brain function is essential.

As already indicated, the main molecule that stores energy is ATP created through a chemical cycle where ADP is converted to ATP by the addition of a phosphate group linked by a high-energy bond. When that energy is required, ATP is converted back to ADP with the release of energy (Figure 9.6).

Cellular respiration is the process of obtaining energy through oxidising nutrient molecules and trapping this energy in ATP. Significantly more energy is made available from nutrients in the presence of oxygen (aerobic respiration) than without it (anaerobic respiration). Aerobic respiration is the process used during normal activity in health.

Maintaining blood glucose level

Gluconeogenesis means 'formation of new glucose' and this is where the other nutrients contribute. Fats (lipids) only contribute a small amount of glucose through glycerol entering the gluconeogenesis pathway with most of the energy coming through acetyl CoA from lipolysis (Figure 9.7). Fats are stored in adipose tissue and broken down into glycerol and fatty acids for release into the blood stream as necessary. Glycerol is used through gluconeogenesis but contributes only three carbon atoms out of approximately 50 in a triglyceride molecule to glucose formation. The fatty acids are converted into acetyl CoA (through beta (β) oxidation) and used for energy production through the CAC. Amino acids are used for formation of structural and functional proteins (i.e. enzymes), which can be broken down back to amino acids. Excess amino acids are broken down: the nitrogen is removed, converted to urea in the liver and excreted in urine through the kidneys. The sugar remaining is converted to glucose or other molecules of the glycolysis or CAC pathways and metabolised for energy.

GO DEEPER

Nutrient metabolism and energy

Figure 9.7 shows how all the different nutrients contribute to the formation of acetyl CoA which contains two carbon atoms. Acetyl CoA enters the Citric Acid Cycle (CAC) (otherwise known as the tricarboxylic acid cycle or the Krebs cycle – after Sir Hans Krebs, the scientist who first described it) through which it loses two carbon atoms. Theoretically, during aerobic metabolism 1 gram of glucose produces 38 molecules of ATP; however, in practice some membrane leakage occurs resulting in about 28-30 ATP molecules per gram of glucose. Figure 9.7 relates to most of this chapter.

Carbohydrates are the major energy source for body function. In the GIT they have already been digested into monosaccharides, i.e. simple sugars (e.g. glucose, fructose) before absorption into the body proper. Surplus glucose is converted into glycogen (a polysaccharide, long chain of monosaccharides) stored in the liver and which is broken down between meals to maintain the blood glucose level. When necessary, protein and fats contribute as energy sources.

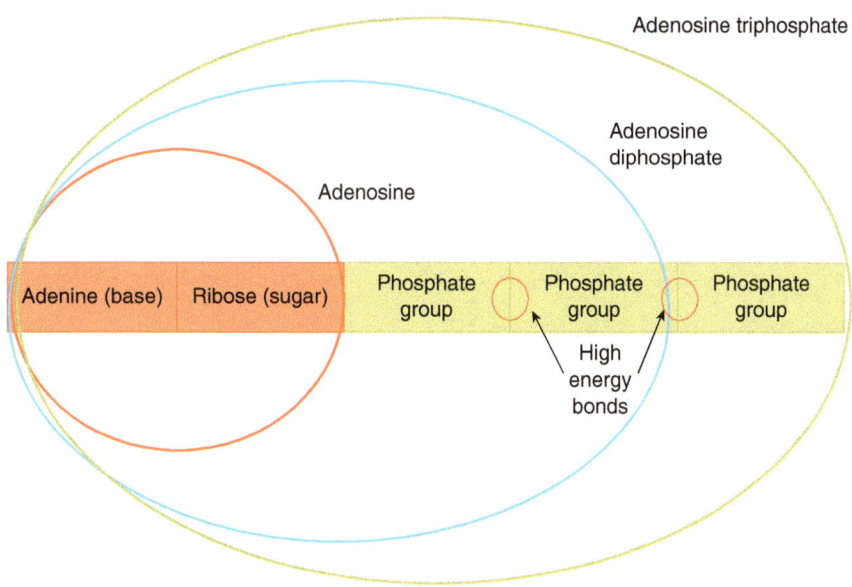

Adenosine = Adenine + Ribose

Adenosine diphosphate = Adenine + Ribose + Phosphate group + Phosphate group

Adenosine triphosphate = Adenine + Ribose + Phosphate group + Phosphate group + Phosphate group

(a) Structure of ATP

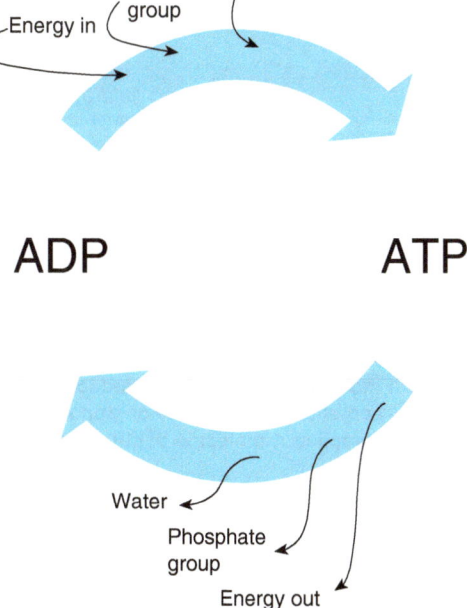

(b) Conversion between ADP and ATP

Figure 9.6 The ADP–ATP cycle

(Continued)

Within the cells, glucose is metabolised through the glycolysis pathway (without oxygen) in the cytoplasm to form pyruvate which, as shown in Figure 9.7, can be converted to acetyl CoA and enters the CAC (which requires oxygen) in the mitochondria to form ATP.

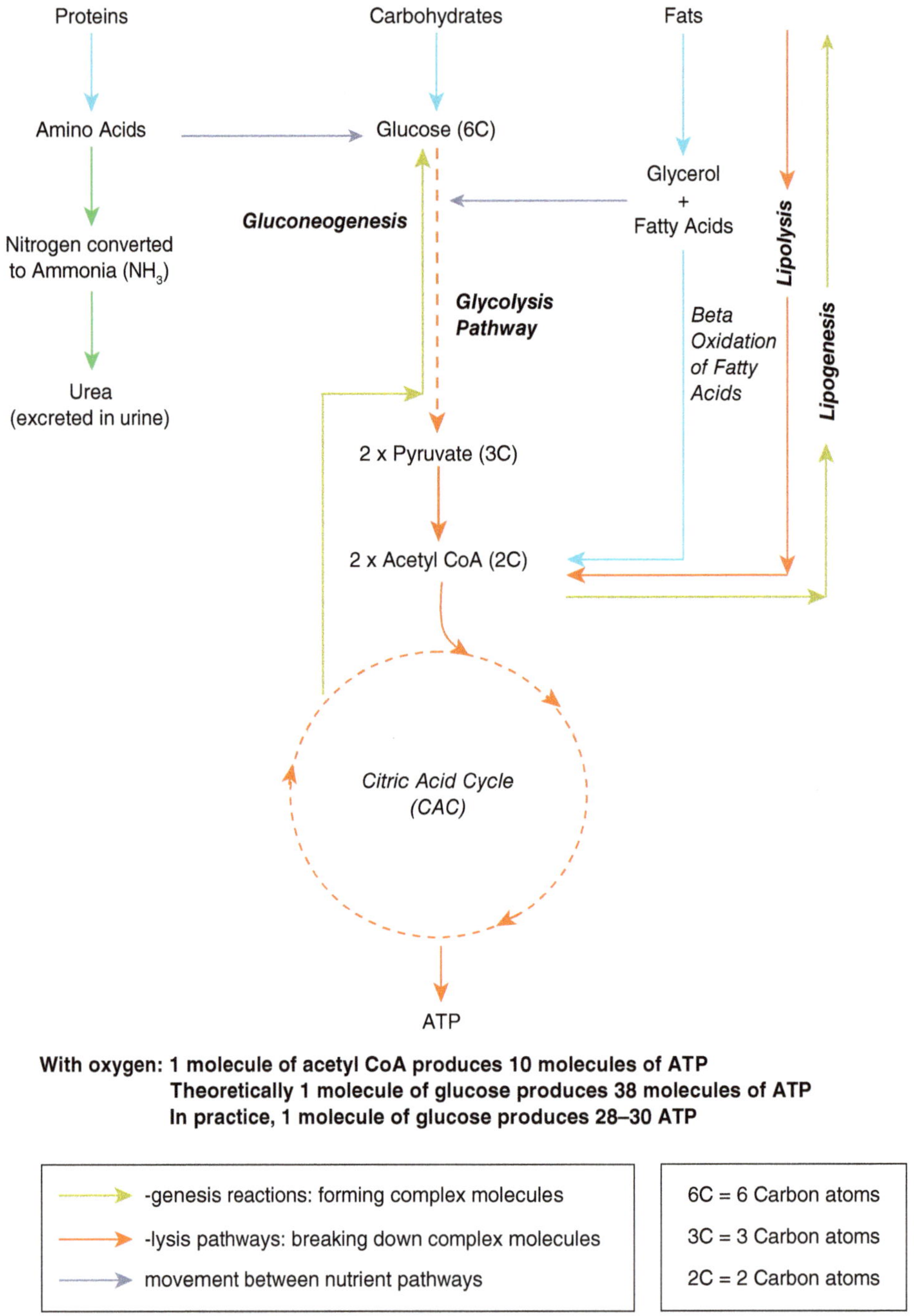

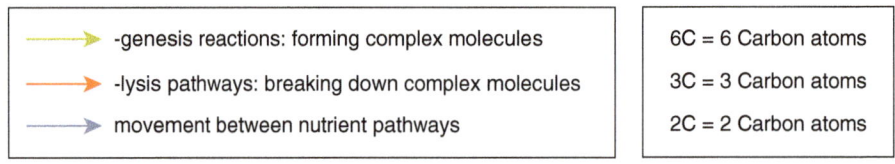

Figure 9.7 Outline of major nutrient metabolism and energy

Metabolism of nutrients in the absence of oxygen (anaerobic metabolism) occurs when the work required is greater than can be produced with the oxygen supplied to the cells. Thus, athletes exercising hard may get a stitch because of the build-up of metabolites due to inadequate oxygen supply. In anaerobic metabolism, one gram of glucose produces only two molecules of ATP.

APPLY

Heart failure and exercise tolerance

People with heart failure or chronic respiratory disease may have difficulty in supplying adequate oxygen to the tissues and there may be some degree of anaerobic with limited aerobic metabolism. This will reduce their ability to undertake physical activity and diminish mental agility.

Regulation of blood glucose level

In most of the world, blood or plasma glucose level is measured in millimoles per litre (mmol/L), although the USA still uses milligrams per decilitre (mg/dL). Normal blood glucose level is about 4 mmol/L (72 mg/dL) but can range to 6.1 mmol/L (110 mg/dL). When we eat a meal the blood glucose level may rise temporarily up to about 7.8 mmol/L (140 mg/dL). However, the measurements vary slightly between different laboratories.

Blood glucose levels are regulated by the action of hormones, mainly pancreatic, but also some others (Chapter 7, Figure 7.12). The pancreas has four cell types which produce different hormones:

- α **(alpha):** glucagon (peptide hormone) raises the blood glucose level when it falls too low by stimulating breakdown of glycogen in the liver into glucose.
- β **(beta):** insulin (peptide hormone) lowers the blood glucose level by promoting glucose absorption from blood into skeletal muscle and fat cells, and enhancing fat storage, as well as stimulating formation of glycogen in the liver.
- δ **(delta):** somatostatin (peptide hormone) (also known as growth-hormone-inhibiting hormone). Regulates the endocrine system: inhibits insulin and glucagon secretion.
- γ **(gamma or PP):** pancreatic polypeptide, which helps to regulate pancreatic function. It reduces appetite. Secretion is initiated by the vagus nerve, and some of the GI hormones.

Insulin and glucagon play major roles in regulating metabolism of carbohydrates, proteins and fats. Insulin is primarily produced in the pancreas and has an anabolic action. Insulin receptors in cell walls trigger a process that leads to transporter proteins in cell membranes to allow glucose from the blood to enter the cell. Insulin also enables glucose storage as glycogen in the liver, storage of fatty acids (as triglycerides) and amino acids in cells and tissues. Glucagon is the major catabolic hormone that, with a number of other hormones, stimulates release of nutrients into the blood. Somatostatin inhibits secretion of both the major hormones (insulin and glucagon) while pancreatic polypeptide also influences glucose metabolism.

GO DEEPER

Pancreatic polypeptide and appetite

Pancreatic polypeptide helps to regulate pancreatic secretions, both endocrine and exocrine, and influences gastrointestinal secretions and glycogen storage in the liver (Lonovics et al., 1981). It has also been

(Continued)

shown to reduce appetite and food intake in most people and secretion is increased after protein intake, and in situations when blood glucose is low such as fasting, exercise and anorexia nervosa (Batterham et al., 2003).

APPLY

Impaired glucose regulation

Both hypoglycaemia (low blood glucose) and hyperglycaemia (high blood glucose) can cause damage. Hypoglycaemia results in inadequate energy available to the brain leading to abnormal behaviour (sometimes mistaken for drunkenness). If prolonged, the person may become unconscious and, if not treated, may die. Hyperglycaemia, as in diabetes mellitus, can result in damage to many body organs leading to, for example, renal failure, blindness or gangrene resulting in leg amputation.

METABOLISM OF NUTRIENTS

Having discussed the interrelationship between the nutrients in providing energy for body function, we are going to look in rather more detail at the specific metabolism of the different nutrients, building on Figure 9.7. In this book we aim to give a reasonable understanding of nutrient metabolism and how the cells work. However, if you want more detail there are numerous suitable books including those listed at the end of the chapter.

GO DEEPER

Citric acid cycle (CAC) and electron transport chain (ETC)

As shown in Figure 9.7, pyruvate (three carbon (C) atoms) is formed through the glycolysis pathway in the cell cytoplasm. It is transferred into mitochondria, converted to acetyl CoA (two carbons) and enters the CAC. The CAC with the Electron Transport Chain (ETC) manage energy production requiring oxygen. Figure 9.8 outlines this and should be looked at along with this description. The cycle begins with acetyl CoA (2C) combining with oxaloacetate (4C) to form citric acid (citrate) (6C). This passes around the CAC through the various molecules indicated. Two carbon dioxide (CO_2) molecules remove the initial two carbon molecules ending back with oxaloacetate. In the process, a number of other molecules NADH (nicotinamide adenine dinucleotide hydride), $FADH_2$ (flavin adenine dinucleotide) and GTP (guanosine triphosphate)) are formed, which contribute to the formation of ATP. GTP is converted directly to ATP, while NADH and $FADH_2$ are dealt with through the ETC within the inner membrane of mitochondria. Electrons from NADH and $FADH_2$ are transferred to oxygen to form water. This takes place in a number of steps with the electrons being passed along the chain of electron carriers ending with the final electron acceptor, oxygen. 'In effect the electrons released from NADH and $FADH_2$ bump down a staircase … with each fall releasing free energy' (Elliot and Elliot, 2009: 183).

The overall oxidation of glucose is shown by the equation:

$$C_6H_{12}O_6 + 6O_2 \rightarrow 6CO_2 + 6H_2O$$

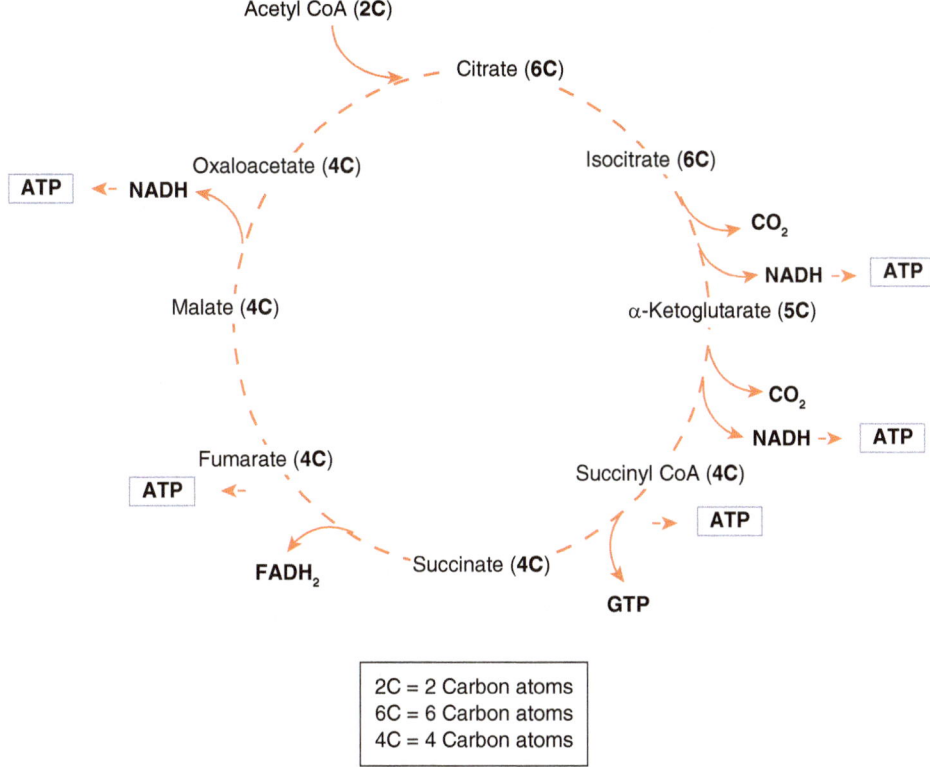

Figure 9.8a The citric acid cycle

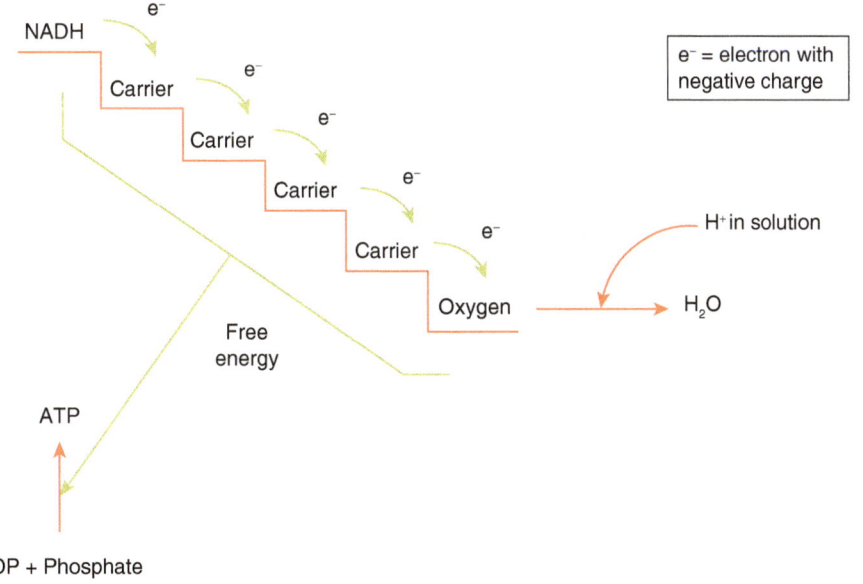

Figure 9.8b The electron transport chain

ACTIVITY 9.6: UNDERSTAND

Watch the following video clip to help develop your understanding of cellular respiration.

This external video can be accessed by **scanning the QR code** with your smart phone or via https://study.sagepub.com/essentialap2e.

CELLULAR RESPIRATION (14:18)

Carbohydrates

Glycogenesis and glycogenolysis

Glucose is the nutrient primarily used for energy by the brain, and also used by other tissues when plenty is available following a meal. During and shortly after a meal, glucose is converted in the liver to glycogen for storage. As the blood glucose level begins to drop before the next meal, glucose is released from glycogen for brain function; other tissues utilise ketones (from fatty acid metabolism) instead. In addition, muscle cells can convert glucose to glycogen, which is stored in the cells, but only for use within the muscle cell.

Figure 9.9 illustrates the structure of sugar, how molecules combine to form glycogen, and the balance between glycogenesis and glycogenolysis.

APPLY

Glycogen store depletion

Under conditions of starvation, glycogen stores are completely used up and protein (i.e. muscle) begins to be used for gluconeogenesis. Glucose levels begin to fall and, under these conditions, brain cells will begin to synthesise the enzymes to enable catabolism of ketones (from fats) for energy. Additionally, neurons can also use lactate as a form of energy (Chapter 5).

Glycolysis

The glycolysis pathway (Figure 9.10) takes place in the cytoplasm of the cell and converts one glucose to two pyruvate molecules in the absence of oxygen. In the process two molecules of ATP and two of NADH are produced. In the absence of oxygen, pyruvate is converted to lactate, enabling athletes to continue using glucose for energy even when the blood is supplying inadequate oxygen. When oxygen is restored, the lactate is reformed into pyruvic acid (pyruvate) and metabolised through the citric acid cycle. If this cannot happen it is transferred to the liver, converted to glucose and then is available for storage or use.

a. Glucose molecule – contains 6 Carbon atoms

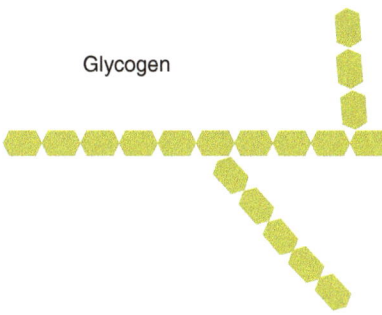

b. Glycogen composed of glucose molecules linked in branching chains through glycogenesis

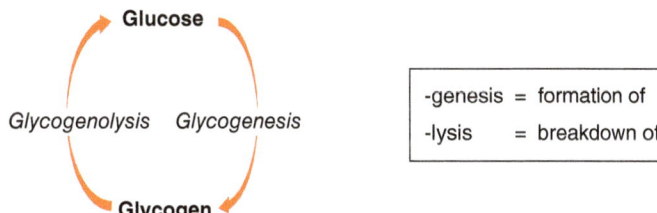

Glycogen

c. Glucose and glycogen balance

Glucose

Glycogenolysis   *Glycogenesis*

Glycogen

-genesis	=	formation of
-lysis	=	breakdown of

Figure 9.9   Glycogenesis and glycogenolysis

ACTIVITY 9.7: UNDERSTAND

Watch the following video clip to help develop your understanding of glycolysis.

This online video can be accessed by **scanning the QR code** with your smart phone or via https://study.sagepub.com/essentialap2e.

GLYCOLYSIS:
THE REACTIONS (5.09)

─── **GO DEEPER** ───

Glycolysis

Figure 9.10 shows the simplified glycolysis pathway.

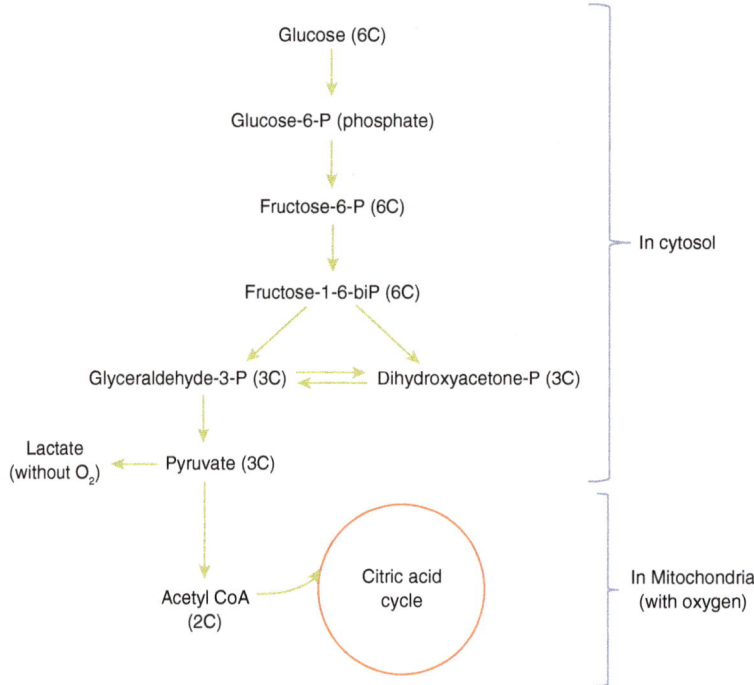

Figure 9.10 Glycolysis – simplified pathway

Gluconeogenesis

This is the formation of new glucose (Figure 9.7). Proteins are broken down into amino acids, the nitrogen is removed and converted into urea for elimination. The remainder of the molecule is converted into glucose or other molecules of the glycolysis pathway or CAC, carried in the blood stream to the brain and, following entry into the brain cells, is metabolised as already described.

Amino acids and proteins

Amino acids

The amino acids which enter the body are used to form structural proteins in the continuous turnover of many body tissues, used in gluconeogenesis, or converted to urea and excreted. Figure 9.11 shows the basic structure of all 20 amino acids, how they combine to form polypeptide chains (i.e. chains of amino acids), and their structure.

Transamination and deamination

Some of the amino acids used in humans are essential (Chapter 8) because they must be present in the diet and cannot be formed from any other source. The non-essential amino acids can be formed by transamination – that is, through transfer of the amino group to another carbohydrate molecule within the cell (Figure 9.12). Amino acids that are not immediately needed for forming proteins in the body are dealt with by deamination in the liver. The nitrogen within the amino acid is removed, converted to ammonia (NH_3) and then into urea which is transported in the blood to the kidneys and excreted in the urine.

Protein structure

We have looked at the structure of the different amino acids and how these combine to form polypeptide chains. Polypeptide chains are formed in ribosomes against the template in messenger RNA, which is formed in the nucleus of the cell, passes through the nuclear pores and combines with the ribosomes. Transfer RNA picks up the particular amino acid coded for by each group of three nucleotides and carries it to the ribosome where it is added to the end of the polypeptide chain being formed. The roles of DNA and RNA are central to protein formation (Chapter 2).

However, this simple chain of amino acids is only the primary structure of proteins. The secondary structure is the shape it folds up into – usually an α-helix, in which the chain twists into a right-handed helix, and the β-pleated sheet (a secondary structure occurring in many proteins that consists of two or more parallel adjacent polypeptide chains arranged in such a way that hydrogen bonds can form between the chains). Different amino acids have varying tendencies to form differing secondary structures. The tertiary structure is determined by the affinity to water of the different parts of the chain. The secondary structure becomes folded up and organised with the hydrophobic (water-repelling) side chains in the centre of the protein molecule or tertiary structure (Figure 9.13a).

Many proteins stop at this point, but some are formed into a quaternary structure when more than one tertiary protein link together to form the final active protein (Figure 9.13b). Haemoglobin is one such protein which consists of four globin chains (two alpha and two beta chains) each with a haem molecule containing iron in the centre.

GO DEEPER

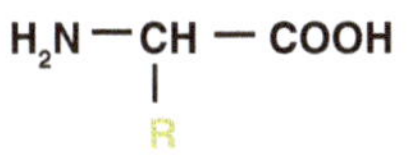

$H_2N-CH-COOH$ with R below

—COOH = Carboxylic Acid functional group
H_2N- = Amino functional group
R = side chain – varies in each amino acid

Figure 9.11a General structure of amino acids

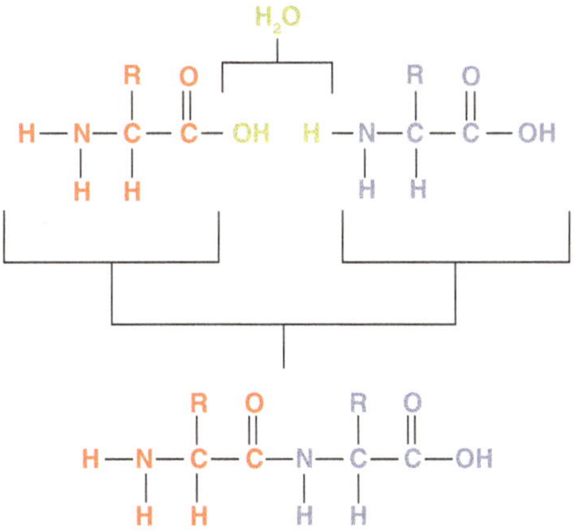

Figure 9.11b Condensation reaction (i.e. water removed from reacting molecules) in formation of polypeptides and proteins

(Continued)

R =	Name	Comments
-H	Glycine	
-CH$_3$	Alanine	
-CH$\big\langle^{CH_3}_{CH_3}$	Valine	Hydrophobic amino acids (have low affinity for water – do not dissolve in it)
-CH$_2$—CH$\big\langle^{CH_3}_{CH_3}$	Leucine	
- CH$\big\langle^{CH_3}_{CH_2-CH_3}$	Isoleucine	
- CH$_2$ — CH$_2$ — S — CH$_3$	Methionine S = Sulphur	Hydrophobic. Involved in methylation
- CH$_2$ —◯	Phenylalanine	Hydrophobic
- CH$_2$ —◯— OH	Tyrosine	
- CH$_2$ ⟨NH ring⟩	Tryptophan	Aromatic side chains (cyclic rings)
- CH$_2$ — COOH	Aspartate	Hydrophilic (have an affinity for water)
- CH$_2$ — CH$_2$ — COOH	Glutamate	
- CH$_2$ — CH$_2$ — CH$_2$ — CH$_2$—NH$_2$	Lysine	
- CH$_2$ — CH$_2$ — CH$_2$ — NH — C$\big\langle^{NH_2}_{NH_2}$	Arginine	
- CH$_2$ ⟨HN NH ring⟩	Histidine	
- CH$_2$ — C$\big\langle^{NH_2}_{O}$	Asparagine	Hydrophilic (but less so than above)
- CH$_2$ — CH$_2$ — C$\big\langle^{NH_2}_{O}$	Glutamine	
- CH$_2$OH	Serine	Weakly hydrophilic
- CH — CH$_3$ | OH	Threonine	
- CH$_2$ — SH	Cysteine	Hydrophobic. Supplies SH groups for certain enzymes
HN — CH — COOH | | CH$_2$ CH$_2$ \\ / CH$_2$	Proline	Unusual structure. Puts kink into polypeptide chains

Figure 9.11c Variation in R side chain

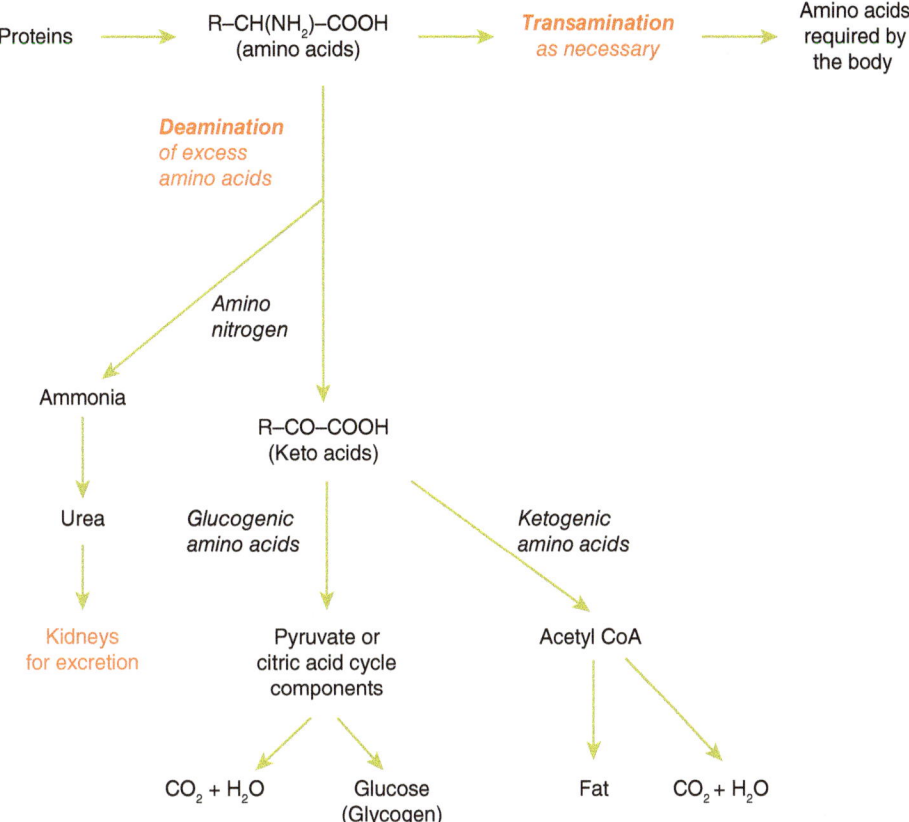

Figure 9.12 Metabolism of amino acids

(a) First three levels of protein structure

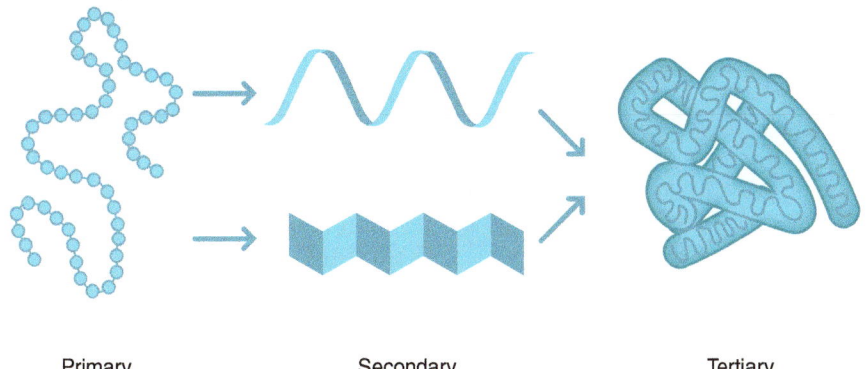

Primary   Secondary   Tertiary

Figure 9.13 (Continued)

(b) Quaternary structure - example haemoglobin

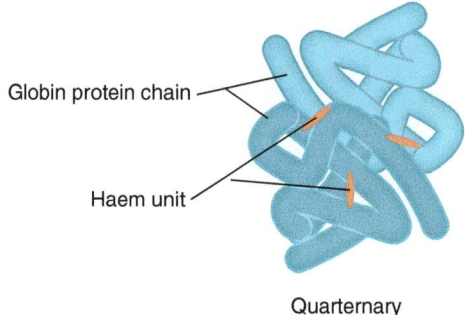

Globin protein chain

Haem unit

Quarternary

Figure 9.13 Four levels of protein structure

Enzymes

Enzymes play a crucial role in regulation of metabolism. They are biological catalysts made of proteins that accelerate, or catalyse, chemical reactions. Catalysts accelerate chemical reactions without themselves being used up. Substrates (molecules at the beginning of the process) are converted into different molecules (products) by the enzyme. You have already come across enzymes involved in digestion in GIT. However, enzymes also exist in the cytosol of every cell of the body and some are positioned within the cell or organelle membrane structure. Almost all metabolic processes in the cell need enzymes in order to occur at rates fast enough to sustain life. Normally, an enzyme will only catalyse one specific reaction and the presence or absence of the enzyme determines the rate of reaction.

Fats (lipids)

There are three groups of lipids and their structures are shown in Figure 9.14.

GO DEEPER

(a) Triglycerides

i. Showing glycerol (in green) and three fatty acids (in purple). The circle indicates how each fatty acid combines with glycerol

ii. Showing glycerol and fatty acids combined as triglyceride. The bottom fatty acid has a double bond in the fatty acid

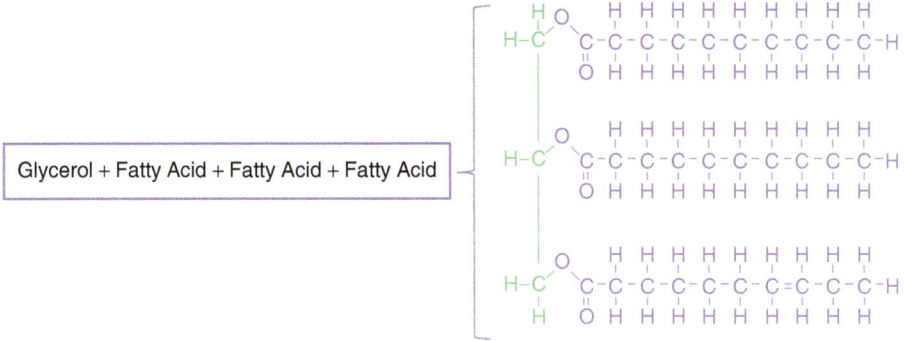

(b) Phospholipids

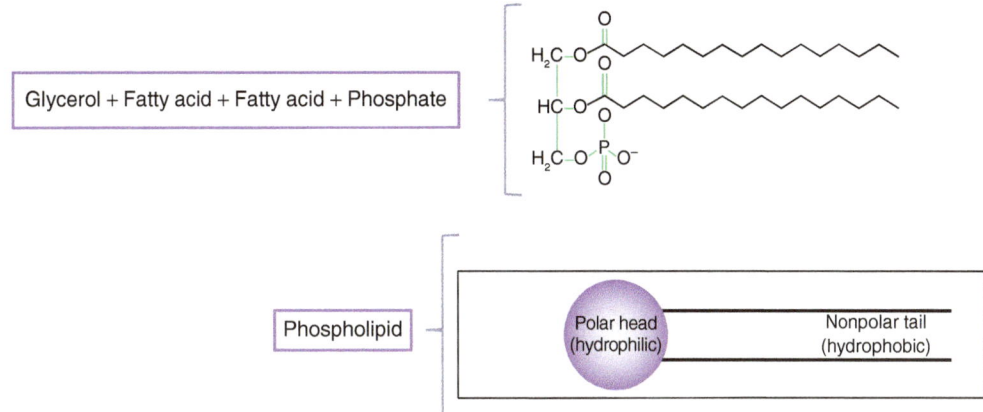

Glycerol + Fatty acid + Fatty acid + Phosphate

Phospholipid

(c) Structure of cell membrane

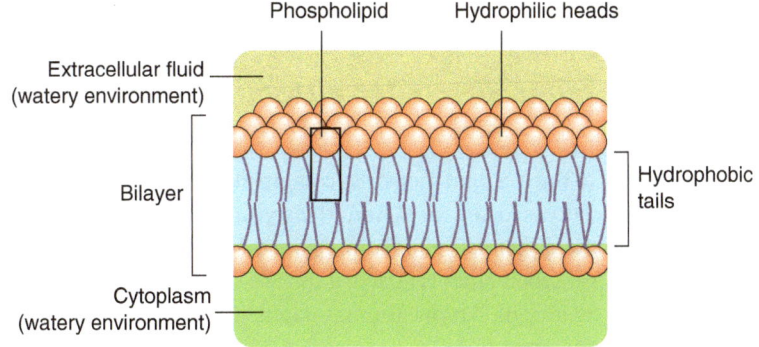

(Continued)

Figure 9.14 (Continued)

(d) Steroids

   i. The steroid nucleus (each number indicates the position of a carbon atom)

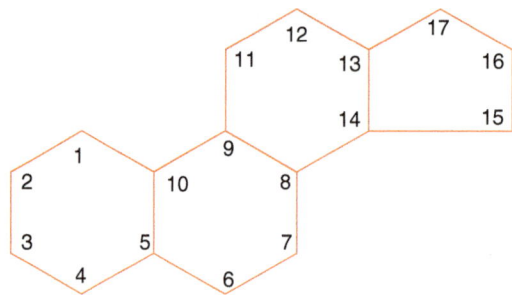

   ii. Steroids of physiological importance (adapted from Pocock and Richards, 2009)

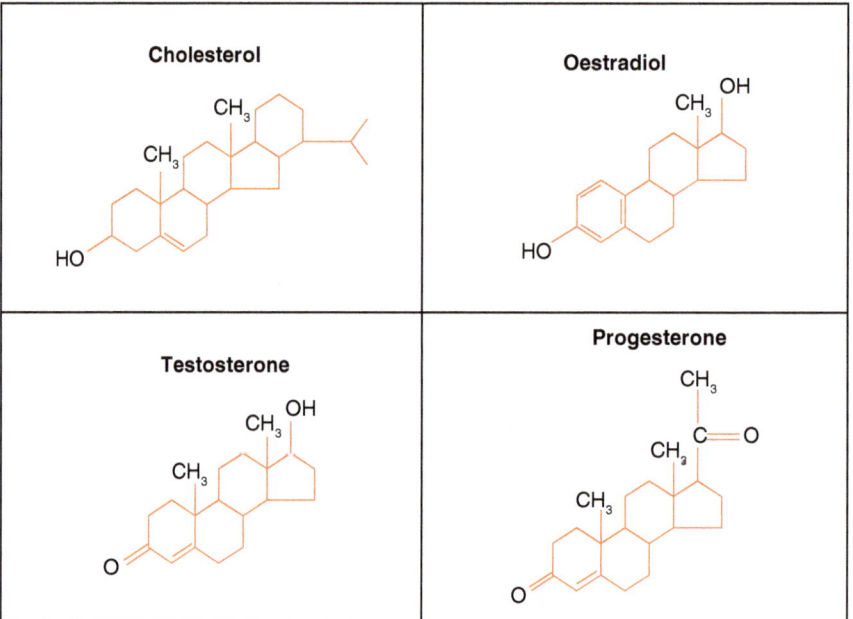

Figure 9.14 (a) Triglycerides (b) Phospholipids (c) Structure of cell membrane (d) Steroids

Triglycerides

These are formed from glycerol (a three carbon molecule) combined with three fatty acids. Figure 9.14a(i) indicates how an OH from the fatty acid and an H from the glycerol (circled) form H_2O (water) leaving the fatty acid combined with the glycerol as shown in Figure 9.14a(ii). Two of the fatty acids in this figure are saturated, that is every carbon atom is attached to those each side by a single bond (with the additional bonds linking with hydrogen), but one is unsaturated, that is it has a double (or it could be a triple) bond between two carbon atoms.

Triglycerides are the major storage form of fat. They are formed from reconstitution of the glycerol and fatty acids absorbed into the lymphatic system from the GIT. Fatty acids can also be built up through lipogenesis from acetyl CoA as indicated in Figure 9.7. When needed, fats are broken down again into the constituent parts and contribute to ATP formation:

- Glycerol is converted to another three carbon molecule and feeds into the glycolysis pathway. It can enter the gluconeogenesis pathway.
- Fatty acids go through the lipolysis cycle and two carbon groups are broken off with each turn of the cycle, converted to acetyl CoA and feed into the CAC.

Phospholipids

There are several types of phospholipids but the commonest are glycerophospholipids. These are similar in structure to triglycerides in that they consist of glycerol with two fatty acid chains, but with a phosphate molecule attached to the third carbon (Figure 9.14b). The phosphate group makes this end of the molecule polar (hydrophilic, i.e. having an affinity for water): the other non-polar end is hydrophobic (i.e. repels water).

Phospholipids play an important role in the structure of cell membranes (Chapter 2, Figure 2.9), with the hydrophilic end on the outside of the membrane and with the hydrophobic end on the inside isolated from the surrounding polar fluids. They are carried round the body in combination with proteins as lipoproteins (Chapter 8).

Steroids

These are an important group of molecules with a common structural framework shown in Figure 9.14 c (i). Cholesterol is the most common steroid molecule and is the precursor for the steroid hormones, some of which are shown in Figure 9.14 c (ii).

Nucleic acids

The structure and function of nucleic acids are considered in Chapter 2 as they are core components in regulating bodily structure and function.

CONCLUSION

Because metabolism is the core of body function, it is important to understand how this occurs and is managed. This chapter has introduced the fate of nutrients once they have entered the internal environment of the body. The importance of glycolysis, the citric acid cycle and the electron transport chain in the production and storage of energy have been considered. The ways in which different nutrients feed into these pathways and are utilised within the body have been described.

The liver plays a vital role in metabolism and its range of activities has been described.

--- GO DEEPER ---

Further reading

Day, H.L. (2005) 'The liver, part 1: anatomy', *Nursing Times*, 101(48): 22.
Day, H.L. (2005) 'The liver, part 2: physiology', *Nursing Times*, 101(49): 22.

- This chapter relates closely with Chapter 2, which considers the cell, and Chapter 8, on how nutrients enter the body. You should try to relate the content of this chapter to the previous ones and bear in mind this material when studying the remainder of this book.

- The constant supply of energy for cell function is essential. Glucose is the major energy source, but fats and proteins can also contribute to energy production. The brain is dependent on glucose supply while other tissues can also use ketones. (Prolonged starvation stimulates the 'turning on' of the enzymes in the brain to enable use of ketones.)

- The major pathways in energy metabolism are the glycolysis pathway, citric acid cycle and electron transport chain.

- Other pathways enable:

  o conversion of nutrients into molecules within the pathways for energy production;

  o conversion of nutrients into storage forms which can be made available for energy as necessary.

- Enzymes are proteins that act as catalysts for the numerous reactions that take place in the body.

- The different lipids in the body play important roles in cell membranes, steroid hormones and other lipid-based structures.

- The liver is very important in storage of various vitamins and minerals, as well as in metabolism and immunity.

REVISE

TEST YOUR KNOWLEDGE

The content of this chapter is important to understand what happens to the working of the body when anything goes wrong with energy production, storage and use. Understanding this chapter will be particularly useful when you begin to study disorders of energy management (e.g. diabetes mellitus, anorexia). To achieve this, the areas to revise are as follows:

1 The functions of the liver

- in relation to nutrient metabolism - carbohydrates, fats and proteins;
- immune function (Chapter 13);
- drugs, toxins and hormone deactivation;
- bile secretion;
- storage.

2 Metabolism for energy, including glycolysis, citric acid cycle and electron transport chain.

3 Regulation of blood glucose level.

4 The major pathways in the metabolism of carbohydrates, fats and proteins.

5 The uses to which the different nutrients are put.

In order to help you revise, consider the following questions, answers for which can be found by visiting **https://study.sagepub.com/essentialap2e**.

Test yourself by revising the chapter first, and then answer these questions without looking at the book. Afterwards compare your answers with the text, including the diagrams, and with the notes you made. Did you miss anything in your notes? Here are the questions:

1   List and briefly explain the functions of the liver.

2   Draw an outline labelled figure showing the anterior and posterior structure of the liver.

3   Describe the blood supply of the liver.

4   Briefly explain catabolism and anabolism.

5   How is energy stored and released in the cells of the body?

6   Outline the major pathways for energy formation.

7   Briefly outline the main pathways of glucose metabolism.

8   Briefly describe the four levels of protein structure.

9   What are the three main groups of lipids and their main functions?

For additional revision resources visit: **https://study.sagepub.com/essentialap2e**.

- Revise key terms with interactive flashcards.

- Test yourself with multiple-choice questions.

- Access the glossary with audio to hear how complex terms are pronounced.

- Explore recommended websites suitable for revision.

REVISE

ACE YOUR ASSESSMENT

REFERENCES

Anderson, J.W., Johnstone, B.M. and Remley, D.T. (1999) 'Breast-feeding and cognitive development: a meta-analysis', *American Journal of Clinical Nutrition*, 70(1): 525–35.

Batterham, R.L., Le Roux, C.W., Cohen, M.A., Park, A.J., Ellis, S.M. et al. (2003) 'Pancreatic polypeptide reduces appetite and food intake in humans', *The Journal of Clinical Endocrinology and Metabolism*, 88(8): 3989–92.

Boutwell, B.B., Beaver, K.M. and Barnes, J.C. (2012) 'Role of breastfeeding in childhood cognitive development: a propensity score matching analysis', *Journal of Paediatrics & Child Health*, 48(9): 840–5.

Cook, N., Shepherd, A., Boore, J. and Dunleavy, S. (2019) *Essentials of Pathophysiology for Nursing Practice*. London: Sage.

De Vita, V.T. Jr, Lawrence, T.S. and Rosenberg, S.A. (eds) (2011) *Cancer: Principles and Practice of Oncology*, 9th edn. Philadelphia: Wolters Kluwer Health/LWW.

Elliot, W.H. and Elliot, D.C. (2009) *Biochemistry and Molecular Biology*, 4th edn. Oxford: Oxford University Press.

Gruszfeld, D. and Socha, P. (2013) 'Early nutrition and health: short- and long-term outcomes', in H. Szajewska and R. Shamir (eds), 'Evidence-based research in pediatric nutrition', *World Review of Nutrition & Dietetics*, 108: 32–9.

Lonovics, J., Devitt, P., Watson, L.C., Rayford, P.L. and Thompson, J.C. (1981) 'Pancreatic polypeptide', *Archives of Surgery*, 116(10): 1256–64.

Pocock, G. and Richards, C. (2009) *The Human Body: An Introduction for the Biomedical and Health Sciences*. Oxford: Oxford University Press.

Quigley, M.A., Hockley, C., Carson, C., Kelly, Y., Renfrew, M.J. et al. (2012) 'Breastfeeding is associated with improved child cognitive development: a population-based cohort study', *The Journal of Pediatrics*, 160(1): 25–32.

Ross, M.H., Kaye, G.I. and Pawlina, W. (2003) *Histology: A Text and Atlas*. New York: Lippincott Williams & Wilkins.

Samols, D., Agrawal, A. and Kushner, I. (2002) 'Acute phase proteins', in J.J. Oppenheim and M. Feldman (eds), *Cytokine Reference On-line*, 1–16. DOI: 10.1006/rwcy.2002.0213.

Soto-Ramirez, N., Alexander, M., Karmaus, W., Yousefi, M., Zhang, H. et al. (2012) 'Breastfeeding is associated with increased lung function at 18 years of age: a cohort study', *European Respiratory Journal*, 39(4): 985–91.

THE RESPIRATORY SYSTEM

GASEOUS EXCHANGE

10

LEARNING OUTCOMES

When you have finished studying this chapter you will be able to:

1. Describe the anatomy of the components of the respiratory system
2. Identify the functions of each component of the respiratory system
3. Describe the phases of respiration
4. Discuss how gases are transported around the body
5. Explain the neurological regulation of respiration

INTRODUCTION

Every cell in the body requires a constant supply of oxygen to undertake their metabolic functions; in doing so they produce carbon dioxide (a waste product) that must be removed. If not removed, carbon dioxide can be toxic to the cells and affect their ability to carry out their functions thereby disturbing homeostasis. The respiratory system works hand in hand with the cardiovascular system (Chapter 12) to maintain homeostasis; the respiratory system draws oxygen from the atmosphere and expels carbon dioxide from the body, while the cardiovascular system transports the blood carrying oxygen to the cells and carbon dioxide out of the cells.

You need to understand how respiration contributes to keeping organs and tissues alive. Therefore, this chapter will explore the physiological responses that occur to ensure exchange of gas from the atmosphere to the cells and tissues and vice versa. The respiratory system has a number of other key functions that will be discussed as we progress through the chapter. As a person-centred nurse it is imperative that you understand the anatomy and physiology of the respiratory system and the contribution it plays in health-related quality of life.

Context

BODIE FAMILY

The respiratory system is vital to how we function physically. This is evident in all members of the Bodie family. At birth Danielle's respiratory system was still immature; approximately one third of her alveoli were developed. The number of alveoli will increase rapidly during the first six months and alveolarisation and secondary septa formation will continue until Danielle is between 18 and 36 months. Further development of her alveoli will continue until she reaches the age of eight, at which stage her alveoli will have reached functional maturity. Matthew (45) is a keen triathlete and his respiratory system will have developed to enable him to participate in this endurance sport. The mature members of the family may start to notice changes in their respiratory systems. For example, Maud (77) and George (84) may find that they become short of breath more readily when out walking due to changes in the structure and function of their lungs and reduced lung capacity. They may also be more susceptible to respiratory tract infections such as pneumonia or bronchitis due to the decrease in the number of alveolar macrophages.

As you work through the chapter, you will identify how this system has an effect on the physical and social aspects of the healthy person and how it can contribute to your practice as a person-centred nurse.

FUNCTIONS OF THE RESPIRATORY SYSTEM

The principle function of the respiratory system is to ensure that the body extracts enough oxygen from the atmosphere and disposes of the excess carbon dioxide. Alongside this, the respiratory system has a number of important functions including:

- **Regulation of blood pH:** carbon dioxide has a direct effect on blood pH. The respiratory system regulates blood pH by changing the levels of carbon dioxide in the blood.
- **Producing chemical mediators:** angiotensin converting enzyme (ACE), a key component in blood pressure regulation (Chapter 12), is produced by the lungs.
- **Sound production:** involved in speaking, singing and non-verbal communication, as air moves through the vocal cords.
- **Providing olfactory sensations (smell):** as molecules are drawn into the superior portions of the nasal cavity the olfactory epithelium sends messages to the central nervous system (Chapter 6).

- **Protection:** microorganisms can be prevented from entering the respiratory system from the atmosphere (by nasal hair) or they can be removed from respiratory surfaces (by coughing and action of the ciliated epithelium).

Structurally the respiratory system is divided into the upper respiratory tract and the lower respiratory tract. It can also be classified functionally into two distinct areas, i.e. conducting zone (carrying gases) and respiratory zone (gaseous exchange). The organs of the respiratory system are: the nose, pharynx, larynx, trachea, two bronchi, bronchioles, two lungs and muscles of breathing, i.e. the intercostal muscles and the diaphragm (Figure 10.1).

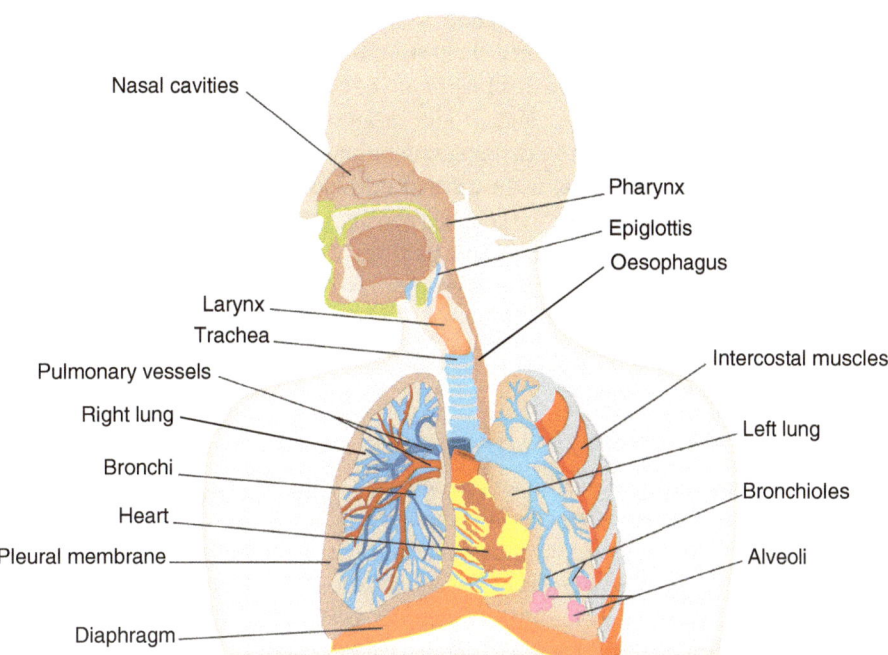

Figure 10.1 Organs of the respiratory system

THE UPPER RESPIRATORY TRACT

Nose

The nose is divided into two portions – the external nose and the nasal cavity – and acts as the primary passage for air entering the respiratory system. The external portion of the nose is that which is found on the face and consists of hyaline cartilage (which allows for slight flexibility of the external nose), nasal bones and extensions of the frontal and maxillary bones. The nostrils (external nares) are found on the under surface of the external nose. Air enters into the vestibule (anterior portion of the nasal cavity) via the flexible tissues of the nose. The epithelium of the vestibule contains coarse hairs that extend across the external nares; large airborne particles get trapped and are thereby prevented from entering the nasal cavity.

The nasal cavity is a large space situated below the nasal bones and above the oral cavity; it is lined with mucous membranes and muscle. The nasal cavity joins the external nose anteriorly and posteriorly and connects with the pharynx via two internal nares (choanae). It is divided into superior and inferior regions. The small superior region contains olfactory epithelium, the sensory organ for smell (Chapter 6). The large inferior region is lined with pseudostratified ciliated columnar epithelium with a large number of goblet cells secreting mucus. The nasal cavity is divided into right and left sides by a perpendicular partition – the nasal septum. The anterior nasal septum consists of septal nasal cartilage and the walls of the nostrils are formed mostly by lateral nasal cartilages (below the nasal bones) and the alar cartilages. The rest of the nasal septum is formed by the vomer, vertical plate of the ethmoid, maxillae and palatine bones.

The hard palate (floor of the nose) is formed by the palatine process of the maxillae and the palatine bone. Ducts (paranasal sinuses and nasolacrimal ducts) open into the nasal cavity. Mucous secretions produced in the associated paranasal sinuses, aided by tears draining through the nasolacrimal ducts, help to keep surfaces of the nasal cavity moist and clean. The paranasal sinuses alongside the nasal cavity act as resonating chambers and are involved in the production of sound as we speak or sing. Hence our voice sounds different when we have a cold or flu.

The walls of the nasal cavity contain three shelf-like projections: the superior, middle and inferior nasal conchae, which extend as far as the nasal septum. The conchae divide the nasal cavity into channels – meatuses (singular meatus). The conchae and meatuses help increase the surface area of the internal nose and prevent dehydration by catching water droplets when we exhale. They also contain a large number of blood vessels; therefore, as the inhaled air passes through the nasal cavity the conchae and meatuses deliver heat and moisture. The conchae contain goblet cells which secrete a large amount of mucus that moistens the air but also traps dust particles. The cilia (hair-like projections) move the dust particles and mucus towards the pharynx where it can be swallowed or spat out, thereby protecting the respiratory system.

Pharynx

The pharynx (commonly known as the throat) is a hollow muscular structure that is shared by both the respiratory (air) and digestive (food) systems. It starts at the internal nares and extends as far as the cricoid cartilage. It lies behind (posterior) the nasal and oral cavities and in front (anterior) of the cervical vertebrae. The wall of the pharynx consists of skeletal muscle that, when relaxed, maintains patency; contraction of the skeletal muscle helps in the process of swallowing (deglutition). The pharynx also acts as a resonating chamber for sound and has a protective function as it houses the lymphoid organs – the tonsils.

The pharynx is divided into three regions (Figure 10.2):

- **The nasopharynx** lies posterior to the nasal cavity and extends to the soft palate, i.e. the posterior portion of the roof of the mouth, which separates the nasopharynx and oropharynx, preventing materials that have been swallowed from entering the nasopharynx or nasal cavity. The uvula is the posterior extension of the soft palate. The walls of the nasopharynx contain five openings: two internal nares, two openings that lead to the auditory tubes (eustachian tubes – air passes through these to equalise the pressure between the atmosphere and the middle ear) and an opening into the oropharynx. The posterior wall contains the pharyngeal tonsil (adenoid), which provides a defence mechanism against infection. The walls of the nasopharynx are lined with pseudostratified ciliated columnar epithelium; the cilia help move the mucus received from the internal nares toward the inferior portion of the pharynx.
- **The oropharynx** extends from the soft palate to the epiglottis. The fauces is the opening from the oral cavity into the oropharynx, hence food, drink and air all pass through here. It is lined with non-keratinised stratified squamous epithelium that protects it from abrasion. The oropharynx contains two sets of tonsils: palatine tonsils and lingual tonsils that protect the respiratory system from infection.
- **The laryngopharynx**, like the oropharynx, is part of both the respiratory and digestive systems and is lined with non-keratinised stratified squamous epithelium. Posteriorly the laryngopharynx extends from the epiglottis to the oesophagus carrying food and fluid to the stomach; anteriorly it conducts air to the larynx.

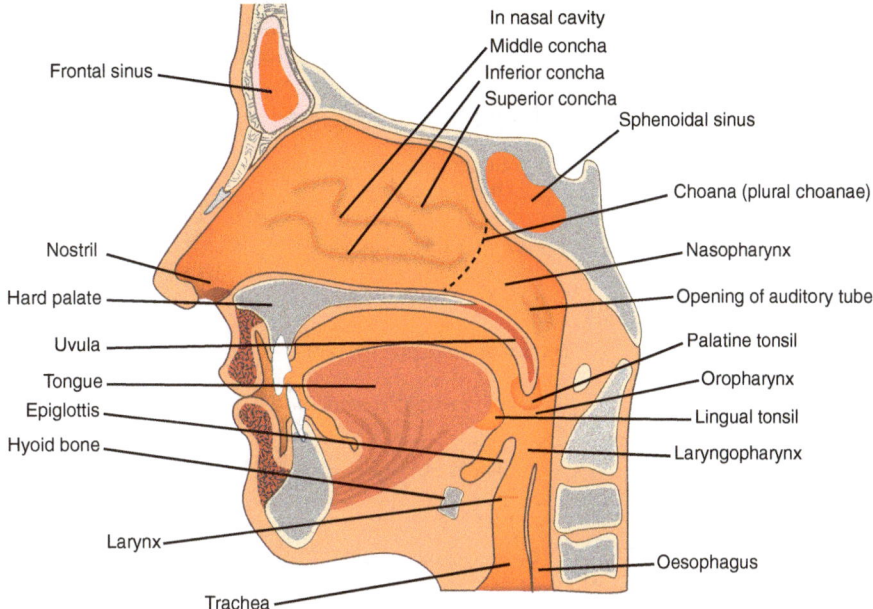

Figure 10.2 Upper respiratory tract

Larynx

The larynx (voice box) connects the laryngopharynx and trachea; it is positioned midline in the neck, anterior to the oesophagus and from the fourth to sixth cervical vertebrae (C4–C6). The wall of the larynx is comprised of nine pieces of cartilage; three occur singly (thyroid, cricoid and epiglottis) with the other six occurring in pairs (arytenoid, corniculate and cuneiform). The cartilages are connected to one another by muscles and ligaments; extrinsic muscles connect the cartilages to other structures in the throat, and intrinsic muscles connect the cartilages to each other (Figure 10.3):

- **Thyroid cartilage** (Adam's apple) is the largest of the cartilages. It is present in both males and females but tends to be larger in males due to the influence of sex hormones during puberty. It is attached to the cricoid cartilage by the cricothyroid ligament.
- **Cricoid cartilage** is the most inferior of the cartilages, forming the base of the larynx. It is attached to the first ring of the trachea by the cricotracheal ligament. The cricoid cartilage is a complete ring of cartilage and acts as an important landmark for emergency airway management. The cartilage rings below the cricoid cartilage are not complete rings (C-shaped), allowing more flexibility in the trachea's movement.
- **The epiglottis** is attached to the thyroid cartilage and projects towards the tongue. The broad portion of the epiglottis is unattached and therefore free to move up and down. The larynx and pharynx rise during swallowing. As the pharynx elevates it widens to receive food and drink: simultaneously the larynx rises and the epiglottis moves down to form a lid over the glottis (two true vocal cords and the space between them). In doing so, food and drink are directed towards the oesophagus and prevented from entering the larynx or airways. If small particles of food, dust, etc., do enter the larynx then a cough reflex usually expels them.
- **Arytenoid cartilages** are responsible for the production of sound; they influence the movement of the mucous membranes (true vocal cords).
- **Corniculate cartilages** are found at the apex of each of the arytenoid cartilages.
- **Cuneiform cartilages** are anterior to the corniculate cartilages and support the vocal cords and lateral aspects of the epiglottis.

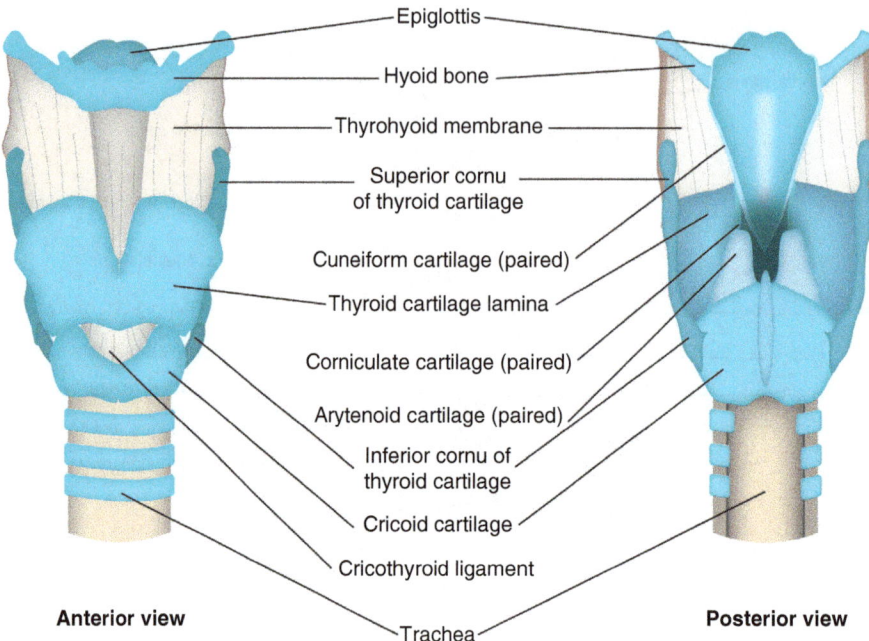

Figure 10.3 Anatomy of the larynx

THE LOWER RESPIRATORY TRACT

Trachea

The trachea (windpipe) is a tubular passageway for air approximately 12 cm long and 2.5 cm in diameter. It is made of dense regular connective tissue and smooth muscle that is reinforced with C-shaped rings of hyaline cartilage which support the trachea anteriorly and laterally. The cartilage protects the trachea and keeps the passageway open for the movement of air. Posteriorly the trachea does not contain cartilage but an elastic ligamentous membrane and bundles of smooth muscle (trachealis muscle). Contraction of the smooth muscle narrows the diameter of the trachea; mucus and foreign objects can be expelled from the trachea during coughing when the smooth muscle is contracted and air moves rapidly through it.

The mucous membrane of the trachea is composed of pseudostratified ciliated columnar epithelium with numerous goblet cells (mucus-producing cells) and functions as a mucociliary escalator. The mucus traps inhaled debris or foreign particles and the cilia propel the mucus towards the pharynx where it is swallowed.

At the inferior end, the trachea divides into two smaller tubes, the right and left primary bronchi, which extend to the right and left lungs respectively. The primary bronchi divide and form the bronchial tree (see lungs below).

Lungs

These are conical-shaped organs located in the thoracic cavity with a wide concave base that rests on the diaphragm and a blunt peak known as the apex which projects above the clavicle. The bronchus, blood vessels, nerves and lymphatic vessels enter at a slit in the mediastinal surface (see mediastinum in glossary) known as the hilum. These structures entering the hilum constitute the root of the lung.

To accommodate the position of the heart, the left lung is smaller than the right and has an indentation called the cardiac impression. The left lung has two lobes (superior and inferior) separated by the oblique fissure. The right lung has three lobes (superior, middle and inferior). The transverse or horizontal fissure separates the upper and middle lobes, while the oblique fissure separates the middle and inferior lobes.

Each lung has a double-layered membrane, the pleural membrane. The parietal pleura is attached to the inside of the thoracic cavity and the visceral pleura is attached to the surface of the lungs. Between the pleura is the pleural cavity containing pleural fluid which reduces the friction between the pleura, allowing them to move freely over each other when breathing. The surface tension of the pleural fluid holds the two layers of the pleura together keeping the chest wall and the lungs in close proximity. The pleura and pleural fluid are also responsible for creating a pressure gradient (lower than atmospheric pressure), which assists in inflation of the lungs (discussed later in this chapter). Along with the pericardium and mediastinum, the pleura compartmentalises the organs of the thoracic cavity thereby preventing the spread of infection from one organ to the others.

Bronchial tree

The bronchial tree is a highly branched system of conducting passages that extends from the primary bronchus to the terminal bronchioles (Figure 10.4). The trachea bifurcates into the right and left primary bronchi. The right primary bronchus is larger in diameter and extends in a more vertical direction than the left. The right primary bronchus subdivides into three secondary (lobar) bronchi and the left primary bronchus subdivides into two secondary (lobar) bronchi. The secondary bronchi then divide into the tertiary (segmental) bronchi – ten in the right lung and eight in the left. Plates of overlapping cartilage provide support to the secondary and tertiary bronchi. The bronchial tree receives its blood supply from the bronchial artery (arising from the aorta), and branches of the pulmonary artery closely follow the bronchial tree as it moves towards the alveoli.

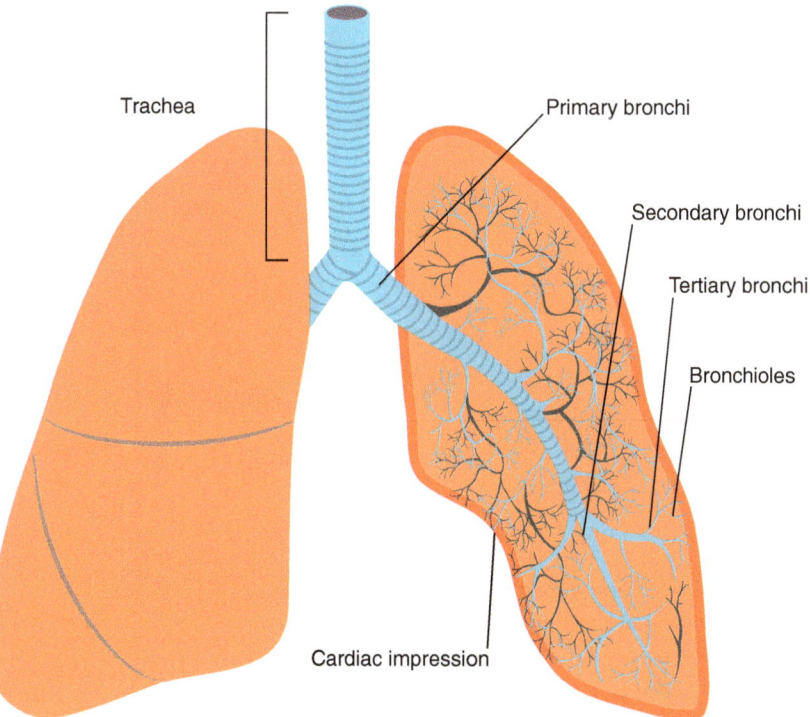

Figure 10.4   Anatomy of the bronchial tree

Bronchioles are small (1 mm diameter or less) continuations of the airway. They contain no cartilage and the epithelial lining becomes cuboidal; cilia, goblet cells and submucosal glands are almost gone. They contain a well-developed layer of smooth muscles that allows them to constrict or dilate (discussed later). Each bronchiole divides into 50–80 terminal bronchioles and they mark the end of the conducting zone. They contain no cartilage, goblet cells or submucosal glands; however, they do contain ciliated epithelium (involved in the mucociliary escalator) to ensure that the terminal bronchioles and alveoli do not become congested. Each terminal bronchiole then gives way to two or more respiratory bronchioles which mark the start of the respiratory zone. They contain small amounts of muscle and are non-ciliated. They further subdivide into two to ten thin-walled passages, the alveolar ducts, that end in grapelike clusters of alveoli, the alveolar sacs (Figure 10.5).

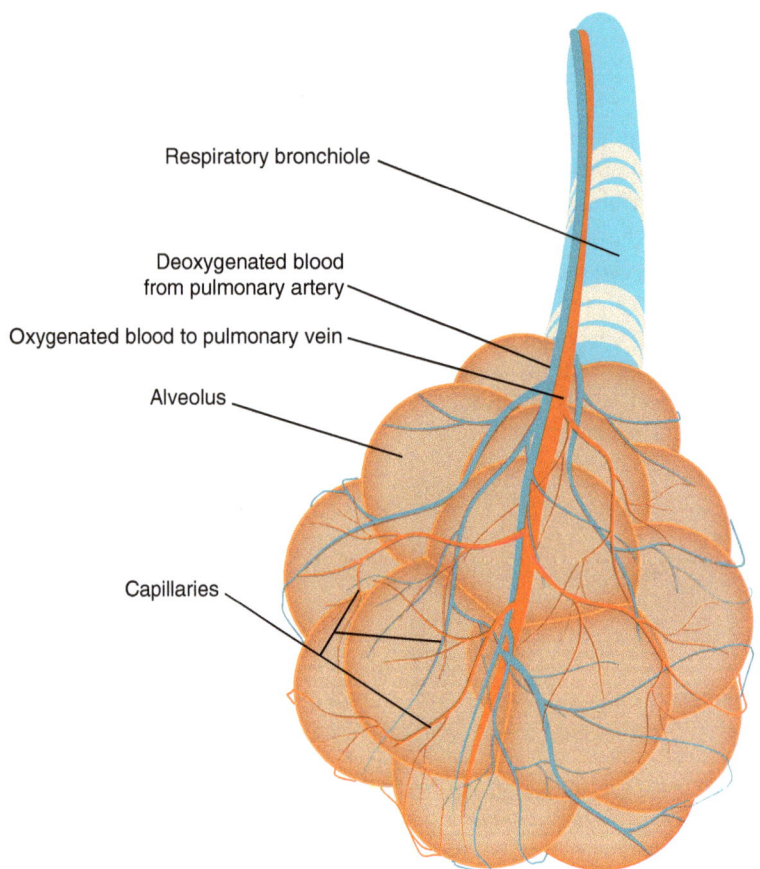

Respiratory bronchiole

Deoxygenated blood
from pulmonary artery

Oxygenated blood to pulmonary vein

Alveolus

Capillaries

Figure 10.5 Alveoli

Alveoli

Each adult lung consists of approximately 150 million alveoli, which provide 70 m^2 of surface area for gaseous exchange to take place. An alveolus is a small cup-shaped pouch lined with simple squamous epithelium and supported by a thin elastic basement membrane. The walls of the alveoli consist of two types of cells: type I alveolar cells (simple squamous epithelial cells) and type II alveolar cells (cuboidal epithelial cells). The thin type I cells are the main site for gaseous exchange. Type II cells are responsible for secreting alveolar fluid, keeping the surface between the air and cells moist. Included in the alveolar fluid is surfactant (a mixture of phospholipids and lipoproteins), which lowers the surface tension of the alveolar fluid thereby maintaining patency of the alveoli (discussed later).

Within the connective tissue and the lumen of the alveoli are the alveolar macrophages (dust cells). These act as the last line of defence against inhaled particles. Large particles (>10 μm in diameter) are usually trapped by the nasal hair or in the mucus of the upper respiratory tract. Smaller particles (2–10 μm in diameter) are usually trapped in the mucus of the bronchi and bronchioles and then removed by the mucociliary escalator. However, particles smaller than 2 μm can enter the alveoli; here they are phagocytosed by the macrophages.

Each alveolus is surrounded by numerous capillaries that consist of a single layer of epithelial cells and a basement membrane. The alveolar and capillary walls together form the respiratory membrane and allow the diffusion of O_2 and CO_2 between the air spaces and the blood.

VENTILATION

The process by which oxygen and carbon dioxide are exchanged between the atmosphere and body cells consists of three distinct phases:

- **Pulmonary ventilation (breathing):** the inhalation (inspiration) and exhalation (expiration) of air involving the exchange of air between the atmosphere and the alveoli.
- **External respiration:** the diffusion of gases between the alveoli and the pulmonary capillaries across the respiratory membrane. Oxygen diffuses from the alveoli to the pulmonary capillaries whereas CO_2 moves in the opposite direction, i.e. from the pulmonary capillaries to the alveoli.
- **Internal respiration:** the diffusion of gases between blood in the systemic capillaries and the tissues. Oxygen diffuses from the systemic capillaries to the tissues and CO_2 moves in the opposite direction, i.e. from the tissues to the systemic capillaries.

--- **UNDERSTAND** ---

It is important to remember that cellular respiration (Chapter 9) is a completely distinct biological concept and not related to the movement of oxygen and carbon dioxide from the external environment to the cell.

In order to understand how ventilation, transport of gases in the blood and the exchange of gases work, we need to first think about the factors that affect breathing. Whilst many of the principles are governed by the gas laws of physics, we also need to look at the muscular actions, alveolar surface tension, lung compliance, airway resistance and pressure gradients that produce this airflow.

Pressure changes

In order for air to pass in and out of the lungs a change in pressure is needed. Just before inspiration the intrapulmonary pressure is equal to atmospheric pressure; air will enter the lungs when the air pressure inside the lungs is less than the air pressure in the atmosphere. Boyle's law states that the pressure of a given quantity of gas is inversely proportional to its volume. Therefore, if you increase the size of the container, the pressure of the gas inside it will decrease. This is important to understand because it is this law that causes the exchange of air between the lungs and the atmosphere. In order for air to flow into the lungs the air pressure in the alveoli must be lower than that of the atmosphere. Therefore, according to Boyle's law, we must increase the size of the thoracic cavity to allow this exchange of air to take place. In order for air to leave the lungs the air pressure inside the lungs must be greater than the air pressure in the atmosphere. Therefore, the size of the thoracic cavity must decrease. The skeletal muscles of the thoracic and abdominal walls change the size of the thoracic cavity thereby altering the volume within. Differences in pressures caused by changes in lung volumes force air into our lungs when we inhale, and out when we exhale.

Inhalation

In order for inhalation to happen, the lungs must expand, increasing lung volume, which in turn decreases the pressure within the lungs below that of the atmosphere, drawing air into the lungs. The diaphragm is a dome-shaped sheet of skeletal muscle that forms the floor of the thoracic cavity that is innervated by the phrenic nerve. During normal quiet breathing, stimulation of the diaphragm by the phrenic nerve causes it to tense and flatten (by approximately 1 cm); this causes enlargement of the thoracic cavity thus reducing the pressure. The external intercostals are responsible for lifting the ribcage up and out and as a result the anteroposterior and lateral diameters of the chest cavity are increased.

During normal quiet inhalation the intrapleural pressure (the pressure between the two pleural layers) is always lower than atmospheric pressure. Just before inhalation this is approximately 4 mmHg less than atmospheric pressure, i.e. atmospheric pressure of 760 mmHg minus 4 mmHg giving an intrapleural pressure of 756 mmHg. As the thoracic cavity increases in volume, intrapleural pressure drops to about 754 mmHg. The parietal pleural and visceral pleura are pulled in all directions as the thoracic cavity increases. This causes the alveolar pressure (pressure inside the alveoli) to drop to 758 mmHg due to the increase in lung volume thereby creating a pressure difference between the alveoli and the atmosphere. As air flows from an area of higher pressure to an area of lower pressure, inhalation occurs (Figure 10.6). Air will continue to flow into the lungs until intrapulmonary pressure is equal to atmospheric pressure. Although the dimensions of the thoracic cavity are only increased minimally, it allows approximately 500 ml of air to flow into the respiratory tract during quiet breathing.

In times of deep forceful inhalations, e.g. during exercise or in some respiratory diseases, accessory muscles of inspiration are also involved in increasing the size of the thoracic cavity. These accessory muscles include: sternocleidomastoid muscles (which run between the mastoid process and the sternum and lift the sternum), the scalene muscles (that lift the first two ribs) and the pectoralis minor muscles (that lift the third to fifth ribs).

Exhalation

When inspiration stops, the phrenic nerves continue to stimulate the diaphragm for a short period of time to prevent the lungs from recoiling too quickly, and thereby making the change from inhalation to exhalation smoother. In normal quiet breathing, inhalation usually lasts about two seconds and exhalation about three seconds. The process of exhalation is also due to a pressure gradient; however, this time it is in the opposite direction. At the end of inhalation, the pressure in the lungs is greater than that of the atmosphere.

Quiet exhalation is a passive process in that no muscular contractions are involved; it results from the elastic recoil of the chest walls and the lungs. It begins as the inspiratory muscles begin to relax. The diaphragm and intercostal muscles relax and the ribs move in and down. This decreases the diameter

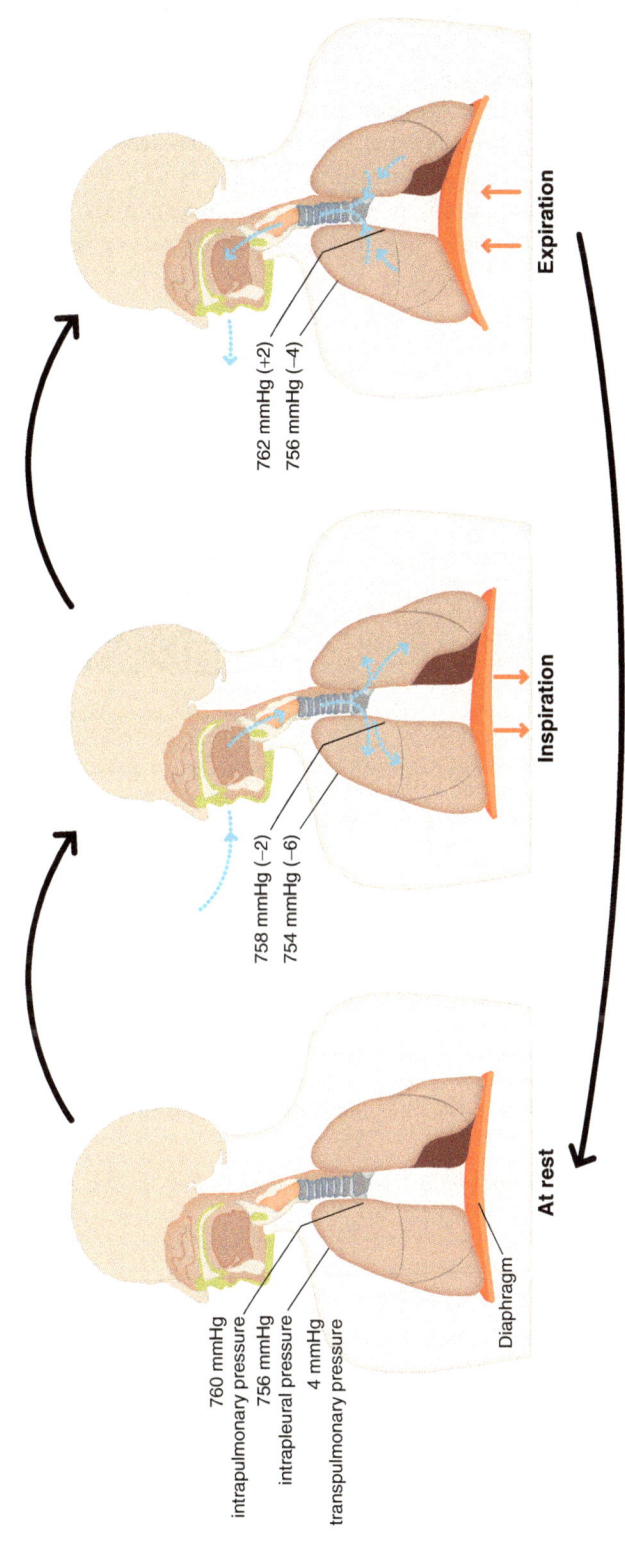

760 mmHg
intrapulmonary pressure

756 mmHg
intrapleural pressure

4 mmHg
transpulmonary pressure

Diaphragm

At rest

758 mmHg (−2)
754 mmHg (−6)

Inspiration

762 mmHg (+2)
756 mmHg (−4)

Expiration

Figure 10.6 Intrapleural pressures during inspiration and expiration

of the thoracic cavity, thereby decreasing lung volume, which in turn will increase alveolar pressure. As alveolar pressure (762 mmHg) rises above atmospheric pressure (760 mmHg), air flows from an area of high pressure to low pressure (Figure 10.6).

Exhalation may become an active process during times of forceful breathing (playing a wind instrument, exercise). During this time your internal intercostal muscles contract, thereby depressing the ribs. Your abdominal muscles also contract, increasing intra-abdominal pressure forcing the viscera and diaphragm upwards and therefore putting pressure on the thoracic cavity. Intrapulmonary pressure increases even higher above atmospheric pressure (20–30 mmHg) causing faster and deeper emptying of the lungs.

ACTIVITY 10.1: UNDERSTAND

Watch the following video clips to help develop your understanding of the mechanics of breathing.

These online videos can be accessed by **scanning the QR codes** with your smart phone or via https://study.sagepub.com/essentialap2e.

BREATHING MECHANISM AND GAS EXCHANGE' (6:32)

3D VIEW OF THE DIAPHRAGM (1:14)

APPLY

Chest auscultation

Chest auscultation (listening to sounds of the lungs) is an important part of respiratory assessment and is used to help diagnose a range of respiratory disorders. It assesses the airflow through the trachea and bronchial tree. Breath sound has three characters: frequency, intensity and quality – these are used to differentiate two similar sounds.

To listen to **normal and abnormal lung sounds**, access the weblink via http://study.sagepub.com/essentialap2e.

How to perform chest auscultation:

1. Auscultation should be done in a quiet room, preferably in a sitting position. If the person cannot assume a sitting posture, assist them in rolling from one side to the other to examine the back.
2. Auscultation should never be done through the clothing. Where possible attempt to warm the diaphragm of the stethoscope before placing on the person's skin.
3. Ask the person to take deep breaths through an open mouth.
4. Using the diaphragm of the stethoscope, start auscultation at the sites identified in Figure 10.7 below.
5. Always compare symmetrical points on each side (Figure 10.7).
6. Listen for the quality of the breath sounds, the intensity of the breath sounds, and the presence of abnormal sounds, e.g. fine and coarse crackles (rales), wheezes (rhonchi), pleural rubs and stridor.

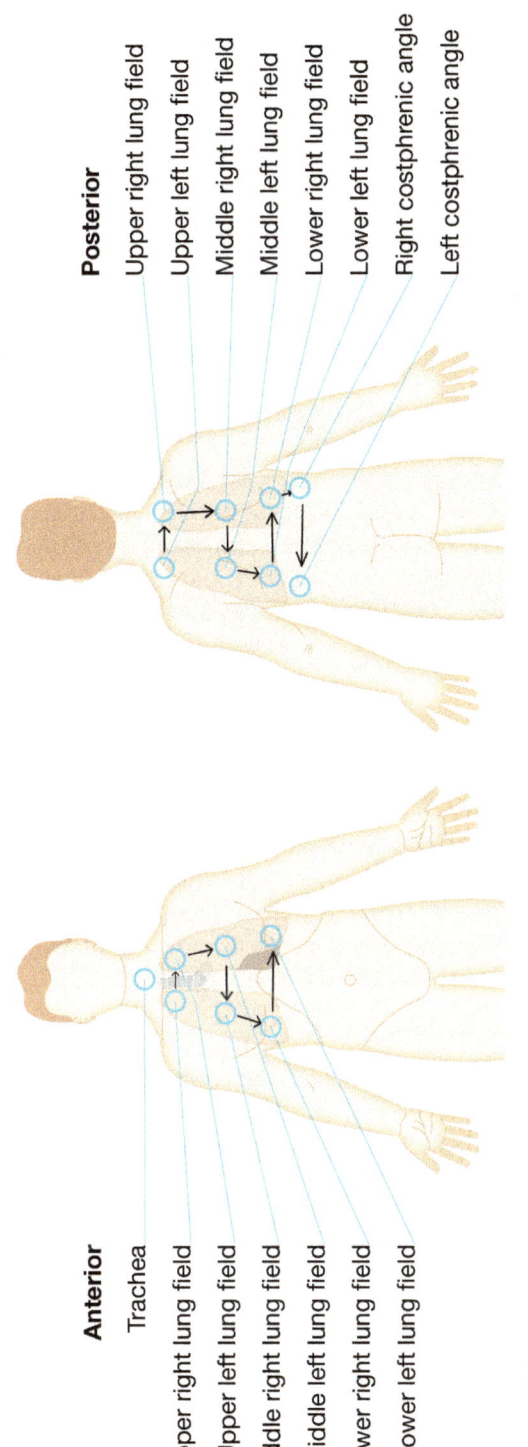

Anterior

Trachea

Upper right lung field

Upper left lung field

Middle right lung field

Middle left lung field

Lower right lung field

Lower left lung field

Posterior

Upper right lung field

Upper left lung field

Middle right lung field

Middle left lung field

Lower right lung field

Lower left lung field

Right costphrenic angle

Left costphrenic angle

Figure 10.7 Auscultation sites

Other factors affecting airflow

As mentioned earlier, there are other factors that can affect the rate of air flow and the ease of inhalation and exhalation. We are going to consider these briefly now.

Surface tension of alveolar fluid

Surface tension is the force exerted by the alveolar fluid that coats the luminal surface of the alveoli. Water molecules are attracted to each other because of the hydrogen bonds, thereby creating surface tension. This tension causes the walls of the alveoli to be drawn inward toward the lumen. If this were allowed to happen the alveoli would collapse on expiration, making reinflation very difficult.

However, as previously mentioned, alveolar fluid secreted from the type II alveolar cells contains surfactant. Surfactant disrupts the hydrogen bonds of water, thereby reducing surface tension.

Lung compliance

This refers to the ease with which the lungs expand. Lungs and chest wall that expand easily are said to have high compliance, whereas lungs and chest wall that resist expansion are said to have low compliance. Lung compliance is related to two factors: elasticity and surface tension. Elastic fibres in the lungs are usually easily stretched and, as already discussed, surfactant reduces surface tension and therefore lungs typically have high compliance. However, lung compliance is reduced by degenerative lung diseases that cause scarring.

Airway resistance

The smooth muscle of the bronchioles allows them to change in size, therefore their diameter governs resistance to airflow. Bronchoconstriction (narrowing of the bronchioles) reduces airflow, whereas bronchodilation (widening of the bronchioles) increases airflow.

Bronchoconstriction can be triggered by airborne irritants (e.g. in asthma), parasympathetic nervous stimulation or release of histamine (the latter may occur in anaphylaxis). Bronchodilation is stimulated by the sympathetic nervous system and release of adrenaline (epinephrine), which binds to β_2 receptors and causes relaxation of the smooth muscle in the bronchial wall.

ACTIVITY 10.2: APPLY

Smoking and health

By now, everyone knows that smoking increases the risk of lung cancer, but you should also know about the other changes which can occur. A number of conditions together known as chronic obstructive pulmonary/airway disease (COPD/COAD) are more likely to occur in smokers. While these vary somewhat, smoking paralyses the cilia of the bronchial tree, thus reducing the effectiveness of the mucociliary escalator in clearing the lungs of mucus. The resulting congestion can bring about infection and inflammation causing scarring of the lung tissue. Over time the tissues become less elastic, and the narrow bronchioles become narrower still. The person may suffer from the following symptoms:

- Productive and persistent cough.
- Frequent chest infections often requiring antibiotics.
- Shortness of breath, which progressively gets worse; occurring initially with exercise, but then occurring with non-strenuous activities, and finally at rest, and eventually needing oxygen therapy.

Use of the accessory muscles of respiration may be visible indicating the effort needed for breathing (Figure 10.8).

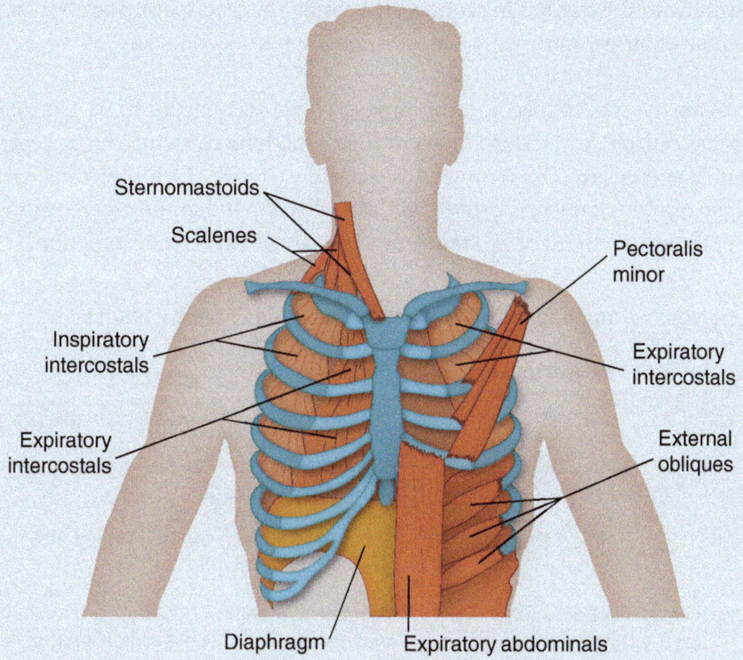

Sternomastoids

Scalenes

Pectoralis minor

Inspiratory intercostals

Expiratory intercostals

Expiratory intercostals

External obliques

Diaphragm   Expiratory abdominals

Figure 10.8   Accessory muscles of respiration

Understanding the long-term effects of smoking will enable you, as a person-centred nurse, to work with people to encourage them not to start or to stop smoking to promote their long-term quality of life.

APPLY

Asthma and drug therapy

Referring back to the Bodie family, Derek Jones has had mild persistent asthma since childhood. Asthma causes transient bronchoconstriction, reducing the airflow into the lungs. Derek has been prescribed a β_2 agonist and a low-dose corticosteroid inhaler to dilate his airways and reduce inflammation. These drugs are inhaled for faster and local action.

BODIE FAMILY

MOVEMENT OF GASES

Lung volumes and capacities

A healthy adult (at rest) will take between 12 and 16 breaths per minute. As identified earlier, each inhalation and exhalation moves approximately 500 ml of air in and out of the lungs. The amount of air that moves with one breath is known as the Tidal Volume (V_T) (measured in millilitres). The amount of air inhaled and exhaled per minute is referred to as the Minute Volume (MV) (measured in litres). This can be calculated by multiplying the tidal volume by the respiratory rate, e.g. 500 ml/breath × 12 breaths/min = 6,000 ml = 6 litres/min. However, not all this air reaches the respiratory zone. Of the air inhaled 70% (350 ml) will reach the respiratory zone and be involved in external respiration, with the rest (30%, i.e. 150 ml) remaining in the conducting zone – the anatomical dead space. The alveolar ventilation rate is the actual volume of air per minute that reaches the respiratory zone, i.e. 350 ml × 12 breaths/min = 4,200 ml/min.

Lung capacities are the sum of specific lung volumes (Table 10.1, Figure 10.9). The total amount of air a person's lungs has capacity to hold is known as their total lung capacity; this is dependent upon the person's age, gender and height. The volume of air exchanged during breathing can be measured using a spirometer and the record is known as a spirogram (Figure 10.9). The spirometer records variables such as rate and depth of breathing, speed of expiration and the rate of oxygen consumption.

Table 10.1 Lung volume and capacities - average values for healthy adults

Measurement	Typical value	Definition
Tidal Volume (V_T)	500 ml	Amount of air inhaled/exhaled in one breath
Inspiratory Reserve Volume (IRV)	Male-3100 ml Female-1900 ml	Amount of air that can be inhaled with maximum effort above tidal inspiration
Expiratory Reserve Volume (ERV)	Male-1200 ml Female-700 ml	Amount of air that can be exhaled with maximum effort above tidal exhalation
Residual Volume (RV)	Male-1200 ml Female-1100 ml	Amount of air left in the lungs after exhalation with maximum effort - keeps alveoli inflated between breaths
Vital Capacity (VC)	Male-4800 ml Female-3100 ml	Amount of air exhaled with maximum effort after maximum inspiration (V_T + IRV + ERV = VC)
Inspiratory Capacity (IC)	Male-3600 ml Female-2400 ml	Maximum amount of air inhaled after a normal exhalation (V_T + IRV = IC)
Functional Residual Capacity (FRC)	Male-2400 ml Female-1800 ml	Amount of air remaining in lungs after exhalation (RV + ERV = FRC)
Total Lung Capacity (TLC)	Male-6000 ml Female-4200 ml	Maximum amount of air the lungs can hold (RV + VC = TLC)

Gaseous exchange

Oxygen and carbon dioxide move between the alveolar air and pulmonary circulation through the process of passive diffusion. Passive diffusion is governed by two gas laws:

- **Dalton's law:** the total pressure of a gas mixture is the sum of the partial pressures of the individual gases.
- **Henry's law:** at the air–water interface, for a given temperature, the amount of gas that dissolves in the water is determined by its solubility in water and its partial pressure in air.

Therefore, according to Henry's law more PO_2 in the alveolar air means that the blood will pick up more. Conversely, as the blood arriving at the alveolus has a higher PCO_2 value than that of air, the blood releases the CO_2 into the alveolar air.

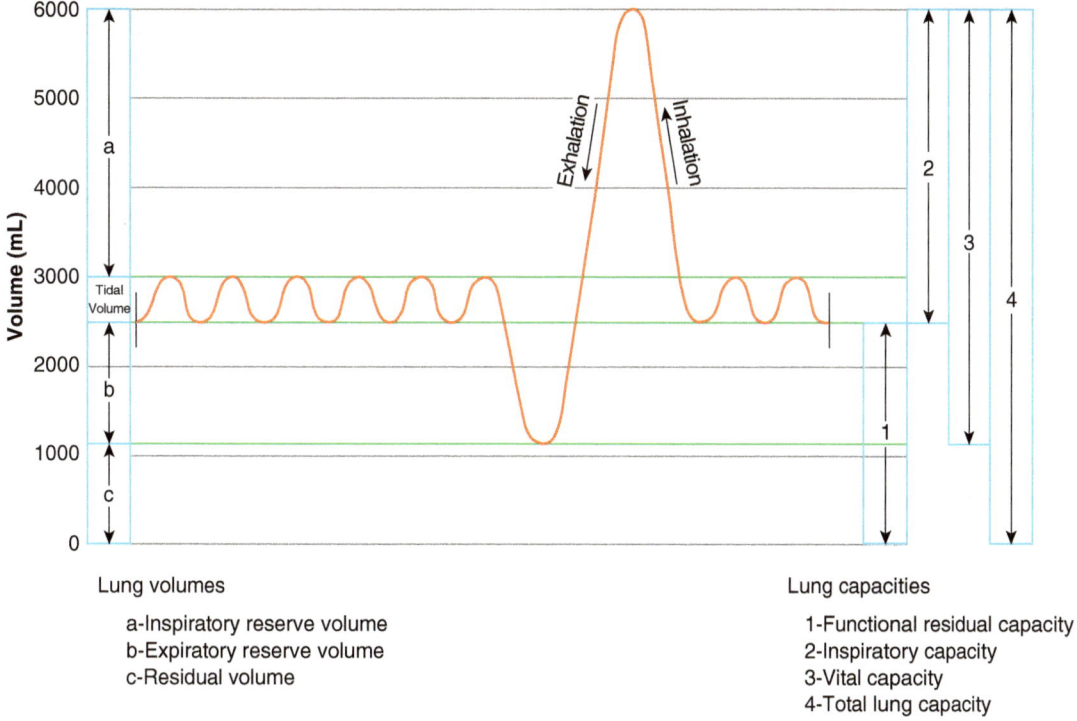

Figure 10.9 Lung volumes and capacities

Lung volumes

    a-Inspiratory reserve volume
    b-Expiratory reserve volume
    c-Residual volume

Lung capacities

    1-Functional residual capacity
    2-Inspiratory capacity
    3-Vital capacity
    4-Total lung capacity

GO DEEPER

Dalton's law

Atmospheric pressure (760 mmHg) is the sum of all gases in the air, i.e. nitrogen (N_2) (78%), oxygen (O_2) (21%), argon (Ar) (0.93%), carbon dioxide (CO_2) (0.04%), variable amounts of water (H_2O) (depending on where you are), plus small quantities of other gases (0.06%).

Each individual gas exerts its own pressure as if no other gases were present; this is known as its partial pressure (PX) (X being the gas). The partial pressure of an individual gas can be determined by multiplying the percentage of that particular gas in the total amount of gas, e.g. partial pressure of O_2 is:

    $PO_2 = 0.21 \times 760$ mmHg (101.3 kPa)* = 159.6 mmHg (21.3 kPa)

Therefore:

    Atmospheric pressure = $PN_2 + PO_2 + PAr + PH_2O + PCO_2 + $ POther gases

These pressures determine the movement of O_2 and CO_2 between the atmosphere and the lungs, between the lungs and the blood, and between the blood and the cells. The gas will move from where there is higher partial pressure to an area of lower partial pressure. For example, alveolar air contains 13.6% O_2 whereas inhaled air has 21% O_2, therefore O_2 will move from the area of higher partial pressure to the area of lower partial pressure.

* Atmospheric pressure at sea level. Respiratory gases in the human body are either measured in mmHg or kPa.

ACTIVITY 10.3: GO DEEPER

To further your understanding, watch this online video about blood gases.

This video clip can be accessed by **scanning the QR code** with your smart phone or via https://study.sagepub.com/essentialap2e.

BLOOD GASES (12:48)

Now that we have a basic understanding of how passive diffusion works, we can look at how the process of gaseous exchange occurs between the alveoli and blood (external respiration) and between the blood and cells (internal respiration).

External respiration

This refers to diffusion of O_2 from the alveoli into pulmonary circulation and the diffusion of CO_2 in the opposite direction. Blood in the pulmonary artery that has been collected from the systemic circulation is low in O_2 and high in CO_2. The partial pressure of O_2 is higher in alveoli, therefore O_2 passively moves out of the alveoli into the circulation and onwards to the left side of the heart. As CO_2 is lower in alveoli than in the pulmonary circulation, CO_2 diffuses into the alveoli ready to be exhaled (Figure 10.10).

Internal respiration

Cells use O_2 to produce adenosine triphosphate (ATP). The concentration of O_2 is always lower in the tissues than in blood. Aerobic respiration produces a molecule of CO_2 for every molecule of O_2 it consumes, therefore the concentration of CO_2 is high in the tissues. As blood flows through the capillaries, O_2 and CO_2 follow the concentration gradient and continually diffuse between blood and tissue. As gases diffuse down the concentration gradients, i.e. from high pressure to low pressure, O_2 will diffuse from the blood into the tissues and CO_2 will diffuse in the opposite direction (Figure 10.10).

Transport of gases

Gas transport refers to the process of carrying gases from the alveoli to the tissues and vice versa, so we need to have an understanding of how O_2 and CO_2 are transported in the blood.

Transport of oxygen

As oxygen does not dissolve easily in water only a small amount (1.5%) is dissolved and therefore transported in plasma. Almost all oxygen molecules (98.5%) bind with haemoglobin in red blood cells (erythrocytes). Each haemoglobin molecule has four iron-containing haem regions, each of which

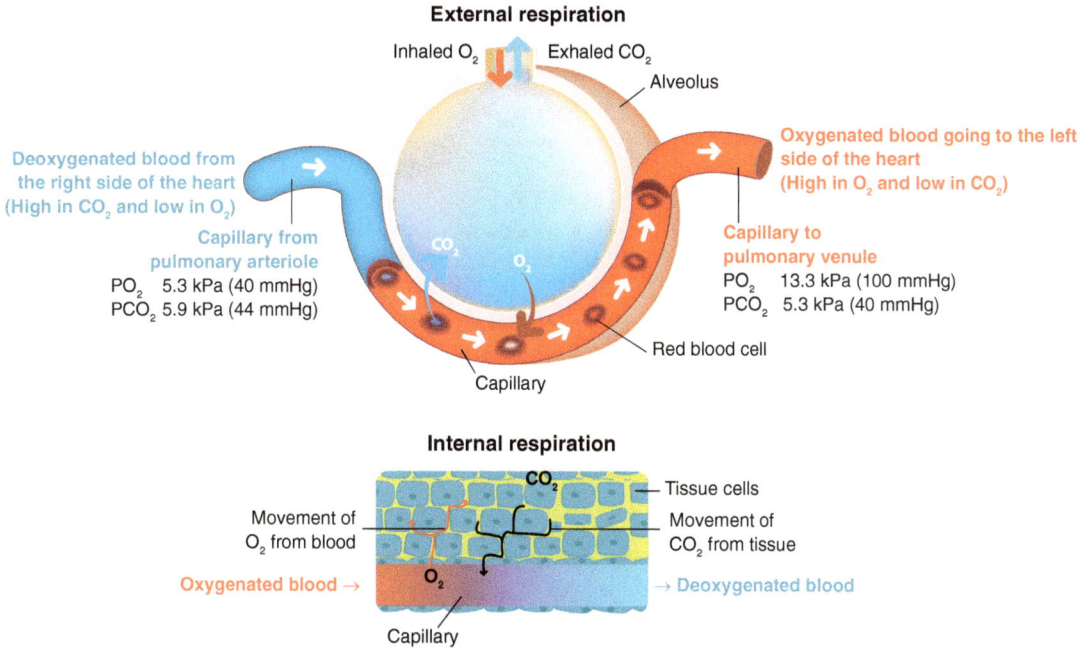

External respiration

Inhaled O_2 Exhaled CO_2

Alveolus

Deoxygenated blood from
the right side of the heart
(High in CO_2 and low in O_2)

Oxygenated blood going to the left
side of the heart
(High in O_2 and low in CO_2)

Capillary from
pulmonary arteriole
PO_2 5.3 kPa (40 mmHg)
PCO_2 5.9 kPa (44 mmHg)

Capillary to
pulmonary venule
PO_2 13.3 kPa (100 mmHg)
PCO_2 5.3 kPa (40 mmHg)

CO_2 O_2

Red blood cell

Capillary

Internal respiration

CO_2 Tissue cells

Movement of
O_2 from blood

Movement of
CO_2 from tissue

Oxygenated blood →

O_2

→ Deoxygenated blood

Capillary

Arterial Blood Gas Test (normal ranges):

Value	Range
pH	7.35–7.45
PaO_2	11–13 kPa (82–100 mmHg)
$PaCO_2$	4.7– 6.0 kPa (35–45 mmHg)
HCO_3^-	22–26 mmol/L
Base Excess	–2 – +2
Saturations (SaO_2)	> 95%

Figure 10.10   External and internal respiration

is capable of binding to one molecule of O_2. Oxygen and haemoglobin bind in an easily reversible reaction to form oxyhaemoglobin. When all four molecules of haem carry an O_2 molecule it is 100% saturated, when three molecules carry an O_2 molecule it is 75% saturated, and if two haem molecules contain an O_2 molecule it is 50% saturated. To enter the tissue cells O_2 must separate from haemoglobin and pass into the plasma and it is the oxygen dissolved in the plasma that diffuses into the tissue cells. The ability of haemoglobin to bind with O_2 at the alveoli and then release it into the peripheral tissues is shown by the oxygen–haemoglobin dissociation curve. The curve is sigmoidal (S-shaped) and describes the relationship between the partial pressure of oxygen (x axis) and oxygen saturation (y axis). As more oxygen molecules bind to the haemoglobin, the affinity becomes greater, and partial pressure increases until the maximum amount that can be bound is reached. As the curve plateaus (at approximately 80 mmHg), increasing levels of PO_2 will not lead to significant increases in saturation. However, at less than 60 mmHg the curve is very steep, therefore small changes in PO_2 will greatly reduce oxygen saturation (Figure 10.11).

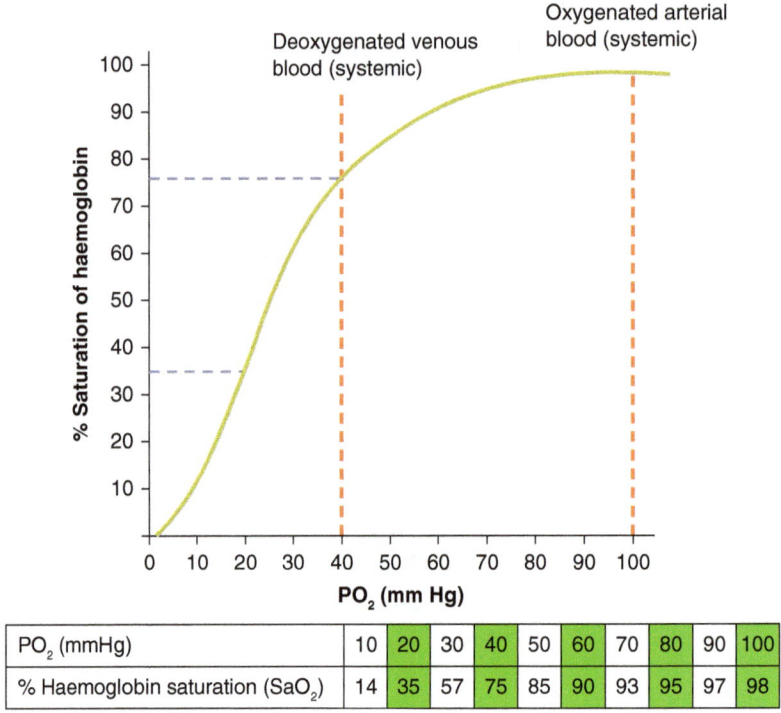

PO$_2$ (mmHg)	10	20	30	40	50	60	70	80	90	100
% Haemoglobin saturation (SaO$_2$)	14	35	57	75	85	90	93	95	97	98

Figure 10.11   Oxygen-haemoglobin dissociation curve

There are a number of factors that promote haemoglobin binding and dissociation including: partial pressure of oxygen, acidity (pH), partial pressure of carbon dioxide, temperature and BPG (see Go Deeper box below). These factors affect the affinity with which oxygen binds with haemoglobin; this can shift the curve to the left (higher affinity) or to the right (lower affinity) (Figure 10.12).

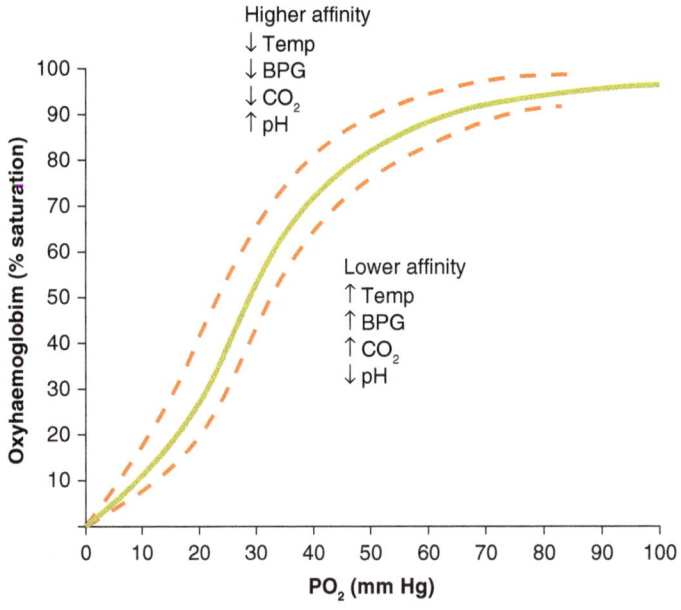

Figure 10.12   Factors affecting affinity of haemoglobin for oxygen

GO DEEPER

Factors affecting the affinity of haemoglobin for oxygen

Partial pressure of O_2: the higher the PO_2 the more O_2 can bind with the haemoglobin. When the PO_2 is high, haemoglobin binds with large amounts of O_2. However, when PO_2 is low, only small amounts of O_2 bind and haemoglobin is only partially saturated. PO_2 is high in the pulmonary capillaries, therefore large amounts of O_2 will bind to the haemoglobin, whereas in the tissue capillaries PO_2 is low, and the O_2 that is dissolved in the plasma will move into the tissue cells.

Acidity (pH): as acidity increases (pH decreases) the affinity of haemoglobin for O_2 decreases (the curve shifts to the right), therefore O_2 unloading is enhanced.

Partial pressure of carbon dioxide: CO_2 can bind with haemoglobin and when it does it has a similar effect to that of pH (the curve shifts to the right). As CO_2 levels rise it causes haemoglobin to release more O_2.

Temperature: as temperature rises so does the amount of oxygen released from haemoglobin. Heat is a by-product of metabolic reactions, therefore more metabolically active cells produce more heat: this promotes the release of O_2 from haemoglobin (the curve shifts to right). Conversely, as body temperature decreases or cellular activity decreases there is less demand for O_2 so it remains bound to haemoglobin (the curve shifts to the left). For every 1° rise in temperature, there is a 10% increase in O_2 consumption.

BPG (2,3-bisphosphoglycerate): formed when red blood cells break down glucose to form ATP. BPG decreases the affinity of O_2 for haemoglobin, and therefore increased O_2 unloading takes place. Increased levels are formed by certain hormones including thyroxine, human growth hormone, adrenaline and testosterone. BPG is also known as DPG (2,3-disphosphoglycerate).

APPLY

Oxygen saturation (SpO$_2$)

As previously discussed, almost all oxygen molecules (98.5%) bind with haemoglobin in red blood cells. Normal readings in a healthy adult range from 94% to 100%.

When a person is in hospital, at times it may be necessary to monitor the oxygen saturation of that person's blood. This is done through the process of pulse oximetry. Pulse oximetry is the technique of measuring oxy-haemoglobin saturation by shining red and infrared light through a peripheral site, such as a finger, toe or nose. Pulse oximetry provides continuous monitoring and may be useful in several clinical situations, e.g. operating and recovery rooms, surgical and medical environments, Intensive Care Units (ICUs) and emergency care.

Although pulse oximetry is a very useful technology, there are situations where you must be careful in applying it:

- Anaemia (not enough functioning haemoglobin).
- Dyes – some surgical procedures, especially in cardiology and urology, call for the injection of dyes into the blood in order to trace blood flow. These dyes affect light transmission through the blood and directly influence the pulse oximeter and lead to wrong readings.
- Hypothermia – the body reduces the heat lost by the skin by constricting the peripheral blood vessels.
- Nail polish – some nail polish and false fingernails may cause false readings.

A study by Milner and Matthews (2012) demonstrated that a significant proportion of pulse oximeters in practice are inaccurate due to not being cleaned or maintained correctly. It is therefore essential that all pulse oximeters undergo periodic calibration and assessment to confirm their accuracy.

Transport of carbon dioxide

Carbon dioxide is transported in three forms:

- 7% is dissolved in plasma. On reaching the lungs it diffuses into alveolar air and is exhaled.
- 23% combines with the protein portion of haemoglobin and plasma proteins to form carbamino-haemoglobin, transporting CO_2 to the lungs for gaseous exchange.
- 70% is transported as bicarbonate ions. As the CO_2 diffuses into the systemic capillaries it reacts with water in the presence of carbonic anhydrase (CA) to form carbonic acid (H_2CO_3). This easily dissociates into hydrogen (H^+) and bicarbonate (HCO_3^-) ions:

$$CO_2 + H_2O \overset{CA}{\leftrightarrow} H_2CO_3 \leftrightarrow H^+ + HCO_3^-$$

Therefore, as CO_2 is picked up by the blood, HCO_3^- accumulates in the red blood cells. Some of the HCO_3^- moves out into the blood plasma, and chloride ions move from the plasma into the red blood cells. This maintains the electrical balance between the blood plasma and red blood cell cytosol. As a result of this, CO_2 is removed from the tissue cells and transported in the blood plasma as HCO_3^-. This reaction is reversed as the blood passes through the pulmonary capillaries and CO_2 is exhaled.

ACTIVITY 10.4: UNDERSTAND

Watch this animated video explaining gaseous exchange.

This online video can be accessed by **scanning the QR code** with your smart phone or via https://study.sagepub.com/essentialap2e.

GASEOUS EXCHANGE (3.08)

CONTROL OF RESPIRATION

Control of breathing depends upon stimuli from the brain to control the skeletal muscles and to act as a coordinating centre to ensure that all the muscles are working together to maintain the rhythmic process. There are a number of neural mechanisms that control breathing: neurons in the medulla oblongata and pons are responsible for the control of unconscious (involuntary) breathing, whereas neurons in the motor cortex of the cerebrum are responsible for voluntary control.

The medulla oblongata contains inspiratory neurons (used during inspiration) and expiratory neurons (used in forced expiration). Nerve fibres travel from these neurons and synapse with neurons in the cervical and thoracic regions. From here, the impulses generated by the inspiratory neurons travel via the phrenic nerves to the diaphragm and the intercostal nerves to the external intercostal muscles. These muscles then contract and inhalation occurs. Inhalation lasts for approximately two seconds, then the inspiratory area becomes inactive, the diaphragm and external intercostal muscles relax allowing for elastic recoil of the lungs and thoracic cage, and passive exhalation occurs.

The expiratory neurons inhibit the inspiratory neurons when there is a need for forced/deeper expiration. Impulses from the expiratory neurons cause contraction of the abdominal and internal intercostal muscles, further decreasing the size of the thoracic cavity during forced expiration.

The pneumotaxic centre located in the upper pons and the apneustic centre in the lower pons are also responsible for regulating ventilation. The pneumotaxic centre sends inhibitory impulses to the inspiratory area, ensuring the lungs do not overfill with air. The duration of inhalation is decreased and the rate of breathing is increased when the pneumotaxic area is active. The apneustic centre sends stimulatory impulses to the inspiratory area, thereby producing a prolonged, deep inhalation.

Voluntary control

We have the ability to control or alter our breathing voluntarily due to the connection between the motor cortex of the frontal lobe of the cerebrum and the respiratory centres. Impulses from the motor cortex travel via the corticospinal tracts to the respiratory neurons in the spinal cord, bypassing the brainstem respiratory centres. The ability to voluntarily control our breathing and hold our breath is limited due to the build-up of CO_2 and hydrogen ions. When CO_2 and hydrogen ions increase to a certain concentration, the inspiratory centre is stimulated and nerve impulses travel to the diaphragm and intercostal muscles and breathing resumes. The limbic system and hypothalamus can also stimulate the respiratory centre; feelings of pain and emotional stimuli can alter respiration, e.g. crying and laughing.

APPLY

Control of respiration and expression

Voluntary control of respiration is necessary to speak and sing. Baby Danielle does not yet have a fully developed nervous system and so is still developing the skills of controlling voluntary respiration alongside learning to speak. Whereas, Thomas loves to sing and so has mastered voluntary breathing control in developing this ability.

BODIE FAMILY

Chemoreceptors and control of respiration

The most potent stimulus for breathing is the pH of body fluids: PCO_2 plays a significant role, and PO_2 is also involved but plays a less significant role. The levels of each of these are detected by chemoreceptors. Chemoreceptors are found in two general locations: centrally (brainstem) and peripherally (carotid and aortic bodies). Peripheral chemoreceptors monitor changes in blood and messages are sent to the respiratory centre via the vagus and glossopharyngeal nerves respectively. Central chemoreceptors primarily monitor the pH of the cerebrospinal fluid (CSF).

Carbon dioxide is lipid soluble and as such easily diffuses into the cells where, in the presence of carbonic anhydrase, it combines with water and forms carbonic acid. As discussed previously, carbonic acid easily dissociates to form hydrogen and bicarbonate ions. Therefore, an increase in CO_2 will increase the number of hydrogen ions; a decrease in CO_2 will decrease the number of hydrogen ions thereby having a direct effect on pH. Chemoreceptors participate in a negative feedback mechanism that maintains the levels of pH (7.35–7.45), PCO_2 (4.7–6.0 kPa or 35–45 mmHg), and PO_2 (>10 kPa or >75 mmHg). With any changes in these parameters (especially pH or PCO_2) from their normal range, chemoreceptors will stimulate the inspiratory area and the rate and depth of breathing will change. An increase in PCO_2 resulting in a decrease in pH will cause the inspiratory area to become increasingly active, and rate and depth of breathing will increase (hyperventilation). This results in excess CO_2 being exhaled and pH returns to normal. With a decrease in PCO_2 the peripheral and central chemoreceptors are not stimulated and stimulatory impulses are not sent to the inspiratory area, resulting in the area setting a moderate rate until CO_2 accumulates and PCO_2 levels rise.

APPLY

Monitoring respiratory rate

BODIE FAMILY

Normal breathing rates will vary depending on the age of the person. It is important that as healthcare providers we are aware of the different rates to help us make decisions regarding a person's respiratory function. An infant like Danielle may take up to 40 breaths per minute: as she moves towards childhood her respiratory rate will slow and may vary between 20 and 40 breaths per minute. As adults, the rate may vary between 12 and 16 breaths per minute. A person over the age of 65 may have a rate between 12 and 18 breaths per minute and someone over the age of 80 may have a rate of 10-30 breaths per minute. Respiratory rate may be affected by many different circumstances, e.g. if the person has an underlying respiratory disorder (asthma, COPD); if they have a chest infection; if they have pyrexia (increased temperature); if they are anxious or in pain. A change in respiratory rate is a significant indicator of a person becoming unwell; it is therefore imperative that the healthcare professional monitors and records it carefully.

APPLY

Respiratory rate and exercise

BODIE FAMILY

Matthew is a keen triathlete competing in many competitions: his respiratory system will have to adjust in response to the intensity and duration of exercise. As the heart pumps more blood to the lungs during exercise, pulmonary perfusion increases. During exercise Matthew's muscles use more O_2 and, in the process, produce more CO_2. Oxygen consumption and pulmonary ventilation increase dramatically to meet the metabolic demands of the body. Neural changes occur due to activation of the limbic system; sensory impulses from the proprioceptors in the muscles and tendons and motor impulses from the primary motor cortex cause an increase in impulses sent to the inspiratory area and therefore increase the rate of breathing.

Due to the nature of the endurance sport Matthew's body produces large amounts of lactate: HCO_3^- ions will buffer H^+ ions (from the lactate) and this reaction releases more CO_2 thereby further increasing PCO_2. The increase in PCO_2 will alter pH detected by the chemoreceptors and in response the rate and depth of breathing increases.

When Matthew finishes his race there will be an abrupt decrease in pulmonary ventilation due to neural changes, followed by a gradual decline to resting level as temperature and blood chemistry levels return to within their normal range.

CONCLUSION

Now that you have read through this chapter and undertaken the activities, you should have a good understanding of how the respiratory system functions. You should have an understanding of how oxygen and carbon dioxide are transported around the body and the movement of gases from the atmosphere to the tissue and cells and vice versa. You should also have an awareness of the other important functions of the respiratory system such as protection, regulation of blood pH, production of chemical mediators, sound production and provision of olfactory sensations – as molecules are drawn into the superior portions of the nasal cavity the olfactory epithelium sends messages to the central nervous system.

GO DEEPER

Further reading

McLafferty, E., Johnstone, C., Hendry, C. and Farley, A. (2013) 'Respiratory system, part 1: pulmonary ventilation', *Nursing Standard*, 27(22): 40-7.

McLafferty, E., Johnstone, C., Hendry, C. and Farley, A. (2013) 'Respiratory system, part 2: gaseous exchange', *Nursing Standard*, 27(23): 35-42.

- Every cell in the body requires a constant supply of oxygen to undertake their metabolic function: in doing so they produce carbon dioxide (a waste product) that must be removed. If not removed it will alter the pH which can be toxic to the cells and affect their ability to carry out its functions, thereby disturbing homeostasis.

- The respiratory system has a number of key functions:

  o Regulation of blood pH – carbon dioxide has a direct effect on blood pH: the respiratory system regulates blood pH by changing the levels of carbon dioxide in the blood.

  o Produces chemical mediators – angiotensin converting enzyme (ACE), a key component in blood pressure regulation (Chapter 11), is produced by the lungs.

  o Sound production – involved in speaking, singing and non-verbal communication, as air moves through the vocal cords.

  o Provides olfactory sensations – as molecules are drawn into the superior portions of the nasal cavity the olfactory epithelium sends messages to the central nervous system.

  o Protection – microorganisms can be prevented from entering the respiratory system from the atmosphere (by nasal hair) or they can be removed from respiratory surfaces (by coughing).

- Structurally the respiratory system is divided into the upper respiratory tract and the lower respiratory tract. It can also be classified functionally into two distinct areas, i.e. conducting zone (carrying gases) and respiratory zone (gas exchange).

- The organs of the respiratory system are: the nose, pharynx, larynx, trachea, two bronchi, bronchioles, alveoli, two lungs and muscles of breathing, i.e. the intercostal muscles and the diaphragm.

- The process by which oxygen and carbon dioxide are exchanged between the atmosphere and body cells consists of three distinct phases:

  o Pulmonary ventilation (breathing).

  o External respiration.

  o Internal respiration.

- Oxygen and carbon dioxide move between the alveolar air and pulmonary circulation through the process of passive diffusion. Passive diffusion is governed by two gas laws:

  o Dalton's law (the total pressure of a gas mixture is the sum of the partial pressure of the individual gases).

  o Henry's law (at the air–water interface, for a given temperature, the amount of gas that dissolves in the water is determined by its solubility in water and its partial pressure in air).

- Gas transport refers to the process of carrying gases (oxygen and carbon dioxide) from the alveoli to the tissues and vice versa.

- Oxygen transport:

  o A small amount is transported in plasma (1.5%).

  o Almost all binds with haemoglobin in red blood cells (98.5%).

- A haemoglobin molecule has four iron-containing haem regions, each binds to one molecule of O_2.

- To enter cells O_2 must separate from haemoglobin.

- O_2 is released into areas where O_2 concentration is low.

- Carbon dioxide is transported in three forms:

  - 7% is dissolved in plasma: on reaching the lungs it diffuses into alveolar air and is exhaled.

  - 23% combines with the protein portion of haemoglobin and plasma proteins to form carbaminohaemoglobin.

  - 70% is transported as bicarbonate ions. As the CO_2 diffuses into the systemic capillaries it reacts with water; in the presence of carbonic anhydrase it forms carbonic acid (H_2CO_3), which easily dissociates into hydrogen (H^+) and bicarbonate (HCO_3^-) ions.

- The depth and rate of ventilation is managed through a control system that uses negative feedback to regulate homeostasis.

  - Centres in the CNS control fundamental respiratory pattern.

  - Motor fibres extend into the spinal cord through the cervical region to the phrenic nerve to the diaphragm.

  - Chemoreceptors are sensitive to chemicals dissolved in body fluids. There are two types: centrally – brainstem, peripherally – carotid and aortic bodies; they respond to CO_2 levels in circulating blood. The presence of hydrogen ions and its effect of decreasing pH stimulate central chemoreceptors and the increased CO_2 triggers ventilation.

REVISE

TEST YOUR KNOWLEDGE

This chapter helps you to understand the structure and functions of the respiratory system, how the respiratory system plays an important role in ensuring that our tissues and cells receive the oxygen they require to function normally and how it excretes the carbon dioxide formed as a waste product. You should also now have an understanding of how gases are transported around the body and how these gases move from the atmosphere to the tissues and cells and vice versa. Revise each section individually and check your understanding of it before you move to the next. Do you understand how alveoli are specially developed to facilitate gaseous exchange? Do you know how O_2 and CO_2 are transported in the body? Can you identify the mechanisms by which respiration is controlled? Challenge your knowledge.

In order to help you revise, consider the following questions, answers for which can be found by visiting **https://study.sagepub.com/essentialap2e**.

Test yourself by revising the chapter first, and then answer these questions without looking at the book. Afterwards compare your answers with the text and with the notes you made. Did you miss anything in your notes? Here are the questions:

1 Outline the functions of the respiratory system.

2   Identify the structures of the upper respiratory tract and describe their functions.

3   Describe the structure and function of the trachea.

4   Explain how alveoli are adapted to ensure that gas exchange takes place.

5   Describe the process of respiration.

6   Explain how the rate and depth of ventilation is controlled.

7   Explain what is meant by the term 'partial pressure' and identify the proportions of gases contained within atmospheric air at sea level.

8   Discuss how oxygen and carbon dioxide are transported.

For additional revision resources visit: **https://study.sagepub.com/essentialap2e**.

- Revise key terms with interactive flashcards.

- Test yourself with multiple-choice questions.

- Access the glossary with audio to hear how complex terms are pronounced.

- Explore recommended websites suitable for revision.

REVISE

ACE YOUR
ASSESSMENT

REFERENCE

Milner, Q.J.W. and Mathews, G.R. (2012) 'An assessment of the accuracy of pulse oximeters', *Anaesthesia*, 67: 396–401.

THE RENAL SYSTEM

FLUID, ELECTROLYTE AND ACID–BASE BALANCE

11

LEARNING OUTCOMES

When you have finished studying this chapter you will be able to:

1. Identify the components of the main fluid compartments in the body and the movement of fluids and electrolytes between these fluid compartments
2. Describe the structure and function of the renal system
3. Discuss the homeostatic control of fluid, electrolyte and acid–base balance within the body

INTRODUCTION

In Chapter 8 you learned how nutrients were obtained through the digestive system and their purposes within the body. In Chapter 1 we identified the main fluid compartments of the body and in this chapter the content of the intracellular fluid (ICF) and extracellular fluid (ECF) within the human body will be considered. The importance of fluid and electrolyte balance and pH (acid–base balance) for normal cell function are considered, particularly in relation to function of nerve and muscle cells. This chapter progresses to examine the structure and functions of the renal system and its role in the regulation of fluid, electrolyte and acid–base balance. The importance of renal function for health will be examined.

Context

At present Danielle does not have conscious control of her bladder; her bladder capacity is between 50 and 70 ml at her age and as it stretches from filling, stretch receptors in the bladder wall initiate reflex contraction of the detrusor (bladder) muscle and relaxation of the internal and external urethral sphincters. However, with maturity of the nervous system (around the age of two), Danielle will develop conscious control and will be able to inhibit relaxation of the external urethral sphincter to prevent micturition.

BODIE FAMILY

Maud has a degree of heart failure following a myocardial infarction (heart attack), which is controlled with medication. One of the early signs of the condition was swelling (oedema) of her ankles although this has now disappeared. George Bodie is not consciously aware of being thirsty, a change in sensation that can occur in ageing. Therefore, it is important that he and his family are aware of the importance of ensuring adequate fluid intake.

Matthew, being a triathlete, is very conscious of the importance of maintaining his fluid and electrolyte balance as he loses these through insensible losses from the skin and lungs in particular during endurance exercise. He compensates for this by consuming fluid and electrolyte drinks.

REGULATION OF WATER AND ELECTROLYTES

Fluids and electrolyte movement

There are four main methods that govern the movement of water and solutes within the body and we will start by explaining these. One of the key concepts is that of osmolality.

─────────────── **UNDERSTAND** ───────────────

Osmolality – Key points

- The concentration of osmotically active particles per 1,000 ml of fluid.
- Monitored by osmoreceptors, specialist cells, found primarily in the hypothalamus.
- Expressed as milliosmoles (mOsm)/L of water.
- Normal blood osmolality: 285-295 mOsm/L water (Kaplan and Kellum, 2010).
- Raised serum osmolality indicates dehydration as solutes are concentrated.
- Low serum osmolality indicates an increase in fluid as solutes are diluted (Maday, 2013).

Osmosis

Osmosis is the movement of water across a semipermeable membrane from a dilute/hypotonic solution (with low osmotic pressure) to a more concentrated/hypertonic solution (which has a higher osmotic pressure). This works to achieve fluid with similar distribution of water and concentration of solute (substance distributed) in the solvent (water in this case) on both sides of the membrane (Figure 11.1a). The pressure exerted in pulling the water across the semipermeable membrane is known as osmotic pressure (Chapter 12). This plays an important part in determining the movement of water and the balance of electrolytes across membranes. Figure 11.1b illustrates what happens to red blood cells when the water concentration (tonicity) differs. The tonicity of a fluid refers to the force exerted by osmotically active particles within that fluid. Hypotonic solutions lose water to an isotonic or hypertonic solution, and hypertonic solutions gain water from isotonic and hypotonic solutions. Some dissolved particles are very osmotically active in that they attract water. Sodium and potassium are very active osmotically and influence water movement in ECF and ICF respectively. Glucose is also osmotically active.

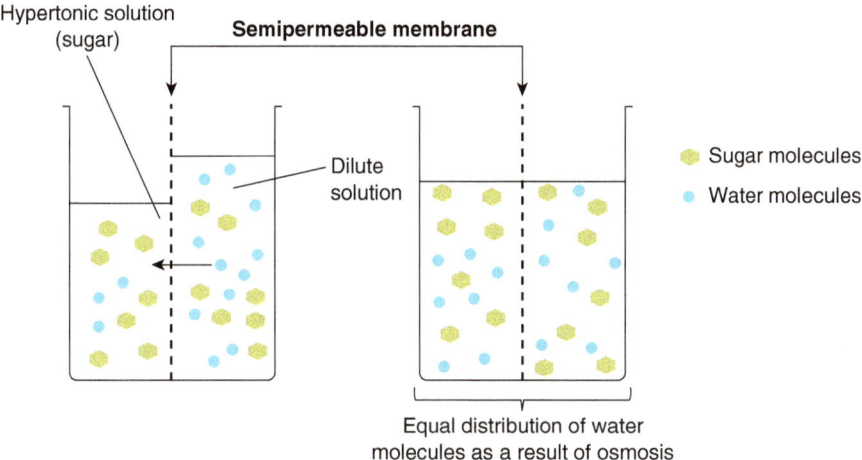

Figure 11.1a Osmosis

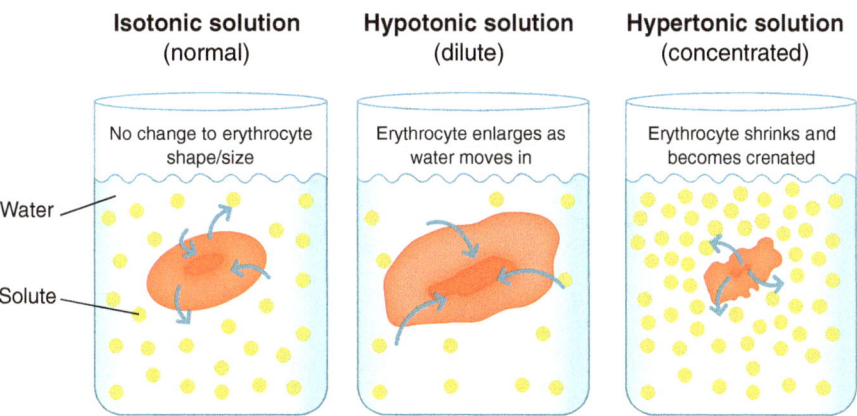

Figure 11.1b Effect of water concentration on red blood cells

Diffusion

This is the movement of ions and molecules from an area of high concentration to an area of low concentration in an attempt to achieve an isotonic balance. It is an important method of achieving a balance of concentration across membranes (Figure 11.2).

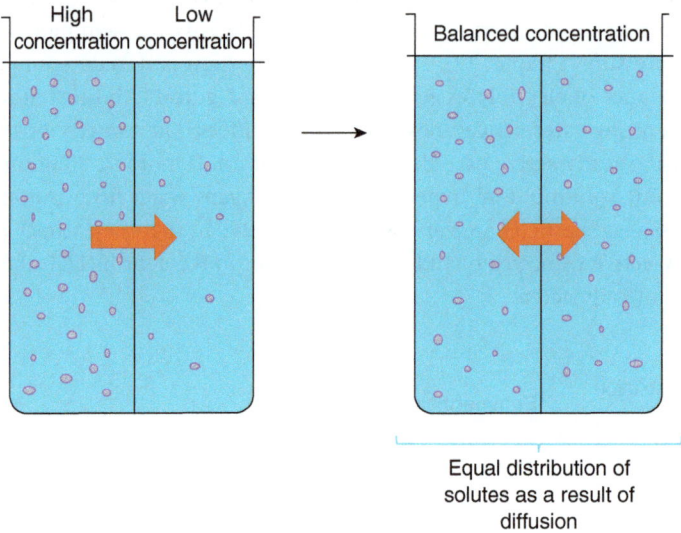

Figure 11.2 Diffusion

Filtration

Filtration involves the movement of water, ions and molecules from an area of higher pressure to an area of lower pressure, usually across a semipermeable membrane, due to the hydrostatic pressure of the fluid. Figure 11.3 illustrates filtration in the glomerulus of the kidney where the pressure in the glomerular

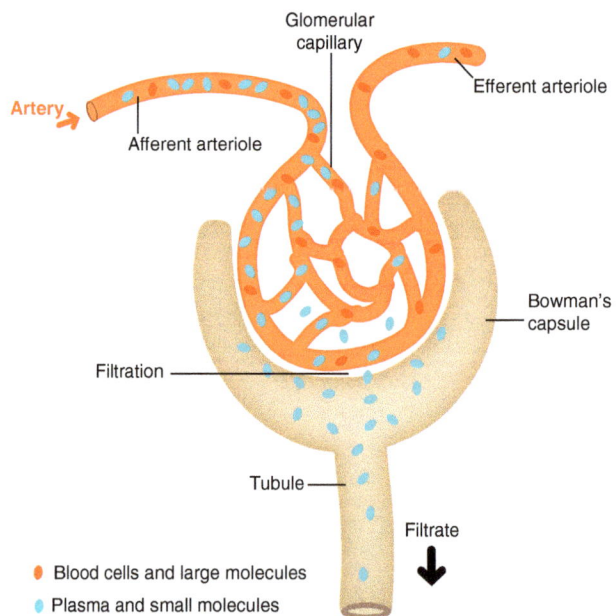

Figure 11.3 Filtration in the glomerulus of the kidney

capillary is greater than in the Bowman's capsule. This forces smaller molecules through the membrane, leaving the cells and larger molecules in the blood circulating through the kidney.

Active transport

This involves the movement of solutes (ions and molecules) across a membrane (usually) against a concentration gradient involving the use of energy (adenosine triphosphate – ATP) and carrier proteins (Figure 11.4). The molecule may be too large to move by diffusion or the higher concentration on the other side of the membrane may prevent diffusion. For example, potassium levels are higher inside a cell than outside; getting more potassium in requires active transport and is crucial to normal cell function. Of these mechanisms, only active transport requires energy.

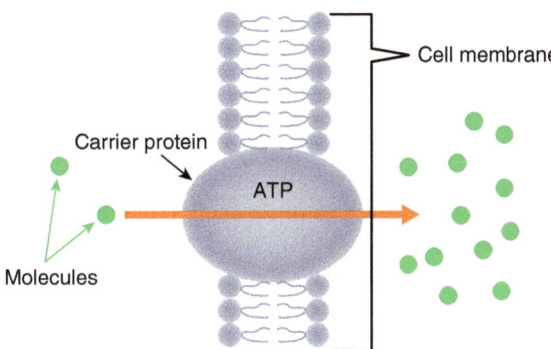

Figure 11.4 Active transport

Water and electrolyte homeostasis

Maintaining water and electrolyte homeostasis is achieved through a complex interplay of the cardiovascular, endocrine, nervous, renal, digestive and respiratory systems. Water balance is not static and is primarily influenced by the following:

- The hypothalamus monitors fluid volume and concentration of the ECF (osmolality) through specialist cells called osmoreceptors. Osmolality tells us the concentration of particles dissolved in the weight of a fluid, providing a representation of water balance and hydration within the body (normal range 285–295 mOsm/kg). The hypothalamus then has a regulatory function in initiating actions to excrete, retain or obtain water and electrolytes through a variety of systems (Chapter 7, Figure 7.10).
- The kidney plays the major role in homeostasis of water and electrolytes through renal excretion and absorption in response to hormonal stimuli and blood volume.
- Fluid moves between fluid compartments as a result of hydrostatic and osmotic pressures and capillary and cell membrane integrity.

Water

Water is essential to life; it provides a transport medium, assists in temperature regulation and performs as a solvent, permitting chemical reactions to occur. Without it, we cannot maintain homeostasis. On average, 60% of our body weight is accounted for by water, with infants having 75% water composition and older adults 55% (Popkin et al., 2010). A healthy adult, under normal conditions, requires approximately 2.5 L of fluid daily to replace fluid lost through faeces, respiration, perspiration (diaphoresis) and urine. The first three of these are referred to as insensible losses as they are not possible to measure in everyday life.

Water balance is maintained by factors that regulate its intake, production and elimination within the body. Some water loss, however, is obligatory including insensible losses from the respiratory system, and sweating through the skin. The faecal content of the large intestine also contains unabsorbed water and is another source of obligatory water loss. Additionally, the blood contains solutes that need to be excreted from the body. The kidneys excrete about 900–1,200 mOsm of solutes daily to maintain homeostasis, using water as the primary component of the solution (urine) to flush these out. Water intake is not solely physiological in origin – there are social aspects to drinking, whether for pleasure or habit. We also generate water through metabolism, known as water of oxidation, producing up to 1 L of water per day, depending on the rate of metabolic activity (Kaplan and Kellum, 2010).

Neuroregulation

A dehydrated person has a low water volume or ECF. The fall in ECF causes water to move from inside cells (intracellular) to outside (extracellular) causing cell shrinkage detected by osmoreceptors in the anterior hypothalamus and preoptic area of the brain (Popkin et al., 2010). The result is thirst, stimulated by a decline in plasma volume of 10–15% or an increase in plasma osmolality by 1–2%. Once a person starts to drink, taste receptors signal to the brain to relay this, quenching thirst even before blood volume increases. The reward sensation of drinking is thought to be a process that enables replacement of water losses before they occur (Popkin et al., 2010). In addition, an increased blood volume triggers the anterior hypothalamus to signal sufficient water intake has been achieved. However, this response is slower as time for absorption is needed. The responsiveness of osmoreceptors diminishes in older adults resulting in a tendency towards dehydration as their thirst mechanism is less responsive (Chapter 7).

Thermoregulation

In Chapter 14 you will discover how the skin regulates heat balance, including through the evaporation of sweat with associated water loss. This depends on various factors, including the temperature of the external environment, layers and type of clothing, levels of activity and air humidity with insensible loss through the skin of 300 to 2,000 ml/hour (Popkin et al., 2010).

Renal regulation

In Chapter 12 you will learn how the renin angiotensin aldosterone system (RAAS) responds to a drop in blood volume to increase water intake. Once osmoreceptors in the hypothalamus detect raised osmolality of the blood it signals the posterior pituitary gland to release antidiuretic hormone (ADH, or vasopressin). ADH release is also influenced by baroreceptors in the cardiopulmonary and peripheral circulation and by angiotensin II (Millard-Stafford et al., 2012). ADH has a direct effect on the renal tubules increasing water reabsorption, leading to an increase in water in the blood. Release of ADH falls away once plasma osmolality rises. Water is reabsorbed in the presence of aldosterone, produced by the adrenal cortex under the influence of angiotensin II. Aldosterone promotes the reabsorption of sodium in the renal tubule, drawing water with it. In contrast to this, atrial natriuretic peptide (ANP) has the opposite action to aldosterone. It is released from the muscle of the atria when blood volume is increased, promoting the excretion of sodium which takes water with it (Chapter 7).

APPLY

Fluid and electrolyte imbalance

If homeostasis of fluid and electrolytes becomes disrupted, the distribution of water and electrolytes can show certain changes. The person may suffer from:

- **Oedema**: swelling of the tissues due to fluid accumulation. Peripheral oedema involves swelling of the feet and ankles, or any dependent parts of the body.
- **Ascites:** swelling of the abdomen due to fluid accumulation. It may occur in liver disease or heart failure, but is also seen in kwashiorkor when the lack of adequate protein in the diet, and the resulting low levels of plasma proteins, lead to a low plasma osmolality and reduced fluid re-entering the blood stream from the interstitial fluid.

GO DEEPER

Ageing and water homeostasis

As people progress into older adulthood, there are age-associated renal changes. Renal mass is reduced by glomerular loss and sclerosis. This impairs the preservation of water and sodium, increasing the likelihood of dehydration and hyponatraemia. As people age, they are more likely to have hypertension, which results in more ANP being released and less ADH and aldosterone. This leads to a further likelihood of water and sodium depletion, particularly considering that any ADH and aldosterone present have less renal mass to stimulate to retain both sodium and water. The thirst response is also dampened in older adults, making behavioural responses to dehydration less efficient. Alongside the possible physical impairment and social isolation, this exacerbates dehydration that has occurred due to physiological changes. Polypharmacy can exacerbate this, particularly if diuretics are prescribed, which may enhance any electrolyte imbalance (El-Sharkawy et al., 2014).

Electrolytes

Electrolytes are positively charged ions (cations, e.g. sodium (Na^+)) or negatively charged ions (anions, e.g. chloride (Cl^-)) dissolved within body fluids. The electrical charge gives rise to their name – when in solution, they conduct electricity. Electrolytes are obtained within the diet and are primarily eliminated from the body by the kidneys, although some can be excreted in sweat. Electrolyte distribution is not equal in all fluid compartments of the body (Figure 11.5 and Table 11.1). This is largely because the cell membrane is selectively permeable, and pumps electrolytes in or out of the cell to achieve the desired concentrations either side of the cell membrane. In addition, different laboratories will often have slightly varying biochemistry results, with the standard results for that laboratory included on the form reporting the results obtained.

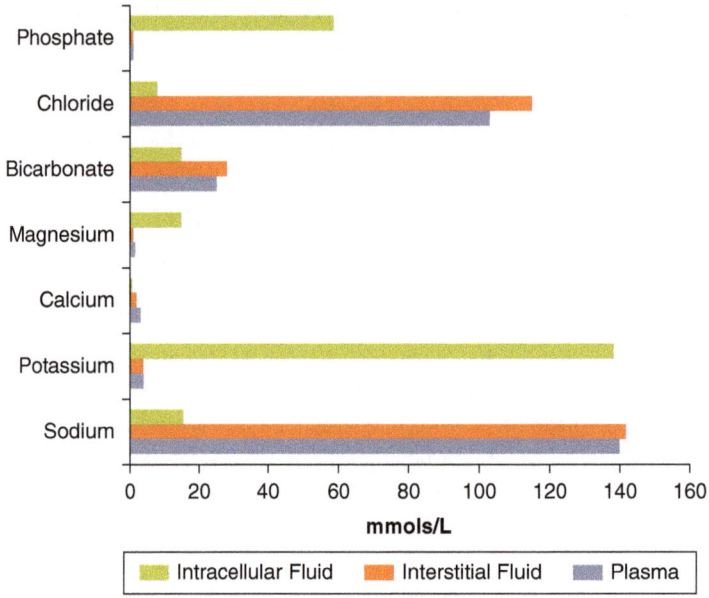

Figure 11.5 Electrolyte distribution in fluid compartments

Table 11.1 Average electrolyte amounts in fluid compartments (in mmols/L)

		Plasma	Extracellular fluid	Intracellular fluid
			Interstitial fluid	
Cations	Sodium	141	143	15
	Potassium	4	4	140
	Calcium	2.5	1.3	0.0001
	Magnesium	1	0.7	15
Anions	Bicarbonate	25	28	15
	Chloride	103	115	8
	Phosphate	1	1	60

Sodium (Na⁺)

Sodium is the main cation in ECF, being a key substance in generating osmotic pressure. With potassium, it plays a vital role in the conduction of nerve signal transmissions (action potentials, Chapter 5). It is therefore important in optimal functioning of the nervous system, including playing a role in the innervation of skeletal and cardiac muscles. An adult needs 1–2 mmol of sodium per kilogram of body weight daily (Kaplan and Kellum, 2010). A 70 kg person should ingest 70–140 mmol daily in diet and fluids. Aldosterone from the adrenal cortex preserves sodium by promoting sodium reabsorption in the renal tubules. Excess sodium is excreted in urine and is enhanced by ANP release from the myocardium (heart muscle).

Potassium (K⁺)

Potassium is the main intracellular cation. It controls cellular osmotic pressure, holding water within the cell and being central to cellular hydration. Low potassium in the ECF results in the cell sacrificing potassium, and therefore some water, into the ECF. Regulation of intracellular and extracellular potassium is central to optimal cell metabolism as potassium has a key role in activating numerous cellular enzymes (Maday, 2013), including those for metabolism of carbohydrates and proteins. Potassium is also essential in conduction of nerve signal transmissions. Renal nephrons play a key role in potassium homeostasis, excreting it in the urine or reabsorbing it within the renal tubule as necessary. An adult needs 0.5 mmol of potassium per kilogram of body weight daily (Kaplan and Kellum, 2010). A 70 kg person therefore requires about 35 mmol daily in their diet and fluids.

APPLY

Exchangeable cations

Sodium and potassium are exchangeable cations. When a person has an increase in potassium intake into the circulation (either from food, fluid or medication), cells will take in extra potassium through active transport and pump out sodium in exchange (Kaplan and Kellum, 2010).

Calcium (Ca²⁺)

Calcium is most abundant in ECF although present as free ions in ICF in very small amounts; calcium can be stored in organelles so the ICF value reads very low. Cells can also pump calcium in when needed. On average, we ingest about 20 mmol of calcium daily, of which two thirds is excreted in faeces. Calcium is also excreted by the kidneys, but 65% of the calcium filtered into the renal tubule is reabsorbed with sodium in the proximal tubule and ascending loop of Henle (Baker and Worthley, 2002). Hormonal regulation of calcium is discussed in Chapter 7. Calcium has an important role in blood clotting (Chapter 12) and it is heavily deposited in bones, which can also release it into the blood stream (Chapter 15). It has a crucial role in neurotransmitter release in neurons (Chapter 5) and also in neuromuscular excitability.

Magnesium (Mg²⁺)

Magnesium is the second most abundant intracellular cation, being essential for hundreds of metabolic processes (Baker and Worthley, 2002), including the formation and use of ATP, i.e. the generation, use and storage of energy, as well as formation of nucleic acids and proteins. Magnesium is also known to keep constant the concentrations of intracellular potassium, calcium and phosphorus. The constancy of these electrolytes is essential for the stability of the conduction system of the heart. Additionally, magnesium has calcium channel-blocking properties for all muscle types, also helping to prevent arrhythmias of the heart (Akhtar et al., 2011). It is absorbed in the small intestine and excreted by renal filtration.

APPLY

Magnesium and the myocardium

BODIE FAMILY

Magnesium is very important in regulating the contraction of the heart. Thus, it is often given to people with unstable heart rhythms. When Maud had her myocardial infarction, she was given a magnesium infusion to help to regulate her heartbeat.

Bicarbonate (HCO_3^-)

Bicarbonate is the second most abundant extracellular anion and an important alkaline buffer used in regulating pH within the body. It also plays an important role in osmotic pressure (Maday, 2013) and can be produced from carbonic acid (from carbon dioxide). The pancreas releases bicarbonate, under the influence of secretin, into the small intestine to neutralise acidic chyme from the stomach. The kidneys are the main regulator of blood HCO_3^- concentrations. The intercalated cells of the renal tubule can either form HCO_3^- and release it into the blood when levels are low or excrete excess HCO_3^- in the urine when blood levels are too high.

Chloride (Cl^-)

The most prevalent ECF anion, chloride plays a role in acid–base balance and production of hydrochloric acid in the stomach. It is taken in the diet and absorbed in the small intestine. Elimination occurs in the renal tubules, where it can be excreted or reabsorbed (Maday, 2013).

Phosphate (PO_4^{3-})

Phosphate is most abundant in ICF. It is important in bone mineralisation and structural components of cells, such as phospholipids, nucleotides and phosphoproteins (Baker and Worthley, 2002). Storage of ATP and transfer of oxygen also depend on phosphate, and it is a buffer in balancing pH. Phosphorus is stored in teeth and bone. Sourced from the diet, phosphorus is excreted at the proximal renal tubule (Maday, 2013). Homeostasis of calcium, magnesium, phosphorus and parathyroid hormone are inter-related, with changes in one impacting on the other.

APPLY

Intravenous fluid therapies

In practice, a person may be unable to eat or drink, or may need supplements of electrolytes in order to facilitate the body to maintain homeostasis. This is often achieved through intravenous fluids. Following assessment of a person's fluid and electrolyte status, usually including venous blood samples, a plan is made to replace any deficit. On average, a person needs around 35 ml/kg/day of fluid (water), or a more accurate calculation would be 4 ml/kg/h for the first 10 kg, 2 ml/kg/h for each of the next 10 kg and 1 ml/kg/h for each kg above 20 kg. The type of fluid chosen will depend on what fluid compartment needs to be hydrated and what electrolytes are required.

A hypotonic fluid will distribute to all of the fluid compartments in the body, proportional to the volume of that compartment. For example, from 1 L of a hypotonic solution administered, two thirds will move into the ICF, 80 ml will remain in the plasma and approximately 53 ml enter the interstitial fluid. Hypotonic solutions are therefore not useful in raising blood volume in someone who is hypovolaemic as most of the fluid will move into the ICF. Hypertonic intravenous solutions are rarely used, and most intravenous fluids are isotonic in nature. In practice, there are two main types of fluids: crystalloids and colloids.

Crystalloids

Crystalloids have a specified balance of water and electrolytes and remain in plasma for a short time, moving easily between fluid compartments. Sodium is normally the main osmotically active solute and crystalloids are primarily for restoring and maintaining fluid and electrolyte homeostasis. The most common crystalloids used are normal saline (0.9% sodium chloride, which is isotonic) and Hartmann's solution (a compound sodium lactate crystalloid solution that is similar in concentration to blood).

Colloids

Colloids have a high molecular weight, preventing them from moving across plasma and cell membranes under normal circumstances. They therefore largely remain in the plasma. As they contain substances that do not dissolve into a true solution, they contribute to osmotic pressure (colloid osmotic pressure). This results in fluid being drawn into the plasma. Blood, albumin and starch solutions, and plasma expanders are all colloids.

RENAL SYSTEM

The renal system is one of the excretory systems of the body and is composed of two kidneys, two ureters (one from each kidney) which transport urine to the bladder, and one urethra enabling excretion of urine from the body (Figure 11.6). The renal system has a number of key functions: excretion (removal of organic wastes from body fluids); elimination (discharge of waste products); and homeostatic regulation (discussed later).

The kidneys

The kidneys are positioned as shown in Figure 11.6, on the posterior wall of the abdomen within the peritoneal cavity and their structure is illustrated in Figure 11.7. They are about 11–14 cm (4.3–5.5 inches) in length, 6 cm (2.4 inches) wide and 4 cm (1.6 inches) thick. Three layers of tissue surround each kidney: the renal capsule, the adipose capsule and the renal fascia. The renal capsule is the innermost layer and consists of a smooth, transparent sheet of irregular connective tissue that is continuous with the outer coat of the ureter: it serves as a barrier against trauma and helps maintain the shape of the kidney. The adipose capsule is the middle layer and is a mass of fatty tissue which surrounds the renal capsule, protecting the kidney from trauma and holding it firmly in place in the abdominal cavity. The renal fascia is the outermost layer consisting of a thin layer of dense irregular connective tissue that anchors the kidney to the surrounding structures and to the posterior abdominal wall.

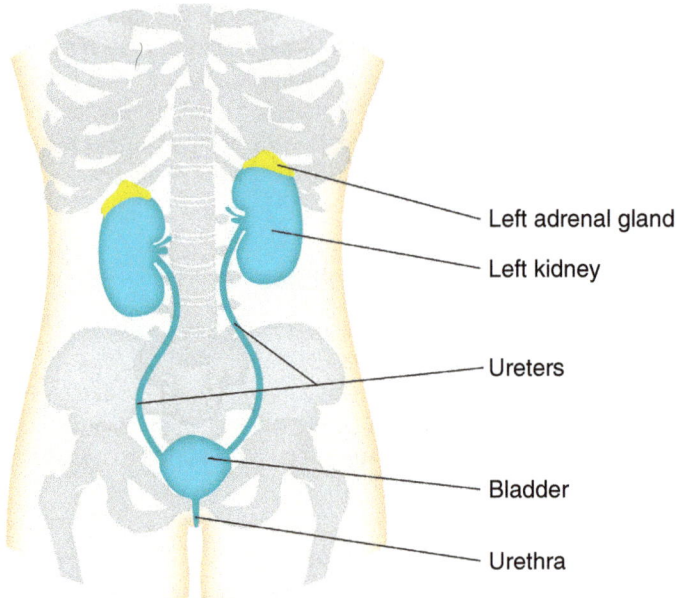

Figure 11.6   Outline of the renal system

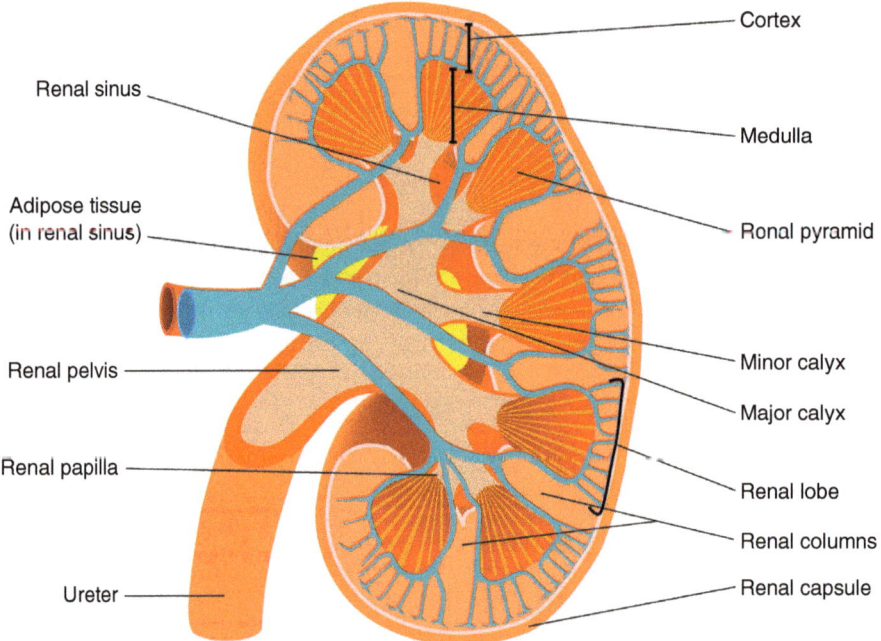

Figure 11.7   Structure of the kidney

There are two distinct regions within the kidney: the renal cortex and the renal medulla. The renal cortex extends from the renal capsule to the bases of the renal pyramids and into spaces between them. The area that extends between the renal pyramids is known as the renal column. Within the renal cortex there are two zones: the cortical zone (outer) and the juxtamedullary zone (inner). The renal medulla is the innermost layer, consisting of pale conical-shaped striations known as renal pyramids (consisting of the base and renal papilla or apex). The base of the pyramid faces the renal cortex and its apex points toward the renal hilum. A renal lobe thereby consists of a renal pyramid, its overlying area of renal cortex and one half of the adjacent renal column. The nephron is the major functional unit of the kidney and filtrate formed here drains into papillary ducts (which extend through the renal papillae) and into minor (8–18) and major (2–3) calyces (cuplike structures). A minor calyx receives urine from the papillary ducts of one renal papilla and delivers it to a major calyx: once the filtrate enters the calyx it becomes known as urine as no further reabsorption can take place.

Homeostatic functions of the kidney

The kidneys play a key role in the maintenance of homeostasis; it contributes to this process by:

- Regulating blood volume and blood pressure by adjusting the volume of water lost in urine; production of renin (Chapter 7).
- Regulating plasma ion concentrations by controlling quantities of sodium, potassium and chloride ions lost in urine. It helps control calcium ion levels through the synthesis of calcitrol (the active form of vitamin D, produced by modification of vitamin D in the liver then the kidney).
- Stabilising blood pH by controlling the loss of hydrogen and bicarbonate ions in urine.
- Conserving valuable nutrients by preventing excretion while excreting organic waste products.
- Assisting the liver to detoxify poisons.

Blood and nerve supply of the kidney

The blood supply to the kidneys is about 20–25% of the cardiac output (1.2–1.25 L/min). There are a number of arteries that deliver blood to the kidney: the renal artery (arising from the abdominal artery); segmental arteries (branches of the renal artery); and interlobar arteries (found between renal pyramids and a branch of the segmental artery). At the base of the pyramid, interlobar arteries branch between the medulla and the cortex and become known as arcuate arteries; divisions of the arcuate arteries produce a series of interlobular arteries (these pass between renal lobes). Interlobular arteries enter the renal cortex and branch off to afferent arterioles. Afferent arterioles divide into a capillary network known as the glomerulus. Glomerular capillaries reunite to form efferent arterioles. The efferent arterioles unite to form peritubular capillaries (surrounding tubular parts of the nephron), peritubular veins unite to form interlobular veins, which unite to form arcuate veins and finally interlobar veins. Blood leaves the kidney via the renal vein through the hilum to the inferior vena cava.

Renal nerves originate in the renal ganglion and pass through the renal plexus, the nervous system supplies both sympathetic impulses and the parasympathetic (vagus) nerve supply. Most of the nerves supplying the kidney are vasomotor nerves and regulate blood flow by vasoconstriction or vasodilation of renal arterioles.

Nephrons

Each nephron consists of a renal corpuscle, a spherical structure consisting of the glomerulus and the glomerular (Bowman's) capsule and the renal tubule, which consists of a long tubular passageway divided into the Proximal Convoluted Tubule (PCT), loop of Henle and the Distal Convoluted Tubule (DCT). Figure 11.8 shows the structure of a nephron. The loop of Henle can vary in length. The various parts of the nephron play different roles in the formation of urine. Each kidney contains about 1.25 million nephrons.

The nephron undertakes different processes in the formation of urine (Figure 11.9):

- Filtration in the renal corpuscle.
- Reabsorption in the proximal convoluted tubule, loop of Henle, distal convoluted tubule and collecting duct.
- Secretion in the distal convoluted tubule.

Renal corpuscle

The renal corpuscle consists of a glomerulus (group of capillaries) within the Bowman's (glomerular) capsule. The blood from the renal artery enters the glomerulus through the afferent arteriole which divides to form the capillaries which recombine to form the efferent arteriole which then passes around the rest of the tubule (Figure 11.9). The Bowman's capsule of epithelial cells acts as a filter of the blood entering through the afferent arteriole.

The juxtaglomerular apparatus is situated in this area and contributes to regulation of fluid and electrolyte balance under the influence of both antidiuretic hormone and aldosterone (Chapter 7, Figure 7.10).

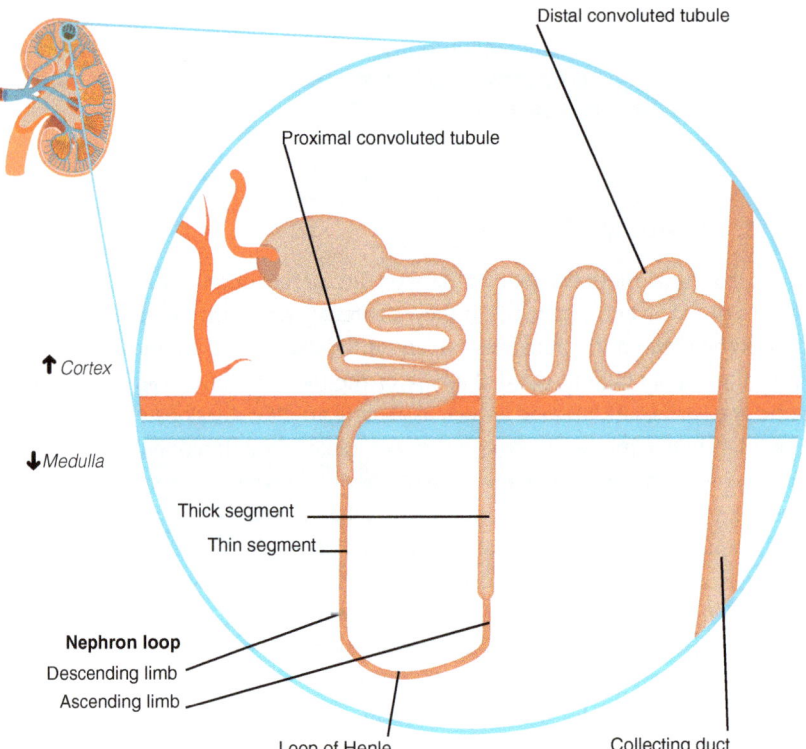

Figure 11.8   **Structure of the nephron**

Proximal convoluted tubule

This segment of the tubule consists of closely fused cuboidal epithelial cells with many mitochondria and a brush border. In a healthy kidney virtually all the nutrients (glucose and amino acids) and most of the electrolytes (sodium, chloride and bicarbonate) are reabsorbed here with the amount of water that retains the osmolality of the filtrate. About two thirds of the filtrate is reabsorbed. This plays no part in electrolyte homeostasis and is known as the obligatory phase of reabsorption.

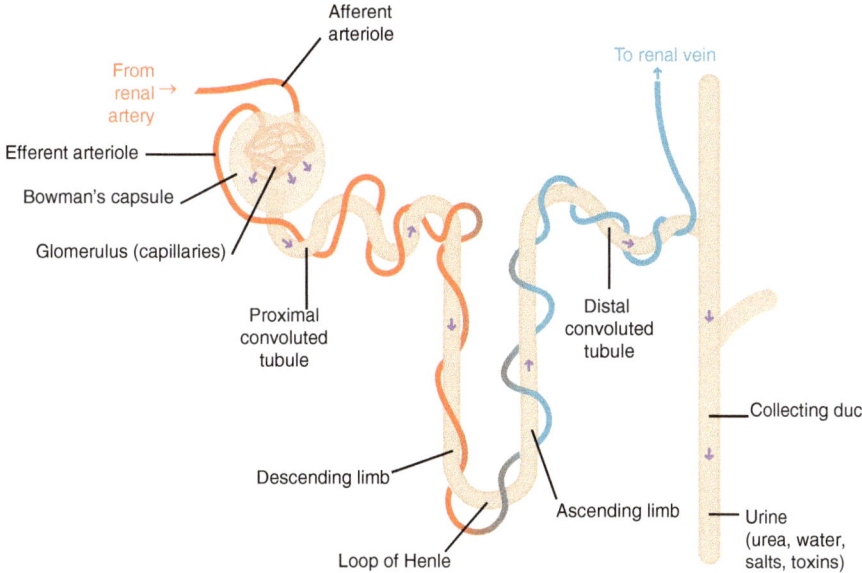

Figure 11.9 Summary of the function of the nephron

Loop of Henle

This section of the tubule dips down into the medulla and back up again. As shown in Figure 11.8 the cells in the cortex are thick (similar cells to the proximal convoluted tubule) but then the cells of the descending limb become thin and permeable. The first part of the ascending limb returning towards the cortex is similar but then the cells become thick with many mitochondria again like the cells of the proximal convoluted tubule.

The loop of Henle creates an osmotic gradient through the medulla (Figure 11.10). The thick section of the ascending loop of Henle moves sodium (as sodium chloride – NaCl) from the filtrate into the interstitial fluid by active transport using ATP. Thus, a large amount of the sodium removed results in filtrate with a low concentration of sodium reaching the distal convoluted tubule. Water is drawn into the interstitial fluid by the high concentration of sodium in the medulla from the descending limb of the loop of Henle and the collecting duct. The water is reabsorbed by capillaries and re-enters the blood circulation. The permeability of the collecting duct is increased by antidiuretic hormone (ADH) thus ensuring that more water is retained in the body and the concentration of the filtrate becomes higher. A fall in ADH means that more water is excreted in the filtrate (resulting in an increase in urinary output).

Distal convoluted tubule

The structure of this is similar to that of the upper part of the ascending limb of the loop of Henle and plays a major role in regulating the electrolyte balance of the body. The RAAS regulates the amount of ions reabsorbed from the distal tubule under the influence of aldosterone (Chapter 7, Figure 7.10). Aldosterone stimulates the reabsorption of sodium into the interstitial fluid (i.e. retention in the body) and secretion of potassium into the tubular filtrate (i.e. loss from the body).

In addition, specialist cells (intercalated cells) in this part of the nephron normally reabsorb the bicarbonate remaining in the filtrate after the reabsorption in the proximal tubule. This leads to excretion of acid urine. These cells actively secrete hydrogen ions into the filtrate and reabsorb bicarbonate and can result in a pH as low as 4 to 4.5.

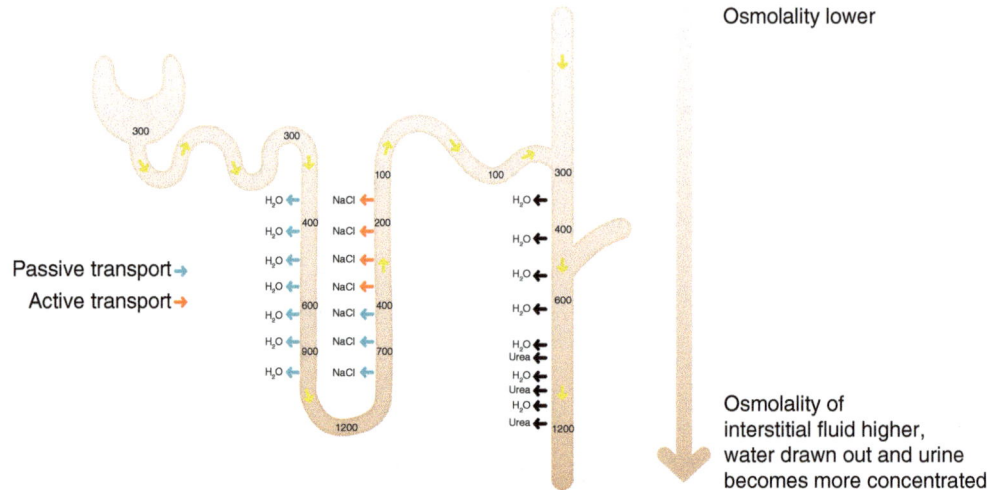

Figure 11.10 The function of the loop of Henle in creating the osmotic gradient

Collecting duct

The collecting duct collects the filtrate from a number of nephrons. The quantity of water reabsorbed into the body through this duct is mainly regulated by ADH. If the body fluids have a high osmolality (greater than 290 mOsm/kg) then ADH secretion is increased and more water reabsorbed leading to highly concentrated urine being excreted. A low osmolality leads to less ADH secretion, less water being reabsorbed and more dilute urine being excreted. The filtrate comes through the collecting ducts via the calyces into the pelvis of the kidney and then the ureter.

Formation of urine

Urine is formed through three physiological processes: filtration, selective reabsorption and secretion.

Filtration

The initial process of urine formation occurs within the renal corpuscle. Blood pressures (hydrostatic pressure and blood colloid osmotic pressure, see Go Deeper Box below) force water and solutes across the wall of the glomerular capillaries and into the capsular space. Solute molecules that are small enough to pass through the filtration membrane are carried by the surrounding water molecules. This process is referred to as glomerular filtration; the glomerular filtrate consists of water, sodium, glucose and waste products. The volume filtered is usually about one fifth of the plasma and known as the Glomerular Filtration Rate (GFR). The GFR is the rate at which the filtrate is formed and is measured in millilitres per minute (normally about 125 ml/minute in adults). The GFR cannot be measured directly but is estimated by examining the rate of excretion of a substance which is readily filtered but neither reabsorbed nor secreted in the nephron, and which does not alter renal function. One substance used is inulin (a plant polysaccharide), which is excreted in direct proportion to its plasma concentration. An easier method for use in clinical practice is to measure creatinine clearance. Creatinine is a normal component in the body, largely retained at a steady level in the blood. It can be filtered from the blood by the kidneys, but it is not reabsorbed or excreted elsewhere.

The filtrate contains electrolytes (sodium, potassium), urea, uric acid and creatinine (waste products of cellular metabolism) but not protein or cells. The filtered blood returns to circulation via the efferent arteriole and eventually to the renal vein to return into the systemic circulation.

GO DEEPER

Filtration pressure

Filtrate is formed when fluid and solutes move from the glomerulus into the glomerular space. In order for this to occur a Filtration Pressure (FP) is required and is the difference in hydrostatic pressure and blood colloid osmotic pressure acting across the glomerular capillaries.

Hydrostatic pressure is the pressure by blood on the surfaces of vessel walls causing fluid to be pushed out of the capillary and into the interstitial fluid. There are two types of hydrostatic pressure involved in the process: Glomerular Hydrostatic Pressure (GHP) (the blood pressure in the glomerular capillaries) and Capsular Hydrostatic Pressure (CHP) (the pressure in the capsular space resulting from resistance to flow along the nephron). CHP pushes water from the filtrate back into the plasma thereby opposing GHP. The Net Hydrostatic Pressure (NHP) is the difference between GHP and CHP (approximately 35 mmHg).

Blood Colloid Osmotic Pressure (BCOP) results from the presence of proteins in a solution. Plasma proteins influence the movement of water between the intravascular space and interstitial compartment. Colloids form colloidal complexes that do not pass through membranes easily due to their size, remaining in the intravascular space. These proteins, primarily albumin, on the plasma side of the membrane give rise to a colloid osmotic pressure gradient that pulls fluids and solutes from interstitial spaces into the intravascular space. BCOP is 25 mmHg.

As stated above, the FP is the difference between the NHP and BCOP and can be found using the following formula:

NHP – BCOP = FP

35 mmHg – 25 mmHg = 10 mmHg

ACTIVITY 11.1: GO DEEPER

To further your understanding, watch this video overview of renal physiology.

This external video link can be accessed by **scanning the QR code** with your smart phone camera or via https://study.sagepub.com/essentialap2e.

RENAL PHYSIOLOGY (10:29)

Selective reabsorption

Selective reabsorption is the movement of fluid and solutes from the tubular system into the peritubular capillaries. Reabsorption occurs via three processes: osmosis, diffusion and active transport. It allows the body to retain fluids, and desired solutes (approximately 99% of the glomerular filtrate) are returned to the blood. Table 11.2 shows the reabsorption that takes place throughout the renal tubule.

Table 11.2 Selective reabsorption throughout the renal tubule

Renal tubule sections		
Proximal convoluted tubule	**Loop of Henle**	**Distal convoluted tubule**
Water 65%	Water 15%	Water 10-15%
Sodium 65%	Sodium 20-30%	Sodium 5%
Chloride 50%	Chloride 35%	Chloride 5%
Potassium 65%	Potassium 20-30%	Calcium (dependent on body's needs)
Glucose 100%	Bicarbonate 10-20%	Urea
Amino acids 100%	Calcium (variable amounts)	
Bicarbonate 80-90%	Magnesium (variable amounts)	
Calcium (variable amounts)		
Magnesium (variable amounts)		
Urea 50%		

Secretion

Secretion is the removal of waste products and substances not required by the body that are extracted from the blood, passed into the convoluted tubules and collecting ducts and excreted from the body in urine. This includes waste products of metabolism (urea, creatinine, ammonium ions), foreign materials such as drugs, H^+ ions or bicarbonate ions as necessary to regulate the pH and some hormones.

Urine

Urine is a yellowish liquid that is approximately 95% water and 5% dissolved solids and gases. It has a pH of 4.5–8.0 (average 6.0) and a Specific Gravity (SG) (amount of dissolved substances in it) of 1.002 (very dilute) to 1.040 (very concentrated). The volume and concentration of urine excreted per day varies according to fluid intake and amount of fluid lost in other ways. Normal composition of urine includes: water, urea, uric acid, creatinine, ammonia, sodium, potassium, chloride, bicarbonate, calcium, magnesium and phosphate. Urine should not normally contain protein, glucose, blood, ketones, white blood cells or casts (solid material moulded within the tubules and consisting of cells or proteins). If any of these are present it may be a sign that the person may have an infection or underlying kidney disease

Acid-base balance

Acid–base balance refers to the presence of acidic compounds, e.g. carbon dioxide, and basic (alkaline) compounds, e.g. bicarbonate. As identified in Chapter 2, cell survival and maintenance of homeostasis is dependent on maintaining pH levels ranging between 7.35 to 7.45 (the average is 7.40). To achieve this there must be the right balance between the number of acidic compounds and the number of basic compounds. Your kidneys and lungs work synergistically to help regulate and maintain pH within this homeostatic range (Figure 11.11).

As metabolic reactions produce an excess of H^+ ions, it is therefore essential that these can be removed from the body. Maintaining H^+ concentration in the blood to an appropriate level (pH 7.35–7.45) is a major homeostatic challenge but is critical to the normal functioning of body cells.

Three major mechanisms are involved in removing and subsequent elimination of H+ body fluids:

1. **Buffer systems** – there are a number of buffer systems that act quickly to temporarily bind excess H+ to make it unavailable. Buffers do not remove H+ from the body but raise pH by converting strong acids and bases to weak acids and bases. The principal buffer systems of the body are the protein buffer system, the carbonic acid-bicarbonate buffer system and the phosphate buffer system.

2. **Exhalation of carbon dioxide** – CO_2 concentration correlates to pH as CO_2 acts as an acid. An increase in CO_2 leads to a decrease in pH (more H+ ions) whereas a decrease in CO_2 leads to an increase in pH (less H+ ions). The rate and depth of breathing controls CO_2 and subsequent pH – increasing the rate and depth means more CO_2 can be exhaled, thereby reducing the level of carbonic acid (H_2CO_3) in blood and raising pH. The opposite happens when the rate and depth of breathing are decreased.

3. **Excretion of hydrogen by the kidney** – this is the slowest mechanism by which H+ is removed from the body. Excretion in urine is the only method of eliminating acids other than H_2CO_3. Cells in the proximal convoluted tubule (PCT) secrete H+ ions into the tubular fluid. In the PCT Na+/H+ **antiporters** secrete H+ as they reabsorb HCO_3^-. Intercalated cells in the collecting duct act in two ways to control H+ levels: (1) excreting excess H+ when pH is too low; and (2) excreting excess HCO_3^- when pH is too high.

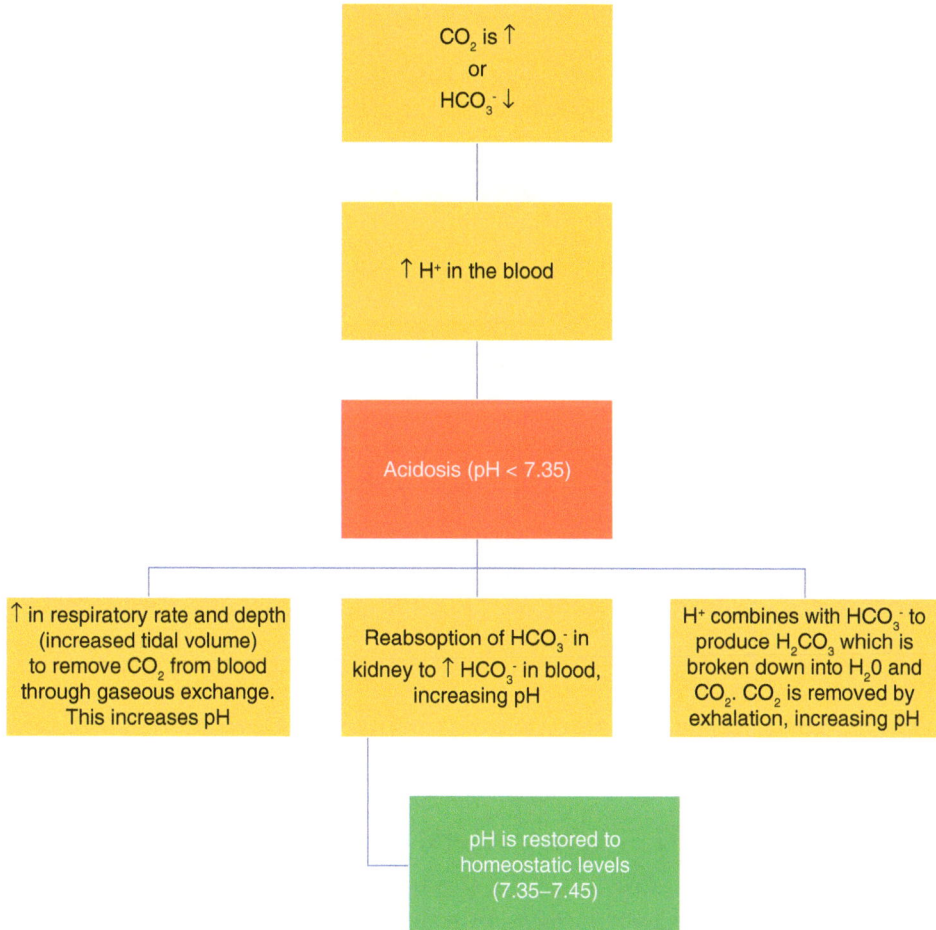

Figure 11.11a   Maintenance of acid base balance – acidosis

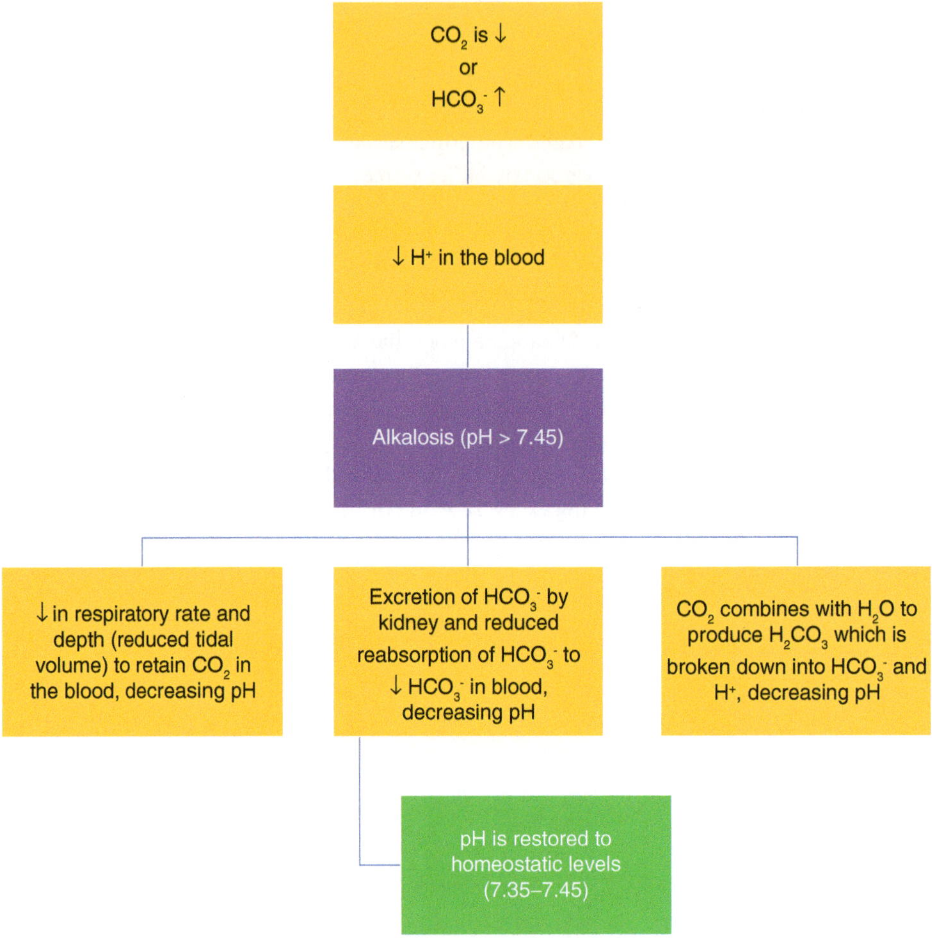

CO$_2$ is ↓
or
HCO$_3^-$ ↑

↓ H$^+$ in the blood

Alkalosis (pH > 7.45)

↓ in respiratory rate and depth (reduced tidal volume) to retain CO$_2$ in the blood, decreasing pH

Excretion of HCO$_3^-$ by kidney and reduced reabsorption of HCO$_3^-$ to ↓ HCO$_3^-$ in blood, decreasing pH

CO$_2$ combines with H$_2$O to produce H$_2$CO$_3$ which is broken down into HCO$_3^-$ and H$^+$, decreasing pH

pH is restored to homeostatic levels (7.35–7.45)

Figures 11.11b Maintenance of acid base balance – alkalosis

Ureters, bladder and urethra

Ureters

The two ureters are tubes of about 30 cm in length which collect urine from the calyces of the kidneys and drain it into the bladder by peristalsis through the action of the circular and longitudinal muscle fibres. This oblique entry minimises reflux of urine back into the ureters as the bladder fills.

--- **APPLY** ---

Vesicoureteral reflux

Vesicoureteral reflux occurs when this oblique entry of the ureter into the bladder is straightened and allows backflow of urine from the bladder into the ureter. This enables bacteria to ascend from the bladder into the kidneys and cause infection.

The bladder

The ureters enter the bladder at the trigone. This is a smooth triangular part of the bladder with the corners identified by the two ureteric openings and the internal urethral orifice (Figure 11.12). When stretched, the receptors in this area signal expansion and, when stretched sufficiently, indicate the need to micturate. This particular part of the bladder is derived from the same tissue as the anterior part of the vagina and responds to oestrogen hormones in the same way (Saez and Martin, 1981).

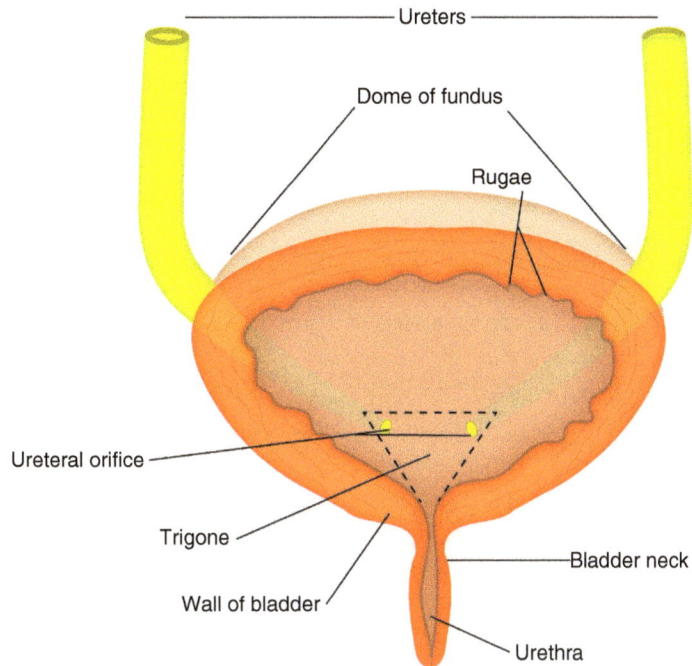

Figure 11.12 The bladder

The main muscle of the bladder is the detrusor, formed of smooth muscle fibres in spiral, longitudinal and circular bundles, which empties the bladder by contraction during micturition. The bladder lining is formed of transitional epithelium which enables considerable expansion to hold a variable volume of urine. When empty, the bladder walls fold into ridges or rugae. As the bladder fills these unfold increasing the volume which can be held. The volume held is very variable but 600–800 ml has been reported, with a point of 150–300 ml at which micturition may occur (Shaw, 2001).

Urethra

The urethra carries urine from the bladder to the external urinary meatus to remove excess fluid from the body. In males, it carries semen as well as urine.

Female urethra

The female urethra is about 4.8–5.1 cm (1.9–2 inches) long and exits the body through the external meatus of striated voluntary muscle between the clitoris and the vagina (Chapter 16). The three layers of the urethra are: muscular, erectile and mucosal, with the muscular layer being continuous with that of the bladder. The proximal two thirds are lined with transitional epithelium (like the bladder) while the final one third has stratified squamous epithelium cells lining it. The opening through the pelvic floor has the external urethral sphincter controlled by the pudendal nerves to control micturition.

Male urethra

The male urethra is significantly longer and more variable in length than the female's ranging from about 15 to 29 cm (6–11.5 inches) (Kohler et al., 2008). The urethra is divided into four parts in men, named after their location (Table 11.3).

Table 11.3 Sections of the male urethra

Region of urethra	Description (Chapter 16)	Epithelium
Pre-prostatic	Passes through wall of bladder Between 0.5 and 1.5 cm long depending on fullness of bladder	Transitional
Prostatic	Passes through prostate gland. Openings: • ejaculatory duct • prostatic ducts	Transitional
Membranous	1 or 2 cm portion passing through external urethral sphincter. Narrowest part of the urethra Bulbourethral glands (posterior to this section but open in spongy urethra)	Pseudostratified columnar
Spongy (or penile)	Runs along the underside of the penis 15-16 cm in length, passes through corpus spongiosum Openings from: urethral gland and bulbourethral glands Lumen runs parallel to penis, except at narrowest point, the external urethral meatus, where it is vertical. Results in spiral stream of urine, which cleans external urethral meatus (lack of equivalent mechanism in female urethra partly explains increased incidence of urinary tract infections in females)	Pseudostratified columnar – proximally, Stratified squamous – distally

Micturition

The micturition reflex is controlled by the sacral segments of the spinal cord while stimuli are passed to the micturition centre in the pons of the hindbrain and to the cerebrum. This allows voluntary control over the autonomic nerves of the micturition reflex.

The two stages of bladder activity are storing (guarding) and voiding. During the storage stage, distension of the bladder occurs as urine collects and when about 150 ml has gathered, the stretching provides a stimulus to the sympathetic nervous supply to the bladder. This inhibits contraction of the detrusor and constricts the internal urethral sphincter around the neck of the urethra, preventing micturition. When the volume of urine in the bladder reaches about 400 ml, a desire to pass urine occurs and is prevented by voluntary contraction of the external urethral sphincter through the pudendal nerves.

At an appropriate time and place we choose to micturate. At this point the neurons in the micturition centre send maximum impulses to the sacral centre and the parasympathetic fibres stimulate contraction of the bladder and relaxation of the internal urethral sphincter. The external urethral sphincter is consciously relaxed enabling micturition to occur.

ACTIVITY 11.2: UNDERSTAND

Watch the following video clip to further your understanding of micturition.

This external video link can be accessed by **scanning the QR code** with your smart phone camera or via https://study.sagepub.com/essentialap2e.

MICTURITION (16:26)

APPLY

Urinary incontinence

Urinary incontinence is defined by the International Continence Society (ICS) (Abrams et al., 2002: 168) as 'a condition in which involuntary loss of urine is a social or hygienic problem'. Urinary incontinence can affect all age groups, males and females. The prevalence is higher in women, particularly older women.

There are various types and causes of urinary incontinence and these include:

- Stress urinary incontinence: this occurs when the urethral sphincter fails to remain closed when there is an increase in abdominal pressure on the bladder, e.g. sneezing, coughing.
- Urge incontinence: this occurs when the detrusor muscle contracts even if the bladder is not full.
- Mixed incontinence: a combination of stress and urge incontinence.
- Overflow incontinence: due to urinary retention that may be caused by obstruction, e.g. a tumour, underactive detrusor muscle or failure of the urethra to open.
- Reflex incontinence: as a result of damage to the spinal cord and loss of sensation of the desire to micturate.
- Enuresis (involuntary loss of urine): nocturnal enuresis refers to urinary incontinence that occurs when the person is asleep. It can affect all age groups and may be related to: a reduced bladder capacity; inability to concentrate urine at night; and difficulty in waking when the bladder becomes full.
- Functional incontinence: this occurs due to a person having an underlying condition, e.g. multiple sclerosis or arthritis that may prevent them from being able to gain access to an appropriate place or at an appropriate time.

As a person-centred nurse it is imperative that we conduct an in-depth assessment of a person's urinary function to highlight any difficulties that they may have so that appropriate interventions can be put into place.

CONCLUSION

The renal system is crucial for the managing of homeostasis of fluid, electrolyte and acid–base balance in the body, which are essential for normal cell function, particularly of the excitable tissues. Understanding homeostasis of these three components of the fluid of the body and its distribution is important for you to be able to provide the necessary care to enable the person for whom you are caring to maintain the appropriate balance of fluid and electrolytes between ICF and ECF. In person-centred nursing, it is necessary to recognise the variation at the different stages of life. The importance of this system is emphasised by the fact that the kidneys are the organs most commonly transplanted.

GO DEEPER

Further reading

Docherty, B. and Foudy, C. (2006) 'Homeostasis. Part 4: fluid balance', *Nursing Times*, 102(17): 22-3.

McLafferty, E., Johnstone, C., Hendry, C. and Farley, A. (2014) 'Fluid and electrolyte balance', *Nursing Standard*, 28(29): 42.

McLafferty, E., Johnstone, C., Hendry, C. and Farley, A. (2014) 'The urinary system', *Nursing Standard*, 28(27): 43-50.

Richardson, M. (2006) 'The urinary system. Part 1 - introduction', *Nursing Times*, 102(40): 26-7.

- The four main methods of movement of fluid and electrolytes are: osmosis, diffusion, filtration, active transport.

- Water and electrolyte homeostasis is achieved by a complex interplay of the cardio-vascular, nervous, endocrine, renal, digestive and respiratory systems.

- Electrolytes are cations (positively charged ions) and anions (negatively charged ions) dissolved in body fluids. The main cations are: Na^+, K^+, Ca^{2+} and Mg^{2+} and the main anions are PO_4^{3-} (phosphate) and HCO_3^- (bicarbonate).

- The renal system is the main system involved in fluid, electrolyte and acid–base balance and consists of two kidneys, a ureter from each kidney running into the bladder and the urethra for elimination. Approximately 25% of the cardiac output passes through the kidneys.

- The renal nephrons are the functional units of the kidneys consisting of the:

  ○ Renal corpuscle, where filtration of the blood occurs.

  ○ Proximal convoluted tubule for obligatory reabsorption.

  ○ Loop of Henle, which creates an osmotic gradient through the medulla thus enabling more water to be reabsorbed back into the body.

  ○ Distal convoluted tubule, which regulates the electrolyte balance of the body.

  ○ Collecting duct, which collects filtrate from a number of nephrons.

- The filtrate from the nephrons passes through the pelvis of the kidney to the ureters, which carry the urine to the bladder where it is stored.

- Micturition is the elimination of urine from the bladder through the urethra by contraction of the detrusor muscle and relaxation of the internal and external urethral sphincters.

- There are two stages of bladder activity:

  ○ Storage phase: under sympathetic stimulation.

  ○ Voiding phase (micturition): under parasympathetic and voluntary nervous control.

The content of this chapter is important to understanding what happens to the workings of the body when anything goes wrong with the homeostatic mechanisms for regulation of fluid, electrolyte and acid–base balance. These are crucial for health, and understanding this chapter is important in the care of any people requiring nursing care. To achieve this, the areas to revise are as follows:

1 The ways in which fluid and electrolytes move between the different fluid compartments of the body.

2 The way in which different systems of the body contribute to this area of homeostasis.

3 The principal cations and anions in the body and their major functions.

4 The overall structure of the renal system.

5 The structure of the nephrons and the function of the different sections.

6 How the bladder stores urine and the involuntary and voluntary control of micturition.

In order to help you revise, consider the following questions, answers for which can be found by visiting **https://study.sagepub.com/essentialap2e**.

Test yourself by revising the chapter first, and then answer these questions without looking at the book. Afterwards compare your answers with the text, including the diagrams, and with the notes you made. Did you miss anything in your notes? Here are the questions:

1   Briefly describe the main functions of the renal system.

2   Name the major structures comprising the renal system.

3   Name the functional units of the kidneys, and briefly describe the role of each part in the formation of urine.

4   Describe how the kidneys maintain acid–base balance.

5   How does the renal system control fluid balance and blood pressure?

6   Describe how micturition is controlled.

REVISE

For additional revision resources visit: **https://study.sagepub.com/essentialap2e**.

ACE YOUR ASSESSMENT

- Revise key terms with interactive flashcards.
- Test yourself with multiple-choice questions.
- Access the glossary with audio to hear how complex terms are pronounced.
- Explore recommended websites suitable for revision.

REFERENCES

Abrams, P., Cardozo, L., Fall, M., Griffiths, D., Rosier, P. et al. (2002) 'The standardisation of terminology of lower urinary tract function: report from the standardisation sub-committee of the International Continence Society', *Neurourology and Urodynamics*, 21(2): 167–78.

Akhtar, M., Ullah, H. and Hamid, M. (2011) 'Magnesium, a drug of diverse use', *Journal of the Pakistan Medical Association*, 61(12): 1220–5.

Baker, S.B. and Worthley, L.I.G. (2002) 'The essentials of calcium, magnesium and phosphate metabolism: part I. Physiology', *Critical Care and Resuscitation*, 4: 301–6.

El-Sharkawy, A.M., Sahota, O., Maughan, R.J. and Lobo, D.N. (2014) 'The pathophysiology of fluid and electrolyte balance in the older adult surgical patient', *Clinical Nutrition*, 33(1): 6–13.

Kaplan, L.J. and Kellum, J.A. (2010) 'Fluids, pH, ions and electrolytes', *Current Opinion in Critical Care*, 16: 323–31.

Kohler, T.S., Yadven, M., Manvar, A., Liu, N. and Monga, M. (2008) 'The length of the male urethra', *International Brazilian Journal of Urology*, 34(4): 451–6.

Maday, K.R. (2013) 'Understanding electrolytes: important diagnostic clues to patient status,' *Journal of the American Academy of Physician Assistants*, 26(1): 26–31.

Millard-Stafford, M., Wendland, D.M., O'Dea, N.K. and Norman, T.L. (2012) 'Thirst and hydration status in everyday life', *Nutrition Reviews*, 70(Suppl. 2): S147–51.

Popkin, B.M., D'Anci, K.E. and Rosenberg, I.H. (2010) 'Water, hydration, and health', *Nutrition Reviews*, 68(8): 439–58.

Saez, S. and Martin, P.M. (1981) 'Evidence of estrogen receptors in the trigone area of human urinary bladder', *Journal of Steroid Biochemistry*, 15: 317–20.

Shaw, D. (2001) 'Volume of a human bladder', in G. Elert (ed.) *The Physics Factbook*. Available at: http://hypertextbook.com/facts (accessed 4 May 2020).

THE CARDIOVASCULAR AND LYMPHATIC SYSTEMS

INTERNAL TRANSPORT

UNDERSTAND: CHAPTER VIDEO

Before working through this chapter, you might find it useful to have an overview of the cardiovascular system. Watch this online video to enhance your understanding.

This video can be accessed by scanning the QR code with your smart phone camera or via https://study.sagepub.com/essentialap2e.

CARDIOVASCULAR
SYSTEM (11:19)

LEARNING OUTCOMES

When you have finished studying this chapter you will be able to:

1. Describe how the components, structure and functions of the cardiovascular and lymphatic systems contribute to homeostasis
2. Describe how blood clotting occurs
3. Understand the principles of the ABO and Rhesus (Rh) blood groups
4. Explain the nervous and endocrine regulation of cardiac activity
5. Discuss what blood pressure is and how it is regulated

INTRODUCTION

Blood and lymph are the key media for transport of substances around the body. The blood and lymphatic vessels are routes for transport of these media. The heart is the pump that provides the force behind the movement of blood and you will learn about the structure and function of the heart. This will enable you to understand the importance of maintaining a healthy heart and how lifestyle activities influence its functioning. As a nurse, you need to comprehend how circulation is fundamental to maintaining organs and tissues. Therefore, this chapter will explore the physiological and behavioural responses that maintain blood volume and perfusion of tissues and organs, including blood clotting, blood vessels changing diameter as needed and renewal of blood cells. Cardiovascular health is a vital consideration across all areas of practice, whether through preventing cardiovascular disease, promoting a healthy diet and active lifestyle, enabling those with cardiovascular disease to maximise their health-related quality of life, or analysing cardiovascular function to identify changed physiological status. Cardiovascular health is central to a fulfilling life, and it is essential for a person-centred nurse to understand its anatomy and physiology.

The Cardiovascular System (CVS) is part of that overall cohesive approach to homeostasis and health. In fact, it is the system that people associate with life – the heartbeat. The heartbeat is the pulse of life: you feel it soar when you are excited, skip a beat when you are smitten, and you feel it bound and increase when you face fear. You also feel it settle when you relax and are calm. Blood is truly the lifeblood: it brings heat to keep you warm, carries oxygen and nutrients to nourish cells and enable them to survive, it delivers hormones to their target for the endocrine system to work, and it carries the soldiers that fight pathogens – the white blood cells. Blood reaches every part of us, and while it brings the fundamentals of life to cells, it also takes away the waste products. Now think about caring for people as a nurse. The heart rate can tell us when the person is under stress, afraid or in pain. Variations in blood pressure may support this. We also know that keeping people nourished is vital to health. As nurses, we often deal with preventing or treating infections in the body. The blood can provide a picture of whether the body is fighting infection, and possibly what type of infection. Alterations to the normal physiology tell you that the body has detected a stressor and is responding to that. This could be an increased need for oxygen and other nutrients, a fall in blood volume, or the preparation to respond to danger.

Context

BODIE FAMILY

Now let us have a look at the Bodie family. Baby Danielle, being young, is growing rapidly. As a result, she has a high metabolic rate due to lots of cell division and development, which requires a continuous supply of oxygen and nutrients, and waste products to be removed efficiently. As a result, her heart rate is much faster than an adult's, particularly as her heart is also developing in size and as a muscle. As her immune system is developing, her blood is circulating the much-needed white blood cells to combat any pathogens that have gained entry to her body. Danielle is also learning to listen to her heart as she develops her intuition; it works with her nervous system to tell her when to anticipate fear, and it also links in with her emotions such as the experience of excitement.

Maud and George have hearts that have been through extensive work. From the moment their hearts started beating in the womb, they have continued to beat their entire lives. During this time, their hearts have endured the stresses of life with them, helping them to battle off illness by increasing the circulation of blood and waste products to their destinations. The elastic fibres in their heart and blood vessels are not as effective as when they were young, which naturally increases their blood pressure. The blood vessels have also endured the pitfalls of particular diets, potentially having fatty deposits laid down on the lining of the vessels, impairing their recoil and narrowing the lumen. However,

Maud, following recovery from her heart attack, and George are active, a key factor in maintaining cardiovascular health.

The other adults of the Bodie family are at stages in their lives that influence their cardiac health. Thomas works as an airline pilot on transatlantic routes experiencing long periods of inactivity. Muscular movement is significant in supporting venous return of blood to the right side of the heart, and so long periods of immobility impair the venous circulation. It is therefore important that Thomas keeps his muscles active during the flight, in the same way that it is important for the passengers on his flight to do so. However, Thomas is from a generation that is aware of the need for physical activity and a balanced, healthy diet. He uses the gym three or four times a week, helping to keep his cardiovascular system healthy. Matthew Bodie is likely to be the person in best cardiovascular health. Although currently unemployed, his job is physically demanding. He is also a triathlete, which involves regular training, during which his cells require more nutrients, including oxygen, which his heart has to circulate effectively and efficiently.

HOMEOSTASIS AND THE CVS

Transport

Before we explore the different components of the CVS, we will briefly consider its key roles in contributing to the maintenance of homeostasis.

Fuel for life

The CVS provides every cell in the body with the nutrients and oxygen needed to sustain its life. The most immediate need is oxygen for cellular respiration, and nutrients are necessary as the energy source for this process. Cells also need to maintain a heat balance to enable chemical processes. Hormones and enzymes regulate biological reactions in the body (Chapters 7 and 9).

Garbage disposal

As cells metabolise and use up nutrients, they produce waste products such as carbon dioxide, ammonia and heat. The cardiovascular system takes these products away to be eliminated through other organs and systems of the body such as the liver, skin, and respiratory and renal systems.

Internal security

The mechanisms for protection from pathogens and injury are relatively ineffective unless they can get to the problem fairly quickly. The CVS enables this by circulating blood cells that fight against pathogens and bringing biological components for blood clotting to the necessary site for haemostasis and healing. For these mechanisms to work and cells to function optimally there has to be the appropriate electrolyte and acid–base balance in the fluid compartments (Chapter 11). By moving fluids and dissolved substances around the body the CVS enables this requirement to be met.

We now need to explore the components of the CVS and how each component is structured and functions. The cardiovascular system is made up of the components in Table 12.1.

Table 12.1 Components of the CVS

Component	Basic role
Blood and lymph	The liquid media of transport
Heart	The pump to move the liquid media around the body at the right pace and force
Blood vessels • Arteries • Veins • Capillaries	To carry blood away from the heart To bring blood back to the heart To allow for the exchange of substances between cells, tissues and the vascular system
Lymph vessels	To bring lymph from between cells into the vascular system

BLOOD

The blood is part of the extracellular fluid (ECF) and is contained within the blood vessels of the body. It consists of:

- **Plasma** (55% of total volume).
- **Blood cells** (45% of total volume).

Plasma

Plasma is a straw-coloured transparent fluid consisting of 90–92% water and a range of dissolved substances including plasma proteins, enzymes, hormones, gases, electrolytes and waste products. These all have distinct roles in maintaining homeostasis.

Plasma proteins are formed in the liver and there are three main types:

Albumins: These are particularly important in creating osmotic pressure (discussed later), drawing fluid back into the capillaries from the interstitial fluid.

Fibrinogen and prothrombin: These are the major proteins involved in blood clotting.

Globulins: Formed either in the liver or in lymph tissue, these proteins are essential for immunity (Chapter 13). Some function as antibodies against infection while others transport some hormones and minerals.

Blood cells

There are three main types of blood cells suspended in the plasma:

1. **Erythrocytes** (or red blood cells).
2. **Leucocytes** (or white blood cells).
3. **Thrombocytes** (or platelets).

The blood cells all originate from pluripotent stem cells which means that they can develop into many different types of cell (Chapter 2). Let us now look at the three main types of blood cells in a little more detail.

Erythrocytes (red blood cells)

Erythrocytes are the most abundant blood cells. They are formed in the red bone marrow in the centre of bones and have a biconcave shape that increases their surface area enabling greater absorption of gases for transport and permitting the cell to get into smaller spaces. Erythrocytes contain the pigment haemoglobin, which transports oxygen and carbon dioxide to and from the cells of the body. Haemoglobin is composed of globin, a protein, and a complex called haem that contains iron. A haemoglobin molecule is a quaternary structured protein containing four globin chains and four haem units as shown in Chapter 9 (Figure 9.13b). Oxygen attaches to haem, thus each haemoglobin molecule carries four oxygen molecules. Each erythrocyte has approximately 280 million haemoglobin molecules, and so can carry four times that amount of oxygen.

In Chapter 2 you discovered the components of a typical cell. Erythrocytes are atypical in that they have no nuclei, ribosomes or mitochondria. As a result, they have a limited lifespan of about 120 days after which they are broken down and recycled. The absence of mitochondria in erythrocytes means that aerobic respiration for energy does not occur and they obtain their energy through anaerobic metabolism of glucose and so do not use the oxygen they carry.

It is important that the body responds to changes in oxygen supply by forming new red blood cells. Erythropoiesis (formation of new erythrocytes) (Figure 12.1) is stimulated through negative feedback. A low level of oxygen in the tissues stimulates release of erythropoietin (EPO) – a hormone from the kidneys. EPO stimulates the red bone marrow to produce more erythrocytes, increasing the amount of oxygen carried.

Sufficient iron, vitamin B_{12} and folic acid in the diet are essential. Additionally, the kidneys and bone marrow need to be in good health to produce and respond to EPO. As erythrocytes age, their cell membrane becomes fragile from wear and tear. When erythrocytes pass through the spleen, they have to pass through narrow channels. The fragile cell membranes of aged erythrocytes lack the necessary flexibility and they become trapped and destroyed with the fragments engulfed by macrophages in the spleen and liver. Haemoglobin is broken into globin, which is recycled into amino acids and haem, further split into iron and biliverdin. The iron is either stored by the liver or reused in the bone marrow to make more haemoglobin, and biliverdin is converted to bilirubin, combined with albumin, removed by the liver and excreted in bile.

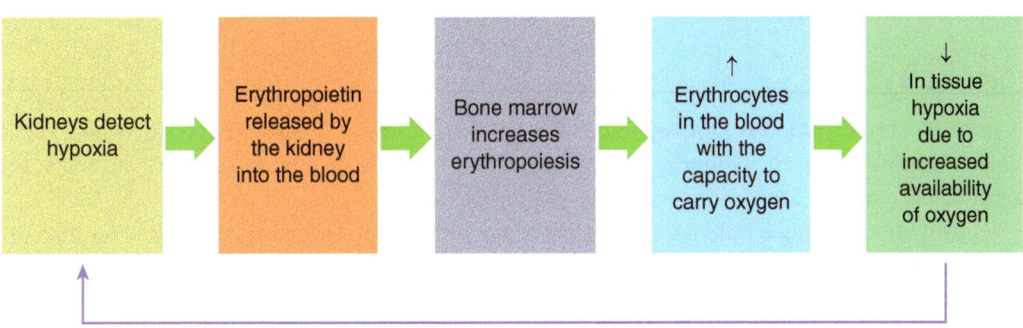

Blocks stimulus until tissue hypoxia reoccurs

Figure 12.1 Erythropoiesis

Blood groups

There are numerous blood groups which are taken into account in ensuring compatible blood transfusions. These blood groups are determined by erythrocytes carrying antigens on their surface. The two major systems are:

- The ABO system.
- The Rhesus (Rh) system.

Combined, these two systems define someone's main blood group and are used in blood grouping to identify potential donor and recipient before blood transfusion. However, before planned transfusion, the relevant blood samples are also cross-matched to check that they are compatible in relation to any of the other considerable number of other (rarer) blood groups which may be present.

APPLY

Antibodies and blood transfusion

People may produce antibodies – immunoglobulins – against the antigens that are not present on their own erythrocytes. While this is not of much significance normally, it becomes important should a person require a blood transfusion and may be important in pregnancy. Should a person receive blood that that has different antigens on the erythrocytes than their own, their immune system will react adversely and this can be fatal.

The ABO system

In the ABO system, blood is grouped under the following antigen groups and develops antibodies. Table 12.3 shows these, the proportion of the population within each group in the UK, and the donor and recipient status of each.

You will see that people in blood group O have no antigens to react with antibodies and thus can donate blood to any other person with ABO blood. Conversely, the people in blood group AB have no antibodies produced and so they can receive blood from anyone (in the ABO system) as they have no antibodies to antigens on donated erythrocytes. The goal is that people should receive blood of their own ABO group whenever possible but in life-threatening situations, group O blood can be safely given to all groups. However, compatibility is more complex than just the ABO system alone and so we must also consider the Rhesus system alongside it.

APPLY

Acute transfusion reaction

An acute transfusion reaction, also called allergic/febrile transfusion reactions, is a reaction occurring during or up to 24 hours after transfusion of blood or components. It can be recognised through these common clinical features:

Hyperthermia (fever)

Flushing

Urticaria (hives)

Rigors

Chills

Tachycardia

Hypertension or hypotension

Collapse

Bone, muscle, chest or abdominal pain

Dyspnoea

Nausea

Malaise

Table 12.2 summarises how these reactions are classified (adapted from SHOT, 2018):

Table 12.2   Classification of acute transfusion reactions

Mild	Moderate	Severe
• Temperature ≥ 38°C and 1-2°C rise on pre-transfusion baseline • Transient flushing urticaria (hives) or rash • Mild febrile/allergic reactions (e.g. localised urticaria, itching)	• Temperature ≥ 39°C or 2°C or more rise on pre-transfusion baseline • Inflammatory signs (myalgia, nausea) • Wheeze with/without flushing urticaria (hives) or rash • Allergic and febrile reactions (one of which must be moderate category, e.g. widespread urticaria, dyspnoea) • Hypotensive episode of 30 mmHg or more during/within one hour after transfusion • Systolic blood pressure < 80 mmHg	• 2°C or more rise on pre-transfusion baseline or temperature ≥ 39°C or rigors and chills • Inflammatory signs (myalgia, nausea) that result in transfusion being stopped • Dyspnoea (bronchospasm, stridor) or circulatory disorder or anaphylaxis • Allergic and febrile reactions (one of which must be severe category, e.g. rigors, collapse, airway oedema) • Clinical hypotension leading to shock

If you suspect an acute transfusion reaction, the first step is to stop the blood transfusion. A rapid clinical assessment of the person is needed, as well as a double-check of their patient ID and the blood compatibility with the person's own blood (Table 12.2). **Serious Hazards of Transfusion (SHOT) UK** provide up-to-date resources on acute transfusion reactions and other blood-related reactions. A link to the website can be accessed via http://study.sagepub.com/essentialap2e

Table 12.3   ABO group, antigen and antibodies, and compatibility

ABO group (% in UK population)	Antigens	Antibodies	Can donate to	Can receive from
A (42%)	A	Anti-B	A AB	A O
B (10%)	B	Anti-A	B AB	B O
AB (4%)	A and B	None	AB	All (universal recipient)
O (44%)	None	Anti-A and Anti-B	All (universal donor)	O

The Rhesus (Rh) system

The Rh system is the second important blood group system, being just as significant as the ABO system (Table 12.4). While there are 50 antigens in this system, D is the one most focused upon as it is highly immunogenic. People are Rh positive (Rh$^+$) if they carry the RhD antigen or Rh negative (Rh$^-$) if they do not. Around 84% of the UK population are Rh$^+$, leaving the remainder as Rh$^-$. Unlike in the ABO system, Rh$^-$ people do not normally carry Rhesus antibodies. However, if they are exposed to Rh+ blood through transfusion or pregnancy they will produce anti-D (anti-Rhesus) antibodies.

Table 12.4   Rh group, antigens, antibodies, donation status

Rh group	Antigens	Antibodies	Can donate to	Can receive from
Rh$^+$	RhD	None	Rh$^+$	Rh$^+$ and Rh$^-$
Rh$^-$	None	None Anti-Rhesus antibodies if exposed to Rh$^+$ blood (transfusion or pregnancy)	Rh$^+$ and Rh$^-$	Rh$^-$

   Table 12.5 considers both the ABO and Rh systems together in determining what blood products a person can receive.

Table 12.5 ABO and rhesus systems and compatibility

ABO group with Rh⁻ factor		Can receive from	ABO group with Rh⁺ factor		Can receive from
	O⁻	O⁺		O⁺	O⁺, O⁻
	A⁻	A⁻, O⁻		A⁺	A⁺, A⁻, O⁺, O⁻
−	B⁻	B⁻, O⁻	+	B⁺	B⁺, B⁻, O⁺, O⁻
	AB⁻	AB⁻, B⁻, A⁻, O⁻		AB⁺	AB⁺, AB⁻, B⁺, B⁻, A⁺, A⁻, O⁺, O⁻

APPLY

Rhesus system and pregnancy

Rh+ people can receive blood that is Rh⁺ or Rh⁻. However, Rh- individuals (particularly women before the menopause) should be given Rh- blood. Rh- people given Rh⁺ blood will create antibodies to the Rhesus factor and if exposed to it again will destroy these blood cells.

In the context of midwifery, a Rh- woman with a Rh⁺ partner may be carrying a Rh⁺ baby. During the first pregnancy this does not usually cause a problem. However, exposure to the baby's Rh⁺ blood during pregnancy or delivery (even in small quantities) will result in the woman forming anti-Rh antibodies (usually RhD antibodies). In future pregnancies with Rh⁺ babies, the antibodies will cross the placental barrier and destroy the baby's blood causing haemolytic anaemia, which may be fatal. To prevent this occurring, rhesus negative women are now given Routine Antenatal Anti-D Prophylaxis (RAADP) (RCOG, 2011). When anti-D is administered to the mother, it causes a fast, non-inflammatory clearance of passive anti-D-coated erythrocytes. This prevents the inflammatory destruction of fetal erythrocytes inducing an immune response (Crowther et al., 2013).

Leucocytes (white blood cells)

Leucocytes are white blood cells and the largest of the blood cells, making up around 1% of blood volume. Their primary function is as part of the immune system providing protection against pathogens and this is discussed in full in Chapter 13. Leucocytes are mostly found in connective and lymphatic tissue, normally only circulating in the blood for a number of hours. There are two basic types of leucocytes:

- **Granulocytes:** have visible granules, which are the organelles, in the cytoplasm.
- **Agranulocytes:** have few or no organelles and so do not have a granular appearance.

These have further subdivisions, which we will explore briefly.

Granulocytes

There are three types of granulocytes:

Neutrophils: Neutrophils engulf bacteria and cellular debris or use chemical agents (lysozyme and peroxidase) to destroy foreign bodies. An increase in neutrophils occurs in acute infections, certain malignant neoplastic diseases (malignant growth or tumours caused by abnormal and uncontrolled cell division) and some other disorders.

Eosinophils: These stain readily with eosin (an acidic dye that reacts to basic (referring to pH) components in a cell). Normally about 1–3% of the total white cell count, eosinophils function in allergic responses and in resisting some infections. They engulf antigen–antibody complexes (complexes formed by the binding of an antibody to an antigen – or immune complex), allergens (substances that cause allergic reactions) and inflammatory chemicals. They can also weaken or kill parasites by secreting chemical agents.

Basophils: These stain blue when exposed to a basic dye (dyes, such as haematoxylin, that react with acidic components in cells) in a laboratory. They normally constitute 1% or less of the total white cell count and assist in the inflammatory response by secreting histamine increasing blood flow by vasodilation. They also secrete heparin, an anticoagulant that increases blood flow.

Agranulocytes

There are two types of agranulocytes:

Lymphocytes: These normally make up about 25% of the total white blood cell count but increase with infection. Lymphocytes are a key defence mechanism for the body as they can destroy cancer cells, cells with a viral infection and foreign cells. There are three types of lymphocytes: B-cells, T-cells and natural killer cells.

- **B-cells** recognise specific antigens and produce antibodies. Exposure to the antigen stimulates rapid cell multiplication, producing large numbers in the plasma.
- **T-cells** secrete immunologically active compounds and assist B-cells. They are cytotoxic and so can destroy foreign cells directly. They can self-regulate other T-cells by preventing overactivity. They also release interleukins, which stimulate other lymphocytes and macrophages (from monocytes).
- **Natural killer cells** provide a rapid response to virally infected cells and respond to tumour cell growth.

Monocytes: These are twice as large as erythrocytes with relatively clear cytoplasm and differentiate into macrophages, large phagocytic cells. They engulf pathogens and dead neutrophils, clear away debris from dead or damaged cells and present antigens to activate other cells of the immune system.

Thrombocytes (platelets)

Thrombocytes, also referred to as platelets, are very small discs without nuclei with a lifespan of eight to 11 days. They are involved in blood clotting, secreting clotting factors (procoagulants), vasoconstricting agents to induce vascular spasm, and then clumping together in platelet plugs. They also dissolve old blood clots, destroy bacteria by phagocytosis and secreting chemical agents, and attract neutrophils and monocytes to infected sites by chemical messengers. They also promote mitosis in fibroblasts and smooth muscle.[1]

Haemostasis

When a blood vessel is damaged, bleeding is stopped through three stages in which platelets are vital:

1. **Vasoconstriction or vascular spasm:** This is the immediate response in haemorrhage. Pain receptors mediate a fast response by stimulating the smooth muscle in blood vessels to constrict to reduce

---

[1]Involuntary, non-striated contractile muscle composed of elongated spindle-shaped cells.

blood flow. Thrombocytes also release serotonin (a neurotransmitter that causes vasoconstriction (narrowing of the lumen of blood vessels through contraction of the smooth muscle within the vessel walls)), which results in a longer lasting vasoconstriction.

2. **Platelet plug formation:** The inner lining (endothelium) of blood vessels is normally very smooth and thrombocytes do not adhere to it. When a vessel is damaged, collagen fibres are exposed and thrombocytes will adhere to collagen using pseudopods (temporary projections of the cell cytoplasm). These also enable them to adhere to other platelets, resulting in a mass accumulation that forms a platelet plug. These pseudopods can contract, bringing the vessel walls closer and forming a temporary seal.

3. **Coagulation (blood clotting):** A complex series of reactions that positive feedback enhances, resulting in blood clot formation. This process needs to be fast with a damaged blood vessel, but should not occur otherwise. Coagulation occurs by two simultaneous pathways (the intrinsic and extrinsic pathways) that merge to a final common pathway (Figure 12.2).

The extrinsic[2] pathway is activated when proteins are released by damaged blood vessels and perivascular tissue (tissue close to/surrounding blood vessels). Factor III combines with factor VII and, in the presence of calcium, activates factor X.

The intrinsic[3] pathway occurs from blood components. It is triggered when thrombocytes release factor XII (through degranulation), leading to activated factor XI, IX and VIII in that order, each triggering the next. Factor VIII activation requires the presence of calcium and platelet factor 3 (PF3). Factor VIII also leads to factor X activation.

The final common pathway starts with the activated factor X from both the intrinsic and extrinsic pathways. This combines with factors III and V in the presence of calcium and PF3 resulting in prothrombin activator. This converts factor II (prothrombin) to thrombin, which converts factor I (fibrinogen) to fibrin. Fibrin reacts with factor XIII and calcium to produce the structural framework of the blood clot. The thrombin speeds this process up by accelerating the production of prothrombin activator by interacting with factor V.

Coagulation depends on a balance between procoagulants and anticoagulants. Procoagulants (e.g. fibrinogen and calcium) promote clot formation while anticoagulants (e.g. heparin and plasmin) cause clot breakdown or fibrinolysis. After the clot has formed and healing of the damaged vessel is underway, fibrinolysis prevents further clot formation.

ACTIVITY 12.2: UNDERSTAND

Watch this video clip which explains the process of haemostasis.

This external video link can be accessed by **scanning the QR code** with your smart phone camera or via https://study.sagepub.com/essentialap2e.

HAEMOSTASIS (9:59)

---

[2]Named the extrinsic pathway as a component from outside the blood begins the process, i.e. the proteins released from damaged tissue and blood vessels.

[3]Named the intrinsic pathway as the components needed for this part of the pathway already exist within the blood.

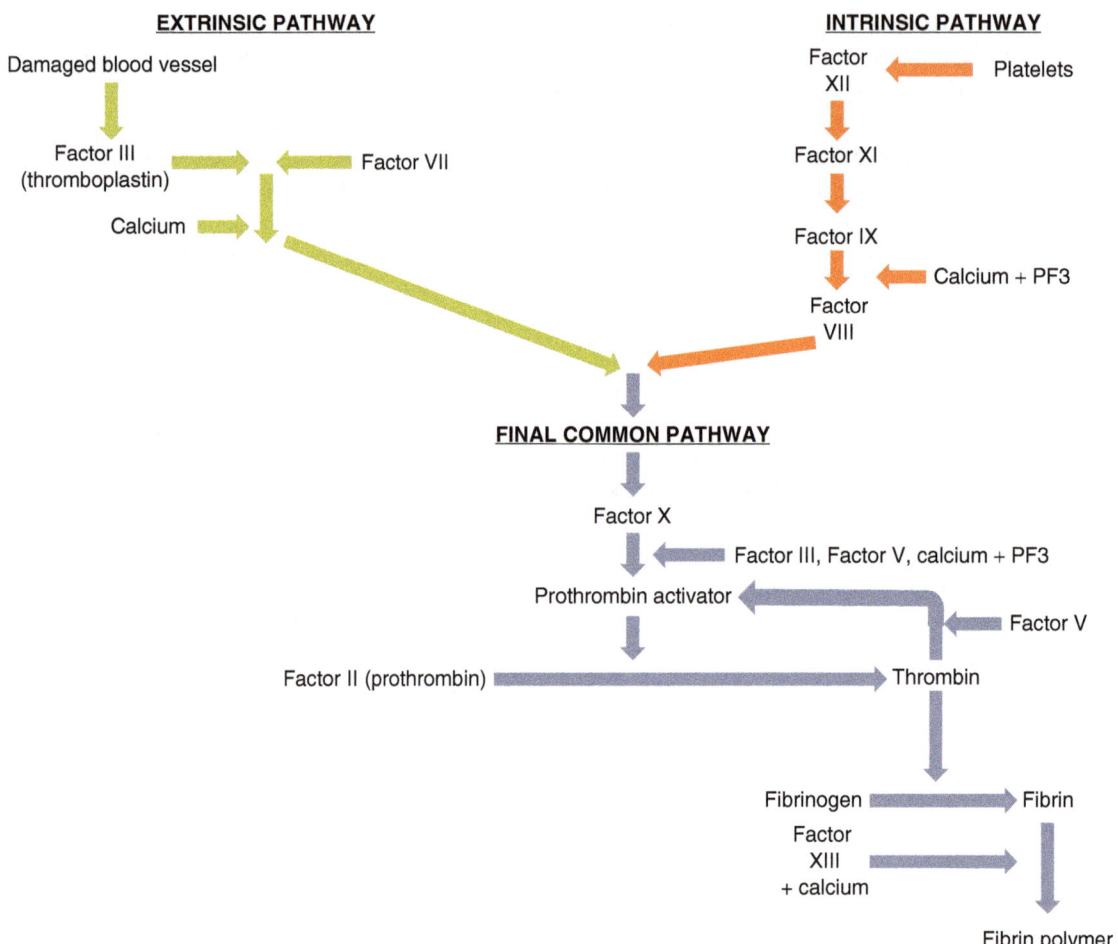

Figure 12.2 Blood clotting

LYMPH

Lymph is the second main medium for transport within the body. Lymph originates from the excess ECF surrounding cells that is not returned to the venous capillaries. Instead it enters the lymphatic capillaries through which lymph moves freely. Lymph is similar in composition to blood, with water containing electrolytes, glucose, fats and leucocytes but normally without erythrocytes and fewer plasma proteins. Lymph from vessels in the arms and legs is clear, whereas lymph containing fatty acids from the intestines is cloudy and is often referred to as chyle (Chapter 8). As lymph moves through the lymphatic vessels, it passes through small, bean-shaped nodes called lymph nodes and is filtered for bacteria, cancer cells and other potentially harmful substances. In terms of leucocytes, lymph primarily contains macrophages and lymphocytes, which help with immunity as described earlier.

So what carries these fluids and what forces move them along? Blood is pumped around the body by the heart, being carried within blood vessels, namely arteries, veins and capillaries. Lymph is carried in lymph vessels and lymph capillaries and eventually returns back to the blood stream (Chapter 13). We will look at the structure of the heart first and then the structure of the vessels carrying blood and lymph.

THE HEART

The heart is the pump that moves blood around your body. Effectively it is two pumps working side by side in one organ, sitting fairly centrally in your chest, with the left lung providing some additional space to the left-hand side (Figure 12.3). Its location in the thoracic cavity and mediastinum,[4] just behind and to the left of the sternum, means that it is well protected by the sternum and ribcage. It is ideally placed to pump blood against gravity to the brain, and gravity assists the flow to the rest of the body. However, the heart needs to pump at sufficient pressure to perfuse (move fluid and nutrients through to cells) all the vital organs and to maintain hydrostatic pressure (see later in chapter) at the arterial end of capillaries.

The right side of the heart pumps blood through the pulmonary circulation (Figure 12.4). The lungs are close by and at the same level, thus this side of the heart has less work to perform than the left. The left side of the heart pumps the blood through the systemic circulation around the whole body. This is further than the lungs, against gravity to the head, and thus the force required is considerably greater. Therefore, the muscle (myocardium) in the left side of the heart is thicker than in the right.

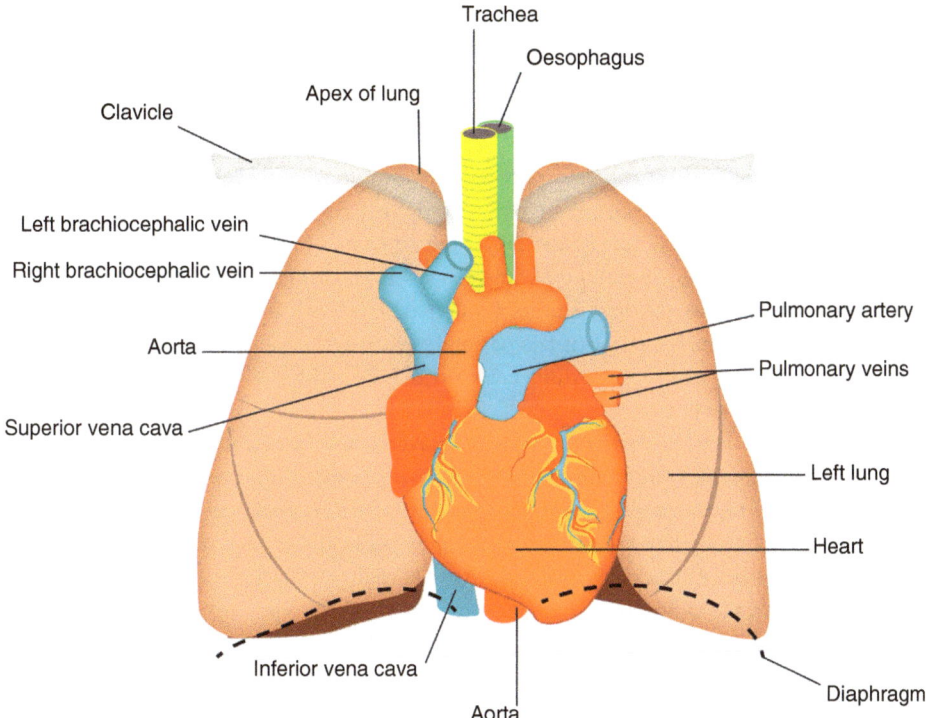

Figure 12.3 Location of the heart

---

[4]The central compartment of the thoracic cavity surrounded by loose connective tissue and containing all the tissues and organs of the chest except the lungs and pleurae.

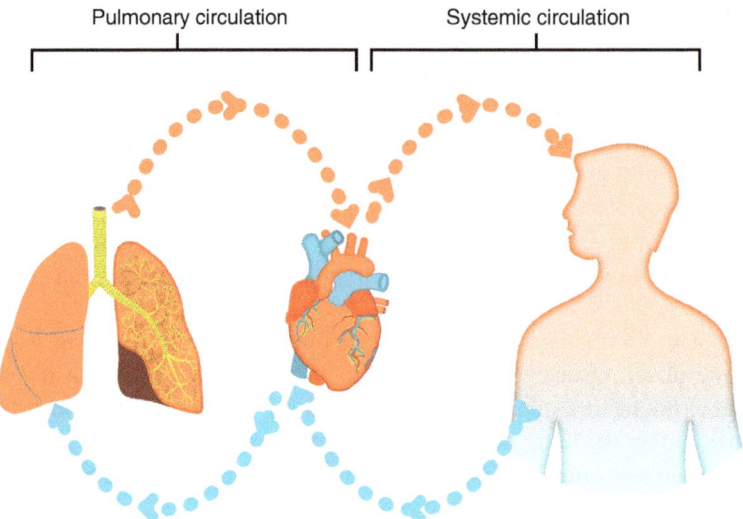

Figure 12.4 Pulmonary and systemic circulations

Structure of the heart

The three layers of the heart support its ability to undertake its life-long function as a pump (Figure 12.5):

1. The pericardium is the outermost layer composed of two sacs, which cover the outside of the heart. It is composed of dense connective fibrous tissue that prevents the heart from overstretching, protecting it from structural damage. The inner layer of the pericardium has two serous membranes similar to the pleural membranes covering the lungs. Serous membranes are double-layered membranes that produce serous fluid (pale, watery fluid) between the two layers to provide lubrication. The outer one is the parietal pericardium and the inner the visceral pericardium (or epicardium) with serous fluid between that permits the membranes to move against each other without friction.

2. The middle layer is the myocardium or muscular layer of the heart consisting of thin filaments of actin and thick filaments of myosin. These fibres pass over each other when the heart beats, shortening the muscle which pumps blood out of the chambers. The thickness of the myocardium varies depending on how strong that area needs to be. For example, the left ventricle (pushing blood round the whole body) has a thicker myocardium than the right ventricle, and the atrial chambers are much thinner than the ventricles.

3. The inner layer is the endocardium which forms the lining of the heart and the heart valves. It consists of a single layer of epithelial cells continuous with the innermost layer of the blood vessels. This ensures that the smooth inner nature of blood vessels continues inside the heart itself. However, the anterior and inferior walls of the right ventricle, the lower two thirds of the septum, and a large amount of the left ventricles are lined by muscle bundles, called the trabeculae carneae, which are not smooth and provide a rougher surface. These muscles are involved in contraction of the heart.

Cardiac muscle is one of the three types of muscle found in the body, the others being smooth and skeletal. It is unique to the heart, striated like skeletal muscle and involuntarily contracts like smooth muscle. Cardiac muscle is innervated through the autonomic nervous system (Chapter 5) but also has its own pacemaker cells, making it autorhythmic. Cells in cardiac muscle contain a large number of mitochondria enabling it to be fatigue resistant.

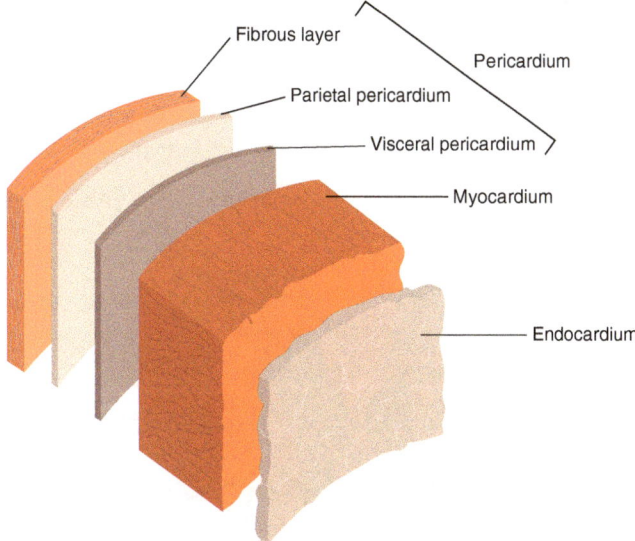

Figure 12.5 Layers of the heart

Chambers of the heart

The heart is comprised of four chambers divided into two sides. Each side of the heart has one atrium and one ventricle, two of each in total (Figure 12.6).

The role of the atria is to pump blood into the larger ventricles from where it is pumped at force out to its destination (Figure 12.6). The right atrium receives blood from the superior and inferior vena cava, which is bringing deoxygenated blood from the body. Blood flows through the tricuspid valve into the right ventricle. From there, it is pumped through the pulmonary valve into the pulmonary trunk, which divides into the right and left pulmonary arteries. From here, deoxygenated blood is taken to the lungs for gaseous exchange through the pulmonary arterioles and capillaries. Capillaries bring the oxygenated blood into pulmonary venules and eventually into the pulmonary veins. These veins empty into the left atrium of the heart where blood is pumped through the bicuspid valve into the left ventricle. From the left ventricle, blood is pumped through the aortic valve into the aorta and from there out to the body through arteries, arterioles and capillaries. As the right side of the heart manages deoxygenated blood and the left side oxygenated blood, it would not be efficient for blood on either side of the heart to mix. As a result, the septum divides the right and left sides of the heart.

Valves of the heart

The atria and ventricles have valves in place to maintain efficiency of the heart. These valves ensure blood flows in one direction, from the atria to the ventricles, while preventing backflow from the ventricles into the atria. The valves between the atria and ventricles are the atrioventricular valves. More specifically, they are called the tricuspid and bicuspid (or mitral – from its shape like a bishop's hat) valves on the right and left sides of the heart respectively (Figure 12.6). The valves at the entrances to the major arteries from the heart are the semilunar valves (half-moon shaped). The pulmonary valve is within the pulmonary artery and the aortic valve in the aorta close to the point where it leaves the ventricles.

When the ventricles relax, the semilunar valves close (to prevent backflow of blood into the heart) and the atrioventricular valves open, allowing blood to move from the atria into the ventricles. The contraction of the atria contributes greatly to this movement of blood. Once the ventricles contract, the atrioventricular valves close and the semilunar valves open, ensuring that blood only moves from the ventricles into the pulmonary trunk and aorta and not back into the atria.

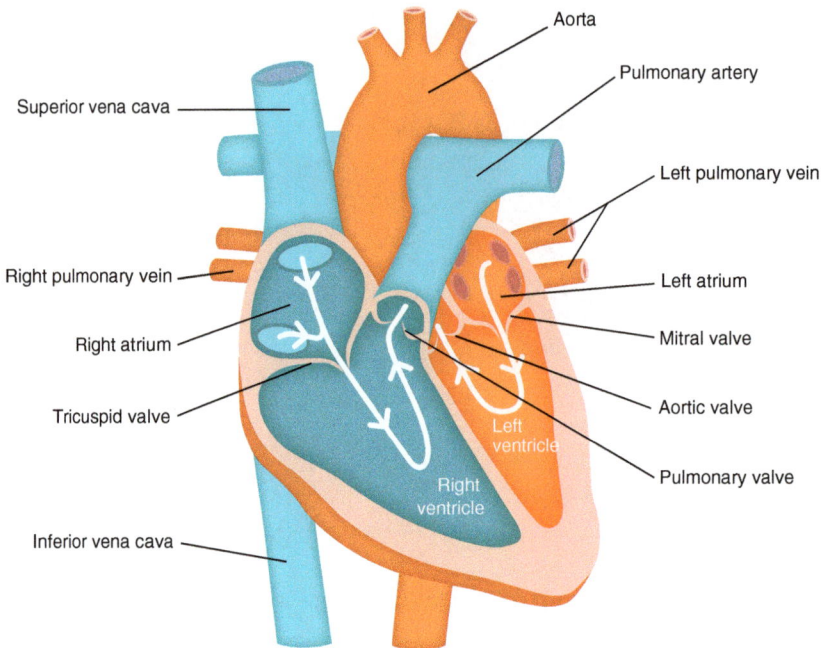

Figure 12.6   Chambers of the heart and blood flow through the heart

ACTIVITY 12.3: UNDERSTAND

Watch this short video clip to see how blood flows through the heart. Study this until you are clear about the blood flow.

This external video link can be accessed by **scanning the QR code** with your smart phone camera or via https://study.sagepub.com/essentialap2e.

HEART CONTRACTION AND
BLOOD FLOW (3:36)

Coronary circulation

We have discussed the two primary circulatory systems, pulmonary and systemic circulations, but there is a third system, the coronary circulation, supplying blood to the heart itself. This is essential as the heart works hard from its initial development in the fetus until the moment of death. Just above the aortic valve lies the opening to the right and left coronary arteries, which supply blood to the heart (Figure 12.7).

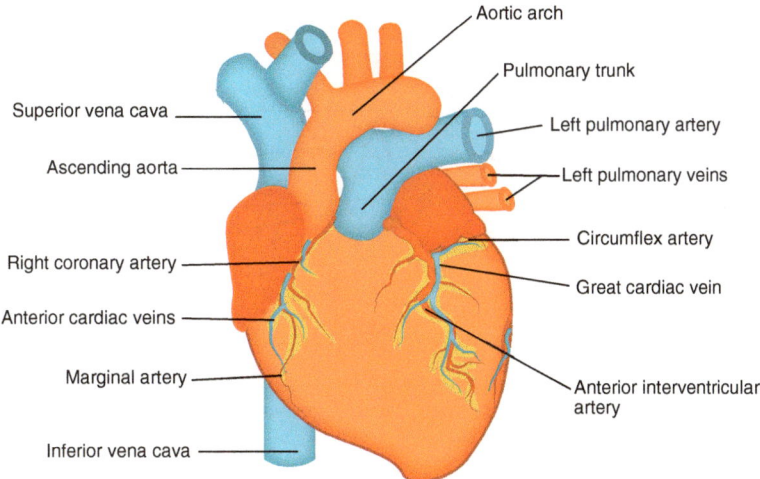

Figure 12.7 Coronary circulation

Approximately 5% of blood coming from the left ventricle moves into these arteries, most for the left ventricle as it requires most oxygen and nutrients being the thickest layer of the myocardium. The coronary arteries spread over the surface of the heart, branching out to provide all the heart with oxygenated blood. Cardiac veins collect the venous blood, with most merging to form the coronary sinus, which empties into the right atrium. The remaining coronary veins empty directly into the right atrium.

CIRCULATION

Now that we have looked at the heart, we need to look at the vessels that blood is pumped through and returned to the heart. We will look at the structure of these vessels first. We also need to consider the circulation of lymph in lymphatic vessels.

There are two distinct parts to the circulation: the systemic circulation and the pulmonary circulation. The pulmonary circulation is the blood supply to the lungs whereas the systemic circulation is the blood supply to the rest of the body. In the pulmonary circulation, pulmonary arteries carry deoxygenated blood from the heart to the lungs to be oxygenated, which is carried back to the heart by the pulmonary veins. In the systemic circulation, arteries carry oxygenated blood from the heart to the rest of the body to be used by the cells, and the deoxygenated blood is carried back to the heart by the veins.

If you find it difficult to remember which vessels leave and enter the heart, then remember the word VEAL – Veins Enter Arteries Leave. This is always the rule.

Blood vessels

There are three main types of blood vessels: arteries carry blood away from the heart (Figure 12.8); capillaries enable exchange of water, nutrients and waste products between the blood and the tissues; and veins carry blood back towards the heart (Figure 12.9). These figures are a useful reference point.

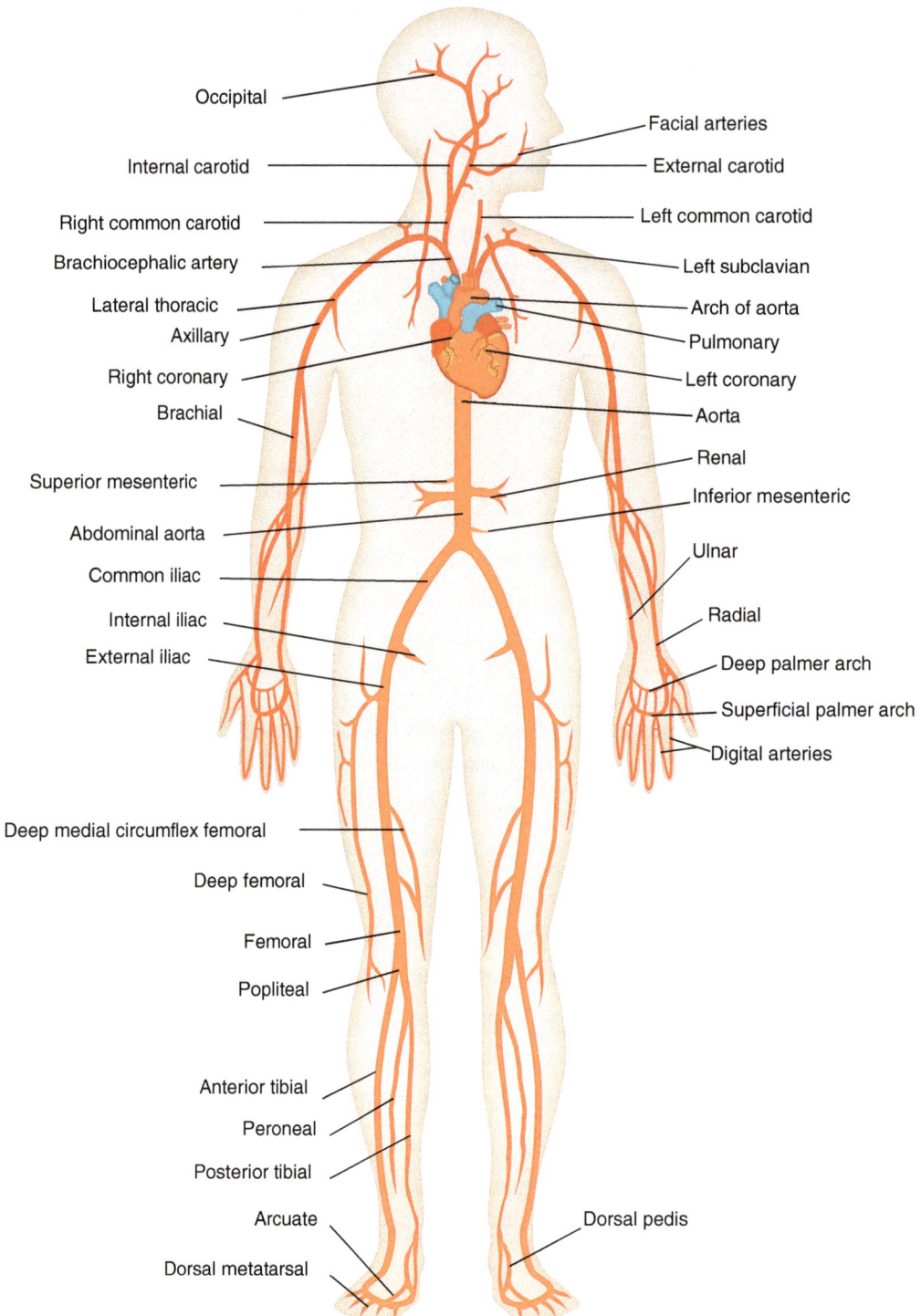

Figure 12.8 Major arteries of the body

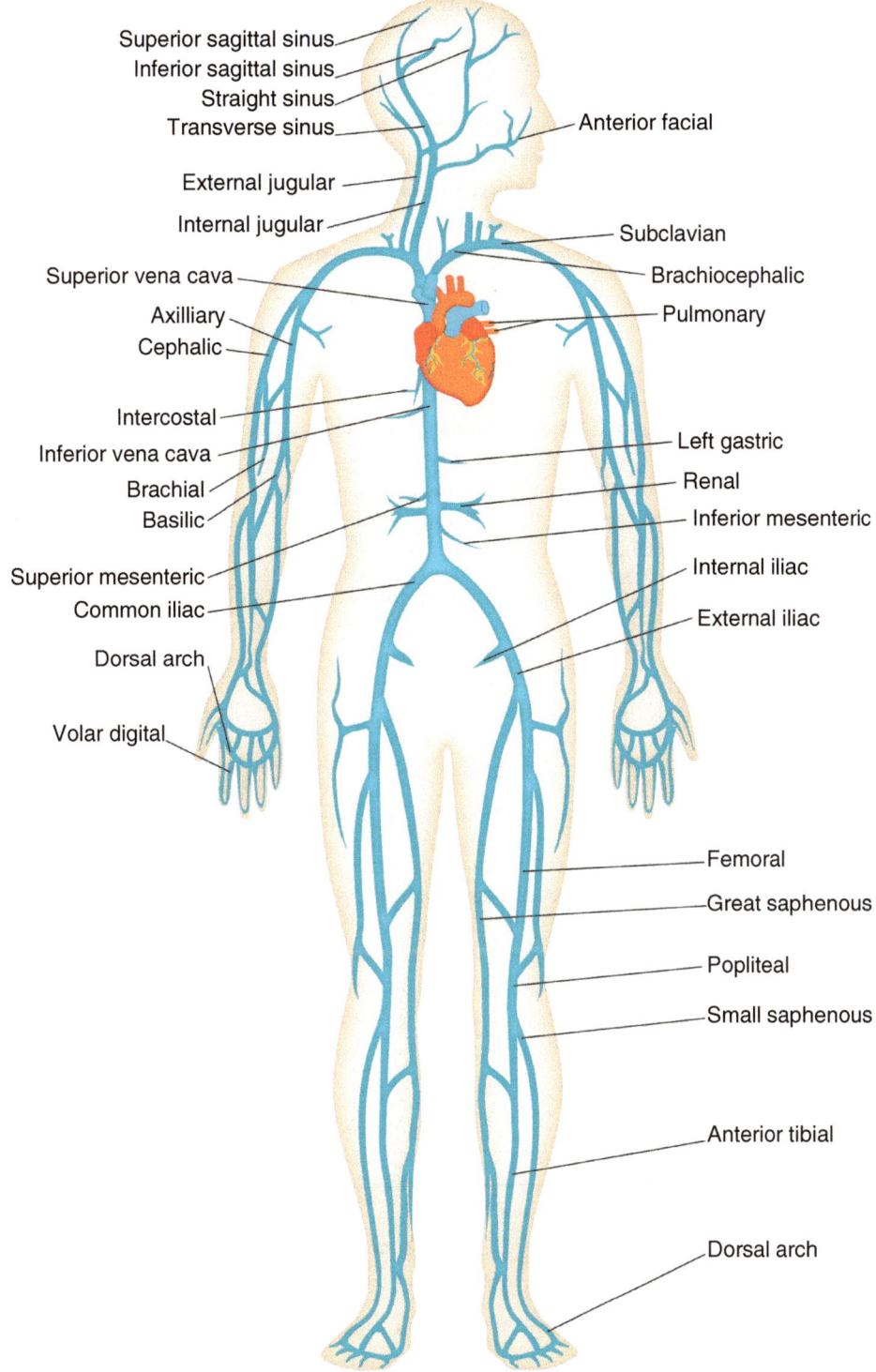

Figure 12.9 Major veins of the body

As blood moves away from the heart the arteries get smaller and eventually divide into arterioles. From the arterioles, blood passes into the smallest blood vessels, the capillaries. Here, fluid and some substances in the blood move into the interstitial fluid under the pressure in the arterioles. These substances are available to nourish cells. The membrane between the blood in capillaries and the interstitial fluid is semipermeable – it will not permit large molecules like plasma proteins to pass through – whereas water and electrolytes move easily. At the venule end of the capillary the pressure is less than in the interstitial fluid and most of the fluid moves back into the venous end of the capillary. The composition of the blood changes as it now carries waste products and less nutrients. Blood is then carried back towards the heart through venules, which merge to become larger and larger veins the closer they become to the heart. This system of blood vessels is a closed system, starting and ending at the heart.

Structure of blood vessels

Arteries and veins have three structural layers similar to all tubular structures (Figure 12.10):

1. An outer fibrous layer (tunica externa/adventitia).
2. A middle layer of muscle and elastic tissues (tunica media).
3. An inner lining of endothelium which is smooth to facilitate flow of blood (tunica intima).

Tunica externa: The tunica externa (tunica adventitia) is the outmost layer consisting of collagen fibres, varying in thickness between vessels. This layer anchors blood vessels to nearby organs, nerves and other blood vessels and provides passage for nerves and small lymphatic and blood vessels. Some blood vessels are very large and require their own blood supply, the vasa vasorum. The other layers are nourished by the blood through diffusion.

Tunica media: This is normally the thickest layer consisting of elastic fibres, collagen and smooth muscle. The smooth muscle allows vasoconstriction and vasodilation thus altering the blood flow and pressure. The elastic fibres allow for expansion and recoil, with collagen providing structural support.

Tunica interna: This is a thin layer of squamous epithelium covering a basement membrane and some fibrous tissue that lines arteries and veins. It is in direct contact with the blood and is smooth to enhance flow and prevent coagulation. This layer is selectively permeable and can secrete vasoactive substances – substances that can alter the diameter of blood vessels by causing them to constrict (get smaller in diameter) or dilate (get larger in diameter) (see later in the chapter).

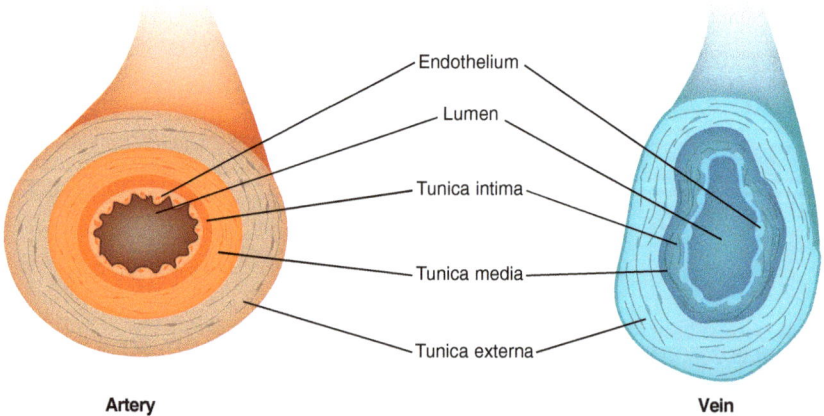

Endothelium
Lumen
Tunica intima
Tunica media
Tunica externa

Artery   Vein

Figure 12.10   Structure of blood vessels

This is the generic structure of veins and arteries, but blood vessels have distinct adaptations that support their role in the CVS.

Arteries and arterioles

The larger the artery (and nearer the heart), the more elastic tissue is present to withstand the pressure of the blood. As arteries get smaller, the muscle tissue increases until, in arterioles, the middle layer is composed entirely of smooth muscle, allowing the walls to dilate and constrict to alter the diameter of the blood vessel. Vasoconstriction decreases the size while vasodilation increases the size of the blood vessel.

Veins and venules

As you learned earlier, capillaries drain blood into venules, which drain into veins, which get larger as they get nearer the heart. The venous circulation is under much lower pressure than the arterial side as it is not under direct pressure from the heart pump. Thus, the tunica media of veins is thinner, which is why veins collapse if cut, while arteries remain open. Some veins have valves to prevent the backflow of blood as it is returning towards the heart (Figure 12.11). Remember, the venous circulation is largely working against gravity in returning blood to the heart and so these valves are necessary.

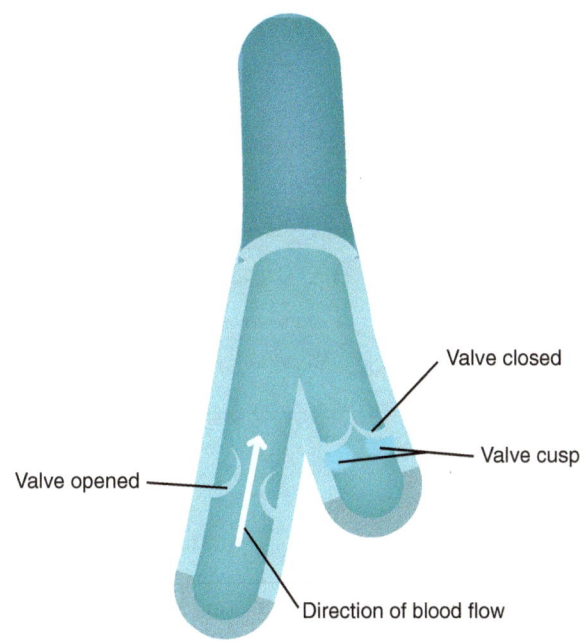

Figure 12.11 Valves in a vein

Blood is moved along by gravity (from the head and neck) and by the skeletal muscle pump with backflow prevented by the valves. This muscle pump is created by the contraction of skeletal muscle, compressing deeper veins and propelling blood towards the heart. In addition, the expansion of the chest by inspiration (breathing in) creates a negative pressure which pulls blood into the chest. The flattening of the diaphragm in inspiration increases intra-abdominal pressure, forcing blood towards the heart.

Capillary circulation

Capillaries are the business end of the circulation, forming a network of fine vessels throughout all tissues of the body and joining arterioles to venules. At the arteriole end, blood is forced through the capillary walls under the influence of blood pressure generated by the heart. Capillaries are made up of only a single layer of endothelial cells, which form a semipermeable membrane that separates the blood and the interstitial fluid compartment (Figure 12.12).

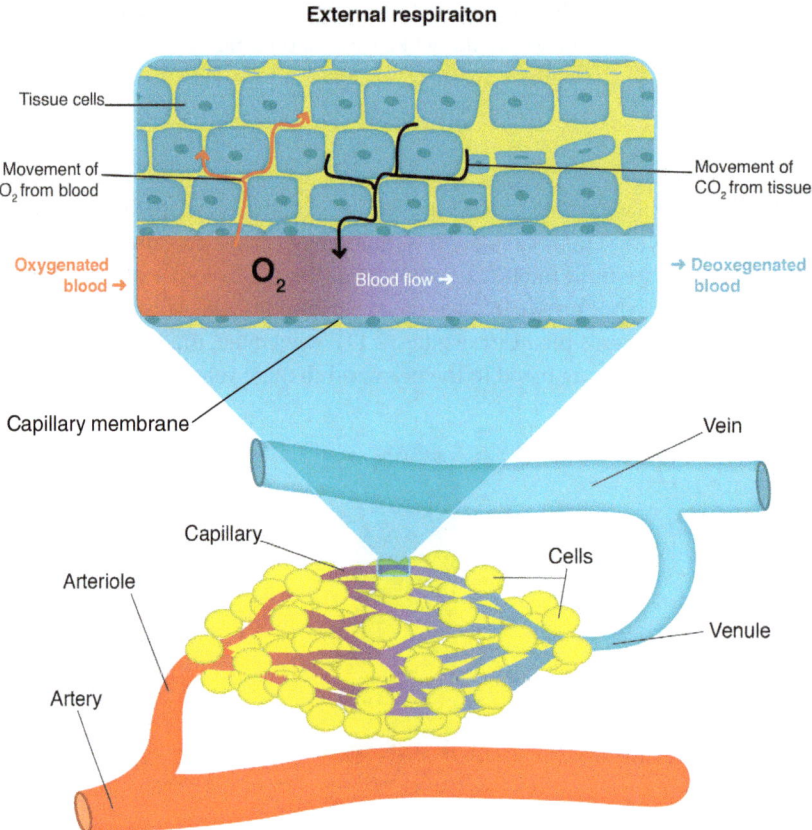

Figure 12.12 Capillary structure

This membrane is selective in what it lets through, permitting water and small molecules to diffuse through the capillary wall into the interstitial fluid and access the cells bathing in it. Large molecules, such as plasma proteins, cannot pass through the wall. Plasma proteins increase the osmolarity of blood as they draw water back into the capillary.

The lymphatic circulation

Lymph flows through the lymphatic vessels and lymph glands that together make up the lymphatic system. Lymphatic vessels have a key role in tissue drainage. The lymphatic system also has a role in the absorption of fat (Chapter 8) and in immunity (Chapter 13). The movement of fluid and small molecules from the circulatory capillaries is largely imbalanced with more fluid moving out of the arterial end of the capillary than is reabsorbed into the venous end, resulting in a net increase in fluid in the interstitial spaces (between the cells). Unless this excess fluid is removed, oedema (excess tissue fluid) forms and this eventually causes significant problems. Lymphatic capillaries absorb the excess fluid, some unused nutrients and waste products and the collected fluid enters larger lymphatic vessels which eventually drain into the right lymphatic duct and thoracic duct. Figure 12.13 shows how the different parts of the body drain back to the circulatory system.

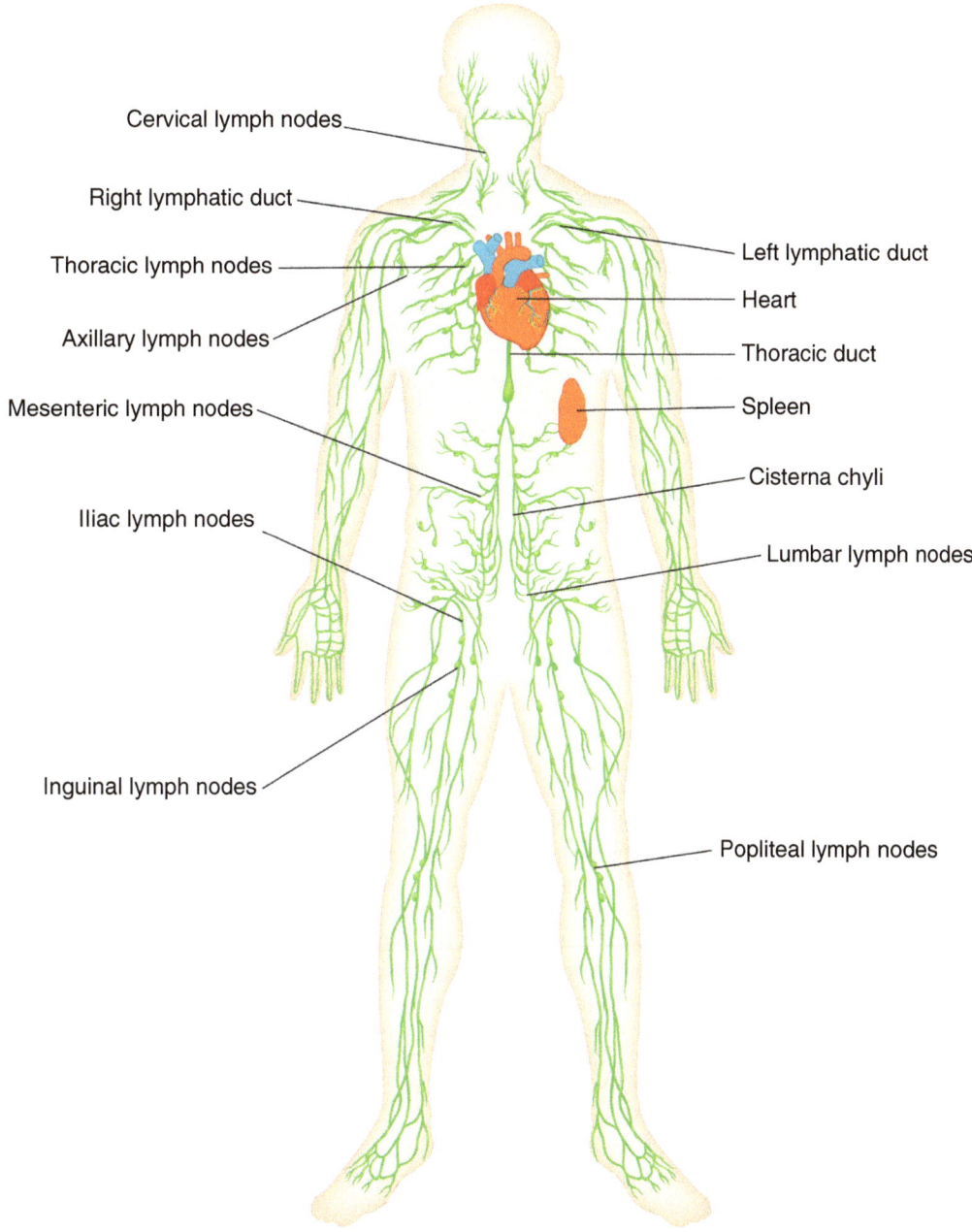

Figure 12.13 Major lymphatic vessels

The thoracic duct returns lymph to the circulatory system by emptying into the left subclavian vein. This lymph largely comes from the legs, pelvic region and abdominal cavity, the left arm and the left-hand side of the thorax, head and neck. The right lymphatic duct drains lymph from the right arm and right-hand side of the thorax, head and neck, returning the lymph to the circulatory system by emptying into the right subclavian vein.

So what propels lymph along lymphatic vessels? Lymphatic vessels have a tubular structure similar to that of blood vessels. The lymphatic vessels also have valves to prevent lymph backflow and the smooth muscle in the vessel walls contracts rhythmically to move lymph along (Figure 12.14). This is assisted by the contraction of adjacent muscles and large arteries.

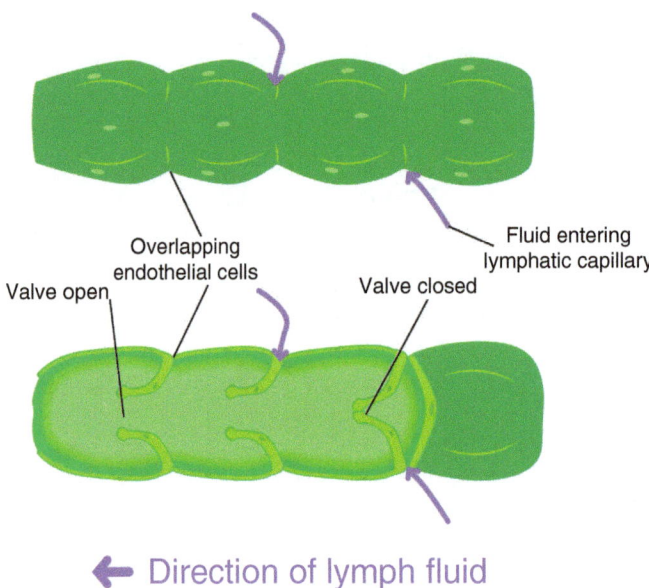

Overlapping endothelial cells

Valve open

Fluid entering lymphatic capillary

Valve closed

← Direction of lymph fluid

Figure 12.14 Lymph vessel

Fluid compartments and movement of fluid

There are two primary fluid compartments in the body, described in Chapter 1: Intracellular Fluid (ICF) and Extracellular Fluid (ECF) separated by membranes. To understand how fluid moves between these compartments, let us look at this a little bit closer.

Capillary membrane: This is freely permeable to water and small molecules such as oxygen, electrolytes and glucose. It is not freely permeable to large molecules such as plasma proteins, retaining them within the plasma.

Cell membrane: This is freely permeable to water but other smaller molecules (including glucose and electrolytes) can only move in by active transport. This means the cell membrane has to actively move them in and out, using energy.

The distinction between what can and cannot pass through these barriers is very important in understanding fluid balance. Fluid moves between these spaces as a result, primarily, of two key forces – hydrostatic pressure and colloid osmotic pressure (Chapter 11). This occurs primarily in the capillaries as this is where the capillary membrane is only one cell thick. These two forces work in opposition to achieve a balance of fluid, and anything that disturbs these will alter the balance of fluid in either compartment.

Hydrostatic pressure: This is the pressure exerted by the blood against the artery wall by the force exerted by the heart pumping. This pressure forces fluid and small molecules out of the capillary at the arterial end.

Osmotic pressure: Osmotic pressure is the pressure exerted by plasma proteins and some electrolytes in the plasma inside the capillaries. In other words, large plasma proteins, like albumin, and electrolytes, like sodium, pull water towards them. They attract and draw water and small molecules back into the capillary at the venous end. Potassium provides osmotic pressure inside the cell, preventing all the water being drawn out of it, as water can move freely across both the capillary and cell membranes.

Hydrostatic pressure is higher at the arterial end of the capillary due to the pumping action of the heart and lower at the venous end of the capillary. Osmotic pressure remains the same along the whole capillary. Thus, fluid and nutrients will move out of the capillary at the arterial end under the higher pressure, and unused nutrients and waste products will move back into the venous end of the capillary as the pressure pushing out is less (Figures 12.15 and 12.16). The net pressure pushing fluid and nutrients out of the capillary at the arterial end is higher than is the net pressure at the venule end, drawing fluid, unused nutrients and waste products back into the capillary. This leaves some fluid behind in the interstitial space. However, the lymphatic capillaries in the interstitial space collect any outstanding fluid. The difference in pressure at the two ends of the capillary is the Net Filtration Pressure (NFP).

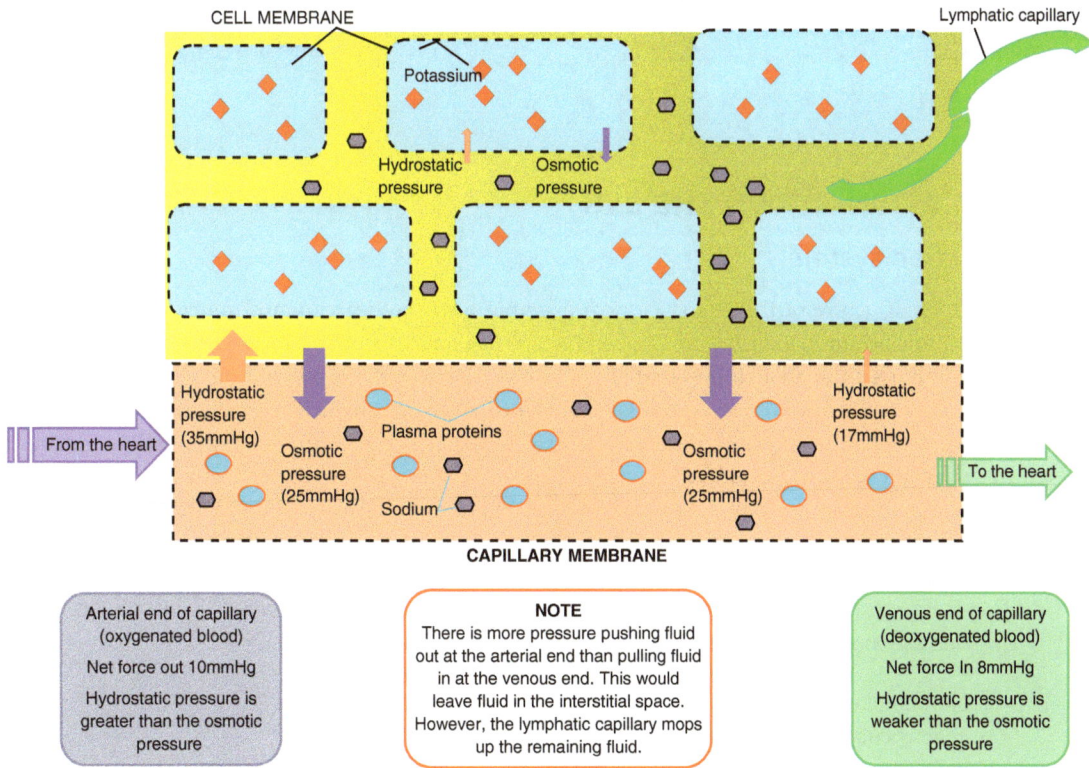

Figure 12.15 Movement of fluid and nutrients between fluid compartments

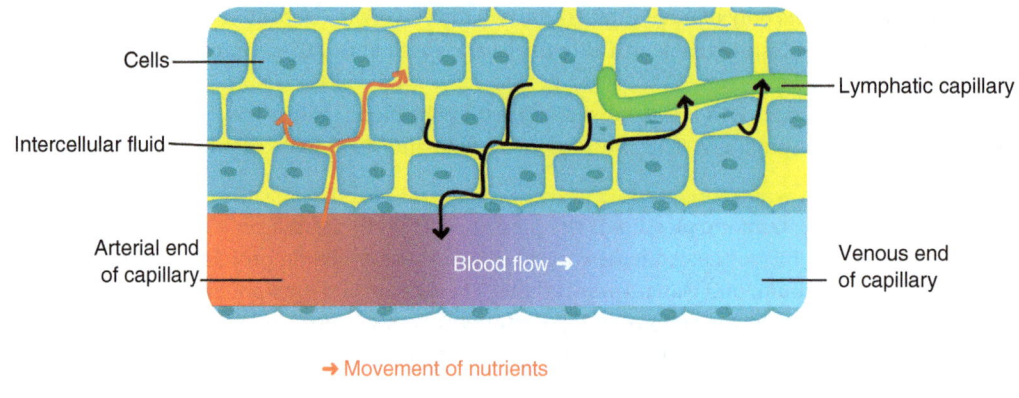

Cells

Intercellular fluid

Arterial end of capillary

Lymphatic capillary

Venous end of capillary

Blood flow ➡

➡ Movement of nutrients

➡ Movement of excess fluid and waste

Figure 12.16 Diffusion of nutrients and waste products between capillaries and cells

ACTIVITY 12.4: APPLY

Hydrostatic and osmotic pressures in practice

Think about how changes in either of these pressures might come about. Take five minutes and think about what would increase or reduce hydrostatic pressure. Similarly, what would increase or reduce osmotic pressure? How would these present in someone?

REGULATION WITHIN THE CVS

Conduction system

Knowing the structure of the heart, we now examine how the heart pumps under the control and regulation of its conduction system. The conduction system initiates and carries an electrical impulse (nervous impulse through action potentials – Chapter 5) across the myocardium, triggering the muscles to contract as it passes. However, it is a controlled electrical impulse to ensure that the pumping action is effective. Additionally, the heart muscle is responsive to hormonal stimulation. Intrinsic contractility is a property of cardiac muscle cells known as autorhythmicity. Autorhythmic cells of the conduction system initiate and distribute electrical impulses to adjacent cells that stimulate the atria and the ventricles to contract. The flow of conduction is unidirectional; it triggers contraction in the direction that blood will flow, from atria to ventricles. Figure 12.17 shows the components of the conduction system and their communication.

The conduction system consists of:

1. Sinoatrial node (SA node).
2. Atrioventricular node (AV node).
3. Atrioventricular bundle (AV bundle)/bundle of His.

4. Right and left bundle branches.
5. Purkinje fibres.

A single cycle of electrical activity moving across the heart and causing contraction is known as the cardiac cycle. In the healthy heart, the pacemaker is the sinoatrial (SA) node, a specialised group of autorhythmic cells in the wall of the right atrium which initiate an impulse that spreads across the atria, triggering the smooth muscle of both atria to contract simultaneously. This pumps blood into the ventricles. The rate of impulse from the SA node is increased by the sympathetic nervous system and reduced by parasympathetic nervous activity (Chapter 5).

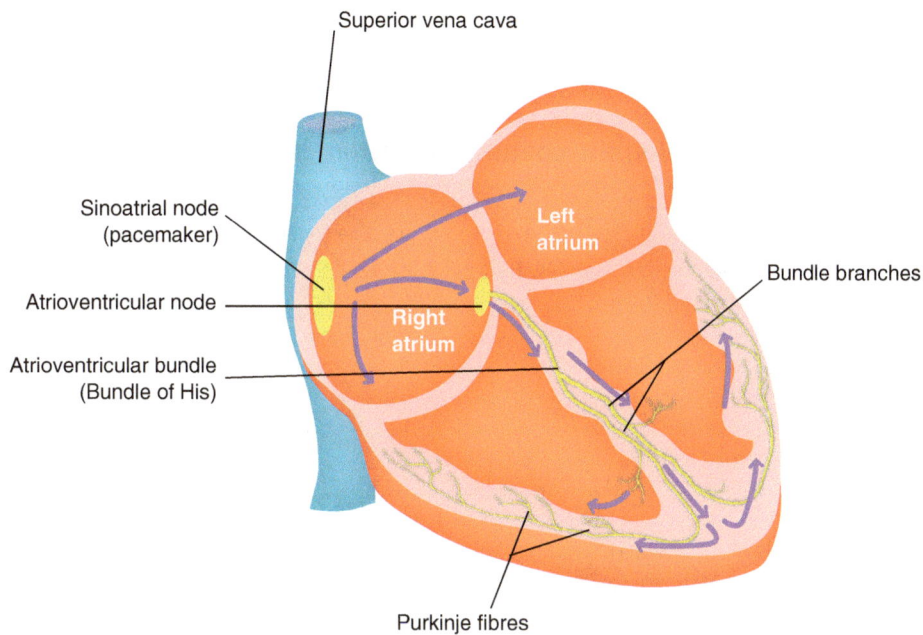

Figure 12.17   Conduction system of the heart

The nervous impulse now reaches a second group of autorhythmic cells in the wall of the septum at the level where atria and ventricles are separated, the atrioventricular (AV) node. There is a short delay at this point to allow the atria to finish contracting and enable the ventricles time to fill with blood from the atria.

Once this delay is over, the nervous impulse is carried from the AV node down to the ventricles through the atrioventricular bundle (AV bundle, often referred to as the bundle of His). This specialised nerve tract divides into right and left bundle branches which pass down the septum dividing out to each ventricle. These subdivide further into the Purkinje fibres to cover more surface area. Purkinje fibres pass over the surface of both ventricles and trigger contraction of the cardiac muscle pumping blood to the lungs and to the rest of the body.

Each cardiac cycle is one complete heartbeat during which the heart contracts (systole) and relaxes (diastole), commonly referred to as the lub-dub of the heart; i.e. you will hear two sounds for every heartbeat. A cardiac cycle normally lasts about 0.8 of a second. A single cardiac cycle consists of systole and diastole of both atria and systole and diastole of both ventricles. The two atria contract while the two ventricles relax then the ventricles contract while the atria relax.

ACTIVITY 12.5: UNDERSTAND

Watch this short video clip to learn more about the conduction system of the heart. Study this until you are clear about the blood flow.

This online video can be accessed by **scanning the QR code** with your smart phone camera or via https://study.sagepub.com/essentialap2e.

CONDUCTION SYSTEM OF THE HEART (4:03)

APPLY

Assessing the conduction system in practice

In Chapter 5 you will have learned how nervous impulses (referred to as action potentials) work in detail. As the heart has a large amount of electrical activity, we are able to measure this through what is known as an electrocardiogram (ECG). Each heartbeat shows a series of connected electrical activity known as the PQRST complex (Figure 12.18).

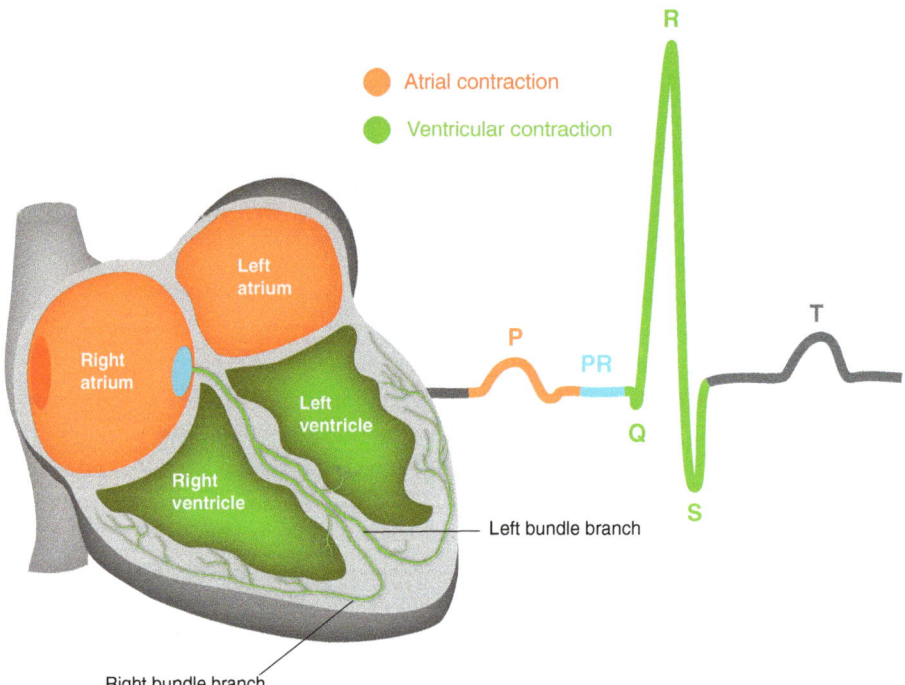

Figure 12.18　Relationship between ECG and cardiac cycle

If you think about it, the atria are much smaller than the ventricles and do not have to pump blood very far. This means there is less electrical activity causing the atria to contract. As the atria contract first, this is the first element of electrical activity seen on the ECG and is the P wave. The slight delay at the AV node means there is a small gap without electrical activity on the ECG between atrial and ventricular activity. The ventricles are much bigger than the atria and more muscular contraction is needed, and there is much more electrical activity visible on the ECG. This is the QRS complex, with the Q representing the electrical activity starting the ventricular contraction, R the peak of the electrical activity in the ventricles, and S the end of the ventricular stimulation. As part of the nervous impulse, there is some small electrical activity as electrolytes move across cell membranes to enable the next conduction cycle to start, the T wave. The ECG allows measurement of the time for each part of the conduction cycle, to measure the heart rate and to look for abnormalities in the conduction cycle process.

ACTIVITY 12.6: APPLY

Maximal heart rate (HR$_{max}$)

Heart rate is an easy method of cardiovascular measurement and is routinely used to assess the response of the heart to and recovery from exercise. The increase seen during exercise is due to the increase in cardiac output and has an upper limit. Maximal heart rate is often interpreted as the upper ceiling for an increase in central cardiovascular function (Robergs and Landwehr, 2002). Physiological evidence demonstrates that the heart's chronotropic response (the change in heart rate over time in response to stimuli), including maximal heart rate, declines with age (Gellish et al., 2007). However, it has been suggested that exercise at an intensity level of 55–65% of maximal heart rate is associated with improved fitness and fewer adverse cardiac events (Atwal et al., 2002).

The estimation of HR$_{max}$ has been largely based on the formula: HR$_{max}$ = 220 – age; however, this has no scientific merit for use in exercise physiology and related fields and validity of the equation has not been established (Tanaka et al., 2001). Following a meta-analysis by Tanaka et al. (2001) a new equation to predict maximal heart rate was formulated. They found that HR$_{max}$ is predicted to a large extent by age alone, independent of gender and physical activity status. Tanaka et al. (2001) established a formula for calculating HR$_{max}$ which was validated by Gellish et al. (2007) through a longitudinal study of 132 participants (male and female).

So how do you predict your HR$_{max}$?

According to Tanaka et al. (2001) the most reliable formula to use is:

$$HR_{max} = 207 - (0.7 \times age)$$

So let's think about Edward Bodie: what's his HR$_{max}$? Edward is 57, therefore using the equation his maximal heart rate would be:

$$HR_{max} = 207 - (0.7 \times 57)$$

$$HR_{max} = 167 \text{ beats per minute}$$

Now, using the equation, calculate what your HR$_{max}$ is.

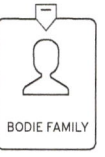

BODIE FAMILY

(Continued)

When do we use HR_{max}?

HR_{max} is used in rehabilitation and disease prevention programmes as a basis to prescribe exercise intensity (Tanaka et al., 2001; ACSM, 2014). It is also used by athletes (elite and non-elite) to design their training programmes to make them more effective in the consumption of oxygen and maintenance of their autonomic nervous system (Plews et al., 2014).

Heart training zones

A number of training zones have been identified by ACSM (2014) and these are used as the basis for many people's training regime:

Healthy heart zone (warm up)

This should be 50-60% of maximum heart rate: the easiest zone and probably the best zone for people just starting a fitness programme. It can also be used as a warm up for more serious walkers. This zone has been shown to help decrease body fat, blood pressure and cholesterol. It also decreases the risk of degenerative diseases and has a low risk of injury. Fats are 85% of food energy burned in this zone.

Fitness zone (fat burning)

This should be 60-70% of maximum heart rate: this zone provides the same benefits as the healthy heart zone, but is more intense and burns more total food energy. The percentage of food energy from fat is still 85%.

Aerobic zone (endurance training)

This should be 70-80% of maximum heart rate: the aerobic zone will improve your cardiovascular and respiratory system and increase the size and strength of your heart. This is the preferred zone if you are training for an endurance event. More food energy is burned with 50% from fat.

Anaerobic zone (performance training)

This should be 80-90% of maximum heart rate: benefits of this zone include improved oxygen consumption (the highest amount of oxygen one can consume during exercise) and thus an improved cardiorespiratory system, and a higher lactate tolerance ability, which means your endurance will improve and you'll be able to fight fatigue better. This is a high-intensity zone burning more food energy, 15% from fat.

Red line (maximum effort)

This is 90-100% of maximum heart rate: although this zone burns the highest amount of food energy, it is very intense. Most people can only stay in this zone for short periods. You should only train in this zone if you have a very high fitness level and have been cleared by a physician to do so.

Recovery heart rate

The recovery heart rate is one that is taken several minutes after exercise. It is taken anywhere between two and ten minutes after exercise. It is taken for 15 seconds and is multiplied by four in order to calculate beats per minute (bpm). The goal is to not exceed 120 bpm.

Task

1. Preferably working with another colleague, assess your baseline pulse rate by taking your radial pulse over one minute. Also count your respiratory rate and check your blood pressure – your colleague can do this for you.
2. Now, exercise for five minutes. This can be sprinting, going up and down a number of flights of stairs, or jogging on the spot.
3. Immediately after the exercise, check your heart rate, respiratory rate and blood pressure.
4. Repeat this check again after five and ten minutes and assess how quickly it took them to return to baseline.
5. Compare the results and evaluate what you think they mean.

Cardiac output and blood pressure

Cardiac output is the volume of blood pumped out of the heart per minute. The amount leaving the right and left sides of the heart has to be the same, otherwise one side will be short of blood and the other congested with too much blood. Cardiac output is calculated by the following equation:

Cardiac output = stroke volume × heart rate

Stroke volume is the blood pumped out of each ventricle per heartbeat, which is approximately 70 ml of blood. The heart rate is the number of heartbeats in one minute, which is approximately 70 beats per minute. For example:

Cardiac output = 70 × 70 = 4,900 ml (almost five litres of blood in one minute)

If the stroke volume or heart rate increases, cardiac output increases. If either drops, then cardiac output drops. The sympathetic nervous system (Chapter 5) or the hormone adrenaline (epinephrine) increase heart rate by raising stroke volume and heart rate. This can occur during fear, anxiety, stress or physical activity.

APPLY

Assessing heart rate

Measuring a person's heart rate is one of the most frequent observations by nurses in practice. This is typically taken at the radial artery as it is easily accessible and reliable, except when the heart rate is irregular. In this situation it should be measured by listening to the heartbeat and counting the rate. The apex beat is when the heart rate is counted by listening to the pulsation of the heart at the apex of the left ventricle at the fifth left intercostal space. It is assessed when there is an irregular heart rate, as not every beat reaches the peripheries and therefore counting the heart rate at the radial pulse in this situation is inaccurate.

(Continued)

Often the radial and apex rate are both counted and compared. The heart rate can also be measured at various other points on the body where arteries are close to the surface, for example at the carotid arteries in the neck, the popliteal arteries (behind the knees), and dorsalis pedal arteries (the prominent arch of the top of the foot between the first and second metatarsal bones). A fast heart rate is referred to as tachycardia and a slow heart rate is bradycardia. The acceptable heart rate ranges (resting heart rate) are generally accepted as:

Table 12.6 Heart rate ranges across the lifespan

Babies aged 0-1 month old:		70-190 beats per minute
Infants aged 1-11 months old:		80-160 beats per minute
Children:		
	1-2 years old:	80-130 beats per minute
	3-4 years old:	80-120 beats per minute
	5-6 years old:	75-115 beats per minute
	7-9 years old:	70-110 beats per minute
	10 years and older:	60-100 beats per minute
Adults:		
	All adults:	60-100 beats per minute

Adults who are very fit and healthy may have a heart rate between 40 and 60 beats per minute and this is not seen as a bradycardia.

In addition to noting the rate of the heart, you should also note the regularity (which may show a problem in the conduction system of the heart) and the amplitude (volume) of the pulsation. A low volume may represent a drop in blood volume or blood pressure, whereas a bounding, full pulse may indicate excess volume in the circulatory system.

Blood pressure is the pressure exerted by the blood against the arterial walls by the pumping force of the heart. This occurs when blood is forced out of the ventricles into arteries that are narrower, encountering resistance as it flows. Blood pressure refers to pressure inside the arterial circulation. Blood pressure is higher in vessels close to the heart than in arteries/arterioles further away. The pressure in the venous circulation is much lower as it is not under direct influence of the heart pumping blood.

Blood pressure is measured in mmHg (millimetres of mercury) and is calculated by the following equation:

Blood pressure = cardiac output × peripheral resistance

You already know how to calculate cardiac output. Peripheral resistance is the resistance to the flow of blood from the wall of the artery vessels. It is mainly determined by contraction or relaxation of the muscle layer of the arterial/arteriolar wall. As the smooth muscle in the arterial/arteriolar walls contracts (vasoconstriction) the diameter of the vessel gets smaller, the amount of blood in contact with the vessel wall increases, and the blood pressure rises. As the smooth muscle in the arterial/arteriolar wall relaxes (vasodilation) the diameter of the vessel gets bigger, the amount of blood in contact with the vessel wall decreases and the blood pressure goes down.

Changes in either cardiac output or peripheral resistance will alter the blood pressure. Control of blood pressure is very important. While it needs to be high enough to force blood through the arterial circulation and to create enough hydrostatic pressure to perfuse the vital organs, if it is too high it will cause trauma to the arterial vessel walls.

Control of blood pressure

There are two control systems in the body that regulate blood pressure:

1. Short-term (fast-acting) control occurs by stimulation from the autonomic division (sympathetic and parasympathetic pathways) of the nervous system (Chapter 5).
2. Long-term (slow-acting) control is achieved through the Renin Angiotensin Aldosterone System (RAAS) of the endocrine system (Chapter 7).

These systems work simultaneously.

Neural control (short-term/fast-acting)

The nervous system detects rises and falls in blood pressure through baroreceptors, which respond to changes in pressure, and chemoreceptors, which respond to changes in the concentration of chemicals. These are found within the CVS and measure changes in pressure and chemical substances such as carbon dioxide and oxygen in the blood. Chemoreceptors largely respond to chemicals affected by respiration (O_2, CO_2 and H^+). High CO_2 concentrations and low O_2 concentrations indicate to the cardiovascular centre, located in the brainstem (Chapter 5), that gaseous exchange is suboptimal. Sympathetic stimulation will relay to the SA node to increase the heart rate (and the rate of breathing to increase gaseous exchange) to compensate. An increase in heart rate will increase cardiac output and therefore blood pressure. In contrast, if O_2 concentration is high or the CO_2 concentration is low then parasympathetic stimulation will relay to the SA node to slow the heart rate, dropping cardiac output and therefore blood pressure. Carbon dioxide and H^+ ions are also vasodilating substances, causing smooth muscle in the arterial/arteriole walls to relax.

Baroreceptors, in the arch of the aorta and carotid sinuses (widening of a carotid artery at its main branch point), also detect rises and falls in blood pressure. A drop in blood pressure is relayed to the cardiovascular centre, which responds by sympathetic stimulation of the SA node, raising the heart rate and causing vasoconstriction of the smooth muscle in the arterial walls (Figure 12.19). Catecholamines, which are hormones/neurotransmitters (such as dopamine, noradrenaline (norepinephrine) and adrenaline (epinephrine)) made by the adrenal glands, the brain and some specialised nerve cells, stimulate arterial vasoconstriction. Both will raise blood pressure as they stimulate alpha-adrenergic and beta-adrenergic receptors, which trigger contraction of the smooth muscle in the arterial/arteriole walls to contract, narrowing the lumen. If the baroreceptors detect that blood pressure is too high, they relay this message to the cardiovascular centre. The cardiovascular centre then relays to the SA node, through parasympathetic stimulation, to slow the heart rate, and signals the smooth muscle in the arterial side of the circulation to vasodilate. Both actions will lower the blood pressure.

Endocrine control (long-term/slow-acting)

The endocrine system has a role in controlling blood pressure through the RAAS. In Chapter 7 you will have discovered that endocrine control is slower than neural control and this is no different for the control of blood pressure. The RAAS primarily responds to low blood pressure and it involves more than one system of the body, which you will discover now as we work through this system. The best way to learn this system is through Figure 12.20 (already introduced in Chapter 7), but you need to be able to understand all the components.

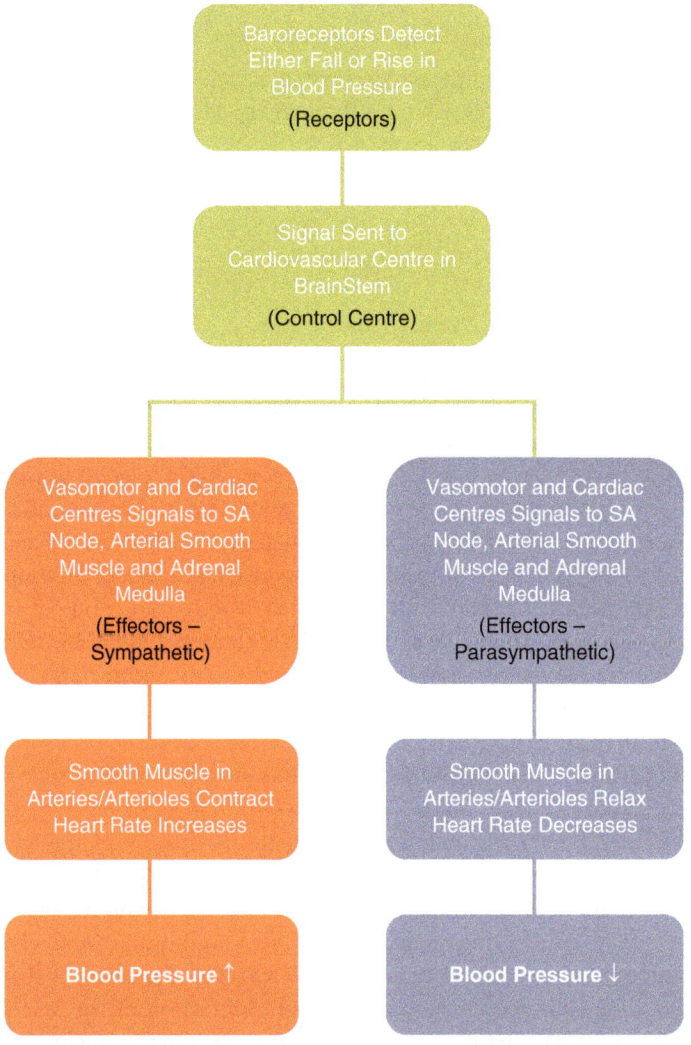

Figure 12.19 Baroreceptor regulation of blood pressure

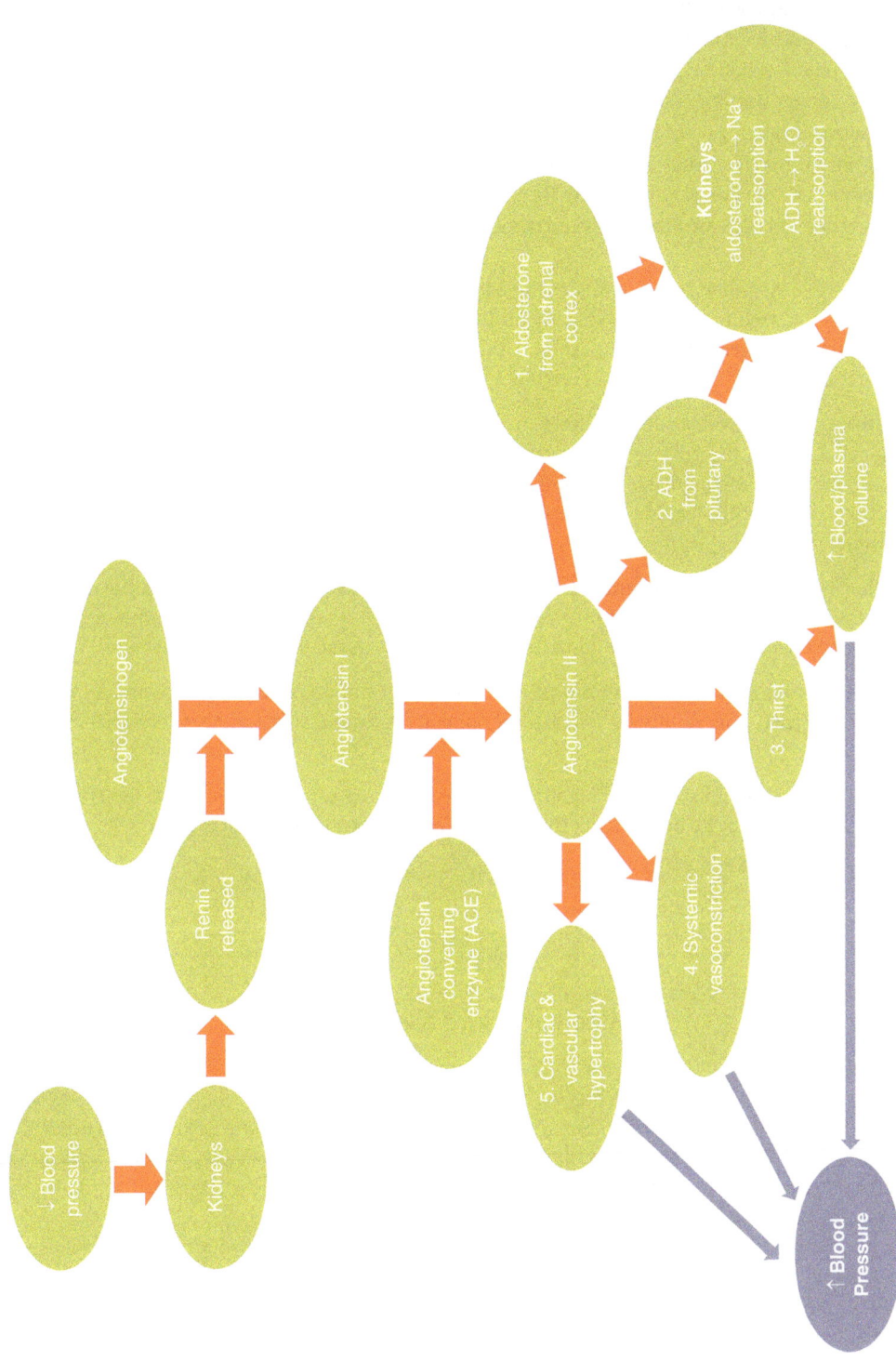

Figure 12.20 Renin angiotensin aldosterone system (RAAS) control of blood pressure

Let's walk through the stages of this system. When blood pressure drops, then blood flow through the glomerular capsule in the nephron of the kidney is reduced. This is detected within the capsule, and results in the release of renin, a hormone, from the afferent arteriole of the nephron. A plasma protein, angiotensinogen, produced by the liver is converted by renin to a variant plasma protein called angiotensin I. Angiotensin I has no biological action and is a precursor to angiotensin II, the active form of this hormone. The lungs and the proximal convoluted tubules in the nephrons of the kidney produce another hormone called Angiotensin Converting Enzyme, or ACE. When ACE comes into contact with angiotensin I, it converts it to the active form, angiotensin II. Angiotensin II then triggers a number of responses that will increase blood pressure through five pathways:

Pathway 1: Angiotensin II stimulates the adrenal cortex to release the hormone aldosterone. Aldosterone is a hormone that works directly on the nephrons in the kidney, triggering the reabsorption of sodium and water back into the blood stream. This increase in volume will increase blood pressure by increasing cardiac output and peripheral vascular resistance.

Pathway 2: Angiotensin II stimulates the release of Antidiuretic Hormone (ADH) (also known as vasopressin) from the posterior pituitary gland. ADH has a direct effect on renal tubules, increasing the permeability of the distal convoluted tubules and the collecting tubules. This increases water reabsorption in both, leading to an increase in blood volume, which leads to an increase in blood pressure.

Pathway 3: Angiotensin II stimulates the thirst centre in the hypothalamus, with the response that the person now feels thirsty. In most circumstances this will result in the person drinking more fluids, leading to an increase in blood volume and subsequently an increase in blood pressure.

Pathway 4: Angiotensin II also directly stimulates the smooth muscles in the arterial circulation to vasoconstrict. This results in increased peripheral vascular resistance, therefore increasing blood pressure.

Pathway 5: Angiotensin II will, on a long-term basis, cause cardiac and vascular hypertrophy. This will raise blood pressure by reducing the elasticity of the heart and arterial walls.

APPLY

Blood pressure assessment

Blood pressure is measured in practice using a sphygmomanometer. When the heart contracts and forces blood out into the (already full) aorta and major arteries, the elastic tissue in their walls enables the vessels to stretch, pushing the blood further through the circulation and causing distension during systole. This is followed by recoil, which spreads throughout the arterial system during diastole. The elastic recoil keeps the blood pressure up to an adequate (but lower) level. This distension of the arteries is what you feel when you are taking someone's pulse.

Blood pressure is measured in millimetres of mercury and is expressed as a fraction. For example, a person's blood pressure may be determined to be 120/80 mmHg. The top figure is the systolic pressure, created during systole (ventricular contraction). Systolic pressure is the maximum pressure exerted by blood against the arterial wall. The lower figure is the diastolic pressure, the pressure exerted during diastole (ventricular relaxation). Diastolic pressure is the lowest pressure exerted by blood against the arterial wall.

Atrial natriuretic peptide

Contrasting with the RAAS is the hormonal action of Atrial Natriuretic Peptide (ANP), a hormone released by the heart when raised blood volume causes stretching of the atrial wall beyond what is expected. ANP works on the proximal convoluted tubules and collecting ducts to reduce the reabsorption of sodium and water. This results in a drop in blood volume as more water is lost from the body through the formation of urine, and thus a fall in blood pressure.

ACTIVITY 12.7: APPLY

Health promotion and education

Considering that the regulation of blood pressure is vital for achieving homeostasis, what do you think the role of the nurse is in preventing cardiovascular disease? Access the following sources of information on the prevention of cardiovascular disease and determine what the priorities are in promoting the health of the public in this regard:

Connolly, S.B., Kotseva, K., Jennings, C., Atrey, A., Jones, J., Brown, A., Bassett, P. and Wood, D.A. (2017) 'Outcomes of an integrated community-based nurse-led cardiovascular disease prevention programme', *Heart*, 103(11): 840–7.

Han, E., Quek, R.Y.C., Tan, S.M., Singh, S.R., Shiraz, F., Gea-Sánchez, M. and Legido-Quigley, H. (2019) 'The role of community-based nursing interventions in improving outcomes for individuals with cardiovascular disease: a systematic review', *International Journal of Nursing Studies*, 100: 103415.

National Institute for Health and Care Excellence (NICE) (2010) *Prevention of Cardiovascular Disease*. London: NICE.

Stewart, J., Manmathan, G. and Wilkinson, P. (2017) 'Primary prevention of cardiovascular disease: a review of contemporary guidance and literature', *Journal of the Royal Society of Medicine Cardiovascular Disease*, 6: 1–9.

Walker, J. (2013) 'Reducing cardiovascular disease risk: cholesterol and diet', *Nursing Standard*, 28(2): 48–55.

World Health Organization (WHO) (2007) *Prevention of Cardiovascular Disease: Guidelines for Assessment and Management of Cardiovascular Risk*. Geneva: WHO.

CONCLUSION

Now that you have read through this chapter and undertaken the activities, you should have a good understanding of how the CVS is structured and functions to achieve optimal homeostasis. Without an effective and efficient CVS, the fundamentals to support life are absent. The CVS provides the fuel for life by circulating nutrients to all cells in the body, transports toxic waste products, and enables the immune system to reach its targets to protect the body. The CVS integrates closely with the nervous system and endocrine system for control and regulation, but also works in association with the renal, respiratory and digestive systems, to name but a few, to maintain homeostasis. You have also learned that the CVS is linked with experiences in life – whether it is a response to fear and the need to protect, enabling people to meet the demands of physical pastimes, or whether it is feeling the physical sensations of excitement and anticipation. Considering all of these factors is central to health-related quality of life, and therefore central to person-centred nursing.

GO DEEPER

Further reading

Farley, A., Hendry, C. and McLafferty, E. (2012) 'Blood components', *Nursing Standard*, 27(13): 35–42.

Farley, A., McLafferty, E. and Hendry, C. (2012) 'The cardiovascular system', *Nursing Standard*, 27(9): 35.

Hurrell, K. (2014) 'Safe administration of blood components', *Nursing Times*, 110(38): 16–19.

Peate, I. (2008) 'Body fluids: components and disorders of lymph', *British Journal of Healthcare Assistants*, 2(3): 115–18.

Peate, I. (2008) 'Body fluids: components and disorders of the blood', *British Journal of Healthcare Assistants*, 2(2): 63–6.

- The CVS has key roles in circulating nutrients, transporting waste products, and enabling white blood cells to target pathogens.

- Blood is composed of nutrient-rich plasma and a variety of blood cells, each with specific functions.

- Lymph has a similar composition to blood except for the absence of erythrocytes and often a high fat content.

- Blood and lymph are transported in tubular vessels with three layers. These vessels make up the lymphatic circulation and the systemic circulation.

- Fluid and nutrients exchange between the systemic circulation and the tissues/cells of the body at the capillaries, under the influence of hydrostatic and osmotic pressures.

- Arterial blood is propelled around the body under the direct influence of the heart, whereas lymph and venous blood rely on muscular contractions, gravity and often the smooth muscle of their vessels.

- The heart is composed of four chambers that make up two pumping systems, the right side pumping blood to be oxygenated in the lungs and the left side pumping oxygenated blood around the rest of the body.

- The heart is composed of three layers of tissue that enable it to function as a pump, provide it with structure and limit its expansion.

- A heartbeat is achieved by the conduction system of the heart, which is regulated by the structure of the related components.

- Blood pressure is controlled neurologically (fast-acting) and through the endocrine system (slow-acting).

- Blood clotting is controlled by two cascading pathways that eventually merge into one final common pathway.

REVISE

TEST YOUR KNOWLEDGE

This chapter covers a lot of material. It is best to break your revision into components when revising. Consider revising in the following sets of information:

- Components of the CVS.

- The composition of blood and lymph, including understanding the variety of cells in both.

- Haemostasis and coagulation.

- The nature of blood and lymph vessels, including their structure.

- The nature of fluid and nutrient exchange at the capillaries.

- The anatomy of the heart and the direction of blood flow through it.

- The conduction system of the heart.

- Regulation of blood pressure.

You may have noticed that this chapter has only one Go Deeper activity. This is because all of the components of this chapter contain fundamental knowledge that underpins your practice as a nurse. When revising this chapter, try to remember the diagrams included. Can you label them? Do you understand the functions of each component? Can you describe the anatomy and can you explain the physiology? Have you undertaken the tasks to engage with deeper learning?

In order to help you revise, consider the following questions, answers for which can be found by visiting https://study.sagepub.com/essentialap2e.

Test yourself by revising the chapter first, and then answer these questions without looking at the book. Afterwards compare your answers with the text and with the notes you made. Did you miss anything in your notes? Here are the questions:

1   What is meant by the circulatory system?

2   What are the functions of blood?

3   What are erythrocytes? Describe their structure and function.

4   Name the five types of white blood cells and identify their roles.

5   A 40-year-old lady is admitted to hospital with acute confusion. She has a cannula in her left arm for fluid administration. In her confusion, she pulls the cannula out. What happens within her body to stop the bleeding and restore homeostasis?

6   Describe the ABO grouping system of blood.

7   Describe how the cardiovascular system contributes to homeostasis.

8   Name the components of the cardiovascular system and describe their main functions.

9   Using the diagram below illustrate the pathway taken by the blood through the heart.

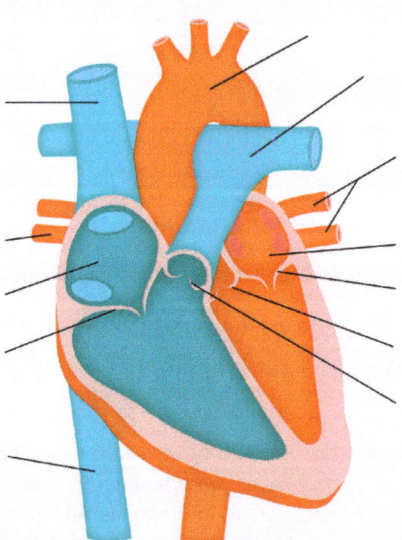

10	What is blood composed of?
11	Where are red blood cells formed?
12	What is the stimulus for erythropoietin production?
13	Describe the three layers of the heart.
14	Label Figure 12.21 identify the components of the conduction system of the heart.
15	Describe how the conduction system in the heart works.
16	What are cardiac output, stroke volume and heart rate?
17	What is blood pressure and how is it created?
18	Describe the neural control of blood pressure.
19	Draw a diagram to represent the endocrine control of blood pressure.

For additional revision resources visit: https://study.sagepub.com/essentialap2e.

REVISE

ACE YOUR ASSESSMENT

- Revise key terms relevant to this chapter with interactive flashcards.
- Test yourself with quizzes and multiple-choice questions.
- Access the glossary with audio to hear how complex terms are pronounced.
- Explore recommended websites suitable for revision.

REFERENCES

ACSM (American College of Sports Medicine) (2014) *ACSM's Guidelines for Exercise Testing and Prescription*, 9th edn. Baltimore: Lippincott Williams & Wilkins.

Atwal, S., Porter, J. and MacDonald, P. (2002) 'Cardiovascular effects of strenuous exercise in adult recreational hockey: the Hockey Heart Study', *CMAJ*, 166(3): 303–7.

Berra, K., Miller, N.H. and Jennings, C. (2011) 'Nurse-based models for cardiovascular disease prevention: from research to clinical practice', *Journal of Cardiovascular Nursing*, 26(4 Suppl): S46–55.

Crowther, C.A., Middleton, P. and McBain, R.D. (2013) 'Anti-D administration in pregnancy for preventing rhesus alloimmunisation', *Cochrane Database of Systematic Reviews*, 2: CD000020.

Gellish, R.L., Goslin, B.R., Olson, R.E., McDonald, A., Russi, G.D. et al. (2007) 'Longitudinal modelling of the relationship between age and maximal heart rate', *Medicine and Science in Sports and Exercise*, 39(5): 822–9.

National Institute for Health and Care Excellence (NICE) (2010) *Prevention of Cardiovascular Disease*. London: NICE.

Plews, D.J., Laursen, P.B., Kilding, A.E. and Buchheit, M. (2014) 'Heart rate variability and training intensity distribution in elite rowers', *International Journal of Sports Physiology and Performance*, 9(6): 1026–32.

RCOG (Royal College of Obstetricians and Gynaecologists) (2011) *The Use of Anti-D Immunoglobulin for Rhesus D Prophylaxis*. Green-top Guideline No. 22. London: RCOG.

Robergs, R.A. and Landwehr, R. (2002) 'The surprising history of the "HRmax = 220 – age" equation', *Journal of Exercise Physiology online*, 5(2): 1–10.

Serious Hazard of Transfusion (SHOT) (2018) *Definitions of current SHOT reporting categories & what to report*. Manchester: SHOT.

Tanaka, H., Monahan, K.D., Douglas, M.S. and Seals, R. (2001) 'Age-predicted maximal heart rate revisited', *Journal of the American College of Cardiology*, 37(1): 153–6.

Walker, J. (2013) 'Reducing cardiovascular disease risk: cholesterol and diet', *Nursing Standard*, 28(2): 48–55.

World Health Organization (WHO) (2007) *Prevention of Cardiovascular Disease: Guidelines for Assessment and Management of Cardiovascular Risk*. Geneva: WHO.

PART 4

SUPPORT AND PROTECTION OF THE INTERNAL ENVIRONMENT

This section of the book focuses on aspects of human structure and function that support the internal environment. It consists of the following three chapters:

- *Chapter 13. The Immune System: Internal Protection*

  This system of the body plays a major role in protection of the body against infection and abnormalities that can arise in body cells, such as cancer. It consists of two sections, the innate and the adaptive immune systems. We share the ancient innate immune system with all other organisms, including plants, fungi, insects and primitive multicellular organisms. It provides a general defence for the body but is short acting. The adaptive immune system has developed in vertebrates and provides a highly specific immunity against particular antigens. Memory cells recognise the antigen to which they have a defence on subsequent meetings and can present a very rapid, long-lasting response.

- *Chapter 14. Skin and Temperature Regulation*

  The skin also plays a major role in protection. It prevents microorganisms gaining ready access to the body, prevents fluid loss and contributes to thermoregulation, thus protecting against hyper- and hypo-pyrexia. It also forms vitamin D, thus contributing to bone health.

- *Chapter 15. The Musculoskeletal System: Support and Movement*

  This system also plays an important role in protection through the physical shelter provided for internal organs by the cavities surrounded by the skeleton. In addition, the marrow of certain bones forms white blood cells which fight infection. The system also enables the body to move rapidly away from potential danger. Furthermore, it makes the red blood cells that are essential for transporting oxygen to the tissues.

THE IMMUNE SYSTEM
INTERNAL PROTECTION

13

UNDERSTAND: CHAPTER VIDEO

Before working through this chapter, you might find it useful to have an overview of the immune system. Watch this video clip to enhance your understanding.

This video can be accessed by **scanning the QR code** with your smart phone camera or via https://study.sagepub.com/essentialap2e.

IMMUNE SYSTEM (9:53)

LEARNING OUTCOMES

When you have finished studying this chapter you will be able to:

1. Describe the different components of the innate immune system and their function in protection of the body
2. Identify the different cell types of the adaptive immune system and their varying functions in protection against infection
3. Explain how the different parts of the immune system interact, and how the nervous and endocrine systems and the gut microbiome influence the function of the endocrine system
4. Discuss how the immune system can be manipulated to aid health
5. Describe how inappropriate functioning of the immune system can interrupt normal functioning of the body

INTRODUCTION

This chapter examines the immune system, which plays the major role in protection of the body against infection and cellular abnormalities. However, the nervous and endocrine systems (Chapters 5 and 7), the skin (Chapter 14), the gastrointestinal tract (Chapter 8) and the microbiome (Chapter 4) all contribute to the development and efficacy of the immune system.

Even single-cell microbes have some degree of protection against infection, while invertebrates and plants have innate immunity, which is also present in vertebrates. However, with the development of vertebrates, a more complex system, adaptive immunity, has also developed. The immune system in vertebrates consists of two major sections: the innate and adaptive systems, which together provide a considerable degree of protection against infection. In this chapter we examine how these systems work and collaborate to maintain the health of the body.

In addition, we will introduce briefly some of the things that can go wrong in relation to the immune system and the effect of these on individuals, and how our understanding of this system is used to promote health and minimise harm.

Context

The immune system plays a crucial role in maintaining health through combating infections and abnormal cells and is important to the Bodie family at all stages of life.

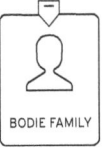
BODIE FAMILY

Danielle was born with some immunity to conditions to which her mother had developed antibodies as they can pass through the placenta from the mother's blood stream. She will continue to obtain antibodies in the breast milk with which she is fed. Her programme of childhood immunisation (see later in chapter) aims to enable her to make her own antibodies against a number of infectious diseases which were once common conditions. Such an immunisation programme not only provides protection against the specific diseases but also provides some protection against other conditions (Kristensen et al., 2000).

As well as routine childhood immunisation, members of the Bodie family who travel to other countries may need immunisation against endemic (i.e. commonly found in an area) conditions. In particular Thomas (the pilot) and the family of Michelle and Kwame who regularly travel to Africa will need specific vaccinations to provide immunity against some conditions not found in this country.

Margaret suffers from hay fever which causes symptoms of eye and nasal irritation and sneezing due to hypersensitivity to pollen during the months when pollen levels are high. While generally considered a minor complaint, it can cause considerable discomfort and disruption to life. However, she is able to manage her symptoms effectively with antihistamines and nasal and eye drops as required. Someone with severe symptoms can be treated with immunotherapy when small doses of the antigen are gradually built up over months or even years allowing tolerance to develop.

The older members of the family will be offered the annual flu vaccine and other vaccines as appropriate (see Table 13.1 in the section on immunisation).

OVERVIEW OF THE IMMUNE SYSTEM

The major organs and tissues involved in immunity include the primary lymphoid organs, i.e. the thymus and bone marrow, and secondary lymphoid organs and tissues, i.e. lymph nodes, the spleen, Waldeyer's ring and lymphoid tissue (mucosa associated lymphoid tissue (MALT) and gut associated lymphoid tissue (GALT)) (Figure 13.1a).

Primary lymphoid organs

Bone marrow is the primary haematopoietic organ in the body, producing all types of blood cells. In terms of immunity, this includes the production and maturation of B-lymphocytes and production of T-lymphocytes, which mature in the thymus. The thymus is an important organ in immunity both before and after birth, into the first few months of infancy. T-lymphocytes, or T cells, mature in the thymus under the influence of a hormone called thymosin; this maturation is necessary for them to perform their immune function.

Secondary lymphoid organs and tissues

The spleen also has an important role in immunity as it contains B- and T-lymphocytes, which collect and present antigens in order for them to be destroyed by phagocytosis. Lymph nodes contain large numbers of lymphocytes, which filter microbes and dead and living phagocytes that have ingested any pathogen or cellular debris. The lymph nodes therefore filter and purify lymph before its return to the systemic circulation. Lymph nodes are located in groups throughout the body but are mostly found in

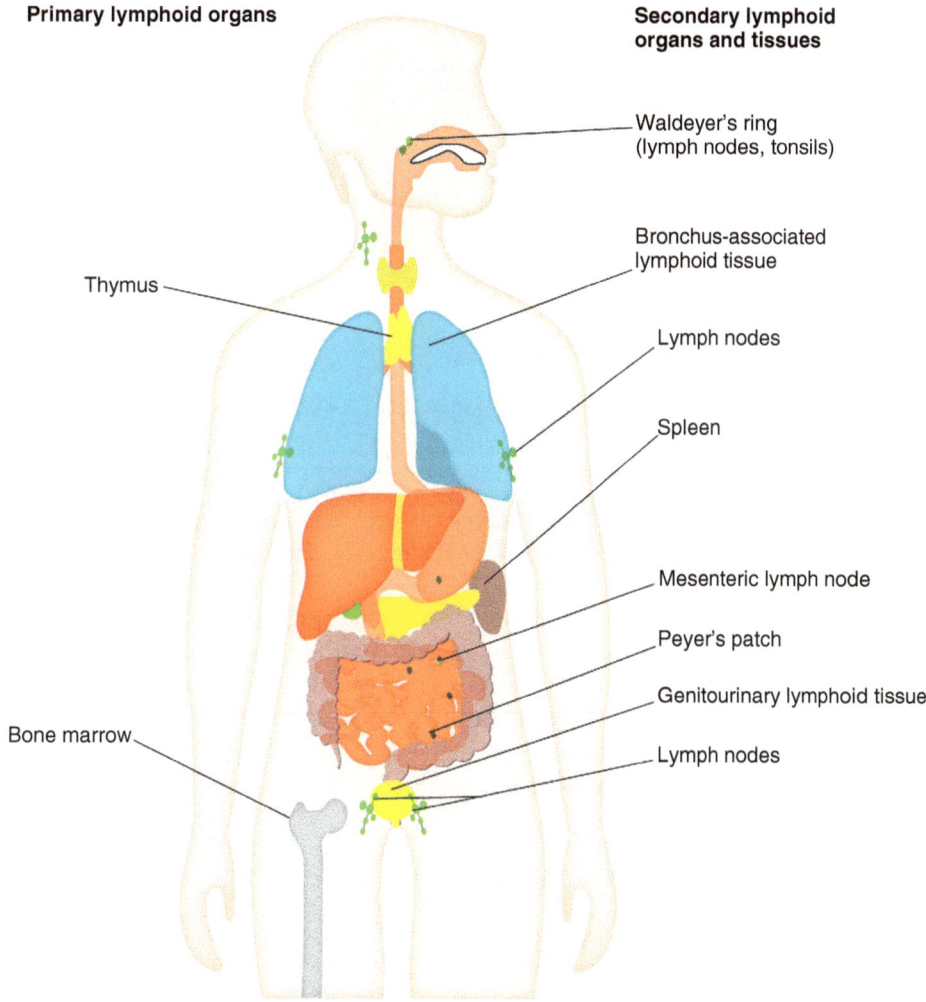

Primary lymphoid organs

Secondary lymphoid organs and tissues

Waldeyer's ring
(lymph nodes, tonsils)

Bronchus-associated
lymphoid tissue

Thymus

Lymph nodes

Spleen

Mesenteric lymph node

Peyer's patch

Genitourinary lymphoid tissue

Bone marrow

Lymph nodes

Figure 13.1a Anatomy of the immune system

areas susceptible to invading pathogens, such as the respiratory and GI tract (Chapter 12, Figure 12.13). Waldeyer's ring (Figure 13.1b) refers to the four tonsillar structures found in the nasopharynx (pharyngeal (or adenoid) and tubal tonsils), oropharynx (palatine tonsils) and posterior third of the tongue (lingual tonsils). MALT and GALT refer to the small concentrations of lymphoid tissue dispersed throughout the submucosal membranes of the GIT (including Peyer's patches and appendix – Chapter 8), nasopharynx, breast, thyroid, lungs and salivary glands.

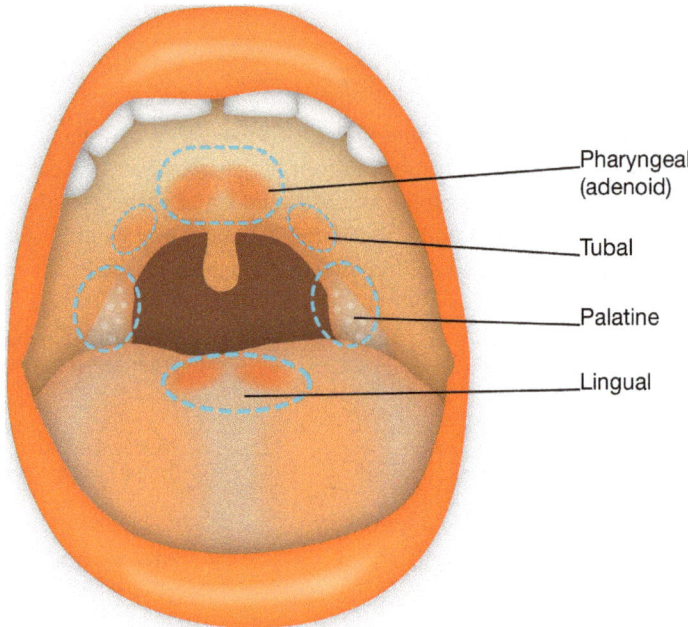

Figure 13.1b   Waldeyer's ring (tonsillar structures)

How the immune system works

The immune system works by recognising pathogens (also called antigens) and foreign cells and differentiating them from the cells of the body. When responding to a pathogen or foreign cell there are two main pathways in immunity (Figure 13.2), both of which we will explore in more depth in this chapter. The first pathway is innate immunity, also referred to as natural or non-specific immunity, which deals with all invasions in the same way and is the most rapid response to a foreign cell or pathogen. The second main pathway is adaptive immunity, also known as acquired or targeted immunity, which reacts to a *specific* antigen/pathogen. This response is tailored and can take a little longer to work as immune cells must either learn about the new antigen/pathogen and produce an antibody to destroy it, or remember the antigen/pathogen from previous exposure and release antibodies to destroy them.

Recognition of non-self

For the immune system to function, it has to be able to differentiate between cells that are part of the individual (host[1]) and those which are not. All cells of the body have a group of proteins on the plasma

---

[1]A host is a living plant or animal, in this case a person, within which a pathogen lives.

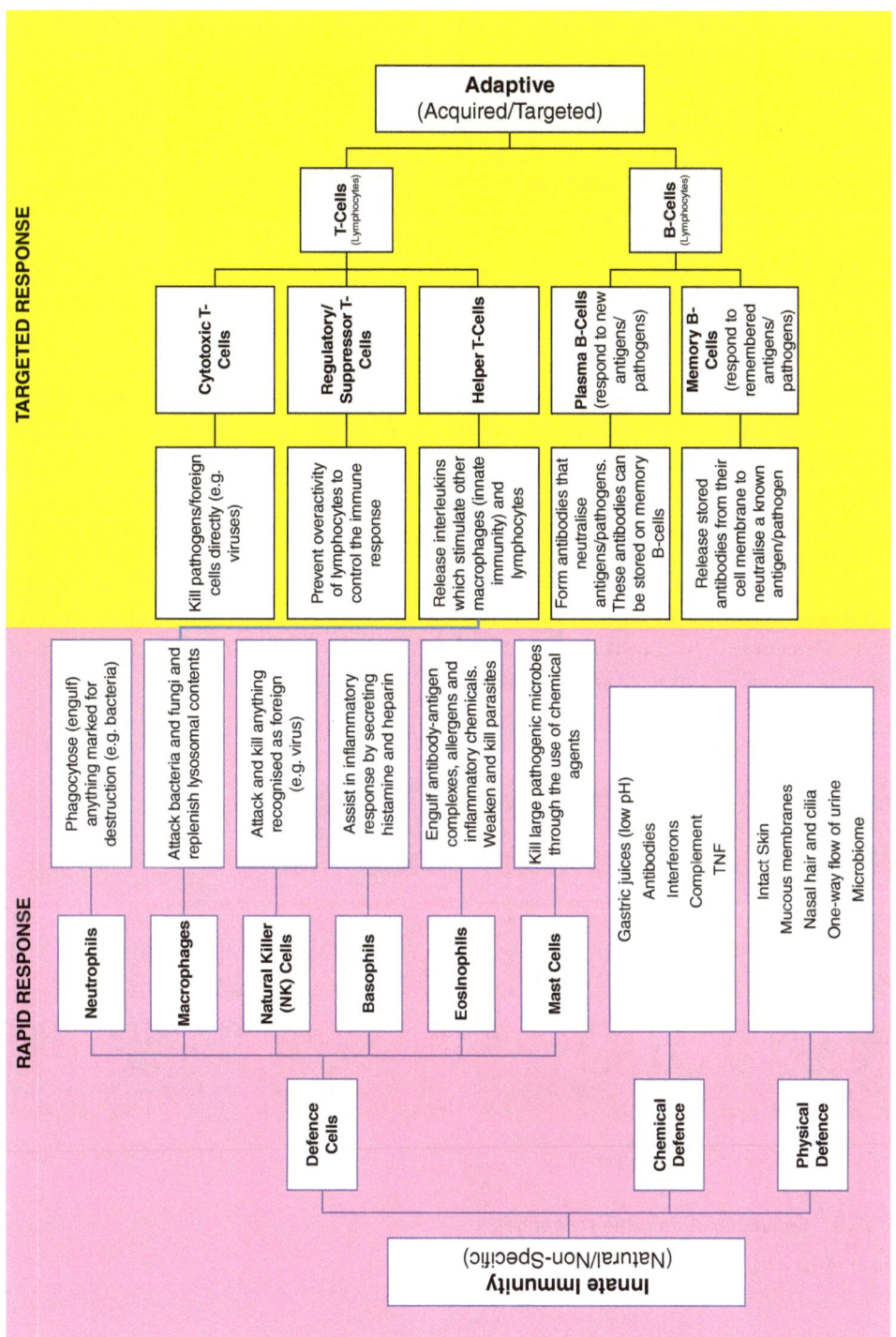

Figure 13.2 Structure of the immune system

membrane that identifies that cell as being part of the individual. During development, the cells of the immune system learn to recognise these (self) as different from those of invading microorganisms (non-self). Recognition of general 'non-self' characteristics is non-specific and is characteristic of the innate immune system. The detection of a particular microorganism from amongst numerous possibilities is specific and is carried out by the adaptive immune system.

The proteins that are recognised as 'self' are known as the Major Histocompatibility Complex (MHC), which in humans is known as the HLA (human leucocyte antigen) system. This is unique to each individual and a near match is a requirement for transplants or blood transfusion (Navarrete, 2000).

Responsiveness of the immune system

The body has a number of mechanisms for protecting it against substances recognised as 'non-self'. These may be invading microorganisms, other harmful substances such as toxins, or abnormal cells (e.g. malignant) that arise in the body and which initiate an immune response. The component which sets off this response is known as an antigen and may be a virus or just specific molecules on the surface of the foreign organism.

The immune response occurs in stages shown in Figure 13.3. In the first stage of the innate response various molecules already present in blood, extracellular fluid (ECF) and secretions either kill or weaken the antigenic organism. Still within innate immunity, the second stage involves recognition of molecules on the antigen and not present on host cells, the Pathogen-Associated Molecular Patterns (PAMPs), and activation of the immune cells in the body. At a later stage, the adaptive immune response is activated and the immune cells deal with the specific antigens. In this chapter the two major aspects of the immune response will be considered in some detail.

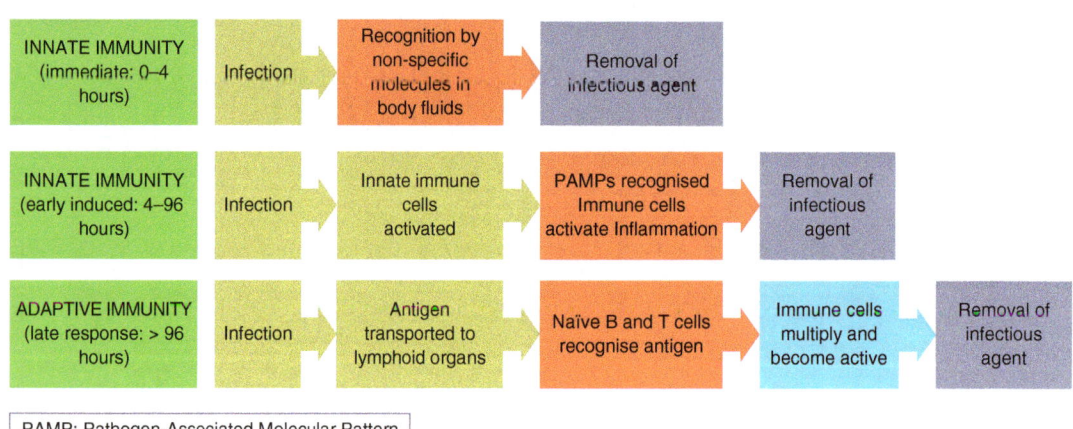

PAMP: Pathogen-Associated Molecular Pattern

Figure 13.3 Sequence of immune responses

Adapted from: Murphy, 2012.

Cytokines and the immune response

The communication system between cells of the immune system is through secretion of cytokines – small proteins that act to pass signals between cells. The cells involved in both the innate and adaptive immune responses both secrete and respond to cytokines, which are released throughout the body (Helbert, 2006) and act both locally and at a distance. The increased cytokine level, primarily from the macrophages, influences various organs of the body including:

- **The liver:** which is stimulated to produce a number of acute-phase proteins, an increase of which raises the viscosity[2] of the plasma. This can be seen by an increased C-reactive protein (CRP) level (which is one of these acute phase proteins – see Chapter 9) indicating infection:
  - The normal level of CRP is less than 10 mg/L.
  - Levels of 10–40 mg/L are found in mild inflammation and viral infections.
  - 40–200 mg/L are found in active inflammation and bacterial infection.
  - >200 mg/L is found in burns and severe bacterial infections.

- **The hypothalamus:** this part of the brain responds by increasing the body temperature. This inhibits replication of viruses and other pathogens, and also indicates infection. Shivering and sweating (rigor) may occur in severe fever.
- **The heart:** a very severe acute-phase response is septic shock. The cytokines increase production of nitric oxide (NO)[3], which reduces cardiac output and increases vasodilation, and thus blood pressure falls, as happens in septicaemia.

INNATE (OR NATURAL) IMMUNITY

This is the first line of defence against pathogens and it reacts much more quickly than the adaptive system. It has three functional characteristics:

1. It reacts rapidly to pathogens.
2. It responds the same way each time it meets pathogens.
3. It starts again on each occasion that it meets a pathogen.

It consists of:

- **Physical and biochemical barriers** of the skin and internal membranes.
- **The inflammatory process.**
- **Cells of the innate immune system:**
  - phagocytes (neutrophils and macrophages);
  - natural killer cells;
  - mast cells;
  - eosinophils.

- **The complement system**, which is activated and destroys pathogens (see below).
- **Interferons** (proteins), which prevent viruses replicating.

The phagocytes and complement system are able to deal with the pathogens living outside the cells, i.e. bacteria and fungi. However, viruses live and reproduce within the cells, and interferons play a major role in combating such infections.

---

[2]Thickness or how resistant a liquid is to flowing.

[3]Nitric oxide (NO) is a hormone that helps cells communicate with each other throughout the body.

Physical and biochemical barriers

Physical barriers play an important role in minimising the entry of pathogens into the body (Figure 13.4). The keratinised external layer of the skin is thick enough to prevent most pathogens entering the body: a break in the skin will allow pathogens to enter and be dealt with by other elements of the immune system. Extensive trauma (e.g. burns) can allow considerable amounts of bacteria to enter, which may overwhelm the body leading to death. The skin merges seamlessly with the membranes lining the organs of the body that link to the external environment, and the respiratory, urinary and gastrointestinal systems. In addition, a number of biochemical substances are formed, which protect against pathogens, and the commensal microbes within the microbiome (Chapter 4) already inhabit sites in the body and prevent pathogens colonising those sites.

The methods of protection in different systems/organs are (Figure 13.4):

- **Skin:** secretions from the sweat and sebaceous glands of the skin contain antibacterial and antifungal chemicals and macrophages (Chapter 14).
- **Respiratory system:** goblet cells secrete mucus, which pathogens stick to and are swept by the ciliated epithelium of the mucosa up to the pharynx or the nose where they are swallowed or sneezed out (Chapter 10).

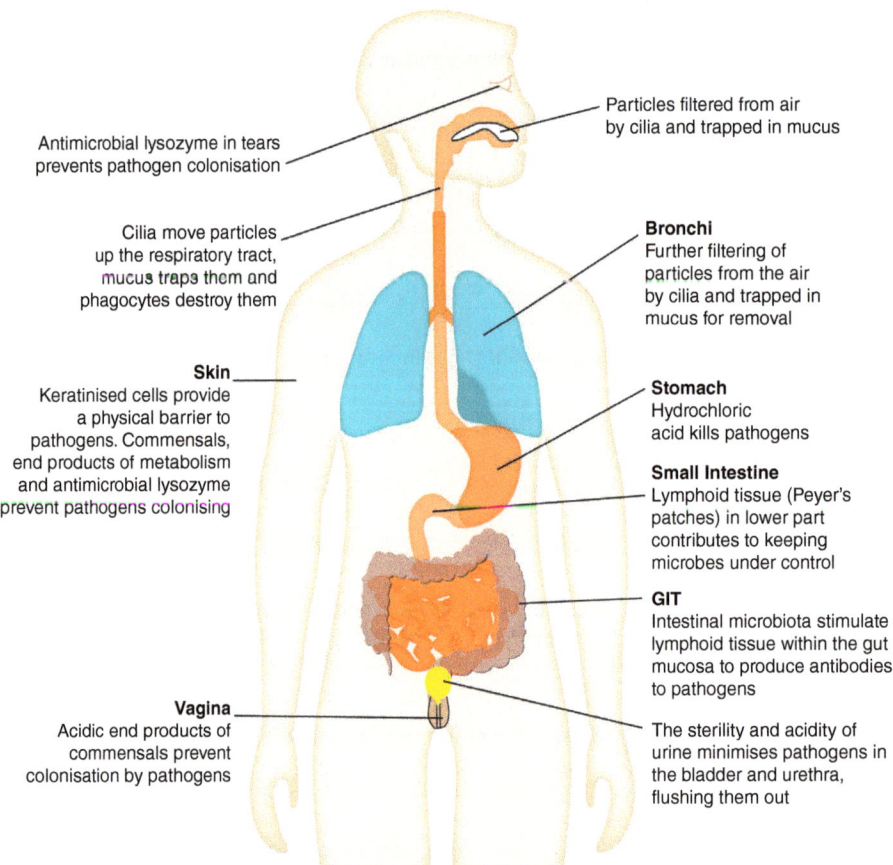

Antimicrobial lysozyme in tears prevents pathogen colonisation

Cilia move particles up the respiratory tract, mucus traps them and phagocytes destroy them

Skin
Keratinised cells provide a physical barrier to pathogens. Commensals, end products of metabolism and antimicrobial lysozyme prevent pathogens colonising

Vagina
Acidic end products of commensals prevent colonisation by pathogens

Particles filtered from air by cilia and trapped in mucus

Bronchi
Further filtering of particles from the air by cilia and trapped in mucus for removal

Stomach
Hydrochloric acid kills pathogens

Small Intestine
Lymphoid tissue (Peyer's patches) in lower part contributes to keeping microbes under control

GIT
Intestinal microbiota stimulate lymphoid tissue within the gut mucosa to produce antibodies to pathogens

The sterility and acidity of urine minimises pathogens in the bladder and urethra, flushing them out

Figure 13.4 Physical and biochemical barriers to infection

- **Renal system:** the flow of urine usually prevents pathogens becoming established. It is often only when renal stones or other blockage occurs or flow of urine is reduced that urinary infections occur.
- **Gastrointestinal tract**: the acid and proteolytic enzymes in the stomach rapidly kill most bacteria entering. However, some reach the colon and compete with the numerous non-pathogenic bacteria normally present (Chapter 8).

The inflammatory process

Inflammation is the body's response to tissue damage. It acts as a protective mechanism to isolate, inactivate and remove the causative agent and/or the damaged tissue. There are a number of catalysts that may trigger an inflammatory response, these include; microbes, physical agents e.g. trauma, extremes of temperature, and chemical agents (e.g. drugs, alcohol, pollutants and antigens).

Inflammation can be divided into two phases:

1. Acute – this is self-limiting i.e. it only continues until the threat is eliminated and lasts approximately 8–10 days from onset of healing.
2. Chronic – this may persist for weeks/months.

Characteristics of the acute and chronic response differ, and each phase involves different biochemical mediators and cells that function together (discussed later). Depending on successful containment of tissue damage and infection, the acute and chronic phases may lead to healing without progression to the granulomatous phase. The granulomatous response occurs if healing has not been initiated. It aims to contain the cause of tissue damage, so it no longer poses any harm to the body.

The acute inflammatory response is a collection of simultaneous processes but can be considered to largely occur in two main phases: vascular and cellular (Figure 13.5).

Vascular phase

In the vascular phase, arterioles and capillaries supplying the damaged area dilate, increasing blood flow and resulting in redness and heat. The permeability of capillaries is increased due to inflammatory mediators from damaged cells, e.g. histamine, and as a result of increased pressure caused by increased blood flow to the area. When this occurs, plasma proteins that are normally confined to the blood stream move into the tissues through widened gaps in the capillary walls. This also increases osmotic pressure in the tissues, drawing fluid in.

Cellular phase

In the cellular phase, leucocytes are chemically attracted to the site of injury when macrophages release cytokines, e.g. interleukins. This results in phagocytic cells, e.g. neutrophils, arriving at the site of injury and engulfing pathogens and cellular debris. This process is essential in wound healing and is described further in Chapter 14.

Changes associated with inflammation are caused by a number of chemical mediators (Figure 13.5). Activated mast cells (a type of white blood cell; specifically, it is a type of granulocyte derived from the myeloid stem cell) initiate synthesis of other inflammatory mediators including, histamine, cytokines, leukotrienes, prostaglandins, platelet activating factor, serotonin and bradykinin (see Go Deeper box for their role).

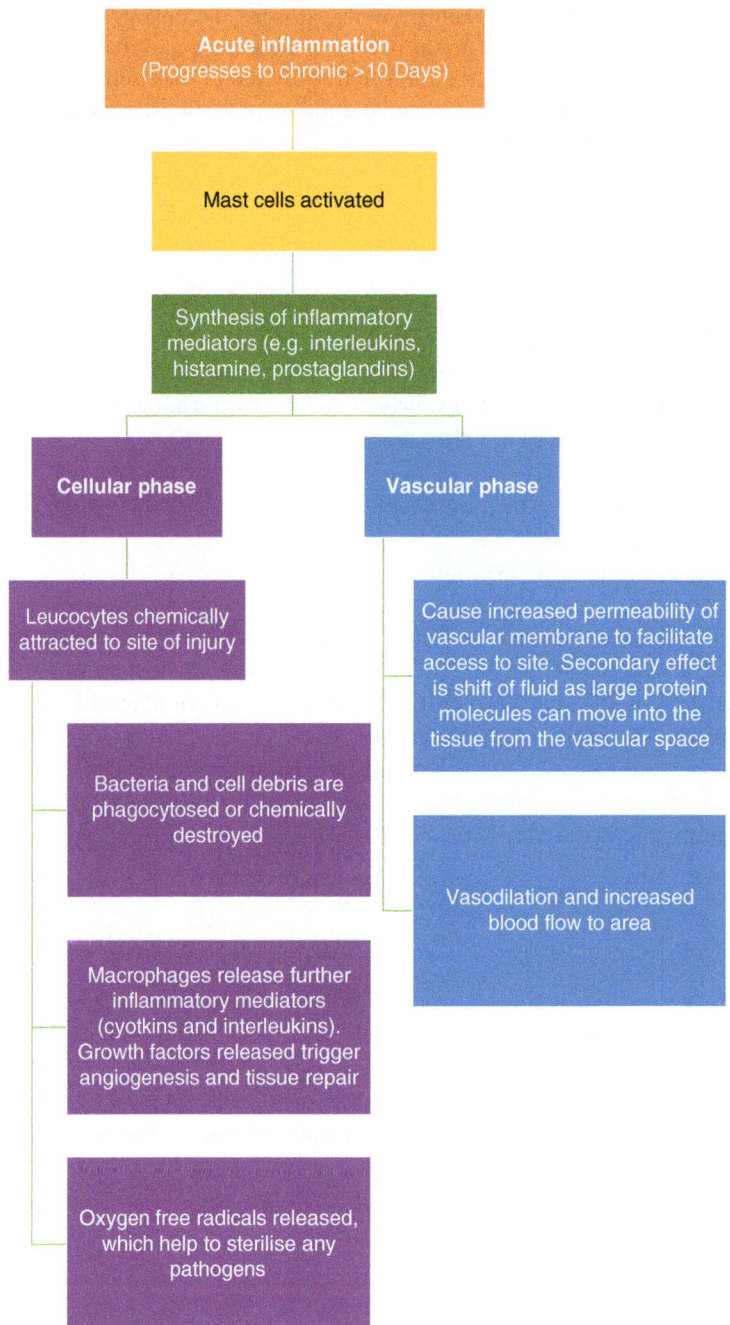

Figure 13.5 Acute inflammatory response

GO DEEPER

Inflammatory mediators

Histamine: increases the permeability of the capillaries to leucocytes and some proteins to allow them to engage pathogens. Acts immediately, causing vasodilation and increased capillary permeability to form exudate.

Cytokines: Cytokine is a general name; other names include lymphokine (cytokines made by lymphocytes), monokine (cytokines made by monocytes), chemokine (cytokines with chemotactic activities), and interleukin (cytokines made by one leucocyte and acting on other leucocytes), Tumor Necrosis Factor (TNF). Proinflammatory cytokines are produced predominantly by activated macrophages and are involved in the up-regulation of inflammatory reactions e.g. interleukin-1 (IL-1) activates macrophages, monocytes and lymphocytes. Interleukin-6 (IL-6) causes hepatocytes to produce C-Reactive Protein (CRP).

Leukotrienes: These are sulphur containing lipids produced from lipoxygenase that initiate histamine-like effects: smooth muscle contraction and increased vascular permeability. Important in later stages of response because they stimulate slower and more prolonged responses than histamine.

Prostaglandins (PG): Cause increased vascular permeability, neutrophil chemotaxis and pain by direct effect on nerves. PGE_1 and PGE_2 (forms of prostaglandin) cause increased vascular permeability and smooth muscle contraction, cyclooxygenase-2 (COX-2) is associated with inflammation.

Platelet-activating factor (PAF): Produced by removal of fatty acids from the plasma membrane phospholipids. Mast cells are major source of PAF although it can also be made by neutrophils, monocytes and endothelial cells. Biological effect is the same as leukotrienes – cause endothelial retraction leading to increased vascular permeability, leucocyte adhesion and platelet activation.

Serotonin: Increase vascular permeability and promotes vasodilation.

Bradykinin: Stimulates unencapsulated sensory nerve endings and is the primary noxious stimulus.

Local and systemic manifestations

The body will produce a number of local and systemic manifestations in response to the inflammatory process. Local manifestations result from the vascular response and subsequent leakage of circulating components into the tissue. Characteristics include:

- Redness (erythema/rubor) and heat – caused by increased blood flow into damaged area.
- Oedema – caused by the shift of protein and fluid into interstitial space.
- Pain – due to increased pressure of fluid on nerves and by local irritation of nerves by chemical mediators e.g. bradykinin.
- Loss/impaired function – due to lack of nutrients for cell or swelling that may interfere mechanically.
- Exudate – results from increased vascular permeability:
  - **Early/mild inflammation**
    - Serous (watery).
    - Few plasma proteins or leucocytes.
  - **Severe/advanced inflammation**
    - Fibrinous (thick and clotted) e.g. pneumonia.
    - High cell and fibrin count.
    - Purulent (suppurative) e.g. cysts, abscesses.
    - Large number of leucocytes.
    - Haemorrhagic.
    - Filled with erythrocytes due to bleeding.

Systemic manifestations occur due to an acute phase response, i.e. a rapid physiologic reaction generated to deal with tissue damage. There are three primary systemic changes:

- Fever (can be beneficial as some organisms are highly sensitive to small changes in temperature; however, can have harmful side effects as it may enhance the host's susceptibility to the effects of the endotoxins associated with gram –ve bacteria):
  - ○ Partially induced by cytokines.
  - ○ Act directly on hypothalamus.
- Leucocytosis (may be accompanied by left shift in ratio of immature to mature neutrophils):
  - ○ Increase in number of circulating white blood cells.
- Plasma protein synthesis:
  - ○ Increased during inflammation (may be pro-inflammatory or anti-inflammatory).
  - ○ Acute phase reactants (maximal circulating levels 10–40 hours after the start of inflammation).
  - ○ IL-1 (indirectly responsible for synthesis of acute phase reactants through the induction of IL-6, which directly stimulates liver cells to synthesis most of these proteins).

Specialised cells

The formation of the cells involved here is discussed under Cells of the Immune System within the section on Adaptive Immunity.

Phagocytes (neutrophils and macrophages)

These are mobile leucocytes (Chapter 2, Figure 2.16 and Chapter 12), which engulf and destroy pathogens. They have special receptors called pattern recognition molecules, which recognise patterns associated with different types of pathogens. For example, some identify the double-stranded RNA, which is found in many viruses at some stage in their life-cycle (but is not produced by mammals). Others recognise certain sugars found on microbes but, again, not in mammals. The phagocytic cells then engulf the microbes by endocytosis (Chapter 2) into a vacuole into which enzymes are secreted that (usually) kill the microbes.

Neutrophils are formed in the bone marrow, migrate into the blood stream and have a short lifespan (days), dying of apoptosis. Monocytes are also formed in the bone marrow but are distributed through the blood stream to the tissues where they stay and differentiate into macrophages (known as microglia in the brain and Kupffer cells in the liver) (Pocock and Richards, 2009).

Natural killer cells

These are formed in the bone marrow from the same lymphoid cell line as B- and T-lymphocytes (discussed later) and circulate in the blood stream. They recognise cells that have been infected by viruses via the altered MHC, which now contains part of the virus protein. They rapidly kill the infected cells by releasing cytotoxic proteins onto the cells, which respond by apoptosis.

Mast cells

These cells develop from the monocytes formed in the bone marrow and deal with the microbes too large to be dealt with by phagocytosis by secreting enzymes from their granules onto the pathogens. These enzymes degrade the microbes and promote inflammation by releasing histamine, thus increasing blood flow to the area.

Eosinophils

These are present in the smallest amounts in the blood stream but play an important role in defence against parasites, allergic responses and tissue inflammation.

Complement system

The complement system is a second major system within innate immunity for dealing with infections and reducing the risk of infection overwhelming the host. It consists of a large number (20+) of plasma proteins, most of which are formed in the liver and function as opsonins to facilitate phagocytosis of bacteria through opsonisation. This is defined as 'the process by which bacteria are altered by opsonins so as to become more readily and more efficiently engulfed by phagocytes' (*The American Heritage Medical Dictionary*, 2012).

GO DEEPER

Complement system pathways

As plasma proteins, these are widely distributed around the body but need to be activated when infective agents enter the body. There are three pathways (classical, lectin and alternative) by which the complement system is activated (Figure 13.6). Different molecules initiate the pathways, but they all result in the same set of actions:

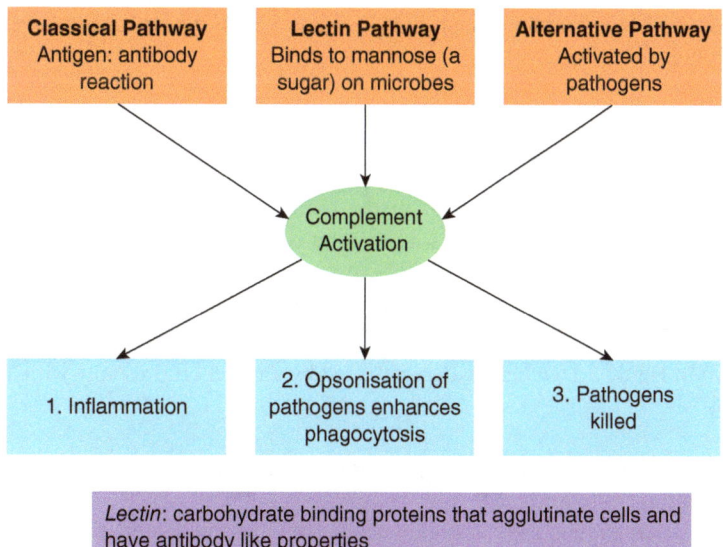

Figure 13.6 Activation of the complement system

(Continued)

1. Complement proteins are activated that then bind to the pathogens, making them ready for phagocytes (carrying the appropriate complement receptors) to engulf them.
2. Fragments of some of the complement proteins attract additional phagocytes to where complement activation has already occurred and activate these phagocytes.
3. The final components of complement proteins also create pores (holes) in the bacterial membrane, damaging and killing them.

Interferons (proteins)

These are proteins secreted by cells of the innate immune system, which can reduce replication of viruses inside the cells of the body. Different interferons have varying roles. Interferons α (alpha) and β (beta) act against replication of viruses within cells by stimulating the production of antiviral proteins in normal cells. Interferon γ (gamma) has little effect against viruses but plays an important role in communication between certain cells of the adaptive system and the macrophages of the innate immune system (Helbert, 2006).

ADAPTIVE IMMUNITY

We have already considered some of the processes involved in innate immunity. However, adaptive and innate immunity interact and the cells involved in both are formed similarly.

Cells of the adaptive immune system

The major pathways in the formation of blood cells emanate from myeloid stem cells (Chapter 2, Figure 2.16 and Chapter 12). The major cells in the adaptive immune system are the lymphocytes of which there are two main types – the B- and T-lymphocytes (Chapter 12). The multipotent stem cells in the bone marrow are the precursors of both types of lymphocytes.

B-lymphocytes

The precursors of the B-lymphocytes remain in the bone marrow and rearrangement of the chains forming the immunoglobulins occurs. If this results in a cell which recognises 'self', the cell normally does not continue its development and undergoes apoptosis while the remaining cells continue their development. Each cell has developed the capacity to respond to a single antigen and matures within the bone marrow. They move to the peripheral lymphoid organs (which include the lymph nodes and the spleen) shown in Figure 2.16 where, when activated by antigen (and usually with T-cell assistance), they develop into:

- **Plasma cells:** that form antibodies as their unique structure determines which are released into the blood stream to react with antigens.
- **Memory cells:** that express their antibody on the cell membrane and are long-lived. When they come into contact with the relevant antigen they divide rapidly and produce antibodies.

B-lymphocytes produce antibodies released into the circulation in what is known as humoral immunity.

T-lymphocytes

The precursors for the T-lymphocytes move from the bone marrow to the thymus gland where they are activated and selected for recognition of non-self-antigens only. They differentiate into different types of T-cell (Murphy, 2012):

- **T-helper cells:** provide essential signals (interleukins) that affect behaviour and activity of other cells, particularly B-lymphocytes and macrophages that enhance their efficiency.
- **T-cytotoxic cells:** directly kill cells infected with viruses or other intracellular pathogens in what is known as cell-mediated immunity.
- **T-regulatory cells:** suppress other lymphocytes' activity and control immune responses.
- **T-memory cells:** these include T-helper and T-cytotoxic cells. These cells do not actively participate in the initial immune response but if they encounter an antigen on a subsequent occasion, they initiate a rapid immune response.

Antibodies

Antibodies (otherwise known as immunoglobulins – Igs) are glycoproteins produced by the lymphocytes, which combine with antigens. They specifically recognise and bind to particular pathogens and aid in their destruction. They are highly specific, that is they bind only with the antigen where they match the binding site, and they combine with varying degrees of affinity (or tightness of binding). The tighter the affinity, the less likely it is that the antigen and antibody will break apart.

The key element about antibody production is the very considerable diversity of antibodies required to match the exceedingly large range of antigens with which the body comes in contact. They are produced from DNA, which has a range of duplicated segments for the different components of antigens (see below). Variation in the antibodies is produced by:

- Random selection from the sets of duplicated segments.
- Variable joining of the segments.
- Mutations that may occur in the different chains.
- Random selection and pairing of the different chains in the antibody.

After this, the particular cell will continue to produce the specific antigen.

Some cells will express antibodies to self-antigens. While in an immature state these will undergo apoptosis. Those with antibodies for non-self-antigens will remain in the body and when they meet their antigen will be stimulated to produce antibodies.

There are five types of antibodies secreted by different cells and with different functions and sites of activity.

ACTIVITY 13.1: UNDERSTAND

Watch the following video to see antibodies explained in more detail.

This external video link can be accessed by **scanning the QR code** with your smart phone or via https://study.sagepub.com/essentialap2e.

ANTIBODIES (7:06)

GO DEEPER

Immunoglobulins

IgG (Imunoglobulin G)

This is found in blood vessels and extracellular fluid and, as the most common type of antibody in the blood, deals with most blood-borne microbes. It activates the complement system by binding to macrophages, thereby enhancing phagocytosis. It also binds to T-cytotoxic cells, destroying infected cells and thrombocytes, aiding the inflammatory process. It is able to cross the placenta and provide immunity to the fetus which is carried into post-delivery life. It is also the main Ig in the secondary immune response (see Figure 13.11, later).

The structure of IgG (Figure 13.7) is the basis for the formation of the other immunoglobulins and consists of two light and two heavy chains. One light and one heavy chain join and the two heavy chains are joined by a disulphide bond (a link involving two sulphur molecules). There are five types of heavy chain but only one occurs in each type of antibody. IgG contains the γ (gamma) heavy chain.

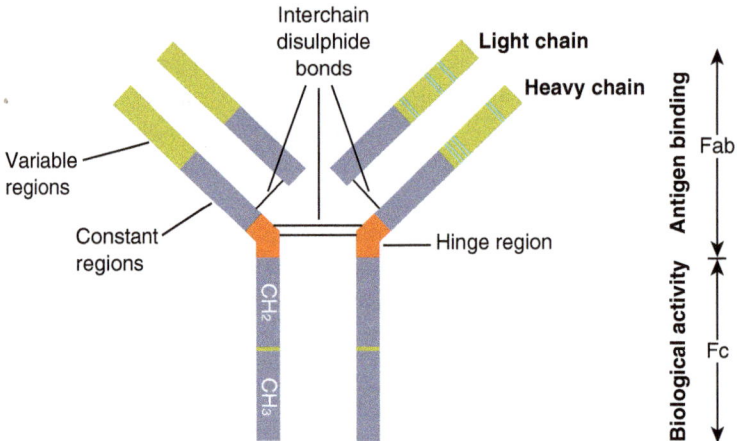

Figure 13.7   **IgG: basic four-chain structure of immunoglobulins**

Adapted from: Section D1, Figure 1, BIOS *Instant Notes: Immunology* Third edition, Peter Lydyard, Alex Whelan, Michael Fanger, Copyright (© 2011), Garland, Taylor & Francis Group. Reproduced by permission of Taylor & Francis Books UK.

IgM

IgM occurs on the surface of B-cells where, as the four-chain complex similar to IgG, it acts as an antigen receptor, agglutinating bacteria in the blood, and promoting lymphocyte activation and suppression. This complex contains the μ (mu) heavy chain. It also occurs in the blood stream as five of these four-chain units joined by disulphide bonds (Figure 13.8). The size of IgM results in this molecule remaining mainly in the blood stream and it is the main immunoglobulin secreted in the immune response. The number of sites that can bind with antigens results in a high level of effectiveness in disposing of microbes.

IgA

IgA, similar in function to IgG, is the main immunoglobulin in secretions both from epithelium lining the respiratory and gastrointestinal tracts, and external secretions (e.g. breast milk, saliva). It contains the heavy α (alpha) chain and exists in two forms (Figure 13.8):

- **Secretory:** the IgA formed in the plasma cells in the Peyer's patches and lymphoid tissue of the epithelium are in the form of dimers, that is two of the four-chain complexes are joined. IgA secretion from these cells, largely in response to the resident microbes in the gut (see Chapter 4), is about 3-4 grams per day, exceeding in quantity all other Ig types (Murphy, 2012). These are the first line of defence against microbes in the epithelial tissues.

- **Circulatory:** in the blood IgA is in the form of a single four-chain molecule produced by cells in the bone marrow. It is present in much smaller quantities.

IgD

This immunoglobulin is present in small amounts in adults and contains the heavy δ (delta) chain (Figure 13.8). It acts as an antigen receptor on B-lymphocytes.

IgE

This is normally present in small quantities in the blood and contains the ε (epsilon) heavy chain. However, it plays an important role in inflammation, worm infection and allergies. In those susceptible, the antigen first primes the IgE-producing mast cells. With future exposures, the antigen stimulates the mast cells to release agents such as histamine which stimulate an allergic response.

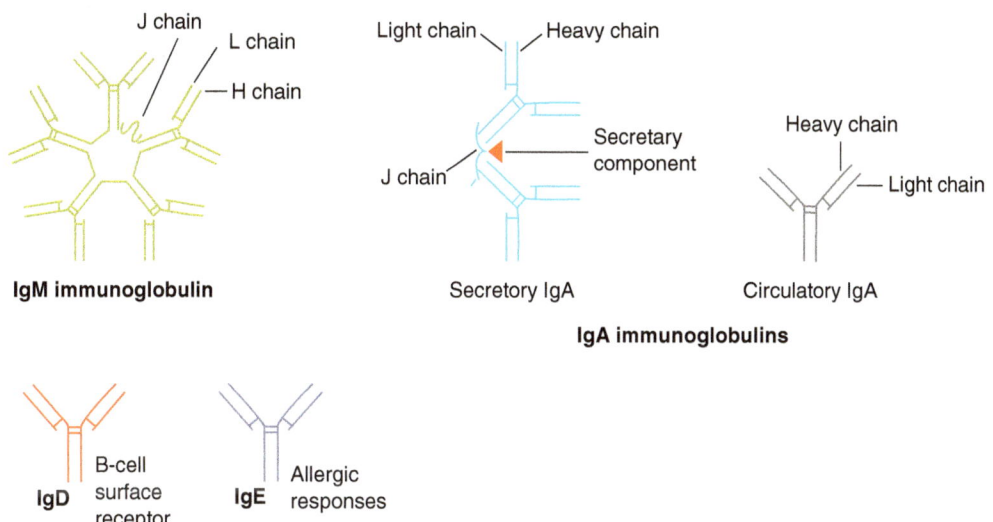

Figure 13.8 Immunoglobulin chain structures: IgM, IgD, IgE, IgA

Adapted from: Section D2, Figure 1, BIOS *Instant Notes: Immunology* Third edition, Peter Lydyard, Alex Whelan, Michael Fanger, Copyright (© 2011), Garland, Taylor & Francis Group. Reproduced by permission of Taylor & Francis Books UK.

APPLY

Immunoglobulins and immunity

Danielle will have received IgM and IgA in the colostrum (the first part of the breast milk to be produced) and the breast milk providing her with temporary immunity to infectious diseases that her mother has had in the past. In addition, Danielle will have received IgG from her mother. IgG easily crosses the

(Continued)

BODIE FAMILY

placenta and has a prominent role in neutralising pathogens and triggering secretion of complement to promote phagocytosis (Waterhouse, 2011). This enhances Danielle's immunity in the first few months of her life.

Intravenous immunoglobulin therapy is used with people who are immunosuppressed or have an autoimmune disorder. These therapies are primarily composed of serum IgG fraction from several thousands of donors, thus it is an expensive blood product (Schwab and Nimmerjahn, 2013). One example of its use is in slowing the progress of multiple sclerosis; intravenous immunoglobulin improves Expanded Disability Status Scale (EDSS) scores and reduces the annual relapse rate (Niimi et al., 2011).

ACTIVITY 13.2: UNDERSTAND

Watch the following video clip to learn about adaptive immunity.

This external video can be accessed by **scanning the QR code** with your smart phone or via https://study.sagepub.com/essentialap2e.

ADAPTIVE IMMUNITY (10:53)

DEVELOPING IMMUNITY

Immunity is the state when an individual has resistance to infection or disease, developed as discussed above through the innate and/or adaptive immune mechanisms. It can be natural or artificial in development and active or passive in nature as shown in Figure 13.9.

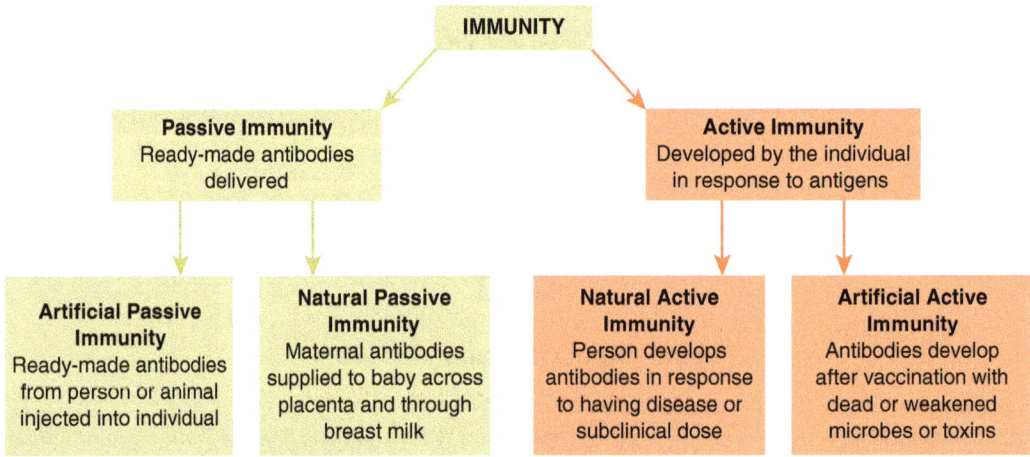

Figure 13.9   Types of immunity

APPLY

Artificial passive immunity

Artificial passive immunity was used in the Ebola outbreak in Africa in 2014. A British nurse who recovered from this frequently fatal disease after being treated with an experimental medicine travelled to America to provide natural antibodies to an American victim of the disease. Initial treatment using these experimental drugs began in 2014, with work on non-human primates demonstrating positive results (Geisbert, 2014). WHO produced interim guidelines for this approach to treatment at that time (WHO, 2014). Whilst some antivirals and antibody therapies have shown some positive effect, trials continue to provide a more definitive picture (CDC, 2019).

Response to antigen

Figure 13.3 showed the three stages in development of immunity involving both innate and adaptive immunity. Here we are looking at the way in which the adaptive immunity components respond to exposure to an antigen. Figure 13.10 shows the four stages of the primary antibody response:

- **Lag:** period between exposure and onset of symptoms. The cells involved are beginning to divide and differentiate.
- **Log phase:** the antibody concentration increases exponentially as cells stimulated by antigen differentiate into antibody-secreting plasma cells.
- **Plateau:** antibody synthesis and decay are balanced, levels are constant.
- **Decline:** antibody decay is greater than formation – levels fall.

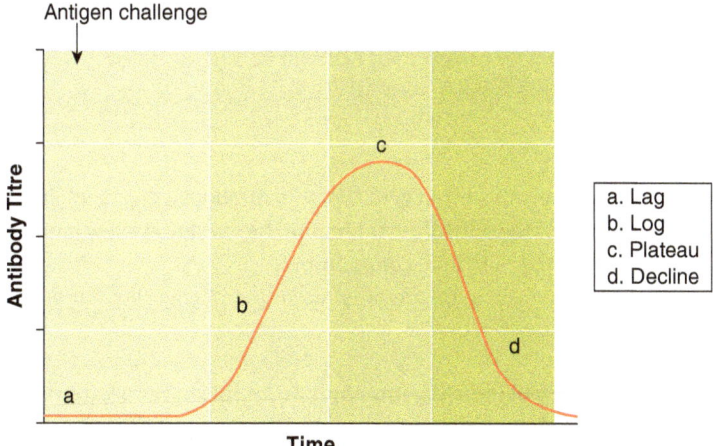

Figure 13.10 The phases of primary antibody response

Very different responses occur with the first and subsequent exposures to an antigen (Figure 13.11) including the following:

- **Timing:** the length of the lag phase is much shorter in the secondary response, while the plateau and decline stages are much longer.

- **Antibody levels:** the levels of antibody are much higher in the plateau stage of the secondary response (note: the scale in Figure 13.11 is logarithmic, i.e. each increment increases by a factor of ten).
- **Antibody type:** in the primary response IgM is most important while in the secondary response IgG is very much more important than IgM.
- **Antibody affinity:** the strength with which the antibody binds to the antigen is usually much greater in the secondary response.

Figure 13.11 clearly shows that the initial exposure to the antigen primes the immune system for a significantly greater response in a subsequent exposure to the same antigen.

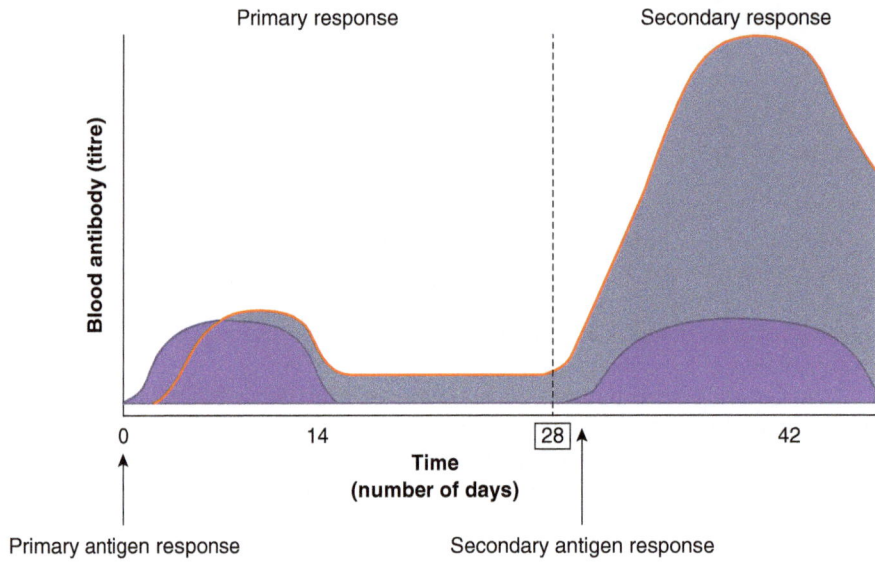

Figure 13.11 Primary and secondary antibody responses

Immunisation

This is an important aspect of public health in general use in developed countries. It is being encouraged in developing countries but one of the difficulties is keeping the vaccine at a low temperature to maintain its viability while being transported to remote communities.

The words immunisation and vaccination are often used interchangeably but, while related, they do not mean exactly the same:

- **Immunisation:** the process of inducing immunity to an infectious organism or agent in an individual (or animal) through vaccination.
- **Vaccination:** the administration of a vaccine which is a biological preparation that improves immunity to a particular disease.

_____ **APPLY** _____

Immunity through vaccination

Danielle (two months) will be starting her vaccination schedule which will continue throughout her life. During her infancy she will receive a number of vaccines including: 5-in-1 (DTaP/IPV/Hib - diphtheria, tetanus, whooping cough, polio and *Haemophilus influenza* type b); pneumococcal conjugate vaccine (PCV); rotavirus; meningitis C; and measles, mumps and rubella (MMR).

George and Maud are receiving the flu vaccine in the Autumn (Fall) of each year to reduce the risk of the disease, and the complications that are more likely in older people. They will probably have been administered the vaccine against pneumonia (pneumococcal polysaccharide vaccine (PPV)) and shingles (herpes zoster vaccine).

Several members of the Bodie family travel abroad and may be exposed to a number of other diseases such as typhoid, cholera or yellow fever. When appropriate, they will take advantage of the vaccinations available through travel clinics or their own general practitioner.

BODIE FAMILY

Vaccination programmes

The vaccination programme within the UK aims to provide protection through promoting immunity to a number of conditions. As well as protecting each individual, if a significant proportion of the population are immunised, the spread of the condition is minimised by breaking the chain of infection thus achieving community or 'herd' immunity. Table 13.1 shows the current UK recommended timetable for protecting the public through immunisation. In addition to the standard list there are additional immunisations available for those at risk.

Vaccinations against measles and tuberculosis (BCG – Bacillus Calmette-Guérin vaccine) both appear to have beneficial effects on child mortality, in addition to the reduction in the specific diseases protected against (Kristensen et al., 2000). However, further research is needed. Occasionally, papers are published with inaccurate results and conclusions: an example is the MMR (combined measles, mumps and rubella) vaccine, which was linked with autism in Wakefield (1999). The paper was disputed by much scientific peer-review and *The Lancet* finally retracted the publication in 2010 (Novella, 2010). In 2019, the World Health Organization highlighted the serious issue of misinformation around vaccinations and the subsequent increase in vaccine hesitancy (Mahase, 2019). The associated reduced uptake has particularly affected more affluent countries. For example, in 2018, there was reduced vaccine confidence in Czech Republic, Finland, Poland and Sweden compared to in 2015 (Larson et al., 2018). In contrast, however, it is *access* to vaccines which is a greater issue globally, particularly in developing countries. For example, in 2018 there were 10 million reported cases of measles and 142,000 deaths related to the disease (WHO, 2019a). In December 2019, this had increased three-fold, with over 30 million cases reported. This increase is attributed, at least in part, to the anti-vax misinformation movement seen globally, a movement that is without evidence to support its agenda. Indeed, some countries that had almost eliminated the disease have now seen a resurgence (WHO, 2019b).

Table 13.1 Vaccination schedule (based on NHS, 2019)

Diseases against which vaccinated	Age of administration
6-in-1 vaccine • diphtheria • hepatitis B • Hib (Haemophilus influenzae type b) • polio • tetanus • whooping cough (pertussis)	Three doses: 8, 12 and 16 weeks old
4-in-1 pre-school booster (diphtheria, tetanus, whooping cough and polio)	3 years and 4 months old
3-in-1 teenage booster (tetanus, diphtheria and polio)	All young people aged 14
Tetanus, diphtheria and polio	13-18 years (around 14 years)

(Continued)

Table 13.1 (Continued)

Diseases against which vaccinated	Age of administration
Pneumococcal disease: Pneumococcal conjugate vaccine (PCV) Pneumococcal polysaccharide vaccine (PPV)	12 weeks and repeated at 1 year 65 years and over
Rotavirus	8 weeks and repeated at 12 weeks
Hib/MenC (Haemophilus influenzae type b (Hib) and meningitis C.)	Single injection given to 1-year-old babies
Meningococcal group B bacteria	8 weeks, 16 weeks and 1 year
MenACWY vaccine (meningococcal bacteria A, C, W and Y)	Children aged 13 to 15
Measles, mumps and rubella (German measles) (MMR)	First dose at 1 year Second dose at 3 years 4 months or soon after
Influenza (children's flu vaccination)	Children aged 2 and 3 (for children born between 1 September 2015 and 31 August 2017) All primary school children Children aged 2-17 with long-term health conditions
Influenza (Adult)	Adults 65 and over People with certain medical conditions Pregnant women Frontline health or social care workers
Human papillomavirus (HPV) (cervical cancer)	Girls and boys aged 12-13 years
Shingles	70 years
Immunisations for those at risk	
Hepatitis B	Four weeks and 12 months for babies born to mothers who have hepatitis B
Tuberculosis	At birth (for children born, or whose parents or grandparents were born, in areas where high numbers of TB cases exist)
Pneumococcal disease	2 until 65 years
Pertussis	From 16 weeks of pregnancy up to 32 weeks

DISORDERS OF THE IMMUNE SYSTEM

Disturbances in the function of the immune system fall into three broad groups, which are introduced here (Figure 13.12).

Hypersensitivity

A number of people suffer from hypersensitivity conditions that often run in families with differing presentations ranging from relatively minor, such as hay fever, to extremely serious if not lethal, such as anaphylactic shock. Atopy is the term for being at a high risk of allergies with raised levels of IgE to specific allergens.

Some individuals move through several presentations of atopy during life. For example, a child may have eczema, followed by hay fever, which may develop in the teens, followed by asthma in middle age.

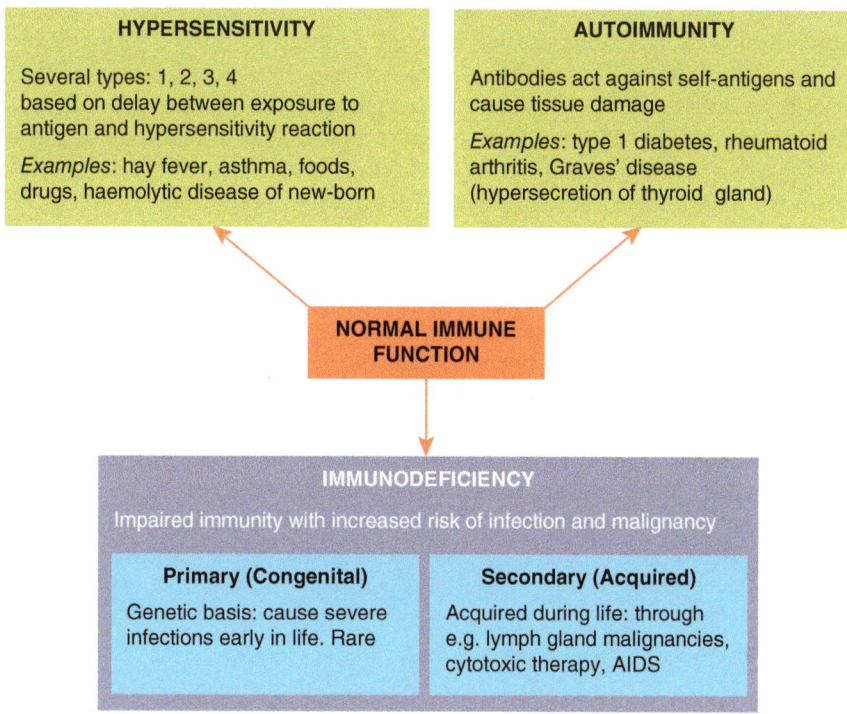

Figure 13.12 Disturbances of immune function

Autoimmunity

The immune system is normally tolerant of self but in autoimmune disorders, tolerance to self-antigens breaks down and components of the immune system recognise and react against components of the body itself. This is due to genetic factors sometimes combined with environmental factors such as infection which break down tolerance.

Reduced immunity (immunodeficiency)

As well as the disturbances given in Figure 13.12, those with these conditions are liable to opportunist infections (Chapter 4), the type of which may give information about the level of immunodeficiency. Malignancies, which may develop are often due to infection, for example, the Epstein–Barr virus can give rise to lymphoma. These types of disorder are discussed in greater detail in books on pathophysiology.

GO DEEPER

HLA linked disorders

Type 1 diabetes is a disorder of glucose regulation characterised by high levels of blood glucose as a result of inadequate or no insulin availability. It is an HLA linked condition, with the risk of developing the condition increased by the HLA-DR3/4 DQ8 genotype in particular (Smyth and Smyth, 2014). This means that while type 1 diabetes is not directly inherited, people may inherit a predisposition to the disorder if they have that specific genotype. This is because HLA-DR3/4 DQ8 appears to increase the risk of an inappropriate immune response to beta cells in the pancreas that are responsible for producing insulin. Other HLA variants are also implicated. Having HLA variants associated with type 1 diabetes does not mean a person will develop it. Indeed, many of the HLA variants are common. Other factors are thought to influence the process such as an environmental triggering event (e.g. mumps, rubella, cytomegalovirus) and/or a T-lymphocyte-mediated hypersensitivity reaction against some beta cell antigen.

An example of a rare autoimmune condition thought to result from a combination of genetic and environmental conditions is narcolepsy – a neurodegenerative disorder resulting in excessive daytime sleepiness, cataplexy (a sudden short episode of muscle weakness), a range of sleep disturbances, and metabolic disturbances. It is caused by the loss of hypocretin (orexin) (a neurotransmitter that regulates arousal, wakefulness and appetite) producing neurons in the lateral hypothalamus (Mahlios et al., 2013). The destruction of these neurons is correlated with HLA (chromosome 6) and an associated autoimmune response thought to be linked with viral infections, such as influenza A(H1N1) infection and vaccine. These can trigger molecular mimicry – when foreign peptides cross-activate self-peptides resulting in those self-peptides being identified as non-self (Kornum et al., 2011; Mahlios et al., 2013).

CONCLUSION

In this chapter we have examined how the two sections of the immune system, the innate and the adaptive systems, are structured and work together to provide protection against infections and malignant cells. We have also reviewed how certain other systems of the body and the microbiome contribute to immunity. In relation to public health, recognising the importance of immunisation will assist you in promoting health in children and adults. Your understanding of the immune system and the introduction to the types of disorders of this system will help you to be a person-centred nurse.

GO DEEPER

Further reading

Hendry, C., Farley, A., McLafferty, E. and Johnstone, C. (2013) 'Function of the immune system', *Nursing Standard*, 27(19): 35–42.

Nigam, Y. and Knight, J. (2008) 'Exploring the anatomy and physiology of ageing. Part 9 – The immune system', *Nursing Times*, 104(47): 58–9.

Weston, D. (2010) 'The pathogenesis of infection and immune response', *British Journal of Nursing*, 19(Supp. 6): S4–11.

An effective immune system is essential for a healthy life and understanding how it functions will enable you to help people in your care. The immune system is not simple and will take some effort to understand. The key points in this chapter you need to learn about are:

- The different components, including the types of cell, comprising the innate immune system and the adaptive immune system, and how they interact in carrying out their function.

- Understanding of the primary and secondary immune response is important when working with people with infectious diseases.

- Recognising how the interaction between the immune system, the nervous and endocrine systems, and the gut microbiome influence the function of the endocrine system.

- Understanding how the immune system is manipulated to aid health through immunisation and the importance of herd immunity in public health.

- The ways in which inappropriate functioning of the immune system, that is hypersensitivity, autoimmunity and reduced immunity (immunodeficiency) can interrupt the normal functioning of the body.

REVISE

This chapter helps you to understand the functions of the immune system. Revise each section individually and check your understanding of each section before you move to the next. Do you understand the types of immunity? Can you identify the specialised cells involved? Challenge your knowledge.

TEST YOUR KNOWLEDGE

In order to help you revise, consider the following questions, answers for which can be found by visiting **https://study.sagepub.com/essentialap2e**.

Test yourself by revising the chapter first, and then answer these questions without looking at the book. Afterwards compare your answers with the text and with the notes you made. Did you miss anything in your notes? Here are the questions:

1 What are the main types and functions of white blood cells?

2 Describe what is meant by innate (non-specific defences) and give examples.

3 A person in your care has an IV cannula in their right arm. When doing your observations you notice that their arm is inflamed. Explain the signs and symptoms you would expect to see, giving a rationale for each.

4 What is acquired immunity?

5 Distinguish between active and passive immunity.

6 John is a 78-year-old gentleman with a history of chronic obstructive pulmonary disease (COPD): he has been advised by his GP to receive the 'flu jab'. Explain how the flu vaccination will protect John from acquiring the flu.

REFERENCES

Centres for Disease Control and Prevention (2019) *Ebola* (Ebola Virus Disease). Atlanta@ CDC. Available at: www.cdc.gov/vhf/ebola/treatment/index.html

Geisbert, T.W. (2014) 'Medical research: Ebola therapy protects severely ill monkeys', *Nature*, doi:10.1038/nature1374 (published online 29 August 2014).

Helbert, M. (2006) *Flesh and Bones of Immunology*. London: Mosby (Elsevier).

Kornum, B.R., Faraco, J. and Mignot, E. (2011) 'Narcolepsy with hypocretin/orexin deficiency, infections and autoimmunity of the brain', *Current Opinion in Neurobiology*, 21(6): 897–903.

Kristensen, I., Aaby, P. and Jensen, H. (2000) 'Routine vaccinations and child survival: follow-up study in Guinea-Bissau, West Africa', *British Medical Journal*, 321: 1435–50.

Larson, H., de Figueiredo, A., Karafillakis, E. and Rawal, M. (2018) *State of vaccine confidence in the EU 2018*. Luxembourg: European Commission.

Mahase, E. (2019) 'Vaccination uptake: access is still biggest barrier, experts warn', *British Medical Journal*, 366: l5576.

Mahlios, J., De la Herran-Arita, A.K. and Mignot, E. (2013) 'The autoimmune basis of narcolepsy', *Current Opinion in Neurobiology*, 23(5): 767–73.

Male, J., Brostoff, J., Roth, D. and Roitt, D. (2012) *Immunology*, 8th edn. Philadelphia: Elsevier Saunders.

Murphy, K. (2012) *Janeway's Immunobiology*, 8th edn. New York: Garland Science (Taylor & Francis).

Navarrete, C.V. (2000) 'The HLA system in blood transfusion', *Best Practice & Research: Clinical Haematology*, 13(4): 511–32.

NHS (2019) *NHS Vaccinations and When to Have Them*. Available at: www.nhs.uk/conditions/vaccinations/nhs-vaccinations-and-when-to-have-them/ (accessed 5 May 2020).

Niimi, N., Kohyama, K., Kamei, S. and Matsumoto, Y. (2011) 'Intravenous immunoglobulin therapy prevents development of autoimmune encephalomyelitis and suppresses activation of matrix metalloproteinases', *Neuropathology*, 31(4): 392–400.

Novella, S. (2010) '*The Lancet* retracts Andrew Wakefield's article', *Science-Based Medicine*, 3 February. Available at: www.sciencebasedmedicine.org/lancet-retracts-wakefield-article/ (accessed 5 May 2020).

Pocock, G. and Richards, C.D. (2009) *The Human Body: An Introduction for the Biomedical and Health Sciences*. Oxford: Oxford University Press.

Schwab, I. and Nimmerjahn, F. (2013) 'Intravenous immunoglobulin therapy: how does IgG modulate the immune system?', *Nature Reviews Immunology*, 13(3): 176–89.

Smyth, D. and Smyth, T. (2014) 'Diabetes and coeliac disease: implementing a gluten-free diet', *Practice Nursing*, 25(6): 292–7.

The American Heritage Medical Dictionary (2012) New York: Houghton Mifflin Company. Available at: http://medical-dictionary.thefreedictionary.com/opsonization (accessed 5 May 2020).

Wakefield, A.J. (1999) 'MMR vaccination and autism', *The Lancet*, 354(9182): 949–50.

Waterhouse, C. (2011) 'Promoting the safe administration of immunoglobulins in neuroscience', *British Journal of Neuroscience Nursing*, 7(3): 529–35.

WHO (World Health Organization) (2014) *Use of Convalescent Whole Blood or Plasma Collected from Patients Recovered from Ebola Virus Disease for Transfusion, as an Empirical Treatment during Outbreaks. Interim Guidance for National Health Authorities and Blood Transfusion Services, Version 1.0*. Geneva: WHO.

WHO (2019a) *New Measles Surveillance Data for 2019*. Geneva: WHO.

WHO (2019b) *Ten Threats to Global Health in 2019*. Geneva: WHO.

SKIN AND TEMPERATURE REGULATION

14

UNDERSTAND: CHAPTER VIDEO

Before working through this chapter, you might find it useful to have an overview of the anatomy and physiology of skin. The following online video clip may enhance your understanding.

This video can be accessed by **scanning the QR code** with your smart phone camera or via https://study.sagepub.com/essentialap2e.

SKIN ANATOMY (12:49)

LEARNING OUTCOMES

When you have finished studying this chapter you will be able to:

1. Identify and explain the roles of the skin and its accessory structures
2. Differentiate between the different layers of the skin and their structures
3. Discuss the impact of ageing on skin
4. Outline the processes of tissue repair
5. Explain how temperature is regulated within the body
6. Explain the impact of hypo- and hyperthermia on homeostasis

INTRODUCTION

As you have learned in Chapter 13, the body has many protective mechanisms. The skin (or integumentary system) protects the body in numerous ways, contributing to homeostasis. The skin's functions in protecting against infection and fluid loss are vital for continued survival. In Chapter 4 we considered the normal flora of the skin, although this will be briefly referred to again here. In this chapter we will discuss the components and functions of the integumentary system and their contribution to homeostasis and protection.

Human cells function optimally at a specific body temperature, and the skin plays a major role in temperature regulation. This will be examined, along with implications for body function of hypo- and hyperthermia. Considering the pivotal role of the skin in maintaining health, changes across the lifespan may lead to psychological or physiological problems that may affect the person's quality of life. Nurses need to consider these factors in the provision of person-centred care and this chapter will focus on these aspects of practice.

You are now working your way through the human body and discovering how each system contributes to maintenance of homeostasis through various mechanisms. The skin is no different: it provides protection for the whole body, but is also central to experiencing life, communicating through touch, experiencing the general senses (Chapter 6); it creates our visual identity, and enables us to express ourselves. From the vulnerability of being born into the world through to our latter years, while skin starts out young and efficient it becomes less effective over time.

As a person-centred nurse, you will need to care for skin, you will use it for therapeutic touch, and it will be your ally in protecting the people you care for from pathogens. This chapter will provide you with the foundations of anatomical and physiological knowledge necessary to underpin and explore your practice.

But this chapter does not stop there. Thermoregulation is a key function of the skin and we will be exploring this in further depth. Metabolism and neurological function all need the body to be within certain temperature parameters to occur effectively. This chapter will enable you to consider your practices and identify risk proactively, underpinning your practice with anatomical and physiological understanding.

Context

Now let us examine the Bodie family. Baby Danielle, at two months, is particularly susceptible to heat loss and to sun damage to her young skin. Her skin is also supporting her developing immune system, keeping pathogens out and dealing with many of those that do get into her skin. Since she was born she has been developing her own skin flora. She is also developing her abilities to sense the environment with her general senses and to touch things, and learning about texture and to respond to the dangers around her. Her skin is one of her main protections. Additionally, as Danielle is developing her nervous system and growing at a rapid rate, temperature control is essential to her optimal development. She is also vulnerable to hypothermia as she has a limited ability to regulate her body temperature. She does, however, have brown adipose tissue to generate heat (Chapter 2).

This is all in contrast to Maud and George. Their skin has thinned, is more susceptible to damage, is less effective as protection and in regulating temperature. They are more vulnerable to cold exposure; their skin is damaged more easily and is slower to heal. However, it still enables them to express and sense the environment, they can still feel each other's embrace, feel their dog cuddle into them, and feel emotion when holding the hand of their great granddaughter. Their skin also tells a story of their lives, having changed shape and weathered due to their bodies' exposure to the external and internal environment. It is essentially a road map of their bodies' experiences of life. Their skin now needs to be treated with more care, to be well hydrated and washed with care. They need to be more attentive to the dangers

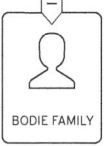
BODIE FAMILY

of injury and the risks of slow healing and of infection. Furthermore, George and Maud are at more risk of hypothermia and hyperthermia than the younger adult members of the family. Older people shiver less effectively to create heat, have less muscle mass to use as fuel for increasing metabolic activity, and their skin functions less well in regulating heat loss and retention. It is the younger adults who can demonstrate the most effective thermoregulation, but only if they are healthy.

The skin of the other adults is probably working optimally as long as they are caring for it and taking a healthy, balanced diet. The temptation in the youngish adults and those in their middle years of life is to excessively expose the skin to ultraviolet radiation from sunlight. This increases the risk of skin cancer, which is rapidly becoming more prevalent, particularly malignant melanoma. Skin cancer is the most common form of cancer for those in their mid-20s and the second most common form in the later teenage years. The choice of wearing sunscreen to protect against this has to be balanced against the risk of vitamin D deficiency from lack of sunlight exposure. Kwame, in particular, is at greater risk of vitamin D deficiency as his black skin, evolved to protect him from the strong sunlight of Africa, may admit insufficient light to stimulate enough vitamin D production in the greyer light of Northern Europe. In the middle years of life, vitamin D deficiency is associated with a number of genetic linked conditions in particular (Rosen, 2011), and so these choices influence the future health of the Bodie family members in these age groups. Margaret Jones, who lives in a location close to the Barrier Reef, is at risk of excessive exposure to the ultraviolet radiation. The remaining members of the Bodie family largely work indoors and in demanding professions, but may expose their skin to ultraviolet radiation damage during holidays. Their working conditions may increase their risk of vitamin D deficiency.

STRUCTURE OF THE SKIN

The skin is the largest organ in the body and has an approximate surface area of 1.8–2 m². It is a multifunctional organ that plays a pivotal role in maintaining homeostasis through various functions:

- Protecting the body from pathogens.
- Regulating body temperature.
- Experiencing the external environment through sensations.
- Absorbing and excreting substances.
- Formation of vitamin D.

First, we will look at the anatomy of the skin.

The skin has two primary layers, the epidermis and the dermis, both with sublayers (Figure 14.1). Some texts refer to a third layer, the hypodermis (thus the alternative name of hypodermic injections for subcutaneous ones), which contains subcutaneous tissue. The skin contains glands (see later), hair and nails. It also contains blood and lymph vessels and is richly supplied with sensory nerve fibres. We will now explore these layers and their accessory structures.

Epidermis

The epidermis is the thin, outermost layer of the skin which is constantly shedding millions of dead cells with new and maturing skin cells moving towards the surface. This process lasts around 30–35 days (Gould, 2012). This layer of the skin is squamous epithelium (Chapter 2) with a high amount of keratin, arranged in layers upon a basal membrane. This basal layer contains melanocytes, i.e. cells that produce melanin, which protect the skin from ultraviolet radiation. Keratin, a protein, and lipids in this layer result in the epidermis being waterproof and tough enough to prevent many pathogens invading and to

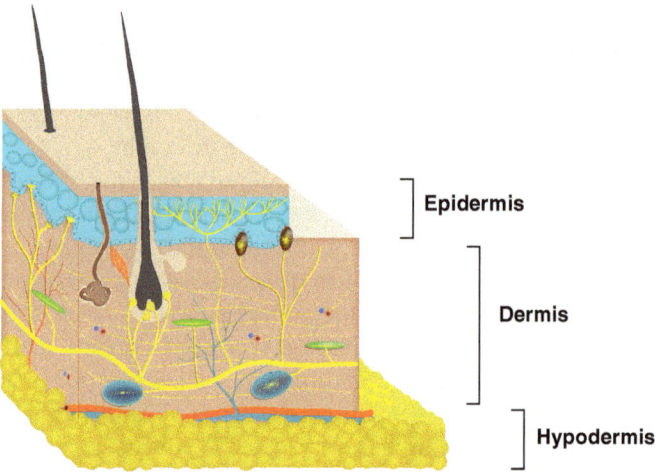

Figure 14.1   Layers of the skin

limit physical damage (Andrews, 2012). The thickness of the epidermis varies throughout the body, being thickest on the areas of the body that have the most contact with the external environment, i.e. the soles of the feet and palms of the hand. The epidermis is avascular, being nourished by blood and lymph supplies in the dermis. Other structures pass through the epidermis from the layers below, including hair and gland ducts. There is a limited supply of nerves to the epidermis, with most sensory fibres being in the dermis below. The outmost surface of the epidermis has fine ridges, or papillae, that help to provide traction and make up your fingerprint pattern.

The epidermis has a number of sublayers which are (from the inside out):

- **Stratum basale** (bottom germative basal membrane): this layer includes stem cells of the epidermis, which mature and differentiate as they move towards the surface.
- **Stratum spinosum:** these cells are 'spiny', which enables them to adhere together and provide structural support to the skin. As cells are pushed further up they produce more keratin which flattens the cells.
- **Stratum granulosum:** within this layer, keratinisation continues. Keratinocytes in this layer and the stratum spinosum also produce glycolipids that spread over the cell, keeping them waterproof. This glycolipid layer, along with the growing distance from a nutrient supply, result in the cells dying.
- **Stratum lucidum:** these cells have clear protoplasm (no organelles) and flattened or no nuclei with melanin to protect against UV light damage. This layer is not present in all areas, for example it is absent from the eyelids.
- **Stratum corneum:** the outermost layer, is heavily keratinised and waterproof. These cells are shed from the epidermis.

As the stratum basale cells divide, they push already formed cells into higher layers. As these cells move into higher layers of the epidermis, they flatten, form keratin, die and are replaced from below. This is the life cycle of the skin cell and is dependent on synchronisation of the following:

- Desquamation, or shedding of surface keratinised cells (the stratum corneum).
- Effective keratinisation of the cells approaching the surface.
- Continual cell division in the stratum basale with newly formed cells being pushed to the surface.

There are four main cell types found in the epidermis (Table 14.1).

Table 14.1 Cells of the epidermis

Cell type	Function
Keratinocytes	These cells produce keratin, a protein that gives skin its strength and flexibility and makes it waterproof
Melanocytes	These cells produce melanin, a pigment that gives skin its colour and also protects it from ultraviolet radiation damage
Merkel cells	These cells are still largely not fully understood but are thought to have a role in touch reception
Langerhans' cells	These cells are the first line of defence in the skin and support the immune system by processing antigens

The dermis

The dermis is more functional than the epidermis and is essential in providing it with nutrients. It is largely comprised of connective tissue with blood and lymph vessels and nerves. This connective tissue is rich in collagen and elastin making this layer structurally sound and elastic, enabling it to stretch and be flexible without being damaged. This thicker layer includes the accessory structures of the skin.

The dermis has two primary sublayers:

- **Papillary layer:** this is made up of fine and loosely arranged collagen fibres and interweaves with the epidermis, connecting it with the dermis.
- **Reticular layer:** this is made up of dense irregular connective tissue. The densely packed collagen fibres make this layer strong, giving the skin its overall strength and elasticity, as well as the structural support for the accessory structures of the skin.

There are four key cells with specific functions in the dermis (Table 14.2).

Table 14.2 Cells of the dermis

Cell type	Function
Fibroblasts	Synthesise the extracellular matrix. The extracellular matrix is a collection of substances outside of cells that provides structural and biochemical support to the surrounding cells. It therefore enables the cells to 'stick' together. These cells have a crucial role to play in wound healing as they produce collagen
Macrophages	These are a form of phagocytes which have a role in both non-specific (innate) and specific (adaptive) defence mechanisms in the body
Adipocytes	Fat cells that make up adipose tissue
Sensory receptors	There are three main functional groups of sensory receptors in the dermis: mechanoreceptors, nociceptors and thermoceptors. Mechanoreceptors respond to mechanical pressure, nociceptors are pain receptors and thermoceptors respond to temperature

The hypodermis

The hypodermis connects the skin to the rest of the body. It is largely composed of adipose tissue and areolar tissue (connective tissue around organs and blood vessels). When it is primarily adipose tissue, it is referred to as subcutaneous fat. The hypodermis is highly vascular and provides an energy store and insulation for the body. Infants and older people tend to have less adipose tissue.

Accessory structures

The skin has numerous accessory structures, largely within the dermis (Figure 14.2).

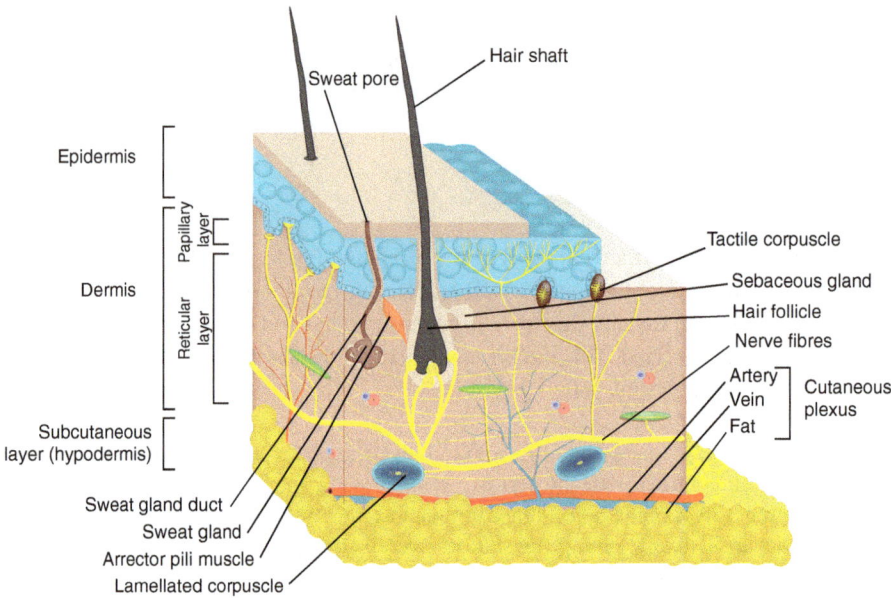

Figure 14.2   Accessory structures of the skin

Blood and lymph vessels

The dermis has a rich supply of blood vessels to provide nutrients to the skin and its accessory structures and to remove waste products. Additionally, blood carries heat and so the constriction and dilation of blood vessels are involved in the regulation of body temperature. Lymph vessels are present within the skin to collect extra fluid and return it to the blood circulation (Chapter 12), and also support immunity.

Sensory nerve endings

Dermal receptors monitor various sensations from the external environment, in particular touch, pressure, pain and temperature. This is central to protecting the body from harm and in experiencing pleasant sensations, which are relayed to the brain through the spinal cord (Chapters 5 and 6).

--- APPLY ---

Pain and burns

In Chapter 6 you will have learned about the different sensory receptors in the skin and how painful stimuli are perceived and processed. In people who have sustained burns, pain sensation at the site is dependent on the thickness of the burn. Burns that are superficial or are partial thickness are more likely to be painful as they result in the exposure of pain receptors in the skin (Butcher, 2011). These can be in large numbers. That pain sensation is exacerbated by any localised oedema and inflammation, which is a natural homeostatic response to trauma. It is commonly believed that full thickness burns will result in no pain for the person as the sensory receptors have been destroyed and there is superficial damage to the neurons. However, the edges of the burn wound are likely to be partial thickness and the surrounding tissue is likely to have a heightened response to pain. This requires pain management to be a consideration in caring for the person with such burns.

Sweat glands and ducts

There are two main types of sweat glands in the skin. These are called eccrine (or merocrine) glands and apocrine glands.

Eccrine (or merocrine) glands are widely distributed and the most numerous. They are abundant on the palms of the hand, soles of the feet and dermis of the forehead. Their ducts open onto the surface of the skin and they produce a clear, watery sweat controlled by the sympathetic nervous system in response to emotion (e.g. fear) or in an attempt to cool the body.

Apocrine glands are located primarily in the armpits and anogenital areas. Their ducts open into hair follicles rather than directly onto the surface of the skin. They become active at puberty and produce sweat similar to that from eccrine glands but also containing fats and proteins making the sweat thicker and cloudy in appearance. Bacteria on the skin digest this sweat, producing odour. They produce pheromones thought to have a role in sexual arousal. This sweat has a limited role in thermoregulation. Specialised apocrine glands produce ear wax (ceruminous glands) and milk (mammary glands).

Sebaceous glands

Sebaceous glands are widely distributed throughout the skin but are not found on the palms of the hands or soles of the feet. Their function is to secrete sebum, an oily substance that softens and lubricates the skin, prevents hair from becoming brittle, reduces water loss from skin and has antimicrobial properties. Sebaceous glands largely secrete sebum into hair follicles. However, some also secrete sebum directly onto the skin.

Hair

Hair is found on most surfaces of the body and its functions are to assist in thermoregulation and to provide mechanical protection against friction. Hair is also linked with arousal stimuli for pleasure. The skin contains hair follicles, within which a cluster of cells is called the bulb. Cell division occurs within the bulb, producing cells that are pushed up and away from the follicle. As they move away from the follicle, they lose their nutritional supply, become keratinised and die. These cells bond together forming hair, known as the root below the surface of the skin and the shaft above the skin. Attached to hair follicles are smooth muscle fibres known as arrector pili, which, on contraction, cause the hair to stand upright trapping air, and insulating the body from heat loss. This occurs in response to cold and fear and is regulated by the sympathetic nervous system.

Nails

Nails are essentially tightly packed, dead, hard keratinised epidermal cells which provide protection for the ends of fingers and toes and facilitate fine motor movements (picking things up, etc.). The nail has a root embedded within the skin and a nail plate. The plate has the nail bed below it, where the cells of the nail germinate. The pale, half-moon shaped area on the nail is the lunula, the visible part of the nail root.

FUNCTIONS OF THE SKIN

We now need to look at the role of the skin in maintaining homeostasis. Figure 14.3 highlights the key functions of the skin, and we now explore these in further detail.

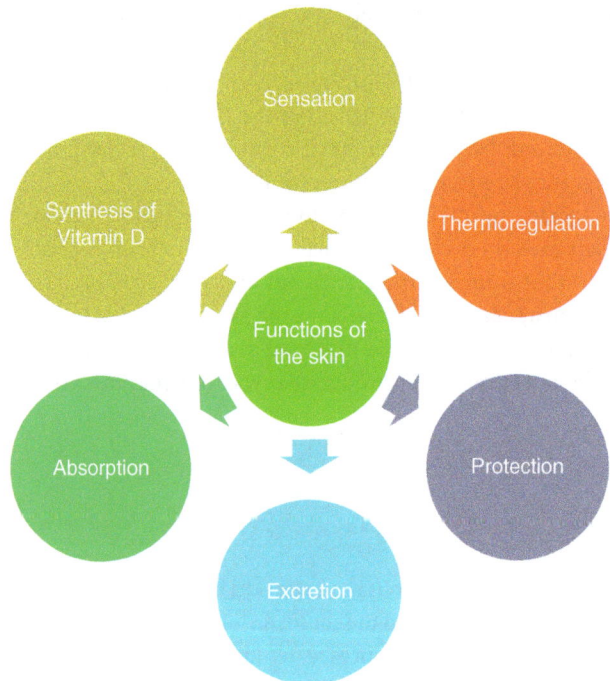

Figure 14.3   Functions of the skin

Sensation

We have already identified that the skin has sensory receptors that detect touch, pressure, pain and temperature. These sensations are important in perceiving the nature of the external environment so that the brain can process these and respond accordingly. This provides some protection, for example rapid removal of your hand from the sensation of excessive heat.

Protection

The skin provides both a mechanical and biochemical barrier to harmful substances. Underlying organs and structures are protected by the skin from physical damage, and the body is protected from fluid loss by the waterproofing by the keratin in cells and the production of oily sebum. The mechanical barrier also prevents pathogens from invading and, should some get past the mechanical barrier, macrophages are present to attack these. Additionally, sebum contains lysozyme, an enzyme that destroys bacteria on the skin. The melanin protects us from harm by UV radiation. Finally, the ability to shed skin cells is a stimulus for the migration of new cells ensuring that skin integrity is maintained. The skin heals after damage, restoring its protection-related functions.

Excretion and absorption

The skin excretes waste products through the production of sweat containing water, sodium, carbon dioxide, ammonia, urea and other aromatic substances. Excessive sweating (diaphoresis) under certain conditions can result in dehydration and hyponatraemia (low serum sodium levels). Someone with impaired renal function often excretes more urea through their sweat.

Some substances can be absorbed through the skin although this is limited by the avascular nature of the epidermis and the solidity of the keratinised cells. When a chemical substance comes into contact with the skin, it may be absorbed into the stratum corneum depending on the concentration or potency of the substance, duration of contact, solubility of the substance and integrity of the skin with which the substance is in contact (Robinson, 2014). While some substances may not get further than the epidermis, some may reach the skin capillaries and thus enter the systemic circulation, exposing the whole body to that environmental substance (Robinson, 2014). While this means that some nutritional substances can be absorbed, it also means that toxic substances can be absorbed, for example heavy metals like lead and mercury.

 APPLY

Topical medication

The ability to absorb medication through the skin is widely used in treating a variety of symptoms and illnesses, and also in health promotion strategies. An example of this is Hannah Jones, who smoked cigarettes in the early part of her career, partly as a strategy to relieve stress. However, she successfully used nicotine replacement patches (Nicotine Replacement Therapy, or NRT) as part of her strategy to give up smoking.

BODIE FAMILY

Synthesis of vitamin D

The skin plays a vital role in the synthesis of vitamin D under the influence of ultraviolet radiation from sunlight. However, it then undergoes further transformation into a more active form through the liver and kidney (Figure 14.4).

As this substance can be obtained from the diet, it meets the criteria for a vitamin. However, in reality it is also a hormone (chemical messenger secreted directly into the blood stream and carried around the body in the blood to attach to their specific receptor and influence the activity of the specific target cells, tissues, organs and systems of the body). While vitamin D has a well-established role in systemic homeostasis, it is also key in regulating the health of the skin itself through:

- Inhibition of cell proliferation.
- Stimulating cell differentiation including that necessary for the epidermis.

- Promoting innate immunity.
- Regulating the hair follicle cycle.
- Suppression of tumour formation (Bickle 2012).

Systemically, vitamin D has fundamental roles to play in (Rosen 2011):

- Mineralisation of bones through calcium absorption.
- Prevention of cardiovascular disease.
- Prevention of diabetes mellitus.
- Prevention of cancer.
- Prevention of immune dysfunction.

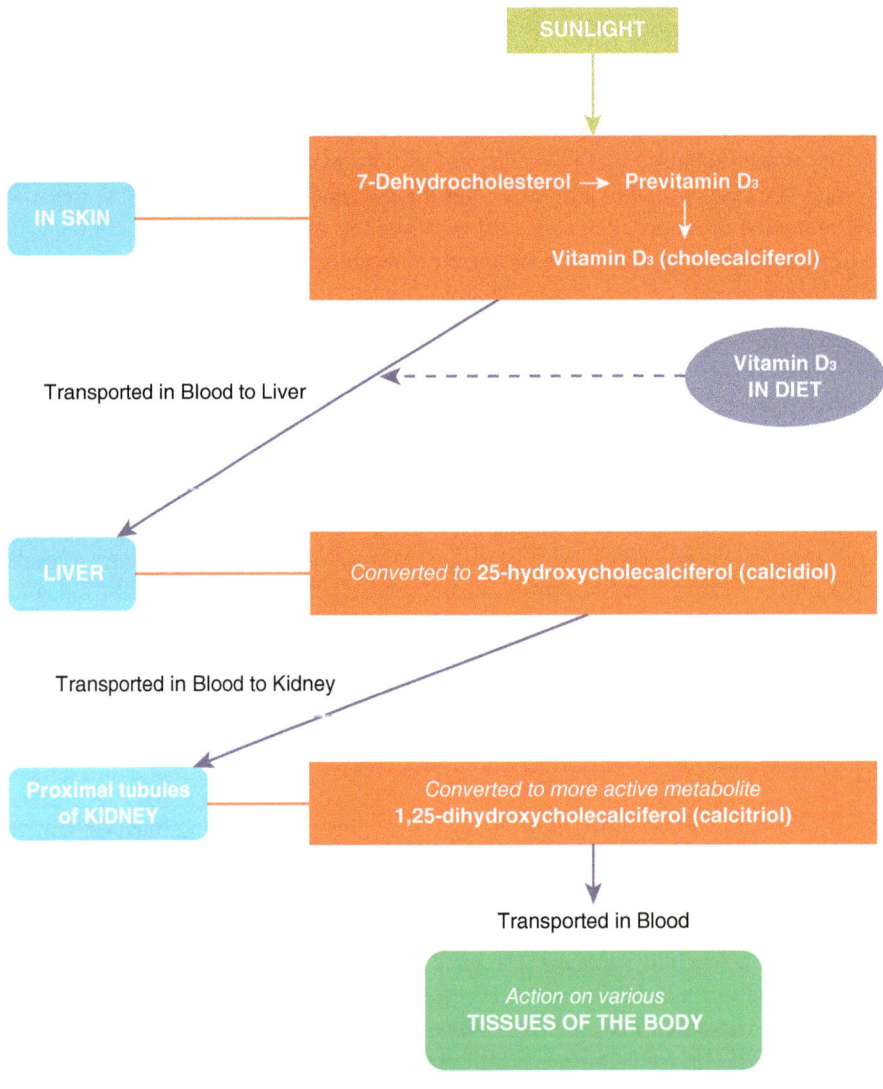

Figure 14.4 Synthesis and activation of vitamin D

Thermoregulation

Later in this chapter we will look at thermoregulation in detail. The skin facilitates the body to release or conserve heat through:

- Altering the diameter of the blood vessels (dilation or constriction) within the skin, thus altering the heat lost through radiation.
- Producing sweat to lose heat through evaporation.
- Using hair to increase or reduce air insulation on the surface of the skin.

THE EFFECTS OF AGEING ON SKIN

From development in the uterus to older age, we see distinct changes in skin appearance as it passes through rapid development and effectiveness in self-renewal to experiencing a continuing decline in cell renewal (Kottner et al., 2013). There are two categories of skin ageing: chronological ageing and photoageing (Dunn and Koo, 2013), both of which are cumulative. The skin, as an organ, is unique as photoageing occurs as a result of sun exposure and pigmentation (Dunn and Koo, 2013). In general, the rest of the body only declines in function as a result of the passing of time, chronological ageing.

Skin ageing can also be viewed as having intrinsic and extrinsic factors (Calleja-Agius et al., 2013). Intrinsic factors are those that are considered genetic. The chromosomes of DNA in cells have DNA sequences at the ends called telomeres. These are reduced in length at each cell division and are thought to govern the lifespan of the cell; they are maintained by an enzyme called telomerase. Skin cells lack this enzyme, resulting in more rapid shortening of telomeres and exposing the cell to oxidative damage. In addition, lower levels of growth hormone as a result of ageing also influence the thickness of the skin. Growth hormone influences the rate of cell proliferation and collagen production in the dermis. As a person ages growth hormone production decreases, reducing cell proliferation and collagen production and therefore decreasing the overall thickness of the skin (Póvoa and Martins Diniz, 2011).

Extrinsic factors are environmental. The biggest contributor here is ultraviolet light which triggers an inflammatory response that increases the production of the enzyme collagenase (Calleja-Agius et al., 2013). Collagenase breaks down the collagen that gives skin its strength and structure. This is why the skin sags and loses shape over time. This is further exaggerated in women as increased collagen damage occurs with post-menopausal oestrogen deficiency (Calleja-Agius et al., 2013). Repeated muscle movements also generate wrinkles in the skin. Sweat and sebum production is markedly reduced, inhibiting thermoregulatory and excretory responses. Kottner et al. (2013) highlight that the cells of the stratum corneum produce less glycolipids resulting in loss of natural moisturising factors that coat the cell membranes and assist in retention of cellular fluid and waterproofing. Calleja-Agius et al. (2013) highlight that the stratum corneum layer does not thin in ageing like the rest of the epidermis, it is just less effective. Another contributing factor to ageing of the skin is the reduction in the dermal–epidermal junction surface area, which increases the fragility of skin as the delivery of nutrients to the epidermis is less efficient.

In general, there is a decline in the thickness and composition of skin. Most cell types are found in lower numbers and microcirculation is reduced. This diminishes perfusion of the skin and impacts on fluid homeostasis resulting in impaired nutrition of the skin. The ability to support thermoregulation is lessened (Bentov and Reed, 2014).

As a result, the structural protection function of skin is compromised, alongside a less efficient ability to heal wounds and a reduced immune response. Older adults are therefore less able to respond to changes in body temperature, less able to keep their skin hydrated, and more likely to incur shear damage to the skin.

ACTIVITY 14.1: APPLY

Caring for skin in practice

A systematic review by Kottner et al. (2013) identified that there are some key choices to make when considering products that are used in caring for the skin. This review identified that:

- Skin dryness and protection can be improved by using products containing syndets (synthetic cleansing substances that surround dirt with small structures that are easily washed away by water) or amphoteric surfactants (additives that are mild, antibacterial and balance out other harsh additives) in comparison to standard soap and by using moisturisers containing humectants (substances that attract water into the epidermis from the dermis and the atmosphere, and also promote desquamation).
- Barrier products that contain occlusives (substances that physically block transepidermal water loss in the stratum corneum) are more effective than standard or no treatment in preventing skin injuries.

Syndets	Amphoteric surfactants	Humectants	Occlusives
• Sodium cocoyl isethionate (the most widely used) • Sulphosuccinates • Alpha olefin sulphonates • Alkyl glyceryl ether sulphonate • Sodium cocoyl monoglyceride sulphate • Betaines	• Coco betaine • Lauryl betaine • Hydroxysultaines	• Propylene glycol • Hyaluronic acid • Glycerin • Sorbitol • Hexylene • Butylene glycol	• Petrolatum • Lanolin • Mineral oil • Silicones (e.g. dimethicone)

Reflection

Consider the skin care you provide in practice.

- What products did you use and were you aware of their ingredients and their potential benefits and drawbacks?
- Did the products contain the beneficial constituents described by Kottner et al. (2013)? (The table above will help you identify these.)
- Do you have a choice in what skin care products you use in practice?
- How can a nurse advocate for a change in skin care products being used within a care setting?

HEALING OF SKIN WOUNDS

Skin is susceptible to damage and injury due to its exposure and contact with the external environment and also when there is a failure of homeostasis. White and Mantovani (2013) refer to injury to the skin as being a concept not limited to trauma, but also resulting from infection, metabolic disturbances, degenerative and ageing changes, and malignant changes. When skin is damaged, it repairs by regeneration and/or fibrosis. Some tissues have the ability to regenerate while others do not.

- **Regeneration** is the ability of the body to replace lost or damaged tissue with the same tissue. In other words, it is a like-for-like situation. Epithelial, bone, connective and smooth muscle tissues can regenerate. Cardiac, nervous and skeletal muscle tissues are limited in this ability.
- **Fibrosis** occurs when there is damaged or lost tissue that is not replaced with the same tissue, but is replaced by fibrotic, or scar, tissue. Scar tissue is made up of collagen fibres, which have a structural but not a functional role, i.e. they hold the tissues together but do not perform the original function.

Types of healing

Depending on the nature of the injury to the skin, wound healing occurs either by primary, secondary or tertiary healing/intention.

Primary healing/intention

Healing by primary intention is considered uncomplicated in that the wound edges are joined together and there is no tissue deficit to remedy. There is usually no infection in the wound and scar tissue formation is minimal. The union of the wound edges occurs in a timely manner though the actual time taken to heal depends on the person's overall health status. For example, in a malnourished person the time taken is extended. In general, the initial union of wound edges takes place over seven–ten days, though the maturation stage of healing can take up to two years.

Secondary healing/intention

Healing by secondary intention occurs when there is a tissue deficit between the wound edges. In this situation the body must fill in and contract the wound area in order to restore tissue integrity. This involves the production of granulation tissue (delicate, pink connective tissue and tiny blood vessels) before scar tissue is formed. Wounds that heal by second intention will always have more granulation tissue in relation to size than a wound healing by first intention. The healing time is longer because of the additional need for tissue growth.

Tertiary healing/intention

Some texts refer to a third type of healing – tertiary intention – which occurs when a wound is intentionally kept open to allow oedema or infection to resolve or to permit removal of exudate. An example is a large abscess which has been drained and packed. Tertiary healing is a form of delayed primary intention and results in more scarring than wounds that heal by primary intention but less than those that heal by secondary intention. The mechanism of healing is similar to that of secondary intention.

Wound healing in practice

Understanding these three types of healing is useful in everyday practice as it can guide wound management. For example, a person with a simple laceration or surgical wound may have this closed by suturing

(Continued)

(with suture material or paper wound closure strips), clipping, or gluing. This enables healing by primary intention.

Where there is a tissue deficit, e.g. a chronic venous leg ulcer, the wound has to be assisted to heal from the edges and the wound bed. Techniques such as Vacuum Assisted Closure (VAC) encourage circulation to the wound and remove excess debris and exudate. This promotes healing by secondary intention.

Sometimes a wound can be deep or tunnelled under the skin and contain excess fluid (exudate). It is important that the deeper tissues heal before the surface closes over, otherwise the wound can break down again or become infected. This is prevented in practice by packing or placing a drain in a wound to intentionally keep it open and remove fluid. This promotes healing by tertiary intention.

Stages/phases of wound healing

Whether a wound heals by primary, secondary or tertiary intention, it will go through a number of stages of healing. Different sources refer to differing numbers of stages of healing, but most agree that there are four key stages (Figure 14.5). They are:

1. Haemostasis.
2. Inflammation.
3. Proliferation (migration, granulation and proliferation).
4. Maturation (and remodelling).

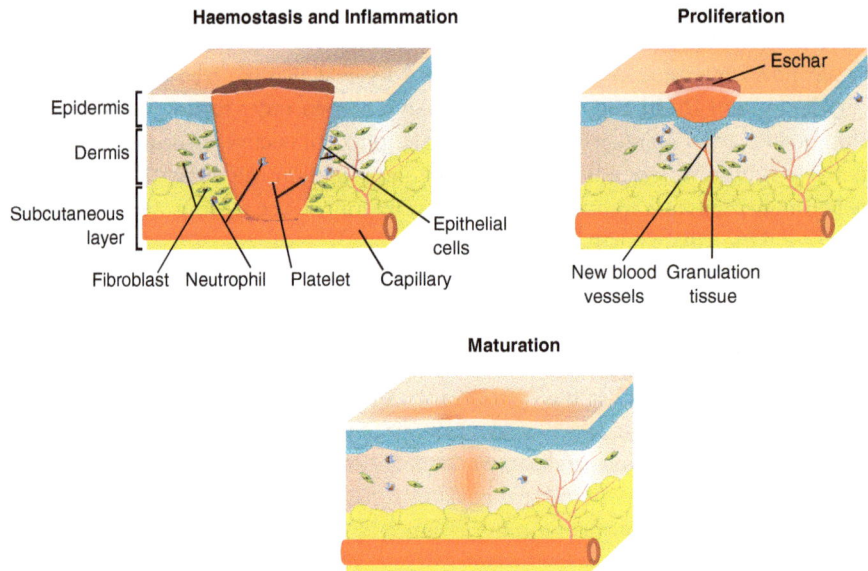

Figure 14.5   Stages of wound healing

Haemostasis

Haemostasis, the first stage of healing, refers to the stopping of haemorrhage at the site of injury by contraction of the smooth muscle in the arterial and arteriole walls, significantly reducing the blood flow. This allows thrombocytes (platelets) to aggregate at the injury site to form a fibrin clot which reduces active bleeding, achieving haemostasis (Chapter 12).

While vasoconstriction at the site of the injury may help with haemostasis, the process leads to a degree of localised tissue hypoxia and acidosis, triggering the release of nitric oxide, adenosine, histamine and other vasoactive substances that counteract this and cause vasodilation (Young and McNaught, 2011). The resulting increased permeability of the vascular membrane allows inflammatory cells access to the injured site, leading to the next stage of healing, inflammation.

Inflammation

The increased blood supply to the site of injury triggers an inflammatory response to prevent infection. White blood cells (neutrophils, macrophages and lymphocytes) are attracted to the site by interleukins and growth hormone. Bacteria and cell debris are phagocytosed and removed from the wound by neutrophils, which normally invade the areas within an hour of injury. Neutrophils release substances toxic to bacteria as well as trapping and destroying them. Neutrophils also release oxygen free radicals (atoms or groups of atoms with one or more unpaired electrons that are unstable and highly reactive) as a by-product, which are antibacterial and help to sterilise the wound by combining with chlorine (Young and McNaught, 2011). Macrophages are also attracted to the wound site and arrive after two to three days. Their growth factors trigger the generation of new blood vessels (angiogenesis) and tissue repair, particularly granulation tissue growth. Around the same time, lymphocytes appear and assist in the generation of new collagen fibres to restructure the damaged tissue. This process continues until all debris and bacteria have been removed.

APPLY

Temperature and wound inflammation

Interleukins (a type of cytokine – a protein released by lymphocytes) cause an increase in the thermoregulatory set-point in the hypothalamus. This is why people may have a raised body temperature when a wound is inflamed.

ACTIVITY 14.2: GO DEEPER

Did you know that oral mucosal wounds heal in an accelerated fashion and display minimal scar tissue formation? Research is now exploring how oral tissue repairs in order to understand whether a clearer understanding of this process could influence skin wound healing.

Go deeper by reading this article:

Glim, J.E., van Egmond, M., Niessen, F.B., Everts, V. and Beelen, R.H. (2013) 'Detrimental dermal wound healing: what can we learn from the oral mucosa?', *Wound Repair & Regeneration*, 21(5): 648-60.

Proliferation

Following the inflammatory phase, the proliferation phase usually begins after three days, taking up to three weeks or more. This phase involves angiogenesis and production of new structural tissue that begins to contract.

New blood vessels begin to form once there is a fibrin–platelet clot and lymphocytes arrive. These blood vessels are weak and permeable, giving an inflamed appearance to the wound, but strengthen over time. Epithelial cells from the edges of the wound move across the wound bed, covering it completely and attaching to the matrix underneath. In a very large wound, this cannot occur across the full wound bed, requiring granulation and contraction to assist this process. The granulation tissue is formed by fibroblasts, numerous at the injury site, laying down structural proteins that produce collagen. This fibrous tissue replaces the fibrin–platelet clot and is referred to as granulation tissue. The fibroblasts then attach to the fibrous tissue and contract, closing the open wound. Some wounds (e.g. burns, gangrene, fungal infections) will form eschar, a piece of dead tissue which is cast off, or surgically removed, before healing can complete.

Maturation (remodelling)

This final phase begins during proliferation, whereby the adapted fibroblasts attach to collagen fibres and contract to pull the wound margins together (or closer together). In this phase, the collagen fibres and proteins are remodelled and realigned along tension lines. Apoptosis of unnecessary cells occurs and the scar tissue continues to strengthen and gradually shrinks. The final tissue is pale in contrast to the pink colour of a newly formed scar. In people with darker skin pigmentation the scar will always remain paler than the skin it replaces.

This phase can take up to two years and strengthens the new tissue. Regenerated tissue will closely resemble what was previously there otherwise it will be scar tissue. The new tissue is rarely as strong as the original, and so the area becomes one that is vulnerable to further injury.

Creating the environment for wound healing

The optimal environment for wound healing is one that facilitates the efficient progression through the stages of wound healing and is outlined in Table 14.3.

Table 14.3 Conditions for optimal wound healing

Condition	Rationale
Absence of necrotic tissue, bioburden and foreign debris	Defective cellular matrix and debris prolong the inflammatory response. They inhibit proliferation as the wound bed is not in optimal condition for epithelialisation and granulation to occur. Bioburden refers to flora that are competing with fibroblasts, neutrophils and macrophages for energy and may result in more non-viable tissue
Free of and protected from infection	High bacterial count will prolong the inflammatory phase, delay proliferation and compete for nutrients needed by fibroblasts, neutrophils and macrophages to function optimally
Absence of dead space	Dead space creates a place for pathogens to survive and thrive, increasing the likelihood of infection. It also promotes seroma formation, a pocket of clear serous fluid, which can cause maceration to the wound bed
Insulated from local hypothermia	Temperature affects the tensile strength of wounds. Cells and enzymes function optimally at normal body temperature. A temperature drop of just 2°C is enough to affect the biological healing process
A balanced moist wound bed	A moist environment prevents tissue dehydration and cell death, it accelerates angiogenesis in the haemostasis phase, increases the breakdown of non-viable tissue and fibrin, and facilitates the interaction of growth factors with target cells
Maintenance of gaseous exchange	Low levels of oxygen interfere with protein synthesis and fibroblast activity, causing a delay in wound healing

ACTIVITY 14.3: APPLY

Wound healing and person-centred nursing

Take some time to think about people in your care with chronic wounds. This means conditions for healing are not optimal.

- How may this person feel?
- What physical challenges may they have and how could this wound impact upon them socially and psychologically?

Nurses need to be able to show compassion for the people they care for and reflecting on these matters enables you to be aware of the challenges people experience.

THERMOREGULATION

Variation in body temperature

The temperature of the internal environment is central to maintaining homeostasis. The temperature the body aims to maintain is set by the hypothalamus and controlled by the thermoregulatory centres in the hypothalamus, the anterior hypothalamus responding to heat and the posterior hypothalamus responding to cold.

The core body temperature (i.e. the temperature centrally within the body – how warm the vital organs are) is normally about 36.8°C but the range for this can be 1°C either side (Peate and Wild, 2012). The shell temperature (at or near the surface of the body) is usually lower than the core temperature and varies with external conditions. Our core temperature shows a circadian rhythm and fluctuates by about 1°C over the period of a day, varies through the menstrual cycle and remains raised during pregnancy (Figure 14.6).

Alterations in body temperature are described as follows:

- Body temperature below 35°C is classified as hypothermia. This can occur as a result of environmental exposure to cold, metabolic disturbances or can be iatrogenic.[1]
- Hyperthermia, or pyrexia is at the opposite end of the spectrum:

  - low-grade pyrexia is from normal temperature to 38.0°C;
  - moderate to high-grade pyrexia is between 38 and 40.0°C (Peate and Wild, 2012);
  - hyperpyrexia is a body temperature above 40.0°C at which serious cellular damage can occur.

If body temperature is too low (hypothermia), metabolism slows and neurological dysfunction occurs. Similarly, with high body temperature (hyperthermia), metabolism is disrupted and neurological dysfunction also occurs (Gomez, 2014).

---

[1]Caused by treatment/intervention by healthcare professionals.

(a) Circadian rhythm

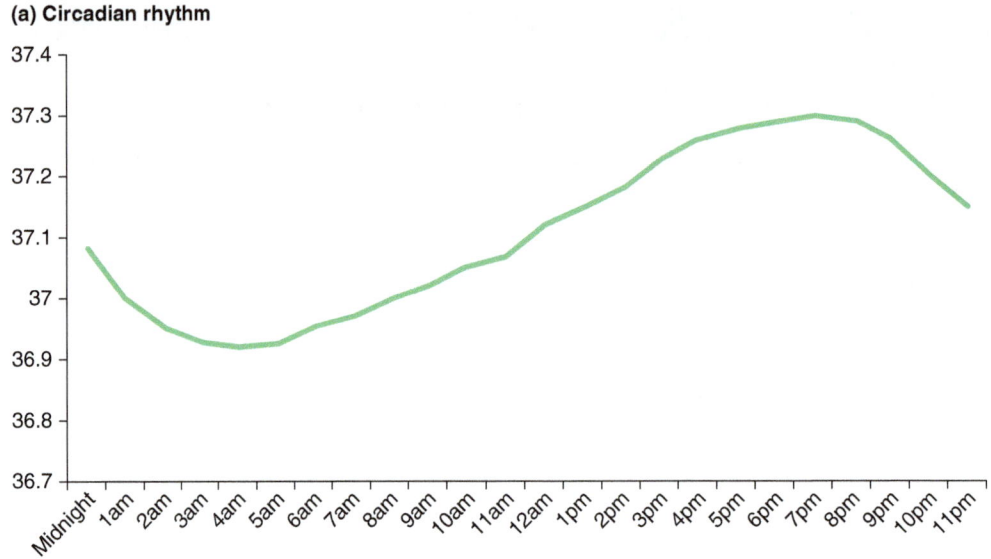

(b) Menstrual cycle vs. pregnancy

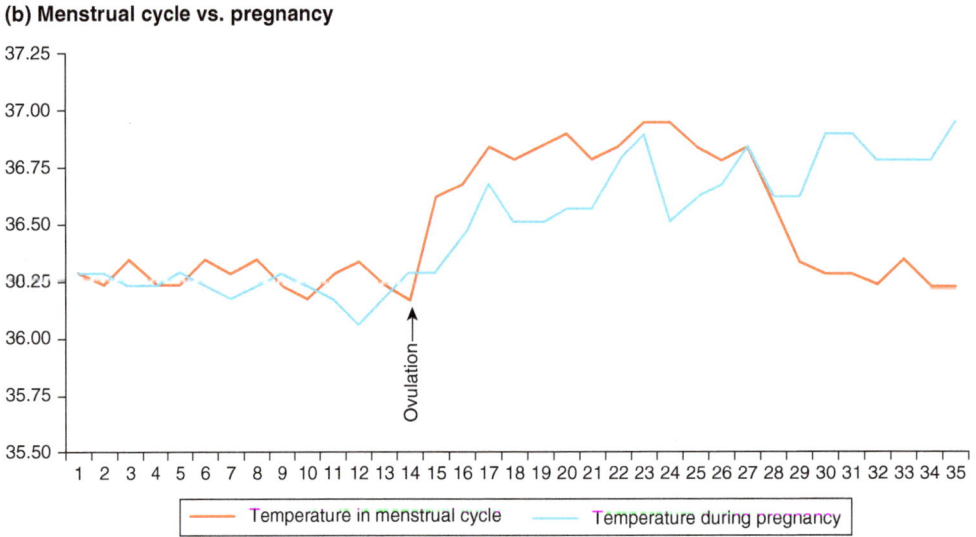

Figure 14.6 Rhythms in temperature regulation

Adapted from: Pocock and Richards (2009).

APPLY

Fever

Fever is a specific circumstance when the body temperature is raised (the person is pyrexial) because the hypothalamic regulating centre has a set-point which is above the normal range, frequently due to infection. When this happens the person shivers and feels cold (they are having a rigor), because the set-point is higher than their actual temperature, and their body is responding by creating additional heat to achieve the new set-point. When their core temperature reaches their new set-point they feel warm again and the mechanisms for temperature regulation work normally again but for the higher set-point. Inhibition of microorganism growth and raised antibody production occurs with a fever and benefits people with some infections (Ganong, 2019).

This is very different from hyperpyrexia caused by heat stroke, other causes of excessive heat production or inadequate thermoregulation when the regulatory mechanisms are no longer functioning. The body temperature can then rise above that compatible with life.

APPLY

Managing fever (pyrexia) in practice

Core body temperature can increase as a result of infection, inflammation or other disease, e.g. hypothalamic injury. In practice, nurses frequently care for people with pyrexia and this care must be evidence-and physiologically informed. Fever is an essential component of the homeostatic response to infection, as raising core body temperature can kill off invading pathogens or reduce their ability to proliferate (Nazarko, 2014). Fever also contributes to the immune response by increasing available antibodies and cytokines, assisting in the activation of a response to invading pathogens by T-cells, neutrophils and macrophages (Nazarko, 2014). Society is very much conditioned to see fever as a negative when it is actually a healthy physiological response. While the concern is that body temperature may rise to adverse levels, this is rarely the case and the body has mechanisms to prevent this. As a result, it can often be more harmful to lower the person's temperature than to allow it to remain high to support the immune response. This, however, needs careful consideration and some critical thinking for every person you care for.

Nurses need to consider how this biological process translates to their practice. Historically it has been the practice to take actions to reduce body temperature when it rises above 38.5°C. It is now advocated that the source of infection should be isolated and treated as the priority, rather than to treat with antipyretics. Nazarko (2014) supports this, highlighting that treating pyrexia with antipyretics will not reduce the length of illness but may prolong it as it impairs the immune response. NICE (2015) support this, identifying that administration of antipyretics for children with pyrexia will not prevent febrile convulsions. They do, however, advocate the administration of paracetamol or ibuprofen (but not together) for distressed children with a pyrexia. In addition, NICE (2015) advocate that tepid sponging should not be used and the child should not be underdressed (i.e. clothes removed) or over-wrapped (have excess clothing on).

Antipyretics may be considered appropriate in cases where the person has a long-term cardiac or respiratory condition. This is because fever can increase metabolism by 25% and such people may not have the physical capacity to support the increased demand for oxygen, nutrients and removal of waste products (Nazarko, 2014).

NICE (2015) provide an algorithm for the management of **feverish illness in children**. To access the algorithm, go to https://study.sagepub.com/essentialap2e.

Control system

The hypothalamus is the control centre with information on body temperature received and actions initiated to adjust the balance between heat production and heat loss to maintain the correct body temperature (Figure 14.7).

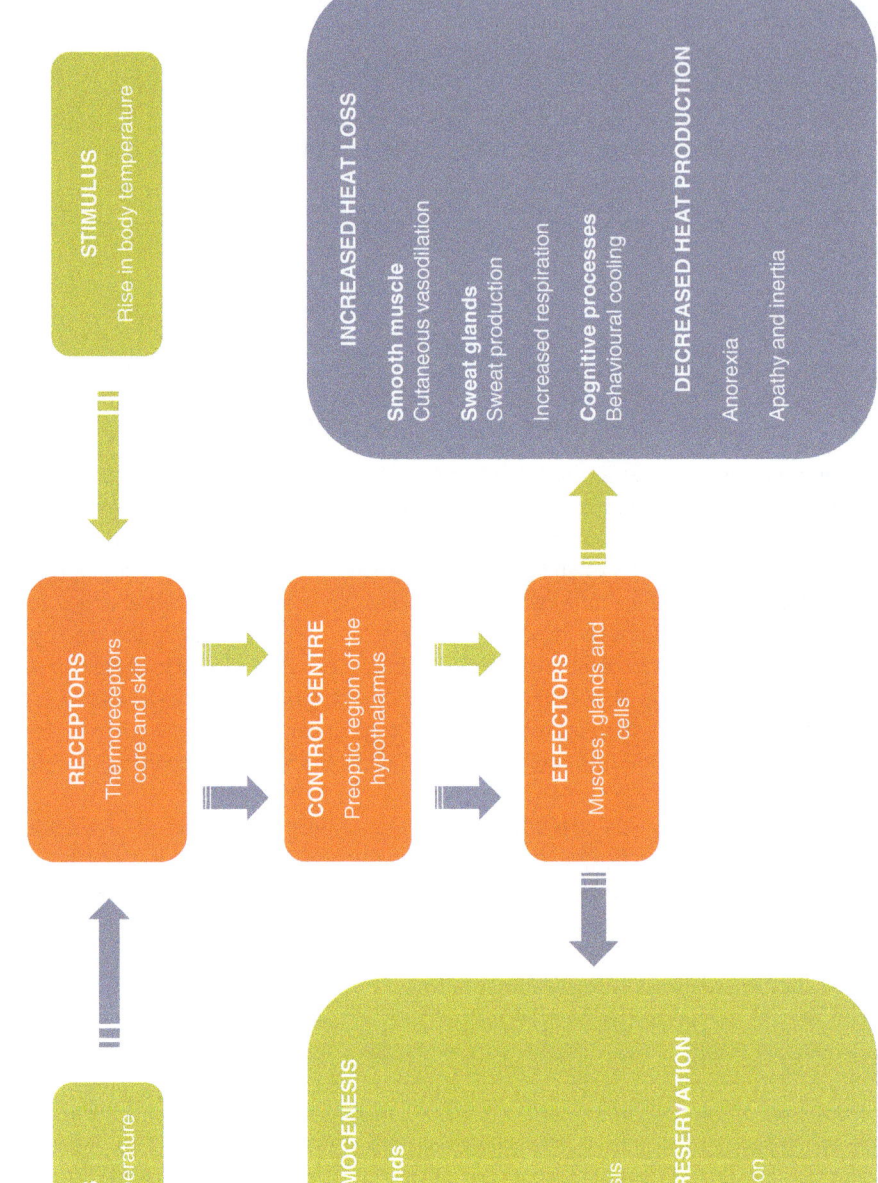

Figure 14.7 Thermoregulation

GO DEEPER

Internal temperature regulation

The primary heat exchange mechanism in the body is through vascular convective heat transfer – i.e. heating of the blood, which then carries heat around the body (Gonzalez-Alonso, 2012).

Core temperature is held within a tight range of 36°C to 37°C. The body responds to changes in temperature through sensory receptors in the skin and internally, primarily in the central pulmonary artery. Neurons in the reticular formation and spinal cord and the vagus nerve all relay signals to the preoptic region of the hypothalamus (Rubia-Rubia et al., 2011; Sund-Levander and Grodzinsky, 2013). From there, signals are sent to either the heat-losing centre in the anterior hypothalamus or to the heat-promoting centre more posterior in the hypothalamus (Rubia-Rubia et al., 2011). These centres then instigate heat loss or heat generation/conservation mechanisms respectively.

Adjusting heat production and loss

There are a number of mechanisms involved in heat production which may be modified up or down to increase or decrease heat production.

Reducing body temperature

There are four key mechanisms for heat loss (Figure 14.8) from the body (Gomez, 2014):

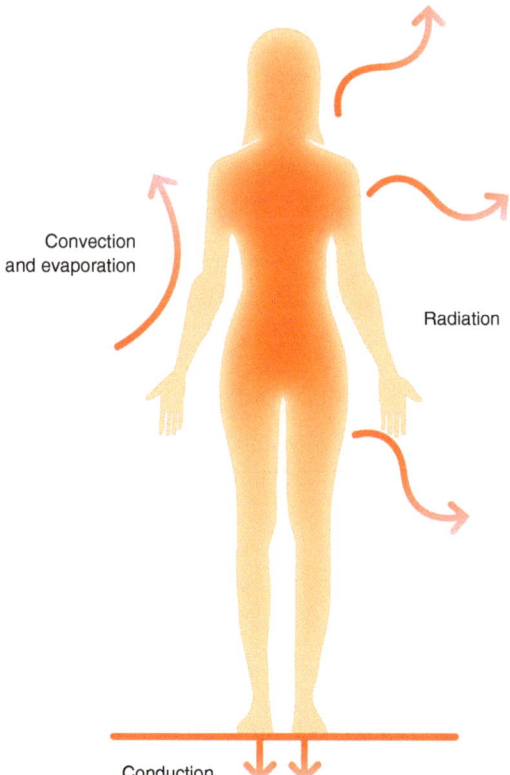

Convection and evaporation

Radiation

Conduction

Figure 14.8 Mechanisms of heat loss

1. **Radiation** refers to the transfer of heat by radiation. This requires no physical contact between surfaces and accounts for 60% of the heat loss from the body at rest. An example is the loss of heat from the body to the colder environment.
2. **Conduction** is heat loss as a result of contact with a surface colder than the body. Some 15% of heat can be lost in this way. Examples include swimming in cold water, lying in cold snow or sitting on a cold surface.
3. **Convection** is a key source of heat loss from the body. Air is heated as it passes over the body. Heated air rises, carrying heat away from the body, and is replaced by cool air. This is a convection current. On average, we lose 5% of heat through this method. An example of heat loss from this method would be being in a cold wind or having a fan blowing cool air onto the body.
4. **Evaporation** describes the conversion of a liquid to vapour by heat resulting in a loss of heat energy, accounting for about 20% of heat loss from the body. This occurs when we sweat and also through expiration of humidified air from the lungs.

In order to reduce body temperature, the body can induce the following:

Cutaneous vasodilation: Vasodilation brings a greater volume of blood close to the surface of the skin, enabling more heat to be lost from the blood to the atmosphere through convection or radiation. The autonomic nervous system regulates this process through innervating the smooth muscle in the arteries and arterioles.

Sweat production: Sweat production is increased when a person is warm, resulting in liquid on the surface of the skin evaporating. This process requires heat energy, resulting in heat loss.

Behaviour: The sensation of feeling warm encourages people to reduce activity, reducing metabolism, and to remove items of clothing to enhance biological heat loss mechanisms. People are also likely to drink cold liquids to enhance cooling by conduction.

Increasing body temperature

Heat is produced through cellular metabolism within the body (Chapter 9), with the liver as one of the most metabolically active organs. The amount of heat created and lost is altered in various ways:

Cutaneous vasoconstriction: Vasoconstriction reduces the amount of circulating blood near the surface of the skin, thus diminishing convective heat loss to the environment. The autonomic nervous system regulates this process through innervating the smooth muscle in the arteries and arterioles and reducing the blood flow.

Piloerection: Arrector pili muscles are stimulated, raising the hairs on the skin and trapping air over the surface of the skin. This provides insulation, which helps to retain and increase body temperature but is minimally effective since people have little skin hair compared to other mammals.

Shivering thermogenesis: Shivering is induced when the first two mechanisms are ineffective. This occurs through a spinal reflex whereby muscle contraction in antagonistic muscle pairs is stimulated, generating heat from metabolic activity. This can increase heat production by 300–400%.

Metabolic thermogenesis: If cold is persistent, then metabolic thermogenesis takes place. This involves autonomic and endocrine stimulation to increase the metabolic rate. The endocrine stimulation comes from release of T_3 and T_4 from the thyroid gland and noradrenaline from the adrenal medulla (Chapter 7). This results in increased appetite to provide nutrients for this increase, although infants and adults working in cold environments often use brown fat which generates heat without ATP synthesis (Chapter 2).

Behaviour: The sensation of feeling cold encourages people to increase activity, increasing metabolism, and to put on additional items of clothing to enhance heat retention. People are also likely to drink warm liquids to transfer heat by conduction and go into the sun to enhance warming by radiation. Exercise is obviously one of the key mechanisms for increasing heat production and many of us will take less exercise when we are very warm. Additionally, we are all aware that if we are cold, having something to eat will usually warm us up through the specific dynamic action of food, which is the amount of energy used to assimilate food into the body.

ACTIVITY 14.4: UNDERSTAND

Watch these two video clips online to help your understanding of temperature homeostasis.

These online videos can be accessed by **scanning the QR codes** with your smart phone or via https://study.sagepub.com/essentialap2e.

The first video is a simple introduction.

This second video is more complicated, but you might it find it helpful in fully understanding this aspect of homeostasis.

BODY TEMPERATURE (3:05)   TEMPERATURE HOMEOSTASIS (10:03)

Thermoregulation through the lifespan

Infants and children have larger surface area to volume ratios than adults, making them more susceptible to rapid heat loss (Hazinski, 2013). Their smaller size also means that their temperature can be manipulated more quickly than that of adults. Below six months, the body cannot shiver effectively to generate heat and so metabolic thermogenesis by brown fat is important in maintaining body temperature. Noradrenaline from the adrenal medulla increases the metabolism of brown fat in these infants, which may be helpful but also leaves them vulnerable to heat loss as they lose this insulating tissue (if it is not replaced) (Hazinski, 2013).

Older people are more prone to both hypothermia and hyperthermia than younger adults. It is well established that older people have lower body temperatures than younger people (Blatteis, 2012). Older people are vulnerable to thermoregulatory issues as they may have a reduced ability to generate heat as a result of:

- A lower lean body mass.
- Impaired mobility – exercise produces heat through metabolic activity and this is reduced.
- Malnutrition secondary to either physical or cognitive impairment.
- Reduced shivering response.

They may also experience increased heat loss as a result of:

- Ineffective cutaneous vasoconstriction.
- Reduced sensation of heat.
- Inability to respond to being cold due to physical or cognitive impairment.
- Exposure to cold because of poor housing, insufficient clothing or inability to access or afford heating.

CONCLUSION

Now that you have read through this chapter and undertaken the activities, you should have a good understanding of how the skin and thermoregulation operate to achieve optimal homeostasis. Thermoregulation is essential for optimal endocrine and neurological function, and the skin is vital to protection, regulation and for experiencing the general senses and communicating through touch and expression. This makes them central to health-related quality of life and protecting people from harm.

─────────────────────── **GO DEEPER** ───────────────────────

Further reading

Garner, A. and Fendius, A. (2010) 'Temperature physiology, assessment and control', *British Journal of Neuroscience Nursing*, 6(8): 397-400.

McLafferty, E., Hendry, C. and Alistair, F. (2012) 'The integumentary system: anatomy, physiology and function of skin', *Nursing Standard*, 27(3): 35-42.

Teller, P. and White, T.K. (2011) 'The physiology of wound healing: injury through maturation', *Perioperative Nursing Clinics*, 6(2): 159-70.

- The skin has key roles in protection, thermoregulation, excretion, absorption, synthesis of vitamin D and sensation.

- The skin is composed of two primary layers, the epidermis and dermis, both of which have sublayers.

- The skin's accessory structures have specific functions that enable the skin to undertake its main roles.

- The skin develops its own microsystem of flora from the day we are born. This flora is largely symbiotic, but can become harmful when we are immunosuppressed.

- Exposure to the environment significantly impacts on skin over time, as does our state of internal health.

- Wounds heal by primary, secondary or tertiary intention and through four key phases.

- Thermoregulation is central to optimal metabolic and neurological functioning.

- The preoptic hypothalamus is the primary control centre for thermoregulation, receiving sensory input from the internal and external environment.

- When a person experiences hypothermia, the body uses cutaneous vasoconstriction, piloerection, shivering thermogenesis, metabolic thermogenesis and behavioural thermogenesis to increase body temperature.

- When a person experiences pyrexia (hyperthermia), the body uses cutaneous vasodilation, sweating and behavioural mechanisms to reduce body temperature.

You have already read about the immune system in Chapter 13, and this chapter builds upon that chapter by looking at the specific functions of the skin in protecting people from harm. However, there is more to skin and to thermoregulation. When revising this chapter, try to remember the diagrams included. Can you label them? Do you understand the functions of each component? Can you describe the anatomy and can you explain the physiology? Have you undertaken the tasks to engage with deeper learning?

In order to help you revise, consider the following questions, answers for which can be found by visiting **https://study.sagepub.com/essentialap2e**.

Test yourself by revising the chapter first, and then answer these questions without looking at the book. Afterwards compare your answers with the text and with the notes you made. Did you miss anything in your notes? Here are the questions:

1   How does an understanding of anatomy and physiology of the skin apply to nursing?

2   How does the body regulate temperature through the integumentary system?

3   A person has a body temperature of 38.7°C. How could a nurse help the body to achieve temperature homeostasis? You must be able to discuss the physiological responses of the nursing actions.

4   What are primary, secondary and tertiary healing?

5   What are the intrinsic and extrinsic factors that affect wound healing?

REVISE

For additional revision resources visit visit: **https://study.sagepub.com/essentialap2e**.

ACE YOUR ASSESSMENT

- Revise key terms with interactive flashcards.
- Test yourself with multiple-choice questions.
- Access the glossary with audio to hear how complex terms are pronounced.
- Explore recommended websites suitable for revision.

REFERENCES

Andrews, H. (2012) 'The fundamental of skin care', *British Journal of Healthcare Assistants*, 6(6): 285–90.

Bentov, I. and Reed, M.J. (2014) 'Anesthesia, microcirculation, and wound repair in ageing', *Anesthesiology*, 120(3): 760–72.

Bickle, D.D. (2012) 'Vitamin D and the skin: physiology and pathophysiology', *Reviews in Endocrine & Metabolic Disorders*, 13(1): 3–19

Blatteis, C.M. (2012) 'Age-dependent changes in temperature regulation: a mini review', *Gerontology*, 58(4): 289–95.

Butcher, M. (2011) 'Meeting the clinical challenges of burns management: a review', *British Journal of Nursing*, 20(15): s44–51.

Calleja-Agius, J., Brincat, M. and Borg, M. (2013) 'Skin connective tissue and ageing', *Best Practice & Research in Clinical Obstetrics & Gynaecology*, 27(5):727–40.

Dunn, J.H. and Koo, J. (2013) 'Psychological stress and skin ageing: a review of possible mechanisms and potential therapies', *Dermatology Online Journal*, 19(6): 18561.

Ganong, W. (2019) *Review of Medical Physiology*, 26th edn. New York: Lange Medical Books/McGraw-Hill.

Glim, J.E., van Egmond, M., Niessen, F.B., Everts, V. and Beelen, R.H. (2013) 'Detrimental dermal wound healing: what can we learn from the oral mucosa?', *Wound Repair & Regeneration*, 21(5): 648–60.

Gomez, C.R. (2014) 'Disorders of body temperature', in M. Aminoff, F. Boller and D. Swaab (eds), *Handbook of Clinical Neurology: Neurologic Aspects of Systemic Disease Part II (Volume 120)*. London: Elsevier, pp. 947–57.

Gonzalez-Alonso, J. (2012) 'Human thermoregulation and the cardiovascular system', *Experimental Physiology*, 97(3): 340–6.

Gould, D. (2012) 'Skin flora: implications for nursing', *Nursing Standard*, 26(33): 48–56.

Hazinski, M.F. (2013) 'Children are different', in M.F. Hazinski (ed.), *Nursing Care of the Critically Ill Child*, 3rd edn. St Louis: Elsevier–Mosby, pp. 1–18.

Kottner, J., Lichterfeld, A. and Blume-Peytavi, U. (2013) 'Maintaining skin integrity in the aged: a systematic review', *British Journal of Dermatology*, 169(3): 528–42.

Nazarko, L. (2014) 'Does paracetamol help or hinder healing in bacterial infections?', *British Journal of Community Nursing*, 19(7): 335–9.

NICE (National Institute for Health and Care Excellence) (2015) *Feverish Illness in Children Overview*. London: NICE.

Peate, I. and Wild, K. (2012) 'Clinical observations 1/6: assessing body temperature', *British Journal of Healthcare Assistants*, 6(5): 215–19.

Póvoa, G. and Martins Diniz, L. (2011) 'Growth hormone system: skin interactions', *Anais Brasileiros de Dermatologia*, 86(6): 1159–65.

Robinson, P.J. (2014) 'Skin', in P. Wexler (ed.), *Encyclopedia of Toxicology*, 3rd edn. London: Academic Press, pp. 283–309.

Rosen, C.J. (2011) 'Vitamin D insufficiency', *The New England Journal of Medicine*, 364(3): 248–54.

Rubia-Rubia, J., Arias, A., Sierra, A. and Aguirre-Jaime, A. (2011) 'Measurement of body temperature in adult patients: comparative study of accuracy, reliability and validity of different devices', *International Journal of Nursing Studies*, 48(7): 872–80.

Sund-Levander, M. and Grodzinsky, E. (2013) 'Assessment of body temperature measurement options', *British Journal of Nursing*, 22(16): 942–50.

White, E.S. and Mantovani, A.R. (2013) 'Inflammation, wound repair, and fibrosis: reassessing the spectrum of tissue injury and resolution', *Journal of Pathology*, 229: 141–4.

Young, A. and McNaught, C.E. (2011) 'The physiology of wound healing', *Surgery*, 29(10): 475–9.

THE MUSCULOSKELETAL SYSTEM

SUPPORT AND MOVEMENT

15

UNDERSTAND: CHAPTER VIDEO

Before working through this chapter, you might find it useful to have an overview of the musculoskeletal system. Watch this video clip to help enhance your understanding.

This video can be accessed by **scanning the QR code** with your smart phone camera or via https://study.sagepub.com/essentialap2e.

MUSCULOSKELETAL SYSTEM
(6:05)

LEARNING OUTCOMES

When you have finished studying this chapter you will be able to:

1. Identify the functions of the musculoskeletal system
2. Describe the structure and function of bone
3. Describe the structure and function of skeletal muscle
4. Explain how muscles and joints work together to facilitate movement and posture
5. Identify types of specific joints
6. Explain the role exercise has in maintaining a healthy musculoskeletal system

INTRODUCTION

This chapter will discuss the role, structure and functions of the musculoskeletal (i.e. skeleton and muscles) system and its contribution to the maintenance of homeostasis. It will also enable you to identify the role of exercise in maintaining a healthy body. This approach facilitates readers in placing their knowledge and understanding within the perspective of the Person-Centred Nursing Framework.

Context

The musculoskeletal system is vital to how we function physically; it provides the framework for your body, creates movements, permits flexibility and provides protection for various organs. The musculo-skeletal system changes throughout the lifespan.

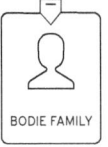
BODIE FAMILY

This is evident in all the members of the Bodie family. Danielle is in the early stages of development: at this stage her skeleton is composed of large amounts of cartilage, but her bones will continue to grow in length and thickness and eventually some of them will fuse to become the adult skeleton. The young adults of the family (Thomas (30), Derek (29), Michelle and Margaret (27), Kwame (28) and Jack (28)), will have formed their own physiques, tall or short, depending on their bone development. Some may be stronger than others with more muscle mass and tone. Their development will have been influenced by environment and genetics and the lifestyle choices they have made.

Matthew (45) is a keen triathlete and his musculoskeletal system will have developed strength and stamina to participate in this endurance sport. The mature members of the family may start to notice changes in their musculoskeletal system. For example, Maud, at 77, is at increased risk of developing osteoporosis (thinning of the bones) due to decreased levels of oestrogen and if she falls may sustain a fracture. Sarah (55) and Hannah (54), going through the menopause, may have reducing oestrogen levels and be at risk of developing osteoporosis in the future. George (84) may find that he is not as strong as he once was due to a decrease in muscle mass and he may also not be as flexible as he once was due to loss of elasticity of skeletal muscle and stiffening of the joints.

As you work through the chapter, you will identify how this system has an effect on the physical and social aspects of the healthy person and how it can contribute to your practice as a person-centred nurse.

THE SKELETAL SYSTEM

The skeletal system provides us with:

- Shape and movement.
- Blood cells.
- A storage area for minerals.
- Protection for some internal organs.

It is mainly composed of bone but also comprises cartilage, ligaments and tendons.

The tissues of the skeleton

Bone

This is highly vascular connective tissue with osteogenic (bone-producing) cells separated by bone matrix. Collagen fibres form between 90 and 95% of the organic part of bone. Bone also includes inorganic crystals

(50–70%) such as calcium phosphate, calcium carbonate and citrate. Although often viewed as dry and brittle, bone is a living tissue with some flexibility allowing it to withstand the everyday pressures of life.

Cartilage

This is more elastic than bone and consists largely of water containing two cell types:

- **Immature chondroblasts**, which secrete the components of the cartilage.
- **Chondrocytes**, which are mature cartilage cells derived from chondroblasts trapped within spaces called lacunae.

Cartilage forms a semi-rigid part of the skeleton and a protective layer at many joint surfaces (Chapter 2). There are three different types of cartilage found in the body:

- **Hyaline (transparent):** covers the ends of synovial joints; connects the ribs to the sternum; forms the larynx and parts of the nose and reinforces the trachea and bronchi.
- **Fibrous:** contains larger amounts of collagen fibre than other types of cartilage making it compressible and able to resist high pressures. It is found in areas of high stress such as the intervertebral discs and knee joints.
- **Elastic:** contains large amounts of elastic fibres providing a high degree of flexibility. It is found in only two places: the external ear and the epiglottis.

Ligaments

These are tough fibrous bands of dense collagen fibres and fibrocytes (spindle-shaped cells) that support internal organs and attach bones to bones, holding them together at joints.

Tendons

These are similar in structure to ligaments and are composed of dense fibrous connective tissue that attaches muscle to bone.

The skeleton

The adult human skeleton contains 206 bones. A child's skeleton is mostly cartilage that through the process of ossification (see later in the chapter) eventually becomes bone. A child has considerably more bones (300–350), many of which fuse together to form the 206 bones of the adult skeleton.

The human skeleton is divided into two parts: the axial skeleton and the appendicular skeleton (Figure 15.1), which are discussed in more detail later in the chapter.

The axial skeleton

This consists of 80 bones including those of: the skull, the vertebral column, the ribs and the sternum. These form the central axis of the body and support the head, neck and torso.

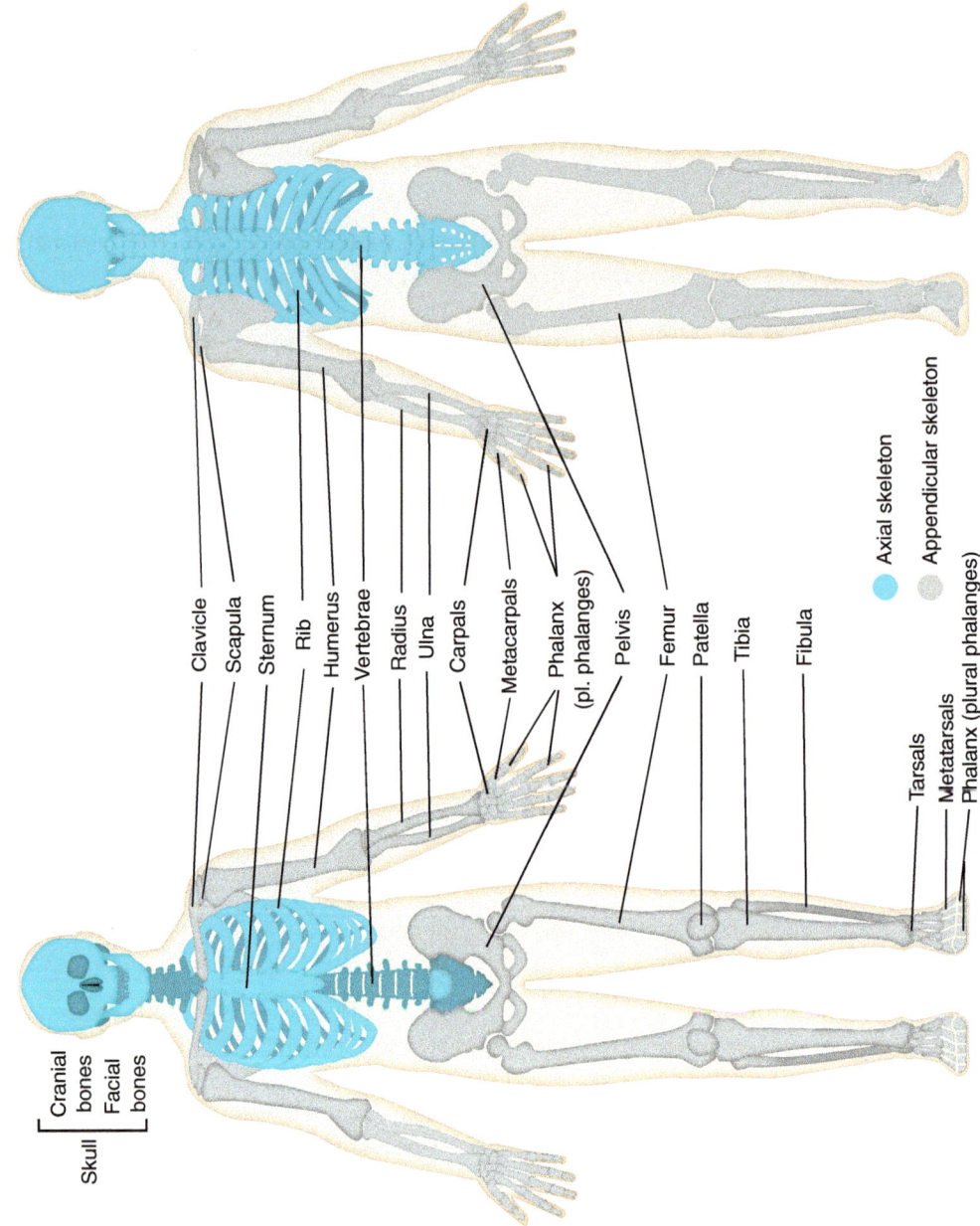

Figure 15.1 The skeleton

Skull — Cranial bones / Facial bones

Clavicle
Scapula
Sternum
Rib
Humerus
Vertebrae
Radius
Ulna
Carpals
Metacarpals
Phalanx (pl. phalanges)
Pelvis
Femur
Patella
Tibia
Fibula

Tarsals
Metatarsals
Phalanx (plural phalanges)

Axial skeleton

Appendicular skeleton

The appendicular skeleton

This consists of 126 bones, those of the arms and legs and those bones that attach them to the axial skeleton, i.e. the shoulder (pectoral) and pelvic girdle.

Bone function

Bones have a number of key functions.

Support

Bones provide structural support by providing a rigid framework to which soft tissues and organs are attached.

Movement

Bones and muscles with joints enable our bodies to move. Tendons and ligaments help determine the direction and the degree of movement of the bone depending on the forces generated by the skeletal muscle. Movements can be very fine/intricate, e.g. threading a needle or they can be gross/obvious, e.g. change in body posture.

Storage

Many essential minerals such as calcium, magnesium and phosphorus are stored in the bones; of these, calcium is the most abundant (about 99%). Bones can release or absorb minerals in response to the body's demands under the influence of hormones (Chapter 7). If there are high levels of calcium in the blood (hypercalcaemia), excess calcium can be deposited in bone. The reverse happens when there are low levels of calcium in the blood (hypocalcaemia); calcium can be released from the bone.

Lipids are also stored within the yellow marrow of the bone; these can be released and used as a source of energy depending on the body's needs.

Protection

Bones provide a rigid structure that gives protection to most of the internal organs and soft tissues within the various cavities of the body:

- **Cranium (skull):** protects the brain.
- **Spinal column of vertebrae:** protect the spinal cord.
- **Thoracic cavity (sternum and ribs):** protects the heart and lungs.
- **Pelvic cavity:** protects the bladder, rectum and anus, and reproductive organs.

Blood cell formation

Bones play an important role in the production of all types of blood cells – red blood cells (erythrocytes), most of the white blood cells (leucocytes) and platelets (thrombocytes) (Figure 2.17). The production of blood cells takes place in the red bone marrow through a process called haemopoiesis. In adults, red bone marrow is only found in certain bones, e.g. the skull, sternum, ribs, pelvis, vertebrae and the ends of the femur and humerus. In the newborn, all bone marrow is red; however, as we age the red bone marrow outside the bones mentioned above changes from red to yellow and stops forming these cells.

Bone structure

The structure of a bone can be discussed in terms of the parts of a long bone. This is one that is longer in length than width, e.g. the femur (thigh bone). A long bone consists of the following parts: diaphysis, epiphysis, metaphysis, articular cartilage, periosteum, medullary cavity and endosteum (Figure 15.2):

- **Diaphysis:** This is the shaft and main portion of the bone. It is long and cylindrical in shape.
- **Epiphyses:** These are the proximal and distal ends of the bone.
- **Metaphyses:** These are between the diaphysis and the epiphyses. Each metaphysis contains the epiphyseal or growth plate, which is made of hyaline cartilage. It is here that the diaphysis can grow in length (discussed later). Once the bone stops growing in length, usually around the age of 18–21, the hyaline cartilage within the epiphyseal plate is replaced by bone to form the epiphyseal line.
- **Articular cartilage:** This is a thin layer of hyaline cartilage that covers the epiphysis of one bone where it forms a joint with another bone. It is responsible for reducing friction and absorbing shocks at freely movable joints (discussed later). Articular cartilage does not contain perichondrium (dense irregular connective tissue which surrounds most cartilage, containing blood vessels and nerves) or blood vessels, therefore if it gets damaged repair is limited.
- **Periosteum:** This is the tough connective tissue membrane that covers the outside of the bone (except where there is articular cartilage). It has two layers: an outer layer of dense irregular connective tissue and an inner layer of osteogenic cells that can develop bone width but not length. Key functions of the periosteum are to:

  - protect and nourish the bone tissue;
  - act as an attachment site for ligaments and tendons;
  - assist in bone repair.

- **Medullary cavity:** This cylindrical cavity within the diaphysis contains blood vessels and bone marrow. It helps to reduce the density and weight of bone whilst the cylindrical shape means it retains strength to withstand the forces exerted upon it.
- **Endosteum:** This single layer of bone-producing cells lines the inner surfaces of the cavities within the bone.

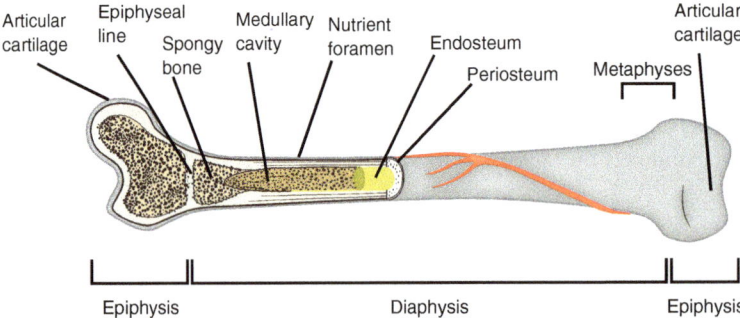

Figure 15.2 Structure of a long bone

Bone composition

Bone is a connective tissue (Chapter 2), and like other connective tissue it contains an abundant extracellular matrix that surrounds the bone-generating cells.

Bone matrix

The matrix consists of: water, collagen fibres and crystallised mineral fibres. Calcium phosphate is the most abundant mineral in the matrix. Crystals of hydroxyapatite are formed by the combination of calcium phosphate and calcium hydroxide. During the formation of the crystals, other minerals such as magnesium, fluoride and sulphate are also deposited in the matrix and as the minerals crystallise, they harden in the process of ossification. The amount and type of crystallised minerals determines the hardness of the bone; however, bone needs to be flexible and this depends on the number of collagen fibres present within the matrix. Calcification only occurs when collagen fibres are present. The mineral salts crystallise initially in the spaces between the collagen fibres. Once the spaces have been filled further crystallisation of the mineral salts occurs around the collagen fibres.

Bone cells

Bone tissue consists of four types of cells: osteogenic cells, osteoblasts, osteocytes and osteoclasts (Figure 15.3).

Osteogenic cells	Osteoblasts
Derived from mesenchymal cells (adult stem cells), they undergo cell division developing into osteoblasts. They are found in the periosteum, endosteum and within the canals that contain the blood vessels.	Bone building cells. They make the bone matrix by synthesising and secreting collagen fibres and other organic components. They also initiate calcification of the matrix.

BONE CELLS

Osteocytes	Osteoclasts
These start as osteoblasts. As they are surrounded by the matrix they are trapped, no longer able to secrete matrix and become osteocytes. Osteocytes are therefore found in mature bone and are the main cell type in bone. Their function is to maintain the daily metabolic function of bone by ensuring exchange of nutrients and waste products with the blood.	Formed by the fusion of approximately 50 monocytes (type of macrophage, white blood cell) and they remove old bone. They are very large, multinucleated and found predominately in the endosteum. Their plasma membrane is folded into deep ruffles and faces the surface of the bone. It secretes powerful lysosomal enzymes and acids that are responsible for dissolving the protein and mineral matrix. This is known as resorption and is part of normal development, maintenance and repair of bone. Removal of old bone is usually aligned to the production of new bone cells by the osteoblasts.

Figure 15.3 Types of bone cells

Categories of bone

Bone is not completely solid, but contains many small spaces. These provide a pathway for blood vessels for exchange of nutrients and waste products and act as a storage area for red bone marrow. The size and number of spaces determine the category of bone – compact or spongy (Figure 15.4). Approximately 80% of the skeleton is compact with the remaining 20% being spongy bone. Approximately 70% of the axial skeleton is spongy bone.

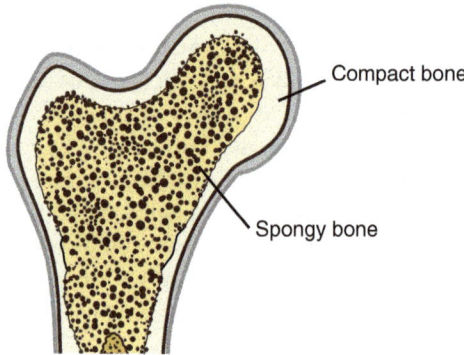

Figure 15.4 Bone tissue

Compact bone

This is the strongest type of bone tissue. It is dense and contains few spaces, thus offering a high degree of protection and support for the bone to withstand the pressures exerted by weight and movement. This type of bone is found beneath the periosteum of all bones and accounts for the majority of the diaphyses of long bones. Figure 15.5 shows the different parts of compact bone.

Compact bone is largely set out in cylinders around central canals through which blood vessels and nerves pass. These canals run longitudinally through the shafts of long bones. The cylindrical structure of this bone around the canals is deposited in overlapping layers, like the layers on an onion. These are called lamellae. In between these layers are tiny gaps called lacunae within which are found osteocytes. The central canal and the innermost lamella are collectively known as an osteon (haversian system) – Figure 15.5.

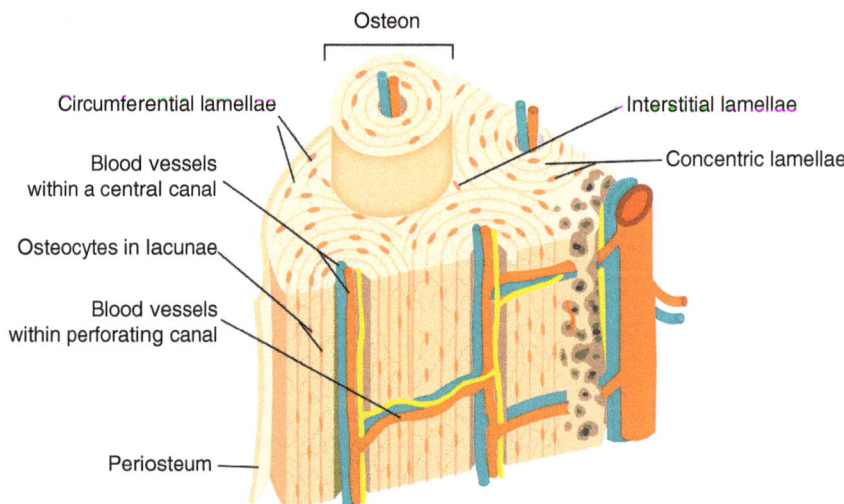

Figure 15.5 Compact bone

Spongy bone

This consists of columns of bone known as trabeculae, indicating why it is also known as trabecular bone tissue (Figure 15.6). Between the columns are spaces filled with red and yellow bone marrow and blood vessels. Spongy bone is always found inside the bone and is protected by a layer of compact bone. It tends to be found in bones that have low levels of stress or where pressures are exerted from a range of directions. The spaces mean that spongy bone is much lighter than compact bone thereby reducing the overall weight of bones, allowing them to move easily when pulled by skeletal muscle.

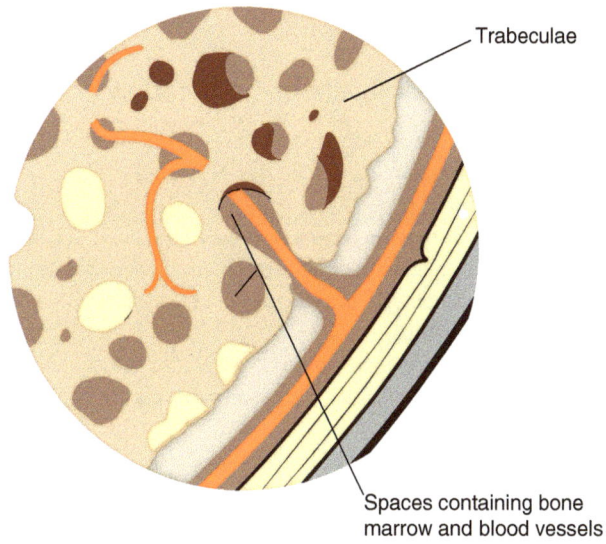

Trabeculae

Spaces containing bone marrow and blood vessels

Figure 15.6 Spongy bone

Blood and nerve supply

Bones receive a rich blood supply particularly where there is a lot of red bone marrow; the blood supply passes from the periosteum to the interior of the bone. Small arteries (periosteal arteries) accompanied by nerves enter the diaphysis through canals (Volkmann's canals) and supply the periosteum and compact bone. A large nutrient artery near the centre of the diaphysis passes through the nutrient foramen (an opening) and enters the medullary cavity; it divides into proximal and distal branches and delivers blood to both compact and spongy bone and the red bone marrow of the diaphysis as far as the epiphyseal line.

Each bone varies in the number of nutrient foramina that it contains, for example the tibia (lower leg bone) has one nutrient foramina whereas the femur (thigh bone) has several. The number of nutrient foramina depends on the bone size and the amount of red bone marrow within it. Metaphyseal and epiphyseal arteries supply blood to the ends of the bones and to the red bone marrow with veins carrying blood away from the bones. There may be one or two nutrient veins accompanying the nutrient artery and exiting the diaphysis. There are numerous metaphyseal and epiphyseal veins that exit with their respective arteries through the epiphyses, and many small periosteal veins drain blood from the periosteum back into the venous system.

Alongside the blood supply is a rich supply of nerves. The periosteum contains many sensory nerve endings, which produce sensations of pain and are sensitive to tension or tearing. This sensitivity explains the severe pain experienced by someone who has sustained a fracture or has a bone tumour.

Bone formation and development

The process of bone formation is called ossification or osteogenesis. During the first few weeks the embryonic skeleton is composed of mesenchymal cells (a special type of undifferentiated tissue) with the general shape of bones. Bone is formed by one of two processes during embryonic and fetal development: intramembranous ossification (occurring within connective tissue) and endochondral ossification (which occurs in cartilage).

GO DEEPER

Intramembranous ossification

Bone formation occurs with the direct conversion of mesenchymal tissue into bone. This occurs in the flat bones of the skull, the lower jaw and the scapula in four stages (Figure 15.7):

1. **The ossification centre:** forms at the site where bone will develop. It occurs when specific chemical messengers cause mesenchymal cells to group together and then differentiate into osteogenic cells and later into osteoblasts. The osteoblasts are responsible for secreting the extracellular matrix until they become surrounded, when it stops.
2. **Calcification:** the osteocytes lie in the lacunae and extend into the canaliculi (tiny canals) radiating in all directions. Over a period of days calcium and other minerals are deposited and the extracellular matrix begins to harden (calcify).
3. **Trabeculae formation:** during formation of the extracellular matrix, trabeculae develop and fuse with one another to produce spongy bone surrounding blood vessels in the connective tissue.

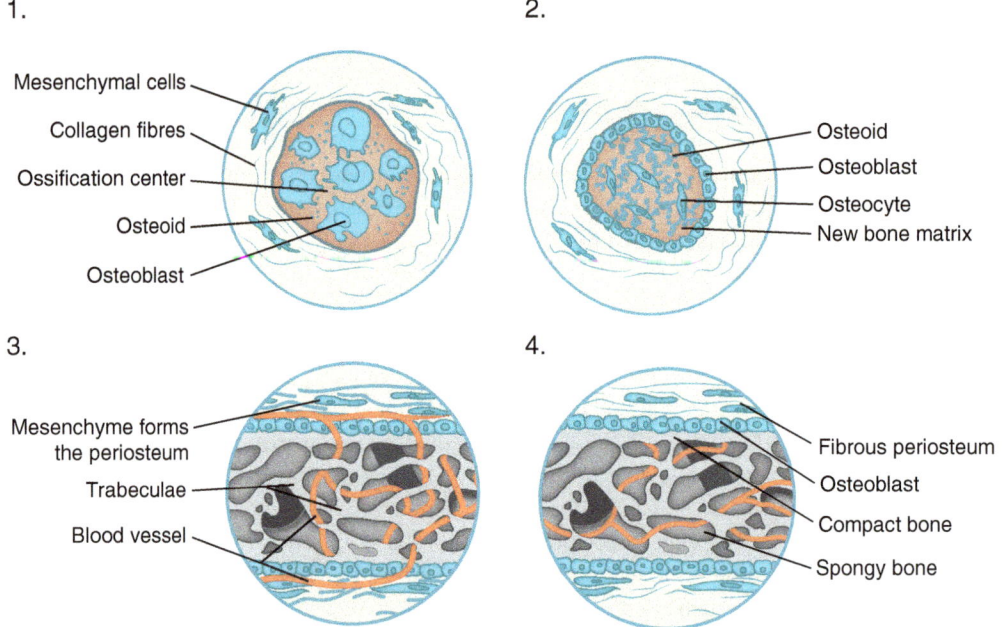

Figure 15.7 Intramembranous ossification

4. **Development of periosteum:** alongside trabeculae formation, the mesenchyme at the periphery of the bone condenses and forms the periosteum. Finally, a thin layer of compact bone replaces the surface layers of spongy bone (the centre remains spongy bone). This newly formed bone undergoes a process of remodelling (discussed later) and eventually becomes the adult shape and size.

Endochondral ossification

This is how most of the bones are formed through replacement of hyaline cartilage with bone. It takes place in six stages (Figure 15.8):

1. **Development of cartilage model:** where new bone is to be formed, specific chemical messengers cause mesenchymal cells to group to resemble the general shape of the future bone. The mesenchymal cells then develop into chondroblasts that secrete a cartilage extracellular matrix producing a hyaline cartilage model. A membrane called the perichondrium forms around the cartilage model.
2. **Growth of the cartilage model:** as more cartilage extracellular matrix is produced chondroblasts become surrounded by the matrix and become chondrocytes. The cartilage model grows in length (interstitial growth) with continued secretion of the cartilage extracellular matrix and cell division of chondroblasts. The cartilage model develops in thickness (appositional growth) as the new chondroblasts continue to secrete cartilage extracellular matrix deposited on the surface of the cartilage. As the cartilage model continues to grow in length and thickness, the chondrocytes in the middle increase in size (hypertrophy) and then deteriorate leaving a cavity within the cartilage. The surrounding matrix begins to calcify. Nutrients cannot diffuse through the calcified matrix and chondrocytes begin to die, leaving spaces which develop into cavities known as lacunae.
3. **Development of the primary ossification centre:** primary ossification continues from the external surface of the bone inwards. As it ossifies a nutrient artery enters the perichondrium and calcifying cartilage through a nutrient foramen into the centre of the cartilage model stimulating osteogenic cells in the perichondrium to differentiate into osteoblasts. Once the perichondrium is producing bone instead of cartilage it becomes known as the periosteum. Periosteal capillaries in the cartilage model grow into the calcified extracellular matrix and form the primary ossification centre where most of the cartilage will be replaced by bone tissue. Spongy bone trabeculae are formed when osteoblasts are deposited and secrete bone matrix over the calcified cartilage.
4. **Development of the medullary cavity:** as the primary ossification centre develops and grows towards the end of the bone some of the new spongy bone is broken down by osteoclasts to form the medullary cavity filled with red bone marrow. Primary ossification forms the diaphysis of the bone which is made of an outer layer of compact bone, lined with spongy bone that surrounds the medullary cavity.
5. **Development of secondary ossification centres:** the epiphyseal artery entering the epiphyses leads to development of secondary ossification centres (usually around the time of birth). The formation of bone in secondary centres is similar to that in the primary centre. The main difference is that spongy bone remains in the centre of the epiphyses, hence no medullary cavities are formed. As opposed to primary ossification, secondary ossification proceeds outwards from the centre of the epiphyses to the external bone surface.
6. **Formation of articular cartilage and the epiphyseal plate:** when the ossification of the epiphyses is completed, the hyaline cartilage covering the epiphyses becomes the articular cartilage and is retained throughout life. The epiphyseal plate (between the diaphysis and the epiphyses) is hyaline cartilage: during childhood it is responsible for growth in length of bones. In adulthood it is replaced by bone.

(Continued)

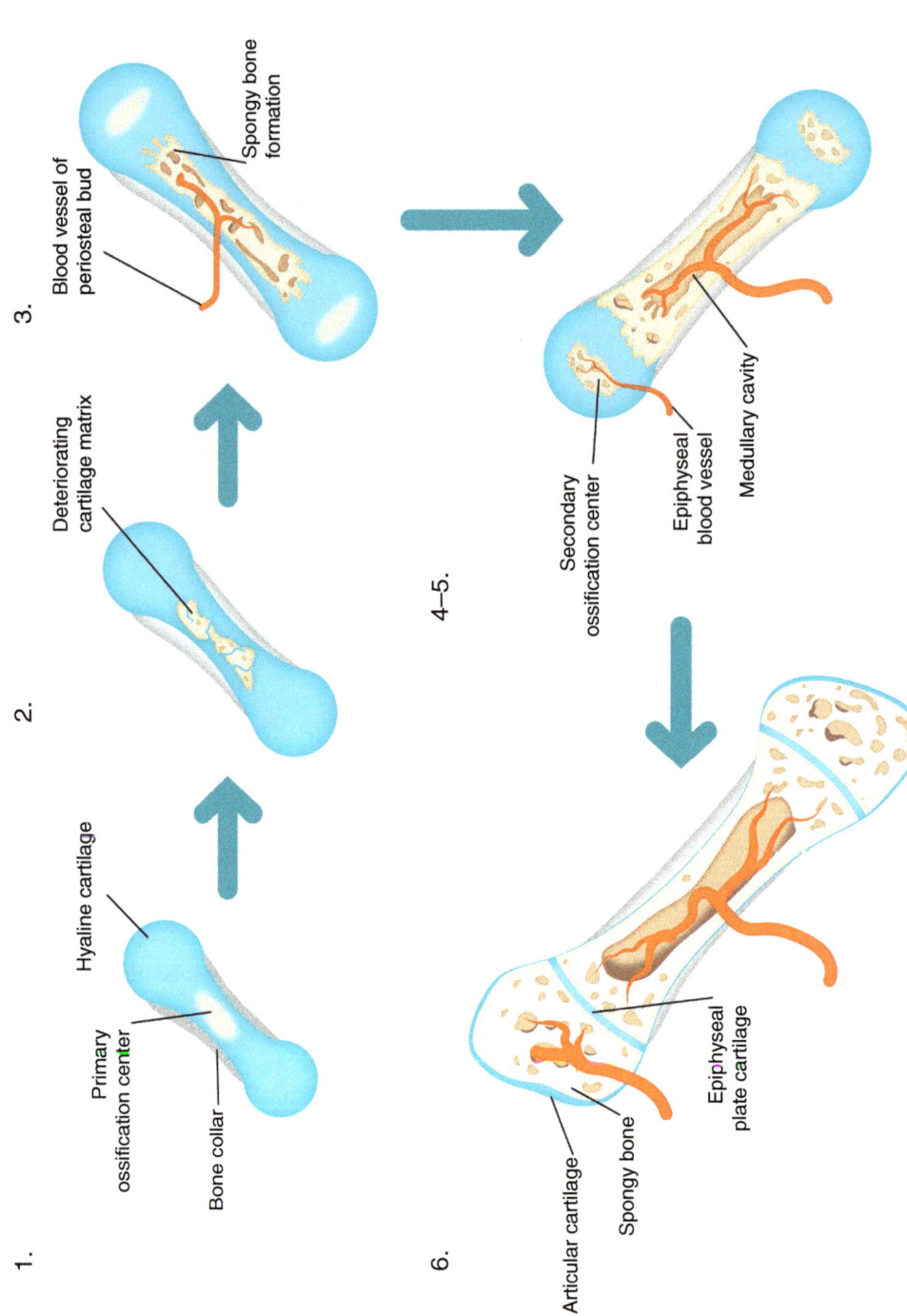

Figure 15.8 Endochondral ossification

APPLY

Osteoporosis

Osteoporosis is a disease of the skeletal system associated with fragile, porous bones with significant reduction in bone mass. Osteopenia refers to a mild to moderate reduction in bone mass below average levels and puts a person at a higher risk of developing osteoporosis. Bone mass is at its peak by the age of 20 years, around the age of 40 years bone mass starts to decrease.

There are a number of factors that may play a role in the development of osteoporosis and these include: genetic factors, hormonal factors, nutrition and level of activity. Increased levels of glucocorticoids (natural or artificial) can also increase the likelihood of developing osteoporosis. Loss of bone mass may also be related to excessive alcohol consumption, caffeine and smoking (Naish and Syndercombe Court, 2014).

Osteoporosis primarily affects middle-aged and older people: it affects more women than men for two main reasons: (1) loss of oestrogen at the menopause means that osteoblast activity and synthesis of bone matrix is greatly reduced, and (2) women's skeletons are lighter with lower bone mass.

As bone mass decreases it increases the likelihood of fractures: these can occur spontaneously due to the mechanical stresses of daily living. The most common fracture due to osteoporosis in persons over the age of 75 years is fracture of the neck of the femur. Alongside fractures, osteoporosis may also cause height loss, bone pain and fractures of the vertebrae, which may lead to kyphosis (excessive outward curvature of the spine, causing hunching of the back). Older women in particular need to ensure that they take enough vitamin D and calcium and enough exercise to reduce the risk of osteoporosis development and fractures.

To help prevent osteoporosis, children (especially young girls) should be encouraged to have an adequate intake of calcium and vitamin D. They should also be encouraged to participate in regular exercise (see the Go Deeper box on The role of exercise in affecting bone strength below).

It is important in person-centred nursing to assess the nutritional requirements of all persons in your care and ensure that they have an adequate dietary intake of calcium. It is also important that the nurse assesses the person's activity levels and encourages exercise to minimise the rate of bone mass loss.

Bone growth

Our bones continue to grow through infancy, childhood, adolescence and early adulthood. Bones continue to grow in length until about the age of 21 but may continue to grow in thickness beyond this time.

Bone length

Bones grow in length from the epiphyseal plate. Two distinct events are involved:

- Interstitial growth of cartilage on the epiphyseal side of the epiphyseal plate.
- Endochondral ossification of the hyaline cartilage on the diaphyseal side of the epiphyseal plate.

Within the epiphyseal plate there are four zones:

- **Zone 1:** resting cartilage. This layer is closest to the epiphysis and contains small chondrocytes that are not actively involved in bone growth. Their role is to attach the epiphyseal plate to the epiphyses and to deliver nutrients to the developing cartilage and store materials, e.g. lipids, glycogen, that are necessary for growth.
- **Zone 2:** proliferating cartilage. Here the chondrocytes are slightly larger and arranged into columns like stacks of coins. The chondrocytes divide and replace dead cells on the diaphyseal side of the epiphyseal plate. These chondrocytes are responsible for secreting the extracellular matrix and produce new cartilage through interstitial growth.

- **Zone 3:** hypertrophic cartilage. The chondrocytes here mature and enlarge in size (hypertrophy) and are arranged into columns with a gradient of maturation. Those chondrocytes nearest the epiphysis are younger and less mature than those cells nearer the diaphysis, which are older and larger.
- **Zone 4:** calcified cartilage. This zone is very thin and contains large dead chondrocytes and calcified extracellular matrix. Osteoclasts help dissolve the calcified matrix and blood vessels, also at this stage osteoblasts from the diaphysis enter the zone. This zone now becomes part of the diaphysis (remember endochondral ossification is the replacement of cartilage with bone) adding overall length to it.

The activities that occur within each of the four zones are the only way that bone can grow in length. Chondrocytes continue to multiply on the epiphyseal side as the bone grows. The older chondrocytes are replaced by new ones and cartilage is replaced by bone on the diaphyseal side of the plate. When adolescence ends, the epiphyseal cartilage cells stop dividing and all remaining cartilage is replaced by bone. The epiphyseal plate closes and fades and creates the epiphyseal line, which marks an end to all growth in bone length.

Bone thickness

Bones only grow in thickness by appositional growth below the periosteum. Periosteal cells at the bone surface differentiate into osteoblasts, which secrete collagen fibres and organic materials to form the bone extracellular matrix. As osteoblasts become surrounded by the extracellular matrix, they develop into osteocytes leading to the formation of ridges on either side of the periosteal blood vessel. The ridges fuse together and form a tunnel enclosing the blood vessels. The periosteum has now become the endosteum. Osteoblasts in the endosteum form concentric lamellae (plates of compact bone) that move towards the centre of the tunnel as bone continues to be deposited. When the tunnel becomes filled with bone it is known as an osteon. As the osteon is forming, new circumferential lamellae are deposited by the osteoblasts under the periosteum resulting in continued thickening of the bone. As more periosteal blood vessels become enclosed this procedure is repeated and growth continues. As new bone is added to the outer surface, bone thickness increases. The diameter of the medullary cavity increases due to bone lining the medullary cavity being destroyed by osteoclasts.

Factors affecting bone growth

There are a number of factors that have an impact on the growth of bone including an adequate intake of vitamins and minerals, and the level of several hormones.

Vitamins and minerals

A number of these are important to bone growth (Table 15.1):

Table 15.1 Vitamins and minerals necessary for bone growth

Vitamins	Role
A	Stimulates the activity of osteoblasts
B_{12}	Needed to produce bone proteins
C	Essential for collagen formation, the building blocks for bone
D	Necessary for calcium absorption
K	Needed to produce bone proteins and facilitates use of calcium for bone growth

Minerals	Role
Calcium and phosphorus	Needed in large quantities to ensure bone growth and provide bone elasticity
Magnesium, fluoride and manganese	Needed in small quantities and help in determining the hardness of bone

Hormones

Several different hormones play a key role in bone growth:

- **Insulin-Like Growth Factors (IGFs)** are the most important hormones needed for bone growth during childhood. They are produced by the liver and by bone tissue in response to human growth hormone (hGH) (Chapter 7). IGFs enhance protein production required for bone growth and stimulate osteoblasts to promote cell division in the epiphyseal plate and periosteum.
- **Thyroid hormones** are needed for growth of all tissues, including cartilage, and stimulate production of osteoblasts.
- **Sex hormones (oestrogen and testosterone)** are responsible for increased activity of osteoblasts and secretion of the extracellular matrix. Both hormones (particularly oestrogen) are responsible for inhibiting growth at the epiphyseal plate and stopping bone length growth. As females have higher levels of oestrogen, cessation in bone length growth occurs earlier, explaining why females are shorter in height than males.

GO DEEPER

The role of exercise in affecting bone strength

Bone tissue can alter its strength due to changes in the mechanical stresses upon it. As more stress is placed, more mineral salts are deposited, and osteoblasts produce more collagen fibres. The main mechanical stresses placed on bone result from the pull of skeletal muscles and gravity. Therefore, high impact stressors such as jumping or weight-lifting exercises will stimulate bone growth more than low impact stressors such as walking. Research has shown that the bones of athletes are noticeably thicker and stronger than those of non-athletes (Andreoli et. al, 2012). Astronauts also lose bone mass by up to 1% per month as they are exposed to microgravity and have no mechanical stresses on bones, leading to a reduction in mineral deposition and number of collagen fibres (Pennline et al., 2014).

Adolescents and young adults should be encouraged to participate in regular weight-bearing exercise to increase their bone mass and density prior to closure of the epiphyseal plate (Nilsson et al., 2014). However, bone mass and density can be increased at any age if weight-bearing activities are carried out (Guadalupe-Grau et al., 2009).

Let's think about Matthew Bodie, a keen triathlete. For him to build strength and endurance to participate in his events, his training incorporates weight training, plyometric exercises (exerting maximum force in as short a time as possible, to increase speed and power), running, swimming and cycling. Although the last two forms of exercise still increase mechanical stresses, they do not have the same impact on bone density and therefore would not increase overall bone mass. The weight training and plyometric exercise would increase mechanical stresses on his bones thereby increasing his overall bone mass and density.

BODIE FAMILY

Bone remodelling

Bones are formed before birth but continually renew themselves throughout an individual's whole life. Bone remodelling is old bone being replaced with new bone: the amount of resorption of bone by the osteoclasts is matched by the new bone created by osteoblasts; the whole process can take between 160–200 days. In areas of high stress, e.g. the hip joint, bone may be replaced more frequently, up to three times per year. During remodelling the osteoclasts attach to the surface of the bone to be resorbed. Once attached, osteoclasts secrete enzymes such as collagenases and lysosomal enzymes. These enzymes attack the organic portion of the bone under their attachment seal. Osteoclasts also secrete acid responsible for dissolving the inorganic salts of the bone matrix. The digested organic substances and the dissolved salts enter the osteoclast, pass through it and are excreted into the extra-cellular space. Osteoblasts migrate into the space created by the osteoclasts and synthesise collagen, osteocalcin (another protein) and other organic substances. The osteoblasts become surrounded by the extracellular matrix and eventually become osteocytes.

Bone repair

A fracture is a break in any bone, classified depending on the severity, shape or position of the fracture line. The process of bone healing takes place in four key steps:

1. **Formation of fracture haematoma:** blood vessels woven throughout the bone are damaged due to the break, blood leaks out of the vessels and a blood clot forms approximately six to eight hours after the fracture. This haematoma destroys nearby bone cells. These dead cells attract macrophages and osteoclasts to the site to remove the dead bone, causing localised swelling and inflammation. Capillaries begin to grow into the haematoma to re-establish a blood supply to the bone. This stage may take several weeks.

2. **Formation of fibrocartilaginous callus:** a procallus is formed when new blood vessels growing into the haematoma start to organise it into granulation tissue. Fibroblasts from the periosteum with osteogenic cells start to enter the procallus, develop into chondroblasts and start producing fibrocarti-lage. A fibrocartilaginous callus of collagen fibres and cartilage closes the gap between the two ends of the broken bone (taking up to three weeks). During the first four to six weeks of fracture healing the fibrocartilaginous callus is very soft and a cast or other support is needed until the callus begins to ossify.

3. **Formation of the bony callus:** osteogenic cells between dead and new bone regions develop into osteoblasts and begin secreting extracellular matrix. They form spongy bone trabeculae, which join the living tissue on either side of the fracture, and the callus is now a bony callus lasting three to four months.

4. **Bone remodelling:** osteoclasts resorb the dead portions of the original fracture. The spongy bone around the periphery of the fracture is replaced by compact bone. Healing of the fracture can some-times be so good that the original fracture line is undetectable by X-ray; however, there is usually evidence of a thickened area on the bone surface that identifies a healed fracture.

ACTIVITY 15.1: APPLY

Identify some of the most common types of fracture using different sources (e.g. the internet, library databases or medical-surgical nursing texts). Can you explain why a bone that has been fractured is stronger after it heals than before the fracture?

ACTIVITY 15.2: UNDERSTAND

Watch this video clip on bones, fractures and repair.

This online video can be accessed by **scanning the QR code** with your smart phone or via https://study.sagepub.com/essentialap2e.

BONE FRACTURE (7:53)

Bone classification

Bones can be classified into six types according to their shape: long, short, flat, irregular, sesamoid and sutural (Figure 15.9):

- **Long bones** are the most common type of bone and are longer than their width. They are mostly compact bone with spongy bone at their centre and ends. They tend to have a slight curvature, which provides strength and the ability to withstand greater pressures. Long bones vary considerably in length and include: the femur (thigh bone, upper leg), tibia and fibula (lower leg), humerus (upper arm), radius and ulna (forearm) and phalanges (fingers and toes).
- **Short bones** are cuboidal in shape and composed primarily of spongy bone with a surface layer of compact bone. They tend to be in areas where there is little movement and include: carpal bones (wrists) and tarsal bones (ankles).
- **Flat bones** are flat, thin and may be slightly curved. They are composed of two layers of compact bone that surround a thin layer of spongy bone. The structure of flat bones means that they provide a high degree of protection and a large area for muscle attachment. These include: the cranial bones (skull), the sternum (breastbone), the ribs and the scapulae (shoulder blades), and the ilium, the largest and uppermost of the bones which form the pelvis.
- **Irregular bones** have complex shapes and do not fit into any of the other categories. They consist of spongy bone with a thin layer of compact bone. Examples include: the vertebrae (back bones), the pelvic bones, some facial bones, e.g. zygomatic bones (cheek bones).
- **Sesamoid bones** develop in certain tendons where there is marked friction, tension and physical stress. The role of sesamoid bones is to protect tendons from excessive wear and tear and they can often change the direction of the pull of the tendon. The number of sesamoid bones may vary in individuals: the only sesamoid bones present in everyone are the two patellae (kneecaps).
- **Sutural bones** are small, irregular-shaped bones that usually appear as additional bones in the natural suture lines of the skull.

Bones of the axial skeleton

As already mentioned, the axial skeleton contains 80 bones and includes: the skull, the spinal column (vertebrae), the ribs and the sternum.

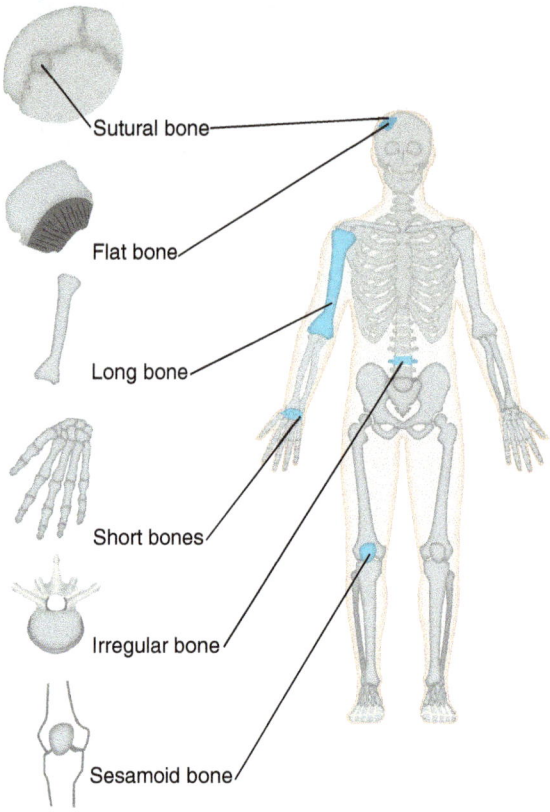

Figure 15.9 Types of bone based on shape

The skull

The skull refers to the bony framework of the head of 22 bones (excluding the auditory ossicles) and is divided into two categories: the cranial bones (8) and the facial bones (14).

- **Cranial bones:** These eight bones form the cranial cavity, which encloses and protects the delicate tissue of the brain. The eight cranial bones are: the frontal bone, two parietal bones, two temporal bones, the occipital bone, the sphenoid bone and the ethmoid bone (Figure 15.10).
- **Facial bones:** These 14 bones form the face and include: two nasal bones, two maxillae (joined to form the upper jaw), two zygomatic bones (cheekbones), the mandible (jawbone), two lacrimal bones, two palatine bones, two inferior nasal conchae and the vomer (Figure 15.11).

The spine (spinal column)

The spine refers to the spinal (or vertebral) column and the mechanical structures that it is composed of, the vertebrae and the ligaments and tendons that connect them.

The spinal column consists of 33 vertebrae, the individual bones of the spine that stack up to complete its bony structure. The vertebrae all have a similar shape, but the distinct sections have structural modifications necessary for movement and protection. Each vertebra is numbered and the number preceded by a letter that indicates its section, e.g. C1 is the first vertebra in the cervical spine.

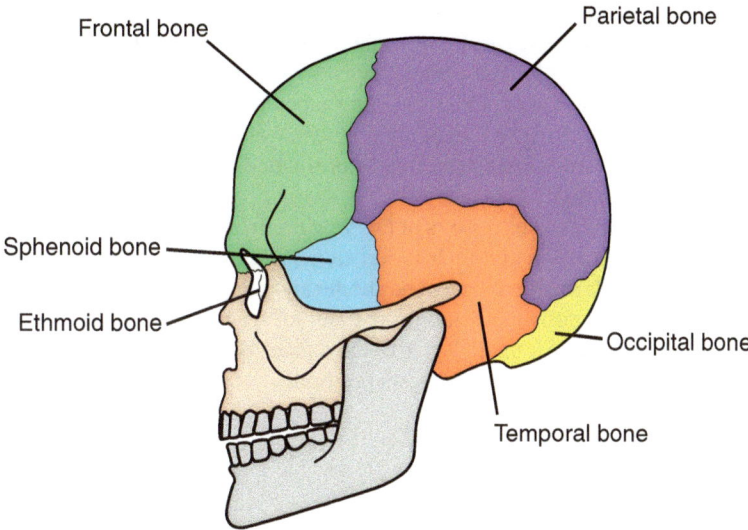

Figure 15.10 The cranial bones

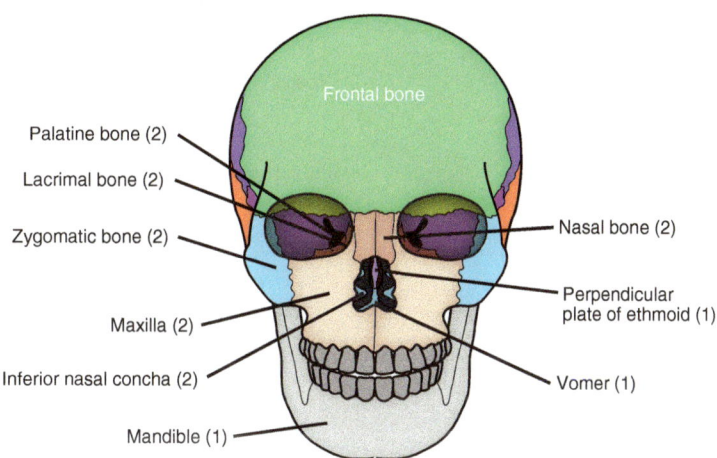

Figure 15.11 The facial bones

A typical vertebra (Figure 15.12) has two main parts – the anterior portion (front) is the vertebral body, which bears the weight. It is spongy bone surrounded by a thicker, harder shell. The vertebral body connects to the posterior section of the vertebra by two bones called the pedicles. These connect with the lamina, leaving a cavity for the spinal cord. The lamina has three body extensions – the spinous process in the middle and the transverse processes at either side. These vary in structure along the spine to permit and restrict movement, and are necessary for attachment of muscles, ligaments and tendons.

Spinal vertebrae

The five sections of spinal vertebrae are (see Figure 15.13):

- **Cervical vertebrae (C1–C7):** The skull rests upon these seven vertebrae at the top of the spine. The first cervical vertebra is the atlas (C1) which, with C2 (the axis), connects the spinal column and skull. The atlas has no vertebral body or spinous process, but the axis has a bony extension (the odontoid peg), which extends up into the gap in the atlas. This structure enables the pivotal movement of the head at the neck. The rest of the cervical vertebrae are typically structured, although the transverse processes are short to facilitate neck movement.

- **Thoracic vertebrae (T1–T12):** There are 12 thoracic vertebrae and they are typically shaped, with T1–T10 having two articular facets, which are cup-shaped to enable connection with ribs.
- **Lumbar vertebrae (L1–L5):** The five lumbar vertebrae are typically shaped and have the largest vertebral bodies as they carry most of the weight of the body. Spinous processes are short and flat, helping to maintain an upright stance.
- **The sacrum (S1–S5):** The sacrum consists of five sacral vertebrae, fused between the ages of 16 to 26. They are atypical in presentation and their fusion allows space for spinal nerves to exit between gaps called foramina. The female sacrum is wider and shorter and more acutely curved than the male sacrum.
- **The coccyx:** The coccyx consists of four vertebrae, fusing in the third decade of life (the 20s). Despite being fused, they have a small degree of movement against the sacrum. These are atypical in formation and the coccyx resembles a cone-shaped bone.

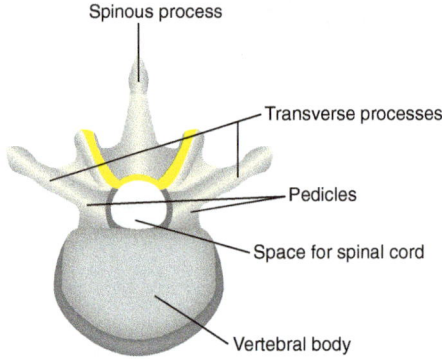

Figure 15.12 Parts of a vertebra

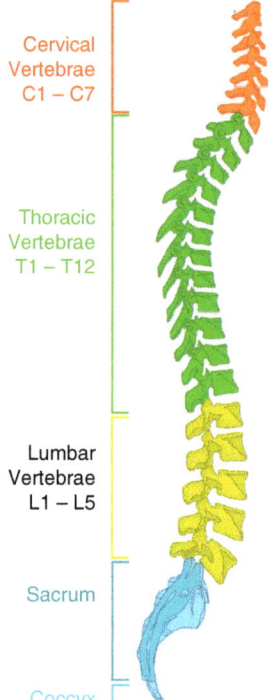

Figure 15.13 Sections of the spinal column

Intervertebral discs

Typically, an intervertebral disc sits sandwiched between the bodies of two vertebrae. Intervertebral discs have a fibrous outer ring and contain a gel-like pulp. When a disc ruptures (or slips), the fibrous outer ring cracks and the gelatinous core can escape. Under physical stress, they can also bulge significantly and press on adjacent spinal nerves, causing pain and other symptoms. Intervertebral discs form fibro-cartilaginous joints, with the vertebrae permitting their slight movement, and also function as shock absorbers within the spine.

Supporting structures

A system of ligaments with tendons and muscles provide the supporting structures that stabilise the vertebrae of the spinal column. Ligaments provide stability during rest and movement and help prevent damage to the cord and vertebrae from excessive movement. Tendons complement ligaments in their function, as they are fibrous tissue that can endure tension through their densely packed collagen fibres. Tendons attach muscle to bone. The muscles of the spine are complex and work together or individually to flex, rotate or extend the spine.

APPLY

When people injure their spine, they may damage the vertebrae, the cord or the supporting structures – or all three! Injuries to the cord can be detected easily by a neurological assessment and through medical imaging (scans such as MRI or CT). Medical imagery also helps to identify injuries to the vertebrae. However, we must not forget about the supporting structures. These can be damaged in situations where the neck has been twisted or moved suddenly and stopped suddenly. These also need to repair themselves.

Thoracic bones

Bones of the thorax protect the organs contained within the thoracic cavity, e.g. the heart and lungs. They include the sternum (breast bone) and ribs (Figure 15.14):

- **The sternum:** consists of three parts – manubrium, body and the xiphoid process.
- **The ribs:** 12 pairs are numbered from superior to inferior. Posteriorly, the ribs connect with the thoracic vertebrae. They are different lengths, 1–7 being the longest and then decreasing in size. Ribs 1–7 are true ribs as they directly attach to the sternum by a strip of hyaline cartilage (costal cartilage). Ribs 8–10 (false ribs) attach indirectly to the sternum. Ribs 11 and 12 are termed floating ribs as they are not attached to the sternum but attached only posteriorly to thoracic vertebrae 11 and 12.

Bones of the appendicular skeleton

The appendicular skeleton consists of 126 bones: the bones of the arms and legs and also those bones that attach them to the axial skeleton, i.e. the pectoral and pelvic girdles.

The pectoral girdle

The pectoral girdle attaches the bones of the upper limbs to the axial skeleton and consists of a clavicle and a scapula on each side (Figure 15.15).

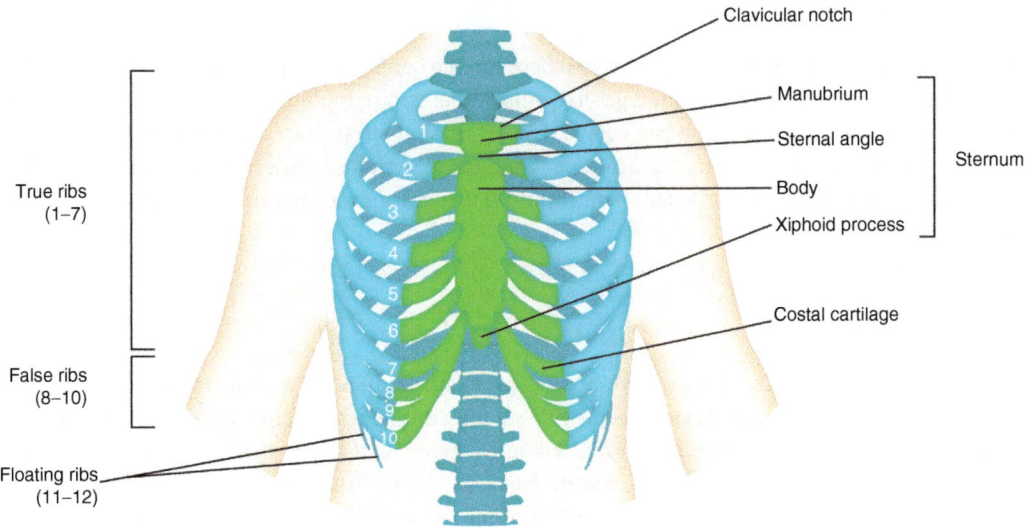

Figure 15.14 Bones of the thorax

- **Clavicles**: commonly known as the collarbones, these lie superior to the first rib on each side and in a horizontal position.
- **Scapulae**: commonly referred to as the shoulder blades, they are situated posteriorly between the second and seventh ribs on each side.

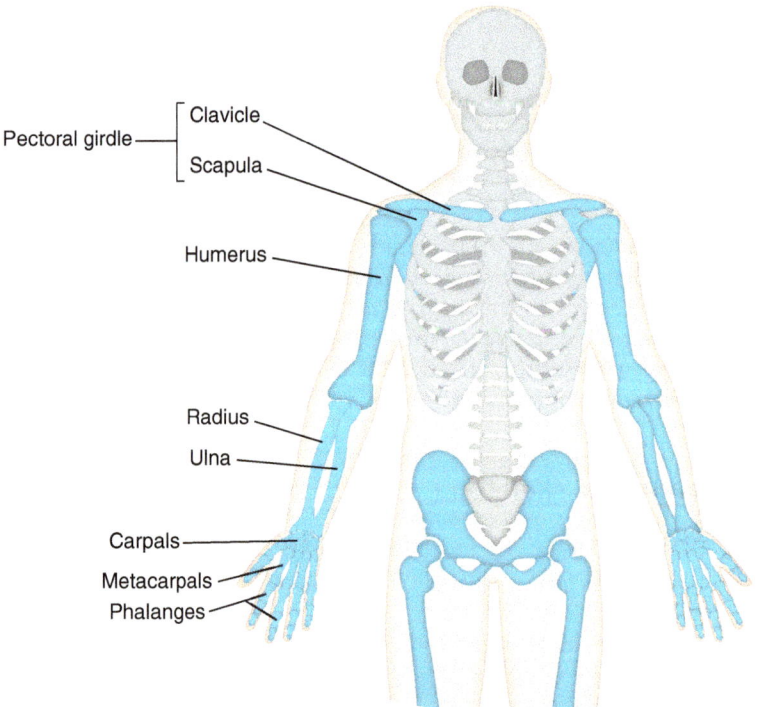

Figure 15.15 Bones of the pectoral girdle and upper extremities

Bones of the upper extremities (arms – upper limbs)

There are 30 bones in each of the upper limbs: humerus, ulna and radius (forearm), eight carpal bones in the wrist, five metacarpals in the palm and 14 phalanges (bones in the digits) (Figure 15.15):

- **Humerus:** This is the largest and longest bone of the upper arm.
- **Ulna:** Located on the medial aspect of the forearm (the same side as your little finger) it is the longest of the forearm bones.
- **Radius:** Located on the lateral aspect of the forearm (the same side as your thumb), this is shorter than the ulna and is wider at the bottom (distal radius) than at the top (proximal radius).The radius and ulna articulate with the humerus to form the elbow joint.
- **Carpal bones:** The wrists consist of eight small bones and are arranged into two rows of four bones including:

  - Row one (proximal row, i.e. closest to the arm) – scaphoid, lunate, triquetrum, pisiform.
  - Row two (distal row, i.e. further away from the arm) – trapezium, trapezoid, capitate and hamate.

- **Metacarpal bones:** These are located in the middle region of your hand (palm). Each of the five metacarpals is divided into proximal (base), medial (shaft) and distal (head) regions and are numbered one to five starting at the thumb. If you clench your fist you will see what are commonly known as 'knuckles', these are the heads of your metacarpal bones.
- **Phalanges:** These are the bones of the digits of the hand. There are two in the thumb and three in each of the other digits.

The pelvic girdle

The pelvic girdle is made up of two pelvic bones (hip bones). In adults each hip bone is made up of three components: ilium, ischium and pubis. In infants and children these are not fused and are separated by cartilage. As a child matures into adulthood, they fuse to form the adult hip bone by around the age of 23. The hip bones join at the front (anterior) at the pubic symphysis and at the back (posterior) with the sacrum; together they form the bony pelvis (Figure 15.16). The role of the bony pelvis is to provide support for the spinal column and the organs of the abdominal and pelvic cavities. In women, the pelvis is shaped to facilitate birth and the joints at front and back of the pelvis tend to soften during pregnancy to permit easier passage of the baby.

Bones of the lower extremities (legs – lower limbs)

Like the upper extremities, each of the lower extremities consists of 30 bones: femur, patella (kneecap), tibia and fibula, seven tarsal bones (ankle), five metatarsals and 14 phalanges (bones of the digits) (Figure 15.17):

- **Femur:** the longest and strongest bone in the body.
- **Patella:** a small triangular sesamoid bone found at the front of the knee joint. The proximal end of the kneecap develops in the tendon of the quadriceps femoris muscle. The patellar ligament attaches the patella to the tibia.
- **Tibia:** commonly referred to as the shin bone, is the stronger of the two bones in the lower leg and is the weight-bearing bone. It articulates with the femur at the proximal end and forms the knee joint.
- **Fibula:** runs parallel to the tibia on the outside of the leg and does not join with the femur but helps to stabilise the ankle joint.

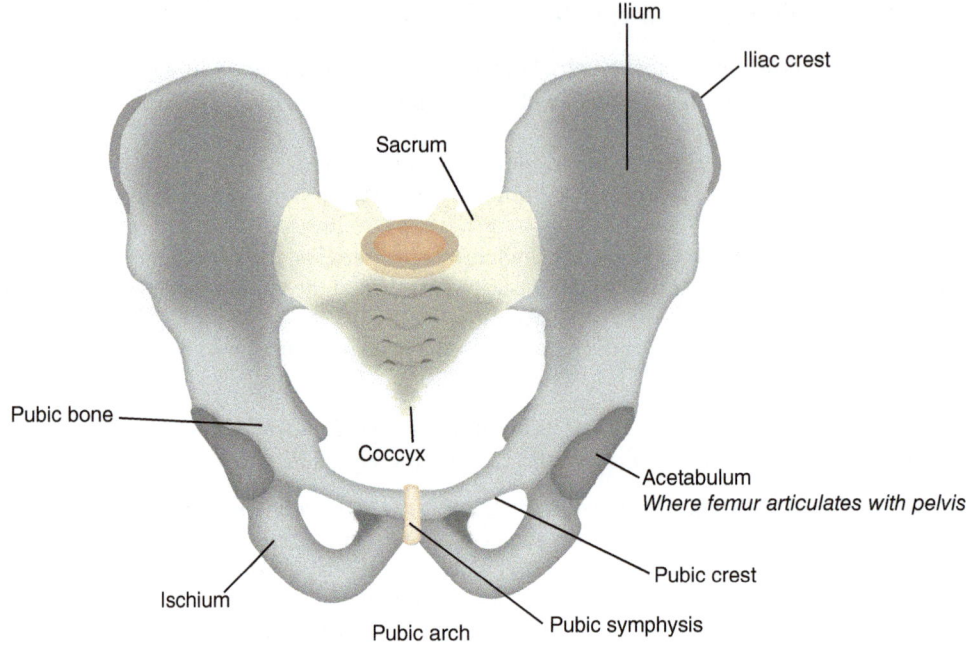

Figure 15.16 The pelvic girdle

- **Tarsal bones:** make up the tarsus or ankle and consist of seven bones: talus, calcaneus, navicular, three cuneiform bones and cuboid.
- **Metatarsal bones:** the region in the middle of the foot and the bones are comparable with the metacarpals, are numbered one to five (starting at your big toe) and divided into proximal, shaft and distal regions.
- **Phalanges:** the bones in the digits of the feet. The great toe (hallux) has two phalanges whereas all the other toes have three.

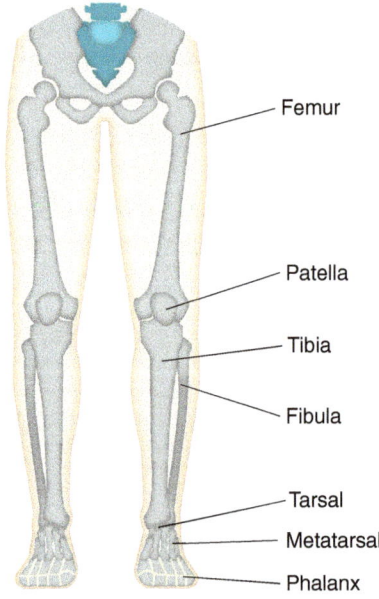

Figure 15.17 Bones of the lower extremities

JOINTS

A joint (or articulation) is the point at which two or more bones meet. They are classified according to the type of connective tissue that connects the bones: there are three types of joints – fibrous, cartilaginous and synovial. The type of connective tissue in joints also relates to the range of movement possible at that joint with an additional classification of:

- **Synarthroses** (fixed or unmovable).
- **Amphiarthroses** (slightly movable).
- **Diarthroses** (freely movable).

We will discuss joints according to the type of connective tissue involved in creating the joint but also identify the movement possible.

Fibrous joints

This type of joint occurs when two bones are attached by fibrous connective tissue, they have no joint cavity and, as they are attached very tightly, there is little or no movement (synarthroses). There are three main types of fibrous joints: suture, syndesmosis and gomphosis:

- **Suture:** This type of fibrous joint is found between the bones of the skull. The tissue between the bones is dense irregular collagenous connective tissue and the periosteum on both the inner and outer surface of the adjacent bones continues over the joint.

APPLY

Newborn babies have two main fontanelles (spaces where the bones have not yet ossified) – anterior and posterior – commonly known as 'soft spots'. These allow the skull to remain flexible during birth and for continued growth of the brain after birth. The posterior fontanelle usually closes within the first two to three months of life while the anterior (larger) one normally closes by the time the child is around 24 months.

- **Syndesmosis:** These joints are found in the lower legs between the distal tibia and fibula (distal tibiofibular joints) and in the forearm between the radius and ulna (radioulnar joint). This type of joint is formed with either a sheet of fibrous tissue (ligament) or a bundle of fibrous tissue (interosseous membrane) between the bones. As these joints have limited movement, functionally they are classified as amphiarthroses.
- **Gomphosis:** This type of joint is formed with collagenous connective tissue. They are specialised joints of pegs fitting into sockets. The only gomphosis joints in the body are those between the teeth and the mandible and maxillae. The connective tissues between the teeth and their sockets are periodontal ligaments and there is limited movement in the joint; however, this may change if the person has gum disease.

Cartilaginous joints

These types of joints are connected by cartilaginous material, in particular hyaline cartilage. As with fibrous joints, they do not have a joint cavity. There are two types of cartilaginous joints: synchondroses (primary) and symphyses (secondary):

- **Synchondroses:** This type of joint is associated with the growth of bones. Hyaline cartilage is involved in bridging the gap. An example of a synchondrosis is the epiphyseal plate that connects the diaphysis and epiphysis. Another example is the joint between the manubrium of the sternum and the first rib. These joints may be temporary and the hyaline cartilage can be replaced with bone. At this stage the joint becomes known as a syntosis or bony joint.
- **Symphyses:** The ends of the bone are covered by hyaline cartilage and fibrocartilage connects the ends of the two bones. They are permanent joints designed for strength and resilience and are found in the midline of the body. The pubic symphysis, which connects the two pelvic bones, is an important example of a symphysis that often becomes more flexible during pregnancy and facilitates the birth.

Synovial joints

This is the most common type of joint in the body. A synovial joint has a distinct characteristic in having a space, the synovial cavity, between the ends of the articulating bones. The large range of movement in synovial joints defines them functionally as diarthroses as they are freely movable. Friction at the articulating surfaces is low because the articular cartilage is elastic and the joint is filled with fluid. Synovial joints all have the same characteristics and contain an articular cartilage, an articular capsule and a joint cavity (Figure 15.18).

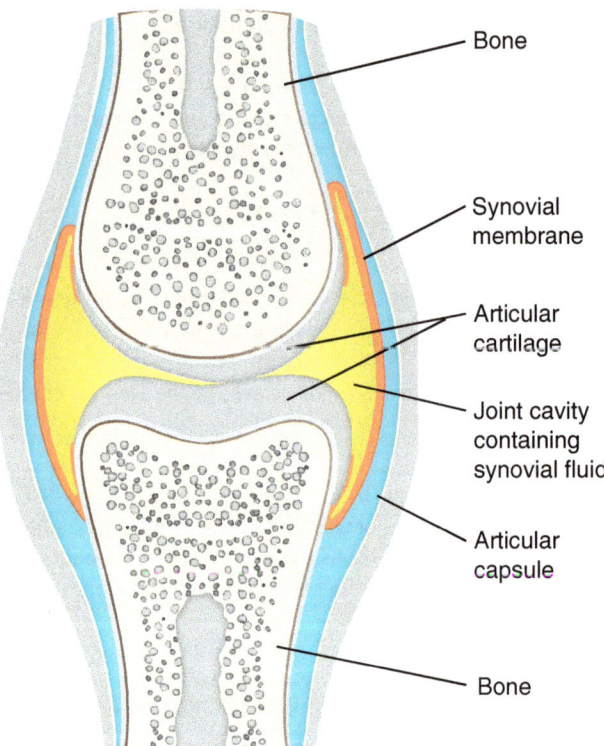

Bone

Synovial membrane

Articular cartilage

Joint cavity containing synovial fluid

Articular capsule

Bone

Figure 15.18   Synovial joint

- **Articular cartilage:** The ends of the articulating bones are covered by a layer of articular cartilage composed of hyaline cartilage. The presence of the hyaline cartilage creates a smooth slippery surface, thereby decreasing the friction and acting as a shock absorber between the ends of the bones.
- **Articular capsule:** The articular capsule is made up of two layers, the outer fibrous membrane and the inner synovial membrane. It acts as a sleeve-like structure to enclose the synovial cavity.

○ *Fibrous membrane*: This is composed of dense irregular connective tissue (predominately collagen fibres) that attaches to the periosteum of the articulating bones. It facilitates a considerable amount of movement whilst providing stability. Some of the fibrous membranes are arranged into parallel bundles as ligaments, which provide additional strength to the joint and prevent it dislocating.

○ *Synovial membrane*: This is composed of areolar connective tissue with elastic fibres enabling the capsule to stretch during movement. The synovial membrane may also contain collections of adipose tissue known as articular fat pads, e.g. the intrapatellar fat pad in the knee.

• **Joint cavity:** this is filled with synovial fluid secreted by the synovial membrane. It is a viscous, clear or pale-yellow fluid containing hyaluronic acid and interstitial fluid. It forms a thin film over the articular capsule surfaces. Its purpose is to lubricate the joint and prevent friction, absorb shocks, provide nutrients and oxygen, and remove waste products and carbon dioxide. Synovial fluid also contains phagocytic cells that digest and remove microbes and debris produced by wear and tear.

In addition, the joint is usually strengthened by ligaments and in some instances may contain articular discs (menisci), labra and fat pads.

Accessory ligaments, articular discs and labra

Many synovial joints contain intracapsular or extracapsular ligaments known as accessory ligaments. Extracapsular ligaments are found outside the articular capsule, e.g. the tibial and fibular collateral ligaments of the knee joint. Intracapsular ligaments are inside the articular capsule but outside the synovial cavity, e.g. anterior and posterior cruciate ligaments of the knee joint. Some synovial joints contain a crescent-shaped pad of fibrocartilage that lies between the ends of the articulating bones; they are attached to the fibrous capsule. These pads are known as articular discs or menisci (meniscus singular).

A fibrocartilaginous lip that extends from the edge of the joint socket, particularly in ball and socket joints, e.g. the hip and shoulder, is known as a labrum (labra plural). The purpose of the labra is to increase the surface area of contact and to deepen the joint socket.

Classification of synovial joints

There are six types of synovial joints: hinge, pivot, ball and socket, saddle, condyloid, and gliding (Table 15.2). They are named after the shape of bones or the type of movement that joint can perform. Movements can be classified according to the form of motion, the direction of movement or the relationship between body parts during the movement. They include: gliding, angular movements, rotation and special movements. Some of the key terms that you may hear associated with types of movement can be seen in Table 15.3.

Table 15.2 Types of synovial joints

Joint type	Movement of joint	Examples
Hinge	Flexion/extension	

(Continued)

Table 15.2   (Continued)

Joint type	Movement of joint	Examples
Pivot	Rotation of one bone around another	Ulna—Radius
Ball and socket	Flexion/extension/adduction/ abduction/internal and external rotation	Hip —Femur
Saddle	Flexion/extension/adduction/ abduction	
Condyloid	Flexion/extension/adduction/ abduction	
Gliding	Gliding movement	

Table 15.3   Types of movements

Action	Definition
Extension	Increases the angle/distance between two bones or parts of body
Flexion	Decreases the angle of a joint
Abduction	Moves away from the midline
Adduction	Moves closer to the midline
Circumduction	A combination of flexion, extension, abduction and adduction
Supination	Rotation of hand and forearm - turns the palms up

Action	Definition
Pronation	Rotation of the hand and forearm - turns the palms down
Plantar flexion	Lowers the foot (points the toes)
Dorsiflexion	Elevates the foot
Rotation	Moves a bone around its longitudinal axis
Inversion	Turning the sole of the foot inwards, outer side of foot on the ground
Eversion	Turning the sole of the foot outwards, inner side of foot on the ground

THE MUSCULAR SYSTEM

There are three types of muscle tissue: skeletal, cardiac and smooth. They share some properties although they vary in terms of location, microscopic anatomy and control by the nervous and endocrine systems (Chapter 2). Muscle tissue is highly specialised and has four major functions: contractility (shortens); excitability (responds to a stimulus and produces action potentials); extensibility (stretches beyond its resting length without damage); and elasticity (returns to its resting length after stretching). Cardiac and smooth muscle have been discussed in Chapters 12 and 8; the purpose in this chapter is to look specifically at skeletal muscle.

Skeletal muscle

Skeletal muscle is the most abundant of the three types of muscle in the body accounting for approximately 40–50% of a person's total body weight. Skeletal muscle is striated (when looked at under a microscope it appears to have light and dark stripes) and is under voluntary control of the nervous system. It has four key functions: production of body movements; maintaining body position; storage and movement; and generation of heat:

- **Body movements:** Skeletal muscle helps produce body movements as it is attached to the bones of the skeleton. Contraction of the skeletal muscle pulls the tendons, which in turn move the skeleton, thereby producing body movements such as walking, running and turning your head.
- **Body position:** Contractions of the skeletal muscle maintain body position and posture by stabilising joints, e.g. sustained contractions of your neck muscles help ensure that your head is maintained upright.
- **Storage and movement:** Skeletal muscle stores 80% of the body's water and is a reservoir for intracellular ions such as potassium. Skeletal muscle contraction also assists in the flow of lymph fluid and returning blood to the heart.
- **Generation of heat:** Contraction of skeletal muscle requires the production and use of adenosine triphosphate (ATP) which is used as energy: three quarters of this energy is released as heat. Skeletal muscle plays an important role in maintaining normal body temperature. When we are cold, involuntary contractions of the skeletal muscles, i.e. shivering, help increase the rate of heat production.

Characteristics of skeletal muscle fibre

Each skeletal muscle is an individual unit comprised of muscle fibres, connective tissue, nerves and blood vessels (Figure 15.19). A layer of dense irregular connective tissue known as fascia supports and surrounds the muscle allowing free movement, and provides an entry and exit route for nerves, blood and lymphatic vessels. Three layers of connecting tissue extend from the fascia to further strengthen and protect the skeletal muscle. These are:

- **Epimysium:** the outermost layer consists of dense irregular tissue and surrounds the entire muscle.
- **Perimysium:** consists of dense irregular tissue and surrounds bundles of muscle fibres (10–100 or more) known as fascicles.
- **Endomysium:** consists of reticular fibres, separates each individual muscle fibre.

All three layers of connective tissue are continuous with the connective tissue that attaches skeletal muscle to bone and other muscles. In some cases, the three layers can extend beyond the muscle and form tendons, which attach the muscle to the bone, e.g. Achilles (calcaneal) tendon that attaches the calf muscle (gastrocnemius) to the heel bone (calcaneus). In other cases, the three layers form an aponeurosis (a flat, broad sheet), e.g. the broad tendinous portion of the oblique and transverse abdominal muscles that attaches to the linea alba.

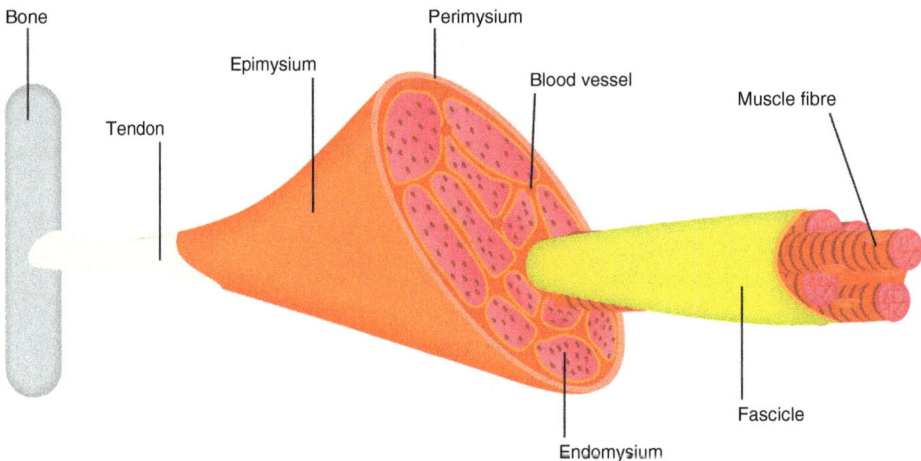

Figure 15.19 Characteristics of skeletal muscle fibre

An artery and one or two veins alongside a nerve extend through the connective tissue to penetrate each skeletal muscle. The capillary beds that surround the muscle fibres have an abundant arterial blood supply and blood is carried from the capillaries by the veins. The skeletal muscle is stimulated to contract by specialised nerve cells called motor neurons (Chapter 5).

ACTIVITY 15.3: UNDERSTAND

Watch this video clip to learn more about the characteristics of skeletal muscle and muscular contraction.

This video clip can be accessed by **scanning the QR code** with your smart phone camera or via https://study.sagepub.com/essentialap2e.

MUSCLE CONTRACTION
PROCESS (7:38)

Skeletal muscle fibre

A single skeletal muscle fibre is a long cylindrical cell that contains multiple nuclei at the periphery of the fibre. A mature muscle fibre could measure up to 10 cm in length and 10–100 µm in diameter. It is striated (striped) due to the arrangement of the different types of filaments. Figure 15.20 shows the different fibres and the striations.

Skeletal muscle is formed during embryonic development by the fusion of mesodermal cells known as myoblasts. Contractile proteins accumulate within the cytoplasm of the myoblasts and are converted to muscle fibres. The number of muscle fibres that you have within your body is determined before birth and remains fairly constant throughout your lifetime. Skeletal muscle has a number of components (Figure 15.20, Table 15.4).

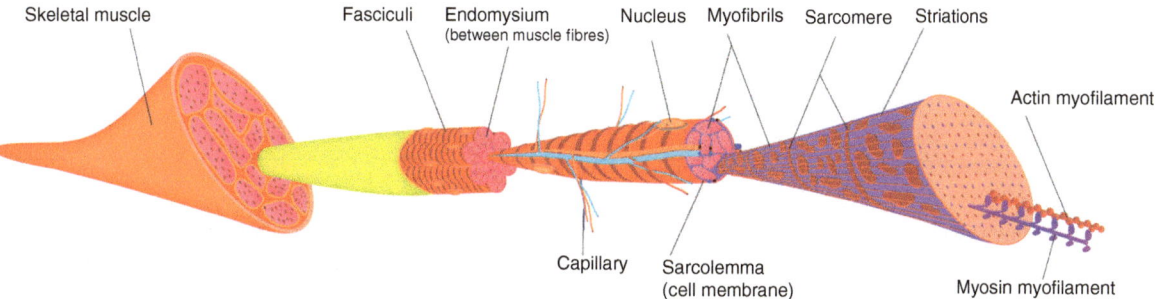

Figure 15.20 Components of skeletal muscle

Table 15.4 Components of skeletal muscle

Component	Description
Sarcolemma	The plasma membrane of the muscle fibre
Transverse T tubules	Enter the sarcolemma from the surface and pass towards the centre. The T tubules are open to the outside of the muscle fibre and are filled with interstitial fluid
Sarcoplasm	The cytoplasm of the muscle fibre. It contains large amounts of glycogen which is used to produce ATP. The sarcoplasm also contains myoglobin, a red-coloured protein found only in muscles and used to bind oxygen. Oxygen is released from the myoglobin when needed by the mitochondria to produce ATP
Myofibrils	Small threadlike structures that are the contractile organelles of the skeletal muscle and extend the entire length of the muscle
Sarcoplasmic Reticulum (SR)	A fluid-filled sac enclosing each of the myofibrils. The SR stores calcium ions when the muscle is relaxed
Terminal cisterns	Open-ended sacs of the SR sit against the sides of the T tubule
Myofilaments	Small protein structures contained within the myofibril. There are two types: thin filaments, composed mainly of actin, and thick filaments, composed mainly of myosin. Both thick and thin filaments are involved in the contractile process. Two further proteins, tropomyosin and troponin, are found in the thin filament. Troponin helps hold tropomyosin in place and tropomyosin prevents actin combining to myosin thereby preventing muscle contraction from occurring
Sarcomeres	Compartments that are the functional unit of the myofibril. They are separated from each other by Z discs and consist of a number of distinct bands and zones. The **A band** extends the entire length of the thick filament and is the darker middle region of the sarcomere. Near the end of the A band thick and thin filaments lie side by side and this area is known as the zone of overlap. The **I band** is found only in thin filaments and is the lighter, less dense region. In the middle of the A band is the narrow area **H zone** that contains only thick filaments. In the middle of the H zone the **M line** is formed by proteins that hold the thick filaments together

Skeletal muscle fibre contraction and relaxation

During skeletal muscle contraction the length of the individual thick and thin filaments does not change despite shortening of the sarcomere. Contraction occurs through a process known as the sliding filament mechanism.

The sliding filament mechanism

At rest there is minimal overlap between the thick and thin filaments and the skeletal muscle is at its longest. Contraction of skeletal muscle occurs when myosin attaches to and stretches along the length of the thin filament at both ends of the sarcomere. This attachment slowly pulls the thin filaments towards the M line. As the thin filaments move inwards the Z discs become closer together and the sarcomere shortens. Shortening of the sarcomere leads to an overall shortening of the muscle fibre resulting in the entire muscle becoming shorter.

 APPLY

BODIE FAMILY

Let's think about the Bodie family. Danielle is an infant but the number of muscle fibres in her body is already determined. Throughout her childhood the muscle fibres will enlarge and she will grow stronger. Other members of the family throughout adulthood can develop the size of their muscles by engaging in exercise.

The contraction cycle

Activity in the motor neuron that supplies the muscle (it may supply many or a few muscle fibres) initiates muscle contraction. Signals produce contraction of the muscle by a process called excitation–contraction coupling and connects the propagation of the action potential (excitation) to the sliding of the filaments (contraction).

Calcium is crucial in muscle contraction. An increase in calcium concentration starts muscle contraction, whereas a decrease in calcium concentration stops muscle contraction. Calcium is stored in the sarcoplasmic reticulum (SR) and released through the calcium-releasing channels on the SR membrane in response to an action potential propagating along the sarcolemma into the T tubules. The calcium flows into the sarcoplasm around the thick and thin filaments and combines with troponin causing it to change shape. This removes tropomyosin from the myosin-binding sites on actin and allows myosin heads to attach to actin, forming cross-bridges and beginning the contraction cycle.

Within the SR membrane active transport pumps use ATP to move calcium from sarcoplasm into the SR. As the muscle action potential moves through the T tubule, calcium channels remain open and calcium continues to flow into the sarcoplasm more quickly than it is removed by the pumps. This continues until action potentials cease through the T tubules and the calcium channels close. Active transport returns calcium back to the SR, decreasing levels of calcium and allowing tropomyosin to cover the myosin-binding sites and leads to muscle relaxation.

The neuromuscular junction

Somatic motor neurons stimulate skeletal muscle fibres. An axon from the motor nerve cell extends to a group of skeletal muscle fibres and muscle action potentials start in the Neuromuscular Junction (NMJ).

The NMJ is the synapse (Chapter 5) between the motor neuron and skeletal muscle fibre separated by a synaptic cleft (small gap). A neurotransmitter (chemical) permits communication between the two cells. The axon terminal (end of the motor neuron) at the NMJ divides into a group of synaptic end bulbs containing synaptic vesicles which contain the neurotransmitter Acetylcholine (ACh). The motor end plate (the sarcolemma opposite the synaptic end bulbs) contains acetylcholine receptors needed for the binding of ACh. A nerve impulse creates a muscle action potential in four steps:

1. **Release of ACh:** the nerve action potential travels along the motor neuron until it reaches the synaptic vesicles, which release ACh into the synaptic cleft.
2. **Activation of ACh receptors:** ACh molecules bind to ACh receptors on the motor end plate and open an ion channel. Sodium flows across the membrane.
3. **Production of muscle action potential:** the flow of sodium into the muscle fibre increases the positive charge and triggers a muscle action potential, which spreads across the sarcolemma and travels down T tubules to the cisternae which release calcium into the sarcoplasm causing muscle contraction.
4. **Termination of ACh activity:** action potential generation stops as ACh is broken down by Acetylcholinesterase (AChE). AChE breaks ACh into acetyl and choline, two products that cannot activate the ACh receptors. The concentration of calcium ions in the sarcoplasm declines, returns to resting levels and muscle contraction ends.

Muscle metabolism

Skeletal muscle fibres' activity can range from high to low levels. During high levels of activity large amounts of ATP are used. ATP is continually produced as the stored ATP only creates enough power for a contraction of a few seconds. ATP is produced from three sources:

1. **Creatine phosphate:** during muscle relaxation excess ATP is produced and diverted to production of creatine phosphate. Creatine kinase (an enzyme) enables the energy from ATP to be transferred to creatine, forming creatine phosphate and ADP (Chapter 9). During muscle contraction levels of ADP increase and creatine kinase transfers the high-energy phosphate group from creatine phosphate back to ADP forming ATP. Excess creatine phosphate is excreted in urine as the waste product creatinine.
2. **Anaerobic respiration:** the most abundant source of energy in muscle fibres is glycogen, used as a source of energy (Chapter 9). During times of increased muscle activity, the production of lactic acid may give you a 'stitch'.
3. **Aerobic respiration:** muscle tissue receives oxygen from haemoglobin or myoglobin and glucose is metabolised to form ATP (Chapter 9).

Muscle fatigue

After a period of prolonged activity, the muscle is unable to maintain its force of contraction – muscle fatigue. It primarily occurs due to changes in the muscle fibre and is thought to occur when there is:

*   inadequate calcium released from the sarcoplasmic reticulum.
*   insufficient oxygen supply.
*   depletion of glycogen.
*   build up of lactic acid.

Muscle tone

A muscle is never totally inactive; sustained contraction of the muscle fibre is maintained by motor neurons. Involving only a few motor neurons means that muscle is relaxed but has tone keeping the muscle firm but not strong enough for movement.

Isotonic and isometric contractions

Contraction of the muscle may be isotonic or isometric: most activities require both types of contraction. During an isotonic contraction, the force of contraction remains unchanged while the length of muscle changes, enabling body movements and movement of objects. There are two types of isotonic contraction:

- **Concentric contraction:** the force generated overcomes the resistance of the object to be moved. The muscle shortens, pulls on the tendon to decrease the joint angle and produces movement, e.g. the lifting phase of a bicep curl.
- **Eccentric contraction:** when the muscle contracts but the length increases; following the bicep curl the weight can be smoothly and controllably lowered. Eccentric contractions are more powerful than concentric contractions, i.e. when lowering a heavy object, you use more strength than lifting it.

In an isometric contraction the force generated does not exceed the resistance of the object to be moved. There is no change in length of the muscle and no movement, but energy is still used. We use isometric contractions continually to resist the downward force of gravity. These contractions help stabilise joints and are essential in maintaining posture.

Types of skeletal muscle fibres

Skeletal muscle varies in composition and functionality. Those containing high levels of myoglobin are 'red muscle fibres', are dark in appearance due to an extensive blood supply and are high in mitochondria. 'White muscle fibres' have low levels of myoglobin and are light in appearance. The rate at which muscle fibres contract and relax, the speed of ATP production, and the rate at which they fatigue vary. Skeletal muscle falls into three main groups based on these characteristics: Slow Oxidative (SO) fibres, Fast Oxidative–Glycolytic (FOG) fibres, and Fast Glycolytic (FG) fibres.

--- **GO DEEPER** ---

Slow Oxidative (SO) fibres: have the smallest diameter and are the least powerful. They contain large amounts of myoglobin and contain a lot of mitochondria. They mainly produce ATP by aerobic respiration. The rate at which ATP is used is relatively slow and the contraction cycle occurs at a slower pace; however, they are less likely to fatigue. These slow twitch fibres are essential for posture and are particularly useful in endurance activities, e.g. running a marathon.

Fast Oxidative-Glycolytic (FOG) fibres: are intermediate in diameter. Similar to the SO fibres, they contain large amounts of myoglobin and have an increased blood supply. They are moderately resistant to fatigue as they produce large quantities of ATP through anaerobic as well as aerobic metabolism. The rate at which ATP is used is three to five times faster than in SO fibres, so their rate of contraction is faster but duration shorter. FOG fibres are used primarily in activities such as walking or sprinting.

Fast Glycolytic (FG) fibres: are the largest in diameter and contain the largest number of myofibrils. They have a low content of myoglobin, relatively few blood capillaries and few mitochondria. They contain large amounts of glycogen and produce ATP through glycolysis. They hydrolyse ATP quickly and generate the most forceful and fastest contraction; however, they fatigue more quickly than SO and FOG fibres. FG fibres are adapted for anaerobic movements of short duration such as weight-lifting. FG fibres increase in size, strength and glycogen content when a person undertakes a strength training programme as they require greater strength for a shorter period of time causing hypertrophy (enlargement) of the muscle.

GO DEEPER

Exercise and skeletal muscle

The number of SO and FG fibres in a person's muscle is genetically determined and accounts for how people perform physically. Those with a high proportion of SO fibres are better at endurance activities such as marathon running compared to those with more FG fibres. Although the overall numbers of skeletal muscle fibres do not change with exercise, the characteristics of fibres already present may change to a degree. If the body is repeatedly exposed to a certain type of training, these lead to changes in the muscle fibre. Aerobic exercises such as running or swimming may lead to changes in some FG fibres with an increase in the number of mitochondria and blood capillaries becoming FOG fibres. There may also be changes in cardiovascular and respiratory systems that enable better delivery of oxygen and nutrients to the tissues; however, they do not increase muscle size.

Alternatively, if the body is repeatedly exposed to exercise that requires short bursts of great strength, e.g. weight-lifting, you will see a gradual change in the number of FG fibres. There is an increase in the number of thick and thin filaments leading to an increase in the muscle size.

Skeletal muscle movement

To produce movement the skeletal muscle exerts a force on tendons, which pull on bones or other structures. The ends of articulating bones are pulled together but do not move equally in response to the contraction, one tends to remain stationary or close to its original position. The attachment of a muscle's tendon to the stationary bone is the origin, and the attachment of the other tendon to the movable bone is the insertion. At a joint there are normally at least two opposing muscles (agonist and antagonist) that produce the movement in opposite directions. An agonist is primarily responsible for producing the action while the antagonist will cause the muscle to move in the opposite direction. For example, when flexing the forearm, the biceps brachii is the agonist and the triceps brachii the antagonist. The reverse happens when the forearm is extending. There are a number of types of movements, which can be seen in Table 15.3.

Principal skeletal muscles

Muscles can be named according to several factors: shape, size, location and number of insertions. The skeletal muscles can be grouped into four areas:

1. Head and neck (Figure 15.21).
2. Upper limbs (Figure 15.22).
3. Thorax and abdomen (Figure 15.23).
4. Lower limbs (Figure 15.24).

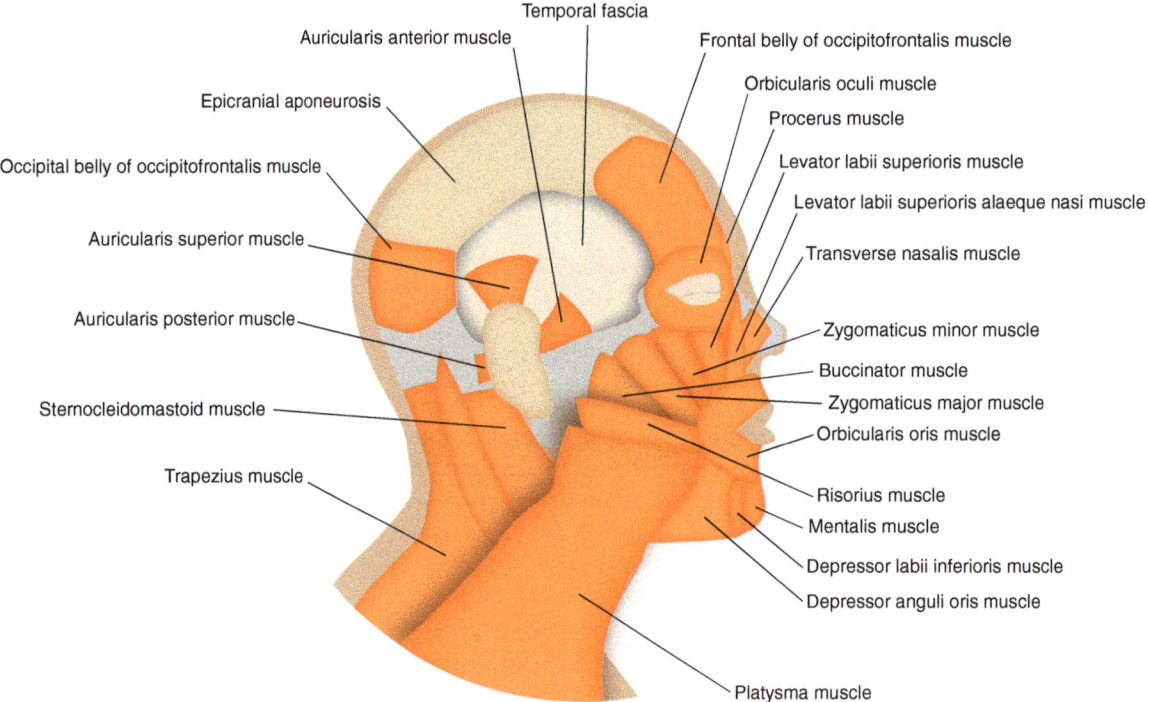

Figure 15.21 Muscles of the head and neck

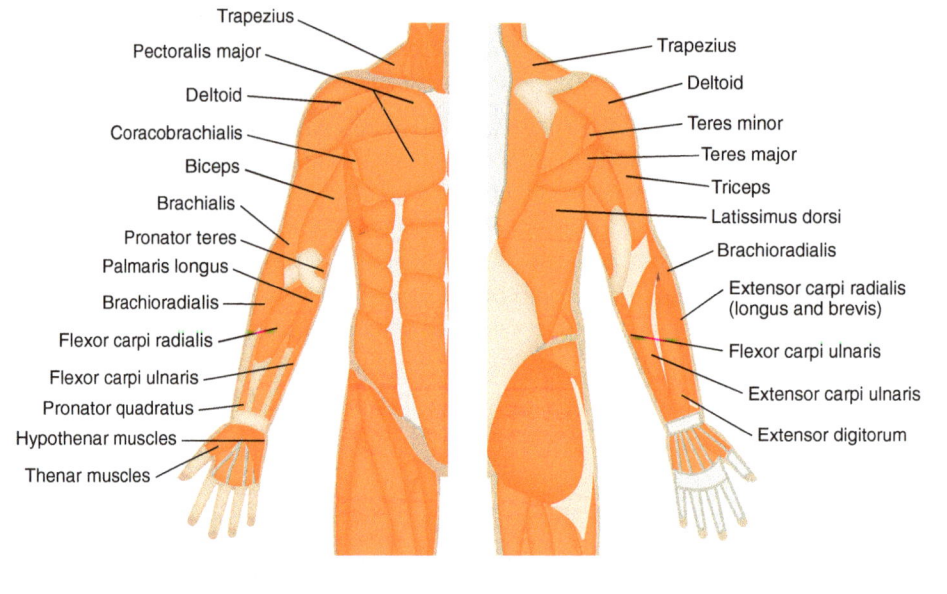

Anterior   **Posterior**

Figure 15.22 Muscles of the upper limbs

(a) Anterior view

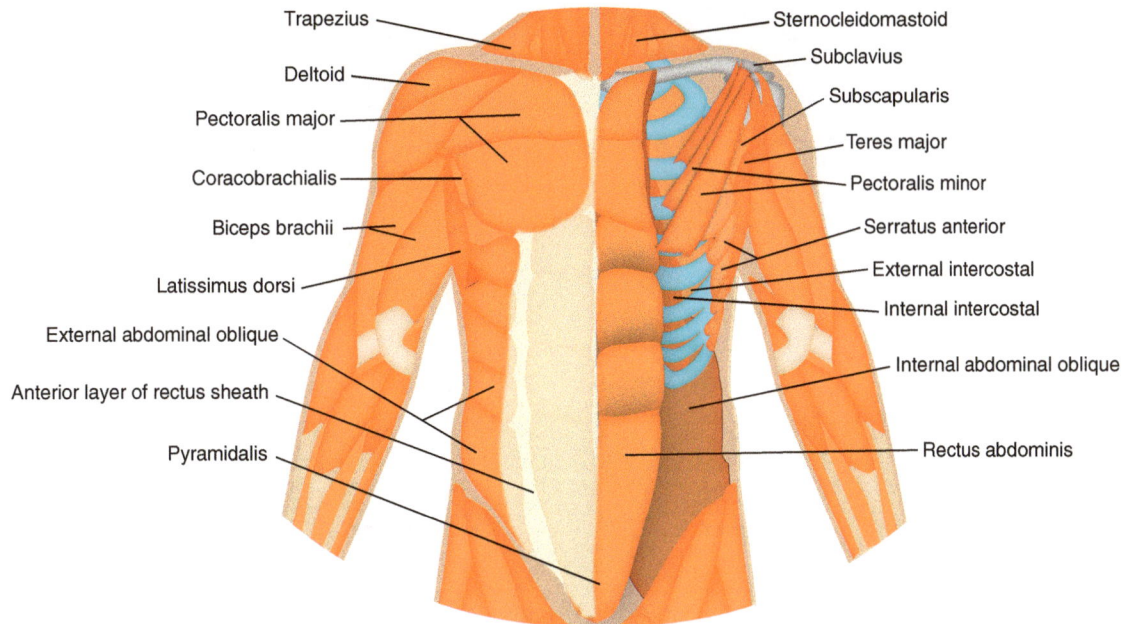

(b) Posterior view

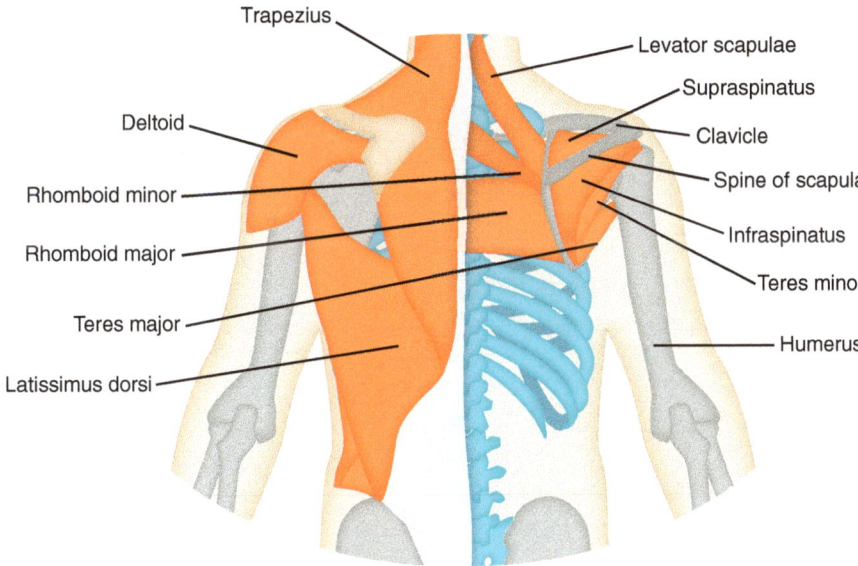

Figure 15.23 Muscles of the thorax and abdomen

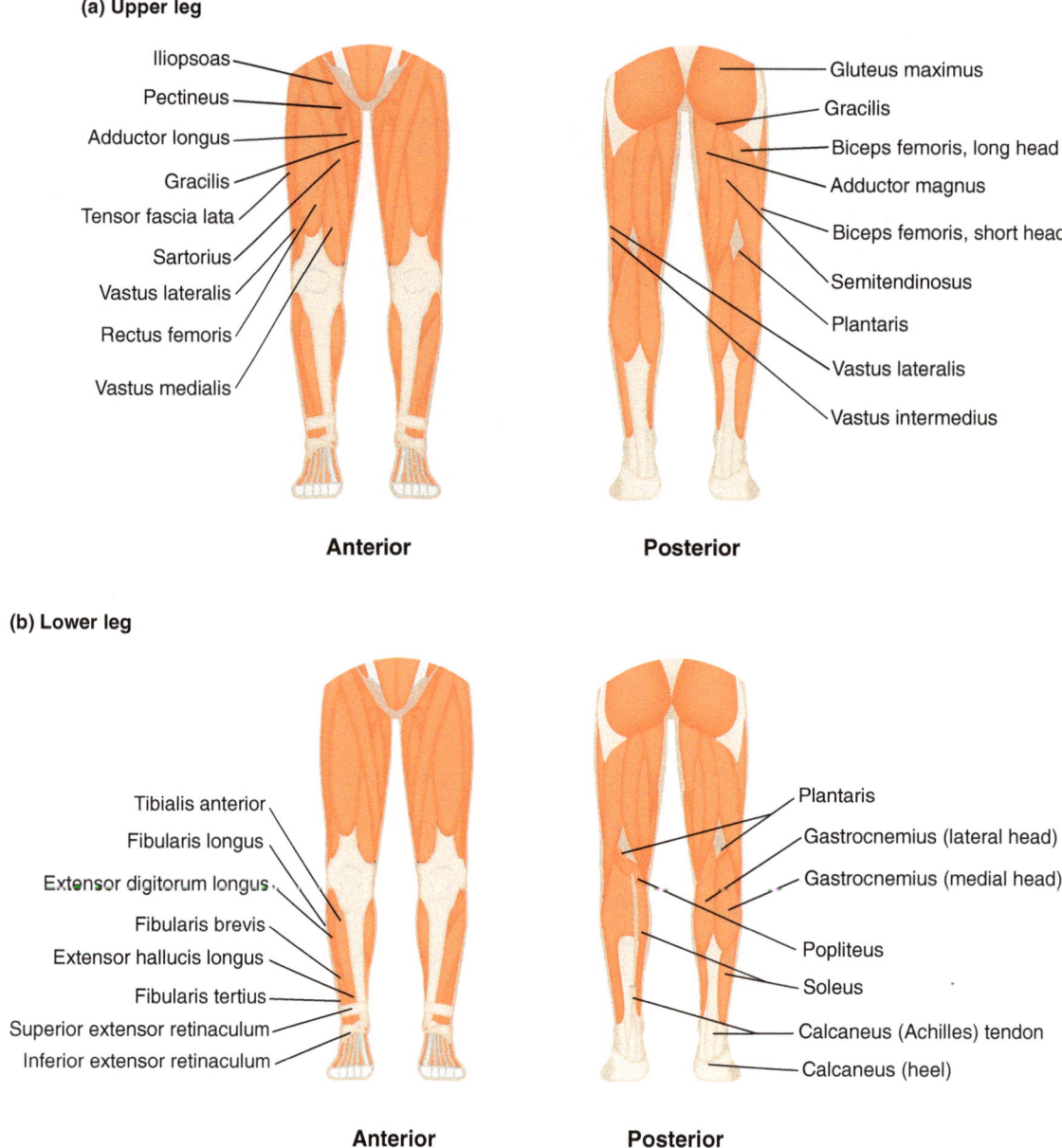

Figure 15.24 Muscles of the lower limbs

WALKING

One of the unique characteristics of mankind is the ability to walk upright on two limbs (bipedalism). While some other current primates can produce a type of upright stance, they do not have the control and anatomical modifications for bipedalism found in our own species.

In walking, as in most body movements, the body moves in such a way that muscles are functioning mainly at or near their resting length, thus enabling maximum strength to be applied. Walking involves the coordination of numerous muscles to provide momentum, sustained balance and minimum exertion for efficient energy use. In normal walking each leg passes through two stages:

- A support phase when the foot is on the ground.
- A swing phase when the foot is off the ground (Ganong, 2019).

With the two legs functioning, the support phases overlap so that periods of double support occur when both feet are on the ground. A comfortable walking speed for young adults is about 80–83.4 m/min (Ganong, 2019; Rose et al., 2006).

APPLY

BODIE FAMILY

It is interesting to think about the development of walking in young children. Danielle is still too young to have started to walk but will begin to learn to walk around the age of one. Her gait will be somewhat stiff, with some flexion of her hips and knees. She will place her feet flat and wide apart giving a wide base for balance. Danielle's early walking movements will see her arms held in abduction and her elbows in extension. This will produce walking in a staccato manner – sudden, short steps. As Danielle matures, the width of her stride will narrow, movements will become smoother and she will begin to use her arms in a reciprocal swing. This smoother walking will result in an increased step length and speed – an adult walking pattern (Kermoian et al., 2006).

Butler et al. (2006) discuss how walking alters during pregnancy and in ageing. During pregnancy, changes in joint flexibility, weight and spinal curvature occur and result in physical discomfort, particularly in the lower back, pelvis, hips and knees. The woman may also be less stable and have a wider stance as a result of an altered centre of gravity.

Older people tend to walk more slowly, take shorter and variable length steps, and have a wider stride (base for balance). In addition, balance tends to become less stable. These changes increase the risk of falls, which become more common with age and are a leading cause of injury in those over 65 years. One of the elements in care of older people is to promote exercise, which minimises these changes. T'ai chi (a Chinese form of exercise based on the martial arts, but fairly slow and gentle) is now undertaken widely in many different countries and has been shown to improve balance and reduce falls in older people (Tsang and Hui-Chan, 2004; Voukelatos et al., 2007).

CONCLUSION

Now that you have read through this chapter and undertaken the activities, you should have a good understanding of how the skeletal and muscular systems operate. You should be able to see how bones and muscles work together to facilitate movement, provide protection and maintain homeostasis through the regulation of minerals.

GO DEEPER

Further reading

Docherty, B. (2007) 'Skeletal system. Part one – bone structure', *Nursing Times*, 103(5): 28-9.
Docherty, B. (2007) 'Skeletal system. Part two – bone growth and healing', *Nursing Times*, 103(6): 28-9.
Docherty, B. (2007) 'Skeletal system. Part three – the axial skeleton', *Nursing Times*, 103(7): 26-7.
Docherty, B. (2007) 'Skeletal system: Part four – the appendicular skeleton', *Nursing Times*, 103(8): 26-7.
Richardson, M. (2006) 'Muscle physiology. Part 1: overview of muscle physiology', *Nursing Times*, 102(47): 28-9.
Richardson, M. (2006) 'Muscle physiology. Part 2: skeletal muscles and muscle fibres', *Nursing Times*, 102(48): 26-7.
Richardson, M. (2006) 'Muscle physiology. Part 3: muscles – the working units', *Nursing Times*, 102(49): 26-7.
Richardson, M. (2006) 'Muscle physiology. Part 4 – movement and muscle problems', *Nursing Times*, 102(50): 26-7.

KEY POINTS

- Bone is a highly vascular connective tissue that has a number of key functions, namely: support, movement, storage, protection and production of blood cells.

- Bone is composed of extracellular bone matrix and four types of bone cells: osteogenic cells, osteoblasts, osteocytes and osteoclasts.

- There are two types of bone: spongy and compact.

- Bone formation takes place during embryonic/fetal development through two processes: intramembranous and endochondral ossification. Bone continues to grow in length until early adulthood and in width throughout the entire lifespan.

- Bone is continually renewing itself through the process of remodelling and can repair itself if it is broken.

- There are five classifications of bones based on their shape: long, short, flat, irregular and sesamoid (the sixth, sutural, is not based on shape).

- There are 206 bones in the adult skeleton. The skeleton is divided into the axial skeleton consisting of the bones of the skull, the spinal column, the ribs and sternum, and the appendicular skeleton consisting of the pectoral and pelvic girdles, bones of the upper extremities and bones of the lower extremities.

- Joints are the point at which two or more bones meet. There are three classifications of joints: fibrous, cartilaginous and synovial. Synovial joints are the commonest type of joint in the body.

- Skeletal muscle accounts for 40-50% of a person's total body weight.

- Skeletal muscle has four key functions: production of body movements, maintaining body position, storage and movement, and heat generation.

- Skeletal muscle fibres are long cylindrical multinucleated cells that are under voluntary control.

- Skeletal muscles contract and relax depending on the levels of calcium and the release of the neurotransmitter acetylcholine.

- There are three types of muscle fibres: slow oxidative, fast oxidative-glycolytic and fast glycolytic.

- Skeletal muscles provide a range of movements including: flexion, extension, adduction, abduction.

- Skeletal muscles can be named according to the shape, size, location and number of insertions. They are grouped in four areas: head and neck, upper limbs, thorax and abdomen, and lower limbs.

REVISE

TEST YOUR KNOWLEDGE

This chapter helps you to understand the functions of the musculoskeletal system, how muscles and joints work together with bones to facilitate movement and posture, and the role exercise plays in maintaining a healthy musculoskeletal system. Revise each section individually and check your understanding of each section before you move to the next. Do you understand how bones are formed? Can you identify the major muscles of the body? Do you know how a bone heals after it is broken? Challenge your knowledge.

In order to help you revise, consider the following questions, answers for which can be found by visiting **https://study.sagepub.com/essentialap2e**.

Test yourself by revising the chapter first, and then answer these questions without looking at the book. Afterwards compare your answers with the text and with the notes you made. Did you miss anything in your notes? Here are the questions:

1   Identify and describe the function of the four types of bone cells.

2   Identify the constituent parts of the axial and appendicular skeleton.

3   List the types of synovial joint found in the human body and explain the role of particular structures common to all types.

4   Skeletal muscle is under control of which division of the nervous system?

5   A 72-year-old lady fell when out walking her dog and sustained a fractured neck of the femur. Describe the stages of bone healing. Include in your answer any nutrients that are necessary to promote the healing process.

6   Explain the steps involved in muscle contraction.

For additional revision resources visit: **https://study.sagepub.com/essentialap2e**.

REVISE

ACE YOUR ASSESSMENT

- Revise key terms relevant to this chapter with interactive flashcards.

- Test yourself with quizzes and multiple-choice questions.

- Access the glossary with audio to hear how complex terms are pronounced.

- Explore recommended websites suitable for revision.

REFERENCES

Andreoli, A., Celi, M., Volpe, S.L., Sorge, R. and Tarantino, U. (2012) 'Long-term effect of exercise on bone mineral density and body composition in post-menopausal ex-elite athletes: a retrospective study', *European Journal of Clinical Nutrition*, 66: 69–74.

Butler. E.E., Druzin, M. and Sullivan, E.V. (2006) 'Gait adaptations in adulthood, pregnancy, aging and alcoholism', in J. Rose and J.G. Gamble (eds), *Human Walking*, 3rd edn. Philadelphia: Lippincott, Williams & Wilkins.

Ganong, W.F. (2019) *Review of Medical Physiology*, 26th edn. New York: McGraw-Hill.

Guadalupe-Grau, A., Fuentes, T., Guerra, B. and Calbert, J.A. (2009) 'Exercise and bone mass in adults', *Sports Medicine*, 39(6): 439–68.

Kermoian, R., Johansen, M.E., Butler, E.E. and Skinner, S. (2006) 'Development of gait', in J. Rose and J.G. Gamble (eds), *Human Walking*, 3rd edn. Philadelphia: Lippincott, Williams & Wilkins.

Naish, J. and Syndercombe Court, D. (eds) (2014) *Medical Sciences*, 2nd edn. Edinburgh: Elsevier.

Nilsson, M., Sundh, D., Ohlsson, C., Karlsson, M., Mellstrom, D. et al. (2014) 'Exercise during growth and young adulthood is independently associated with cortical bone size and strength in old Swedish men', *Journal of Bone and Mineral Research*, 29(8): 1795–804.

Pennline, J.A., Mulugeta, L., Lewandowski, B.E., Thompson, W.K. and Sibonga, J.D. (2014) 'The Digital Astronaut Project Remodeling Model'. Human Research Program Investigators' Workshop; 12–13 February, Galveston, Texas.

Rose, J., Morgan, D.W. and Gamble, J.G. (2006) 'Energetics of walking', in J. Rose and J.G. Gamble (eds), *Human Walking*, 3rd edn. Philadelphia: Lippincott, Williams & Wilkins.

Tsang, W.W. and Hui-Chan, C.W. (2004) 'Effect of 4- and 8-wk intensive Tai Chi training on balance control in the elderly', *Medicine & Science in Sports & Exercise*, 36(4): 648–57.

Voukelatos, A., Cumming, R.G., Lord, S.R. and Rissel, C. (2007) 'A randomized, controlled trial of tai chi for the prevention of falls: the Central Sydney Tai Chi Trial', *Journal of American Geriatrics Society*, 55(8): 1185–91.

PART 5

THE NEXT GENERATION

This section of the book focuses on the fundamental feature of all living things, that of reproduction in order for the human race to continue and evolve. It also considers how the body changes as it grows, develops and ages across the lifespan. It consists of the following two chapters:

- *Chapter 16. The Reproductive Systems*

  This chapter examines the structure, function and development of the male and female reproductive systems. It also examines how gametes are formed, conception and maternal changes during pregnancy and delivery. It relates to Chapter 2 (cell division) and Chapter 3, which discusses genetic makeup. It also looks at reproduction in the wider context of human relationships.

- *Chapter 17. Development through the Lifespan*

  This chapter discusses the changes that occur throughout the lifespan from the beginning of life through to older age. While it follows on from the preceding chapter on reproduction, it has links across all chapters as ageing and development result in changes in all systems of the body.

THE REPRODUCTIVE SYSTEMS

16

UNDERSTAND: CHAPTER VIDEO

Before working through this chapter, you might find it useful to have an overview of the reproductive systems. Watch this online video to enhance your understanding.

This video link can be accessed by **scanning the QR code** with your smart phone camera or via https://study.sagepub.com/essentialap2e.

REPRODUCTIVE SYSTEMS
(12:20)

LEARNING OUTCOMES

When you have finished studying this chapter you will be able to:

1. Describe the anatomy of the male and female reproductive systems and how these develop from the fetal structures into the sexually differentiated structures of the mature male and female
2. Describe the endocrine regulation of function of the reproductive systems, pregnancy, birth and lactation
3. Recognise the interaction between the physiological and psychological/emotional aspects of the interaction between male and female in achieving conception
4. Understand the way in which the mother adapts to the demands of pregnancy and lactation

INTRODUCTION

Reproduction is fundamental to the biological need for sustaining life as well as the personal need to create and nurture. This chapter will focus on human reproductive biology, through conception to birth and afterwards. You will learn about the structure and function of the male and female reproductive systems with reference to their development through life, their function in conception, maternal changes during pregnancy and delivery, and nurturing of the infant after birth. This chapter enables you to understand how human diversity is achieved through the reproductive process and identifies the relationship with reproductive health in the wider context of a healthy person. It must be noted that those of you aiming to become midwives will need to study parts of this content in greater detail than provided here.

Context

Within the Bodie family, reproduction has obviously been important in producing the next generation of their family as four generations now exist. Maud and George had three children. With the first delivery, labour was long and some assistance was required: forceps were applied to aid delivery. Maud then needed sutures to repair her perineum. Her other two children were delivered normally, but she was admitted to hospital for her deliveries because her first had needed forceps.

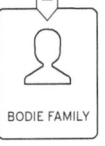

BODIE FAMILY

Hannah's second pregnancy was of twin girls, identified as identical twins at birth although they have taken different paths in life and have a number of characteristics where they differ (see epigenetic factors in Chapter 3). Michelle has recently had her first baby with Kwame, Danielle. She was quite a big baby at eight pounds 13 ounces (4 kg) and her delivery was facilitated by using a ventouse or vacuum extractor. This consists of a cap that fits on the baby's head and is attached to a suction device and is used to apply gentle traction to the baby's head to facilitate delivery. It is something that was not available when Maud was having her babies. It leaves a small swelling, and sometimes some bruising, on the baby's head but this disappears quite soon. Danielle is still being breast-fed.

SEX DIFFERENTIATION

Sex refers to the biological composition of male and female, in the historically accepted sense. Gender is the social expression of the sense of being male or female, for example, and is largely socially, culturally and psychologically determined. This element will be discussed later in this chapter.

The sex differences between male and female begin at conception when the female gamete (an ovum) is fertilised by a male gamete (a spermatozoon – plural spermatozoa, otherwise known as sperm). Normally each of these gametes contains 23 chromosomes – 22 autosomes and one sex chromosome formed through meiosis (Chapter 2). The ovum contains an X chromosome and the sperm either an X or Y. The zygote (fertilised ovum) contains 23 pairs of chromosomes, 22 pairs of autosomes (one of each from each parent) and one sex chromosome from each parent. The two sex chromosomes in the gamete are either an XX or an XY. A zygote with an XX pair will normally develop into a female, while one with an XY will normally develop into a male. However, hormonal influence also determines sex: the developing embryo, regardless of genetic sex, can potentially develop male or female reproductive systems (Mawhinney and Mariotti, 2013).

The Y chromosome plays a vital role in determining sex. It contains a gene (SRY – Sex-Determining Region of the Y) that codes for the production of Testis-Determining Factor (TDF), initiating the differentiation into male sex organs, most notably the testis. Lack of this gene permits female sex organs to develop.

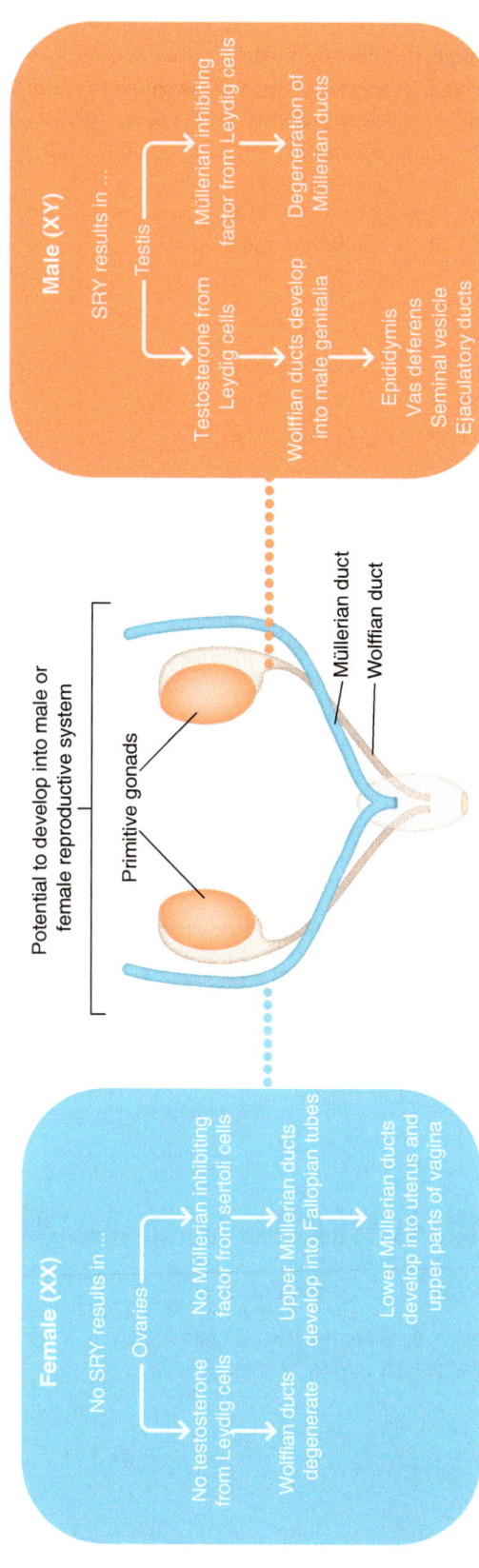

Male (XY)

SRY results in …

Testis

Testosterone from Leydig cells → Wolffian ducts develop into male genitalia → Epididymis
Vas deferens
Seminal vesicle
Ejaculatory ducts

Mullerian inhibiting factor from Leydig cells → Degeneration of Müllerian ducts

Potential to develop into male or female reproductive system

Primitive gonads

Müllerian duct

Wolffian duct

Female (XX)

No SRY results in …

Ovaries

No Mullerian inhibiting factor from sertoli cells → Upper Müllerian ducts develop into Fallopian tubes → Lower Mullerian ducts develop into uterus and upper parts of vagina

No testosterone from Leydig cells → Wolffian ducts degenerate

Figure 16.1   Sex differentiation

The embryo has genital (gonadal) ridges from which either ovaries or testes develop after about five to six weeks (Figure 16.1). Wolffian (mesonephric) ducts develop into the male reproductive tract, eventually forming the epididymis, vas deferens and seminal vesicle. Müllerian (paramesonephric) ducts differentiate into the female reproductive tracts, forming the oviduct (Fallopian tube), uterus and upper portion of the vagina (Mawhinney and Mariotti, 2013).

In summary, male sex anatomy results from having a Y sex chromosome and the presence of androgens (male sex hormones). Female sex anatomy results from two X sex chromosomes and the absence of androgens. The organs of the male and female reproductive systems can be identified in three groups (Table 16.1).

Table 16.1   Male and female reproductive organs

	Male	**Female**
Gonads (gamete and hormone producing organs)	Testes	Ovaries
Internal genitalia	Epididymis Vas deferens Accessory glands: • Seminal vesicles • Prostate gland • Bulbourethral glands	Fallopian (uterine) tubes Uterus Uterine cervix Vagina
External genitalia	Scrotum Penis	Mons pubis Pudendal cleft Labia majora Labia minora Bartholin's glands Clitoris

ACTIVITY 16.1: UNDERSTAND

Watch this online video explaining sex differentiation.

This video clip can be accessed by **scanning the QR code** with your smart phone or via https://study.sagepub.com/essentialap2e.

SEX DIFFERENTIATION (12:04)

STRUCTURE AND FUNCTION OF MALE REPRODUCTIVE SYSTEMS

Development of male genitalia

During the third to fifth months, the cells in the testes differentiate into three main types of cell:

- **Spermatogonia:** which are the germ cells for sperm formation.
- **Leydig cells:** outside the seminiferous tubules and which produce testosterone.
- **Sertoli cells:** which produce anti-Müllerian hormone and act as the 'nurse' cells for the germ cells (fetal spermatogonia).

Male external genitalia develop during the third and fourth months of gestation and the fetus continues to grow, develop and differentiate.

Testosterone produced from Leydig cells in the primitive testes furthers the formation of the male reproductive system through further differentiation of Wolffian ducts and common primordial genital tissue (Mawhinney and Mariotti, 2013) (Figure 16.2). Testosterone and Müllerian-Inhibiting Factor (MIF) secreted from immature Sertoli cells both inhibit differentiation of the Müllerian ducts into the female reproductive system.

Twelve weeks after conception, male genitalia are formed. The testes begin to descend through the inguinal canal between weeks six and ten, entering the scrotum by around week 28. The labelled components of the male reproductive system are shown in Figure 16.3 while the cross-sectional Figure 16.4 illustrates the relationships between them in the body.

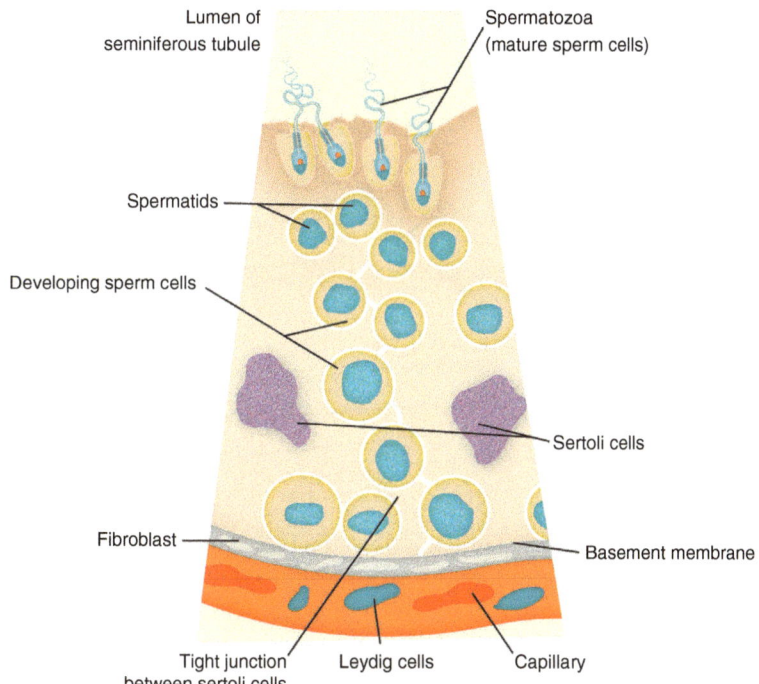

Figure 16.2 Cross-section of the seminiferous tubule

The functions of the male reproductive system are:

- **Testes:** To form the male gametes (sperm) and male hormones (testosterone).
- **Spermatic ducts and accessory glands:** To carry the sperm through the tubules for activation.
- **Penis:** To penetrate the female and deposit sperm within the female reproductive system.

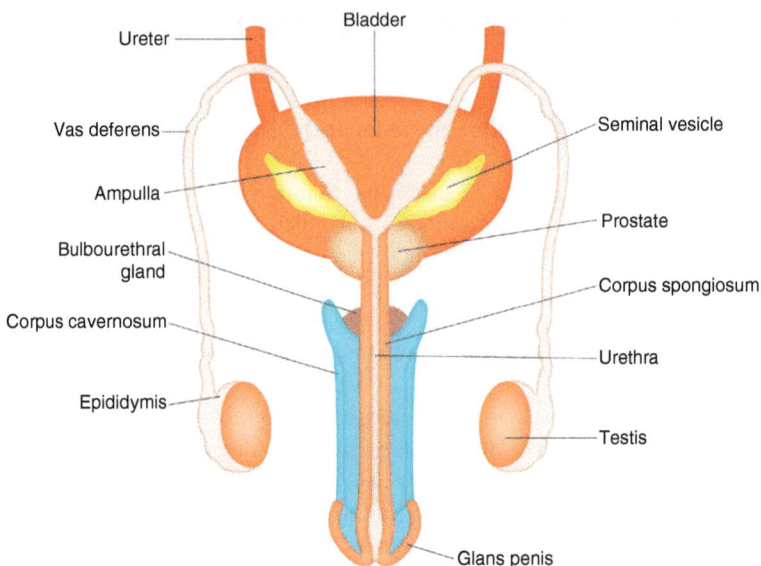

Figure 16.3 Diagram of the male reproductive system (posterior view)

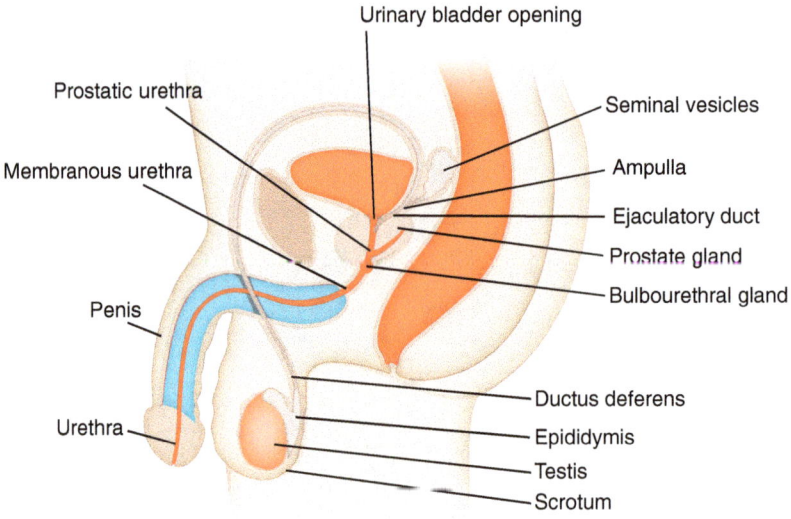

Figure 16.4 Cross-section of the male reproductive system

Testes

The testes are the reproductive glands which form the sperm and are oval in shape, 4–5 cm (2 inches) in length and 2.5 cm (1 inch) in diameter (Figure 16.5). The testes initially form within the abdomen but usually descend into the scrotum through the inguinal canal by the time of birth (Hughes and Acerini, 2008). They are suspended in the scrotum by the spermatic cord and the scrotum hangs outside the body, maintaining the testes at a temperature 1–2°C lower than core body temperature – the optimal temperature for sperm production. Their position in the scrotum also ensures they are not exposed to intra-abdominal pressure causing sperm to be squeezed out of the testes before maturation. Passing through the spermatic cord, the testicular artery, vein and nerve supply enter the testes.

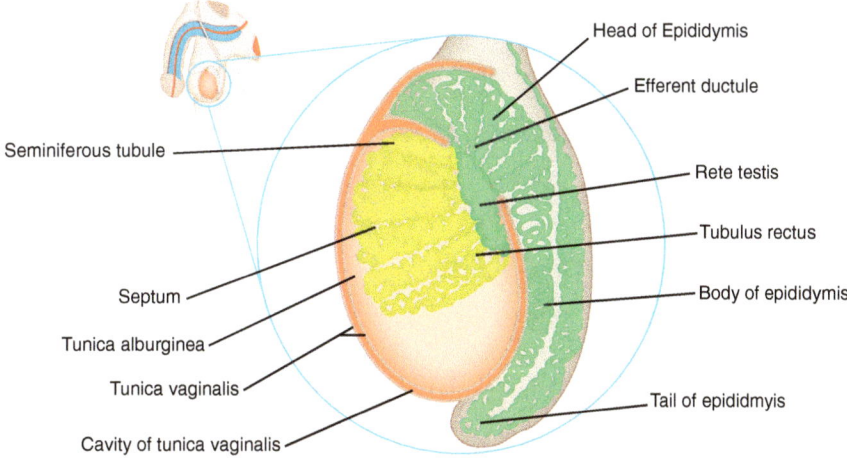

Figure 16.5 A testis

Figure 16.5 shows the three layers of tissue within the testis:

* **Tunica vaginalis:** covers the anterior and lateral surfaces of the testes and is derived from the peritoneum with descent of the testis. The fluid within prevents friction within the scrotum.
* **Tunica albuginea:** is fibrous tissue covering and creating partitions between seminiferous tubules. It also connects to the tunica vaginalis and epididymis.
* **Tunica vasculosa:** This innermost layer of connective tissue contains the blood supply to the testis.

APPLY

Cryptorchidism

Undescended testicle is a relatively common condition known as cryptorchidism (unilateral or bilateral) when a boy is born without both testes in the scrotum. These will usually descend to the normal position within three to six months and cause no problems. However, cryptorchidism should be treated if this does not happen as infertility can occur later in life if the testes are not positioned correctly.

Seminiferous tubules and spermatogenesis

Each testis contains about 200–300 lobules of which each contains 1–4 seminiferous tubules, which produce sperm (spermatozoa) from germinal epithelial cells. Spermatozoa are continuously produced after puberty and approximately 200 million sperm are contained in each ejaculation.

Sperm are tiny cells that have three main sections (Figure 16.6):

1. **The head** with the nucleus containing 23 chromosomes and the acrosome (a cap) which contains enzymes to penetrate the ovum for fertilisation.
2. **The mid-piece** largely containing mitochondria for energy.
3. **The tail** (flagellum) which propels the sperm through the female reproductive tract.

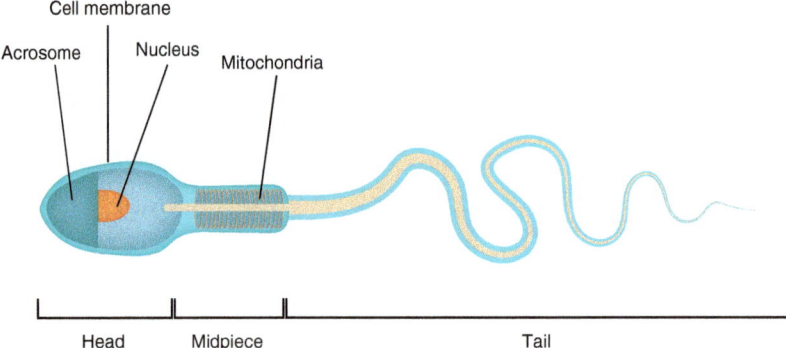

Figure 16.6 A sperm cell

Sertoli cells

Sertoli (sustentacular) cells support maturation by providing nutrients to germ cells and removing their waste material (Figure 16.2). They are activated by follicle stimulating hormone (FSH) produced by the pituitary gland, and by testosterone. Sertoli cells also produce inhibin, a peptide hormone that inhibits the production of FSH and increases spermatogenesis. Sertoli cells are stacked tightly together, forming the blood–testis barrier, which prevents large molecules gaining access to germ cells. This is important, as germ cells only contain half the chromosomes of most cells, making them genetically different from the cells of the body as a whole, and so they would be attacked by the immune system if this blood–testis barrier did not exist.

Sperm cells do not swim before ejaculation but Sertoli cells bath them in fluid which moves them through the seminiferous tubules to the rete testis. This is a network of tubules in the testis (Figure 16.5) that carries sperm to the efferent ducts. In this area, the sperm become concentrated and other fluids are reabsorbed. Infertility may result if this is not carried out effectively.

Testosterone secretion

Testosterone, the male sex hormone, is secreted by the Leydig cells (Figure 16.2) in the testes and has a number of different functions:

- Promotes male sex organ development.
- Influences libido and sexual behaviour.
- Stimulates metabolism, protein synthesis and muscle growth.
- Stimulates the growth of pubic hair, chest hair and reduces scalp hair.
- Increases bone calcium retention and strength.

While it is the major male hormone, it is also secreted in small amounts in women and influences sexual arousal.

Spermatic ducts and accessory glands

The spermatic ducts take sperm from each testis to the urethra in their journey to leave the body during ejaculation. This starts in the epididymis, continues with the ductus (vas) deferens, and ends with the ejaculatory duct at the seminal vesicles where the union with the urethra occurs.

The male reproductive system has three accessory glands: the seminal vesicles, prostate gland and bulbourethral glands.

Epididymis

This six-metre coiled and convoluted tube on the posterior border of the testis receives sperm entering the head of the epididymis from efferent ducts. The spermatozoa are still immature and incapable of fertilising ova at this stage, but travel through the length of the epididymis to its tail over 20 days. Here, they are stored, protected and remain fertile for 40–60 days. Once infertile, they disintegrate and are reabsorbed by the epididymis.

Ductus (vas) deferens

The tail of the epididymis straightens, turns 180 degrees and becomes the ductus deferens, a 40–45 cm (16 inch) long tube that ascends out of the scrotum into the pelvic cavity, via the inguinal canal. Its smooth muscle uses peristalsis (Chapter 8) to move the sperm. The ductus deferens terminates behind the bladder, widens and joins the duct of the seminal vesicles to become the ejaculatory duct, a short 2 cm (1 inch) duct that passes through the prostate gland and merges with the urethra.

APPLY

Testicular cancer

Testicular cancer is one of the most common forms of cancer in younger men, usually affecting them between the ages of 15 and 49. In order to promote early detection, men are encouraged to perform testicular self-examination (TSE) monthly. Jack's father has a history of testicular cancer and so Jack has been vigilant in performing TSE monthly since he was a teenager.

BODIE FAMILY

Accessory glands

The positions of these three sets of glands are shown in Figures 16.3 and 16.4. Together they contribute the bulk of the fluid of the ejaculate by secreting different liquids into the spermatic duct with the sperm.

Seminal vesicles

These two glands are blind-ended tubular glands of smooth muscle lined by stratified columnar epithelial cells. They each merge with the vas deferens on the same side to form the ejaculatory duct; the two ejaculatory ducts merge with the urethra. Secretions, stimulated by testosterone, are stored in the lumen of the seminal vesicles. During ejaculation, fluid from the seminal vesicles is expelled and mixes with that from the testes and the other accessory glands to form semen.

Secretions from the seminal vesicles:

- Comprise 60% of semen.
- Are alkaline and neutralise the acidic pH of the vaginal tract during sexual intercourse.

Prostate gland

The circular prostate gland, about 4 cm (1–2 inches) in diameter, surrounds the urethra and ejaculatory duct adjacent to the bladder. It contains between 30 and 50 tubuloacinar glands inside a single fibrous capsule with 20 pores to empty prostatic fluid into the urethra. The smooth muscles of the prostate help expel semen during ejaculation. Prostatic fluid is the first part of the ejaculate containing most of the sperm:

- It comprises around 30% of semen.
- It appears to enhance motility, survival and protection of genetic material compared with sperm in fluid mainly from the seminal vesicles.

Bulbourethral glands

These produce 10% of seminal fluid to provide lubrication for the head of the penis and protect spermatozoa by alkaline pH neutralising any acidity of the urethra from residual urine.

Penis

The penis has two functions: it is the passage for urine and is also the male sexual organ specialised to deliver sperm into the vagina. The urethra carries urine from the bladder and also transports seminal fluid from the ejaculatory duct through the penis into the vagina.

The penis is a tubular organ which anatomically consists of two halves: the root and the shaft and glans; the different components are labelled in Figure 16.7. The root is the internal attachment to the body wall and the shaft and glans make up the externally visible components of the penis. The urethra exits at the glans through the urethral meatus.

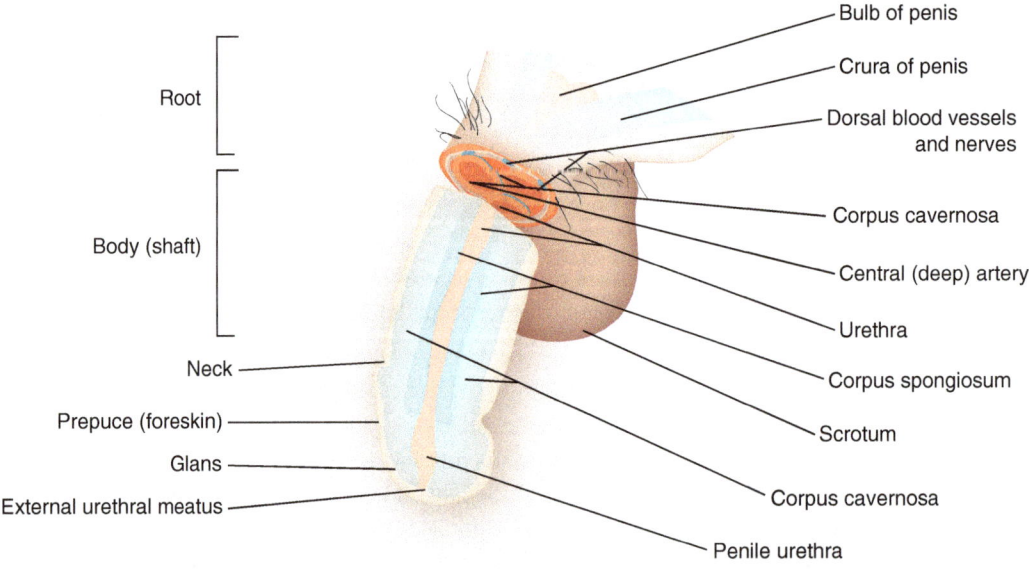

Figure 16.7   The penis

The loose skin surrounding the shaft permits expansion as an erection occurs during arousal. The foreskin (prepuce) is the skin over the glans where it is attached by a fold of tissue known as the frenulum. The glans and enclosing skin produce a waxy substance, smegma, from sebaceous glands.

The main body of the penis has three cylindrical bodies of erectile tissue surrounded by a fibrous sheath (tunica albuginea) and separated by a median septum. On the ventral surface, with the central penile urethra, is the corpus spongiosum, which expands at the distal end of the penis, filling the glans. Above this are the two corpus cavernosa. Erectile tissue is highly vascular and contains lacunae (tiny blood sinuses) that fill with blood during sexual arousal, expanding the penis. The lacunae are separated by connective tissue and smooth trabecular muscle, the latter providing muscle tone in the erect penis (Figure 16.8). Within the penile root, the corpus spongiosum attaches to the perineal membrane (surrounding the muscle of the perineum) at the bulb and the corpora cavernosa diverge (crura) and attach to the pubic arch.

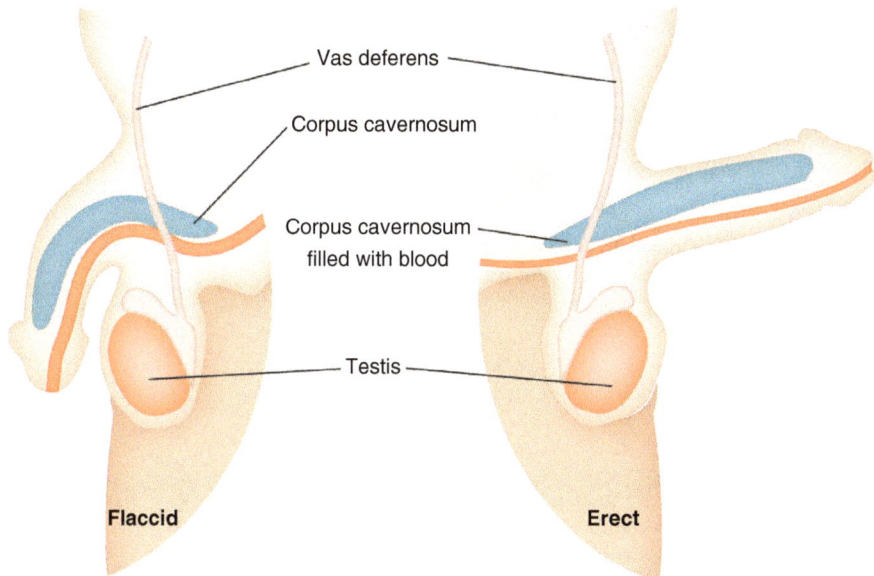

Figure 16.8 The flaccid and erect penis

ACTIVITY 16.2: UNDERSTAND

Watch this online video providing an overview of the male reproductive system.

This external video link can be accessed by **scanning the QR code** with your smart phone camera or via https://study.sagepub.com/essentialap2e.

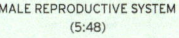

MALE REPRODUCTIVE SYSTEM
(5:48)

Endocrine regulation of male sex organs

Testicular activity is regulated by the endocrine activity of the hypothalamus and pituitary gland (Figure 16.9). The gonadotrophins from the hypothalamus stimulate the secretion of the two pituitary hormones, which act on male and female sexual organs – the follicle stimulating hormone (FSH) and luteinising hormone (LH). In the male, FSH acts on the Sertoli cells and stimulates spermatogenesis and secretion of the hormone inhibin, which acts through negative feedback to reduce FSH secretion from the anterior pituitary. FSH is thus maintained at a relatively constant level.

LH (which used to be known as Interstitial Cell-Stimulating Hormone – ICSH) stimulates the Leydig cells to secrete testosterone. Through negative feedback, testosterone inhibits secretion of both the hypothalamic and pituitary hormones, which initially cause testosterone secretion. Testosterone carries out a number of functions:

- Initiates the development of the secondary sexual characteristics at puberty (see below).
- Maintains function of the accessory reproductive organs.
- Promotes the sex drive by its action on the brain.
- Ensures normal sperm production by its action on the Sertoli cells.

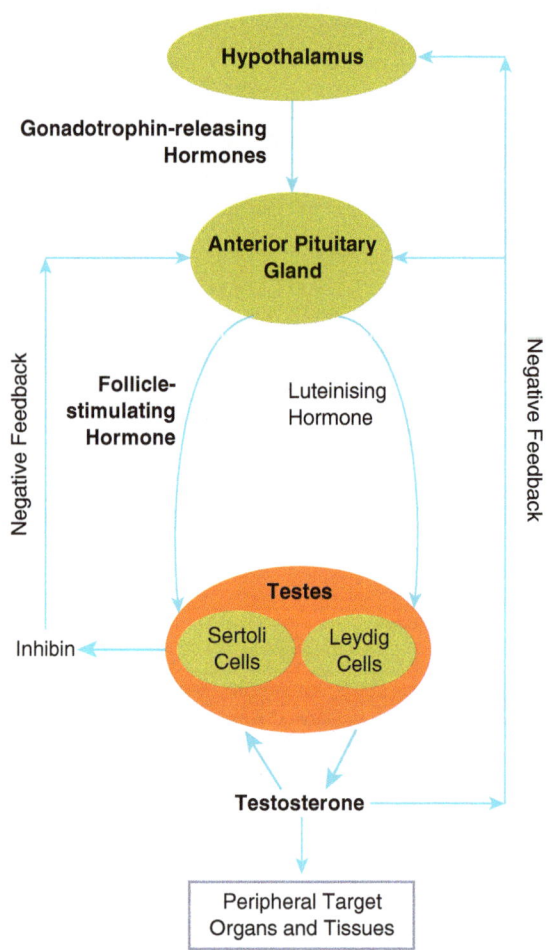

Figure 16.9 Endocrine regulation of the male sex organs

THE FEMALE REPRODUCTIVE SYSTEM

The female reproductive system develops from the Müllerian ducts (Figure 16.1). Parts of these fuse to form the uterus and vagina and the fused medial walls are resorbed from below upwards. Occasionally, this resorption does not fully occur and a bicornuate uterus, which has two 'horns' separated by a septum, results. The vagina develops from a rod of epithelial cells and the vagina forms as the centre of this rod breaks down. Development is complete by about the 20th week of gestation. The female reproductive organs are mainly internal and are shown in Figures 16.10 and 16.11. Figure 16.12 illustrates the female external genitalia.

The functions of the female reproductive system are to:

- Produce the female gametes, the ova.
- Transport the ovum along the Fallopian tube where it is fertilised.
- Protect and nurture the developing embryo and fetus until ready for birth.
- Deliver the baby safely.

Changes in the mother during pregnancy enable her to carry out these functions and prepare her for nurturing the baby during the early stage of life.

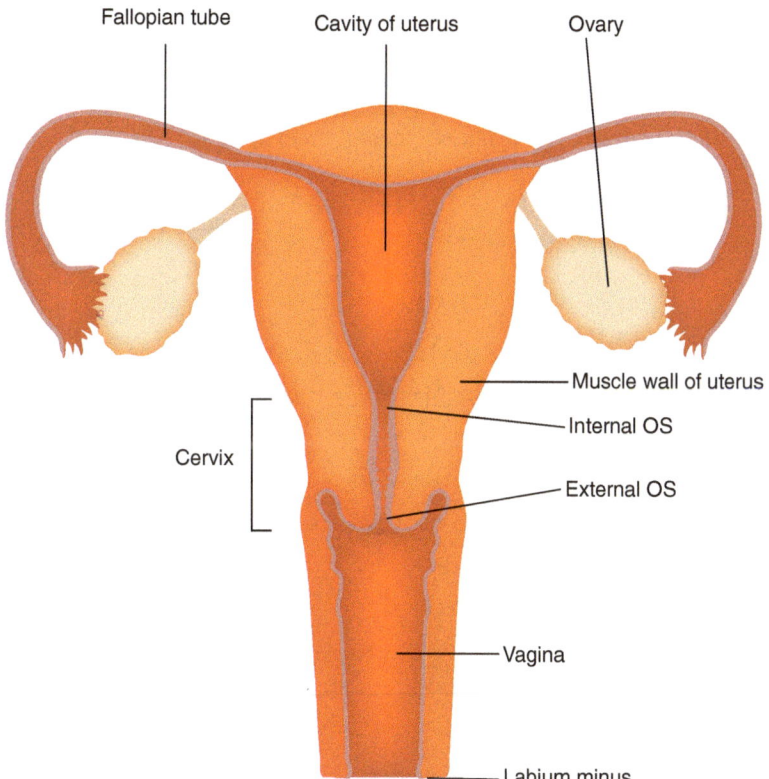

Figure 16.10 Anatomy of the female reproductive system (anterior view)

Ovaries and fallopian tubes

The ovaries in the female produce the female gamete and sex hormones (discussed later). Primordial ova are present in the ovary at the birth of a baby girl, but development is halted at meiotic stage 1 (Chapter 2,

Figure 2.13). At puberty (Chapter 17) further development takes place to form the mature ovum. Figure 16.13 shows maturation of the ovum followed by ovulation and development of the corpus luteum. Ovarian follicles and corpora lutea (singular corpus luteum) both play an important role in the endocrine function of the ovaries.

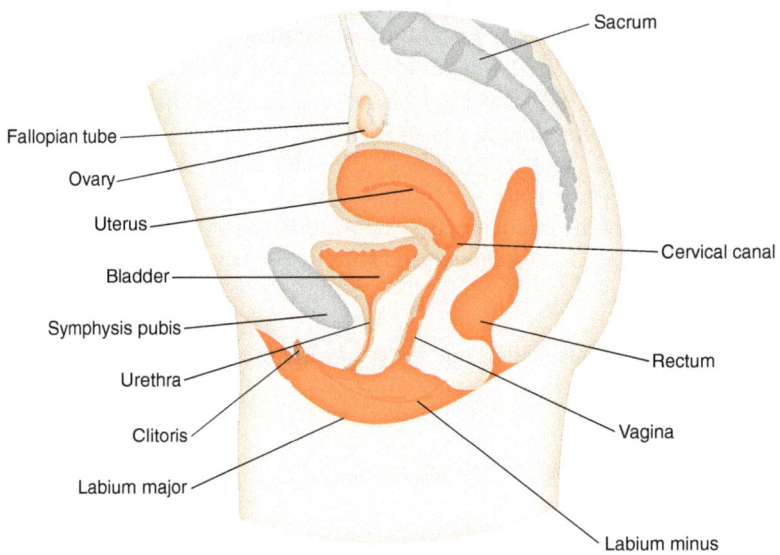

Figure 16.11   A sagittal section of the female reproductive system

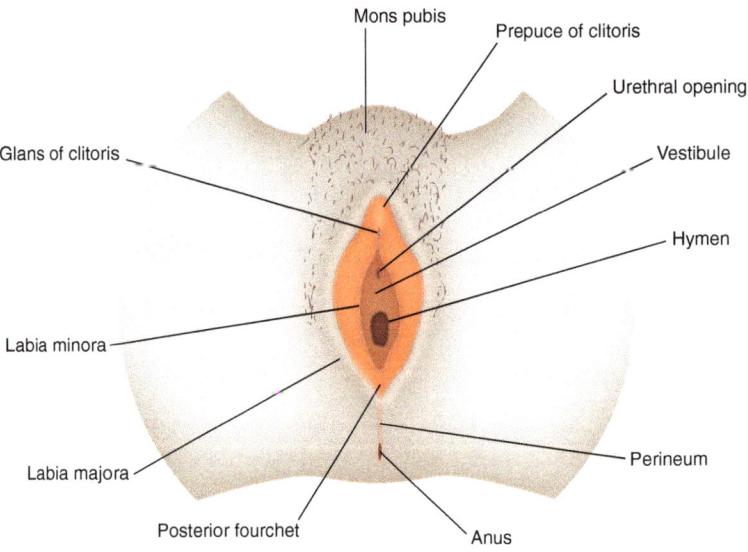

Figure 16.12   The external female genitalia (vulva)

Chapter 2 has considered meiosis with four daughter cells, one ovum and three polar bodies (Figure 2.16). At birth all the ova that the female will ever have are present in the ovary and these diminish in number through her life. Very few of these ova are actually released. Figure 16.14 shows how the numbers of non-growing follicles (NGF) diminish with age in women (Hansen et al., 2008).

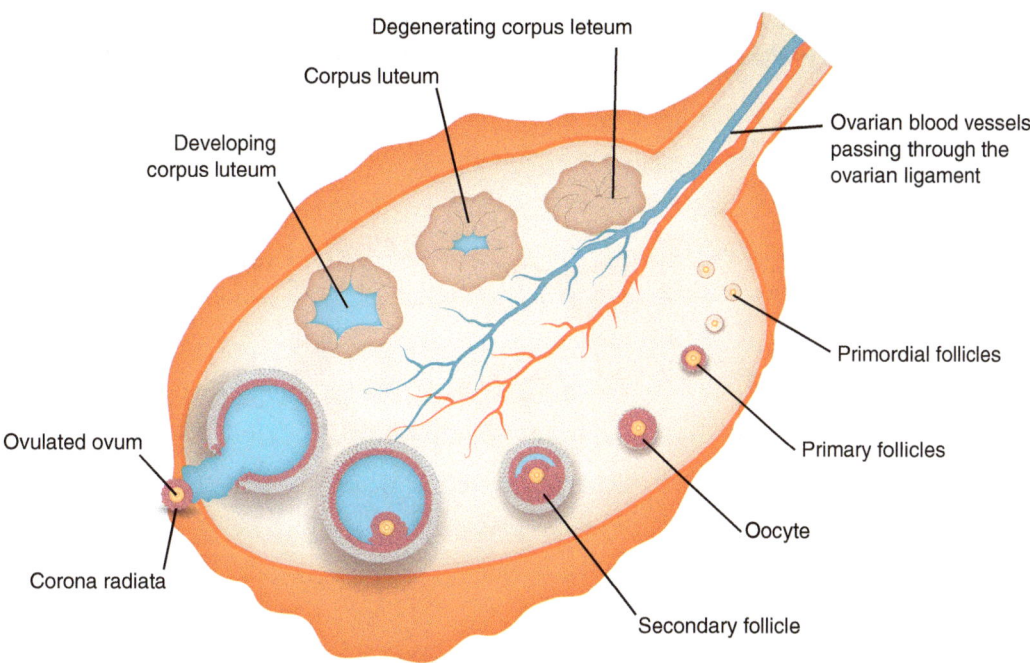

Figure 16.13 Ovary illustrating development of ovum and corpus luteum

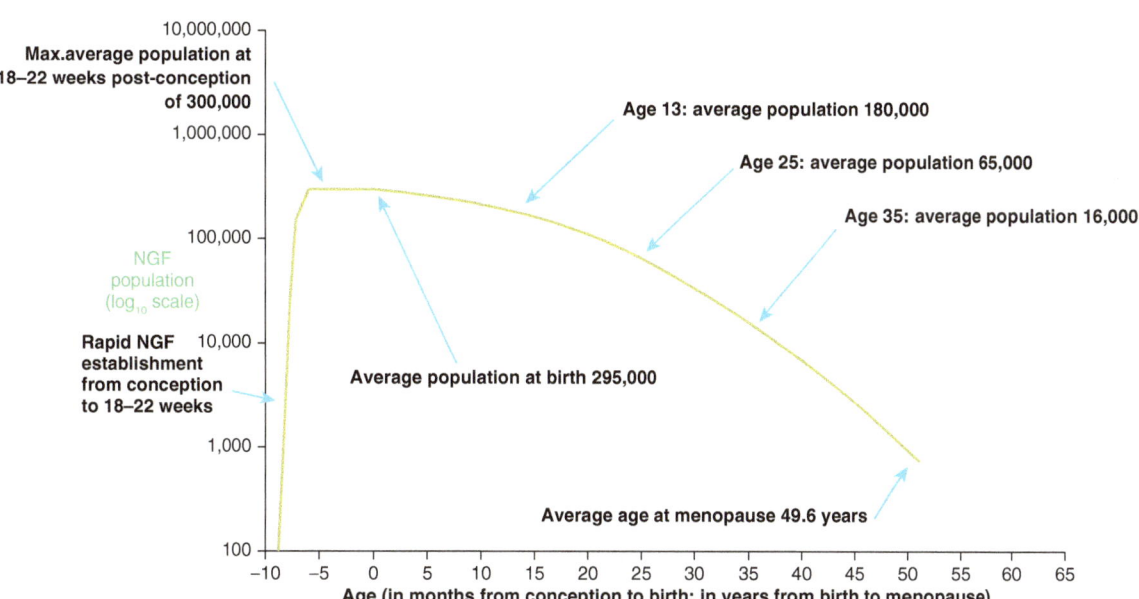

Figure 16.14 Numbers of follicles in the ovaries from conception to the menopause

Source: Adapted from Wallace, W.H.B. and, Kelsey, T.W. (2010) 'Human ovarian reserve from conception to the menopause', PLoS ONE, 5(1): e8772. doi:10.1371/journal.pone.0008772. ©22010 Wallace, Kelsey. Redrawn and reproduced under Open Access Creative Commons License.

The ovum is wafted into the Fallopian tube (oviduct, or salpinges) by the action of the fimbriae at the end of the Fallopian tube and is then moved along it to the uterus by the action of the ciliated epithelium lining (Figure 16.15).

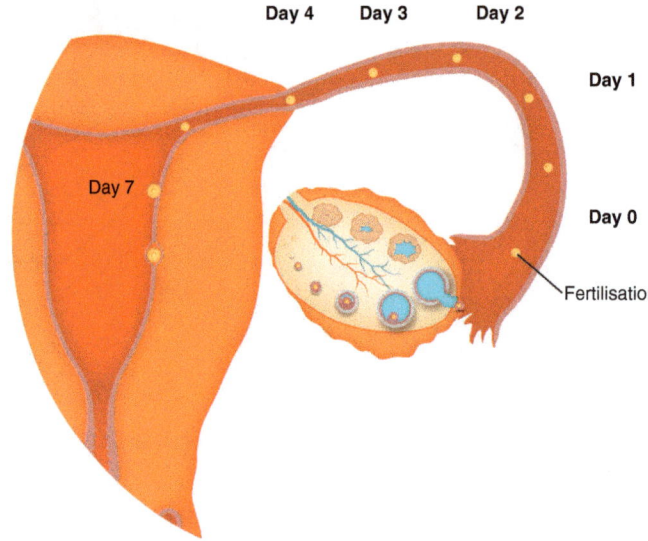

Figure 16.15   Ovary and Fallopian tube showing movement of ovum to uterus

Uterus

The uterus (Figures 16.10 and 16.11) has a thick wall of non-striated (or smooth) muscle and is lined with glandular endometrium. Figure 16.15 shows the fertilised ovum developing into a blastocyst, or bunch of cells, which embeds into the endometrium to become an embryo and moves through the stages of development (Chapter 17). The embryo is nourished initially from the tissue fluid of the endometrium, later through the placental blood supply.

The base of the uterus is thickened into a ring of muscle and fibrous tissue with a central passage – the cervix – through which the sperm enter the uterus, and menstrual fluid and babies leave. This expands to allow the baby to pass through at the end of pregnancy

Vagina

The vagina is the fibromuscular tube which runs from the cervix of the uterus to the opening to the external genitalia (Figure 16.12). It is a self-cleaning organ with healthy microbiota that flow towards the exit minimising the risk of infection.

The vulva (external female genitalia)

The vulva (Figure 16.12) consists of the labia majora (outer lips) and labia minora (inner lips), mons pubis, clitoris, vulval vestibule and associated glands, external urethral orifice and opening of the vagina.

The vulva plays an important role in sexual activity with its extensive nervous supply enabling pleasure when stimulated.

Labia majora and minora

The labia majora are lip-like, often plump, structures either side of the vulva composed of skin and adipose tissue, which start from the pudendal cleft through the base of the mons pubis and come together behind the vaginal opening just in front of the perineum. The perineum is the area between the anus and the vaginal opening. During puberty the outer surface of the labia majora grows pubic hair, while the inner surface is smooth, hairless, and resembles a mucous membrane. Both surfaces contain numerous oil and sweat glands. The nerve supply is less extensive than in the rest of the vulva.

The labia minora are two hairless folds of skin between the labia majora, and the surfaces are similar to mucous membrane. They protect the vestibule, urethra and vagina with the upper front part protecting the clitoris. The nerve supply is extensive. The clitoris is the female equivalent of the penis in men, contains some erectile tissue and responds to sexual arousal by enlarging and becoming firm. It is surrounded and protected by the clitoral hood formed by an extension of the labia minora. It has been estimated that the glans or head of the clitoris contains 8,000 sensory nerve endings (Di Marino, 2014).

APPLY

Midwifery care of the perineum

The perineum is important in midwifery during labour and delivery. The vaginal opening has to stretch to enable the baby's head (the part of the baby with the biggest circumference) to exit and this includes stretching of the perineum, which can be damaged through tearing or by controlled cutting to increase the size of the vaginal opening. Tearing of the perineum is categorised in Table 16.2.

Midwifery care aims to minimise perineal damage. An episiotomy is a surgical incision under local anaesthetic of the perineum carried out to increase the size of the vaginal opening to facilitate the birth of the baby. It reduces the damage to the perineum and is readily repaired by suturing after delivery. It is amongst the commonest procedures carried out on women and has been shown to reduce obstetric and anal sphincter rupture (Räisänen et al., 2011). In addition, women can be taught to undertake digital perineal massage during the last five weeks or so of pregnancy to stretch the perineum; this reduces the number of episiotomies and, thus, suturing (Beckmann and Stock, 2013).

Table 16.2 Classification of perineal trauma (NHS Quality Improvement Scotland, 2008)

Degree	Trauma
First	Damage to the skin only
Second	Damage to the perineal muscles but not the anus
Third: 3a 3b 3c	Perineal injury including damage to the anal sphincter Damage to external anal sphincter of less than 50% thickness Damage to external anal sphincter of more than 50% thickness Internal anal sphincter also torn
Fourth	Perineal injury including both external and internal anal sphincters and anal epithelium

Vulval vestibule and associated organs

Figure 16.12 shows the vestibule and organs associated with it. It is delineated by the labia minora to either side with the clitoris in front and the joining of the two labia minora behind.

The urethra and vagina open into the vestibule. The urinary meatus (external urethral opening) is about 2.5 cm (1 inch) behind the clitoris in front of the vagina. The opening of the vagina (introitus) is surrounded or partially covered by the hymen, a membrane which is usually broken with intercourse and in some cultures used as an indication of virginity. However, it can also be broken by physical activity and is, thus, not a valid indicator of sexual activity.

There are also a number of glands that open into the vestibule around the lower end of the urethra and the anterior wall of the vagina. These include the greater and lesser vestibular glands, including Bartholin's glands that are to the right and left and slightly posterior to the vaginal opening, which they lubricate with their mucous secretions. Infection of this gland can occur and blocking of the duct results in a Bartholin's cyst. The tissues surrounding these glands are stimulated with sexual activity and swell with blood.

Endocrine regulation of female sex function

The major difference in endocrine function and, thus, regulation between the male and female is that the male system continues its function smoothly while the female system works to a monthly (menstrual) cycle during the reproductive years of a woman's life.

GO DEEPER

Energy consumption and the endometrium

It appears that the energy use involved in a combination of resorption and shedding of the endometrium and then building it up again, is less than the energy required for maintaining it in a metabolically active state at all times. It is proposed that six days' worth of food is saved over four menstrual cycles. Humans (and chimpanzees) are thought to undergo visible menstruation because they have large uteri with more endometrium than can be reabsorbed, unlike other mammals (Strassmann, 1996).

Figure 16.16 shows the changes that occur in the pituitary hormone secretion, ovarian hormones, and the endometrium of the uterus through the menstrual cycle. This is numbered from the first day of menstruation at which point hormonal levels are at their lowest level and the endometrium drops to its thinnest.

The FSH and LH from the anterior pituitary reach a peak at about the mid-point of the cycle. As the ovarian follicle develops it begins to secrete oestrogen, reaching its peak just before ovulation occurs, stimulating growth of the endometrium so that it is ready to welcome a fertilised ovum. After ovulation the follicle turns into the corpus luteum, which is a temporary endocrine structure that secretes fairly high levels of progesterone and medium levels of oestrogen and inhibin (although this is not shown in the figure). Inhibin provides negative feedback on gonadotrophin hormones, thus reducing FSH secretion. A new corpus luteum is formed during each menstrual cycle.

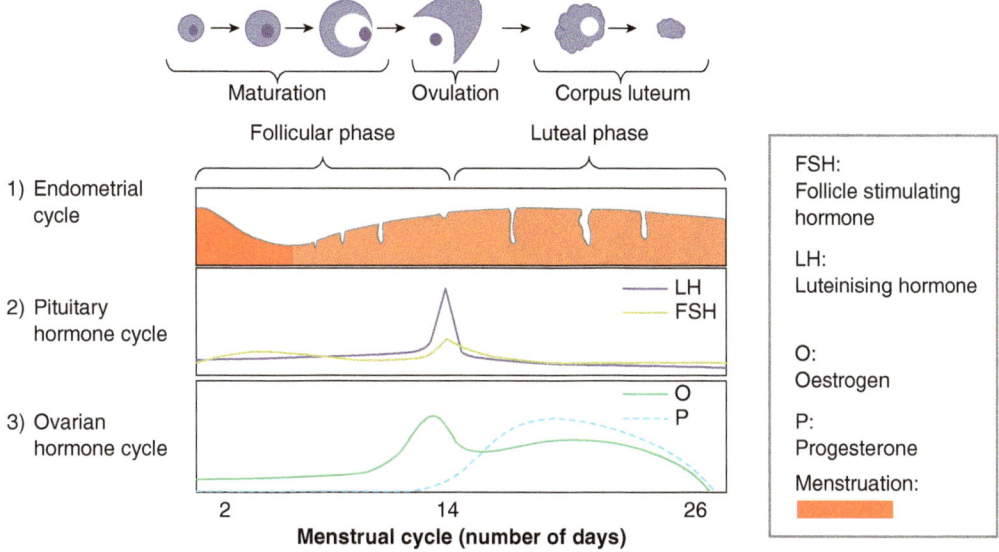

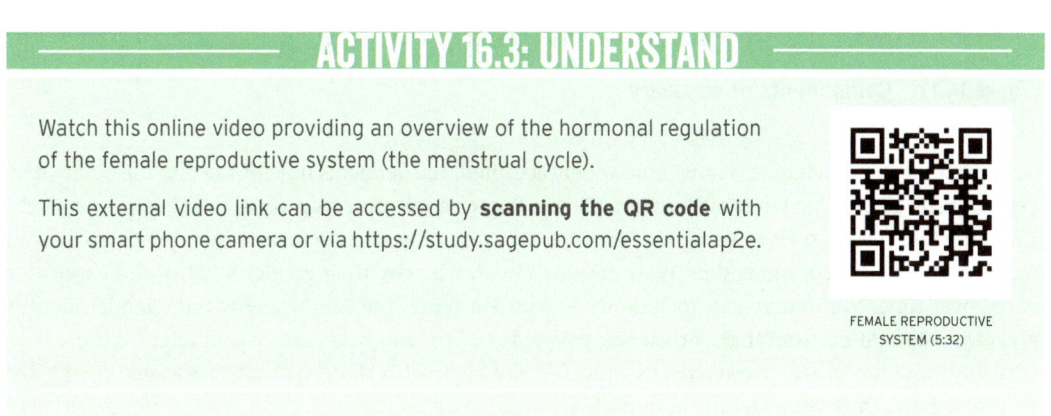

Figure 16.16 Endocrine regulation of the menstrual cycle

SEXUALITY IN PERSON-CENTRED NURSING

Sexuality is the interrelationship between the biological, sociocultural and psychological factors that influence the ways in which we experience and express ourselves as sexual beings and which need to be taken into account in providing sensitive person-centred care. It is clear from Figure 16.17 that sexuality is a broad concept that is multifactorial and interrelational, as well as biological in nature.

Sexuality involves interplay between the reproductive system and the limbic system of the brain in terms of emotion and relationships, as well as cognition and perception. Gender role, i.e. how people behave as men or women (as society perceives them), is moderated by the environment in terms of what is socially and culturally acceptable. It is closely associated, but not always aligned, with gender identity – the sense of being male or female.

Gender identity, however, is considered to be biological, determined prenatally and largely consistent with gender anatomy. In males, gender identity is influenced by levels of androgens, specifically testosterone (Gooren, 2006). This may or may not be further aligned with sexual orientation, the romantic or sexual attraction to another. Again, sexual orientation is considered to be determined prenatally by endocrine conditions. For example, homosexuality is correlated with atypical endocrine conditions during embryonic development (Balthazart, 2011).

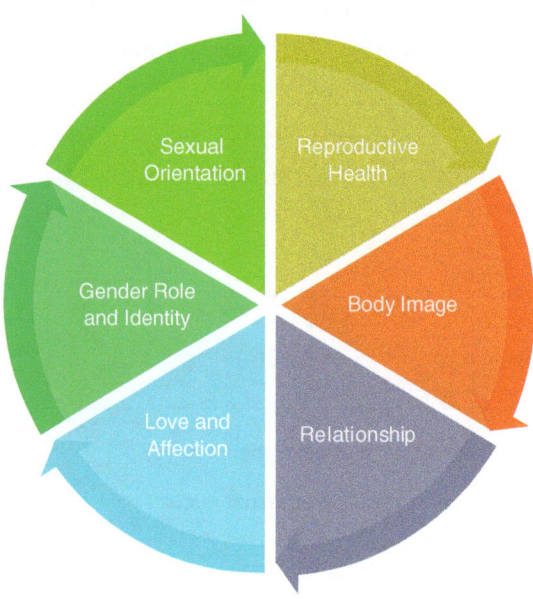

Figure 16.17   Components of sexuality

In contemporary society, it is now more widely accepted that gender is not confined to the binary concepts of male and female but is rather more diverse. Some people now identify as **non-binary**, whereby they associate their gender as being located on a spectrum of gender identities that are not exclusively male or female. Indeed, some identify as **gender fluid**, whereby their gender identity and expression varies over time. Non-binary can include those who are **transgender**, whereby their gender identity and expression differs from that of their sex assigned at birth, and **intersex**, which refers to those born with any variation in sex characteristics which do not align with those typically associated with being male or female. Those who identify as non-binary may have a preference for not being referred to as male or female when pronouns are being used (e.g. he, her, his, she), but prefer the singular use of they, their and them, among others. In being person-centred, establishing personal preferences in a respectful way will facilitate a more effective therapeutic relationship.

GO DEEPER

Biological determinants of sexual orientation

Swaab (2004) and Rahman (2005) provide further links to the biological determinants of sexual orientation, reporting the existence of DNA markers on the X chromosome for orientation. Ellis et al. (2008) provide more supporting evidence, identifying a potential link in genes on chromosomes 1 and 9. They report that heterosexual males and females exhibited statistically identical frequencies of the A blood type, but that homosexual men have a low incidence and homosexual women a high incidence of this blood type. Additionally, they report unusually high proportions of homosexual men and women being Rh negative. It is true to say that the biological determinants of sexuality are still not fully understood and their expression is significantly influenced by culture, society and life experiences.

APPLY

Sexuality and person-centred care

Person-centred nursing can be considered as encompassing holistic care which is underpinned by an 'acceptance that health is determined and defined by interrelated social, psychological and biological factors' (Wynne et al., 1997, cited by Earle, 2001) and this should include sexuality. Earle (2001) discusses the way in which disabled people are often infantilised by society and the role of the nurse in treating them as adults in the area of sexuality. She presents a continuum of activities in relation to facilitated sex (Table 16.3). Different countries deal with this issue in differing ways and it is an issue that can cause difficulties for nurses. Francis (1998) discusses how she helped two disabled people gain sexual satisfaction.

Sexual activity

Sexual activity is a normal part of human living with two major purposes. The primary purpose is reproduction – to maintain the species. However, as important in enhancing the quality of life, is the pleasure obtained by both males and females in participating in sexual activity. In providing person-centred care we need to be able to take account of individuals' needs and wishes in providing health promotion, and relevant education to safely enjoy sexual activity.

Table 16.3 Continuum of facilitated sex

The continuum	The role of the nurse: examples
Providing accessible information, advice and services	Arranging for information to be available in Braille, large print and audio-tape
Fostering an environment which allows intimacy	Acceptance and acknowledgement of patient's sexual needs
Offering and observing need for privacy	Closing doors, providing curtains
Encouraging and enabling social interaction	Arranging suitable transportation
The procurement of sexual goods	Purchasing or arranging the purchase of pornographic magazines
Arranging paid-for sexual services	Assistance with arranging, or information on how to arrange, paid-for sex; willingness to discuss this as an option for the patient
Facilitation of sexual intercourse with another party	Undressing, or helping to undress, patient
Facilitation of masturbation	Assisting patient with positioning and technique
Sexual surrogacy	Assistance with arranging, or information on how to arrange a sexual surrogate

Adapted from: Earle, S. (2001) 'Disability, facilitated sex and the role of the nurse', *Journal of Advanced Nursing*, 36(3): 433-40. Reprinted with the kind permission of John Wiley & Sons.

Sexual activity begins with a range of behaviours that can initiate arousal. Three modes of sexual arousal can result in an erection in males: psychogenic; reflexogenic; and REM (Rapid Eye Movement) sleep induced (Dean and Lue, 2005) (Table 16.4). Once aroused, neurotransmitters are released from nerve terminals in the corpora cavernosa, which trigger the process to induce an erection (Dean and Lue, 2005) (Figure 16.18).

Table 16.4   Modes of arousal leading to an erection

Psychogenic	Reflexogenic	REM sleep induced
This occurs following audiovisual stimuli or sexual thought/fantasy and a resulting emotional reaction. These primarily lead to activation of the bilateral inferior temporal and frontal cortex and the limbic system (insula and cingulate gyrus). Signals are then sent through the spinal cord to spinal nerves (T11-L2 and S2-S4) to the penis via the cavernous nerve where the erectile process is activated	Stimulation of the male genitalia results in a spinal reflex arc response known as the bulbocavernosus reflex. Sensory input is sent to the spinal erection centres (T11-L2 and S2-S4), with signals ascending to the cerebral cortex (sensory perception) and autonomic signals relaying to the cavernous nerve where the erectile process is activated	REM sleep induces increased activity in the pontine reticular formation, activating cholinergic neurons in the lateral pontine tegmentum. This is thought to initiate signals to the spinal erection centres to induce an erection

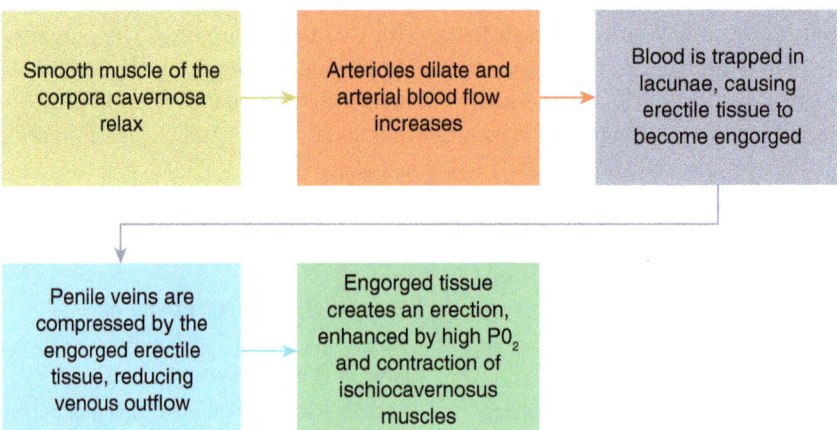

Figure 16.18   Physiology of an erection

During arousal and sex, there are various stages of physical response. Researchers have identified four stages of sexual response in women and men: arousal, plateau, orgasm and resolution (Table 16.5) (Rowland and Gutierrez, 2017). There is some debate about whether males achieve orgasm more quickly than women, but general opinion seems to be that they do.

Table 16.5   Stages of sexual response

	Male	Female

Excitement or arousal	• Physical or psychological stimulation → erection
	• Blood flows into the three spongy areas (corpora) in penis
	• Penis grows in size
	• Scrotum becomes tighter and testes drawn up towards the body
		• Dilation of blood vessels →
		  o Fluid secreted from vaginal walls - lubrication
		  o External genitalia (clitoris, vaginal opening, labia) become swollen
		  o Flushing, particularly on chest and neck
		• Top of vagina expands
		• Pulse and respiration rates increase, blood pressure rises

	Male	Female
Plateau	• Glans (head) of penis widens • Blood vessels in and around penis fill, penis grows <50% larger • Testes continue to rise • Pulse and blood pressure rise, breathing gets faster, thighs and buttocks tighten	• Increased blood flow causes swelling and firmness of bottom third of vagina • Breasts may increase in size by <25% • Blood flow round nipple increases → looks less erect • Nearing orgasm, clitoris pulls back against pubic bone and seems to disappear - continuous stimulation needed
Orgasm (and male ejaculation)	• Contractions in pelvic floor and ductus deferens move sperm, and seminal vesicles and prostate gland add fluid to sperm • Semen consists of 5% sperm and 95% fluid • Contractions are part of orgasm, and are followed by ejaculation of semen out of penis	• Rhythmic contractions of genital muscles during orgasm with intense and pleasurable release of tension • Most do not have recovery period, may have another orgasm if stimulated again • Not all women have orgasm on every occasion, foreplay is important in reaching orgasm
Resolution	• Recovery phase: penis and testes return to normal size • Breathing heavily and fast, pulse rate high, sweating • Recovery period of minutes to hours before another orgasm - gets longer with increasing age	• Body slowly returns to normal state • Swelling goes down • Pulse, blood pressure, respiration return to normal levels

Intercourse

Intercourse occurs when the male inserts his erect penis into the women's vagina (Figure 16.19). Arousal continues in both sexes to orgasm and ejaculation of semen into the vagina. Orgasm is a psychophysiological response; it is connected with rhythmic bodily contractions, erogenous stimulation, and the

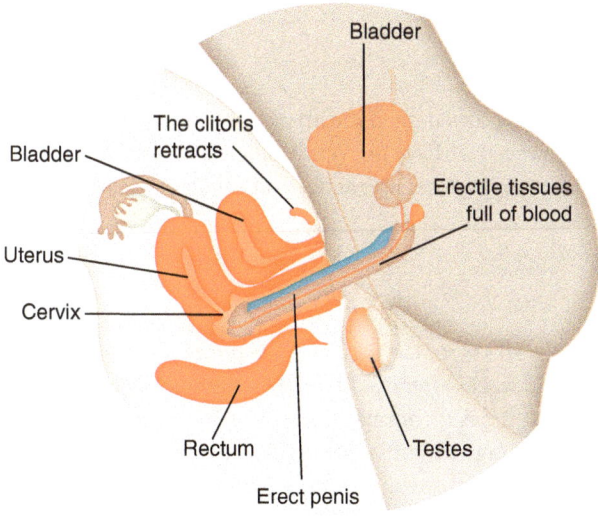

Figure 16.19 Male–female intercourse

emotional response to the sexual experience. Orgasm provides the pleasurable, whole body sensation that makes sexual experiences attractive to people (Llanes et al., 2013). Contractions of the pelvic and vaginal muscles help the sperm to enter the uterus from where some will enter the Fallopian tube and one will fertilise an ovum. The coordination of simultaneous orgasm and ejaculation in the male is central to effective reproduction, while achievement of orgasm in the woman is not essential.

An erection subsides (detumescence) when the arousal stimulus ends, allowing the smooth muscle of the corpora cavernosa to contract, reducing arterial flow. This initially slowly reduces pressure in the erectile tissue and venous outflow is slowly restored. This is followed by a rapid reduction in pressure and full restoration of venous return (Dean and Lue, 2005). Comparable, though less obvious, changes also occur in women.

--- **GO DEEPER** ---

Physiology of ejaculation

The glans of the penis contains encapsulated receptors (Krause-finger corpuscles) and free nerve endings that are stimulated by sensations in the penile shaft, perineum and testes (Giuliano and Clement, 2005). Along with psychogenic stimuli, these peripheral sensory inputs relay to the spinal cord where spinal thoracolumbar and lumbosacral nuclei are innervated, leading to efferent stimulation of the accessory sexual glands, smooth muscles of the spermatic ducts, and striated muscle of the urethral sphincter. The intensity of the stimuli results in ejaculation through these pathways.

Ejaculation occurs in two phases: emission and expulsion. In emission, an autonomic reflex contraction occurs in the spermatic ducts, seminal vesicles and prostate, resulting in semen being collected in the prostatic urethra (Llanes et al., 2013). The urethral bulb expands, increasing intra-urethral pressure. Some seminal fluid is released prior to expulsion, particularly from the bulbourethral glands and prostate. In expulsion, the striated sphincter muscle of the urethra relaxes and semen is propelled along the urethra by the cyclic contraction of the urethral smooth muscles and the muscles at the base of the penis (Llanes et al., 2013). A refractory period occurs after ejaculation, preventing any further ejaculation for minutes to hours. Ejaculation can occur without orgasm, but usually the two coincide.

MATERNAL CHANGES

If fertilisation of the ovum has occurred, then the blastocyst embeds in the endometrium and develops as described in Chapter 17. However, we have not yet considered how the mother adapts to what is a radical change in her functioning to support her pregnancy and enable the baby to develop.

Pregnancy

Pregnancy usually lasts 38 weeks from the date of conception which, in someone with a regular four-weekly menstrual cycle, is 40 weeks after the first day of her last menstrual period. During that period the uterus grows to about 20 times its normal size. Normally it is about the fourth month before the mother can feel her baby moving. Towards the end of pregnancy, the baby normally takes up a head down position with the back towards the side and front of the woman's body and the head drops into the pelvis (Figure 16.20) ready for a cephalic presentation birth – the most common position for delivery.

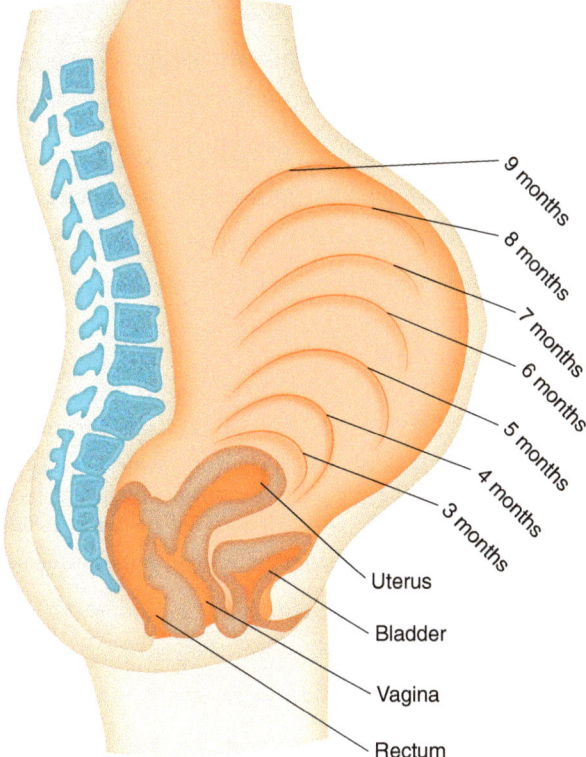

9 months
8 months
7 months
6 months
5 months
4 months
3 months

Uterus

Bladder

Vagina

Rectum

Figure 16.20   Growth of the uterus and position after the fetal head drops into the pelvis

A number of normal physiological changes occur in the mother to provide the baby with adequate nutrients and oxygen for normal development, and protection from potentially harmful substances. A number of different bodily systems in the mother are altered.

Endocrine system

Secretion of a number of hormones changes during pregnancy to maintain the pregnancy and adjust maternal structure and function.

- Progesterone and oestrogen levels rise throughout, suppressing the hypothalamic and menstrual cycles. The placenta takes over the secretion of oestrogen. The placenta also begins to secrete Human Chorionic Gonadotrophin (hCG), which maintains the corpus luteum in producing progesterone, a function later taken over by the placenta. The progesterone relaxes smooth muscle. (hCG excretion levels in the urine are measured to confirm pregnancy.)
- Secretion of prolactin from the pituitary gland during pregnancy prepares the breasts for breast-feeding.
- Parathyroid hormone increases and raises the levels of calcium in the body by its action on the gut (increased absorption) and kidneys (increased reabsorption).
- Adrenal hormones (cortisol and aldosterone) are also increased.
- Human Placental Lactogen (hPL) from the placenta stimulates fat metabolism by the woman, so that glucose is conserved for use by the fetus. It can decrease the sensitivity of maternal tissue to insulin, sometimes resulting in gestational diabetes.

Immune system

The baby contains genes from both parents and thus has a different genetic makeup from his/her mother and theoretically is susceptible to immune attack. However, maternal immune tolerance prevents an attack by the mother's immune system against the fetus and placenta. In particular, the placenta acts as a barrier between the fetus and the mother's immune system. However, some immunoglobulin G molecules can pass the placenta from the mother giving the fetus some immunity against disease, which lasts for some time after delivery.

Physical changes

During pregnancy there is virtually always an increase in weight due to growth of the fetus and placenta, the amniotic fluid surrounding the fetus, increased breast size and an increase in water and fat retention. The increasing size of the uterus causes significant changes in the shape and posture of the woman's body (Figure 16.21). These lead to a change in centre of mass, resulting in alterations in posture, gait, balance and an increased risk of falls. The hormones oestrogen and relaxin are associated with remodelling of soft tissues, ligaments and cartilage. The sacroiliac and symphysis pubis joints stretch and become wider to facilitate birth of the baby.

Cardiovascular adaptation

During the pregnancy the capacity for supplying substances for growth to and removal of waste products from the fetus must increase. Thus, plasma and blood volume increase and cardiac output rises due to increased heart rate and stroke volume. Blood pressure initially falls (weeks 12–26) due to the action of relaxin causing smooth muscle relaxation and vasodilation, and then returns to pre-pregnancy levels by week 36. Increased erythropoietin secretion increases red blood cell numbers. White cell count also increases.

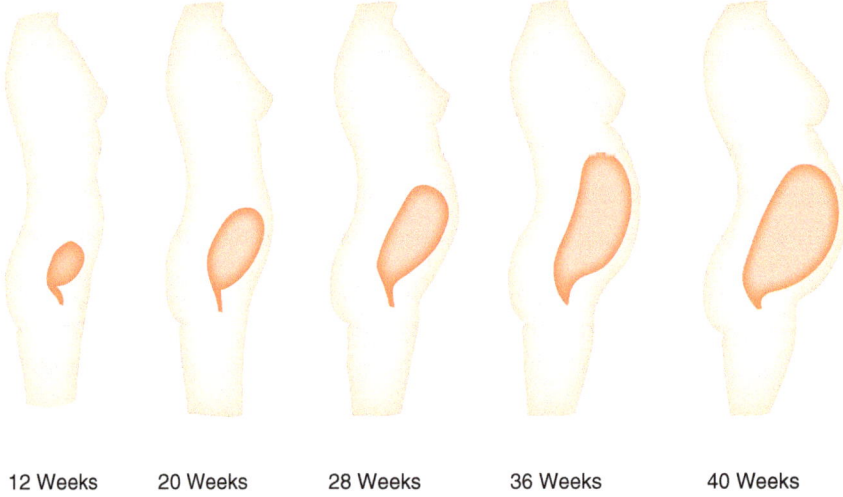

12 Weeks   20 Weeks   28 Weeks   36 Weeks   40 Weeks

Figure 16.21 Physical changes during pregnancy

APPLY

Blood coagulation and pregnancy

Blood becomes more likely to coagulate during pregnancy due to increased production of coagulation factors by the liver. This leads to increased risk of Venous Thromboembolism (VTE), particularly following labour. The enlarging uterus restricts venous return and lymphatic drainage from the legs and fluid can accumulate in the feet causing oedema.

Homeostasis

During pregnancy the body needs increased nutrients for the growth of the baby and developments within the mother. On average, a weight gain of 9.1–13.6 kg (20–30 lbs) occurs. Additional protein is one major requirement for development of the muscular tissues of the uterus to facilitate delivery, and for breast development for lactation and breast-feeding. Vitamin supplements are recommended during pregnancy.

APPLY

Nutritional issues

Folic acid (a B vitamin) is recommended before conception and in early pregnancy to reduce the incidence of neural tube defects. These are conditions in which the spinal cord and spine do not develop normally and result in severe disabilities. Insulin resistance can occur during pregnancy resulting in gestational diabetes. Morning sickness (nausea and vomiting) is relatively common during the first trimester (one third of pregnancy) and sometimes beyond. Hyperemesis gravidarum is a severe condition in which persistent severe vomiting can lead to weight loss and dehydration. Acid reflux due to prolonged gastric emptying and constipation due to decreased colonic activity often also occur.

Delivery

How labour is initiated is still not entirely clear, but it appears that interaction between the fetus, the mother and the placenta are all involved, with phases modulated by hormonal, biochemical, mechanical and immunological factors (Vidaeff and Ramin, 2008).

Normal labour

In the most common form of delivery, the crown of the head, which has the smallest diameter in the fetus, comes first in what is known as a vertex presentation (Figure 16.22), which begins with the head of the baby descending into the pelvis.

Uterine contractions begin as short and weak with long intervals between and get stronger, longer and closer together as labour progresses. The contractions are said to feel like menstrual cramps but become stronger and more painful during labour, occurring at about three-minute intervals as delivery approaches. The key element is the head pressing against the cervix of the uterus, which gradually stretches it open until it is completely effaced and the cervix is said to be fully dilated. The head then begins to come out, turns to allow the shoulders to pass through and then the baby is delivered. The head is only allowed to come out slowly to enable the perineum to stretch.

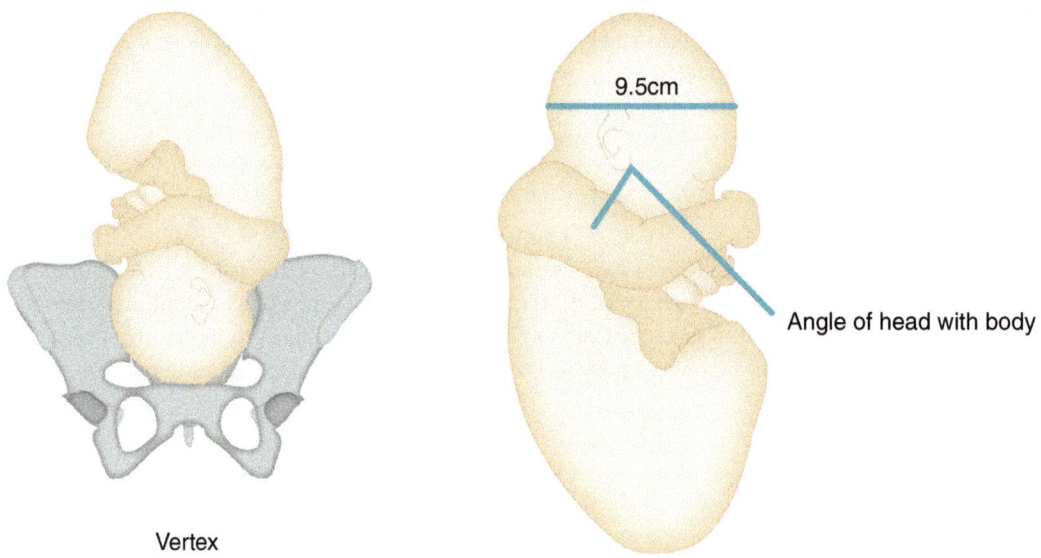

Vertex

a. Smallest diameter of fetal head entering pelvis

9.5cm

Angle of head with body

b. Vertex - smallest diameter of fetal head

Figure 16.22   Vertex presentation of the fetus

Labour occurs in three stages:

- **First stage:** this lasts from the beginning of proper contractions until the cervix is fully dilated. Its length is variable, an average of eight hours in a first pregnancy, but shorter, sometimes much shorter, in subsequent labours. Usually the 'waters break' during this stage when the membranes around the baby lining the uterus break and the amniotic fluid gushes out.
- **Second stage:** this lasts from full dilation of the cervix until delivery of the baby. During this stage the woman feels the need to push the baby out. If the woman is having difficulty in expelling the baby, an episiotomy may be performed.
- **Third stage:** this lasts from delivery of the baby to delivery of the placenta and membranes. These must be checked to ensure that they are complete as any left in the uterus can result in post-partum haemorrhage. Completion of the third stage is often facilitated by the administration of oxytocin.

Hormonal control of labour and delivery

Oxytocin is a hormone produced by the paraventricular and supraoptic nuclei of the hypothalamus, which passes down nerve fibres and is stored in the posterior pituitary gland before being released into the blood stream. It plays an important role in reproduction and intimacy as well as having other functions discussed in Chapter 7. In relation to childbirth, oxytocin plays an important role through its action on the uterus and, after delivery, its action on the breasts.

Towards the end of pregnancy, the raised levels of oestrogen increase the numbers of oxytocin receptors in the uterus. Once labour starts, contractions cause cervical dilation, which stimulates prostaglandin secretion and sends nerve impulses back to the brain. Positive feedback increases oxytocin and causes yet more contractions (Figure 16.23). After delivery of the baby the presence of oxytocin stimulates uterine contractions and facilitates delivery of the placenta. It also promotes bonding by the mother with the baby.

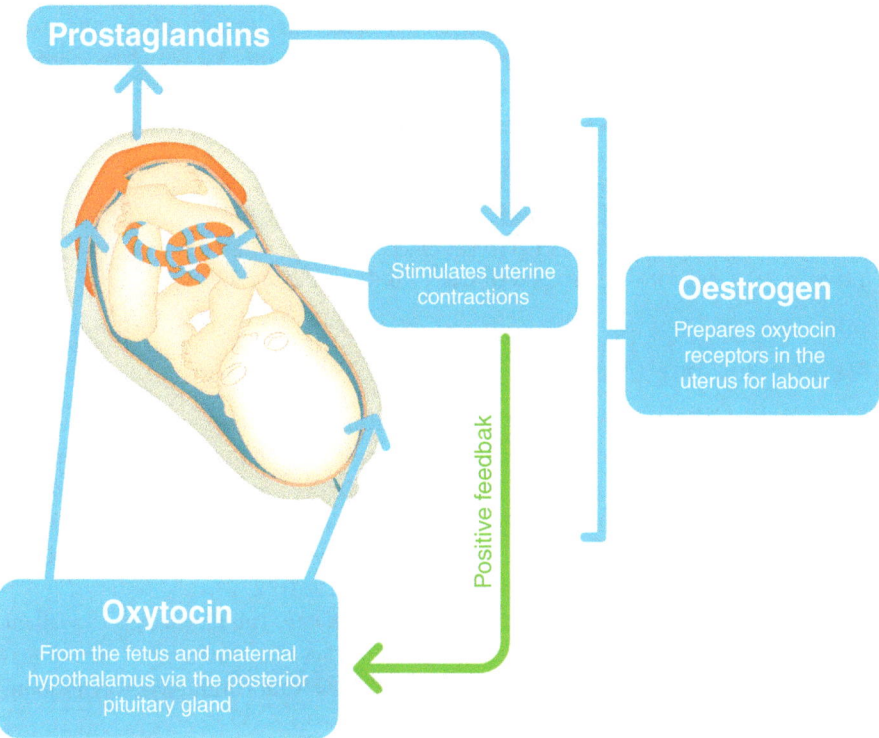

Figure 16.23 Oxytocin in labour and delivery

ACTIVITY 16.4: UNDERSTAND

Watch this useful online video to further your understanding of the endocrine regulation of pregnancy, hormones, and giving birth.

This video can be accessed by **scanning the QR code** with your smart phone camera or via https://study.sagepub.com/essentialap2e.

HORMONAL REGULATION
(10:14)

Post-partum changes

Two main issues are relevant post-natally:

- Returning to the non-pregnant state.
- Infant feeding and nurturing.

Return to non-pregnant state

The post-partum period (or puerperium) is usually considered to last for six weeks following delivery. It is the period during which the mother's body returns to its usual state, including hormone levels and size and condition of the uterus. The cervical os (opening of the cervix: internal – into the uterus; external – into the vagina) returns to the non-pregnant tight opening. A vaginal discharge called lochia lasts up to about 60 days but more commonly to 33 days; it contains blood, mucus and uterine tissue and moves through three stages (Oppenheimer et al., 1986):

- **Lochia rubra (red):** Three to five days.
- **Lochia serosa (brownish/pink):** Up to ten days.
- **Lochia alba (white/yellow):** Second through third to sixth week.

Infant feeding and nurturing

Under endocrine control during the second and third trimester of pregnancy, the breasts enlarge in preparation for breast-feeding the infant. Figure 16.24 shows the structure of the female breast with the lactiferous sinuses where milk is stored prior to feeding. Exclusive breast-feeding is now recommended (WHO/UNICEF, 2003) for the first six months of life (in certain countries) and reduces the incidence of gastrointestinal and respiratory infections. It also inhibits restoration of menstruation and enhances weight loss in the breast-feeding mothers (Kramer and Kakuma, 2012).

The milk 'comes in' about the second or third day after delivery. The first secretion available for the baby is colostrum which contains more white cells and antibodies (particularly IgA) than in mature milk and therefore provides protection against infection for the baby. Colostrum gives way to mature breast milk over about two weeks.

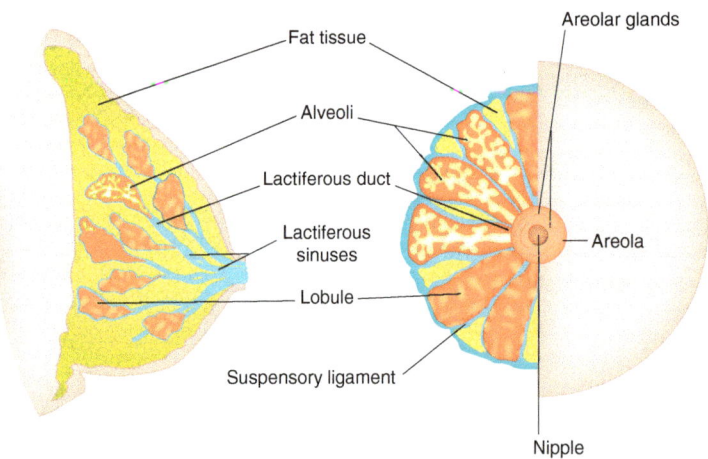

Figure 16.24 Breast structure

GO DEEPER

Body fat and development

Body fat distribution is different in men and women, with women having more gluteofemoral fat and less abdominal and visceral fat, and thus a lower Waist–Hip Ratio (WHR). A relationship between the WHR and the breast-fed baby's cognitive development has been identified. It appears that the fat in these areas contains the Long Chain Polyunsaturated Fatty Acids (LCPUFAs), especially DHA (omega-3 Docosahexaenoic Acid), that are essential for neurodevelopment. About 60-80% of the LCPUFAs needed for brain development come from this area of fat storage (Lassek and Gaulin, 2008), which is not used until late pregnancy and during lactation when brain development is at its maximum. Loss of fat from the gluteofemoral region occurs with each pregnancy and with breast-feeding.

Endocrine regulation of lactation

After delivery of the placenta the levels of oestrogen and progesterone fall rapidly: the fall in oestrogen initiates milk production. The major hormones involved in lactation and breast-feeding are prolactin and oxytocin. Prolactin has a wide number of effects but, in this context, during pregnancy it stimulates growth of the mammary glands (i.e. the breasts) in preparation for lactation. Following delivery, the levels fall, but then secretion of prolactin is stimulated by the baby suckling on the nipple and promoting milk production.

Oxytocin stimulates contraction of the smooth muscle that lines the lactiferous ducts of the breast and squeezes the milk out of the nipple for the infant. The oxytocin stimulates expression of milk into the lactiferous sinuses and the baby's mouth. The positive feedback from the baby's suckling occurs through nerve impulses sent to the hypothalamus, which stimulates prolactin release from the anterior pituitary gland and oxytocin release from the posterior pituitary gland. These hormones maintain lactation and increase the volume of milk to ensure adequate nutrition for the infant.

APPLY

Breast-feeding

Michelle is still breast-feeding Danielle who is making good progress in weight and development. Initially Michelle had some difficulty with breast-feeding and early on had a couple of days when her breasts were engorged (i.e. swollen with breast milk) because of an imbalance between the milk taken by the baby and the amount produced by the mother. This settled within a few days when Danielle settled into a routine of feeding regularly and well.

BODIE FAMILY

Nurturing

The big question in child development is related to the relationship between nature (innate qualities including genetic makeup) and nurture (environment and experiences) in determining individual differences in behaviour and physical state. In Chapter 3 we considered the genetic makeup and its importance, with some conditions and physical properties of the individual being entirely or mainly due to one's genes (for example Huntington's disease). It is clear that nurture is also of critical importance in development of the nervous system including IQ (Intelligence Quotient) and behaviour. It is now abundantly clear that it is the interaction between nature and nurture that is important (Plomin, 2005).

APPLY

Nature and nurture

Bhattarcharjee's (2015) article on nature and nurture in *National Geographic* discusses the importance of nature and nurture. It emphasises that 'a baby's brain needs love to develop'.

To access the URL link to this article, go to https://study.sagepub.com/essentialap2e.

CONCLUSION

In this chapter we have examined how the reproductive systems of men and women work to produce the next generation. The genetic determination of structural sex by the XX or XY chromosomes (female or male gender respectively) is central although it does not necessarily determine gender identity. Knowledge of the structure and function of the male and female reproductive systems is important in understanding the contribution of each to the joining of the gametes from the two parents to form the zygote. The changes that occur in the pregnant woman enable the normal growth and development of the embryo to fetus, and the normal delivery of the baby and provision of optimum nutrition. Recognition of the sexual nature of men and women will help you to understand the importance of recognising the support needed during different stages of life.

GO DEEPER

Further reading

Ayaz, S. (2013) 'Sexuality and nursing process: a literature review', *Sexuality and Disability*, 31(1): 3-12.

Lingen-Stallard, A. (2009) 'Normal labour and birth', *Journal of Paramedic Practice*, 1(11): 446-53.

McLafferty, E., Johnstone, C., Hendry, C. and Farley, A. (2014) 'Male and female reproductive systems and associated conditions', *Nursing Standard*, 28(36): 37.

1   The male and female reproductive systems both develop from the same fetal struc-tures, which become differentiated into male under the influence of testosterone, or continue to develop into the female structures.

2   The endocrine system plays a central role in development of the fully functional reproductive systems. It is also central in the maintenance of pregnancy, delivery of the baby and enabling lactation.

3   In order to maintain the pregnancy and prepare for lactation, the pregnant woman undergoes a number of physiological changes.

4   Interactions between physiological and psychological/emotional aspects of the peo-ple concerned are important for quality of life in relation to sexual activity.

5   The quality of nurturing during early life is central to the development of individuals.

Revision of this chapter (as of all the others) is important as reproduction influences such a large part of our lives. Understanding the anatomy and physiology is important but equally the relationship aspects are central to a good quality of life. There are eight key areas to revise:

1   Sex differentiation.

2   Structure and function of the male reproductive system.

3   Endocrine regulation of the male sex organs.

4   Structure and function of the female reproductive system.

5   Endocrine regulation of the female sex organs.

6   Stages in the sexual response.

7   Maternal changes and hormonal control of pregnancy and delivery.

8   Post-partum changes and infant feeding and nutrition.

In order to help you revise, consider the following questions, answers for which can be found by visiting https://study.sagepub.com/essentialap2e.

Test yourself by revising the chapter first, and then answering these questions without looking at the book. Some answers may need knowledge from different parts of the chapter. Afterwards compare your answers with the text, the figures and the notes you made. Did you miss anything? Here are the questions:

1   Label the diagram of the male reproductive system (Figure 16.25).

2   Draw a diagram to show the endocrine regulation of the male sex organs.

3   Label the diagram of the female reproductive system (Figure 16.26).

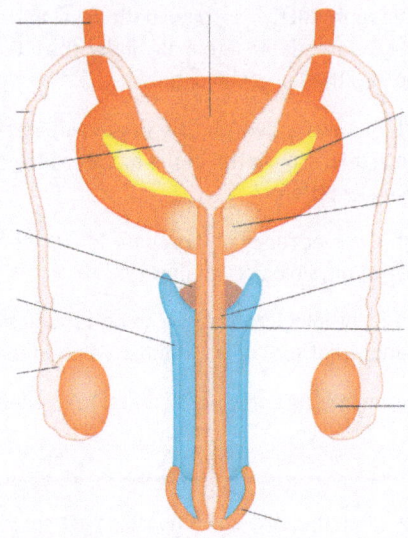

Figure 16.25

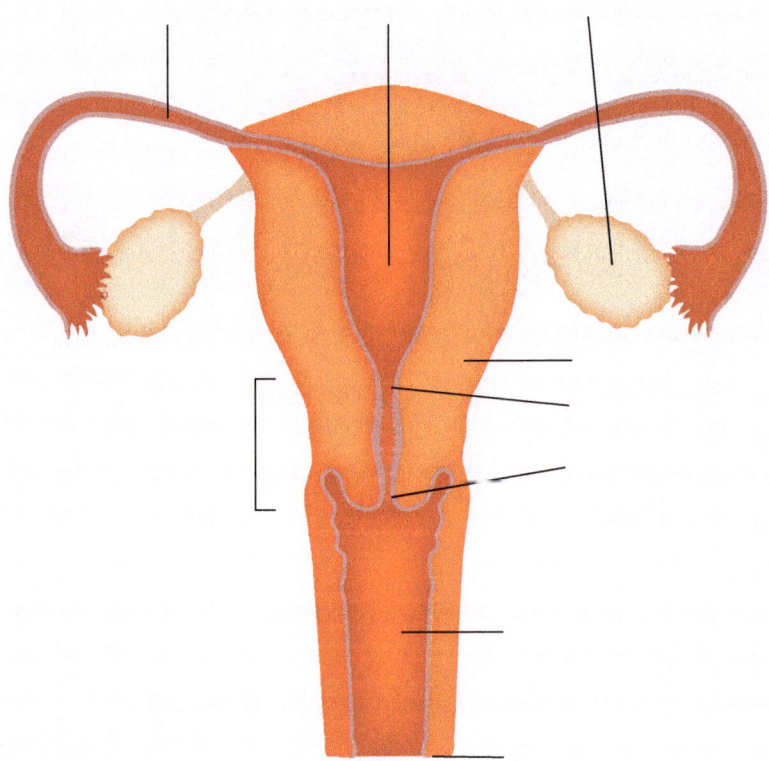

Figure 16.26

4 Outline the endocrine changes during the menstrual cycle and the effects on the endometrium.

5 Outline the stages of the sexual response.

6 Briefly describe the maternal changes during pregnancy.

7 What changes occur in the woman following delivery?

For additional revision resources visit: https://study.sagepub.com/essentialap2e.

REVISE

ACE YOUR
ASSESSMENT

- Revise key terms with interactive flashcards.

- Test yourself with multiple-choice questions.

- Access the glossary with audio to hear how complex terms are pronounced.

- Explore recommended websites suitable for revision.

REFERENCES

Balthazart, J. (2011) 'Minireview: hormones and human sexual orientation', *Endocrinology*, 152(8): 2937–47.

Beckmann, M.M. and Stock, O.M. (2013) 'Antenatal perineal massage for reducing perineal trauma', *Cochrane Database of Systematic Reviews*, Issue 4. Published online: 30 April (DOI: 10.1002/14651858. CD005123.pub3) (accessed 5 May 2020).

Bhattacharjee, Y. (2015) 'Baby brains: the first year', *National Geographic*, January. http://ngm. nationalgeographic.com/2015/01/baby-brains/bhattacharjee-text (accessed 5 May 2020).

Dean, R.C. and Lue, T.F. (2005) 'Physiology of penile erection and pathophysiology of erectile dysfunction', *The Urologic Clinics of North America*, 32(4): 379–95.

Di Marino, V. (2014) *Anatomic Study of the Clitoris and the Bulbo-Clitoral Organ* (Google eBook). Springer.

Earle, S. (2001) 'Disability, facilitated sex and the role of the nurse', *Journal of Advanced Nursing*, 36(3): 433–40.

Ellis, L., Ficek, C., Burke, D. and Das, S. (2008) 'Eye color, hair color, blood type, and the rhesus factor: exploring possible genetic links to sexual orientation', *Archives of Sexual Behavior*, 37(1): 145–9.

Francis, H. (1998) 'Hot potatoes: the agony and the ecstasy', *Nursing Times*, 94(24): 34–6.

Giuliano, F. and Clement, P. (2005) 'Neuroanatomy and physiology of ejaculation', *Annual Review of Sex Research*, 16(1): 190–216.

Gooren, L. (2006) 'The biology of human psychosexual differentiation', *Hormones and Behavior*, 50(4): 589–601.

Hansen, K.R., Knowlton, N.S., Thyer, A.C., Charleston, J.S., Soules, M.R. et al. (2008) 'A new model of reproductive aging: the decline in ovarian non-growing follicle number from birth to menopause', *Human Reproduction*, 23(3): 699–708.

Hughes, I.A. and Acerini, C.L. (2008) 'Factors controlling testis descent', *European Journal of Endocrinology*, 159(supplement): 75–82.

Kramer, M.S. and Kakuma, R. (2012) 'Optimal duration of exclusive breastfeeding' *Cochrane Database of Systematic Reviews Issue 8*. Art. No.: CD003517 (DOI: 10.1002/14651858.CD003517.pub2).

Lassek, W. and Gaulin S. (2008) 'Waist–hip ratio and cognitive ability: is gluteofemoral fat a privileged store of neurodevelopmental resources?', *Evolution and Human Behavior*, 29(1): 26–34.

Llanes, L.L., Ballester, G.L.A. and Elías-Calle, L.C. (2013) 'Ejaculation and sexual pleasure in males: a complex relation with multiple determinants', *Revista Sexología y Sociedad*, 19(1). Available at: www.revsexologiaysociedad.sld.cu/index.php/sexologiaysociedad/article/view/236/297 (accessed 23 July 2020).

Mawhinney, M. and Mariotti, A. (2013) 'Physiology, pathology and pharmacology of the male reproductive system', *Periodontology* 2000, 61: 232–51.

NHS Quality Improvement Scotland (2008) *Perineal repair after childbirth: a procedure standards tool to support practice development*. Edinburgh: NHS Quality Improvement Scotland.

Oppenheimer, L.W., Sherriff, E.A., Goodman, J.D., Shah, D. and James, C.E. (1986) 'The duration of lochia', *British Journal of Obstetrics and Gynaecology*, 93(7): 754–7.

Plomin, R. (2005) 'Nature and nurture: genetic and environmental influences on behavior', *The Annals of the American Academy of Political and Social Science*, 600 (1): 86–98.

Rahman, Q. (2005) 'The neurodevelopment of human sexual orientation', *Neuroscience & Biobehavioral Reviews*, 29(7): 1057–66.

Räisänen, S., Vehviläinen-Julkunen, K., Gissler, M. and Heinonen, S. (2011) 'High episiotomy rate protects from obstetric anal sphincter ruptures: a birth register-study on delivery intervention policies in Finland', *Scandinavian Journal of Public Health*, 39(5): 457–63.

Rowland, D. and Gutierrez, B. R. (2017) 'Phases of the sexual response cycle', *Psychology Faculty Publications*. *62*. Available at: https://scholar.valpo.edu/psych_fac_pub/62 (accessed 23 July 2020).

Strassmann, B.I. (1996) 'The evolution of endometrial cycles and menstruation', *The Quarterly Review of Biology*, 71(2): 181–220.

Swaab, D.F. (2004) 'Sexual differentiation of the human brain: relevance for gender identity, transsexualism and sexual orientation', *Gynecological Endocrinology*, 19(6): 301–12.

Vidaeff, A.C. and Ramin, S.M. (2008) 'Potential biochemical events associated with initiation of labor', *Current Medicinal Chemistry*, 15: 614–19.

Wallace, W.H.B. and, Kelsey, T.W. (2010) 'Human ovarian reserve from conception to the menopause', *PLoS ONE*, 5(1): e8772. (DOI:10.1371/journal.pone.0008772).

WHO/UNICEF (World Health Organization/United Nations International Children's Fund) (2003) *Global Strategy for Infant and Young Child Feeding*. Geneva: WHO.

DEVELOPMENT THROUGH THE LIFESPAN

---------- **UNDERSTAND** ----------

This chapter considers changes in physiology across the lifespan. It will be beneficial to have studied each chapter thoroughly before reading this one.

---------- **LEARNING OUTCOMES** ----------

When you have finished studying this chapter you will be able to:

1. Outline the processes involved in prenatal development leading to embryo and placental formation and fetal development
2. Describe the major changes in transition to independent life
3. Identify major characteristics of growth and development in childhood
4. Differentiate between puberty and adolescence and specify the changes that take place during puberty
5. Identify the stages of adulthood and ageing
6. Specify the main changes which occur in the different systems of the body through adulthood

INTRODUCTION

Throughout life, from fertilisation of the egg to death, changes occur in the way in which the human body functions and thus in the requirements needed to maintain health. In order to meet the needs of individuals throughout their lifespan, you need to understand the stages in development including growth in size and increase in functional ability through childhood to adulthood, through activity in midlife, to diminution in height and reduced vitality in old age. Figure 17.1 demonstrates the changes in physical appearance throughout life.

Figure 17.1 Physical appearances throughout life

During these stages in life the person-centred care needed by individuals will change as their abilities to care for themselves first increase and later diminish.

Context

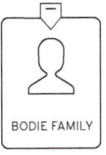

BODIE FAMILY

The Bodie family represent between them the changes that occur throughout life. Danielle has grown from the zygote through embryonic and fetal life and, following birth, the baby is growing and developing rapidly during the early stage of her independent existence. She will progress through childhood and adolescence to adulthood. The adults in the family will have bodies that are fully developed with the younger adults at their peak of activity and the older adults moving towards, or in, the later stages of life. Through this chapter we will be using examples of changes occurring among the members of this family.

PRENATAL DEVELOPMENT

Pregnancy is divided into three trimesters, each three months in duration. The first main stage of development in life is that between fertilisation of the egg from the female by the sperm from the male, and birth. It is described in two parts. Embryonic development is the first eight weeks, after which fetal development lasts until birth. The male and female reproductive systems and their roles in creating the following generation have been considered in Chapter 16.

Critical windows in development are when susceptibility to damage is greatest. All organs and tissues (except adipose tissue) begin developing during the embryonic period and are thus at the highest risk of damage. Care of the mother during this stage, and before the pregnancy begins, is important to ensure the best possible state of health and, thus, optimal conditions for the development of the embryo and fetus.

Folic acid

Folic acid supplements before and during the early stages of pregnancy reduce the risk of neural tube defects (i.e. spina bifida and anencephaly) or incomplete formation of the spinal cord and the brain. Public health measures to increase folic acid supplementation have been successful in reducing these conditions. However, it appears that obese women and those taking certain medications need a higher dosage of folic acid around conception (Talaulikar and Arulkumaran, 2013).

In earlier chapters we have examined the two types of cell division and the different types of cells formed. In the last chapter, Figure 16.15 illustrates roughly the first seven days of development:

- the fertilisation of the ovum.
- cell division as it is moving along the Fallopian tube.
- formation of a morula (a dense ball of about 16 cells).
- development into a blastocyst – a ball of cells with a fluid-filled cavity and an inner cell mass.
- implantation of the blastocyst into the uterine wall.

Programmed cell death (apoptosis)

We have looked at cell division as a major process in development, but we also need to consider apoptosis as another major process in moving through the stages of development. This removes cells and tissues no longer required and permits remodelling. An example of apoptosis is removal of the tissue that joins the fingers during early development; this is genetically programmed and enables growth and development of the five separate digits on each hand. In addition, this process may be initiated by some external stresses such as hypoxia, nutrient deprivation, viral infection and damage to the cell membrane, which result in cell suicide (Figure 17.2). A range of factors can cause intracellular apoptotic signals to be released by a damaged cell and lead to the following stages:

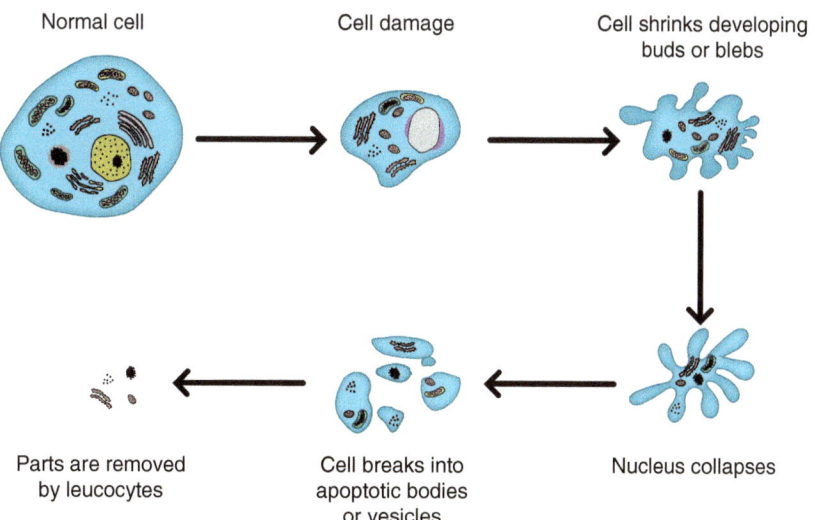

Normal cell   Cell damage   Cell shrinks developing buds or blebs

Parts are removed by leucocytes   Cell breaks into apoptotic bodies or vesicles   Nucleus collapses

Figure 17.2 Apoptosis

1. Cell shrinkage and rounding as the cytoskeleton[1] breaks down.
2. The cytoplasm appears dense, and organelles become tightly packed.
3. Chromatin (the material of which chromosomes are composed) condenses into compact patches against the nuclear membrane.
4. The nuclear membrane breaks and the DNA degrades and breaks into numerous units.
5. The cell membrane develops irregular buds or blebs.
6. The cell breaks apart into apoptotic bodies or vesicles, which are then phagocytosed (Ralston and Shaw, 2008).

ACTIVITY 17.1: UNDERSTAND

Watch this online video explaining cell fate – division, senescence and death.

This video can be accessed by **scanning the QR code** with your smart phone or via https://study.sagepub.com/essentialap2e.

CELL FATE (8:01)

Embryogenesis

During the first eight weeks of development, growth is rapid and the basic structure of the human body is laid down. Fertilisation of the ovum by the sperm to form the zygote with a complete chromosome count of 46 takes place in the Fallopian tube and cell division begins. The small solid bunch of cells (morula) enters the uterus at about three days and develops a fluid-filled space (the blastocyst) (Figure 17.3). This embeds in the uterine wall at about seven days. The mucosal cells of the uterine wall react by enlarging and becoming filled with glycogen for the initial nourishment of the embryonic cells.

Here we are examining in brief the remaining processes involved in the first eight weeks after which the embryo is known as a fetus. Following implantation, gastrulation occurs when the three germ layers (Figure 17.4) of the embryo form and the embryo develops a head-to-tail and front-to-back orientation.

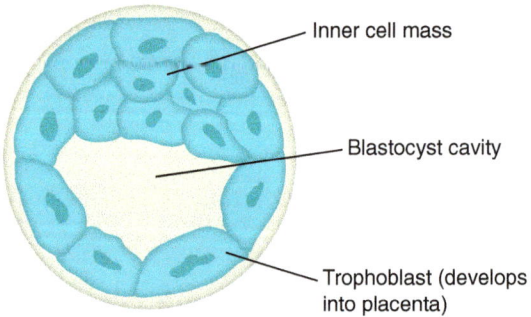

Inner cell mass

Blastocyst cavity

Trophoblast (develops into placenta)

Figure 17.3 Blastocyst

---

[1]A dynamic network of fibres made of protein microfilaments and microtubules to form an internal framework of the cell, allowing movement.

During embryogenesis, a coordinated range of changes occur in expression of genes, cell growth and movement, and differentiation into the different tissues indicated in Figure 17.4. By the end of this period of development, the external features of the fetus and most of the organs are recognisable.

Early development is of two sets of structures: the three germ layers of cells that form the embryo; and the support tissues from which develop the placenta and umbilical cord, which nourish the embryo, and the amniotic sac that surrounds and protects it. During embryonic development, the basic structure of the body is laid down and the heart starts beating at week four.

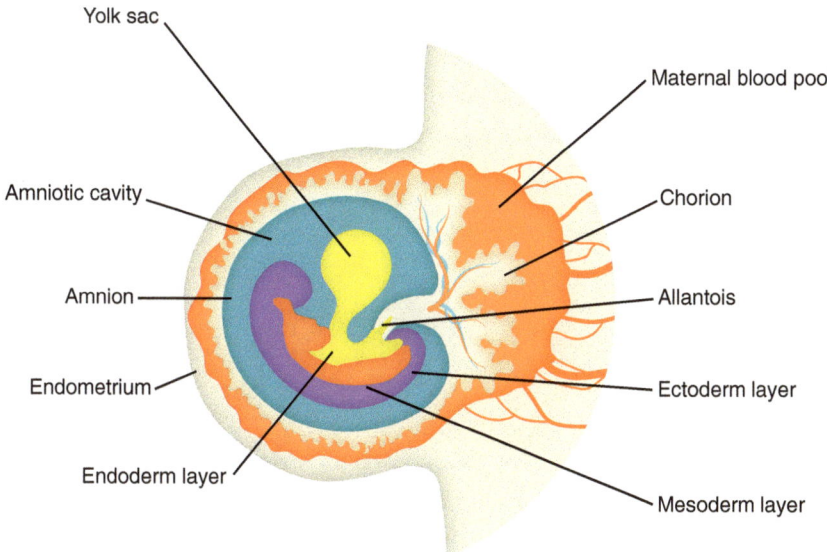

Figure 17.4   Gastrulation stage: three germ layers

APPLY

Healthy embryonic development

The embryo is particularly susceptible to external influences, which may result in major abnormalities or in susceptibility to disease in later life. Many babies whose mothers were prescribed Thalidomide during the first trimester for morning sickness were born with major abnormalities including missing or shortened limbs. This has led to a more rigorous testing of drugs before they are introduced. Substances that cross the placenta and cause harm to the developing embryo are called teratogens.

The impact of the Dutch Hunger Winter when the population were subjected to famine during late 1944 to April 1945 demonstrated the importance of good nutrition for development, particularly during the first trimester. Poor nutrition during this period had negative consequences on both physical and mental health into adulthood (Roseboom et al., 2011).

The placenta

The placenta is mainly established by the fourth week of pregnancy, although it continues to develop through pregnancy. It is formed by both embryonic and maternal cells and the blood vessels of the two separate individuals are close but not linked to each other, permitting exchange of substances, but minimal cells, between the two circulations. During intrauterine life, the fetus obtains nutrients and oxygen from the mother and waste products are eliminated through the placenta into the maternal circulation from which they are excreted. Figure 17.5 is a representation of a section of the placenta at the end of the first trimester.

The placenta is also an endocrine organ, producing human chorionic gonadotrophin during early pregnancy (Chapter 16).

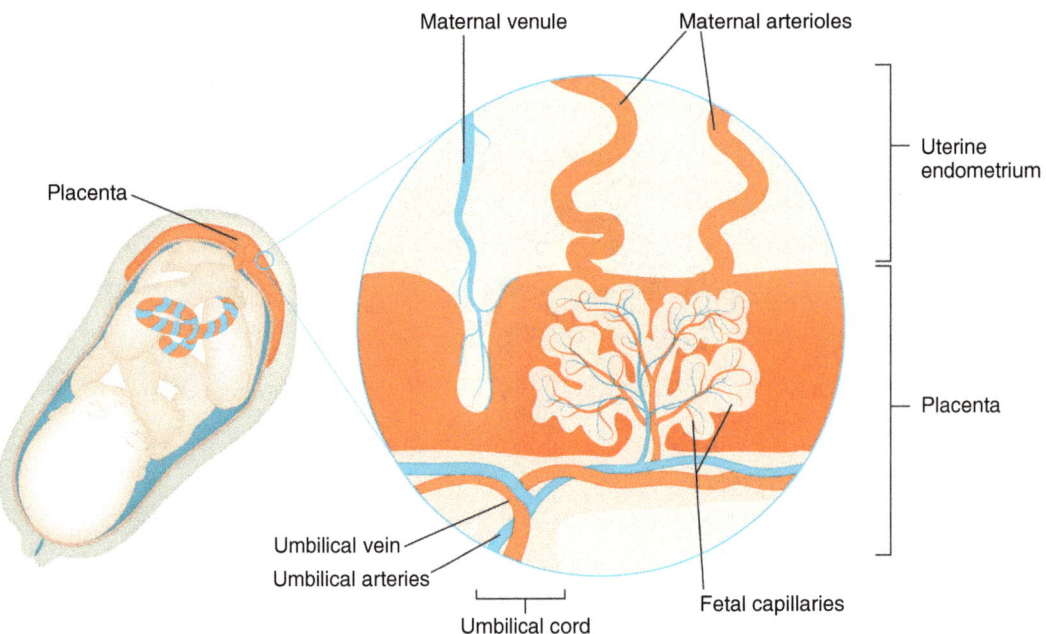

Figure 17.5   The placenta

The fetus

During fetal development, rapid growth and differentiation occur; the organs and systems develop in size and complexity to achieve readiness for beginning life after birth. Table 17.1 summarises development through fetal life; adipose tissue (the one tissue not developed during embryonic development) is laid down during the later stages of gestation. Figure 17.6 includes several stages of fetal development starting with the embryo at the beginning of the fetal stage of development. The small size of the limbs and the large head are clearly visible. This figure also shows later stages during gestation.

Table 17.1   Fetal development

Weeks	Size (week)	Development
9-12	2.2 cm (9)	Intestinal coils form, urine formation begins, red blood cells formed in liver, external genitalia by week 12
13-16	8.5 cm (14)	Rapid growth, muscle and bones develop (bones begin to ossify), liver and pancreas form secretions
17-20	13.5 cm (20)	Growth slows but continues, muscle develops, eyebrows, lashes and nails appear

Weeks	Size (week)	Development
21-24	27 cm (22)	Lanugo (soft hair) covers body, bone marrow begins to form cells, lower airways begin to develop, fat storage begins
25-28	30 cm (26)	Rapid brain development, some control over body functions, respiratory system allows gas exchange
29-32	33 cm (30)	Increase in body fat, bones developed but still pliable, stores iron, calcium and phosphorus, breathing movements but lungs not mature
33-36	40+ cm (36)	Body fat increases, weight may be over 2 kg, fingernails reach fingertips

Adapted from: British Nutrition Foundation (ed.) *Nutrition and Development: Short- and Long-Term Consequences for Health*. Reproduced with permission from Wiley © John Wiley & Sons 2013.

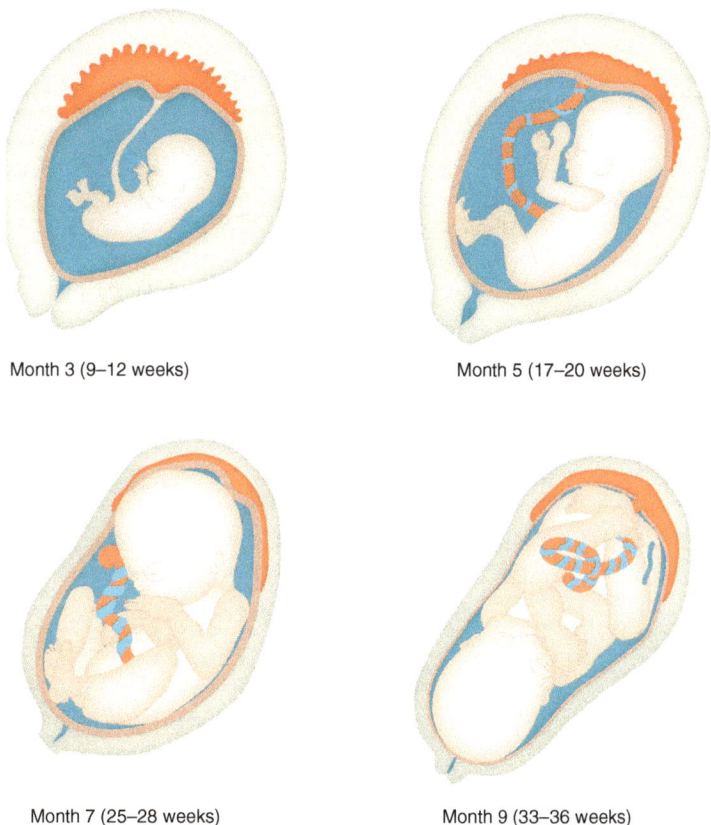

Month 3 (9–12 weeks) Month 5 (17–20 weeks)

Month 7 (25–28 weeks) Month 9 (33–36 weeks)

Figure 17.6 The developing baby through pregnancy

EARLY LIFE

Transition from intrauterine to independent life

During fetal life, the developing individual is dependent on the mother for all its requirements. While its systems are developing during the pregnancy, they have to make significant adaptations after delivery. In essence, the placenta undertakes the functions of many body systems. At birth, the respiratory and circulatory systems have to adapt immediately to the external environment, while the gastro-intestinal tract, sense organs and immune system adapt over hours and days following parturition (giving birth).

In utero, the lungs are non-functional and the heart and circulation enable most of the blood to bypass the lungs. At birth, changes must occur immediately to enable the lungs to inspire air and provide oxygen to the body. As the baby takes its first breath the lungs expand and changes occur in the circulation to enable adequate blood flow to the lungs. Figure 17.7 shows the changes that occur in the heart to stop blood passing straight from the right to left atrium, thus preventing it bypassing the lungs. Figure 17.8 indicates the changes that occur in the heart and in the wider circulation as the infant achieves the adult-type circulation.

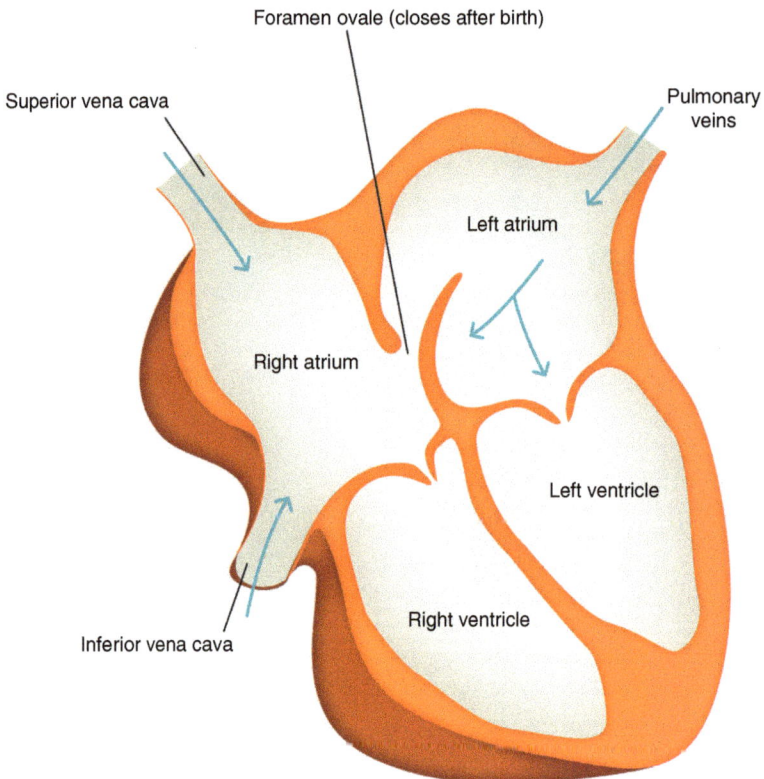

Figure 17.7 Cardiac changes following birth

Circadian rhythm

One of the important factors in the way in which the body works is circadian rhythm. Most animal cells have a rhythm in their function of about 24 hours (i.e. circadian), which is normally maintained at 24 hours by the day–night light cycle. If deprived of light, it tends to drift out of phase, becoming either shorter or longer. It is thought to be regulated by photoreceptors in the eyes, the hypothalamic suprachiasmatic nuclei and the pineal gland (Chapter 7) including the pineal hormone, melatonin (Warren and Cassone, 1995).

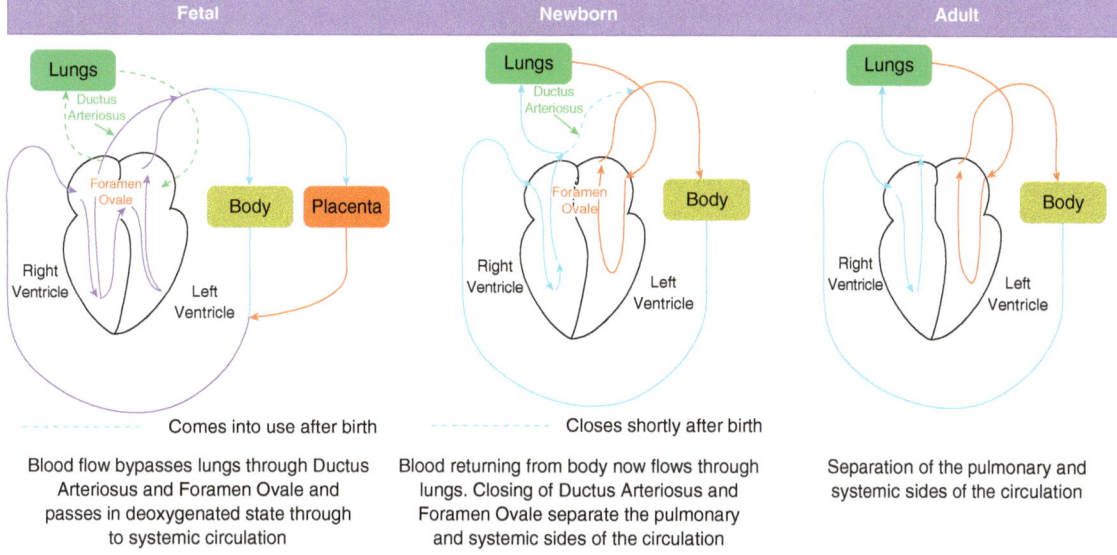

Figure 17.8 Changes in the circulation after birth

Development of circadian rhythm in bodily function is important for normal development. This begins during the last ten weeks of fetal life in response to environmental cues and maternal interaction and physiological functions (Mirmiran et al., 2003).

Growth

Following birth, there are various stages in developing to adulthood. During this period body growth is the major characteristic with girls slightly ahead of boys in their growth. There are a range of charts available for monitoring development at different stages through reviewing various characteristics such as height and weight, and stages of puberty (Table 17.2). However, different tissues have their maximum growth at different times in life with a definite steady growth in the size, but also the synapses, of the brain during childhood.

Infancy is from birth to about 24 months and, after some initial weight loss (usually 5–10% of birth weight), is a period of rapid growth (a continuation of fetal growth) compared to other stages in life, with height reaching about half of the adult value. However, during this same period the brain achieves 80% of its adult size.

From infancy to adolescence, growth in height, body systems and weight all occur along with increasing motor skills and coordination. While the brain reaches adult weight some time before the early teens (British Nutrition Foundation, 2013) it is still developing synapses and being remodelled until about 25 years in men.

Table 17.2 Growth charts (RCPCH, 2015)

Early years charts - including:
0-4 years
Neonatal and Infant Close Monitoring (NICM) (0-2 years)
Personal Child Health Record (PCHR), or 'red book', charts
Education and training materials
School age charts - including:
2-18 years
Childhood and Puberty Close Monitoring (CPCM) (2-20 years)
Down syndrome (0-18 years)
Body Mass Index (BMI) (2-20 years)
Personal Child Health Record (PCHR), or 'red book', charts

APPLY

Monitoring child growth

BODIE FAMILY

Monitoring child growth is an important aspect of assessing the health status of children and the use of growth charts enables the health visitor or other health professional to judge the child's development against the norm for the age. Danielle will be undergoing regular height and weight checks.

WHO has developed child stature and weight charts which are adapted for different ethnic groups, ages and countries: the Royal College of Paediatrics and Child Health have produced a number of videos for healthcare professionals on measuring weight, length/height and head circumference (RCPCH, 2015). They also publish the UK-WHO charts and we suggest that you look at the range of charts available at the Child Health website.

The URL link to the **UK-WHO growth charts** can be accessed via the companion website https://study.sagepub.com/essentialap2e.

APPLY

Vitamin D and sunlight exposure

BODIE FAMILY

A balanced diet is essential for normal growth and this includes ensuring adequate vitamin D. In recent years there has been an emphasis on limiting exposure of children and adults to the ultraviolet (UV) rays of the sun in order to limit the development of various types of skin cancer. The use of sun-block cream and keeping the body covered has been encouraged, and the advice followed. However, this has had repercussions.

A lack of adequate vitamin D is now being identified in many parts of the world (Holick, 2006). While those living in Northern latitudes (for example, the UK, and particularly the more Northern parts of the country) are developing vitamin D deficiencies, some from hot climates are also affected because they wear clothing that covers the skin. The risk of developing a number of different conditions (e.g. some common cancers, autoimmune diseases, hypertension, and infectious diseases) is increased with vitamin D deficiency (Holick and Chen, 2008). Those at all ages in the Bodie family need to consider taking vitamin D.

A number of different hormones are associated with growth at different times until adult proportions are reached (Table 17.3). There is some evidence that secretion of melatonin by the pineal gland inhibits puberty and that a considerable drop of melatonin initiates the changes of puberty (Yun et al., 2004). Chapter 7 discusses the detail of the endocrine system.

Table 17.3 Hormones in human growth at different ages

Thyroid hormones	Starts high, remains constant until around the age of four, declines steadily until age 20
Growth hormone	Increases rapidly from birth to around age of two, remains constant to age 16, declines steadily to age 20
Androgens and Oestrogens	Increases steadily from around the age of nine, peaks at around the age of 15 and declines until around the age of 20

Adapted from: Barrett et al., *Ganong's Review of Medical History*, 2010.

PUBERTY AND ADOLESCENCE

Adolescence is the stage in life when the child turns into an adult, identified by WHO (2014) as between the ages of ten and 19. It has been described as 'the period of physical, cognitive, and social maturation between childhood and adulthood' (Blakemore et al., 2010: 926). Providing more detail, the Canadian Paediatric Society (2003) has stated that 'adolescence begins with the onset of physiologically normal puberty and ends when an adult identity and behaviour are accepted' (p. 577). With an increasing period for education, with individuals taking part in all the activities of student life, the end of adolescence may not occur until the early or mid-20s. However, the major focus here is on the physiological changes occurring, known as puberty.

Puberty

Puberty begins with varying hormone levels causing changes in the reproductive systems to achieve functional reproductive capacity and development of secondary sexual characteristics. It starts with hormonal signals from the hypothalamus and pituitary gland to the gonads (ovaries in girls, testes in boys). These hormonal changes stimulate libido (the desire for sexual activity) and enhance development of all organs of the child to achieve the adult body. The stages in puberty were originally described by Tanner in the 1960s (Marshall and Tanner, 1969, 1970) and details are given in Table 17.4 and illustrated in Figure 17.9. Menarche (the beginning of menstruation) is the major landmark for girls around age 12–13, while the first ejaculation at about 13 is the key event for boys.

Puberty usually starts earlier in girls than boys, with girls beginning puberty at 10–11 years old and completing it by 15–17, and boys beginning it at 11–12 and completing it by 16–17. While there has been some debate about whether puberty is occurring earlier now than in the middle of the last century, it has been concluded that menarche and breast development is occurring earlier (Euling et al., 2008). Cole (2003) has summarised and reviewed the changes in child development over the years. He identified that how children develop over the years has been changing since the mid-19th century, occurring earlier. He reports that adult height has stabilised, but weight is increasing with a rise in obesity. He also states that in Northern Europe the age at menarche has also stabilised at about 13 years.

Table 17.4   The immature and mature male and female bodies

	Tanner stage	Changes	Usual age
Genitals (male)	1	Prepubertal (testicular volume less than 1.5 ml; small penis of 3 cm or less)	age 9 and younger
	2	Testicular volume between 1.6 and 6 ml; skin on scrotum thins, reddens and enlarges; penis length unchanged	9–11
	3	Testicular volume between 6 and 12 ml; scrotum enlarges further; penis begins to lengthen to about 6 cm	11–12.5
	4	Testicular volume between 12 and 20 ml; scrotum enlarges further and darkens; penis increases in length (to 10 cm) and circumference	12.5–14
	5	Testicular volume greater than 20 ml; adult scrotum; penis 15 cm in length	14+
Breasts (female)	1	No glandular tissue: areola follows the skin contours of the chest (prepubertal)	age 10 and younger
	2	Breast bud forms, with small area of surrounding glandular tissue; areola begins to widen	10–11.5
	3	Breast begins to become more elevated, and extends beyond the borders of the areola, which continues to widen but remains in contour with surrounding breast	11.5–13
	4	Increased breast size and elevation; areola and papilla form a secondary mound projecting from the contour of the surrounding breast	13–15
	5	Breast reaches final adult size; areola returns to contour of the surrounding breast, with a projecting central papilla	15+
Pubic hair (male and female)	1	No pubic hair at all (prepubertal state)	age 10 and younger
	2	Small amount of long, downy hair with slight pigmentation at the base of the penis and scrotum (males) or on the labia majora (females)	10–11.5
	3	Hair becomes coarser and curlier, and begins to extend laterally	11.5–13
	4	Adult-like hair quality, extending across pubis but sparing medial thighs	13–15
	5	Hair extends to medial surface of the thighs	15+

Table information taken from W. A. Marshall, J. M. Tanner, 1969 and W. A. Marshall, J. M. Tanner, 1970.

GO DEEPER

Puberty triggers and delays in males

To trigger puberty, gonadotrophin-releasing hormone (GnRH) (from the hypothalamus) levels increase as a result of decreased negative feedback sensitivity and other factors. GnRH triggers the pituitary gland to release follicle stimulating hormone (FSH) and luteinising hormone (LH). FSH matures germ cells and stimulates spermatocytes' meiosis. LH stimulates testosterone production in Leydig cells.

However, puberty can be delayed in some people, with Gamma-Aminobutyric Acid (GABA) preventing GnRH release. Without GnRH, FSH and LH are not produced. These are needed for the production of testosterone and maturation of germ cells (spermatogenesis).

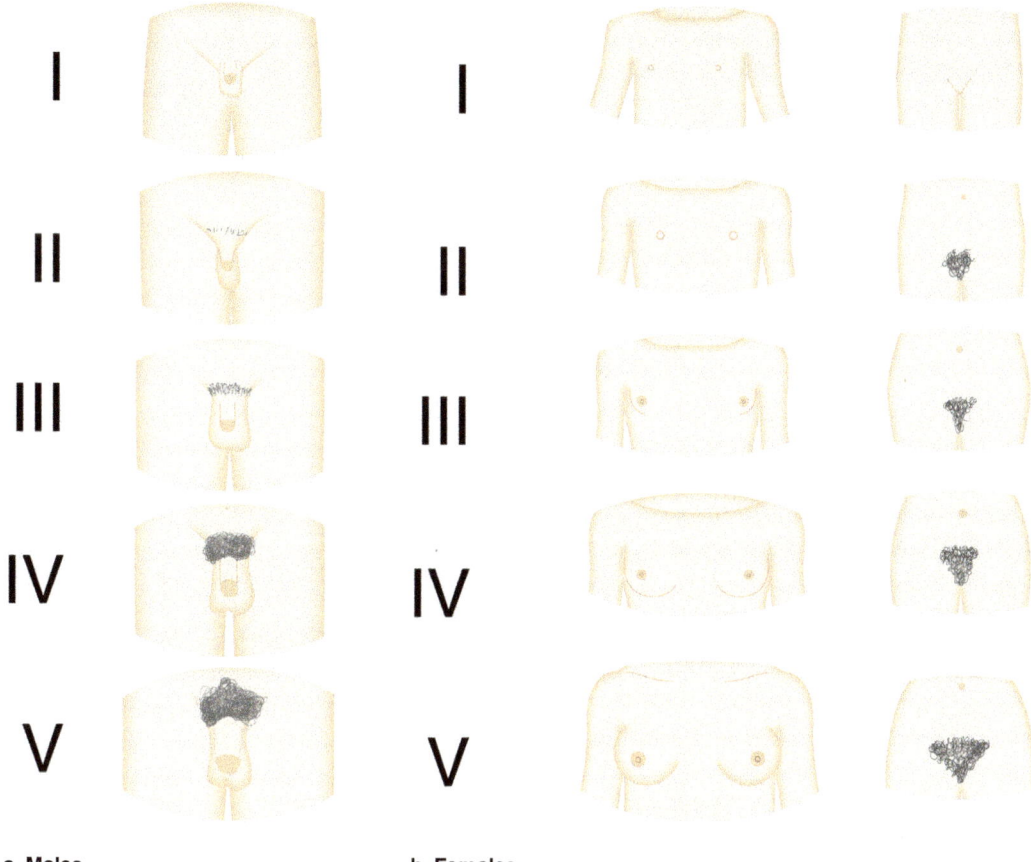

a. Males

b. Females

Figure 17.9   Illustration of the Tanner scale for males and females

Source: Reproduced under the Creative Commons 3.0 Attribution-ShareAlike License CC BY-SA 3.0 Tanner Scale Male: http://goo.gl/7cxTLM. Tanner Scale Female: http://goo.gl/haB9Cb, ©Michal Komorniczak, 2009.

APPLY

Puberty and obesity

Public health issues are related to early puberty in that it has been linked with childhood obesity. In girls, early menarche is correlated with adult obesity, type 2 diabetes and breast cancer (Ahmed et al., 2009).

The physical changes that occur are about the child's body developing into the man or woman with changes in function, as well as growth, of many of the organs and tissues of the body. The secondary sex characteristics have already been indicated in Table 17.4. Growth increases during the first half of puberty, with girls reaching their full height earlier than boys while skeletal changes result in the broad shoulders of young men, and the broad hips of women designed to facilitate childbirth (Figure 17.10). Growth of body hair and enlargement of the Adam's apple (angle of the thyroid cartilage in front of the larynx) in young men result in the characteristic male body and deeper voice. Figure 17.10 shows the male and female bodies at the end of puberty.

Adolescence

This stage in development occurs somewhat earlier in girls than boys, enabling boys, in general, to reach a greater height. The definition is problematic, but it is generally agreed to begin with the first signs of puberty (i.e. the process of achieving reproductive maturity) and to last until adulthood is reached, which may be in the early 20s.

--- **GO DEEPER** ---

Brain development

One of the key issues of adolescence is brain development. It appears that parts of the limbic system (concerned with reward) develop more rapidly than the prefrontal cortex (control mechanisms), resulting in a tendency to increased risk-taking in adolescents (Casey et al., 2008). In addition, it has been demonstrated that a 'phase shift' in circadian rhythm towards increased evening activity occurs with puberty (Crowley et al., 2007). Associated with the rising times required for school attendance this can result in reduced sleep in adolescents, with susceptible individuals exhibiting disturbed behaviour, psychological problems and poor academic performance (Carskadon et al., 1998).

During adolescence and early adulthood, changes in the brain continue particularly in the frontal and parietal cortices. Following the increase in grey matter during puberty, there is a decrease in grey matter after puberty, as 'synaptic pruning' occurs with refinement of the brain structure and enhancement of executive function. This has been described as:

> the capacity that allows us to control and coordinate our thoughts and behaviour. ... selective attention, decision-making, voluntary response inhibition and working memory ... these ... [have] a role in cognitive control, for example filtering out unimportant information, holding in mind a plan to carry out in the future and inhibiting impulses. (Blakemore and Choudhury, 2006: 301)

ADULTHOOD

Changes throughout life

Adulthood can be considered as having three stages:

- **Early adulthood:** about 20 to 40 years.
- **Middle adulthood:** about 40 to 65 years.
- **Late adulthood:** about 65 until death.

While there is considerable variation between individuals, optimum physical functioning is achieved in early adulthood with noticeable decline beginning in middle adulthood (Boyd and Bee, 2014). During early and middle adulthood individuals need to be fit to undertake their work, bring up their children and participate in family life. They need adequate nutrition, exercise and leisure. Having a lifestyle that enables the person concerned to live within their individual circadian rhythm is beneficial to health.

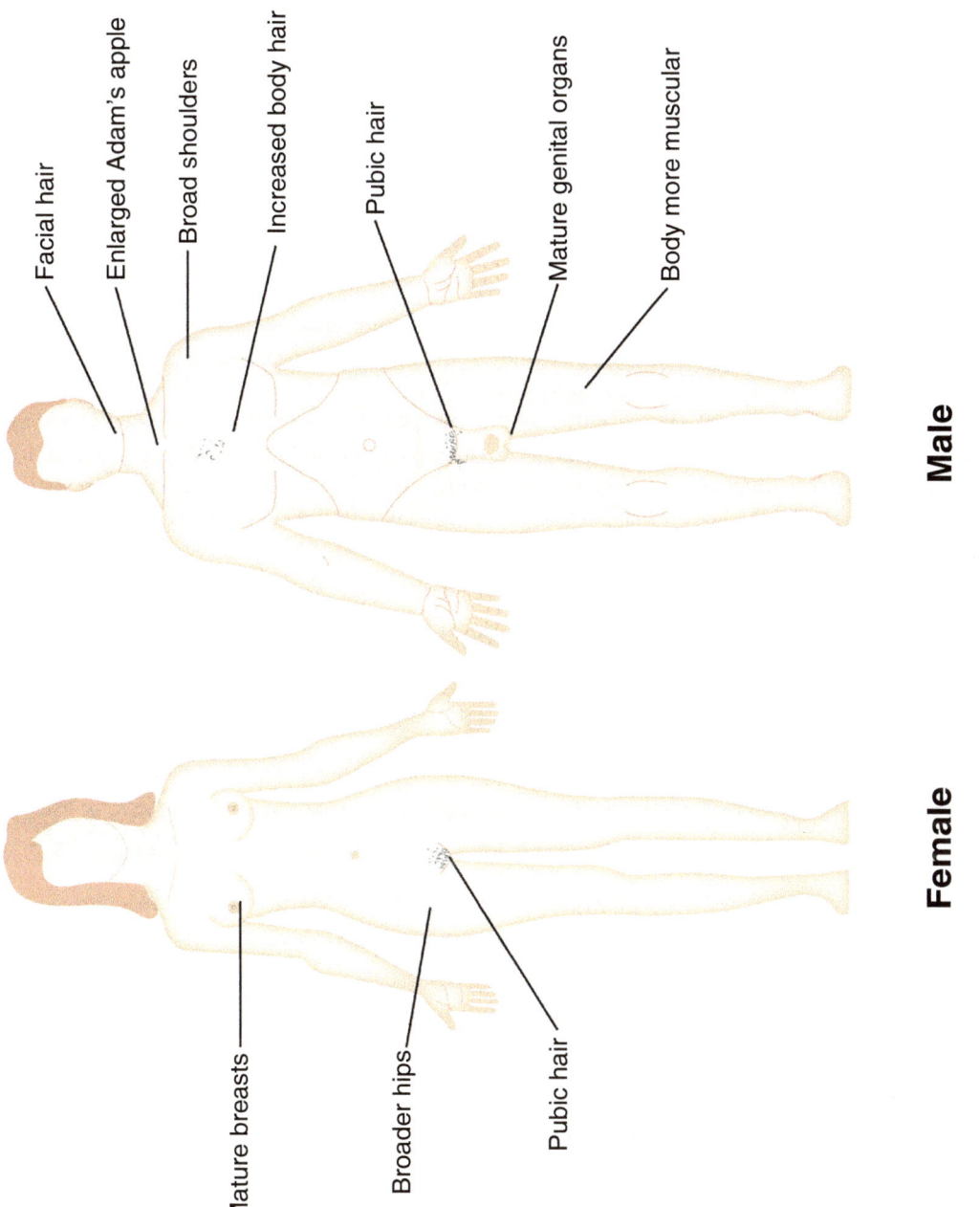

Facial hair

Enlarged Adam's apple

Broad shoulders

Increased body hair

Pubic hair

Mature genital organs

Body more muscular

Male

Mature breasts

Broader hips

Pubic hair

Female

Figure 17.10   The mature male and female body

GO DEEPER

Circadian rhythms and health

BODIE FAMILY

Thomas undertakes long-haul flights as an airline pilot and has to try to manage activity and sleep to fit the pattern of his work and maintain his ability to function. A study by Petrelli et al. (2006) reported that such pilots often report considerable fatigue because of 'irregular sleep schedules, long duty days, night flying, and multiple time zone changes' (p. 1347). Within this study, on flights with several stages, at the end of a stage, it was the overall amount of sleep in the previous 24 hours (including napping on board) that was related to self-reported fatigue rather than the timing of sleep. However, commercial airlines take the amount of work and rest, not work and sleep, into account when planning duty rosters. It is emphasised that pilots themselves are responsible for managing their own rest.

In relation to physical functioning, Table 17.5 provides an indication of the changes throughout life. In person-centred nursing, you must take account of the physical and functional differences between individuals, including those related to healthy ageing.

It is commonly considered that cognitive abilities also decline in later life. However, it is proposed (Ramscar et al., 2014) that the apparent changes in cognitive ability in healthy older people result from the wider experience from which they must draw and discriminate to achieve the correct answers in psychological tests. This takes longer than in younger individuals with less knowledge from which to draw. Therefore, slower response times may not represent a decline in cognitive function.

ACTIVITY 17.2: UNDERSTAND

Watch this online video to gain an understanding of how the brain changes with ageing.

This video can be accessed by **scanning the QR code** with your smart phone or via https://study.sagepub.com/essentialap2e.

COGNITIVE FUNCTION (5:12)

Table 17.5 Physical changes throughout life

System/ function	Age at which begins to be visible	Change
Vision	Mid-40s	Thickening of lens, loss of accommodation power, poorer near vision, increased sensitivity to glare
Hearing	50-60	Very high and very low tones more difficult to hear
Smell	About 40	Detection and discrimination between different smells declines
Muscles	About 50	Loss of muscle tissue, particularly fibres for bursts of strength or speed
Bones	Mid-30s (women)	Osteoporosis (loss of calcium) and osteoarthritis (degeneration of joint cartilage and underlying bone); more obvious after 60

System/ function	Age at which begins to be visible	Change
Heart and lungs	35-40	Little change at rest but, e.g. cardiac output or aerobic capacity, show age changes during work or exercise
Immune	Throughout life	Diminished ability to produce antibodies, decline in efficiency of immune system
Reproductive	Mid-30s (women)	Increased reproductive risk and reduced fertility; menopause from 40-60, hot flushes, irregular periods
	Early 40s (men)	Gradual decline in viable sperm, very gradual decline in testosterone from early adulthood
Cellular elasticity	Gradual	Loss of elasticity in most cells – skin, muscle, tendon, blood vessel cells decline faster with sunlight exposure
Skin	About 40	Increased wrinkles (loss of elasticity), loss of efficiency of oil-secreting glands
Hair	Variable	Becomes thinner, may become grey
Height	About 40	Compression of spinal discs → loss of height of 2.5-5 cm (1-2 inches) by 80 years
Weight	Variable	Tendency to increase weight in middle adulthood, but lose weight late in life. However, rate of overweight/obesity in population very variable and increasing

Reproduced from Boyd, D. and Bee, H. *Lifespan Development*, Global edition, ©2019. Printed and electronically reproduced by permission of Pearson Education Inc, New York.

Changes in the reproductive systems

Men and women's reproductive systems show very different patterns of change in their activity with ageing.

Female

With increasing years there is a decline in fertility with fewer fertile ova being released until the menopause when reproductive life ends. During these later years of reproductive life there is also an increase in miscarriages and other obstetric difficulties (Lubna and Santoro, 2003).

Women come to a clear end of their reproductive life at the menopause when menstruation ends, usually between 45 and 55 years (the average in the UK is 51 years) although in some this may happen earlier. However, the actual time of the menopause is when the ovaries reduce and then cease functioning with changes in hormone secretion, which may take place over a few years. The reduction of oestrogens is considered to cause osteoporosis in older women due to loss of calcium from the bone, sometimes resulting in compression fractures of the vertebrae and bowing of the spine with loss of height (Old and Calvert, 2004). The most obvious signs of the menopause are changes in the pattern of menstruation: the frequency and amount of blood lost may change from the pattern normal for the individual. A year without periods is usually taken as implying one year post-menopause.

 APPLY

Menopause

Women vary in the presence and severity of menopausal symptoms, which might last two to five years and may include some of the following:

(Continued)

- Reduced sex drive.
- Vaginal dryness and itching, and pain or discomfort during sex.
- Hot flushes.
- Problems with sleeping, sometimes associated with night sweats.
- Urinary tract infections.
- Changes in mood and tiredness, sometimes depression and/or anxiety.
- Headaches.
- Palpitations (when heartbeats become noticeable).

BODIE FAMILY

Sarah Bodie (55) and Hannah Jones (54) are both currently passing through the menopause and compare notes on how they are managing their symptoms. Sarah has recently stopped taking Hormone Replacement Therapy (HRT) tablets after two-and-a-half years during which her symptoms, particularly the hot flushes and vaginal irritation, have been minimised. She is coping well now. Hannah is still using HRT patches which are effective in controlling her symptoms.

Male

Men can continue to produce sperm and can inseminate a fertile woman producing a healthy child until late in life (e.g. the tenth decade). However, the volume of semen, and the quality of sperm (motility and genetic makeup) diminish with ageing: there is some evidence for chromosomal abnormalities becoming commoner in older men (Kühnert and Nieschlag, 2004). In addition, there is evidence of a decline in testosterone secretion from 40 onwards which can influence a number of physiological functions (Feldman et al., 2002). Nevertheless, in many men endocrine function is adequate to maintain fertility (Hermann et al., 2000).

APPLY

Ageing and sperm production

With age, there can be a reduction in the production of spermatozoa. This is thought to be related to either a loss of germ cells from DNA damage and mutations, or damage to the Sertoli cells that nourish them (Paul and Robaire, 2013). It may also be related to a reduction in testosterone production. As a result, an increase in paternal age is correlated with an increased risk of complex multigene diseases in children (Paul and Robaire, 2013). As males age, the refractory period after ejaculation also increases (Llanes et al., 2013). This can often result in a sense of a loss of virility.

Ageing

The abilities of people in late adulthood decline but the rate of decline varies with physical and mental activity, genetic makeup and environmental conditions. Many in their 60s and early 70s are fit and capable of looking after themselves, but some are much less able. With increasing age elderly people become more frail and debilitated. However, some, and increasingly more of them, live in good health into their 100s.

Older adults are often studied in three subgroups (varying somewhat with different authors); one such grouping is: 'young-old, 65–74 years; middle-old, 75–84 years; and oldest-old, ≥85 years' (Zizza et al., 2009: 481). In all these age groups there can be considerable variation in physical and mental health status. However, in general, the following descriptions apply:

- **Young-old group:** are generally fit. They have probably been working until fairly recently and may indeed still be using their expertise. Many will be taking part in varying types of exercise such as golf, t'ai chi, Pilates, and are possibly involved with a range of local activities.

- **Middle-old group:** are getting less active but some can still participate in physical exercise although perhaps less vigorously. Many will be involved in a range of activities, some of which are likely to be social in nature, maintaining their interactions with others. Some may be becoming somewhat forgetful. Some may need a limited amount of support

---

APPLY

BODIE FAMILY

Managing health in older age

Maud (77) and George (84) come within the middle-old group. Although Maud was diagnosed with heart failure when she was 74 her condition is under good control with drugs (digoxin to improve heart function and warfarin to thin her blood). Both of them are still active, taking the dog for walks and participating in various activities.

- **Oldest-old:** many in this group will be becoming frailer, both physically and mentally, and are likely to need some support in maintaining their well-being.

APPLY

Person-centred nursing and ageing

With increasing numbers of older people living into their 80s plus, retirement is becoming a much more important and long-lasting part of the life-cycle. Regular physical exercise is recognised as an important factor in maintaining mental health (Etgen et al., 2010). George and Maud (the great-grandparents) have joined U3A (University of the Third Age), which offers a wide range of activities that deal with health and fitness, learning, and social activities run by the members for the members. Many people following retirement take advantage of such organisations as U3A to enjoy the activities, but also to meet others. The aim of person-centred nursing at this stage of life is to help individuals to enjoy healthy fulfilled lives, not just long ones.

Eventually at the end of life, bodily functions decline until the person is no longer able to do the things that he or she perceives as central to life being worth living. The role of the health professional is not to keep them alive at any price but to support the person concerned in making the choices that will enable him or her to achieve an end to life which meets their desires (Gawande, 2014).

As an individual moves to the end of life, there will be diminished function in all systems of the body resulting in:

- Fatigue and weakness.
- Reduced blood flow through the skin which increases the risk of pressure sores (decubitus ulcers).
- Loss of appetite and reduced fluid intake leading to dehydration.
- Cardiac failure/dysfunction.
- Renal failure.
- Nervous system dysfunction.

At the end of life most people gradually lose consciousness, become comatose, and die peacefully as breathing ceases and the heart stops beating.

CONCLUSION

This chapter has briefly reviewed the development of the body from fertilisation to death. In providing person-centred care it is important to understand the stage of life that the individual is in and provide care appropriate to their needs at that time.

GO DEEPER

Further reading

Levin, R. (2015) 'Sexuality of the ageing female – the underlying physiology', *Sexual & Relationship Therapy*, 30(1): 25-36.

Maguire, S.L. and Slater, B.M. (2013) 'Physiology: physiology of ageing', *Anaesthesia & Intensive Care Medicine*, 14(Regional Anaesthesia): 310-12.

Navaratnarajah, A. and Jackson, S.H. (2013) 'Medicine in older adults: the physiology of ageing', *Medicine*, 41(Medicine in Older Adults): 5-8.

1. The first eight weeks of pregnancy is the stage of embryonic development during which all organs and tissues (except adipose tissue) begin development and are, thus, at greatest risk of damage from environmental or toxic factors. Preconceptual care is important to minimise congenital abnormalities.

2. During early life the importance of nutrition and care of the infant is crucial for normal growth and development. Monitoring of these processes takes place during early life and adolescence to ensure appropriate interventions can be provided if necessary.

3. Puberty is a period of substantial change as children turn into adults going through the anatomical and physiological changes for the functions necessary for reproduction. This is a part of adolescence but equally important is the achievement of adult identity and behaviour.

4. Adulthood is the major part of life when work and bringing up a family are central concerns. In the early part of this stage of life most people are healthy and able to undertake these roles effectively. Later in life tissues, organs and systems begin to decline in strength and function.

5. Understanding the physiological changes which occur with ageing enables person-centred nursing to be planned to provide the necessary care and support for individuals.

In revising the content of this chapter, it is important to be aware of the underpinning principles. Start your revision by reviewing both person-centred nursing at the beginning of the book and homeostasis in Chapter 1, before moving on to this chapter.

In order to help you revise, consider the following questions, answers for which can be found by visiting **https://study.sagepub.com/essentialap2e**.

Test yourself by revising the chapter first, and then answer these questions without looking at the book. Afterwards compare your answers with the text and with the notes you made. Did you miss anything in your notes? Here are the questions:

1. At what stage of prenatal development is the embryo most susceptible to damage? Name one substance recommended to minimise such damage.

2. What is apoptosis and why is it important in development?

3. What do you understand by embryogenesis and what are the major structures created?

4. What are the functions of the placenta?

5. Briefly describe the cardiovascular changes that occur in the transition from intra-uterine to independent life.

6. Which tissues grow most rapidly in the first few years of life?

7   What do you understand by the terms puberty and adolescence?

8   What are the three stages of adulthood and of ageing?

9   Outline the changes in the reproductive systems of men and women that occur as they get older.

10 Outline the physiological changes which commonly occur as someone is dying.

REVISE

For additional revision resources visit: **https://study.sagepub.com/essentialap2e**.

ACE YOUR ASSESSMENT

- Revise key terms with interactive flashcards.
- Test yourself with multiple-choice questions.
- Access the glossary with audio to hear how complex terms are pronounced.
- Explore recommended websites suitable for revision.

REFERENCES

Ahmed, M.A., Ong, K.K. and Dunger, D.B. (2009). 'Review: childhood obesity and the timing of puberty', *Trends in Endocrinology & Metabolism*, 20(5): 237–42.

Barrett, K.E., Barman, S.M., Boitano, S. and Brooks, H.L. (2010) *Ganongs Review of Medical Physiology*, 23rd edition. New York: McGraw Hill Medical.

Blakemore, S-J. and Choudhury, S. (2006) 'Development of the adolescent brain: implications for executive function and social cognition', *Journal of Child Psychology and Psychiatry*, 296–312.

Blakemore, S.J., Burnett, S. and Dahl, R.E. (2010) 'The role of puberty in the developing adolescent brain', *Human Brain Mapping*, 31(6): 926–33.

Boyd, D.R. and Bee, H.L. (2019) *Lifespan Development*, Global edn. Essex: Pearson.

British Nutrition Foundation (2013) *Nutrition and Development: Short- and Long-Term Consequences for Health*. Chichester: Wiley-Blackwell.

Canadian Paediatric Society (2003) 'Age limits and adolescents', *Paediatrics and Child Health*, 8(9): 577.

Carskadon, M.A., Wolfson, A.R., Acebo, C., Tzischinsky, O. and Seifer, R. (1998) 'Adolescent sleep patterns, circadian timing, and sleepiness at a transition to early school days', *Sleep*, 21(8): 871–81.

Casey, B.J., Getz, S. and Galvan, A. (2008) 'The adolescent brain', *Developmental Review*, 28: 62–77.

Cole. T.J. (2003) 'The secular trend in human physical growth: a biological view', *Economics & Human Biology*, 1(2): 161–8.

Crowley, S.F., Acebo, C. and Carskadon, M.A. (2007) 'Sleep, circadian rhythms, and delayed phase in adolescence', *Sleep Medicine*, 8(6): 602–12.

Etgen, T., Sander, D., Huntgeburth, U., Poppert, H., Förstl, H. et al. (2010) 'Physical activity and incident cognitive impairment in elderly persons', *JAMA Internal Medicine (previously Archives of Internal Medicine)*, 170(2): 186–93.

Euling, S.Y., Herman-Giddens, M.E., Lee, P.A. et al. (2008) 'Examination of US puberty-timing data from 1940 to 1994 for secular trends: panel findings', *Pediatrics*, 121(suppl. 3): S172–191.

Feldman, H.A., Longcope, C., Derby, C.A., Johannes, C.B., Araujo, A.B. et al. (2002) 'Age trends in the level of serum testosterone and other hormones in middle-aged men: longitudinal results from the Massachusetts Male Aging Study', *The Journal of Clinical Endocrinology & Metabolism*, 87(2): 589–98.

Gawande, A. (2014) *Being Mortal: Illness, Medicine and What Matters in the End*. London: Wellcome Collection, Profile Books.

Hermann, M., Untergasser, G., Rumpold, H. and Berger, P. (2000) 'Aging of the male reproductive system', *Experimental Gerontology*, 35(9–10): 1267–79.

Holick, M.F. (2006) 'Resurrection of vitamin D deficiency and rickets', *The Journal of Clinical Investigation*, 116(8): 2062–72.

Holick, M.F. and Chen, T.C. (2008) 'Vitamin D deficiency: a worldwide problem with health consequences', *The American Journal of Clinical Nutrition*, 87(suppl):1080S–6S.

Kühnert, B. and Nieschlag, E. (2004) 'Reproductive functions of the ageing male', *Human Reproduction Update*, 10(4): 327–39.

Llanes, L.L., Ballester, G.L.A. and Elías-Calle, L.C. (2013) 'Ejaculation and sexual pleasure in males: a complex relation with multiple determinants', *Revista Sexología y Sociedad*, 19(1). Available at: www.revsexologiaysociedad.sld.cu/index.php/sexologiaysociedad/article/view/236/297 (accessed 23 July 2020).

Lubna, P. and Santoro, N. (2003) 'Age-related decline in fertility', *Endocrinology and Metabolism Clinics*, 32(3): 669–88.

Marshall, W.A. and Tanner, J.M. (1969) 'Variations in pattern of pubertal changes in girls', *Archives of Disease in Childhood*, 44(235): 291–303.

Marshall, W.A. and Tanner, J.M. (1970) 'Variations in pattern of pubertal changes in boys', *Archives of Disease in Childhood*, 45(239): 13–23.

Mirmiran, M., Maas, Y.G.H. and Ariagno, R.L. (2003) 'Development of fetal and neonatal sleep and circadian rhythms', *Sleep Medicine Reviews*, 7(4): 321–34.

Old, J.L. and Calvert, C. (2004) 'Vertebral compression fractures in the elderly', *American Family Physician*, 69(1): 111–16.

Paul, C. and Robaire, B. (2013) 'Ageing of the male germ line', *Nature Reviews Urology*, 10(4): 227–34.

Petrelli, R.M., Roach, G.D., Dawson, D. and Lamond, N. (2006) 'The sleep, subjective fatigue, and sustained attention of commercial airline pilots during an international pattern', *Chronobiology International*, 23(6): 1347–62.

Ralston, A. and Shaw, K. (2008) 'Gene expression regulates cell differentiation', *Nature Education*, 1(1): 127.

Ramscar, M., Hendrix, P., Shaoul, C., Milin, P. and Baayen, H. (2014) 'The myth of cognitive decline: non-linear dynamics of lifelong learning', *Topics in Cognitive Science*, 6(1): 5–42.

RCPCH (Royal College of Paediatrics and Child Health) (2015) *UK–WHO growth charts, 0–18 years*. London: RCPCH. Available at: www.rcpch.ac.uk/child-health/research-projects/uk-who-growth-charts/uk-growth-chart-resources-2-18-years/school-age (accessed 5 May 2020).

Roseboom, T.J., Painter, R.C., van Abeelen, A.F.M., Veenendaal, M.V.E. and de Rooij, S.R. (2011) 'Hungry in the womb: what are the consequences? Lessons from the Dutch famine', *Maturitas*, 70: 141–5.

Talaulikar, V. and Arulkumaran, S. (2013) 'Folic acid in pregnancy', *Obstetrics, Gynaecology and Reproductive Medicine*, 23(9): 286–8.

Warren, W.S. and Cassone, V.M. (1995) 'The pineal gland: photoreception and coupling of behavioral, metabolic, and cardiovascular circadian outputs', *Journal of Biological Rhythms*, 10(1): 64–79.

WHO (World Health Organization) (2014) *Maternal, Newborn, Child and Adolescent Health: Adolescent Development*. Geneva: WHO.

Yun, J.A., Bazar, K.A. and Lee, P.Y. (2004) 'Pineal attrition, loss of cognitive plasticity, and onset of puberty during the teen years: is it a modern maladaptation exposed by evolutionary displacement?', *Medical Hypotheses*, 63(6): 939–50.

Zizza, C.A., Ellison, K.J. and Wernette, C.M. (2009) 'Total water intakes of community-living middle-old and oldest-old adults', *The Journals of Gerontology. Series A, Biological Sciences and Medical Sciences*, 64A(4): 481–86.

GLOSSARY

A

Abduction movement away from the midline.

Absorption the process by which one substance absorbs or is absorbed (taken in or assimilated) by another.

Absorption (intestinal) movement of nutrients across the intestinal wall into the blood and lymphatic circulation.

Accommodation change in shape of the lens of the eye.

Achondroplasia a genetic disorder characterised by abnormally slow conversion of cartilage to bone during bone development, resulting in short stature (causing disproportionate body structure: normal size truck, short limbs).

Acidosis when blood pH is less than 7.35.

Acinar cells cells of the pancreas that secrete pancreatic juice.

Action potential the change in electrical potential in the membrane of a neuron to send an electrical impulse along its length.

Active transport the movement of solutes (ions and molecules) across a membrane (usually) against a concentration gradient involving the use of energy (ATP) and carrier proteins.

Adaptive immunity a specific immune response consisting of antibody responses and cell-mediated responses where each pathogen is 'remembered' by a signature antibody.

Adduction movement closer to the midline.

Adenoids (tonsils) a mass of enlarged lymphatic tissue.

Adenosine Triphosphate (ATP) the energy store of the cell used to power cellular activities.

Adipocytes fat cells that make up adipose tissue.

Adipose tissue tissue composed of adipocytes, specialist cells that store fat. There are two types; white and brown. Brown adipose tissue is used in infants to generate heat. White adipose tissue is more prevalent in adults and is hormonal, secreting hormones such as leptin that reduce appetite and fat storage.

Adolescence the stage in life when the child turns into an adult (normally between the ages of ten and 19).

Aerobic   relates to an organism that exists with, or is dependent on, oxygen.

Agranulocytes   leucocytes that have few or no organelles and so do not have a granular appearance.

Albumin   the main plasma protein, which generates plasma colloidal osmotic pressure and serves as a transport protein.

Aldosterone   a hormone produced by the adrenal cortex under the influence of angiotensin II that promotes the reabsorption of sodium in the renal tubule, drawing water with it.

Alkalosis   when blood pH is greater than 7.45.

Allele   alternative form of the same gene located at the corresponding position on homologous chromosomes.

Allergen   substance that can cause an allergic reaction.

Alpha-adrenergic receptors   receptors that, when stimulated by catecholamines, constrict blood vessels, contract uterine muscles, relax intestinal muscles and dilate pupils.

Alveolar ducts   thin-walled air passages that are subdivisions of bronchioles and lead to alveoli of the lungs.

Alveolar fluid   fluid lining the alveoli.

Alveolar ventilation rate   the actual volume of air per minute that reaches the respiratory zone.

Alveoli   small cup-shaped pouches/air sacs at the terminal end of alveolar ducts where gaseous exchange takes place.

Amino acid   organic compounds containing nitrogen that combine to form proteins.

Amniotic fluid   fluid within the amniotic sac surrounding the fetus/embryo in the uterus.

Amphiarthroses   joint with limited movement.

Ampulla   sac containing hair cells embedded in the cupula.

Amygdala   part of the limbic system that contributes to storage of emotional experiences as memories and regulates emotional learning.

Anabolism   the building up or synthesising of large and complex molecules. This process requires energy.

Anaerobic   relates to an organism that exists without, or is not dependent on, oxygen.

Anal canal   terminal end of the gastrointestinal tract.

Androgens   male sex hormones.

Angiogenesis   the formation of new blood vessels.

Angiotensin I   a biological inactive plasma protein converted to angiotensin II by angiotensin converting enzyme (ACE).

Angiotensin II   a hormone that brings about a number of responses designed to increase blood volume and blood pressure.

Angiotensin Converting Enzyme (ACE)   a hormone produced by the lungs and proximal convoluted tubules in the nephrons of the kidney that converts angiotensin I to angiotensin II.

Angiotensinogen a plasma protein produced by the liver and converted by renin to angiotensin I.

Anterolateral pathways sensory pathways for carrying temperature, pain and coarse touch sensations to the brain.

Antibodies glycoprotein immunoglobulins (Igs), produced by lymphocytes, that combine with antigens to inactivate them by changing the antigen's chemical composition, render them immobile or prevent them from penetrating cells.

Antidiuretic Hormone (ADH) a hormone produced by the hypothalamus and released by the pituitary gland. It results in water reabsorption in the renal tubule to retain/increase water levels in the body.

Antigen any substance foreign to the body that evokes an immune response.

Antigen–antibody complex complex formed by the binding of an antibody to an antigen – or immune complex.

Antiporter an integral membrane transport protein that simultaneously transports two different molecules, in opposite directions, across the membrane.

Anus opening at the terminal end of the gastrointestinal tract through which faeces is eliminated.

Aperture opening, hole or gap.

Apneustic centre area in the lower pons of the brain responsible for regulating ventilation (with the pneumotaxic centre).

Apocrine gland sweat gland with ducts that open into hair follicles rather than directly onto the surface of the skin and produce sweat containing fats and proteins. Specialised apocrine glands produce ear wax (ceruminous glands) and milk (mammary glands).

Apoptosis programmed cell death.

Appendicular skeleton bones of the arms and legs and those bones that attach them to the axial skeleton.

Appendix (vermiform) a tube of about 9 cm in length attached to the caecum. It has a large amount of lymphoid tissue.

Arachnoid mater lining of the brain separated from the dura mater by the dural space and from the pia mater below by the subarachnoid space. It contains large blood vessels and arachnoid villi for reabsorption of CSF.

Archaea microorganisms similar to bacteria in size and structure but genetically and functionally dissimilar from them. They often exist in extreme environmental conditions.

Arrector pili smooth muscles fibres attached to hair follicles.

Arteries blood vessels that carry blood away from the heart.

Arterioles small arteries.

Articular capsule a sleeve-like structure that encloses the synovial cavity.

Articular cartilage thin layer of hyaline cartilage that covers the epiphysis of one bone where it forms a joint with another bone.

Articulation *see* Joint.

Ascending (afferent) neurons neurons that carry information up the spinal cord to the brain for processing.

Astrocytes   neuroglia that secure neurons to their blood supply, form the blood–brain barrier, and regulate the external chemical environment of neurons by removing excess ions and promote re-uptake of neurotransmitters released during synaptic transmission.

Atoms   the basic building blocks of matter.

Atopy   being at a high risk of allergies with raised levels of IgE to specific allergens.

Atrial Natriuretic Peptide (ANP)   a hormone released from the muscle of the atria of the heart when blood volume is increased, promoting the excretion of sodium in the renal tubule, taking water with it.

Atrioventricular (AV) bundle   specialised nerve tract that divides into right and left bundle branches that pass down the septum of the heart dividing out to each ventricle to distribute the electrical (cardiac) impulse.

Atrioventricular (AV) node   specialised autorhythmic cells in the wall of the septum of the heart that delay the electrical (cardiac) impulse temporarily before conducting it down to the ventricles through the atrioventricular bundle.

Atrioventricular valves   valves between the atria and ventricles.

Atrium   upper chamber of the heart, of which there are two, which receive blood into the heart and pump it into the ventricles. The right atrium receives deoxygenated blood from the veins of the systemic circulation, the left atrium oxygenated blood from the pulmonary vein.

Auditory tube   *see* Eustachian tube.

Auricle (pinna)   fleshy part of the ear.

Autocrine   relating to a hormone that has effect only on the cells which produce it.

Autonomic nervous system   a division of the nervous system that contains cranial and spinal nerves that subconsciously control visceral organs (including those of the circulatory, digestive and respiratory systems) through regulating involuntary activities via its sympathetic and parasympathetic subdivisions.

Autorhythmic cells   cells of the conduction system that initiate and distribute electrical impulses to adjacent cells, which stimulate heart muscle to contract.

Autosome   any chromosome that is not a sex chromosome.

Avascular   referring to the absence of blood vessels.

Axial skeleton   the bones of the skull, vertebral column, ribs and sternum forming the central axis of the body and supporting the head, neck and torso.

Axon   long projection in a neuron that carries action potentials.

Axon terminal   end(s) of an axon.

B

Bacteria   plural of bacterium – a single-celled spherical, spiral, or rod-shaped microorganism that appear singly or in groups or chains, existing independently or parasitically.

Bacteriophages   a virus that infects and replicates within a bacterium.

Baroreceptors   sensory receptors that respond to changes in pressure.

Basal cells (taste) stem cells located at the periphery of taste buds that mature into taste cells to replace those that have died.

Basal ganglia *see* Basal nuclei.

Basal metabolic rate the rate at which energy is used while at rest to maintain vital organ functions.

Basal nuclei (basal ganglia) nuclei (groups of nerve cell bodies) of grey matter buried within the white matter.

Basophils granulocytes that assist in the inflammatory response by secreting histamine and heparin to increase blood flow by vasodilation and thin the blood.

B-cells lymphocytes that recognise specific antigens and produce antibodies.

Beta-adrenergic receptors receptors that, when stimulated by catecholamines, increase heart rate, stimulate cardiac contraction, dilate bronchi, dilate blood vessels and relax the uterus.

Bile liquid secreted by the liver containing water, salts, mucus, cholesterol, lecithin and bilirubin.

Bilirubin a breakdown product of haemoglobin converted from biliverdin.

Biliverdin an intermediate substance produced in the degradation of haemoglobin that is converted to bilirubin.

Blastocyst stage in embryonic development when the morula has developed a fluid-filled space and then embeds into the uterine wall at about day seven of pregnancy.

Blood part of the ECF contained within blood vessels that consists of plasma and blood cells.

Blood–Brain Barrier (BBB) a structural and chemical barrier that strictly regulates substances passing from the circulatory system into the nervous system.

Blood pressure the pressure exerted by the blood against the arterial walls by the pumping force of the heart = cardiac output × peripheral resistance.

Blood vessels the tubes forming the circulation through which the blood is transported around the body. Arteries carry blood away from the heart, veins carry blood back to the heart, and the fine capillaries join the two and permit the fluid in the blood to enter the ECF surrounding the cells.

B-lymphocyte type of lymphocyte that differentiates into plasma cells that synthesise antibodies to react with specific antigens that stimulate them.

Bolus a small rounded mass of a substance (e.g. chewed food).

Bone remodelling when bone is reshaped by replacing old bone with new.

Bowman's capsule a cup-shaped sac at the beginning of a nephron that collects filtrate from the glomerulus which it encloses.

Brachial plexus network of nerves in the shoulder carrying neurons from the spine to the arms.

Bronchi the divisions of the trachea at the inferior end.

Bronchial tree a highly branched system of conducting passages that extends from the primary bronchus to the terminal bronchioles.

Bronchioles small continuations of the airway, derived from bronchi.

Bulbourethral glands glands that produce 10% of seminal fluid.

C

Caecum   first part of the large intestine.

Calcification   accumulation of calcium salts in a body tissue.

Callus   bony healing tissue formed around the ends of a fracture.

Calyces   chambers of the kidney through which urine passes.

Capillaries   small blood vessels that link arterioles to venules and enable exchange of water, nutrients and waste products between the blood and the tissues.

Carbohydrates   macronutrients that are a key source of energy for the body.

Cardiac cycle   one complete heartbeat during which the heart contracts (systole) and relaxes (diastole).

Cardiac output   the volume of blood pumped out of the heart per minute = stroke volume × heart rate.

Carpal bones   eight small bones of the wrist.

Carrier protein   *see* Transport protein.

Cartilage   a strong, flexible and semi-rigid connective tissue.

Cartilaginous joint   type of joint connected by cartilaginous material.

Catabolism   the breaking down of substances to provide energy and raw materials for anabolism.

Catalyst   a substance that accelerates chemical reactions without themselves being used up.

Catecholamines   hormones/neurotransmitters (such as dopamine, noradrenaline (norepinephrine), and adrenaline (epinephrine)) made by the adrenal glands, the brain and some specialised nerve cells that stimulate alpha-adrenergic and beta-adrenergic receptors.

Cauda equina   nerves resembling a horse's tail that extend down to the sacrum.

Cell differentiation   the process by which a less specialised cell becomes more specialised.

Cell/plasma membrane   double-layered lipid membrane including proteins that controls movement of substances into and out of the cell thus determining composition of cytoplasm (cell contents excluding organelles) – ions and nutrients.

Cellular respiration   metabolic processes that occur in cells to create energy and waste from nutrients.

Central Nervous System (CNS)   the division of the nervous system made up of the brain and spinal cord.

Centrioles   two cylinders in the centrosome at right angles to each other.

Centrosome   core of the cell cytoskeleton from which single filaments radiate and enable movement of vesicles and organelles.

Cerebellomedullary cistern   CSF filled space that links the ventricles with the subarachnoid space.

Cerebellum   part of the brain that coordinates the muscles of the body, regulates muscle tone and posture, and has an important role in cognition.

Cerebral aqueduct   connection between the third ventricle and the fourth ventricle.

Cerebrospinal Fluid (CSF)   fluid created in the choroid plexus that protects and nourishes the CNS.

Cerebrum   part of the brain consisting of the cerebral cortex (grey matter), underlying white matter, the basal nuclei and the limbic system.

Cervix   thickened ring of muscle and fibrous tissue at the base of the uterus with a central passage.

Chemical senses   the sensations of smell (olfaction) and taste (gustation).

Chemoreceptors   sensory receptors that respond to changes in the concentration of chemicals.

Cholesterol   most common steroid molecule and the precursor for the steroid hormones.

Chondroblasts   immature chondrocyte cells that, when mature, secrete the components of cartilage.

Chondrocytes   mature cartilage cells derived from chondroblasts.

Choroid   heavily pigmented layer of the eye that sits behind the retina, prevents light scattering within the eye and also nourishes the retina.

Choroid plexus   networks of blood vessels in each ventricle of the brain that produce CSF.

Chromatids   duplicated chromosomes formed during cell division.

Chromosome   structures in the nucleus composed of DNA tightly coiled many times around proteins. They are responsible for genetic expression.

Chyle   a milky bodily fluid consisting of lymph and emulsified fats, or free fatty acids (FFAs). It is formed in the small intestine during digestion of fatty foods, and taken up by lymph vessels specifically known as lacteals.

Chylomicrons   type of lipoprotein created in the small intestine to aid absorption.

Chyme   partly digested food mixed with digestive enzymes.

Cilia   fine hair-like projections from certain cells that sweep in unison to move particles and mucus away.

Ciliary body   a continuation of the choroid containing the ciliary muscle and ciliary processes.

Ciliary muscle   muscle that encircles the lens of the eye to change its shape to focus light for vision.

Ciliary processes   part of the ciliary body that secretes aqueous humour.

Cingulate gyrus   part of the limbic system thought to integrate emotion and sensory experiences.

Circadian rhythm   the inherent biological clock related to biological processes that occur regularly at about 24-hour intervals.

Circumduction   a circular movement – a combination of flexion, extension, abduction and adduction.

Clavicle   the collarbone.

Clitoris   female equivalent of the penis in men, containing erectile tissue and responds to sexual arousal by enlarging and becoming firm.

Coagulation   a complex series of reactions in which positive feedback enhances the action of clotting factors, resulting in blood clot formation.

Cochlea   spiral cavity of the inner ear containing the organ of Corti for hearing.

Codon   a triplet of nucleotides in the mRNA chain that codes for a specific amino acid in the synthesis of a protein molecule.

Cognition   the mental process of acquiring and interpreting knowledge, including perception, intuition and reasoning.

Collagen   the primary structural protein in connective tissue.

Collagenase   an enzyme that breaks down collagen.

Collecting duct   a duct in the kidneys that collects the filtrate from a number of nephrons.

Colon   largest section of the large intestine between the caecum and the rectum.

Colonisation   usually of microbes, to become established in a new environment.

Colostrum   first secretion of breast milk which contains more white cells and antibodies (particularly IgA) than mature milk.

Commensal   normal symbiotic flora living on or within the body, deriving benefit without harming or benefiting the host.

Compact bone   strong, dense bone tissue.

Complement system   a second major system within innate immunity consisting of a large number (20+) of plasma proteins that facilitate phagocytosis of bacteria through opsonisation.

Compounds   chemicals made from atoms of different elements joined by chemical bonds.

Concentric lamellae   plates of compact bone.

Conchae   long, narrow curled shelves of bone.

Conducting zone   parts of the airway that deliver inhaled and exhaled air to and from the lungs but in which no gaseous exchange takes place.

Conduction   transfer of heat as a result of contact with a surface colder or hotter than the body.

Cones   photoreceptors sensitive to blue, green and red light and which allow colour vision and are important in day vision.

Conjunctiva   a thin, transparent protective membrane that coats the anterior surface of the eye, except for the cornea, and the inner surface of the eyelid.

Connective tissue   structural tissues that provide structural support for the organs of the body.

Consciousness   a state of explicit awareness dependent on both biological arousal in the brain and the processing of experiences (perception).

Constriction   narrowing of the pupil of the eye.

Conus medullaris   narrowing of the spinal cord into a conical shape towards the end.

Convection   heat loss from air being heated as it passes over the body, rises and is replaced by cool air.

Core body temperature   the temperature centrally within the body – how warm the vital organs are.

Coronary circulation   blood supply to the heart.

Corpus callosum   major tract of nerve fibres joining the two sides of the brain.

Corpus luteum   a temporary endocrine structure in the ovary developed from the follicle after ovulation that secretes fairly high levels of progesterone and medium levels of oestrogen and inhibin.

Corticospinal tracts   carries information from the cerebral cortex to the spinal cord.

Cranial nerves   12 pairs of nerves that pass directly from the brain through openings in the skull.

C-Reactive Protein (CRP)   a plasma protein which increases in level in response to inflammation.

Creatinine   a waste molecule generated from muscle metabolism.

Creatinine clearance   the rate at which creatinine is cleared from the blood by the kidney.

Cribriform foramina   pores in the ethmoid bone of the skull that allow the axons of olfactory neurons to pass through to the olfactory bulb.

Crystallisation   the formation of crystals.

CT   Computerised Tomography, or computerised axial tomography (CAT) – a scan that creates a digital image of structures inside the body from a series of x-rays (ionising radiation).

Cupula   a gelatinous membrane in the ampullae of the semi-circular canals of the ear.

Cytokines   small proteins which act to pass signals between cells.

Cytoplasm   a clear substance that consists of all of the contents outside of the nucleus of a membrane-bound cell.

Cytoskeleton   a dynamic network of fibres made of protein microfilaments and microtubules to form an internal framework of a cell, allowing movement.

Cytosol   the aqueous component of the cytoplasm of a cell, within which various organelles and substances are suspended.

Cytotoxic   toxic to living cells.

D

Deamination   the process in which the nitrogen part of amino acids are removed and excreted if there is an excess of protein intake.

Dentine   hard, porous material that makes up the bulk of the tooth.

Deoxyribonucleic Acid (DNA)   a nucleic acid molecule consisting of two polynucleotide chains in the form of a double helix that plays a central role in protein synthesis and is responsible for the transmission of hereditary characteristics.

Depolarisation   reversal of the resting potential in the neuronal cell membrane when stimulated.

Dermatome   an area of skin supplied by a single spinal nerve.

Dermis   layer of the skin below the epidermis.

Descending (efferent) neurons   neurons that carry instructions down the spinal cord from the brain to the muscles and glands of the body.

Desquamation   shedding of surface keratinised cells from the stratum corneum of the skin.

Detrusor   main muscle of the bladder formed of smooth muscle fibres in spiral, longitudinal and circular bundles that empties the bladder with contraction.

Detumescence   the subsiding of an erection.

Diaphoresis   excessive sweating.

Diaphragm a dome-shaped sheet of skeletal muscle that forms the floor of the thoracic cavity innervated by the phrenic nerve and used as one of the primary muscles of respiration.

Diaphysis the shaft and main portion of a bone.

Diarthrosis freely moveable joint.

Diastole relaxation of the heart muscle following a heartbeat.

Diencephalon part of the brain composed of the thalamus, hypothalamus and epithalamus.

Diffusion the movement of ions and molecules from an area of high concentration to an area of low concentration in an attempt to achieve an isotonic balance.

Digestion breaking down food into individual nutrients both mechanically and chemically.

Diploid two complete sets of chromosomes.

Disaccharides carbohydrates composed of two monosaccharides (simple sugar molecules).

Dorsiflexion movement that elevates the foot.

Duct a channel/tube leading from an exocrine gland or organ.

Ductus (vas) deferens a 40–45 cm (16 inch) long tube that ascends out of the scrotum into the pelvic cavity, via the inguinal canal, terminating behind the bladder where it widens and joins the duct of the seminal vesicles.

Duodenum first part of the small intestine connecting the stomach to the jejunum.

Dura mater outermost layer of the brain and spinal cord that prevents friction against the skull, and is a lining that retains CSF within the CNS.

Dynamic equilibrium maintenance of the body position in response to sudden movements, perceived through rotational movement.

E

Eccrine (merocrine) gland most common sweat gland in the skin with ducts that open onto the surface of the skin and produce a clear, watery sweat.

Ejaculation the action of expelling semen from the body.

Ejaculatory duct a short 2 cm (1 inch) duct that passes through the prostate gland and merges with the urethra.

Elastin an elastic, fibrous glycoprotein in connective tissue.

Electrolyte a substance that dissociates into ions when dissolved in a solution and attains the ability to conduct electricity.

Element a pure chemical substance consisting of a single type of atom.

Elimination the act of removing waste products from the body.

Embryo the collection of cells that has developed from the fertilised egg before all the major organs have developed and differentiated fully. The embryo becomes a fetus eight weeks after fertilisation.

Enamel hard, bone-like structure that protects the tooth.

Endocardium innermost layer of the heart muscle forming the lining of the heart and heart valves.

Endocrine glands ductless glands that release hormones directly into the blood stream.

Endocytosis energy-using process by which cells absorb molecules (such as proteins) by engulfing them by means of a coated vacuole or vesicle.

Endolymph fluid within the cochlear duct (scala media) of the ear.

Endometrium the mucous membrane lining the uterus that thickens during the menstrual cycle for possible implantation of an embryo.

Endoplasmic reticulum a cell-wide network of membrane that provides a surface on which lipids and proteins can be formed and transported round the cell. There are two types: smooth and rough.

Endosteum single layer of bone-producing cells lining the inner surfaces of the cavities within bone.

Endotoxin toxic substance bound to the wall of bacteria.

Enteroreceptors sensory receptors in the internal environment.

Enzyme biological catalyst of protein.

Eosinophils granulocytes that function in allergic responses and in resisting some infections by engulfing antigen–antibody complexes, allergens and inflammatory chemicals and can also weaken or kill parasites by secreting chemical agents.

Ependymal cells neuroglia that create and secrete CSF and circulate it by cilia activity.

Epidermis the thin, outermost layer of the skin.

Epididymis a tightly coiled mass of thin tubes that carries sperm from the testes to the ductus deferens in the male reproductive system. Sperm mature as they pass through the epididymis.

Epigenetics the study of changes caused by modification of gene expression rather than alteration of the genetic code itself.

Epiglottis a thin, cartilaginous flap that covers the glottis during swallowing, preventing swallowed substances entering the larynx.

Epiphyses the proximal and distal ends of a bone.

Epithalamus relay centre composed primarily of the pineal gland and the habenula.

Epithelial cells cells of the epithelium.

Epithelial tissue tissue that covers the body and lines cavities, organs and glands.

Epithelium a membranous cellular tissue that covers the surface or lines a tube or cavity of body structure.

Equilibrium the sense that helps maintain balance and awareness of orientation in space (proprioception).

Erythrocytes red blood cells – formed in the red bone marrow and containing haemoglobin, which transports oxygen and some carbon dioxide to and from the cells of the body.

Erythropoiesis formation of new erythrocytes.

Erythropoietin (EPO) a hormone produced by the kidneys that stimulates the red bone marrow to produce more erythrocytes.

Eukaryotes   these include both animal and vegetable, and single- and multiple-celled organisms with more complex cells that have a nuclear membrane surrounding the characteristic defining chromosomes.

Eustachian (auditory) tube   tube that connects the nasopharynx and tympanic cavity and opens to allow air to move in to equalise air pressure on either side of the tympanic membrane.

Evaporation   the conversion of a liquid to vapour resulting in a loss of heat energy.

Eversion   turning the sole of the foot outwards, inner side of the foot on the ground.

Exocrine glands   glands that release secretions through ducts to the surface of an organ or into a cavity.

Exocytosis   the release of cellular substances by fusing the vesicular membrane with the plasma membrane and subsequently expelling the substances out of the cell.

Extension   movement that increases the angle/distance between two bones or parts of the body.

External respiration   the diffusion of gases between the alveoli and the pulmonary capillaries across the respiratory membrane.

Exteroreceptors   sensory receptors located in the skin which monitor the external environment.

Extracellular Fluid (ECF)   fluid outside of cells made up of plasma (the fluid component of blood) and interstitial/intercellular fluid (fluid between cells).

Extracellular matrix   collection of substances outside of cells that provides structural and biochemical support to the surrounding cells, enabling them to 'stick' together. These cells have a crucial role to play in wound healing as they produce collagen.

Extrinsic   originating outside or/external to.

F

Facultative aerobes   microorganisms that can grow if oxygen is available or not.

Falciform ligament   ligament dividing the liver into the left and right lobes.

Fallopian tube   a pair of tubes along which the ovum travels from the ovaries to the uterus.

Fat   a macronutrient composed of glycerol and fatty acid combinations.

Fauces   the opening from the oral cavity into the oropharynx.

Femur   the longest and strongest bone in the body located in the upper leg.

Fertilisation   the fusion of two gametes (sex cells): an ovum and a spermatozoon.

Fetus   a human embryo is called a fetus from the end of the second month of pregnancy until birth. At this stage, differentiation becomes more distinct and vital organs develop further and become active.

Fibre   a non-starch polysaccharide (or cellulose) composed of oligosaccharides and polysaccharides.

Fibrinogen   a glycoprotein involved in blood clotting.

Fibroblasts   cells that synthesise the extracellular matrix.

Fibrosis   when lost or damaged tissue is not replaced with the same tissue, but is replaced by fibrotic, or scar, tissue.

Fibrous joint type of joint when two bones are attached by fibrous connective tissue.

Fibula bone running parallel to the tibia on the outside of the lower leg.

Filtration the movement of water, ions and molecules from an area of higher pressure to an area of lower pressure, usually across a semipermeable membrane, due to the hydrostatic pressure of the fluid.

Fissure A normal groove/furrow that separates an organ into lobes or parts.

Flexion movement that decreases the angle of a joint.

Flora the microorganisms (algae, bacteria or fungi) living in or on the body.

Follicle a secretory cavity, sac or gland.

Follicular cells these are also known as thyroid epithelial cells and are located in the thyroid gland where they produce and secrete thyroid hormones, thyroxine (T_4) and triiodothyronine (T_3).

Foramen magnum opening in the base of the skull through which the spinal cord passes.

Foramen of Monro connection between the third ventricle and the lateral ventricles.

Forebrain largest section of the brain consisting of the diencephalon and the cerebrum.

Foreskin (prepuce) the skin over the penile glans.

Fovea centralis located near the optic disc, this is an area of high-density cones in the retina of the eye and is responsible for detailed images.

Fracture a break in any bone.

Free nerve endings general sensory receptors that respond to temperature changes of heat and cold, as well as pain.

Frenulum fold of tissue where the foreskin attaches to the penile glans.

Fungi any of a wide variety of single or multi-cell organisms that reproduce by spores, including the mushrooms, moulds, yeasts, and mildews.

G

Gallbladder small sack located at the underside of the liver that stores and concentrates bile.

Gamete a mature reproductive cell (ovum or spermatozoon) having a single set of chromosomes.

Ganglion a cluster of neurons or neuronal cell bodies in the peripheral nervous system.

Gaseous exchange the process that occurs in alveoli where oxygen is extracted from inhaled air into the blood stream and carbon dioxide is extracted from the blood for elimination through exhalation.

Gastrointestinal Tract (GIT) a hollow tube that runs from the mouth to the anus with the purpose of moving nutrients from the external environment into the internal environment and eliminating ingested waste products.

Gender fluid when a person's gender identity and expression varies over time.

Gender identity the sense of being male or female.

Gender role   how people behave as men or women (as society perceives them).

General senses   the sensations of pain, temperature and touch (pressure, vibration and proprioception).

Genes   section of DNA that codes for a specific polypeptide chain.

Genetics   refers to specific individual genes and how they are inherited.

Genomics   refers to the complete genetic makeup of an individual and how it interacts with environmental and lifestyle factors.

Genotype   the overall structure of an individual's genetic makeup.

Gestation   the process or period of development from conception to birth.

Ghrelin   the 'hunger hormone' produced by cells in the stomach, which enhances the sensation of feeling hungry before an expected meal.

Glans   the rounded head of a penis.

Globulins   proteins essential for immunity with some functioning as antibodies against infection while others transport some hormones and minerals.

Glomerular Filtration Rate (GFR)   the rate at which the filtrate is formed in the Bowman's capsule of the kidneys, measured in millilitres per minute.

Glomerulus   Tiny tufts of capillaries which carry blood within the kidneys.

Glucagon   a peptide hormone that raises blood glucose level when it falls too low by stimulating breakdown of glycogen in the liver into glucose.

Gluconeogenesis   formation of new glucose from protein and glycerol (from fat).

Glycerol   a three carbon molecule produced by hydrolysis of fats.

Glycogen   carbohydrate store in the body.

Glycogenesis   the conversion of glucose to glycogen for storage in the liver.

Glycogenolysis   the process by which glycogen is broken down into glucose to provide an immediate energy source.

Glycolipid   a molecule consisting of a carbohydrate plus a lipid.

Glycolysis   breakdown of glucose by enzymes, releasing energy and pyruvate.

Glycoprotein   a molecule consisting of a carbohydrate plus a protein.

Goblet cells   cells that produce mucus, found in the respiratory and gastrointestinal tracts.

Golgi body   a cell organelle composed of stacks of flattened sacs of membrane, which receives proteins and lipids from the endoplasmic reticulum and packages them into secretory vesicles. These are stored and moved to the plasma membrane when needed.

Golgi tendon organs   general sensory receptors that detect tension as part of proprioception.

Gomphosis   type of fibrous joint formed with collagenous connective tissue (those between the teeth and the mandible and maxillae).

Granulation   the formation of new connective tissue.

Granulocytes leucocytes that have visible granules, the organelles, in the cytoplasm.

Grey matter grouping of nerve cell bodies.

Gustation sensation of taste.

Gyri folds in the white matter of the cerebral cortex in between sulci.

H

Habenula a relay from the limbic system in the brain, and deals with sleep, stress, pain, and reinforcement processing.

Haematoma localised collection of blood outside blood vessels.

Haemoglobin within red blood cells, a protein composed of globin and haem and containing four globin chains and four haem units enabling each haemoglobin molecule to carry four oxygen molecules.

Haemopoiesis production of blood cells.

Haemostasis the process by which bleeding is stopped.

Haemostasis (wound healing) the first stage of healing, referring to the stopping of haemorrhage at the site of injury by contraction of the smooth muscle in the arterial and arteriole walls, significantly reducing the blood flow.

Hair receptors general sensory receptors that interpret touch by movement of hair (root hair plexus).

Half-life the time required for any specified property, e.g. the concentration of a hormone or drug in the body, to decrease by half.

Haploid one complete set of chromosomes present in gametes.

Haustrations small segmented pouches of the colon that give it a segmented appearance and maximise surface area for absorption of water and minerals.

Hearing the ability to interpret sound.

Heart rate the number of heartbeats in one minute.

Helix upper rim of the auricle.

Helminths multi-cellular parasitic worms.

Hemianopia blindness in one half of the visual field.

Hepatic artery branch of the coeliac artery that delivers a rich supply of oxygenated blood to the liver.

Hepatic vein vein carrying blood from the liver to the vena cava.

Hepatocytes functional cells of the liver.

Hepatopancreatic ampulla junction of the pancreatic and bile ducts as they open into the duodenum.

Heterozygous a person in whom the cells contain two different alleles of a gene in a pair.

High Density Lipoproteins (HDLs) type of lipoprotein that mops up cholesterol in the blood transporting it to the liver for elimination in bile.

Hilum   the point where structures such as blood vessels and nerves enter an organ.

Hip bone   unit bone of the pelvic girdle made up of the ilium, ischium and pubis bones.

Hippocampus   part of the limbic system primarily associated with formation of memory, and organising sensory and cognitive experiences for storage.

Homeostasis   regulation of the internal environment in order that a level of consistency is maintained necessary for the cells and organs of the body to operate optimally.

Homologous   pair of autosomes, one from each parent, similar in length, gene position and centromere location.

Homozygous   having identical pairs of genes for any given hereditary characteristics.

Hormones   chemical messengers secreted directly into the blood stream and carried around the body in the blood to attach to their specific receptor and influence the activity of the specific target cells, tissues, organs and systems of the body.

Human Leucocyte Antigen (HLA)   *see* Major Histocompatibility Complex (MHC).

Humerus   the largest and longest bone of the upper arm.

Hyaline cartilage   transparent cartilage.

Hydrolysis   when a water molecule is added to a substance which usually splits into two parts.

Hydrophilic   having an affinity for water; readily absorbed or dissolved in water.

Hydrophobic   the tendency to repel or not mix with water.

Hydrostatic pressure   the pressure exerted by the blood against the artery wall by the force exerted by the heart pumping.

Hymen   a membrane which covers or partly covers the opening of the vagina (until broken by physical activity or intercourse).

Hyperpolarisation   A change in cell membrane potential that makes it more negative as a result of excess potassium leaving the cell. The opposite of a depolarisation, it prevents an action potential for a short period (refractory period) by increasing the stimulus threshold necessary to trigger an action potential.

Hyperthermia (pyrexia)   raised body temperature above 38°C.

Hypertrophy   enlargement of an organ/tissue from the increase in size of its cells.

Hypodermis   lowest layer of the skin that connects it to the rest of the body.

Hypothalamus   several nuclei and tracts of axons below the thalamus that control the autonomic nervous system, neuro-endocrine system and limbic system.

Hypothermia   body temperature below 35°C.

I

Ileum   terminal part of the small intestine between the jejunum and the first part of the large intestine – the caecum.

Immunisation   the process of inducing immunity to an infectious organism or agent in an individual or animal through vaccination.

Immunodeficiency   immune system's ability to fight infectious disease is compromised or absent resulting in increased liability to opportunist infections.

Immunoglobulins   class of proteins that function as antibodies.

Infancy   period of time from birth to about 24 months.

Inflammation   an increased blood supply to a site of injury that triggers an inflammatory response to prevent infection. Second stage of wound healing.

Ingestion   the taking in of food into the digestive system.

Innate immunity   the rapid first line of defence against pathogens that responds the same way each time it encounters a pathogen.

Insula   part of the limbic system (insular lobe) that regulates thermosensation, nociception, somatosensation, viscerosensation, gustation.

Insulin   a peptide hormone that lowers blood glucose level by promoting glucose absorption from blood into skeletal muscle and fat cells, and enhancing fat storage. Insulin receptors in cell walls trigger a process that leads to transporter proteins in cell membranes to allow glucose from the blood to enter the cell.

Insulin-Like Growth Factors (IGFs)   (the main one being IGF-1). These have some similarities in structure to insulin and are formed mainly in the liver under the stimulus of growth hormone. They promote growth of most tissues in the body especially bone growth during childhood.

Intercostal muscles   muscles running along and between the ribs that move the chest wall in respiration.

Interferons   proteins which prevent viruses replicating.

Internal capsule   major tract of sensory and motor fibres, which carry information to and from the cerebral cortex.

Internal respiration   the diffusion of gases between blood in the systemic capillaries and the tissues.

Interneurons   neurons that act as connections between descending and ascending neurons. Also called association neurons.

Intersex   refers to those born with any variation in sex characteristics that do not align with those typically associated with being male or female.

Intervertebral discs   discs that enable vertebrae to bind together and also function as shock absorbent material.

Intracellular Fluid (ICF)   fluid inside cells in which organelles and other substances are suspended or dissolved.

Intracrine   relating to a hormone in which activation of the hormone occurs within the cell where it was created and binds with nuclear receptors to modify function of that cell.

Intrinsic   originating within or naturally inherent to the body.

Inversion   turning the sole of the foot inwards, outer side of the foot on the ground.

Ion channels   pores in the cell membrane that enable movement of ions and thus determine the concentration of ions within and outside of the cell.

Iris   the coloured part of the eye around the pupil that regulates the amount of light entering the eye.

Islet cells (islets of Langerhans)   cells of the pancreas that secrete the hormones insulin and glucagon.

Isometric contraction   muscle contraction where there is no change in length of the muscle and no movement, but energy is still used and muscle tension increased.

Isotonic contraction   muscle contraction where the length of muscle changes enabling body movements and movement of objects, but there is little change in muscle tension.

J

Jejunum   middle part of the small intestine between the duodenum and ileum.

Joint (articulation)   the point at which two or more bones meet.

Joint cavity   cavity filled with synovial fluid secreted by the synovial membrane.

Junk DNA   areas of DNA that do not appear to code for any genes.

Juxtacrine   relating to a hormone that has effect only on receptors in the immediate neighbourhood of its secretion.

Juxtaglomerular apparatus   specialised cells near the glomerulus in the kidney that stimulate the secretion of the adrenal hormone aldosterone and play a major role in renal autoregulation.

K

Karyotype   appearance of the chromosomes in a cell, with reference to their number, size, shape, etc.

Keratin   a hard, fibrous protein that is found in skin cells and hair.

Keratinocytes   cells that produce keratin.

Krause end bulbs   general sensory receptors that detect light touch and texture.

Kupffer cells   macrophages in the sinusoids of the liver that provide a barrier preventing antigens and bacteria absorbed in the gut entering into the systemic circulation and also degrade haemoglobin into haem and globin molecules.

L

Labia majora   plump, lip-like structures either side of the vulva.

Labia minora   two hairless folds of skin between the labia majora.

Labyrinth   a series of cavities within the temporal bone that contain the main organs of balance and hearing.

Lacrimal apparatus   a system in the eye that is composed of lacrimal glands and ducts that manage the flow of tears.

Lacrimal canals   ducts that carry tears away from the eye.

Lacrimal ducts ducts that carry tears from the lacrimal gland to the surface of the eye.

Lacrimal fluid (tears) lubricant for the eye made by lacrimal glands containing water, salts, mucus and lysozyme.

Lactate an ester or salt of lactic acid, the final by-product of anaerobic glycolysis.

Lactation production/secretion of breast milk by the mammary glands.

Lacteals lymphatic vessels of the small intestine.

Lacunae cavities in bone created by chondrocytes.

Lamellated corpuscles general sensory receptors that detect deep pressure, stretch, vibration and tickle sensations.

Langerhans cells cells that are the first line of defence in the skin and support the immune system by processing antigens.

Larynx the hollow muscular organ part of the air passage to the lungs housing the vocal cords. Otherwise known as the voice box.

Lateral geniculate nucleus relay centre in the thalamus for the visual pathway.

Lens (eye) lens suspended by suspensory ligaments to the ciliary body within aqueous humour, used to focus light appropriately.

Leptin the 'satiety hormone' that regulates how much fat is stored in the body by adjusting the sensation of hunger and the amount of energy expended.

Leucocytes white blood cells that provide protection against pathogens. They are mostly found in connective and lymphatic tissue, normally only circulating in the blood for a number of hours.

Leydig cells cells outside the seminiferous tubules that produce testosterone.

Libido the desire for sexual activity.

Ligament a small band of tough, fibrous connective tissue that joins two bones or cartilages or holds together a joint.

Limbic system complex system of neurons and networks in the brain concerned with instinct, mood, emotion and memory.

Lipogenesis the metabolic formation of fat from acetyl-coA and glycerol.

Lipolysis breakdown of lipids by hydrolysis to release fatty acids.

Lipoproteins types of fat that are triglycerides and cholesterol units combined with proteins and phospholipids.

Lochia vaginal discharge lasting approximately 33–60 days postpartum.

Long Chain Fatty Acid (LCFA) a fatty acid with 16 or more carbons.

Low Density Lipoprotein (LDL) type of lipoprotein largely composed of cholesterol – cells use enzymes to break them down to release the cholesterol.

Lumbosacral plexus nervous plexus in the lumbar region of the body formed from the first four lumbar nerves (L1-L4) and from part of the last thoracic nerve. Neurons extend to the legs.

Lung compliance the ease with which the lungs expand.

Lunula   the pale, half-moon shaped area on the nail.

Lymph   the fluid in lymph vessels formed from excess ECF surrounding cells and similar in composition to blood, with water containing electrolytes, glucose, fats, and leucocytes but normally without erythrocytes and fewer plasma proteins.

Lymph node   small, bean-shaped nodes that filter lymph for bacteria, cancer cells, and other potentially harmful substances.

Lymphatic system   collects excess ECF and returns it to the systemic circulation.

Lymphatic vessels   tubular structures similar to blood vessels that carry lymph.

Lymphocytes   agranulocytes that can destroy cancer cells, cells with a viral infection and foreign cells. There are three types of lymphocytes: B-cells, T-cells and natural killer cells.

Lysosome   a cell organelle containing enzymes that break down unneeded large molecules, which are recycled or excreted from cells.

Lysozyme   a protective bactericidal enzyme.

M

Macrophage   a form of phagocyte that has a role in both non-specific (innate) and specific (adaptive) defence mechanisms in the body – mature monocyte.

Maculae   patches of hair and support cells in saccules and utricles of the vestibule. Hair cells are sensory receptors and support cells secrete the otolithic membrane.

Major Histocompatibility Complex (MHC)   proteins that are recognised as 'self', which in humans are known as the HLA (Human Leucocyte Antigen).

Mamillary bodies   part of the limbic system thought to be a relay centre with a distinct role in memory operations.

Mast cells   cells developed from monocytes that deal with microbes too large to be dealt with by phagocytosis by secreting enzymes from granules onto the pathogens.

Maturation (remodelling)   final phase of wound healing that begins during proliferation, whereby adapted fibroblasts attach to collagen fibres and contract to pull the wound margins together (or closer together).

Meatus   a passage or opening leading to the interior of the body.

Mechanoreceptors   sensory receptors that respond to mechanical pressure.

Medial lemniscal pathways   sensory pathways for discriminative touch, vibration and proprioception sensations to the brain.

Mediastinum   the central compartment of the thoracic cavity surrounded by loose connective tissue, which contains all the tissues and organs of the chest except the lungs and pleurae.

Medulla oblongata   part of the brain that links the brain with the spinal cord and contains nuclei that act as relay stations for sensory information coming from the spinal cord. Also the location where corticospinal and some sensory pathways cross over.

Medullary cavity   cylindrical cavity within the diaphysis of bones containing blood vessels and bone marrow.

Meiosis the division that occurs to form the gametes – spermatozoa or ova – in preparation for fertilisation and formation of the zygote which develops into the fetus.

Melanin a pigment in the skin that protects it from ultraviolet radiation.

Melanocytes cells that produce melanin.

Melatonin hormone produced by the pineal gland that contributes to the regulation of the sleep–wake cycle.

Menarche the first occurrence of menstruation.

Meninges three layers of connective tissue that surround the brain and spinal cord.

Menopause the ceasing of menstruation.

Menstruation monthly shedding of the endometrium.

Merkel cells general sensory receptors that detect the tactile sensations of light touch, texture, shapes and edges.

Mesenchymal cells a type of stem cell that develop into vessels (lymphatic and circulatory) and connective tissue.

Metabolic rate amount of energy used.

Metabolism all of the organic and chemical reactions in the body, using all of the nutrients that the body takes in to create molecules for structure, chemical reactions, and for energy.

Metabolites a substance produced during or from taking part in metabolism.

Metacarpal bones bones in the middle region of the hand (palm).

Metaphysis the portion of the bone between the diaphysis and the epiphysis.

Metatarsal bones bones in the region in the middle of the foot.

Microbiome 'the ecological community of commensal, symbiotic, and pathogenic microorganisms that literally share our body space and have been all but ignored as determinants of health and disease' (Lederberg and McCray, 2001: 8).[1]

Microbiota the microorganisms of a particular location.

Microflora community of microorganisms (algae, fungi and bacteria) that live in or on another living organism (e.g. the body).

Microglia neuroglia that are specialised macrophages capable of phagocytosis, protecting neurons from pathogens.

Microorganism a microscopic living organism such as a bacterium, virus or fungus.

Microvilli microscopic cellular membrane projections that increase the surface area of cells. Found on the surface of the small intestine.

Micturition the act of passing urine.

Midbrain part of the brain that links the thalamus, hypothalamus, pons and medulla oblongata.

---

[1]Lederberg, J. and McCray, A.T. (2001) ''Ome sweet 'omics – a genealogical treasury of words', *The Scientist*, 15(7): 8.

Mineral inorganic micronutrient needed for a variety of reasons in the body, largely to do with structure, fluid balance, nervous and muscular activity and blood clotting.

Minute volume the amount of air inhaled and exhaled per minute.

Mitochondria cellular organelle known as the 'power house'. ATP is created for storage of energy which is released when required.

Mitosis cell division that results in two genetically identical daughter cells.

Mixture molecules of elements and compounds mixed together, without chemical bonds.

Molecule a group of two or more atoms held together by chemical bonds.

Monocytes granulocytes that engulf pathogens and dead neutrophils, clear away debris from dead or damaged cells and present antigens to activate other cells of the immune system.

Monosaccharides simple carbohydrates (glucose, galactose and fructose) that are easily digested and absorbed in the body.

Morula small solid bunch of embryonic cells (days 4–5 after fertilisation) when nearing entry to the uterus.

Motor division/system division of the PNS that carries nervous impulses sent from the CNS to cells/organs/muscles to initiate responses.

MRI Magnetic Resonance Imaging – a type of imaging scan that uses magnetic fields and radio waves to produce detailed images of the inside of the body.

mRNA the template for each amino acid in the polypeptide chain in the formation of a protein.

Mucosa mucous membrane or a lining of a part of the body that secretes mucus.

Mucus a viscous slippery secretion produced by mucous membranes. Being rich in mucins, its function is to moisten and protect.

Muscle a collection of fibrous tissue that can contract, produce movement or maintain posture.

Muscle spindles general sensory receptors that detect muscle stretch for proprioception.

Myelin sheath an insulating layer around the axon of a neuron that prevents passive movement of ions across the cell membrane, increasing the speed of conduction of an action potential.

Myocardium middle layer of the heart muscle consisting of thin filaments of actin and thick filaments of myosin.

Myofibrils small threadlike structures that are the contractile organelles of the skeletal muscle.

Myofilaments small protein structures contained within myofibrils.

Myoglobin a red-coloured protein found only in muscles used to bind oxygen.

N

Nasal septum a thin piece of cartilage and bone inside the nose separating it into the right and left sides (nostrils/nares).

Nasopharynx cavity posterior to the nasal cavity and extending to the soft palate.

Natural killer cells lymphocytes that provide a rapid response to virally infected cells and respond to tumour cell growth.

Negative feedback control system in the body that moves the stimulus back towards normal (such as body temperature when it is raised or lowered).

Neurofibrils (neurofilaments) filaments in neurons that contribute to the transport of cellular material and facilitate axon movement and growth.

Neurogenesis growth of new neuronal tissue.

Neuroglia non-neuronal cells that support and protect neurons.

Neuromuscular junction the synapse between a motor neuron and skeletal muscle fibre separated by a synaptic cleft.

Neuron (nerve cell) specialised cell of the nervous system that transmits nervous impulses (action potentials).

Neurotransmitter chemical messenger that transmits nerve signals across a synapse.

Neutrophils granulocytes that engulf bacteria and cellular debris or use chemical agents (lysozyme and peroxidase) to destroy foreign bodies.

Nociception perception of pain.

Nociceptors sensory receptors that respond to pain stimuli.

Nodes of Ranvier gaps between areas of myelination on the axon where ions can easily flow into the ECF.

Non-binary when a person associates their gender as being located on a spectrum of gender identities that are not exclusively male or female (the binary forms).

Nuclei brain structures composed of compact clusters of neurons.

Nucleolus contained within the nucleus of the cell, this structure is involved in the formation of ribosomes.

Nucleus cellular organelle that contains the genetic information (DNA) within chromosomes. It provides the template (RNA) for protein formation.

Nutrient foramina the external opening for the entrance of blood vessels in a bone.

Nutrients substances that are ingested, digested, absorbed and metabolised to maintain homeostasis through their roles in structure and function within the body, primarily carbohydrates, proteins, fats, minerals and vitamins.

O

Obligate aerobes microorganisms that can only grow in the presence of oxygen.

Oedema excess tissue fluid.

Oesophagus a muscular tube that connects the oropharynx to the stomach, passing through the diaphragm.

Oestrogen   one of several steroid hormones, largely secreted by the ovaries and placenta.

Olfaction   sense of smell.

Olfactory epithelium   specialised tissue inside the nasal cavity involved in smell.

Olfactory glands   glands in the olfactory mucosa that produce mucus to lubricate the surface of the olfactory epithelium.

Olfactory mucosa   found within the nasal cavity, this section of the nasal mucosa contains the sensory nerve endings for smell.

Oligodendrocyte   a type of neuroglial cell that produces myelin in the central nervous system.

Oligosaccharides   carbohydrates composed of approximately three to nine monosaccharides and used to make glycoproteins and glycolipids.

Opportunists   microbes that are normally harmless within the body but can cause disease when the body is immunocompromised or the microbe is in the wrong place in the body.

Opsonisation   the process by which a pathogen is marked for destruction by phagocytosis.

Optic chiasm (or chiasma)   point on the visual nerve pathways where half the fibres go to the lateral geniculate nucleus on the same side of the brain and the other half crosses over to the opposite side.

Optic disc   a convergence of the retinal ganglion cells to form the optic nerve.

Organ   a differentiated structure with specific function(s) made up of organised cells and tissues.

Organ of Corti   structure in the cochlea that converts sound vibrations into nervous impulses through the long, stiff hairs called stereocilia.

Organelle   a cellular structure with a specific function (small cell organ).

Orgasm   the pleasurable, whole body sensation that makes sexual experiences attractive to people.

Oropharynx   the part of the pharynx below the soft palate and above the epiglottis that is continuous with the mouth and that extends from the soft palate to the epiglottis.

Osmolality   the concentration of osmotically active particles per 1,000 ml of fluid.

Osmoreceptors   a group of specialised sensory neurons in the hypothalamus that are stimulated by increased osmolality of the extracellular fluid.

Osmosis   the movement of water across a semipermeable membrane from a dilute/hypotonic solution (with low osmotic pressure) to a more concentrated/hypertonic solution (which has a higher osmotic pressure).

Osmotic pressure   the pressure exerted by plasma proteins and some electrolytes in the plasma inside the capillaries, pulling water and small molecules towards them. Includes the pressure exerted in pulling water across a semipermeable membrane.

Ossicles   tiny bones in the ear that connect with the eardrum, transferring and magnifying sound vibration from the auditory canal to the inner ear.

Ossification   the formation of bone.

Osteoblasts   bone-building cells that synthesise and secrete collagen fibres and other organic components and also initiate calcification of bone matrix.

Osteoclasts bone cells formed from fused monocytes that secrete powerful lysosomal enzymes and acids for dissolving the protein and mineral matrix.

Osteocytes bone cells, derived from osteoblasts, found in mature bone, and the main cell type in bone.

Osteogenesis (ossification) formation of bone.

Osteogenic bone producing.

Osteogenic cells bone cells derived from mesenchymal cells (adult stem cells) that develop into osteoblasts.

Osteon chief structural unit of compact bone.

Otolithic membrane a thick gelatinous glycoprotein layer in the vestibule that detects movement through gravitational pull.

Ova *see* Ovum.

Oviduct *see* Fallopian tube.

Ovulation the point in the menstrual cycle when a mature ovarian follicle releases an ovum (female gamete), which moves down the Fallopian tube towards the uterus.

Ovum the female gamete. Plural ova.

Oxidation the loss of electrons or an increase in oxidation state by a molecule, atom or ion.

P

Palpebrae the eyelids.

Palpebral fissure the gap between open eyelids.

Pancreatic juice an alkaline liquid secreted by the pancreas containing sodium bicarbonate, inactive proteolytic enzymes and a number of active enzymes.

Paneth cells endocrine cells in the crypts of the intestinal epithelium.

Papillary layer upper layer of the dermis made up of fine and loosely arranged collagen fibres that interweaves with the epidermis.

Paracrine relating to a hormone that has effect only in the neighbourhood of its secretion.

Parahippocampal gyrus structure of the limbic system that works with the hippocampus in processing declarative memory.

Parasympathetic nervous system a division of the autonomic nervous system which reduces the activity of many of the systems of the body, but increases digestion, absorption and storage of nutrients.

Passive diffusion the passive movement of a solute from an area of high concentration to an area of lower concentration until the concentration of the solute is uniform throughout and reaches equilibrium.

Patella a small triangular bone found at the front of the knee joint (kneecap).

Pathogen an agent causing disease or illness to its host.

Pathogenic   ability to cause or produce disease.

Perfuse   to move fluid and nutrients through to cells of organs/tissues.

Pelvic girdle   unit of two pelvic bones (hip bones) and the sacrum and coccyx.

Pericardium   outermost layer of the heart muscle that encloses it and is composed of two sacs – the parietal pericardium and visceral pericardium (or epicardium).

Perichondrium   dense irregular connective tissue which surrounds most cartilage, containing blood vessels and nerves.

Perilymph   fluid within the scala tympani and scala vestibule.

Perineum   area between the anus and the scrotum or vulva.

Periosteum   the tough connective tissue membrane that covers the outside of the bone (except where there is articular cartilage).

Peripheral Nervous System (PNS)   all of the nervous system outside the CNS.

Peripheral resistance   the resistance to the flow of blood from the wall of the arterial vessels.

Peristalsis   the ripple-like waves created by the relaxation and contraction of muscle in a tubular structure that results in the movement of material along its length.

Peritoneum   the serosa of the small and large intestine composed of two serous membranes – the parietal, lining the walls of the abdomen; and visceral, covering the abdominal organs.

Peyer's patches   aggregations of lymph tissue within the lower part of the ileum.

pH   a chemical term that indicates the level of acidity/alkalinity in a solution and is determined by the amount of hydrogen ions (H^+) in a solution.

Phagocytes   mobile white cells which engulf and destroy pathogens.

Phagocytosis   process by which a cell (e.g. white blood cell) engulfs a pathogen, other cells, cell debris or foreign particles.

Phalanges   bones of the digits of the hand and feet.

Pharynx   the section of the gastrointestinal tract that extends from the mouth and nasal cavities to the larynx.

Phenotype   the observable expression of an individual's characteristics resulting from the interaction of its genotype with the environment.

Phospholipids   type of fats composed of one glycerol molecule, two fatty acids and a phosphate group.

Photoageing   ageing of the skin as a result of sun exposure and pigmentation.

Photoreceptors   sensory receptors that respond to light as a stimulus.

Pia mater   a thin, impermeable fibrous membrane adjacent to the brain and spinal cord ensuring that CSF remains within the subarachnoid space. Blood vessels pass through the pia mater to the brain and spinal cord.

Pineal gland   gland that secretes serotonin during the day and melatonin at night, in regulating the sleep–wake cycle.

Pinna *see* Auricle.

Pituitary gland a small pea-sized extension below the hypothalamus that links the nervous and endocrine systems.

Pituitary stalk the connection between the hypothalamus and the posterior pituitary gland.

Placenta the vascular organ formed in the uterus during pregnancy from maternal and embryonic tissues to deliver oxygen and nutrients to the fetus and move fetal waste products into the maternal circulation.

Plantar flexion movement that lowers the foot (points the toes).

Plasma the colourless fluid part of blood or lymph in which cells are suspended, primarily composed of water and plasma proteins, enzymes, hormones, gases, electrolytes and waste products.

Plasticity the adaptability of an organism to changes in its environment or differences between its various habitats.

Pleural cavity thin, fluid-filled space between the two layers of the pleural membrane.

Pleural membrane the serous membrane lining the lungs.

Pneumotaxic centre area in the upper pons responsible for regulating ventilation (with the apneustic centre).

Polypeptides amino acids combined together by peptide bonds.

Polysaccharides carbohydrates composed of many monosaccharides, usually more than ten, and often used as a form of energy storage, such as glycogen in the liver and muscle cells.

Pons collection of nerve fibres connecting the two hemispheres of the cerebellum, brain and the spinal cord. It also acts as a relay station, including with the cranial nerves, and has a role in regulating breathing.

Portal fissure fissure in the liver that allows for entry and exit of blood vessels, lymph vessels, nerves and bile ducts.

Portal system circulation system within which the blood from one capillary network flows into another through a vein without first having gone through the heart.

Portal vein vein that delivers nutrient-rich deoxygenated blood from the stomach, pancreas and small and large intestine to the liver.

Positive feedback control system that enhances the original stimulus (such as blood clotting).

Post-partum the period immediately after birth of a child (six weeks normally).

Prebiotic an ingested substance not digested in the small intestine that beneficially affects the host by selectively stimulating the growth and/or activity of one or a limited number of bacteria that can improve the host's health.

Primary intention uncomplicated healing of the skin where wound edges are joined together and there is no tissue deficit to remedy.

Probiotics live microorganisms that confer a health benefit on the host.

Prokaryotes single-celled organisms of relatively simple structure without a cell nucleus surrounding the chromosomes and which include bacteria.

Proliferation rapid reproduction of a cell or organisms such as microbes.

Proliferation (phase) stage of wound healing that involves angiogenesis and production of new structural tissue that begins to contract.

Pronation rotation of the hand and forearm – turns the palms down.

Proprioceptors sensory receptors that determine the relative position of parts of the body.

Propulsion the movement of food along the full length of the gastrointestinal tract.

Prostate gland a circular gland, about 4 cm (1–2 inches) in diameter, that surrounds the urethra and ejaculatory duct adjacent to the bladder and produces prostatic fluid.

Prostatic fluid the first part of the ejaculate containing most of the sperm.

Protein nitrogenous organic compound macronutrient composed of combined polypeptides.

Prothrombin a protein that is converted into thrombin for blood clotting – the precursor to thrombin.

Protozoa a diverse group of mostly motile unicellular eukaryotic organisms (including amoebas, flagellates, ciliates and sporozoans) that mostly live parasitically.

Pseudopods temporary projections of the cell cytoplasm.

Puberty the process that occurs when changing hormone levels cause development in the reproductive systems to achieve functional reproductive capacity and development of secondary sexual characteristics.

Puerperium *see* Post-partum.

Pulmonary circulation carries the blood between the heart and the lungs permitting exchange of blood gases.

Pulmonary ventilation (breathing) the inhalation (inspiration) and exhalation (expiration) of air involving the exchange of air between the atmosphere and the alveoli.

Pulp core layer of the tooth containing a rich supply of blood and nerves entering from the root.

Pupil the dark circular gap in the iris.

Purkinje fibres subdivisions of the AV bundle that pass over the surface of both ventricles, triggering their contraction.

Pyloric sphincter ring of smooth muscle around the opening of the stomach into the duodenum.

R

Radiation the transfer of heat as waves or particles passing through space.

Radius shorter of the forearm bones.

Receptor a structural component of a sensory neuron, or other cell, that has the ability to respond to a sensory stimulus.

Rectum part of the large intestine, which stores faeces until elimination through the anus.

Redox reactions   when oxidation and reduction reactions occur in parallel.

Reduction   the gain of electrons or a decrease in oxidation state by a molecule, atom or ion.

Reflex arc   nerve pathway involved in a reflex action.

Refraction   bending of light by the lens and cornea.

Refractory period   period that follows repolarisation after a nervous impulse, for a time the neuron fails to respond to a stimulus of threshold intensity.

Regeneration   the ability of the body to replace lost or damaged tissue with the same tissue.

Renin   hormone from the afferent arteriole of the nephron that converts angiotensinogen to angiotensin I.

Repolarisation   restoration of the polarised state of a neuronal cell membrane after a nerve impulse.

Respiratory zone   part of the airway in which gas exchange takes place.

Reticular Activating System (RAS)   area of the brain composed of a number of nuclei that connect throughout the forebrain, midbrain and hindbrain and controls arousal mechanisms used in maintaining consciousness and wakefulness states essential for selective attention and purposeful responses.

Reticular formation   a core of nerve cell bodies that extend from the spinal cord up through the medulla, pons and midbrain to the hypothalamus and thalamus and with connections to the cerebral cortex involved in regulating skeletal muscle tone, autonomic control of the cardiovascular and respiratory systems, and somatic and visceral sensations.

Reticular layer   layer of the dermis below the papillary layer made up of dense irregular connective tissue.

Retina   a thin transparent membrane attached within the eye that absorbs stray light and is rich in sensory receptors and neural tissue.

Ribonucleic Acid (RNA)   a nucleic acid molecule that provides a code for the formation of proteins in collaboration with ribosomes.

Ribosomes   cellular organelles known as the protein factories. When loose in cytoplasm, they form proteins for use within the cell.

Right lymphatic duct   drains lymph from the right arm and right-hand side of the thorax, head and neck, emptying it into the right subclavian vein.

Rods   photoreceptors responsible for night vision, only seeing black, shades of grey and white.

Rotation   movement of a bone around its longitudinal axis.

Rough endoplasmic reticulum   a type of endoplasmic reticulum combined with ribosomes that is involved in protein (enzymes and hormones) formation for storage and secretion from the cell.

Ruffini corpuscles   general sensory receptors that detect heavy touch, pressure and stretching.

Rugae   series of ridges produced by folding of the wall of an organ such as the stomach and bladder.

S

Saccule   one of the two otolith organs located in the vestibule in the ear.

Saliva   a slightly acidic liquid composed of water, enzymes, hormones, antibodies and antimicrobial substances.

Salpinges   *see* Fallopian tube.

Saltatory conduction   electrical activity jumps between Nodes of Ranvier in the myelin surrounding the nerve.

Sarcolemma   the plasma membrane of the muscle fibre.

Sarcomeres   compartments that are the functional unit of the myofibril.

Sarcoplasm   the cytoplasm of the muscle fibre.

Sarcoplasmic Reticulum (SR)   a type of smoother endoplasmic reticulum – fluid-filled sac enclosing each myofibril that controls calcium ion concentration in muscle cells.

Satellite cells   neuroglia that regulate the external chemical environment of neurons, particularly calcium ions.

Satiety   the opposite of hunger.

Scapula   the shoulder blade.

Schwann cell   a type of neuroglial cell that produces myelin in the peripheral nervous system.

Sclera   the white, tough outer layer of the eye composed of connective collagen fibres and contains blood vessels and nerves, providing shape and rigidity to the eye.

Scleral venous sinus   sinus in the sclera that drains the aqueous humour of the eye.

Sebaceous glands   glands in the skin that produce sebum.

Sebum   an oily substance that softens and lubricates the skin, prevents hair from becoming brittle, reduces water loss from skin and has antimicrobial properties.

Second messenger   an intracellular signalling molecule that is released by a cell in response to exposure to extracellular signalling molecule, e.g. a hormone.

Secondary intention   healing that occurs when there is a tissue deficit between the wound edges requiring a process to fill in and contract the wound area in order to restore tissue integrity.

Segmentation   the mixing of the contents of the gut to promote digestion.

Semi-circular canals   three bony canals in the inner ear filled with endolymph, involved in control of balance.

Semilunar valves   valves at the entrances to the major arteries from the heart.

Seminal fluid   semen – fluid that contains sperm and other fluids produced in the male reproductive system and expelled during ejaculation.

Seminal vesicles   two blind-ended tubular glands that merge with the vas deferens on the same side to form the ejaculatory duct.

Sensory division/system   division of the PNS that carries sensory information through afferent neurons to the CNS for processing.

Sensory receptors   organs/receptors that convert energy from one form into electrical energy which is transmitted through sensory nerve fibres.

Septum pellucidum   part of the limbic system related to regulating rage, pleasure and mood.

Serosa   the outermost layer of lining of a tubular structure that is fibrous and elastic.

Serotonin   a neurotransmitter that has a role in memory function, mood and behaviour regulation (anxiety, depression, experience of pain, behaviour, cognition and perception), causes vasoconstriction, and influences gut motility.

Serous membrane   double-layered membrane that produces serous fluid (pale, watery fluid) between the two layers to provide lubrication.

Sertoli cells   cells essential for testis formation and spermatogenesis and that produce anti-Müllerian hormone and act as the 'nurse' cells for the germ cells.

Sexual orientation   the romantic and/or sexual attraction to another.

Sexuality   the interrelationship between the biological, sociocultural and psychological factors that influence the ways in which we experience and express ourselves as sexual beings.

Shell temperature   the temperature at or near the surface of the body.

Short Chain Fatty Acids (SCFA)   a fatty acid with less than eight carbons.

Sight   *see* Vision.

Sinoatrial (SA) node   specialised group of autorhythmic cells in the wall of the right atrium that initiate an impulse that spreads across the atria causing them to contract simultaneously. Known as the heart's natural pacemaker.

Sinus   a cavity within a bone or other tissue.

Sinusoids   blood vessels whose walls are incomplete, being lined with highly permeable endothelium.

Skeletal muscle fibre   a long cylindrical cell that contains multiple nuclei at the periphery of the fibre.

Smegma   a waxy substance produced by sebaceous glands in the glans of the penis.

Smooth endoplasmic reticulum   a type of endoplasmic reticulum in the cell that is involved in steroid (lipid) and carbohydrate metabolism and detoxification.

Somatic nervous system (motor)   division of the motor system responsible for voluntary (conscious) control of body movements through stimulating contraction of skeletal muscles.

Somatosensation   sensations originating mainly in the skin including proprioception, touch and temperature.

Sound   audible vibration of molecules.

Spermatic ducts   tubes that take sperm from each testis to the urethra.

Spermatogonia   germ cells for sperm formation.

Spermatozoa   *see* Spermatozoon.

Spermatozoon   sperm, the male gamete. Plural spermatozoa.

Spinal cord   the neural tissue encased within the spine which runs from the medulla to the level of the first or second lumbar vertebrae.

Spinal nerves   31 pairs of nerves that carry motor, sensory and autonomic signals between the spinal cord and the body.

Spine   the spinal/vertebral column and the mechanical structures that it is composed of: 33 vertebrae and the ligaments and tendons that connect them.

Spinocerebellar tracts   sensory pathways for sensations for muscles and tendons (stretch) to the cerebellum to coordinate skeletal muscle movement.

Spongy bone   softer, honeycomb-like bone tissue on the inside of bones with spaces containing red and yellow bone marrow and blood vessels.

SRY   Sex-determining Region of the Y (chromosome) – gene that codes for the production of Testis-Determining Factor (TDF).

Stapedius muscle   tiny muscle of the ear that dampens large vibrations of the stapes that can occur from loud noises. This protects the oval window and decreases sensitivity of hearing.

Static equilibrium   maintenance of balanced body position relative to the force of gravity.

Stem cells   newly formed cells that have the potential to differentiate into any type of body cell.

Sterile   free from bacteria or other living microorganisms.

Sternum   breast bone.

Steroid hormones   there are a number of different steroid hormones, all formed from cholesterol, manufactured in the liver and acquired from the diet. The hormones are formed in several different endocrine organs of the body.

Stomach   bean-shaped sac designed to house food for digestion.

Stratum basale   base layer of the epidermis that has cells which mature and differentiate as they move towards the surface.

Stratum corneum   outermost layer of the epidermis where cells are heavily keratinised and water-proof.

Stratum granulosum   epidermal layer of the skin above the stratum spinosum where keratinisation continues and glycolipids are produced making the cells waterproof.

Stratum lucidum   epidermal layer of the skin above the stratum granulosum where cells have clear protoplasm (no organelles), and flattened or no nuclei.

Stratum spinosum   epidermal layer of the skin above the stratum basale that provides structural support to the skin and produces more keratin.

Stroke volume   the volume of blood pumped out of each ventricle per heartbeat.

Sulci   fissures that divide the brain into lobes.

Supination   rotation of the hand and forearm – turns the palms up.

Supporting cells (taste)   cells that lie between taste cells but do not contain taste receptors.

Surfactant   a fluid mixture of phospholipids and lipoproteins that lowers the surface tension of alveolar fluid to maintain the patency of alveoli.

Suture   type of fibrous joint found between the bones of the skull.

Sweat   waste product secreted by sweat glands that contains water, sodium, carbon dioxide, ammonia, urea and other aromatic substances.

Symbiotic   a relationship between different species where both organisms benefit from the presence of the other.

Sympathetic nervous system   a division of the autonomic nervous system that prepares the body for activity (the fright, flight or fight response) through modifying activity of the different systems of the body to increase the supply of nutrients and oxygen for the different activities.

Symphysis   type of cartilaginous joint that forms permanent joints designed for strength and resilience.

Synapse   a specialised intercellular site between two communicating neurons where rapid, highly localised transmission of chemical and electrical signals occurs across a minute gap.

Synaptic cleft   narrow extracellular gap that separates the pre- and post-synaptic membranes of communicating neurons.

Synarthrosis   fixed or unmovable joint.

Synchondrosis   type of cartilaginous joint associated with the growth of bones.

Syndesmosis   type of fibrous joint found in the lower legs between the distal tibia and fibula and in the forearm between the radius and ulna.

Synovial fluid   a viscous, clear or pale yellow fluid containing hyaluronic acid and interstitial fluid that forms a thin film over articular capsule surfaces.

Synovial joint   joint with a distinct characteristic of having a space, the synovial cavity filled with synovial fluid, between the ends of the articulating bones. A high degree of mobility.

Synovial membrane   membrane that produces synovial fluid.

Systemic circulation   the blood supply to all of the body except the lungs.

Systole   contraction of the heart muscle during a heartbeat.

T

Tactile corpuscules (tactile discs)   general sensory receptors that detect the tactile sensations of light touch, texture, shapes and edges.

Tarsal bones   bones of the ankle.

Tarsal glands   glands in the eyelid that produce a protective oily lubricant that helps prevent water/tear evaporation from the eye.

Taste (gustatory) cells   cells that contain taste hairs that lead into a taste pore with receptors that synapse with a neuron, but are not neurons themselves.

T-cells   lymphocytes that secrete immunologically active compounds and assist B-cells, they are cytotoxic and so can destroy foreign cells directly and can self-regulate other T-cells by preventing over activity. They also release interleukins which stimulate other lymphocytes and macrophages (from monocytes).

T-cytotoxic cells a T-lymphocyte that is antigen-specific and able to source and kill specific types of virus-infected cells.

TDF Testis-Determining Factor – hormone that initiates the differentiation into male sex organs.

Telomerase enzyme that can lengthen telomeres and result in continued division, preventing them from progressively shortening during successive rounds of chromosome replication.

Telomere a short length of specific DNA that acts as a buffer against damage and shortening of the chromosome end at cell division.

Tendons dense fibrous connective tissue that attaches muscle to bone.

Tensor tympani muscle tiny muscle connected to the ossicles that limits movement and increases tension on the eardrum to prevent damage to the inner ear from loud noises.

Terminal cisterns open-ended sacs of the sarcoplasmic reticulum that sit against the sides of T tubules.

Tertiary intention healing that occurs when a wound is intentionally kept open to allow oedema or infection to resolve or to permit removal of exudate.

Thalamus relay centre for nervous impulses moving to and from the cerebrum.

T-helper cells a T-lymphocyte that assists other cells in the immune response by recognising antigens and secreting cytokines to activate T- and B-cells.

Thermoceptors sensory receptors that respond to temperature.

Thermogenesis the production of heat.

Thermoregulation the regulation of body temperature.

Thermosensation perception of temperature.

Thoracic duct lymph vessel that returns lymph collected from legs, pelvic region and abdominal cavity, the left arm and the left-hand side of the thorax, head and neck to the circulatory system by emptying into the left subclavian vein.

Thrombin an enzyme in the blood that has a role in coagulation by converting fibrinogen to fibrin.

Thrombocytes (or platelets) very small discs without nuclei that secrete clotting factors (procoagulants), vasoconstricting agents to induce vascular spasm, and clump together in platelet plugs. They also dissolve old blood clots, destroy bacteria by phagocytosis, secrete chemical agents, attract neutrophils and monocytes to infected sites by chemical messengers and promote mitosis in fibroblasts and smooth muscle.

Tibia the shin bone – the stronger of the two bones in the lower leg.

Tidal volume the amount of air that moves with one breath.

T-lymphocyte type of lymphocyte that has a central role in cell-mediated immunity.

Tonicity (of a fluid) the force exerted by osmotically active particles within a fluid.

Trabeculae supporting columns of connective tissue within an organ/structure.

Trachea a tubular passageway of connective tissue and smooth muscle reinforced with C-shaped rings of hyaline cartilage extending from the larynx to the bronchial tubes to bring air to and from the lungs.

Transamination   the transfer of the nitrogen element from one amino acid molecule to another molecule to form a needed amino acid.

Transcription   the first step of gene expression where a particular segment of DNA is copied into RNA.

Transcytosis   transport of large molecules by capturing them in vesicles, carrying them across a cell, and exporting them out the other side.

Transfer protein   a cell membrane protein that facilitates the transport of a substance across that membrane.

Transgender   when someone's gender identity and expression differ from that of their sex assigned at birth.

Transients   microorganisms that get incorporated briefly into the microbiota from the diet or environment but do not remain long and are unable to colonise the body.

Transport protein   a protein that transports materials within an organism.

Transverse T tubules   small tubules that run transversely through striated muscle fibres and through which action potentials are transmitted from the sarcoplasm into the fibre.

T-regulatory cells   a T-lymphocyte that suppresses other lymphocytes' activity and controls immune responses.

Triglycerides   type of fat composed of one molecule of glycerol joined with three fatty acid molecules.

Trimester   a three-month period of pregnancy.

Tunica externa   outer fibrous layer of blood vessels.

Tunica interna   inner, smooth layer of blood vessels.

Tunica media   middle muscular layer of blood vessels.

Tympanic cavity   air-filled cavity behind the eardrum that connects to the nasopharynx by the Eustachian tube (or auditory tube).

Tympanic membrane   a thin semi-transparent partition between the auditory canal and the middle ear.

U

Ulna   the longest of the forearm bones.

Umbilical cord   the vascular cord that connects an embryo or fetus to the placenta.

Urea   formed in the liver from amino acids, this is the primary nitrogenous waste product of protein catabolism (breakdown).

Ureters   pair of tubes which collect urine from the calyces of the kidneys and drain it into the bladder by peristalsis.

Urethra   tube that carries urine from the bladder (and semen in men) to the external urinary meatus for elimination from the body.

Urinary meatus   external urethral opening.

Uterus   the womb.

Utricle   one of the two otolith organs located in the vestibule of the ear.

Uvula   small, fleshy, conical body projecting downward from the middle of the soft palate.

V

Vaccination   the administration of a vaccine, which is a biological preparation that improves immunity to a particular disease.

Vagina   the fibromuscular tube that runs from the cervix of the uterus to the opening to the external genitalia.

Valency   the combining power of an element determined by the number of electrons in the outer shell. Elements in the same group of the periodic table have the same valency.

Vascular tunic   section of the eye made up of the choroid, ciliary body and iris.

Vasoconstriction   narrowing of the lumen of blood vessels through contraction of the smooth muscle within the vessel walls.

Vasodilation   widening of the lumen of blood vessels through relaxation of the smooth muscle within the vessel walls.

Veins   blood vessels that carry blood back towards the heart.

Ventricle (brain)   chambers of the brain filled with CSF.

Ventricle (heart)   lower, larger chamber of the heart, of which there are two, that receive blood from the atria for pumping out around the body. The right ventricle pumps blood to the lungs and the left out to the rest of the body.

Venules   small veins.

Vertebrae   small bones forming the spine.

Vertex presentation   The vertex presentation is when the baby's chin is tucked down towards its chest, so that the vertex of the skull is the leading part entering the mother's pelvis. This position gives the smallest diameter of the head for delivery.

Very Low Density Lipoproteins (VLDLs)   precursors to LDLs, these lipoproteins have their triglycerides removed in adipocytes and become LDLs.

Vesicle   a cellular organelle comprising of a fluid enclosed by a lipid bilayer membrane, within which substances can be stored for secretion.

Vestibular apparatus   organ of balance consisting of the vestibule and the semicircular canals of the ear.

Vestibule (ear)   one of the two major parts of the vestibular apparatus – consists of a pair of membranous sacs, the saccule and utricle.

Vestibule (nasal)   anterior portion of the nasal cavity.

Vestibule (vulvar)   part of the vulva between the labia minora into which the urethra and vagina open.

Villi minute finger-shaped processes of the mucous membrane of the small intestine.

Virus any of a wide variety of mostly pathogenic microorganisms consisting of a single nucleic acid chain surrounded by a protein coat, capable of replication only within a living cell.

Viscerosensation sensations originating in internal organs including pain, palpitations and spasms.

Vision the conversion of electromagnetic radiation to electrochemical energy in order to produce images of our surroundings.

Vitamin organic micronutrient required in small quantities necessary for metabolism.

Vitreous humour the clear, gel-like substance filling the eyeball behind the lens.

Volkmann's canals small channels in the bone that transmit blood vessels from the periosteum into the bone.

Vulva external genitals of the female.

W

Water of oxidation water created through metabolic reactions.

White matter grouping of nerve cell axons.

Z

Zona fasciculata layer of the adrenal gland that secretes glucocorticoid hormones, including cortisol.

Zona glomerulosa layer of the adrenal gland that secretes aldosterone.

Zona reticularis layer of the adrenal gland that produces androgens.

Zygote the name given to the fertilised egg cell formed by the fusion of a sperm cell and an ovum.

APPENDIX 1

INTRODUCTORY SCIENCE

INTRODUCTION

To understand the way in which the human body works you need to know something about the chemistry underpinning structures, metabolism and homeostasis.

ATOMIC AND MOLECULAR STRUCTURE

All human tissues are formed of matter with the same basic structure as the rest of our material world. Matter is made of elements and compounds, singly or in mixtures (Figure A1).

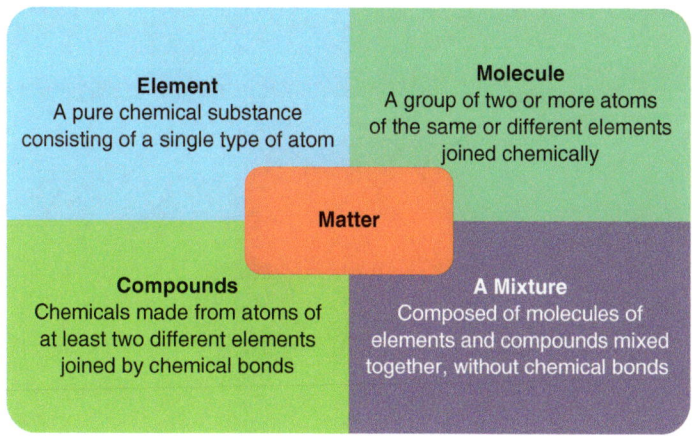

Element
A pure chemical substance consisting of a single type of atom

Molecule
A group of two or more atoms of the same or different elements joined chemically

Matter

Compounds
Chemicals made from atoms of at least two different elements joined by chemical bonds

A Mixture
Composed of molecules of elements and compounds mixed together, without chemical bonds

Figure A1 Matter

Atoms

Atoms are the basic building blocks of matter. Atoms consist of three types of subatomic particles: protons, neutrons and electrons. Each atom is a bit like the solar system and consists of: a nucleus containing a specific number of protons and a number of neutrons, and electrons orbiting around the nucleus (Figure A2).

Every atom of a specific element has the same number of protons in the nucleus and number of electrons in orbit: this is the atomic number. The number of neutrons can vary somewhat in different isotopes (versions) of the same element and the combined weight of protons and neutrons is the atomic mass. Different isotopes of the same element have the same atomic number but different atomic mass (or atomic weight).

The electrons orbit in a number of shells around the nucleus and Figure A3 shows the maximum number of electrons that can inhabit each shell. With increasing atomic number, one additional proton and electron is added. The shells of electrons build up from the centre and the number in the outer shell determines the valency, i.e. the combining power of an element or number of bonds it can make with another element (see Chemical Bonds below).

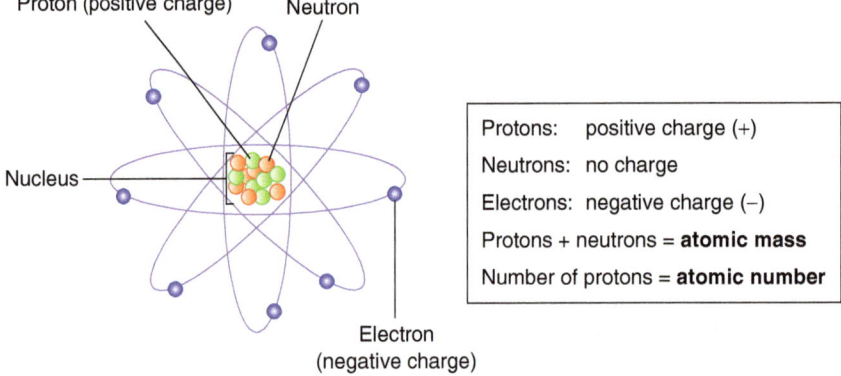

Figure A2 Atomic structure

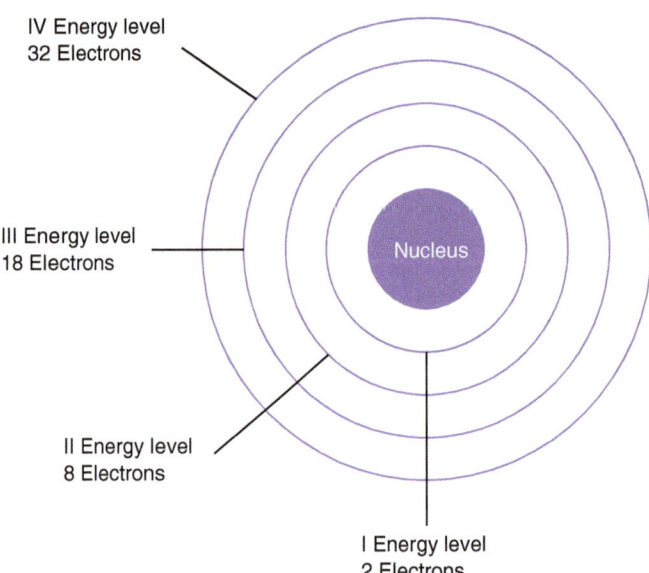

Figure A3 Electron shells with maximum number of electrons in each of the first four shells

You might find it interesting to watch this external video on atomic structure.

This video can be accessed by **scanning the QR code** with your smart phone camera or via https://study.sagepub.com/essentialap2e

ATOM STRUCTURE (2:13)

Periodic Table of the Elements

The original tables of elements that were produced were based on atomic weight. However, in 1869 a Russian chemist, Dmitri Mendeleev (1834–1907) described what is now known as the Periodic Table of the Elements (Figure A4). He had worked extensively on the similarities between elements but is said to have written:

> I saw in a dream a table where all elements fell into place as required. Awakening, I immediately wrote it down on a piece of paper, only in one place did a correction later seem necessary. (Kedrov, 1966/67)

Figure A4 Periodic Table of the Elements

Thus far, 118 elements have been identified, although only 98 of these are naturally existing. Arranged in order of atomic number, at regular intervals elements resemble each other and Mendeleev arranged them in columns or periods (Figure A4). In group 1 you can see a series of seven elements that are similar in chemical and physical properties and have the same valency. Each element is identified by one letter or two letters (e.g. H = hydrogen; He = helium). In his Table Mendeleev left spaces where a number of elements should fit but had not yet been identified, although they were later.

You might find it interesting to watch this external video on acids and bases.

This video can be accessed by **scanning the QR code** with your smart phone camera or via https://study.sagepub.com/essentialap2e

ACIDS, BASES AND PH (5:47)

Molecular structure

A molecular formula shows the number of atoms in a particular molecule. For example, the formula for glucose is $C_6H_{12}O_6$, i.e. it contains

6 Carbon atoms, 12 Hydrogen atoms, 6 Oxygen atoms

CHEMICAL BONDS

Chemical bonding follows the octet rule (or the duet rule for small elements). The octet rule says that in order to be stable, an atom needs eight electrons in its outer shell, or two electrons for the small atoms of H, He, B (boron), Li (lithium) and sometimes Be (beryllium). Stability of the atom is always the aim and electrons can be lost or gained to achieve this. There are four main types of chemical bonds.

Ionic bonds

Ionic bonds form when two atoms have a large difference in electronegativity (i.e. an atom's ability to attract an electron). An ionic bond is created by the transfer of one electron from the outer shell of an atom to the outer shell of another atom:

- **Cation:** the atom that loses an electron (carrying a negative charge) and thus becomes positively charged.
- **Anion:** the atom that gains an electron and thus becomes negatively charged.

The positive and negative charges on a cation and anion are attracted to each other and form an ionic bond. Figure A5 shows ionic bonding between sodium and chlorine.

Sodium (Na from the Latin natrium) has 11 electrons around it, one in its outside shell. Chlorine (Cl) has 17 electrons with seven in its outside shell. An atom which has an electric charge (positive or negative) is known as an ion. The Na becomes Na$^+$ (a sodium ion) as it donates an electron to the chlorine

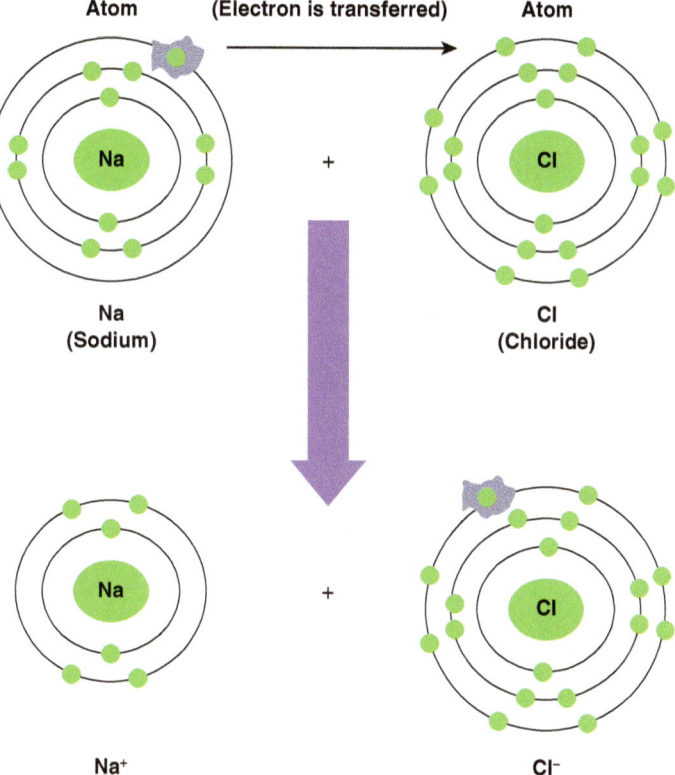

Figure A5 Ionic bonding

and is left with eight electrons in its outside shell. The Cl becomes Cl⁻ (a chloride ion) and now also has eight electrons in its outside shell. The cation and anion are attracted to each other in an ionic bond. In this particular example, each atom has a valency of one. This bond is often formed between a metal (e.g. Na) and a salt (e.g. Cl or bicarbonate).

In a dry condition, NaCl is in solid crystalline form (salt). However, when dissolved in a suitable solvent (e.g. water) it ionises into two charged ions ($NaCl \rightarrow Na^+ + Cl^-$) and is known as an electrolyte. Intracellular Fluid (ICF) and Extracellular Fluid (ECF) are solutions of electrolytes with different concentrations of various ions. The positive and negative symbols indicate that the substance is ionic due to chemical dissociation and the ions are not directly linked.

This is the major form of bonding among the electrolytes in the ICF and ECF.

Covalent bonds

These form when two atoms have a small difference in electronegativity and are often formed between similar atoms, e.g. metal to metal or non-metal to non-metal. Covalent bonds occur when the two atoms share electrons, usually leading to a strong bond. Figure A6 illustrates covalent bonding between carbon (C) and hydrogen (H) to form methane. The carbon has six electrons, two in the inner shell and four in the outer, thus needing four more electrons to achieve stability. Each of the four H atoms has one electron and needs a second to achieve stability. Each hydrogen atom shares its electron with the carbon (achieving eight electrons in the C outside shell) and the carbon shares one electron with each hydrogen atom (achieving two electrons in the H shell). Thus, all atoms have achieved stability.

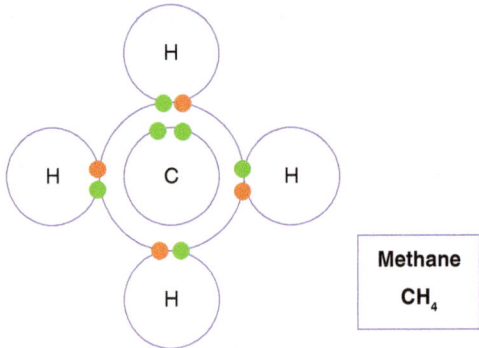

Figure A6   Covalent bonding

This and the next type are the commonest forms of bonding in the nutrients and materials comprising the structure of the human body.

Polar covalent bonds

Polar covalent bonds fall between ionic and covalent bonds: although they are covalent, they also have significant ionic properties. The two elements have a moderate difference in electronegativity and one atom becomes slightly negative and the other slightly positive, although not enough to make ions. They are considered slightly positive or slightly negative. Figure A7 illustrates this slight electric charge in a water molecule.

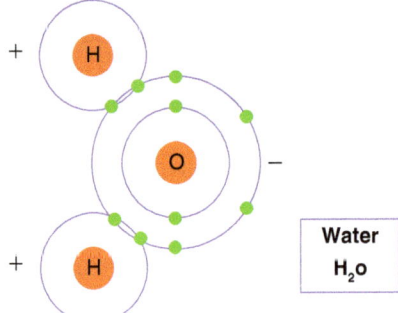

Figure A7   Polar covalent bonds

Hydrogen bonds

Hydrogen bonds are rather different, in that they only occur between hydrogen and a few other atoms (oxygen (O), nitrogen (N) or fluorine (F)). Figure A7 shows the positive and negative poles in water due to the polar covalent bonding. Figure A8 illustrates the positive hydrogen attracted to the negative pole of another water molecule. This bonding throughout water increases the cohesion of the liquid and results in a higher boiling point than would otherwise occur.

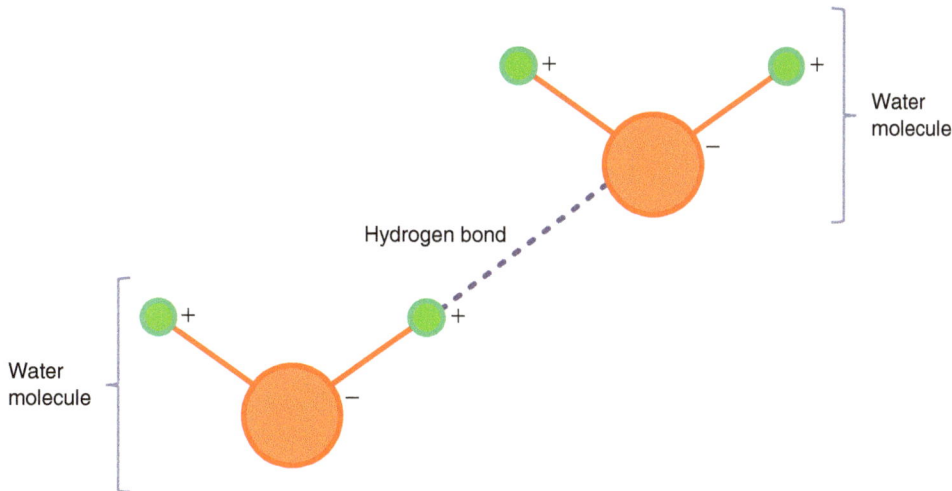

Figure A8 Hydrogen bond

pH (ACID-BASE BALANCE)

Acid–base balance or pH is an important aspect of homeostasis as human cells function most effectively at a pH of about 7.4 – slightly alkaline. To understand the concept of pH, you first need to know the structure of water and how it behaves (Figure A7). It dissociates (breaks up) into two ions:

- One hydrogen ion created by the loss of its one electron (negatively charged) leaving it with an overall positive charge. This electron transfers to the hydroxyl ion.
- One hydroxyl ion is created by the oxygen atom combined with one hydrogen atom, and with the extra electron from the hydrogen ion leaving it negatively charged.

$H_2O \rightarrow H^+ + OH^-$

Water molecule → Hydrogen ion + Hydroxyl ion

pH is a chemical term that indicates the level of acidity/alkalinity in a solution and is determined by the amount of hydrogen ions (H^+) in a solution. The pH scale runs from 0 to 14 with 0 being strongly acidic (i.e. high levels of hydrogen ions) and 14 strongly alkaline (or basic) (either low levels of hydrogen ions or high levels of hydroxyl ions). The neutral point is 7, i.e. pure water (Figure A9). Figure A9 shows the pH of different components in the body and some examples for comparison.

Variation in the pH of a solution is minimised by buffers or buffer systems. A buffer is able to hold or release H^+ as necessary to minimise the change in pH. The main buffer system in the body is the carbonic acid–bicarbonate system although proteins in the erythrocytes also absorb H^+. The carbonic acid–bicarbonate buffer system interacts in solution as shown in Figure A10. Note the reversible reactions, essential for the buffer system to function.

At the normal blood pH of 7.4 the ratio of bicarbonate (HCO_3^-) to carbonic acid (H_2CO_3) is 1:20. The lungs and kidneys are both involved in regulating the overall balance.

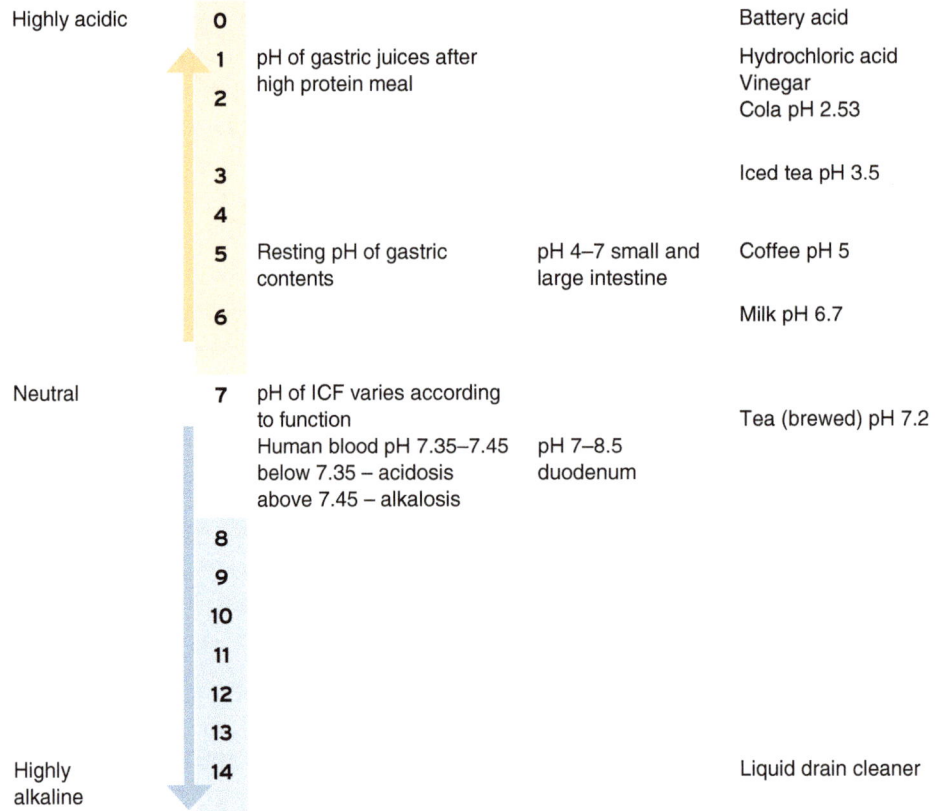

Figure A9 The pH scale

Carbon dioxide + Water ⟷ Carbonic acid ⟷ Bicarbonate ion + Hydrogen ion

$$CO_2 + H_2O \longleftrightarrow H_2CO_3 \longleftrightarrow HCO_3^- + H^+$$

This equation is reversible and so allows the body to regulate pH in a number of ways. Additional free H^+ ions in the blood will lower its pH, making it more acidic. The body then has two main options:

1. Excrete hydrogen ions (H^+) ions in the urine via the kidneys. This will raise pH.
2. Bind hydrogen ions (H^+) with bicarbonate ions (HCO_3^-) which will become carbonic acid (H_2CO_3) which dissociates into water (H_2O) and Carbon dioxide (CO_2). Carbon dioxide can then be eliminated from the body via the lungs, which raises pH.

If the blood is too alkaline (pH > 7.45), then bicarbonate ions (HCO_3^-) can be excreted in the urine via the kidneys.

Figure A10 Carbonic acid–bicarbonate buffer system

CHEMICAL REACTIONS

There are a number of types of chemical reactions in which molecules take part including synthetic ones (in which more complex substances are formed) and ones in which complex molecules are broken down into simpler molecules. In a chemical reaction only the substances there to start with (the reactants) can be present in the end products. No atoms are formed or destroyed. The reactants meet, chemical bonds are broken, and atoms become rearranged and create new bonds to make the end products.

Chemical equations identify clearly the reactions which occur. The example below is a simple one:

HCl	+	NaOH	→	NaCl	+	H_2O
Hydrogen chloride	+	Sodium hydroxide	→	Sodium chloride	+	Water
(hydrochloric acid)		(caustic soda)		(common salt)		

Some chemical reactions are reversible. The two sides of the equation are in balance and the reversible nature of the reaction is shown by the arrows pointing in both directions, e.g.:

$$AB \rightleftharpoons A + B$$

Reaction rate: work, energy and catalysts

Chemical reactions occur at a rate determined by the concentration of the chemicals involved and the temperature. Chemical reactions involve energy transformations essential for the work of the body and the major currency is ATP (Adenosine Triphosphate, Chapter 9), which stores energy in high-energy bonds. The definition of work in physics is the product of a force and the distance over which it is applied. Work is carried out when a force is applied to an object which is moved through a distance. However, the work in living things is considered as energy transformations, including:

- **Synthetic work:** formation of the complex molecules necessary for cell division and life including DNA (Chapter 2). This requires ATP.
- **Electrical work:** carriage of nervous impulses through an action potential is dependent on maintaining an electrical potential difference across cell membranes. This also requires ATP.
- **Mechanical work:** movement caused by contraction of our muscles also needs ATP.

While some reactions occur spontaneously, many need additional assistance to occur at a useful rate and ATP is the main source of additional energy. In addition, many reactions in the body require a catalyst: a substance that, in small amounts, increases the rate and lessens the energy used in a chemical reaction without permanent change itself. In human physiology these are mainly enzymes, proteins formed within the cells of the body from the DNA template (Chapter 9). Enzymes are often named after the substance on which they act with -ase at the end: e.g. lipase is an enzyme that breaks down lipids.

Balanced equations

As atoms are neither created nor destroyed in everyday life, a chemical reaction must include the same number and type of atoms on each side of the equation, i.e. in the reactants and the products. The metabolism of glucose illustrates this (Table A1).

Table A1 Balancing an equation

Glucose	+2	Oxygen	→	Carbon dioxide	+	Water
$C_6H_{12}O_6$	+	O_2	→	CO_2	+	H_2O

In the equation above, the subscript numbers specify the numbers of that particular atom present and you need to have the same number on each side of the equation. Working this out:

Glucose:	C_6 indicates 6 Carbon atoms		6 Carbon atoms are in 6 CO_2
	H_{12} indicates 12 Hydrogen atoms		12 H atoms are in 6 water (H_2O) molecules, 2 in each
	O_6 indicates 6 Oxygen atoms		Oxygen needed = 12 O in CO_2 (2 in each CO_2) + 6 in H_2O
			Total: 12+6=18 O atoms

6 O atoms are in glucose, 12 more are needed – 6 O_2 molecules

The overall balanced equation is:	$C_6H_{12}O_6 + 6O_2 \rightarrow 6CO_2 + 6H_2O$

Oxidation, reduction, hydrolysis reactions

You need to know something about certain types of reactions.

Oxidation is the loss of electrons or an increase in oxidation state by a molecule, atom or ion.

Reduction is the gain of electrons or a decrease in oxidation state by a molecule, atom or ion.

GO DEEPER

The term oxidation state is used as an indication of the charge that the atom/molecule would have if all bonds were ionic (i.e. involving the movement of electrons) which does not happen in covalent bonds.

Oxidation and reduction reactions occur in parallel and are known as redox reactions. Simple redox reactions are the oxidation of carbon to carbon dioxide (CO_2), or the reduction of carbon by hydrogen to form methane (CH_4).

Hydrolysis is when a water molecule is added to a substance which usually splits into two parts (Figure A11).

Figure A11 Hydrolysis

ORGANIC CHEMISTRY

This is the branch of chemistry concerned with carbon and is the most important in the context of human biology. The structures of the molecules concerned are relatively easy to understand. They are based on the valencies of a relatively small number of elements given in Table A2.

Table A2 Major elements in organic chemistry in biology

Element	Symbol (common molecular form)	Atomic number	Electrons in outer shell	Valency	Main properties
Carbon	C	6	4	4	Non-metal, second element in human body
Oxygen	O (O_2)	8	6	2	Colourless gas, 21% of air, first element in human body
Hydrogen	H (H_2)	1	1	1	Colourless highly flammable gas
Nitrogen	N (N_2)	7	5	3 or 5	Colourless gas, 78% of air
Sulphur	S	16	6	2	Bright yellow crystalline solid at room temperature
Phosphorus	P	15	5	3 or 5	Highly reactive, found in compounds

Carbon has four electrons in its outer shell and the orbits of these give it a three-dimensional (pyramidal) shape (see Figure A12). Carbon-based molecules can be shown two-dimensionally as in Figure A13 and can have single, double or triple bonds between the carbon atoms. When a bond greater than one is present, the substance is unsaturated. Cyclic compounds also occur with alternate double and single bonds forming a circle.

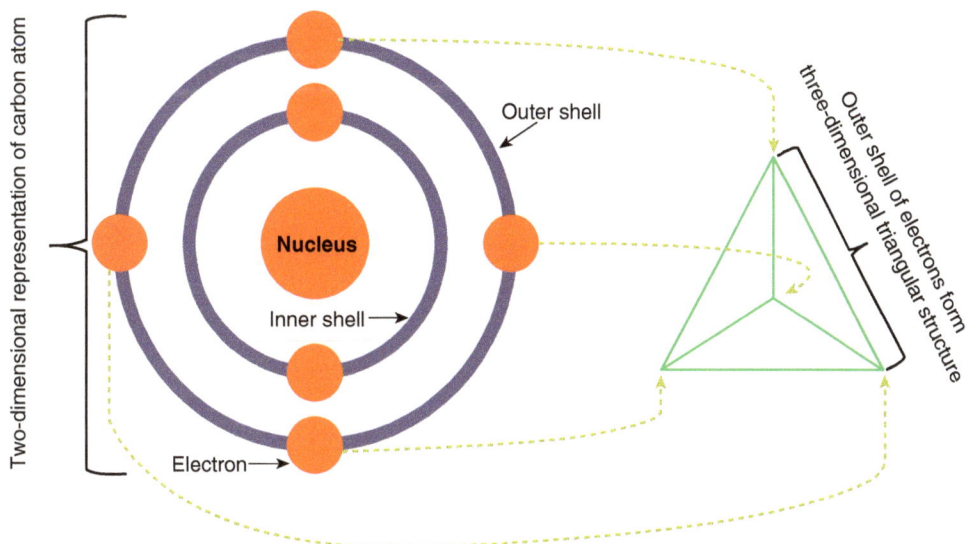

Figure A12 The three-dimensional shape of carbon

Carbon can form long chains as in fatty acids (Chapter 9) and can have a number of different side chains.

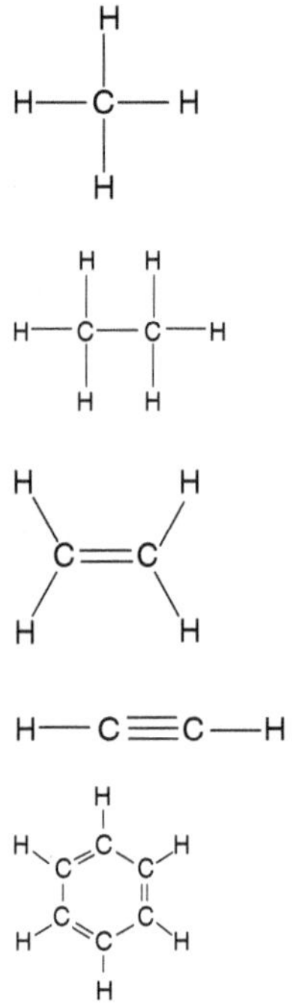

Methane. CH_4
Single bonds

Ethane. C_2H_6
Single bond between carbon atoms

Ethylene C_2H_4 $H_2C=CH_2$
Double bond between carbon atoms

Acetylene C_2H_2 $HC\equiv CH$
Triple bond between carbon atoms

Benzene C_6H_6
Alternate double and single bonds

Figure A13 Carbon bonds

REFERENCE

Kedrov, B.M. (1966/67) 'On the question of the psychology of scientific creativity (on the occasion of the discovery by D.I. Mendeleev of the Periodic Law)', *Soviet Psychology*, 5(2): 18–37. Accessed from *The Soviet Review: A Journal of Translations*. Summer 1967, Vol. VIII, No. 2.

APPENDIX 2

UNITS AND NUMERALS

In this section we clarify two issues:

- The international units used in measurement in science.
- The Roman numerals used in some numbering systems in science and in many monuments to famous scientists.

SI (SYSTÈME INTERNATIONAL) UNITS

History

Many different systems of measurement have been used around the world. Historically some of these were based on parts of the body – such as thumb length. However, as you can imagine, this leads to many problems, for example not everyone's thumb is the same length. Therefore, it was necessary to develop a system to help overcome these problems. Work initiated in 1948 resulted in the launch in 1960 of a standardised system of measurement, the Système International or SI system. It is derived from the metre-kilogram-second (MKS) system rather than the centimetre-gram-second (CGS) system and is still evolving. Imperial units (for example: feet and yards; pounds and stones; fluid ounces, pints and gallons) are still used in a number of English-speaking countries alongside SI units which are now used almost universally in science, technology, engineering, etc. While the UK now mainly uses SI units, the USA still uses imperial (known as United States customary) units quite widely. In 1999, confusion at NASA in conversion between US and SI measures resulted in the loss of a space probe to Mars (Stephenson et al., 1999).

The SI system is based on seven standardised units of measurement plus a number of derived units. Amendments to this evolving system are managed and agreed internationally through three organisations derived from the Metre Convention of 1875 and still extant:

- **General Conference on Weights and Measures** (Conférence générale des poids et mesures or **CGPM**) meets every four to six years with representatives from all member states. It receives reports from CIPM (below) and endorses new developments in SI. On 1 January 2015, there were 55 Member States and 41 Associates of the CGPM.

- **International Committee for Weights and Measures** (Comité international des poids et mesures or **CIPM**) of 18 scientists of high repute meets annually at the BIPM (see below) and advises the CGPM on administrative and technical matters.
- **International Bureau of Weights and Measures** (Bureau international des poids et mesures or **BIPM**), the international metrology centre based at Sèvres in France helps to ensure uniformity of SI weights and measures around the world by undertaking a range of activities to support the CGPM and CIPM.

SI and related units

The SI and associated units considered here relate to all areas of science, but the major focus will be on those used within the biological sciences. The seven base SI units are specified in Table A3 and both names and symbols are written in lower case except for Ampere and Kelvin which are both named after people. From the seven basic units, other units are derived for a variety of purposes (e.g. Joule (J) the SI unit of work or energy; Pascal (Pa) the SI unit of pressure). There are also a number of units accepted for use with the SI units although they are not officially part of this system (BIPM, 2006). The litre is an important one of these units. It is the metric unit of volume and SI accepts it as 'a special name for the cubic decimetre' and defines it as exactly 1 cubic decimetre, 1,000 cubic centimetres, or 0.001 cubic metre.

Table A3   Base SI units

Category	Name	Abbrev.	Current Specification
Length	metre	m	Distance travelled by light in vacuum during 1 / 299792458 of a second: most precise measurement
			An international standard metre of 90% plutonium and 10% iridium is held at the BIPM at Sèvres
Mass	kilogram	kg	Mass is the amount of matter in an object and is judged against the international kilogram of 90% plutonium and 10% iridium in the form of a cylinder, diameter and height about 39 mm held at the BIPM
			(Mass is not the same as weight. Weight is determined by the pull of gravity on mass and varies with the distance from the centre of the earth although are interchangeable at most points of the earth)
Time	second	s	This is scientifically defined in relation to radiation corresponding to the transition between two states of the caesium 133 atom at 0°C
			More usually, it is part of the day which is divided into 24 hours, each hour divided into 60 minutes, each minute divided into 60 seconds. A second is 1 / (24 × 60 × 60) of the day
Electric current	Ampere	A	Less relevant in biology, electric current is a flow of electric charge often carried by moving electrons in a wire. It can also be carried by ions in an electrolyte or along nerve fibres
Temperature	Kelvin	K	The Kelvin, unit of thermodynamic temperature, is the fraction 1/273.16 of the thermodynamic temperature of the triple point of water (i.e. point at which water is solid, liquid and gas in equilibrium)
			A degree Celsius (equal to Centigrade, still sometimes used) is equivalent to a degree Kelvin. 0°K is known as absolute zero and is the point at which no atoms are moving and it can get no colder
			273°K is equivalent to 0°C and is the freezing point of water

Category	Name	Abbrev.	Current Specification
Amount of substance	mole	mol	The mole is the amount of material with as many elementary particles as there are atoms in 0.012 kg of carbon 12
			The elementary particles are atoms, molecules, ions, electrons, other particles, or specified groups of such particles
Luminous intensity	candela	cd	A measure of light emitted

Prefixes for SI units

SI units of measurement must be able to work across a very wide range and this is achieved by the use of prefixes. When a quantity reaches 1,000 × its units it is shortened to kilo units (e.g. think of weight; if someone weighs 70,000 grams, we write this as 70 kilograms (kg)). When a quantity is only one thousandth of its unit, we write this as milli units. We often measure small lengths in millimetres (mm). When a quantity is only one millionth of its unit, we write this as micro units (μ or mcg). Some drugs are given in microgram quantities (e.g. digoxin may be prescribed as 250 μg or 250 mcg). The SI system allows sizes to be made bigger or smaller by the use of appropriate prefixes (Table A4).

Table A4   Prefixes with SI units

Prefix	Fraction	Decimal	Example
kilo (k)	one thousand	1000	kilogram (kg)
deca (da)	ten	10	
deci (d)	one tenth	0.1	
centi (c)	one hundredth	0.01	centimetre (cm)
milli (m)	one thousandth	0.001	millilitre (ml)
micro (μ)	one millionth	0.000001	microgram (μg)
nano (n)	one thousand millionth	0.000000001	nanogram (ng)

ROMAN NUMERALS

Roman numerals are used to a considerable extent in science and it is important that you understand these and are able to translate these numbers into the normal numbers used today. Roman numerals are still used in a number of specific situations, including the cranial nerves (Table 5.6 in Chapter 5), and are based on the following seven symbols:

I	1
V	5
X	10
L	50
C	100
D	500
M	1000

There is no zero in this system, so the ordering of the letters determines the value of the overall number. It works as follows: the number is read from the left in order of value, with the largest value at the left. However, *subtractive notation* is used to reduce the number of symbols used (see c and d). Some examples make this system clearer:

Arabic Numeral	Roman Numeral	Explanation
2	II	I + I (1 + 1)
4	IV	V – I (5 – 1)
99	IC	C – I (100 – 1)
1066	MLXVI	M + L + X + V + I (1000 + 50 + 10 + 5 + 1)
2016	MMXVI	M + M + X + V + I (1000 + 1000 + 10 + 5 + 1)

REFERENCES

BIPM (2006, updated in 2014) *SI Brochure: The International System of Units (SI)*, 8th edn. Sèvres, France: International Bureau of Weights and Measures (Bureau international des poids et mesures). www.bipm.org/en/publications/si-brochure/chapter4.html (accessed 29 March 2015).

Stephenson, A.G., LaPiana, L.S., Mulville, D.R., Rutledge, P.J., Bauer, F.H. et al. (1999). *Mars Climate Orbiter Mishap Investigation Board Phase I Report*. NASA, 10 November. Available at: ftp://ftp.hq.nasa.gov/pub/pao/reports/1999/MCO_report.pdf (accessed 27 March 2015).

APPENDIX 3

DESCRIPTORS OF THE BODY

In learning about the anatomy and physiology of the people for whom you will be caring, you also need to understand the different terms used to identify positions and movements of the different parts of the body. This section covers the relevant terms.

The anatomical position is the upright stance of the body with the face forward, the arms at the side, and the palms of the hands facing forward. It is used when describing the relation of body parts to one another, as in Figure A14. Table A5 shows the terms used to indicate the relative positions of the organs and parts of the body.

Table A5 Terms of position within the body

Ventral/anterior	Front of the body	Dorsal/posterior	Back of the body
Superior	Above	Inferior	Below
Medial	Towards the midline	Lateral	Away from the midline
Proximal	Closer to point of origin or body	Distal	Further from point of origin or body
Superficial	Closer to body surface	Deep	Further from body surface

The body can be divided by three lines:

- Horizontal: into superior (upper/above) and inferior (lower/below).
- Sagittal: into right and left halves either side of the median line.
- Coronal: into anterior (front half) and posterior (back half).

The sections of the abdomen are described by a number of terms (Figure A15). The internal organs of the body are distributed through two main cavities, dorsal (posterior/back) and ventral (anterior/front) with subdivisions (Figure A16). The differing movements of the body are specified in Chapter 15, Musculo-skeletal System.

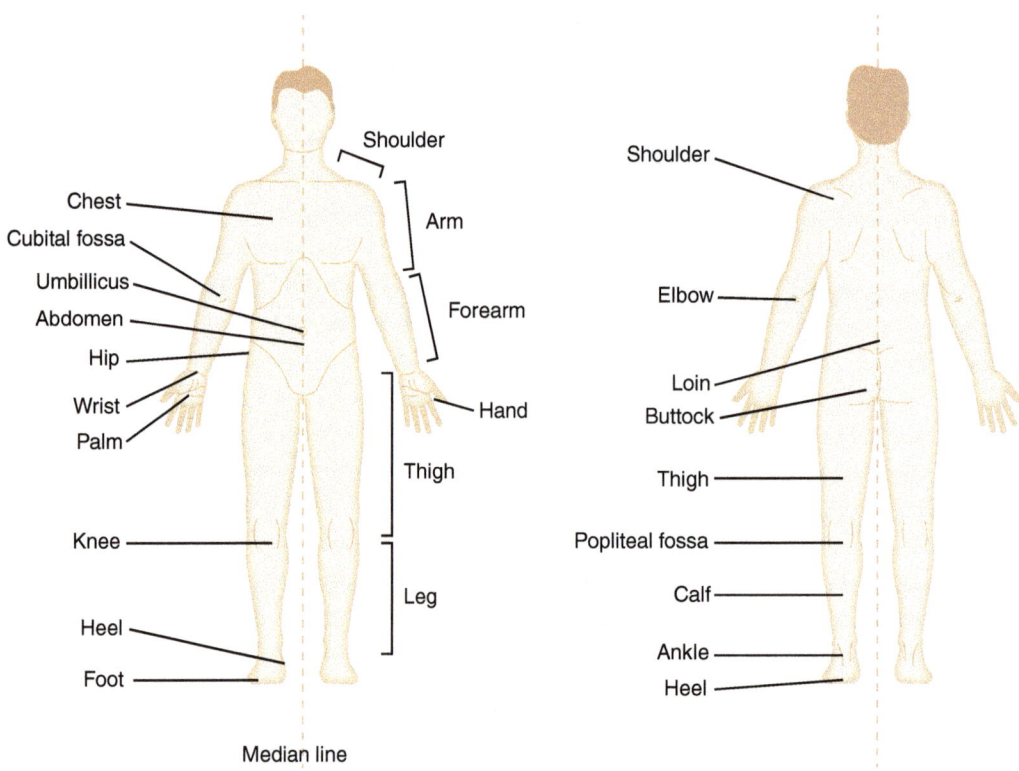

Figure A14a Anatomical names

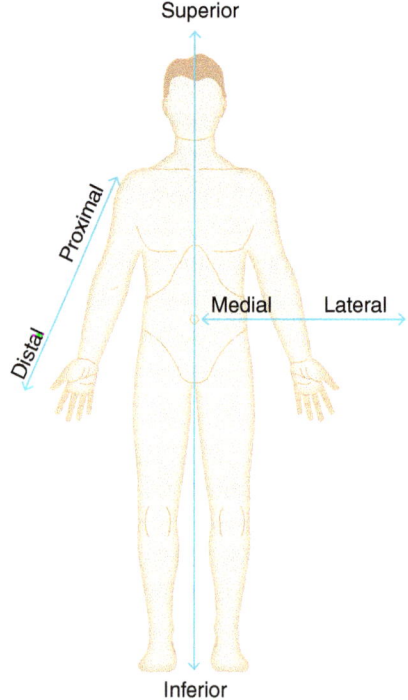

Figure A14b Anatomical directions

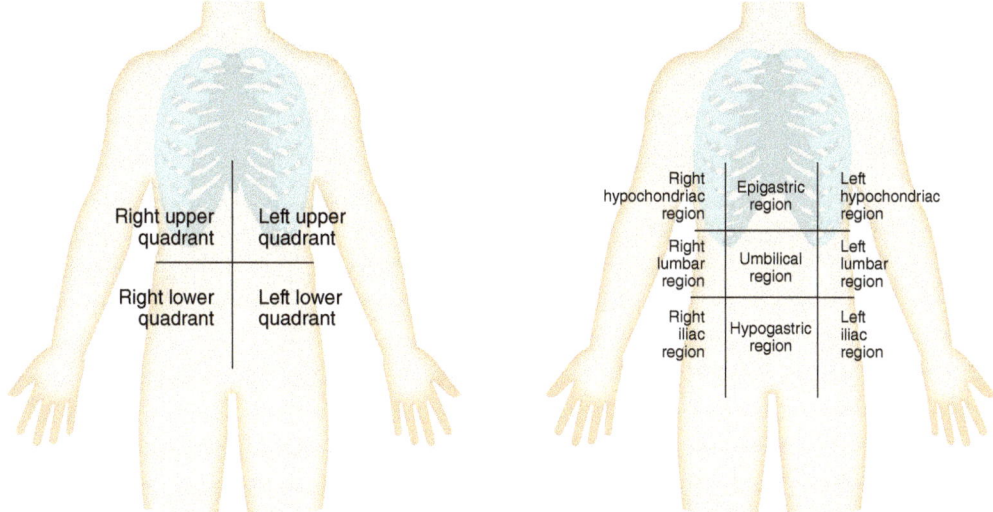

The four abdominal quadrants

The nine regions of the abdominal cavity

Figure A15   The quadrants and regions of the abdomen

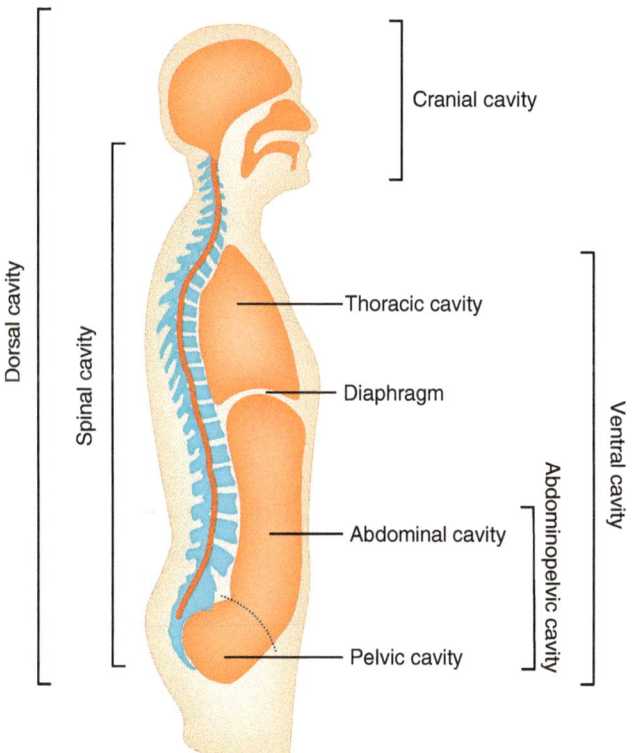

Figure A16   Body cavities and subdivisions

INDEX

CPSIA information can be obtained
at www.ICGtesting.com
Printed in the USA
BVHW010043250522
637570BV00018B/1

9 781526 460325